P9-CKB-233

MARINE BIOLOGY

AN ECOLOGICAL APPROACH

Fifth Edition

James W. Nybakken
Moss Landing Marine Laboratories

An imprint of Addison Wesley Longman, Inc.
San Francisco Boston New York
Capetown Hong Kong London Madrid Mexico City
Montreal Munich Paris Singapore Sydney Tokyo Toronto

Acquisitions Editor: Elizabeth Fogarty
Project Editor: Heather Dutton
Assistant Editor: Chriscelle Merquillo
Managing Editor: Wendy Earl
Design Manager: Bradley Burch
Production Editor: Scott Hitchcock
Text Designers: Yvo Riezebos and Electronic Publishing Services Inc., N.Y.C.
Cover Designer: Yvo Riezebos
Cover Photo: Norbent Wu / Stone
Art Studio: Electronic Publishing Services Inc., N.Y.C.
Photo Researcher: Clare Maxwell
Manufacturing Coordinator: Vivian McDougal
Project Coordination and Electronic Page Makeup: Electronic Publishing Services Inc., N.Y.C.
Marketing Manager: Josh Frost

Library of Congress Cataloging-in-Publication Data
Nybakken, James Willard.
Marine biology : an ecological approach / James W. Nybakken.--5th ed.
p. cm.
Includes bibliographical references.
ISBN 0-321-03076-1
1. Marine biology. 2. Marine ecology. I. Title.

QH91 .N9 2001
577.7--dc21 00-058603

ISBN 0-321-03076-1

5 6 7 8 9 10 - QWT - 04 03

1301 Sansome Street
San Francisco, CA 94111

Chapter opening photo credits: • Chapter 1: Kim Westerskov/Stone • Chapter 2 Peter Parks/Mo Yung Productions/Norbert Wu • Chapter 3: Norbert Wu • Chapter 4: courtesy of the author • Chapter 5: Gerry Ellis/Minden Pictures • Chapter 6: Darrell Gulin/DRK Photo • Chapter 7: David Scharf/Peter Arnold, Inc. • Chapter 8:Carr Clifton/Minden Pictures • Chapter 9: Fred Bavendam/Minden Pictures • Chapter 10: Eiichi Kurasawa • Chapter 11: Ben Osborne/Stone

CONTENTS

(*Chapter 2 continues*)

(Chapter 7 continues)

(Chapter 7 continued)

PREFACE

In recent years, there has been an increasing interest in the field of marine biology at the undergraduate level in our two-year community colleges, as well as in four-year colleges and universities. That interest apparently arises from the growing awareness and concern at all levels of society of the importance of the world's oceans as sources of food, reservoirs of minerals, major suppliers of oxygen, regulators of climates, and the ultimate dumping ground for the mounting burden of human waste materials. This concern has been popularized and brought into focus by the various international disputes on fishing rights, the whale problem, international law of the seas conferences, and numerous television programs and books about marine life.

As a result of this heightened public awareness, marine biology and marine ecology courses have attracted a wide spectrum of students with varying backgrounds who desire a basic understanding of the biological processes that operate in the oceans.

It is important to understand that in the oceans, just as on land, there are scientific principles that govern the organization and perpetuation of organisms and associations. Although those principles operate somewhat differently in the ocean than on land because of the physical-chemical properties of water, they can be readily explained and may be understood by those with a minimum background in other sciences and in mathematics.

The texts in marine biology available to serve these courses appeared to me to lack an ecological approach to the entire marine environment; they tended to emphasize specific areas, habitats, and organisms to the exclusion of broader concepts and processes. I was, thus, stimulated to write this book as an introduction to marine biology, emphasizing the ecological principles governing marine life throughout the world, not as a purely taxonomic or regional approach, so it would be useful in all parts of the world. I have, furthermore, purposely downgraded any discussion of pure oceanography, except as it bears on organization of the associations or communities, in order to increase coverage of the biology. The sequence of topics in this book closely follows the sequence in my upper division undergraduate course in marine ecology, which I taught for many years at the Moss Landing Marine Laboratories.

This text is designed for the undergraduate student in marine biology. It presumes a certain minimum background in basic concepts of chemistry, physics, and biology, but no more than would be obtained from introductory college courses in each of these fields. Familiarity with the major invertebrate phyla is helpful. Some acquaintance with the basic ecological concepts is also helpful, but if lacking, they may be obtained from Chapter 1. Although the text is generally aimed for an undergraduate majors marine biology course, I have tried to imbue this book with sufficient rigor and detail that it will also be useful in upper-division undergraduate courses in marine ecology and biological oceanography. To this latter end, I have included text citations to the primary literature.

This text stresses ecological processes and adaptations that structure marine associations and permit their persistence through time. It is not a guide to the local fauna and flora. It is presumed that familiarity with this, if necessary, will be obtained in laboratory and field trips in which the instructor will provide the taxonomic expertise.

Depending on the time available in the course, this book includes perhaps more material than can be adequately covered in a single quarter or semester. In that case, the instructor may choose to concentrate on those chapters or areas of most concern in his or her geographical region. For example, instructors in temperate zones may wish to deemphasize coral reefs and spend less time on deep-sea biology and more on the intertidal, to which they have direct access. Suggestions for use of the book under varying course conditions are given in the Instructor's Manual. However, I hope the comprehensive nature of this book means that whatever instructional choices are made, information will be available from the book.

CHANGES IN THE FIFTH EDITION

The changes made in this edition maintain the rigorous treatment of all subjects established in the fourth edition and primarily reflect new knowledge and understanding about marine ecology.

New subjects in this edition include a section on plate tectonics and continental drift, a comparison of marine and terrestrial biodiversity, a large section comparing ocean and terrestrial productivity, a discussion of how sea birds find prey using olfaction, the killer whale and sea otter interaction in Alaska, the importance of upwelling along the Pacific coast to community structure in the intertidal, the importance of small temperature changes to keystone predation, data on the importance of coral reefs as carbon sinks, succession and stability of coral reefs in time and space, recent discoveries on the diseases of coral reefs, collapse of Pacific coast salmon fishery, extensive coverage of the changes in ecosystems due to overfishing, a large section on the effects of fishing gear on communities and life histories of the target species, the importance of bycatch and ghost fishing to marine communities, the occurrence of dead zones in nearshore waters and significance of the increase in diseases in the oceans.

Topics that have been substantially upgraded or changed from the fourth edition include a newly revised section on the microbial loop and carbon flux in the oceans, the use of satellites to calculate net primary productivity, the use of bio-optical methods to measure photosynthesis, additional data on the importance of marine mammals to ecosystems, new information on seasonality of reproduction in the deep sea and the incidence of planktotrophy, lecithotrophy and direct development, updates on number of species, endemism and biogeography of the deep sea hydrothermal vents and seeps, additional data on the sea ice communities, newer data on the role of recruitment and post-recruitment processes on infaunal communities in soft sediments, the role of gametophytes in kelp forest ecology, new information on the role of sea urchin recruitment and disease in creating the barren grounds in Atlantic Canada, many more examples of predation, herbivory, competition and physical factors acting to structure the Atlantic intertidal communities, additional adaptations against water loss in marsh plants, a complete rewrite of marsh zonation and its causes plus comparison between the New England and southern Atlantic salt marshes, new information on competition among corals and between corals and other space occupying invertebrates, and a new revised discussion of the role of iron in promoting algal growth in the ocean.

This edition initiates full four color illustrations throughout the text thus increasing the value of the illustrations as well as placing them next to the text rather than is a set of separate color inserts.

In this edition, I have again included additional information and examples that point out the controversies in marine ecology, because I believe that it is important for the student to learn that not only are there many unsolved problems in the field of marine ecology, but also that there are disagreements among scientists about how these various communities and systems are organized. In this latter connection, I think it is particularly important for the student to recognize that studies elucidating the mechanisms ordering a given community in a given geographical area likely cannot be extrapolated as a "universal" explanation for the organization of communities in similar environmental settings throughout the world.

One other comment deserves mentioning here and should be kept in mind when using this text. Considering the all pervasive presence of humans in all parts of the planet, except possibly the deep sea, it is unlikely that there are any pristine natural habitats left and we are in fact studying and describing habitats that are in some way disturbed.

The first editions of this text were perceived by many as distinctly biased toward the Pacific coast, even though I felt that I had made an effort to avoid that bias. I had striven to remedy that in the second through the fourth editions and have continued in that direction in this edition with more coverage and examples from the Atlantic and Gulf coasts, as well as other areas. I think it is important to point out in this context that most of the East Coast of the United States south of New England, as well as the Gulf coast, consists of sedimentary environments, such as open sand beaches, protected sand beaches, muddy shores, and large estuaries and marshes. Rocky shores are rare. The Pacific coast, by contrast, consists mainly of rocky shores, with few estuaries and other sedimentary environments. I hope, therefore, that I will be forgiven for continuing to use Pacific coast examples in discussions of the rocky shores and commended for expanding a discussion of the dominant Atlantic and Gulf coast sedimentary environments. As part of an effort to make this book more useful globally, I have also included in this edition more examples and comparisons with communities in areas outside the North American continent.

Some who have used this book in more advanced classes have suggested that I remove the very basic introductory ecology material from Chapter 1 because their students have had it already. I have chosen not to do so in deference to those who teach more basic courses without major prerequisites and who feel that their students need it. I can suggest only that those whose classes are more advanced simply ignore it. Many users of this book have suggested that it would benefit from the inclusion of summaries of the chapters. In this edition I have added to the end of each chapter a section called Summary of Key Concepts which, I hope, will answer that request. Finally, the glossary has been updated to include all bold face terms and concepts in the text.

INSTRUCTOR'S MATERIALS

As with the fourth edition, an Instructor's Manual with Test Bank is available to text adopters. The material has been updated with problems reflecting the updated coverage. An art CD-ROM for instructors includes all of the figures from the text in Powerpoint. The fifth edition also includes a text specific Companion Web site.

ACKNOWLEDGMENTS

As the author of this textbook, I am indebted to a large number of marine scientists whose research work is the foundation on which this book is based and who are too numerous to mention in their entirety here. A few of this vast number are mentioned in the text and listed in the end-of-chapter bibliographies. I repeat here my thanks to all those colleagues and friends mentioned in the prefaces to the first, second, third and fourth editions who helped me by reviewing chapters, and who contributed illustrations for the preceding editions.

In addition to those colleagues who helped me in the first four editions of this book, I would like to thank the many who have helped with this edition by providing me with constructive criticisms of the fourth edition and pointing out errors. I am particularly indebted here to my Moss Landing colleagues Nicholas Welschmeyer, James Harvey, Greg Cailliet, and William Broenkow. And thank you as well to the following colleagues who reviewed this fifth edition manuscript: Ellen Baker (Santa Monica College: Santa Monica, CA); Joe Britton (Texas Christian University: Fort Worth, TX); Harold Cones (Christopher Newport University: Newport News, VA); Horst Felbeck (University of California at San Diego: Marine Biology Research Division: La Jolla, CA); David J. Fox (University of Tennessee: Knoxville, TN); Walter Goldberg (Florida International University: Miami, FL); Chuck Holliday (Lafayette College: Easton, PA); Osmund Holm-Hansen (Scripps Institute of Oceanography: La Jolla, CA); Dave Krupp (Windward Community College: Kaneohe, HI); Beth Pearson Lowe (Palomar College: San Marcos, CA); Bryan Ness (Pacific Union College: Angwin, CA); Brian Palenik (Scripps Institute of Oceanography: La Jolla, CA); Paul Renaud (University of Connecticut: Groton, CT); Dave Richard (Rollins College: Winter Park, FL); Richard Rosenblatt (Scripps Institute of Oceanography: La Jolla, CA); Erik P. Scully (Towson University: Towson, MD); James Small (Rollins College: Winter Park, FL).

I would also like to thank Diane Nelson, John Heine, Roger Seapy, Robert Higgins and Ron Shimek for additional photos for this edition. Finally, I would like to thank the staff of Benjamin Cummings and particularly Elizabeth Fogarty and Heather Dutton for their help and encouragement in the process of this revision.

Because of the text's broad scope and necessary limitations in length, I have often been forced to treat certain complex or poorly understood processes through generalizations that some may consider seriously oversimplified. Undoubtedly, some errors and omissions may remain. Therefore, please feel free to pass on to me any potential errors, omissions, or questions of interpretation, as well as any other comments and suggestions. All observations will aid in revising future editions.

James W. Nybakken
Moss Landing Marine Laboratories
P.O. Box 450
Moss Landing, CA 95039-0450
nybakken@mlml.calstate.edu

INTRODUCTION TO THE MARINE ENVIRONMENT

About 71% of the surface of this planet is covered by salt water. The water depth averages 3.8 km, a volume of 1,370 × 10⁶ km³. Since life exists throughout this immense volume, the oceans constitute the single largest repository of organisms on the planet. These organisms include representatives of virtually all phyla and are tremendously varied. All, however, are subject to the properties of the seawater that surrounds them, and many features common to these plants and animals result from adaptations to the watery medium and its movements. Before we can consider the major associations or assemblages of organisms, we need to examine briefly the physical and chemical conditions of seawater and aspects of its motion (oceanography), and some basic ecological principles and terms. These principles and terms will be central to understanding the processes discussed in the remainder of this text. Finally, we will make some comparisons between aquatic and terrestrial ecosystems to point out some fundamental differences in organization.

PROPERTIES OF WATER

Water is the substance that surrounds all marine organisms. It composes the bulk of the bodies of marine plants and animals, and it is the medium in which various chemical reactions take place, both inside and outside living organisms.

Chemical Composition

Pure water is a very simple chemical compound composed of two atoms of hydrogen (H) joined to one oxygen (O) atom. Expressed symbolically, it is H_2O. The hydrogen atoms are bonded to the oxygen asymmetrically such that the two hydrogens are at one end of the molecule and the oxygen is at the other. The bonding between the hydrogen atoms and the oxygen is via shared electrons, each hydrogen sharing its single electron with the oxygen. In this manner, oxygen receives the two electrons needed to complete its outer electron shell, and each hydrogen the one needed for its outer shell. However, the oxygen atom tends to draw the electrons furnished by the hydrogen atom closer to its nucleus. This creates a slight negative charge at the oxygen end of the molecule, while the removal of the electrons away from the hydrogens results in a slight positive charge at that end. This electrical separation creates a polar molecule.

The polar nature of the water molecule means that the hydrogen end, which is positive, will attract the negative, or oxygen, end of other water molecules. This gives rise to weak bonds, called **hydrogen bonds**, between adjacent water molecules. These bonds are only about 6% as strong as the covalent bonds between the hydrogen and oxygen in the water molecule itself, and they are easily broken and reformed (Figure 1.1). It is this hydrogen bonding

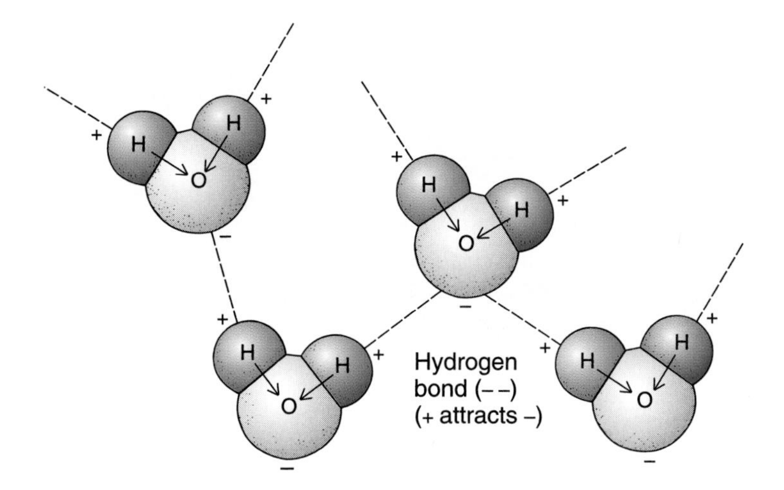

FIGURE 1.1 Diagrammatic representation of a series of water molecules, indicating the polar nature of the molecules and the hydrogen bonding.

between adjacent water molecules and the polarity of the water molecule that is responsible for many of the unique chemical and physical properties of water. Indeed, if water were not polar and did not form such bonds with adjacent molecules, it would be a gas, not a liquid, at room temperature, and its freezing point would be lower than temperatures found over most of the earth's surface. Under such conditions, life as we know it on earth would be impossible.

Physical and Chemical Properties

Because of the hydrogen bonding, water tends to stick firmly to itself, resisting external forces that would break these bonds. This is called **cohesion**. At air-water interfaces, the strength of cohesion forms a "skin" over the water surface strong enough to support small objects. This phenomenon is called **surface tension**. The surface tension of water is the highest of all common liquids and permits the existence of associations of organisms either suspended below it or moving over the top. Cohesion is also responsible for the **viscosity** of water. Viscosity is a property of a material that indicates the force necessary to separate the molecules and allow passage of an object through the liquid. Viscosity of seawater is altered by temperature and salinity. It increases with a decrease in temperature or an increase in salinity. This resistance to flow or movement is important in the sinking rate of objects (pp. 49–54) and in problems of movement in the water by animals (see pp. 100–104).

Another set of properties concerns the effects of heat. Water, in order to evaporate, must absorb more heat in comparison with other liquids of simple molecules. This means that the **heat of vaporization** of water is high—the highest of most common substances. This is a direct result of the strength of the hydrogen bonding between molecules, which must be broken to allow the escape of a molecule. Because of the high heat of vaporization, water evaporates slowly at earth's atmospheric temperatures and absorbs considerable heat, leading to atmospheric cooling. Similarly, this high heat of vaporization means a high boiling point (100°C), with the result that most water is in a liquid phase on earth rather than a gas. Related to this is the **latent heat of fusion**, which is the amount of heat gained per unit mass when a substance changes from a solid to a liquid, or lost when it changes from liquid to solid. Water, again, has the highest value of most common compounds. Melting ice, therefore, takes up large quantities of heat. When ice is formed, large quantities of heat are given off.

The high values for both the heat of vaporization and latent heat of fusion mean that it takes more heat to change the temperature of a given quantity of water than virtually any other common substance. Water also has a high **heat capacity**. This means that water is a strong buffer against both rising and falling temperatures and, therefore, moderates the climate. It also means that the range of temperatures experienced within any body of water is less than that in air.

Water has a peculiar *density-temperature* relationship. Most liquids become more dense as they are cooled. If cooled until they become solid, the solid phase of such liquids is more dense than the liquid phase. This is not true for water. Pure water becomes more dense as it is cooled until it reaches 4°C. Cooling below this temperature decreases the density, and when freezing occurs, there is a marked decrease

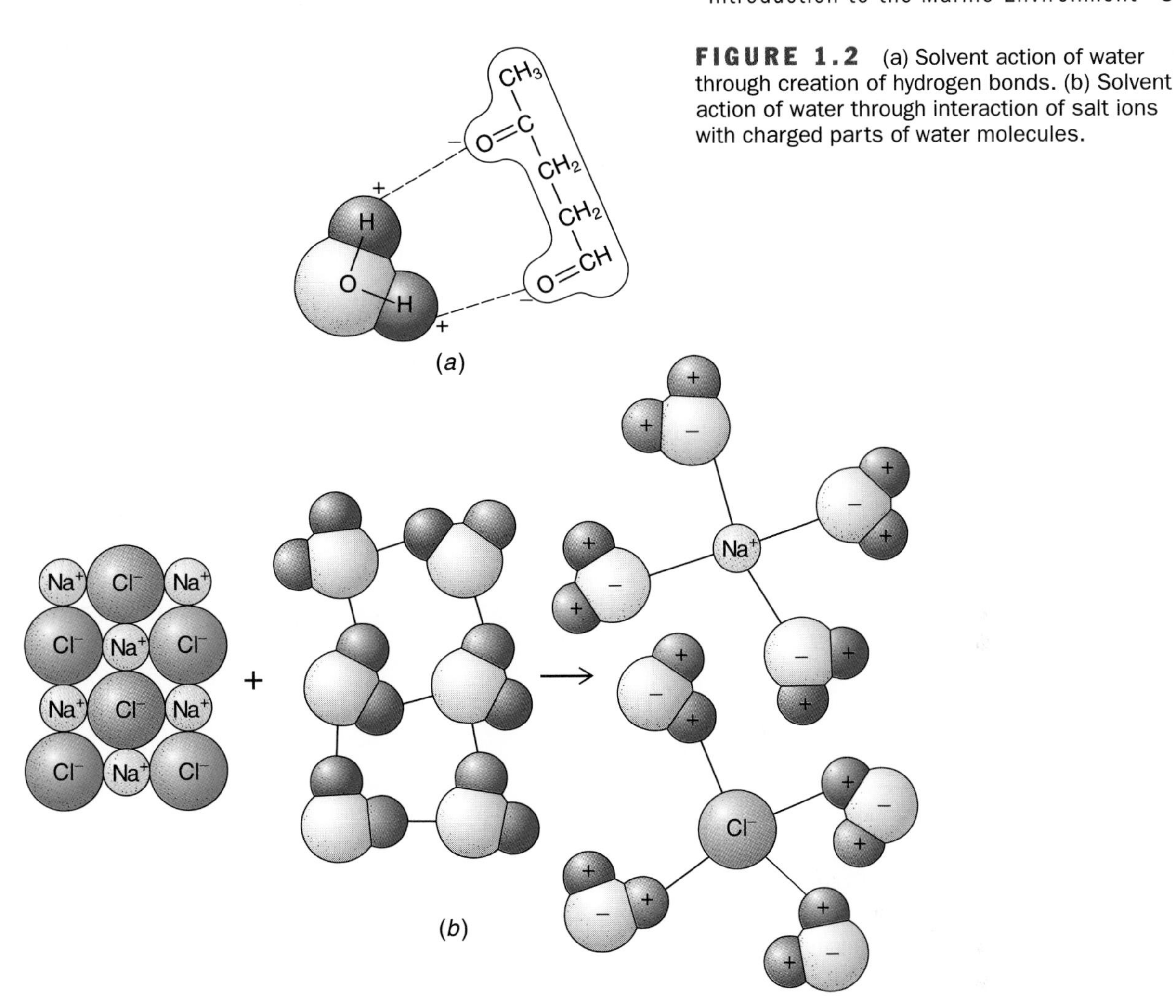

FIGURE 1.2 (a) Solvent action of water through creation of hydrogen bonds. (b) Solvent action of water through interaction of salt ions with charged parts of water molecules.

in density. Therefore, ice is lighter than water and floats upon it. This property is of utmost importance to life in the oceans; otherwise, major volumes of the oceans would be uninhabitable as large blocks of ice. Different water masses in the world's oceans have different densities due to their different temperatures and salinities. Water masses may sink or rise through other water masses depending on their relative densities and may also spread out and flow horizontally. Such behavior is common in the world's oceans, and such density differences are responsible for deep ocean currents.

The chemical properties of water that are of importance are those concerned with the solvent capacities (Figure 1.2). Water is almost a universal solvent, with the ability to dissolve more substances than any other liquid. This is because the solvent action is of two types: one depends on the polar character of the molecule and the other on the hydrogen bonding. Various nonpolar organic and inorganic compounds containing oxygen atoms or hydrogen atoms bonded to either oxygen or nitrogen atoms are held in solution by hydrogen bonding. The polar dissolving action of water for various salts depends on the interaction of the salt ions with the charges on the water molecule. When a salt is dissolved in water, it breaks into its component ions; thus, common table salt, NaCl, when put into water, dissociates into Na^+ and Cl^- ions. It is through interaction of the charges on the salt ions and the water molecules that salts are held in solution.

Still other chemical properties of water depend on the breaking of the strong bonds between the hydrogen and the oxygen in the water molecule. In any volume of water, a few of the water molecules have separated the charges so completely that the molecule breaks into two charged parts, H^+ (hydrogen ion) and OH^- (hydroxyl ion). Here, one hydrogen

TABLE 1.1

Some Properties of Water

Property	*Compared to other substances*
Surface tension	The highest of all common liquids
Conduction of heat	The highest of all common liquids except mercury
Viscosity	Relatively low viscosity for a liquid (decreases with increasing temperature)
Latent heat of vaporization: The quantity of heat gained or lost per unit mass by a substance changing from a liquid to a gas or a gas to a liquid phase without an increase in temperature (cal/g)	The highest of all common substances
Heat capacity: The quantity of heat required to raise the temperature of 1g of a substance 1°C (cal/g/°C)	The highest of all common solids and liquids
Density: Mass per unit volume (g/cm3 or g/ml)	Density determined by (1) temperature, (2) salinity, (3) pressure, in that order; the temperature of maximum density for pure water is 4°C; for seawater the freezing point decreases with increasing salinity
Dissolving ability	Dissolves more substances in greater quantities than any other common liquid

Source: After Ingmanson and Wallace, 1973.

atom moves away, leaving its electron with the bonding electrons on the oxygen atom, making the free hydrogen positive and leaving the OH negative. Certain substances, when they dissolve in water, do so by reacting with these ions. An example is carbon dioxide (CO_2), which reacts as follows:

$$CO_2 + HOH \rightleftharpoons H_2CO_3$$

These properties are all summarized, with some additional ones, in Table 1.1.

Seawater

Seawater is water in which are dissolved a variety of solids and gases. A 1,000-g sample of seawater will contain about 35 g of dissolved compounds, collectively called salts. In other words, 96.5% of seawater is water and 3.5% dissolved substances. The total amount of dissolved material is termed the **salinity**. Marine biologists and oceanographers in the past usually expressed salinity in terms of parts per thousand, abbreviated as ‰, but the new term is **practical salinity units** or **psu**. Thus, if a typical seawater sample has 35 g of dissolved compounds in 1,000 g, it has a salinity of 35 psu (35‰).

Dissolved substances include inorganic salts, organic compounds derived from living organisms, and dissolved gases. The greatest fraction of the dissolved material is composed of inorganic salts present as ions. Six inorganic ions make up 99.28% by weight of the solid matter. They are chloride, sodium, sulfur (as sulfate), magnesium, calcium, and potassium ions (Table 1.2). These can be considered the major ions. Four minor ions add an additional 0.71% by weight so that ten ions together make up 99.99% by weight of the dissolved substances.

The salinity of various parts of the open ocean away from coastal areas varies within a narrow range, usually 34–37 psu, and averages about 35 psu. The differences in salinity are due to differences in evaporation and precipitation. Higher salinities occur in subtropical oceans, where there is a high evaporation rate, and lower values in temperate oceans, where there is less evaporation. Equatorial salinities are low because of precipitation in the intertropical convergence. In inshore areas and partially enclosed seas, the salinity is more variable and may be near 0 psu where large rivers discharge fresh water, to near 40 psu in the Red Sea and Persian Gulf.

Even though variations in salinity from near 0 to 40 psu exist, the ratios among the most abundant ions remain virtually constant. This is important; it means that concentrations of nearly all ions in a given sample of water can be determined by measuring only one. This fundamental relationship is the basis for the measurement of salinity in seawater. Salinity can be established by measuring a single property, which may be the chlorinity (chloride ion concentration), for example, or the electrical conductivity or refractive index, which also depend on the salt content.

Among the remaining 0.01% of dissolved substances in seawater are several inorganic salts that are crucial to marine organisms. Included here are the inorganic nutrients, phosphate and nitrate, required by plants to synthesize organic material in photosynthesis, and silicon dioxide, required by diatoms and radiolarians to construct their skeletons. In contrast to most ions, nitrate (NO_3^-) and phosphate (PO_4^{3-}) do not exist in a constant ratio with other elements or ions and tend to be in short supply in surface waters, varying in abundance as a result of biological activity (see Chapter 2 for discussion). Supplies of these essential nutrients may often become limiting to plant growth in some cases (see pp. 62–66).

Other substances essential to life processes exist as trace amounts and include such elements as iron, manganese, cobalt, and copper. Of particular interest is iron. According to Martin et al. (1994), in some parts of the world's oceans where nitrate and phosphate are abundant, plant production is limited by a lack of trace amounts of iron in the water column (see p. 489). Certain organic compounds, such as vitamins, are also present in minute amounts, but little is known of their variations.

The salt content of seawater has a definite effect on its properties. The maximum density of pure water occurs at 4°C, but in all salt water with a salinity above 24 psu, the density continues to increase to the freezing point. Because of the salt content, the freezing point is also reduced from 0°C, the amount being a function of the salinity. For seawater of 35 psu, the freezing point is −1.9°C. Upon freezing, however, the salts are excluded and the density decreases; thus, ice floats on the surface. The importance of this density increase is that very cold and very dense surface water can be formed and sink to the bottom of the ocean basins, where it forms the environment for deep-sea organisms and, more importantly, creates a fairly vigorous circulation that aerates the deep ocean.

TABLE 1.2

Major and Minor Constituents of 34.8 psu Seawater

Ion	*Percent by Weight*
A. Major	
Chloride (Cl^-)	55.04
Sodium (Na^+)	30.61
Sulfate (SO_4^{2-})	7.68
Magnesium (Mg^{2+})	3.69
Calcium (Ca^{2+})	1.16
Potassium (K^+)	1.10
Subtotal	99.28
B. Minor	
Bicarbonate (HCO_3^-)	0.41
Bromide (Br^-)	0.19
Boric acid (H_3BO_3)	0.07
Strontium (Sr^{2+})	0.04
Subtotal	0.71
Total	99.99

Source: From Deep *Sea Biology: Developments and Perspectives* by N. B. Marshall, © 1980 Garland STPM Press.

Two gases dissolved in seawater are of metabolic importance: oxygen and carbon dioxide. The solubility of gases in seawater is a function of temperature; the lower the temperature, the greater the solubility. Therefore, the colder the water, the more oxygen it can hold. Even so, the solubility of gases in water is not great. At 0°C, 35 psu seawater contains about 8 ml of O_2 per liter, whereas air contains 210 ml/liter. At 20°C, 35 psu seawater contains only 5.4 ml/liter. The reason that deep ocean water does not become anoxic (devoid of oxygen) through biological activity as it flows across the ocean floor is that, when water that is to become the deep ocean water sinks from the surface, it is so cold that it has a maximum amount of oxygen, more than can be consumed by the limited populations of animals in deep water. Oxygen is not distributed uniformly with depth in the ocean. A typical vertical profile of the oxygen content shows a maximum amount in the upper 10–20 m, where photosynthetic activity by plants and diffusion from the atmosphere often leads to supersaturation (Figure 1.3). With increasing depth, the oxygen content declines, reaching a minimum value somewhere between 200 and 1,000 m in open ocean waters. This **oxygen minimum zone** may have oxygen values that approach zero in some areas. Below this zone, oxygen values increase somewhat with depth, but they usually do not reach surface values (Figure 1.3). The occurrence of the oxygen minimum zone is usually attributed to biological activity depleting the oxygen,

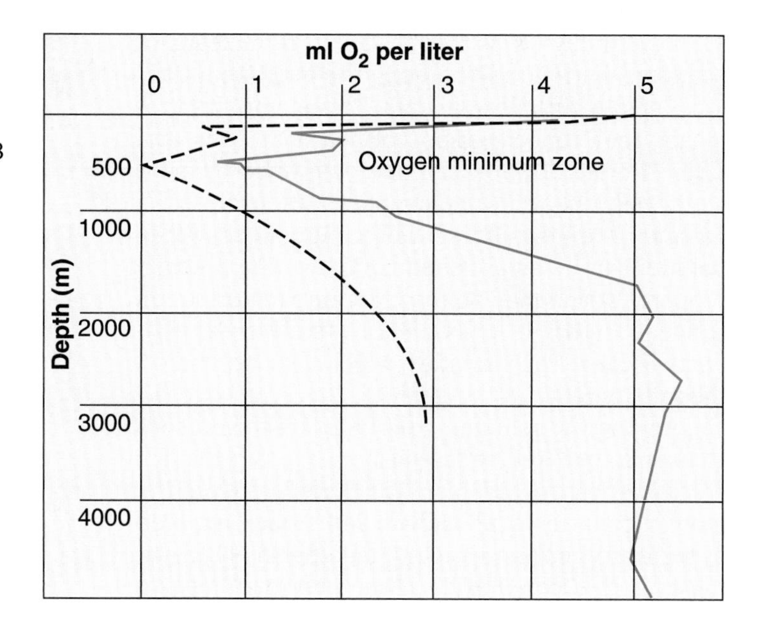

FIGURE 1.3 Change in dissolved oxygen with depth in the eastern tropical Pacific Ocean (dashed line) and the tropical Atlantic Ocean (solid line). (Modified from *Oceanology: An Introduction*, D. E. Ingmanson & W.J. Wallace, © 1973 Wadsworth Publishing Company.)

coupled with the absence of photosynthetic activity and contact with the atmosphere to permit renewal. The oxygen increase below the oxygen minimum zone results from the influx of the cold, oxygen-rich waters that originally sank at high latitudes.

The solubility of carbon dioxide is somewhat different from that of oxygen, since it reacts chemically in the water. Carbon dioxide is abundant in seawater, which has a considerable capacity to absorb the gas. This is because carbon dioxide, upon entering seawater, reacts with the water to produce carbonic acid, as follows:

$$CO_2 + HOH \rightleftharpoons H_2CO_3$$

Carbonic acid further dissociates into a hydrogen ion and a bicarbonate ion:

$$H_2CO_3 \rightleftharpoons H^+ + HCO_3^-$$

Bicarbonate may further dissociate into another hydrogen ion and a carbonate ion:

$$HCO_3^- \rightleftharpoons H^+ + CO_3^{2-}$$

The major reservoir of carbon dioxide in the ocean is the bicarbonate ion, as can be seen from its place as one of the most abundant minor ions in seawater (see Table 1.2). In contrast to oxygen, it is more abundant in seawater than in air. Therefore, availability of carbon dioxide is not limiting to plant growth in seawater.

The carbon dioxide–carbonic acid–bicarbonate system is a complex chemical system that tends to stay in equilibrium. If CO_2 gas is removed from seawater, the equilibrium will be disturbed, and carbonic acid and bicarbonate will shift to the left in the above equations, until more CO_2 is produced and a new equilibrium is set up.

The reactions shown result in the production or absorption of free hydrogen ions (H^+). The abundance of hydrogen ions in solution is a measure of **acidity**. More H^+ ions mean a more acid solution, and fewer H^+ ions mean a more basic or alkaline solution. **Alkaline** (or basic) solutions are those that have large numbers of OH^- ions and few H^+ ions. Acidity and alkalinity are measured on a logarithmic scale of 1–14 pH units. The higher the concentration of H^+ ions on this scale, the lower the pH value; hence, low pH values indicate acid conditions. Conversely, high pH values indicate low H^+ concentrations and high OH^- concentration. The neutral point is pH 7, where equal numbers of both ions occur (Figure 1.4).

Although pure water is neutral in pH because dissociation of the water molecule produces equal numbers of H^+ and OH^- ions, the presence of CO_2 and the strongly alkaline ions—sodium, potassium, and calcium—in ocean waters tends to change this. Seawater is slightly alkaline, usually in the pH range of 7.5–8.4. At the pH of ocean water, the carbon dioxide–carbonic acid–bicarbonate system functions as a

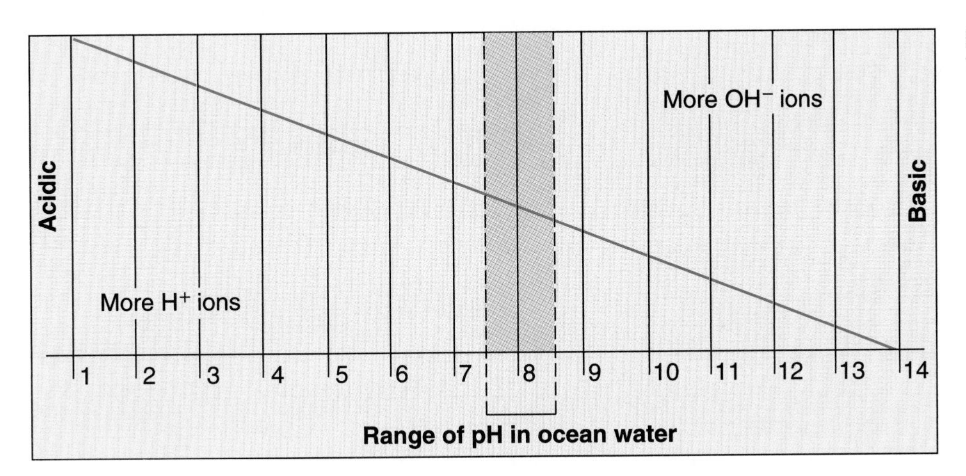

FIGURE 1.4
The pH scale.

buffer to keep the pH of seawater within a narrow range. It does this by absorbing H^+ ions in the water when they are in excess and producing more when they are in short supply. This is accomplished by shifting the reactions shown previously to the right when there are too few H^+ ions, producing more bicarbonate ions and carbonate ions, and shifting the reaction to the left when there are too many H^+ ions, producing more undissociated carbonic acid, carbon dioxide, and bicarbonate ions.

BASIC OCEANOGRAPHY

This is not a book on oceanography, which is the study of all aspects of the physics, chemistry, geology, and biology of the sea. We are concerned only with giving a basic understanding of life in the sea and how it is organized and persists. To understand the ecology of various marine associations, it is necessary to know something about the structure and motion of the ocean water masses.

Geography and Geomorphology of the Oceans

The three major oceans are connected to each other, converging in the area around the Antarctic continent. The world ocean has been separated for convenience into four major divisions—Pacific, Atlantic, Indian, and Arctic—in order of decreasing size. Projecting from, or partially cut off from, these larger oceans are smaller marginal seas, such as the Mediterranean, Caribbean, Baltic, Bering, South China, and Okhotsk. The locations of the major oceans and seas are given in Figure 1.5.

The oceans are not equally distributed over the earth. Oceans cover more than 80% of the Southern Hemisphere but only 61% of the Northern Hemisphere, where most of the earth's landmasses occur.

On the margins of the major landmasses, the ocean is very shallow, overlying an underwater extension of the continent called the **continental shelf**. Forming only 7–8% of the total ocean area, the continental shelf slopes gently from shore to depths of 100–200 m. The shelf extends offshore for up to 400 km off eastern Canada but extends only a few kilometers offshore along most of the Pacific coast of North America. At the outer edge of the shelf, there is an abrupt steepening of the bottom to become the **continental slope**. The continental slope descends precipitously to depths of 3–5 km. At these depths, the bottom becomes the flat, extensive, sediment-covered **abyssal plain**. Such plains monotonously cover the floors of vast areas of the oceans at depths between 3–5 km (Figure 1.6).

Abyssal plains are broken in several places by various submarine ridges. This mid-ocean ridge system is an extensive, contiguous chain that has been found in all oceans. The best known is the mid-Atlantic ridge, which bisects the Atlantic Ocean into east and west basins and runs from Iceland into the South Atlantic, where it links with a similar ridge in the Indian Ocean. Occasionally, the ridges break the surface to form islands. The Azores, Ascension, and Tristan da Cunha are islands formed by the mid-Atlantic ridge. These extensive ridge systems mark the boundaries of the various crustal plates of the earth and are sites of volcanic activity.

In certain areas, the abyssal plains are cut by deep, narrow troughs called **trenches**. Most of these trenches lie in an arc bordering the islands

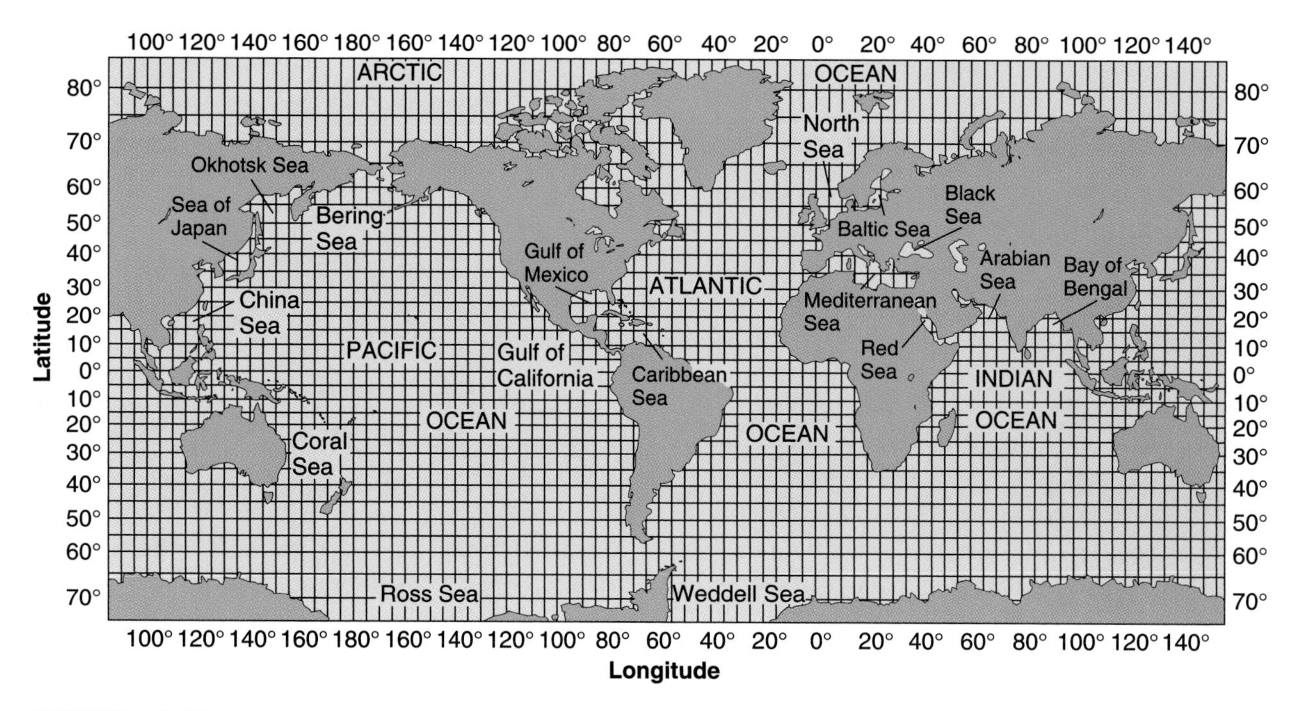

FIGURE 1.5 Major oceans and seas of the world.

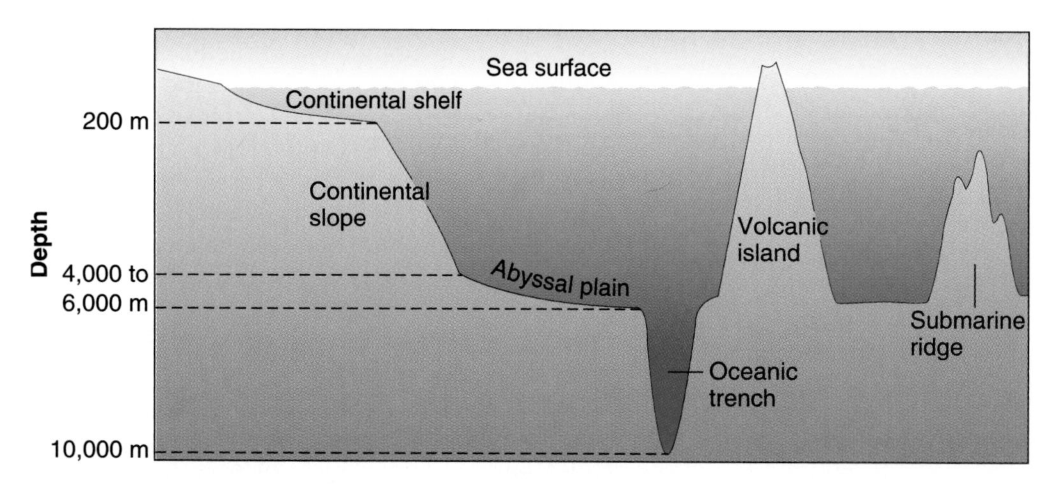

FIGURE 1.6 Diagrammatic cross section of an ocean basin, showing the various geographic features. (Not to scale; large vertical exaggeration.)

and continents in the Pacific Ocean. The trenches have depths from 7,000 to more than 11,000 m. The deepest area known is the 11,022-m Challenger Deep in the Marianas Trench.

Finally, there may be isolated islands and submarine **seamounts** formed by isolated volcanic action. Such mountains, as opposed to the ridges, rise individually from the abyssal plain.

Plate Tectonics

In the 1960s, when new evidence became available about the ocean floor, it became apparent to scientists that the earth's crust was divided into a number of large units called **plates** and that these plates were bounded by the aforementioned ridge and trench systems. The sizes of the plates vary, but most of the planet is covered by seven major plates (Figure 1.7). The plates

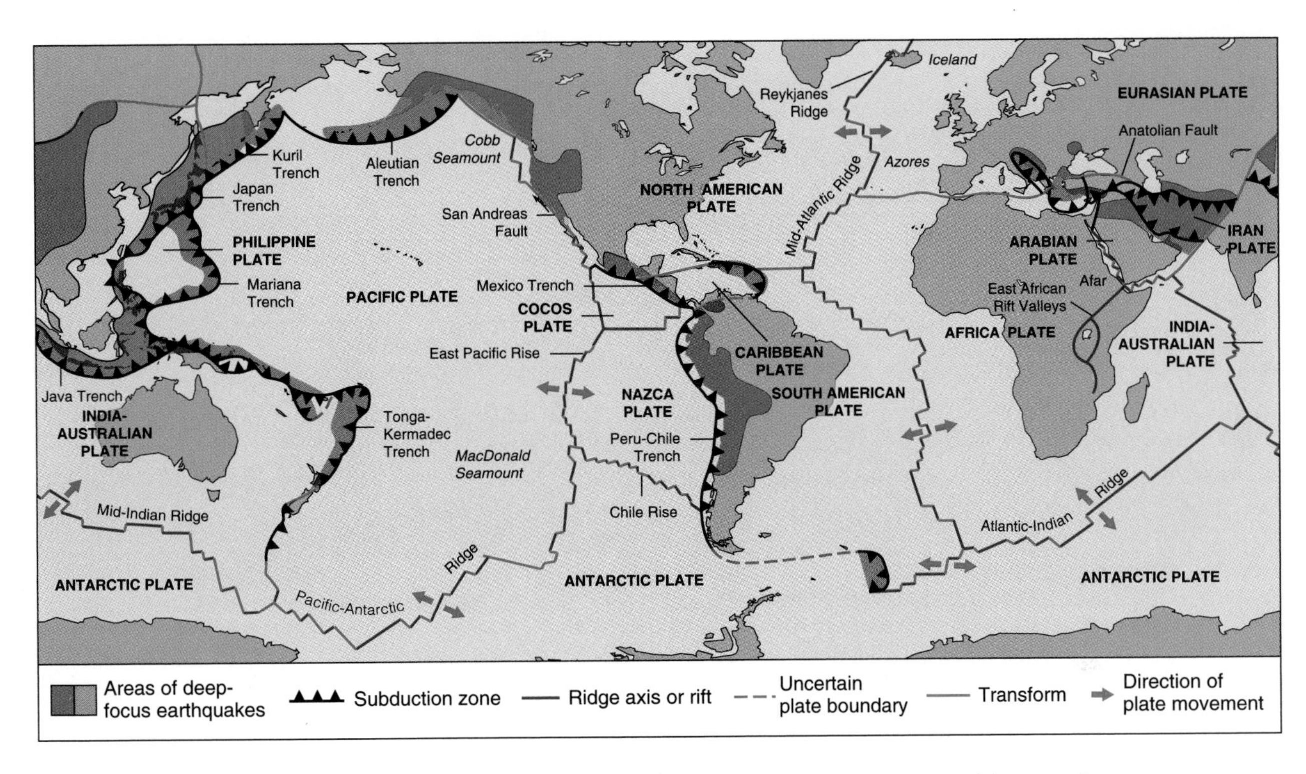

FIGURE 1.7 Location and boundaries of the earth's plate boundaries, mid-ocean ridges, and trenches. Arrows indicate direction of plate movement. (From *Scientific American,* Vol. 226, No. 5, May 1972. Reproduced with permission from the estate of Mary and Dan Todd.)

themselves are rigid and float on the underlying mantle. The major plates all encompass both continental and oceanic crust. The plates are not static but move slowly relative to each other. The movement of these plates over geologic time is responsible for the differing positions of the continents through time, a phenomenon called **continental drift**.

The plates move because the oceanic ridges are centers of volcanic activity where new crustal material is formed and added to the crust. As this happens the plates move laterally, in opposite directions, causing **seafloor spreading**. The opposite of this spreading phenomenon occurs in the trench systems, where the margin of one plate dives beneath the margin of the adjoining plate, a process called **subduction**. Here the subducting crust is melted into the mantle layer again (Figure 1.8). Both the oceanic ridge system and the subduction zones of the trenches are the sites of volcanic and seismic activity.

The plate boundaries, particularly the oceanic ridge systems, are significant in marine ecology because they are weak spots in the earth's crust where the mantle pushes up, producing cracks and fissures. Water seeps into these and is heated to very high temperatures by the mantle material. The heated water subsequently emerges again, carrying with it many reduced chemical compounds in what are called **hydrothermal vents**. These hydrothermal vents are the sites of unique communities of deep-sea organisms that were first discovered in 1977 and are discussed in detail in Chapter 4.

Temperature and Vertical Stratification

Temperature is a measure of the energy of molecular motion. In the world oceans it varies, moving north to south with changes in latitude and vertically with depth. Temperature is a singularly important factor governing the life processes and the distribution of organisms. Vital life processes, collectively termed **metabolism**, function only within a relatively narrow range of temperatures, usually between 0 and 40°C. However, a few organisms can tolerate temperatures somewhat above and below these limits. These include the cyanobacteria living in 85°C hot springs. Within the temperature range in which life processes operate, metabolism is temperature dependent. In general, for organisms that do not regulate their internal temperatures, metabolic processes are increased by a factor of 2 for each 10°C rise in tem-

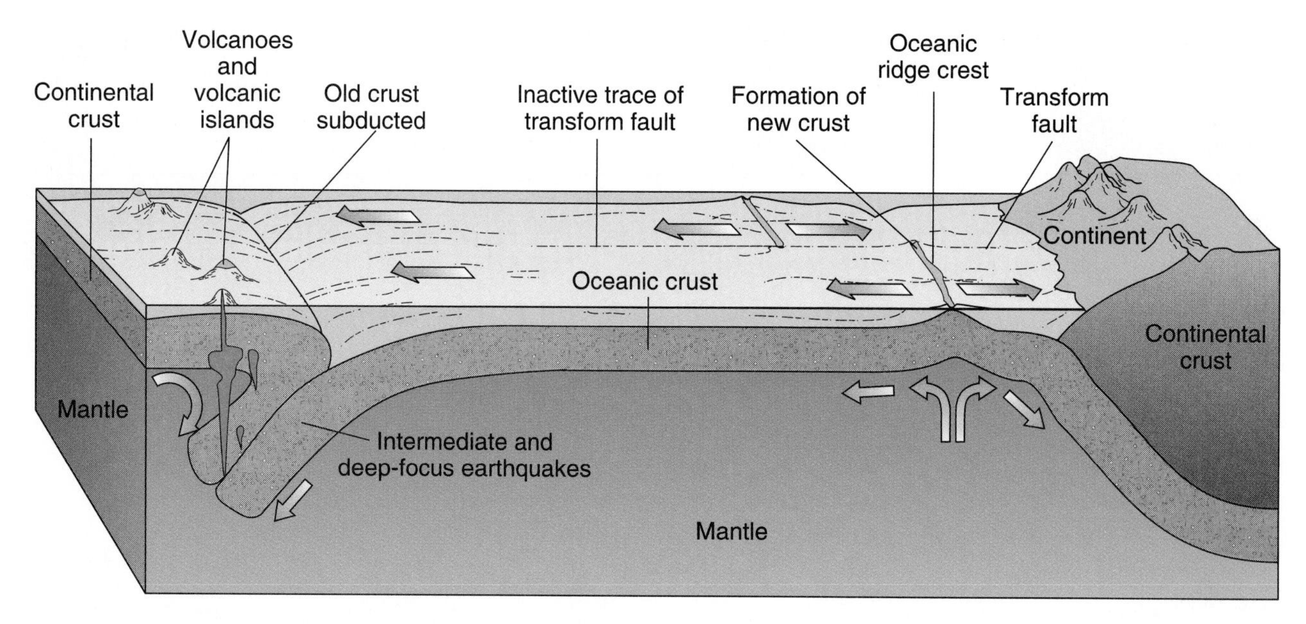

FIGURE 1.8 Diagrammatic representation of the ocean and continental crusts, showing the formation of new crust at the oceanic ridges and the destruction of the oceanic crust in the subduction zones. (Modified from *Marine Biology*, Levinton, Oxford University Press)

perature. With the exception of the marine birds and mammals, and some fishes, marine organisms are **poikilothermic** or **ectothermic**, meaning that their body temperatures vary with that of the surrounding water mass. **Homeothermic** or **endothermic** birds and mammals have the ability to regulate their own internal temperature, regardless of the temperature of the water mass. Since most marine organisms are poikilothermic and since the seawater temperatures vary latitudinally, the distribution of marine organisms follows closely the geographical differences in ocean temperatures.

On the basis of surface ocean temperatures and overall organism distribution, four major biogeographical zones can be established: **polar, cold temperate, warm temperate (subtropical)**, and **tropical (equatorial)** (Figure 1.9). Transition zones between these areas also exist. Boundaries may vary somewhat with season, so these zones are not absolute.

Temperature in the oceans also has a marked variation with depth. Surface waters in the tropical regions are warm, 20–30°C, throughout the year, and temperate zone surface waters are warm in the summer.

Below the warm surface water, the temperature begins to fall, and over a narrow depth range of 50–300 m it undergoes a very rapid decline. The depth zone of most rapid temperature decline is the **thermocline**. Below the thermocline the temperature continues to decrease with depth, but at a much slower rate, so that the water mass below the thermocline is nearly isothermal all the way to the bottom. The thermocline is a persistent feature of tropical waters and occurs in temperate waters in the summer months. Thermoclines are absent in polar waters. In winter, thermoclines extend through a more narrow depth range over the relatively shallow waters of the continental shelf than those formed in the open ocean. The importance of this zone is discussed in Chapter 2.

Temperature also has an effect on the density of seawater. Warm seawater is less dense than cold seawater of the same salinity. Density is also a function of salinity, increasing salinity causing increasing density. However, the range of temperatures found throughout the world's oceans is greater than the range of salinities. Temperature is, therefore, more important in affecting the density. The rapid change in temperature that produces the thermocline also means the density of seawater changes rapidly over the same depth range. Such a zone of rapid density change is called a **pycnocline**.

Water Masses and Circulation

As a result of different temperatures and salinities and their effects on density, the seawater of the world's oceans can be separated into different water

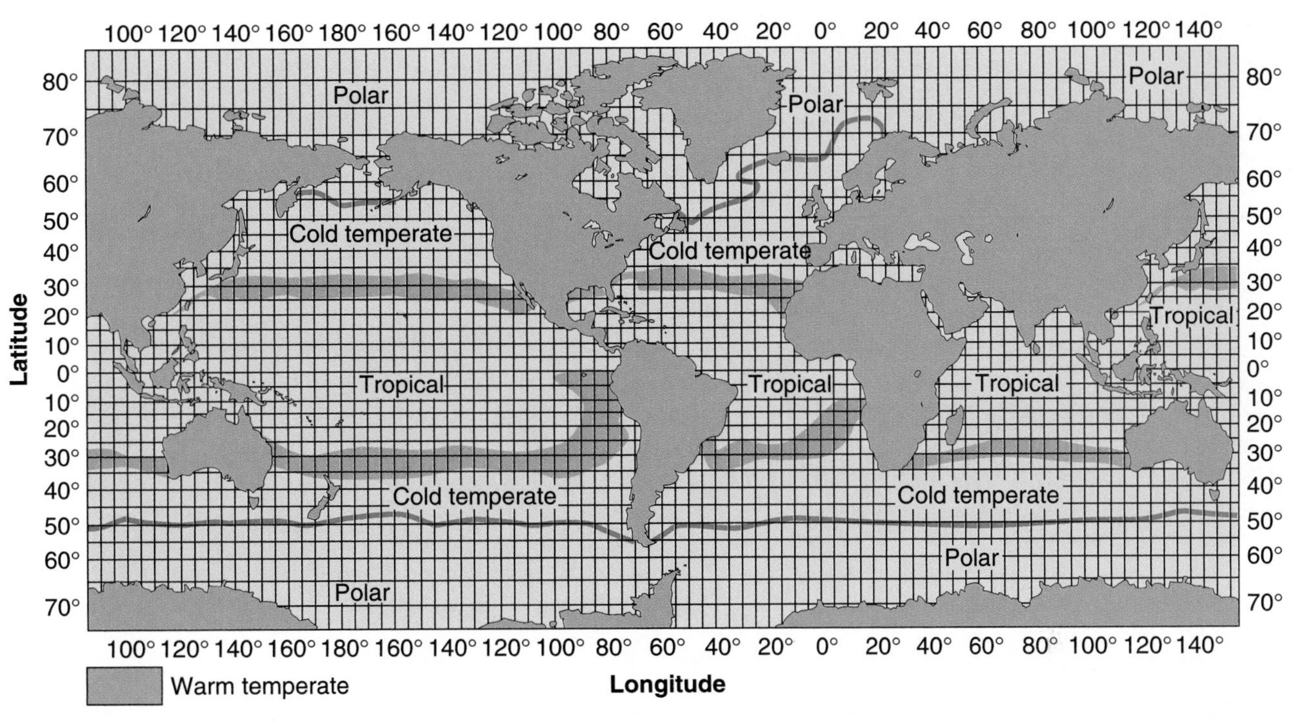

FIGURE 1.9 Major biogeographical regions of the world's oceans, based on temperature.

masses. The **upper water mass** (surface water mass) of the oceans includes all well-mixed water above the thermocline. Below the thermocline are several **deep water masses** extending to the bottom.

The upper water mass of the oceans is in constant motion. The motion is induced primarily by the action of winds blowing across the surface of the water. These winds produce two kinds of motion, **waves** and **currents**. Waves range in size from ripples only a few centimeters in height to storm waves, which may tower as high as 30 m. Other than height, waves are further characterized by **wavelength**, which is the horizontal distance between the tops or crests of successive waves (Figure 1.10). The **period** of a wave is the time required for two successive wave crests to pass a fixed point. In addition to wind, waves may be generated by earthquakes, volcanic explosions, and underwater landslides, which create the destructive waves known as **tsunamis**. The attraction of the moon and sun produces the waves known as **tides** (see Chapter 6).

Wind wave height in the open ocean is dependent on wind **speed**, the distance or **fetch** over which the wind blows, and the **duration** that the wind blows.

All waves behave similarly. Once generated, the waves move outward and away from the center of origin. This repeated horizontal progression results in very little horizontal transport of the individual water molecules. The water molecules transcribe a circle, moving upward and to one side as a wave crest approaches and to the other side and down to near their original position as the wave crest passes by. While the wave form and the energy are transported horizontally, the water remains nearly stationary (Figure 1.10). The passage of waves generates movement, not only in the surface water molecules, but also in water down to a depth approximately equal to one-half the wavelength. With each depth interval below the surface equal to one-ninth the wavelength, the orbits followed by the water particles diminish by one-half. By the time a depth of half a wavelength is reached, the movement is almost imperceptible.

As waves enter shallow water and begin to encounter the bottom effects, they slow their forward motion and the wavelength decreases. As a result, they begin to increase in height and become steeper. At a point where the water depth is 1.3 times the height of the wave, it will "break," releasing the energy onto the shore. Waves are of greatest importance biologically in shallow water and the surf zone, as discussed in Chapter 6.

Currents are water movements that result in the horizontal transport of water. The major ocean current systems are produced by a few major wind belts that

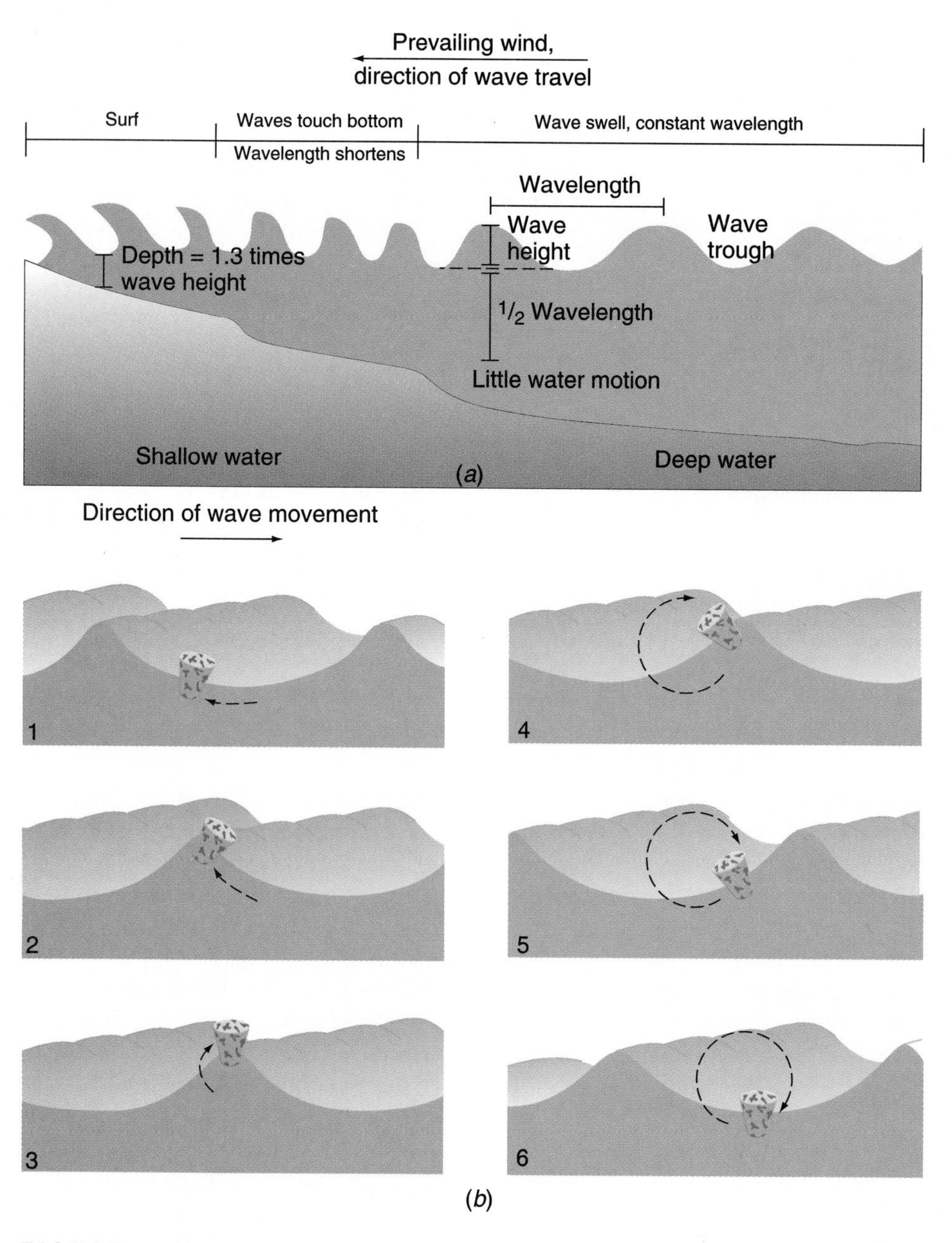

FIGURE 1.10 (a) Features of waves and change of wave form as a wave enters shallow water. (b) Cork floating in the water demonstrates the passage of a wave, indicating that the water itself does not move. As the wave passes, the cork transcribes an imaginary circle (dashed line).

succeed each other latitudinally around the world and where the winds are steady and persistent in direction. These winds are caused by differential heating of the atmospheric air masses aided by the **Coriolis effect**. The backbone of the system are the northeast trade winds blowing from northeast to southwest between the equator and 30°N latitude and the southeast trades in similar position south of the equator, moving air from southeast to northwest. Between 30° and 60°N and S latitudes, the westerlies blow from the southwest to the northeast in the Northern Hemisphere and to the southeast in the Southern Hemisphere. In the polar regions, cold air flows toward the equator as the polar easterlies (Figure 1.11).

These winds set the surface waters into motion, producing the slow, horizontally moving currents that

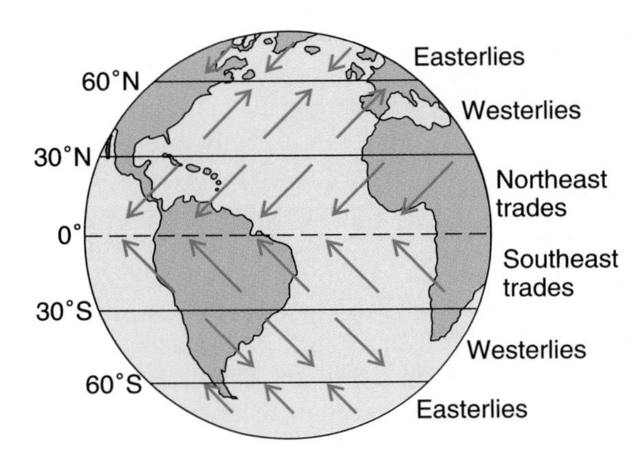

FIGURE 1.11 The main wind belts of the earth and their prevailing direction of motion (arrows).

can transport huge volumes of water across vast distances in the oceans. Such currents influence the distribution of marine organisms and also lead to the displacement of biogeographical zones through the transport of warm water into colder regions, and vice versa. The major currents of the world's oceans are shown in Figure 1.12.

A comparison of Figures 1.11 and 1.12 will show that the ocean currents do not flow parallel to wind direction. They are deflected into roughly circular **gyres** that move clockwise in the Northern Hemisphere and counterclockwise in the Southern Hemisphere. The deflections and gyres are the result of the Coriolis effect. The Coriolis effect is the result of the rotation of the earth on its axis. The spinning of the planet imparts a deflection to moving water, displacing it to the right in the Northern Hemisphere and to the left in the Southern Hemisphere. Because the rotation of the earth is from west to east and because the deflection of the currents created by the trade winds imparts a water movement at the equator parallel to the equator moving from east to west, the net water movement is from east to west, piling up water on the western side of ocean basins. As water builds up on the western sides, it meets the continental or island chain landmasses and is deflected north or south as continental boundary currents. These boundary currents, in turn, moving poleward, fall under the influence of the westerly winds. The westerlies impart more energy to the currents and drive them in an easterly direction, eventually crossing the ocean basins to return water to the eastern side of the basin. Continental landmasses on the eastern side deflect the moving water toward the equator, which completes the pattern. These huge circular current patterns, called gyres, are found in all major ocean basins (see Figure 1.12).

These global currents are surface currents. How deeply do they affect the water column? This can be

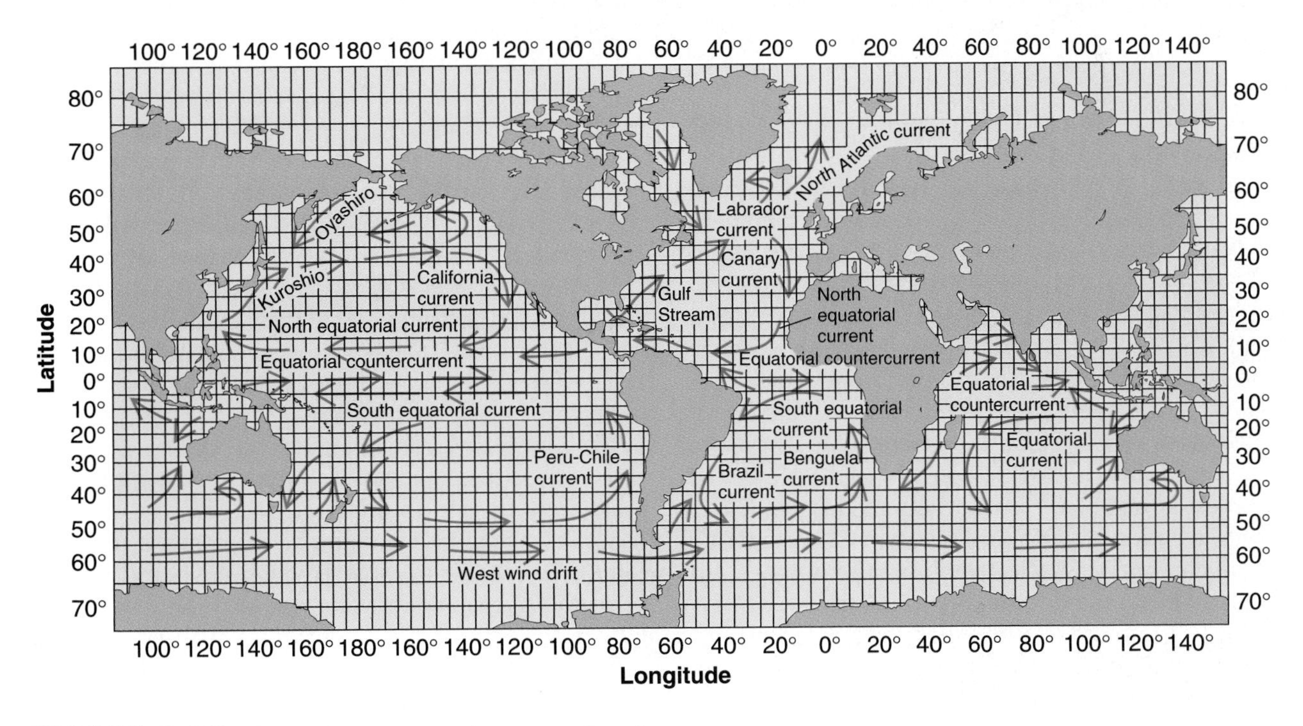

FIGURE 1.12 The major ocean surface current systems.

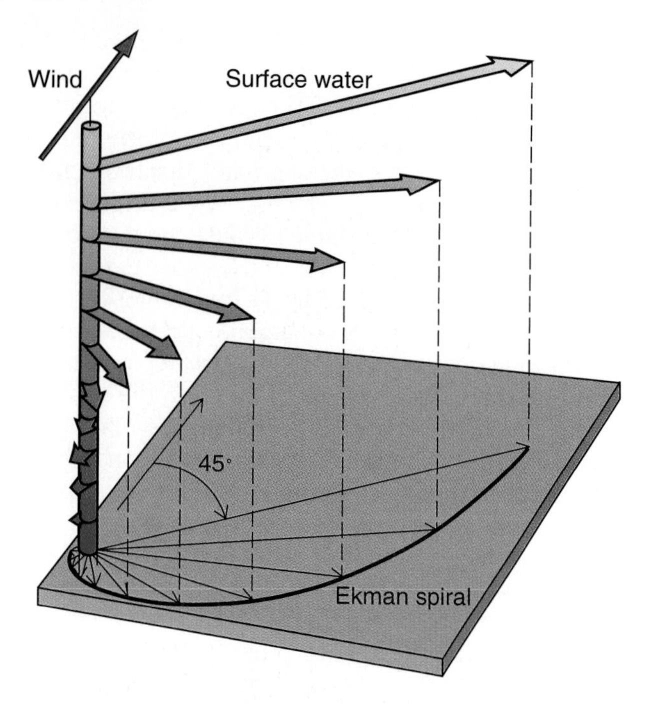

FIGURE 1.13 The Ekman spiral. The wind-driven surface current moves at an angle of 45° to the direction of the wind, to the right in the Northern Hemisphere, to the left in the Southern Hemisphere. Successively deeper water layers are deflected even further with respect to those immediately above them and move at slower speeds. Net water movement is at 90° to the wind. (Modified from *Oceanography for Meteorologists* by Sverdrop.)

estimated by noting what happens to deeper layers of water when wind stress causes the surface water to move. As previously noted, because of the Coriolis effect, the direction of the water movement is deflected from that of the initiating wind. The energy of the wind is passed down through the water column, and it sets each successively deeper water layer into motion. Each layer, however, receives a decreasing amount of energy, and therefore its velocity is lower. At the same time, due to the Coriolis effect, each layer set in motion is deflected with respect to the one immediately above it. The result is the **Ekman spiral** of current directions and velocities from the surface downward (Figure 1.13). The depth at which wind stress fails to impart motion varies, but it is roughly limited to the upper 100 m.

In certain areas and under certain conditions, the wind-induced lateral movements of the water may also bring about a vertical circulation or upwelling of water. Along the eastern margins of ocean basins, for example, the wind-driven surface currents along the continental margins flow toward the equator. At the same time, the Coriolis effect tends to push these surface waters offshore. This water is then replaced by deeper water transported vertically along sloping surfaces to the surface. Similarly, along the equator the two equatorial currents flowing west are deflected away to the right north of the equator and left to the south. To replace this water, subsurface water upwells to the surface (Figure 1.14). Such wind-induced upwelling is limited to the upper 200 m of the water column.

Movement in the deep water mass is quite different from that in the surface mass. The deep water mass is isolated from the wind; therefore, its motion cannot be dependent on it. Movement in the deep water mass does, however, result from changes occurring in water at the surface. Seawater increases in density with a decrease in temperature and an increase in salinity. When seawater increases in density, it sinks. Therefore, to move water into the deep basin of the oceans, it is necessary to increase its density at the surface. This is accomplished in two ways. Warm water from the tropics or subtropics is high in salinity due to evaporation. In the northern Atlantic Ocean, such warm saline water is transported out of the tropics by the Gulf Stream. In the region of Iceland and Greenland, it meets the very cold waters of the Labrador Current moving south. Cooling at the surface increases the density of this highly saline water, and it sinks to form North Atlantic deep water. Similarly, warm water moving south in Antarctic seas loses heat to the atmosphere, which causes the sinking of water masses. A final, very high density water mass is produced in the Weddell Sea in Antarctica, where very cold water becomes more saline when winter freezing occurs. This mass sinks to become the bottom water of most ocean basins.

Since these dense waters are all cold and produced at the surface, they contain large amounts of oxygen, which is then transported to the depths. Without this oxygen, deep water would be anoxic.

These deep and bottom water masses move very slowly north and south to form the deep water of all ocean basins. Many hundreds of years are required for these waters to move through the ocean basins. North Atlantic deep water, for example, may have been away from surface contact for hundreds of years from the time it sank near Iceland until the time it again surfaced in the Antarctic region. It is, however,

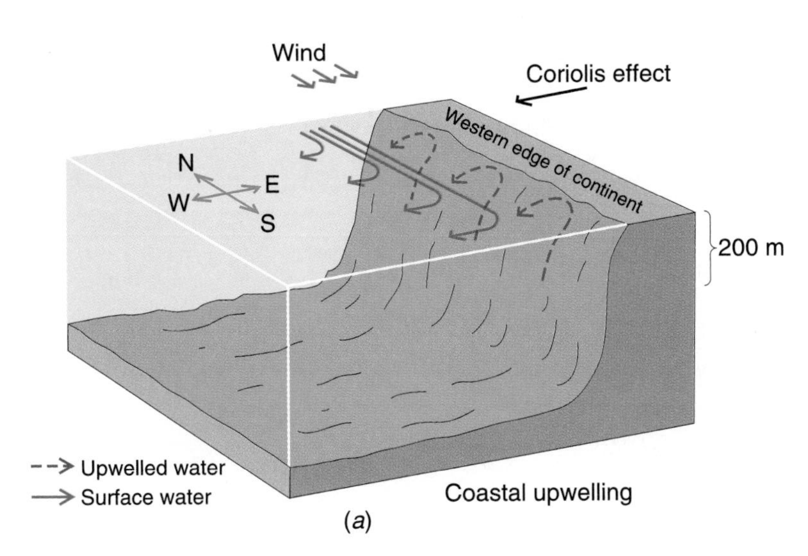

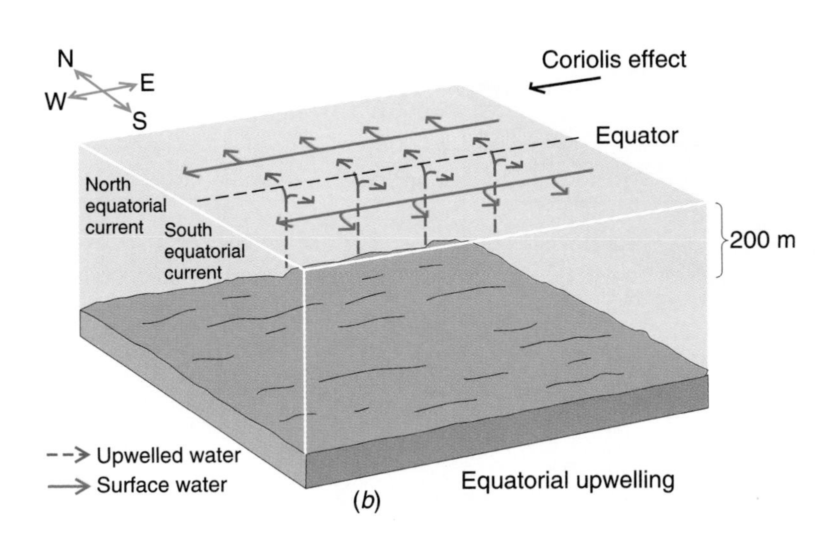

FIGURE 1.14 (a) Coastal upwelling in the Northern Hemisphere. Along the western margins of continents, the wind and Coriolis effect act to move water offshore as indicated by the blue arrows. This water is replaced by water moving up from depths of about 200 meters (dashed arrows). (b) Equatorial upwelling. Along the equator, the Coriolis effect acts on the westward-flowing currents, pulling water north in the Northern Hemisphere and south in the Southern Hemisphere (blue arrows). This is replaced by cool water moving up from the 200-meter depths (dashed arrows).

a major contributor to the productivity of Antarctic seas, because when it surfaces, it brings with it large amounts of inorganic nutrients (NO_3^-, PO_4^{3-}) accumulated during the many years beneath the photosynthetic zone. A general outline of these deep water masses and movement is given in Figure 1.15.

SOME ECOLOGICAL PRINCIPLES

Because this is a text that emphasizes habitats and the ecology of marine organisms and because certain users of this book may lack a background in basic ecology, it is necessary at this point to cover briefly some basic ecological concepts and terms.

Terms and Definitions

Ecology is the science that treats the spectrum of interrelationships existing between organisms and their environments and among groups of organisms. It is important to realize initially that living things do not exist as isolated individuals or groups of individuals. All organisms interact with their own species, other species, and the physical and chemical environments that surround them. In this interactive process, the organisms have effects on each other

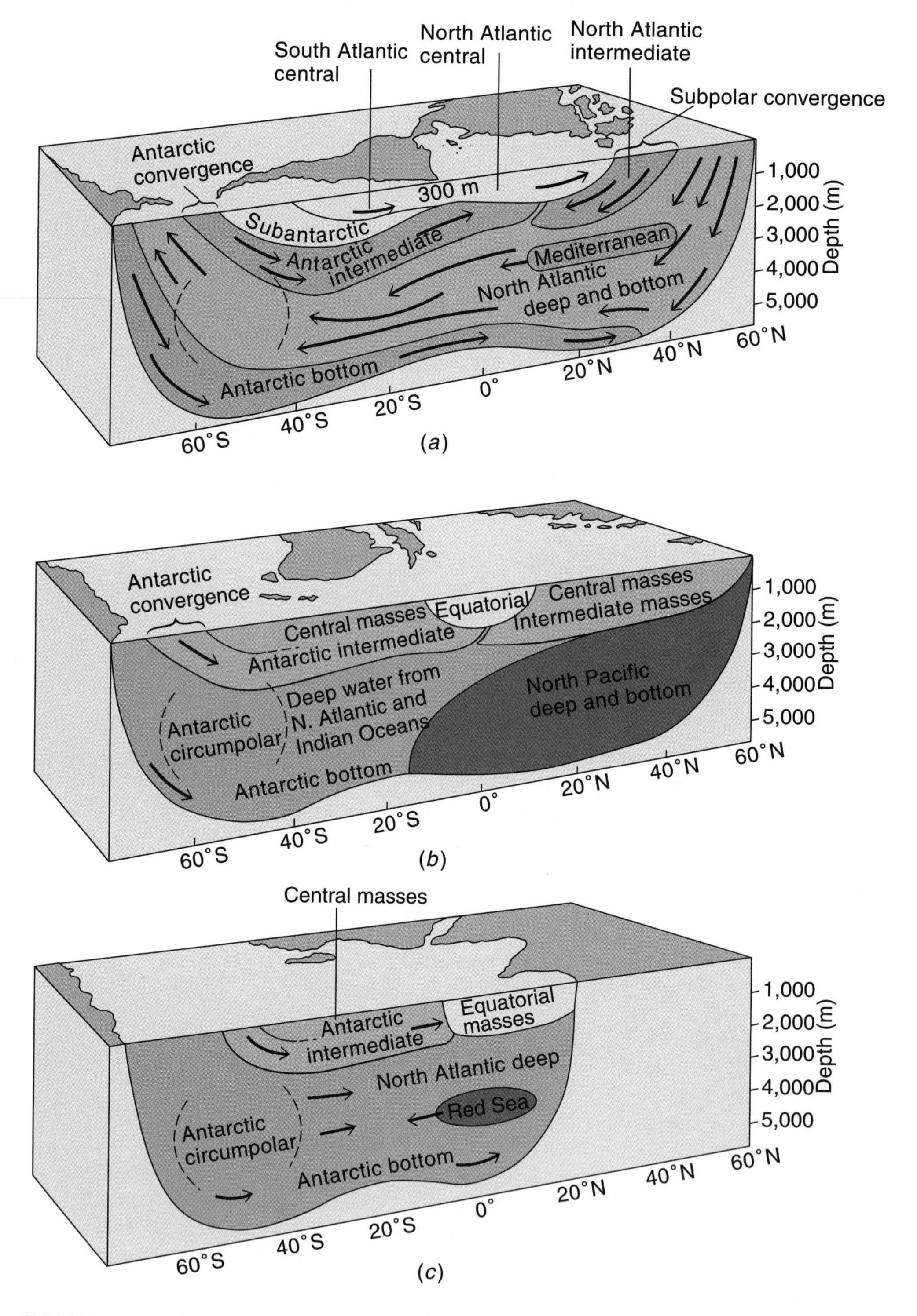

FIGURE 1.15 Subsurface water masses and circulation patterns in the three major oceans. (a) Atlantic Ocean. (b) Pacific Ocean. (c) Indian Ocean. (From *Oceanology: An Introduction*, D. E. Ingmanson & W.J. Wallace, © 1973 Wadsworth Publishing Company.)

and on the surrounding environment. Similarly, the different factors of the environment affect the activities of the organisms. The various organisms and the environmental factors may be further organized into various levels, each of which is somewhat broader than the preceding. A *species* is a natural group of actually or potentially interbreeding individuals reproductively isolated from other such groups. All the individuals of a given species in an area constitute a *population*. Several different species populations that tend to occur together in a particular geographical area constitute an ecological *community*. A community or a series of communities and the surrounding physical and chemical environment constitute an *ecosystem*. Ecosystems are the most complex entities and have many interactive components. They are so large and complicated that ecologists tend to study them by concentrating on their component parts, such as various communities or populations. Ecosystems, however, may be considered on a large or small scale, depending on the number of communities included and the dimensions of the surrounding nonliving environment. On the largest scale, it is possible to consider the earth as a single ecosystem comprising all the various terrestrial, freshwater, and marine communities. At the other end of the scale, it is possible to consider a tide pool or freshwater pond as an ecosystem. Both have the biological and nonbiological components the definition requires.

The oceans of the world may also be considered a single ecosystem in which a series of communities is influenced by and, in turn, influences the physical and chemical factors of the surrounding seawater. This large ecosystem may be divided into smaller sections in which the physical and chemical parameters have differential effects on the populations of organisms, changing the composition and adaptations of organisms within that area. For example, the physical factors acting on a rocky seashore, such as waves and movements of the bottom, are not those that affect small organisms floating in the open ocean. Both areas, therefore, contain different organisms and different adaptations. It is the aim of this book to give an understanding of the functioning of the total marine ecosystem by considering the functioning of these major subsections that form the basis of the subsequent chapters.

Ecosystem Components

An ecosystem is a functional unit of variable size composed of living and nonliving parts that interact. The component parts and the whole system function through a sequence of operations involving energy and the transfer of energy. With few exceptions, the original energy source is the sun. Energy from the sun is captured by the **autotrophic** component, the green plants and algae. Energy captured is stored in the chemical bonds of organic materials in the plants, which are the "food" that drives the **heterotrophic** component of the system. Heterotrophic organisms include all other life forms that obtain their energy through consumption of the autotrophic organisms, through consumption of organisms that have ingested the autotrophs, or through the absorption of dissolved organic matter from the environment. Such an arrangement of autotrophs and succeeding levels of heterotrophs is called a **trophic structure**, in which each successive level is called a **trophic level.** Trophic structure is a characteristic feature of all ecosystems. The first trophic level is the autotrophic or producer level, where the energy is initially captured and stored in organic compounds. As the energy is passed from level to level in such a system, the majority of it is lost through heat and metabolic use by the organisms. The amount lost at each transfer is variable, but substantial, usually ranging between 80 and 95%. The system thus becomes self-limiting in that, at a certain point, not enough energy remains to be passed on to sustain another level. This is visualized as either an energy pyramid or a trophic pyramid (Figure 1.16). In such a structure, the trophic level that consumes the photosynthetic organisms (autotrophs, first level) includes animals called **herbivores**. Herbivores are consumed by **carnivores**, which are eaten by still other, often larger, carnivores. All levels above the second consist of carnivores or **omnivores** (animals that eat both plant and animal material). Within each trophic level or population, the amount of living material at any instant in time is the **standing crop**.

The final component of the trophic structure of an ecosystem consists of the **decomposers**. These are organisms, chiefly bacteria, that break down the complex organic molecules of dead organisms, releasing simple molecules usable again by the autotrophs or by those heterotrophs that can absorb organic molecules from the environment. Decomposers act on every trophic level.

The necessary abiotic components of the trophic structure of an ecosystem are an energy source, a nutrient source, and a water source. Photosynthetic organisms cannot fix energy and produce complex

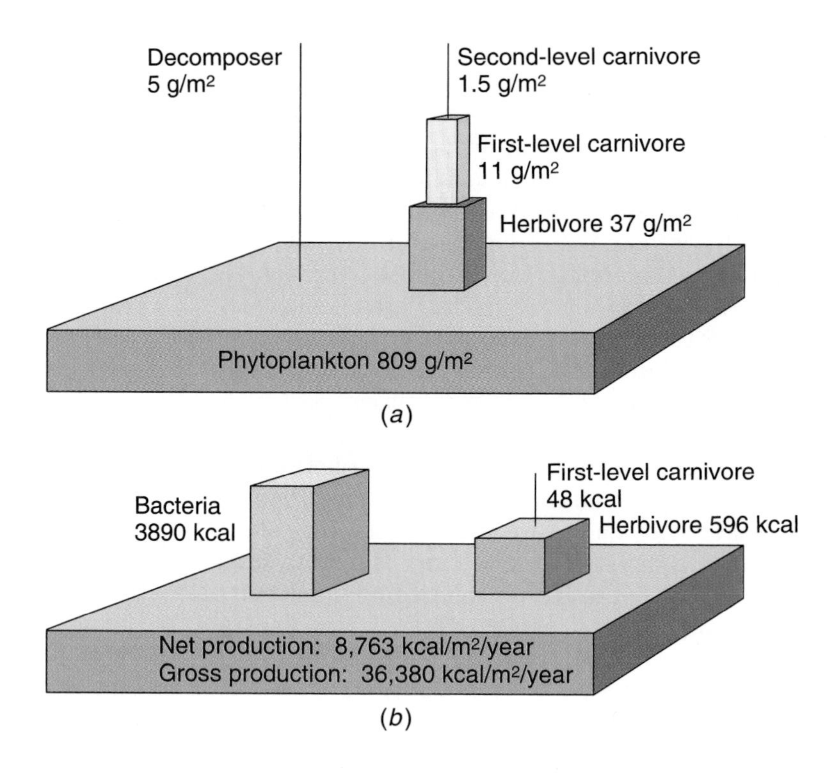

FIGURE 1.16 Trophic pyramids. (a) Biomass pyramid. (b) Energy pyramid. The thickness of the bars indicates the relative amounts. (From *Ecology & Field Biology*, 3/e R. L. Smith, 1980. Copyright © 1980 Benjamin/Cummings. Reprinted by permission of Addison Wesley Educational Publishers.)

organic molecules without sunlight for energy and without a series of inorganic nutrients, of which nitrates and phosphates are the most important.

Since ecosystems are composed of from one to many communities, each composed of many populations of producers, consumers, and decomposers, the transfer of energy through an ecosystem may follow several pathways. Each pathway that transfers energy from a given photosynthetic source through a given series of consumers is called a **food chain**. The combination of all food chains in a given community or ecosystem is called a food web. The **food web** is thus a summary of all the pathways by which energy moves from one level to another through a community or ecosystem.

Biogeochemical Cycles

Among many of the chemical elements and compounds in ecosystems there is a cycling back and forth between organisms and the physical environment. Such repeated transfers are called **biogeochemical cycles**. A few of these cycles involve chemical compounds vital to the continued maintenance of life in the ecosystem and are extremely important. Perhaps the most significant are the cycles involving carbon, nitrogen, and phosphorus (Figure 1.17).

In all these cycles there is a major reservoir or pool of the element, from which the element is continually moving in and out as it passes through organisms. Each cycle also contains a sink, into which a certain amount of the chemical passes and from which it is not recycled in the normal course of events. Over long periods of time, the loss to the sink may become limiting, unless the sink can be tapped again. This latter situation usually develops through geological action that moves the sink into an area where organisms, erosion, or other factors release the elements. Biogeochemical cycles tend, finally, to have self-regulating feedback mechanisms that keep the cycles in equilibrium.

In the carbon cycle, the reservoir is in the form of CO_2, which exists in water in the carbonic acid–bicarbonate–carbonate system (Figure 1.17a). It is fixed into organic compounds by photosynthetic organisms; transferred to animals through herbivory, absorption, and predation; and returned to the reservoir via respiration and bacterial action. A nonrenewable loss occurs when heavy carbonate materials, such as shells, are deposited in deep oceans as calcareous sediments.

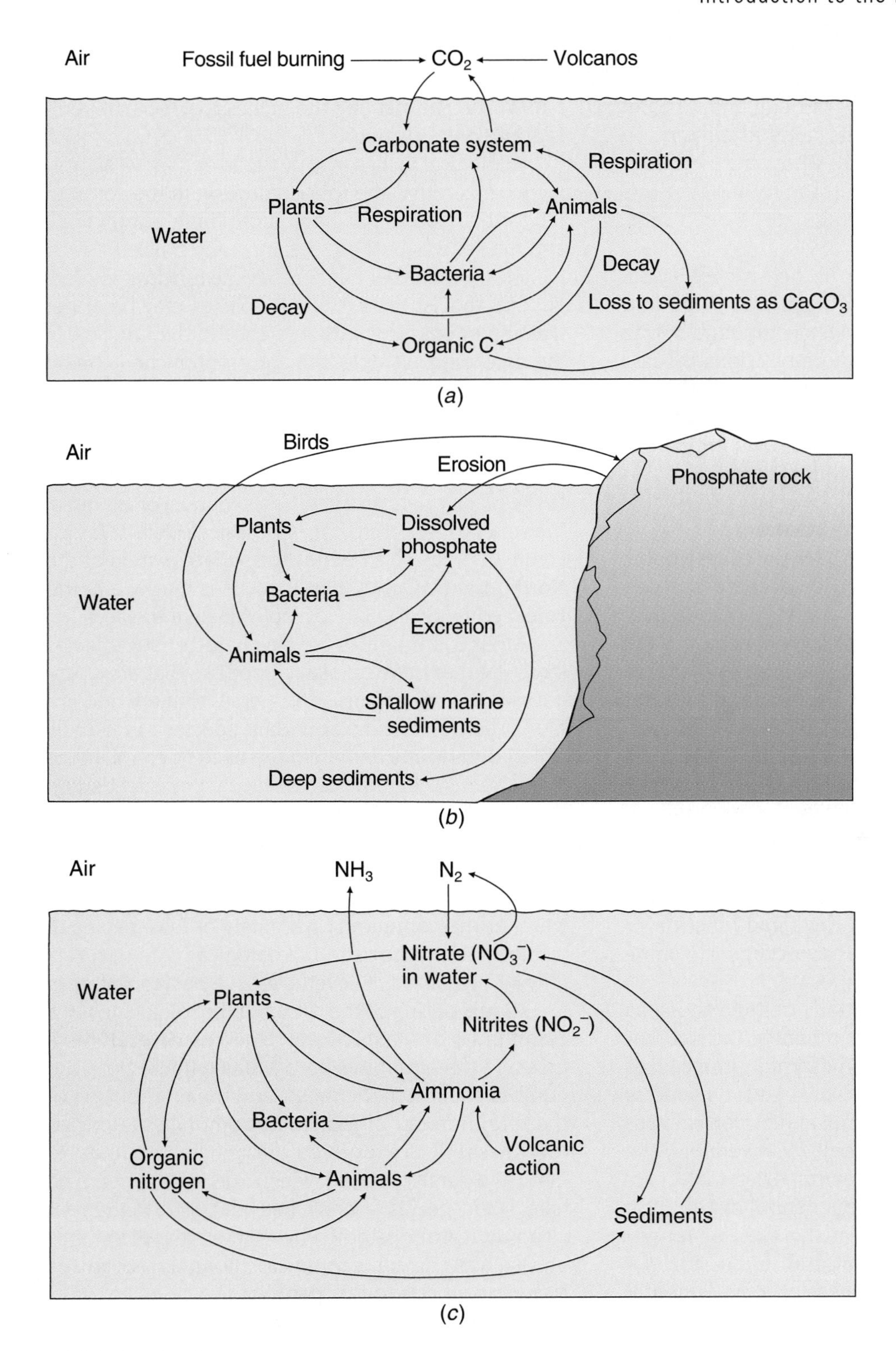

FIGURE 1.17 Biogeochemical cycles in the sea. (a) The carbon cycle. (b) The phosphorus cycle. (c) The nitrogen cycle.

In the phosphorus cycle, the major reservoir is in phosphate rock. Here, erosion brings the chemical into the water, where it cycles through animals and plants and is returned to the general circulation through decay and excretion. Phosphorus is lost when it sinks to the bottom in the form of various skeletal materials. These accumulate in deep sediments (Figure 1.17b).

Air is the major reservoir in the nitrogen cycle. Nitrogen gas is not used by most organisms and must be converted into a nitrogen compound before it can enter the cycle. This is normally done by bacteria and/or algae. Useful nitrogen compounds may also be produced through volcanic activity. Once in compounds, the nitrogen cycles through organisms with some loss to deep sediments (Figure 1.17c).

Biotic Structure of Ecosystems

Communities and ecosystems have the same trophic levels throughout the world. However, the species constituting each level differ among geographical areas, and in some areas, each level may have more or fewer species than another. This implies that ecologically equivalent species replace one another geographically and that the species structure of communities is variable, even if the trophic structure is not. For example, seagrass beds in the tropics may consist of several species of grasses, whereas those in the temperate zone consist of but one. They are different species, but ecologically fulfill the same role as autotrophs. Species that are ecological equivalents are those that perform the same function or role in a given community. They occupy the same ecological niche.

A **niche** in ecology is usually defined today as the role of an organism in a community. The fact that the term niche has had other meanings often causes confusion. A niche can be established for a species for virtually every environmental and biological variable affecting a species. For any given variable, the range within which the species can exist and reproduce is a part of its niche. The potential distribution of a species along all such niche axes or ranges would then be the **fundamental niche** for the species. This theoretical concept of a fundamental niche is difficult to comprehend because it requires us to think in a large number of dimensions simultaneously, whereas we are used to only three. Because we cannot completely determine the fundamental niche of any organism, we tend to restrict our discussion of the niche to only a few variables. We speak then of niches with respect to food, space, breeding sites, and so forth. These are **realized niches** and represent the actual distribution of the species in the real world. They will be the niches to which the term is applied in this book.

Ecological niches may be broad or narrow. Narrow niches mean that the role or function in the community or ecosystem has been more finely subdivided, and the species is, therefore, more specialized. Broad niches, on the other hand, mean the function is more generalized. At the extremes, species may be either *specialists* or *generalists* with reference to their niches. In the preceding example, the autotroph niche is broad in the temperate zone and occupied by a single generalist grass; in the tropics, several more specialized grasses subdivide the niche.

In contrast to niche, an ecological **habitat** refers to the place where an organism is found. For example, the small copepod crustacean *Calanus finmarchicus* (see Figure 2.5) has as a habitat the surface waters of the North Atlantic Ocean, but its niche is that of a herbivore feeding on the microscopic algae in the area.

Most communities have a characteristic species structure that consists of a few species that are abundant and a larger number of species that are uncommon. The numerically abundant species are usually called **dominants** and are often used to characterize a community. We thus can speak of a mussel bed on the rocky shore reflecting the dominance by the mussels *Mytilus edulis* or *Mytilus californianus*. This pattern of few common and many uncommon or rare species seems to be true at each trophic level. It is also true whether the community has many or few species in total. Species structure in ecological communities may be measured in several ways. **Species richness** is a simple listing of the total number of species in a community or trophic level. **Species diversity** is a measure that combines into a single figure both the number of species (richness) and the distribution of the total number of individuals among the species (evenness). It is expressed through various mathematical diversity indices, which need not concern us here. High species diversity has been used in the past as an indicator of stable environments and communities, but current ecological thought recognizes many exceptions to this paradigm.

There are currently two major schools of thought in ecology that address how high species diversity is produced and maintained in communities. One school, the equilibrium school, suggests that the species composition of communities is usually in a state of equilibrium and that high diversity is due to a large number of habitats and/or finely divided, narrow niches maintained by various feedback mechanisms (niche diversification) and the stable nature of the physical environment that allows the equilibrium to be achieved. The other school, the nonequilibrium or intermediate disturbance school of Connell (1978) and others, says that communities and species are rarely in a state of equilibrium and that high species diversity is maintained through continual or gradual environmental change and periodic disturbance. This promotes a changing species composition through the presence of many species that are not highly specialized. In this model, communities that experience frequent disturbances that kill most or all the resident species would tend to be colonized by only those few species that have the ability to attain maturity and reproduce quickly before the next catastrophic event. At the other end of the spectrum, those areas that experience infrequent, small disturbances at long time intervals would also tend to have low diversity. In this case, it is because the most efficient competitors have the time to eliminate most or all of the other species and take over the community. Only when the disturbances are intermediate does diversity increase. This is because enough time is available to permit establishment of a variety of slower-growing and -reproducing species, and the interval between disturbance events is short enough to avoid competitive exclusion. It is probably not possible at our current level of knowledge to decide whether one or the other or both of these theories is correct for all the various marine communities, but the evidence that has been gathered thus far seems to favor the intermediate disturbance theory for at least some communities.

Each species in a community has certain tolerances with respect to all environmental factors. If the limit of tolerance is exceeded in some area by a given factor (say, temperature), the species will be absent. Similarly, each species requires a certain minimum amount of various materials. If the concentrations of these necessities, such as nitrate, fall below the minimum, the species disappears. More important, if any *single factor* exceeds the tolerance level or if any single necessary substance is reduced below the minimum, the species will be eliminated. This is true even if all the other factors and substances are favorable. This is known as **Leibig's law of the minimum**.

Since these *limiting factors* vary for each species and since communities are defined on the basis of recurrent groups of species, it follows that the boundaries between communities, representing changes in various environmental factors, are also not sharp. The result is areas of transition between adjacent communities where species gradually drop out and others come in. Such boundary zones are called **ecotones**. Compared with communities that usually change rather slowly in time and space, ecotones change rapidly.

Communities are not static units. They change in structure and composition with season and over longer periods of time. Some communities tend to change in an orderly fashion over periods of many years until they reach a stage that perpetuates itself indefinitely as long as the climate does not change or there is no disturbance. In such a sequence, each community modifies the environment, making it suitable for the next community. This orderly process of community change controlled through modification of the physical environment is called **ecological succession** and the terminal, persistent community the **climax**. The intermediate communities are called **seres**.

The scenario just described is the classical or **facilitation model** of succession as originally developed for terrestrial plant communities by Clements (1936). A second model for succession, the **inhibition model**, was proposed by Egler (!954), and it differs from the classical in that no species is competitively superior, and whichever species gets to a site first holds it against later settlers. In this model, succession is not an orderly, predictable process, and there is no climatic climax. Here succession proceeds from short-lived to long-lived species. A third model, proposed by Connell and Slayter (1977), called the **tolerance model**, is intermediate between the other two models. In this model, early colonizing species are not necessary, and any species can start succession. Community change then occurs as more tolerant or more competitively superior species prevail.

In the marine environment, ecological succession seems to follow the second or third model. In bottom communities following a disturbance, the area is usually reoccupied by organisms that happen to have their propagules in the water and available to settle. The first occupants may change, depending on the time of year that the disturbance occurs. Furthermore, the initial occupiers often do not modify the environment to make it unsuitable for them and more suitable for subsequent organisms. In fact, they may even prevent the occupation of the area by other, later successional forms by various means and thus delay the climax community development. Some examples of these kinds of succession are given in Chapters 5 and 6. Finally, succession probably does not occur in the open waters among pelagic communities.

Ecological Control and Regulation

Populations, communities, and ecosystems are all regulated by various factors. The major controlling factors for ecosystems and communities are energy, the physical factors collectively termed climate or environment, and the interaction among various species that compose the systems. Ultimately, nearly all systems on earth are limited by the amount of energy available from the sun. Tolerance limits of various species to such abiotic factors as temperature, light, nutrients, and salinity also limit the extent of populations and communities in the oceans. The final aspects of regulation and control considered here are those dealing with the interaction among populations that keep populations within limits.

Biologists have long known that all species possess the reproductive potential to produce much larger populations than are observed under natural conditions and that if population explosions occur, they are quickly reduced. What are the biological factors that exert control? They are grouped under competition, predation, and parasitism and disease.

Competition is an ecological term referring to the interaction among organisms for a necessary resource that exists in short supply. Competition may be intraspecific (among individuals of the same species) or interspecific (among individuals of different species). Competition may be for any number of things but usually is limited to items, such as light, food, nutrients, water, and space. In a competitive interaction, either the competitors manage to share the limited resource or one excludes the other. In the first case, both individuals are hampered, which inhibits their growth, development, and reproduction; thus, the numbers of both organisms are limited. In the second case, one individual is eliminated, again controlling population. Interspecific competition is usually between two closely related species and has led to the **competitive exclusion principle**, which says no two species with exactly the same requirements can coexist in the same place at the same time; that is, complete competitors cannot coexist. As population numbers increase, competition usually increases, because the limited resource becomes more scarce. This increased competition increases the stress on animals and plants and absorbs energy otherwise used for reproduction, limiting populations.

Predation can be defined as the consumption of one species by another. The animal that is the consumer is the predator; the victim, the prey. A special case concerns those animals that consume autotrophic (photosynthetic) organisms. They are called herbivores. A **grazer** is an animal that feeds on plants or sessile animals. Predators and herbivores vary considerably in their ability to regulate the numbers of organisms they consume. In some cases, the predator may be the most important factor in regulating numbers of a prey species. In other cases, a predator may have little effect on a prey population. In the first case, the removal of the predator will have a marked effect on the prey population, causing it to increase dramatically. In the latter case, predator removal has little effect on the prey population. We shall see examples of both these situations in the following chapters. A special case of predation concerns those predators that have a profound effect, not only on their prey population, but also on the entire community of which they are a part. In these cases, the removal or depletion of the numbers of the predator causes great changes in the presence and abundance of many species in the community, most of which are not the prey of the predator. As a result, the entire community structure may be changed. Such predators are called **key industry** or **keystone species**.

Parasitism and disease are the final biological controls on populations. As with predation, they may exert strong control, or they may have little effect. **Parasites** are organisms living in or on other organisms from which they derive nourishment and shelter. Many, perhaps all, marine organisms have parasites, but we know considerably less about the role of parasites and disease in regulating populations of marine organisms than we do for terrestrial organisms. Perhaps the best-documented change induced in a marine community by a disease is the loss of the eelgrass beds in the Atlantic Ocean in the 1930s (see pp.216-217).

LARVAE AND LARVAL ECOLOGY

A very large number of marine organisms, both fish and various invertebrate phyla, inhabiting both pelagic and benthic systems, produce larvae in their life cycles. **Larvae** are independent, morphologically different stages that develop from fertilized eggs and that must undergo a profound change before assuming adult features. They are almost always consider-

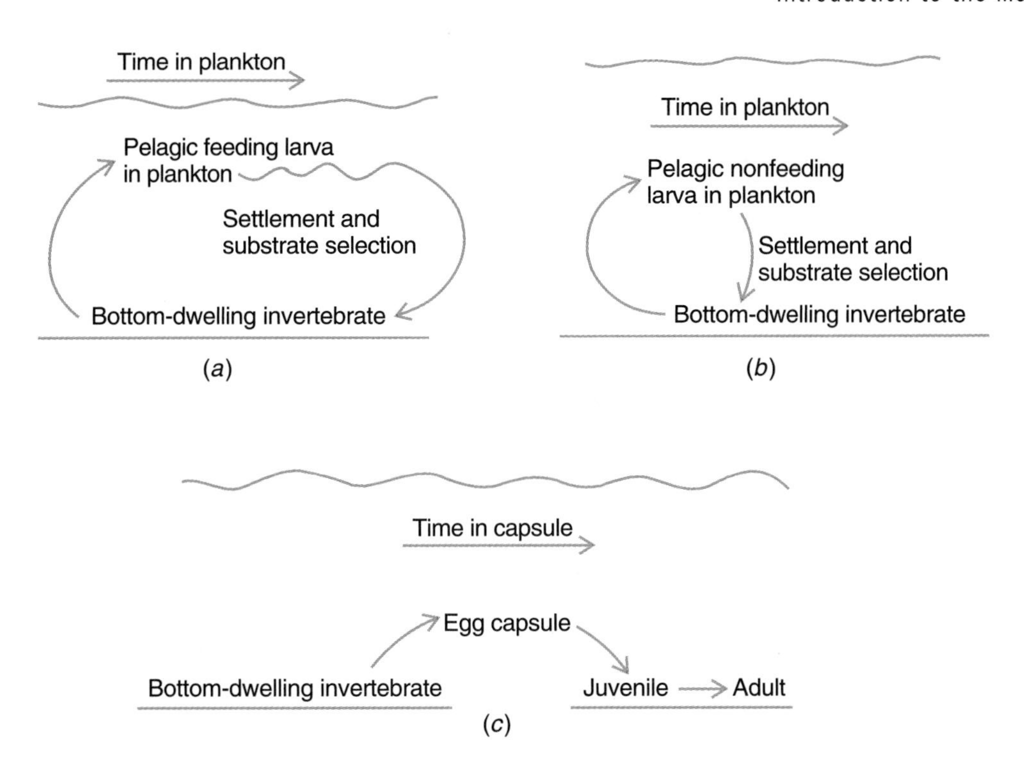

FIGURE 1.18 Representations of the three patterns of development found among benthic invertebrates. (a) Planktotrophic. (b) Lecithotrophic. (c) Nonpelagic.

ably smaller than the adult stage of the species. Larvae and their ecological characteristics are important in the establishment and maintenance of many marine communities and associations. This is because, as Chia (1988) notes, in any given marine habitat the distribution and abundance of benthic invertebrates are maintained by four factors: larval recruitment, migration, asexual reproduction, and mortality. Most benthic communities are composed of species that reproduce by producing various larval types that undergo a free-swimming stage in the water column before metamorphosing into benthic adults. Even many organisms that pass their adult lives in the water column also produce larvae. Therefore, an understanding of larval ecology is central to understanding how such communities persist. We discuss the basic features of larval ecology here, leaving the specific role for larvae in each larger community to be covered in the appropriate chapter.

Larval Types and Strategies

As noted by Vance (1973), there are three possible paths that development may take in a benthic invertebrate with a given amount of energy available with which to produce young. One way is to produce a great many very small eggs. Such eggs hatch quickly into larvae that are free-swimming in the plankton. Because so little yolk is put into each egg, the larvae are dependent for nutrition on the plankton and are called **planktotrophic larvae**. A second course is to produce fewer eggs and endow each with somewhat more energy in the form of yolk. Such eggs hatch into larvae that, because of their yolk reserve, do not feed in the plankton and spend less time in the plankton before settling. Such larvae use the plankton phase mainly for dispersal and are called **lecithotrophic larvae** (Figure 1.18).

A final option is to dispense with the larval phase altogether. In this course, the adult produces a few eggs, each with a very large amount of yolk. Such eggs undergo a long development without additional energy sources. They pass through the larval stages in the egg and hatch as juveniles. There is no free-swimming larval stage. This is **nonpelagic** or **direct development**. Such hatchlings are called **nonpelagic larvae** or **juveniles**.

There are two major assumptions of the Vance model. One is that mortality in the plankton is constant over time and is not related to size of the larvae; the other is that egg size is inversely related to time spent in the plankton.

Each of the three developmental pathways noted by Vance has both advantages and disadvantages. In the case of the planktotrophic larvae, the advantages are that a very large number of young can be produced for a given amount of available energy, and wide dispersal is assured through the long time spent in the plankton. The disadvantages are that the larvae depend for nutrition on the plankton, a notoriously unpredictable situation, and that long residence in the plankton increases the chances that a predator will consume the larvae. The chances of being consumed by a predator in the plankton are extremely high, and the pelagic zone is thus a very dangerous habitat.

Lecithotrophic larvae have the advantage of spending less time in the plankton and thus have less chance of being consumed. They also are not dependent on the vagaries of the plankton for food. The disadvantage of this strategy is that, because of the greater amount of energy placed into each egg, fewer eggs and larvae can be produced per unit of available energy. A shorter time in the plankton also means less dispersal ability. A lecithotrophic larva also presents a larger target for visual predators in the plankton.

The nonpelagic development has the advantage of reducing the planktonic mortality to zero. In this case, however, the disadvantages are that only a few eggs can be produced because of the great amount of energy each takes and that dispersal is nil.

Since all three types of development exist, it can be assumed that under certain conditions one type or another is favored. Even though we are currently ignorant of all the reasons that some areas and conditions favor one larval type over another, it is possible to suggest some that are valid.

In polar waters, nonpelagic development is common, even though productivity is reduced and confined to a narrow summer peak (p. 68) and it takes a long time for benthic animals to obtain sufficient energy to reproduce. Such development avoids the problem of precise timing to hit the very short polar bloom period, where failure would mean the death of most, if not all, larvae, and it ensures the highest survival rate of the few eggs that can be produced.

Lecithotrophic larvae are common under conditions somewhat similar to those in polar waters. Lecithotrophy is preferred over nonpelagic development if dispersal is of prime concern to the survival of the organism in question. Lecithotrophy is also more energy efficient than planktotrophy when planktonic mortality is high. Both lecithotrophy and nonpelagic development were thought to be the most common mode of development in deep-sea organisms by Thorson (1950), but recent work has suggested that planktotrophy is also common and in some taxa may be the dominant form of development.

The advantages of planktotrophy are that eggs are cheap to produce, can be produced in large numbers, and have high dispersal rates. The disadvantages are primarily the great attrition rate suffered while in the plankton and the unpredictable plankton food source. Planktotrophic larvae, therefore, should be common where the plankton food is predictable over long periods, where planktonic mortality is low, where dispersal is of importance, and where development time is short. The area that fulfills most of these criteria is the tropics, and it is there that we find most forms with planktotrophic larvae.

Indeed, it was Thorson (1950) who first suggested that there was a latitudinal gradient in the proportions of the three larval types. From his studies, primarily in cold temperate and Arctic waters, Thorson suggested that planktotrophy was most common among tropical marine organisms, and lecithotrophy and nonpelagic development became more prevalent as one approached polar latitudes. Planktotrophy also decreased and lecithotrophy and nonpelagic development increased in moving from shallow waters to the deep sea.

The problem with the simple Thorson models is that there is no nice fit to them if one studies a variety of marine invertebrates. Strathmann (1977, 1978) has clarified the problems by demonstrating that the predicted effect of various hypotheses about larvae and larval development is confounded when looking at the real world. For example, species with a feeding larval form (planktotrophic) very often have low survival and recruitment most years, but occasionally, they have a very successful year. One would expect that an organism that produces such larvae would be long-lived as an adult and breed many times (iteroparous) simply because in most years the young would not survive, and it would take several years of producing young to have a successful recruitment. However, many short-lived animals that breed only once (semelparous) have feeding planktotrophic larvae.

Similarly, it has been suggested that small animals more often are brooders, producing a few eggs, because they simply are not big enough to allocate a large amount of energy to produce numerous eggs. To ensure the survival of the few, they brood them through to a more advanced state. This seems illogical because if brooding ensures a higher percentage of survival of young, why shouldn't all animals, large and small, do it? Certainly, the larger animals would gain by getting a higher return of juveniles per energy unit input. In other words, there must be some other compensating advantage to sticking with the "risky" planktotrophic development other than simple survival of the young. One such advantage might be dispersal, since large animals are less likely to be moved over long distances by water currents and rafting than are small animals. This is particularly true if the large adults are also sedentary. However, both animals that are permanent members of the plankton and those that as adults are sedentary or sessile on the bottom have planktotrophic larvae. Since adults in the water column are as dispersible as the larvae, this would negate the argument for dispersibility. One reason larger animals may not brood is that they may simply lack the physical space to brood a large set of eggs or be unable to aerate the very large egg mass properly. Strathmann (1985) has even suggested that planktotrophy may be an adaptation to prevent suffocation of eggs in large egg masses, since eggs must have oxygen and there are physical constraints on how egg masses are constructed so oxygen can reach interior eggs.

In summary, whereas we have several hypotheses as to why organisms may have a particular larval developmental type, we can find many exceptions. Together these suggest there are many factors involved in selection for a given larval type, they do not act similarly for all the animals in a given habitat, and there may be morphological or phylogenetic constraints on the development of a particular type.

Obrebski (1979) has also offered a model for the existence of different larval types. This model suggests that the type of larva—planktotrophic, lecithotrophic, or nonpelagic—should be predicted by the relative abundance of suitable substrates for settlement. In this model, those benthic organisms colonizing abundant habitat patches would have less need for long-lived planktotrophic larvae, whereas those inhabiting rare patches would benefit from having long-lived planktotrophic larvae in order to search out the rare habitat type.

Vance's models have also been challenged by Underwood (1974), who has shown that surveys of the existing data reveal many discrepancies to the predicted modes of reproduction. This has led Underwood to conclude that other factors are of importance and that energy considerations alone are inadequate to explain the different strategies.

We find ourselves in the somewhat awkward position of having several different hypotheses for the existence and perpetuation of the different larval types but with the real data not seeming to fit any model completely and with several larval types coexisting in the same area. This must strongly suggest that other factors are also at work in determining larval development type. Why should this be? We can only suggest that for each species there is a necessity to balance all the factors, including energy allocation, dispersal, relative abundance of microhabitat space, longevity of adults, competitive ability of adults, and undoubtedly others, against the need for larger numbers of young. The type of reproduction that has evolved is that which balances all of these factors with survival. This is reinforced by the finding of certain benthic infaunal species that can alter their reproductive strategy among the three developmental patterns depending on prevailing conditions. An example is the common polychaete worm *Capitella capitata*, which Grassle and Grassle (1974) have shown can produce either planktonic or benthic larvae depending on the conditions.

Larval Ecology and Community Establishment

Many, if not most, marine benthic communities are composed primarily of species that have a free-swimming larval stage. The exception to this rule appears to be the aforementioned polar waters. Since over larger areas the benthic communities seem to persist over time (see Chapter 5) but are dependent for this continued existence on the settlement of larvae from the plankton, and since the adults live only a short time, how is this stability maintained? What mechanisms are there that will act to ensure that the "right" larvae will settle and mature to keep the community going? A partial answer is found in the characteristics of larvae.

Since the pioneering work of D. P. Wilson (1952), many laboratory studies on larvae have disclosed that larval settlement and metamorphosis are responses on the part of the larvae to complex and highly specific environmental stimuli, including light, gravity, and fluid movement. The most influential clues, however, appear to be chemical, emanating from biological sources. Most larvae have preferences as to the areas in which they will settle. Larvae do not just settle out of the water column on whatever substrate is available when the time comes for them to metamorphose into adults. Larvae have the ability to "test" the substrate, probably responding to certain physiochemical factors; if the substrate

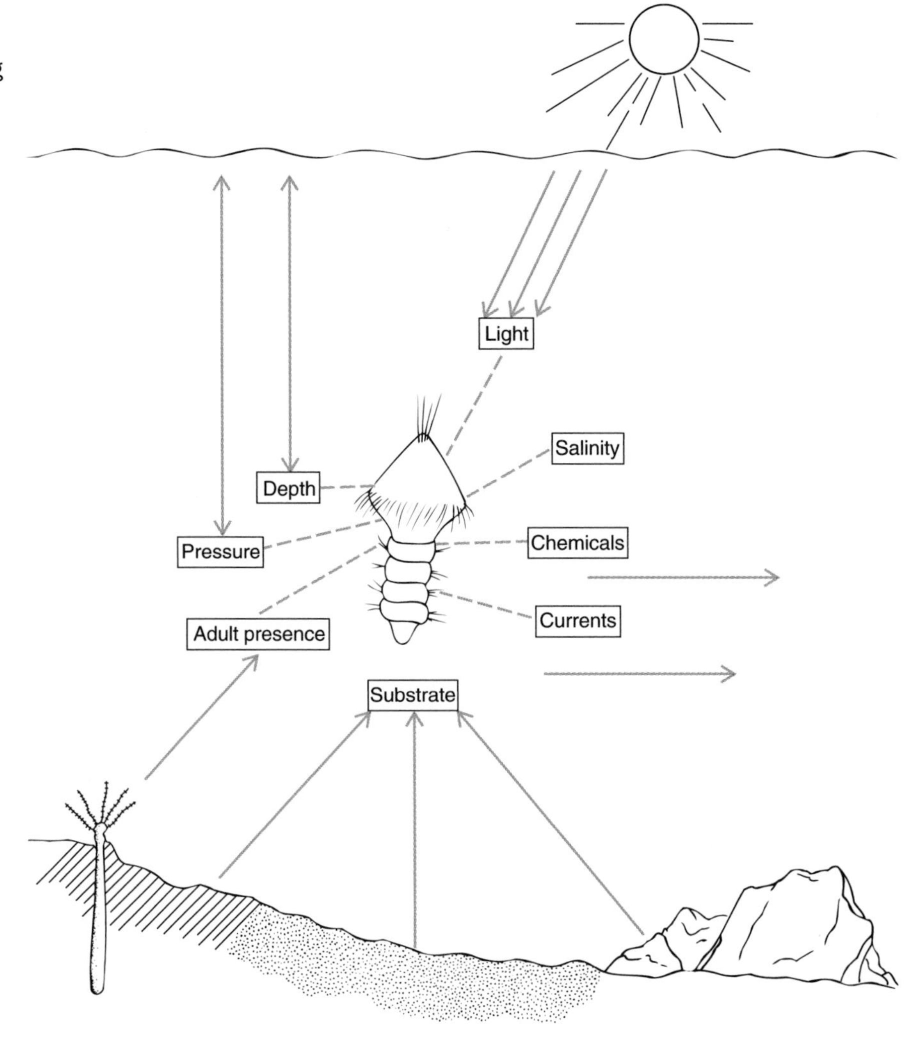

FIGURE 1.19 Various factors of the environment to which larvae respond in selecting a suitable site for settling.

is unsuitable, they do not settle out or metamorphose. This means that certain substrates will always be attractive to certain larvae and repellent to others. Larvae also respond to the presence or absence of adults of their own species. Many larvae preferentially settle where the adults are living. They are, as Crisp and Meadows (1962) have shown, attracted to the area by a chemical or **pheromone** secreted by the adult. Since the presence of the adult means that the area is suitable for habitation, this mechanism ensures survival of the young. It also ensures that the community will persist.

The larvae of many invertebrates also can delay their own metamorphosis for a certain period of time if they cannot find suitable substrates at the time they are supposed to settle. This delay of metamorphosis is a finite period. After a certain time, metamorphosis will take place even over unsuitable substrates. The ability to delay settlement, however, is yet another factor that can be employed to ensure that the larvae will settle in the proper place.

Larvae, as Thorson (1966) has noted, also respond to other physicochemical factors, such as light, pressure, and salinity. Many free-floating larvae are positively phototactic in the early stages of their larval life. This condition ensures that they will reside in the upper, faster-moving waters where dispersal is greatest. Later, as time for settlement approaches, they become negatively phototactic and migrate toward the bottom. Some larvae are very sensitive to light and pressure and inhabit only certain levels in the water column, areas with a definite light regime and pressure. Segregation of larvae into various layers in the water column also means that wherever a given layer impinges on the bottom, it will have only a certain complement of larvae available for settlement, a further way of ensuring a nonrandom bottom community (Figure 1.19).

These various studies designed to determine the mechanisms that control larval settlement and metamorphosis have tended to strongly favor the hypothesis of active habitat selection by the larvae. Most of these studies, however, have been done in the laboratory in still water and have not considered the alternative hypothesis that larvae may be passively deposited on the seafloor by physical processes. Butman (1987) has also suggested that various hydrodynamic processes are also active in determining where larvae settle.

Finally, different species have different times of reproduction and, hence, different times of the year during which larvae are in the water column. As we shall see, this may also have an effect on the associations that develop.

Despite the considerable abilities of larvae to discriminate among substrates and select places to settle, there is often considerable variability from year to year in both species abundances and composition within bottom infaunal communities. Samples taken close to each other often show many differences in species composition and abundance. If the larvae have such selective powers, why is there such variability? The answer to this lies in the life histories of the various invertebrates composing the community, their interaction with each other and with the physical environment, and the effects of predators and competition.

Life History Strategies

According to MacArthur (1960), it is possible to recognize two quite different types of life history patterns among organisms in any habitat. These are introduced here and will be referred to in subsequent chapters. One type can be called **opportunistic** or "*r* selected." Opportunistic marine species are characterized by having short life spans, rapid development to reproductive maturity, many reproductive periods per year, larvae present in the water during much or all of the year, and high death rates. Usually, they are small animals and often are sedentary or **sessile** (fixed in place). The second type of life history can be termed **equilibrium** or "K selected." Equilibrium species are those that have long life spans, relatively long development time to reach reproductive maturity, one or more reproductive periods per year, and low death rates. Usually, these species are somewhat larger than opportunists, and they are often mobile (Table 1.3).

The characteristics of opportunistic and equilibrium species define the two ends of a continuum. It must be remembered that although some species will have all characteristics of an opportunistic species and others those of equilibrium species, still others will fall somewhere in the middle and have varying mixtures of these features. For purposes of discussion, however, it is easier to consider the ends of the spectrum.

TABLE 1.3

Summary of the Characteristics of Opportunistic and Equilibrium Species

Features	*Equilibrium* (K)	*Opportunistic* (*r*)
1.Reproduction periods	Few per year	Many per year
2. Development	Slow	Rapid
3. Death rate	Low	High
4. Recruitment	Low	High
5.Colonizing time	Late	Early
6. Adult size	Generally large	Generally small
7. Mobility	High	Low

Opportunistic species are favored where the substrate is subjected to frequent disturbance from some particular agent. In the case of shallow-water soft bottoms, the common disturbance agent is waves or other water motion that stirs up the bottom or carries away the upper layer of sediment. Another disturbance may be the rapid deposition of sediment on the bottom. In certain cases, biological activity may also disturb the bottom, for instance when large fish, such as rays, dig into the bottom. When such a disturbance occurs, it usually kills the resident organisms. The result is an open area, which can be resettled by organisms. Because of their attribute of having larvae in the water most of the time, opportunists quickly settle these open areas. Furthermore, if disturbances occur frequently, the opportunists will have the advantage because they can quickly mature and reproduce again before being destroyed by the next disturbance.

Equilibrium species, on the other hand, tend to inhabit areas that are not subject to frequent disturbance, allowing them to complete their life cycles. Too frequent disturbance would mean destruction before they could reach reproductive maturity; hence, no larvae would be present in the water column to settle the area again.

Whenever the bottom is subject to frequent disturbance, the benthic communities are usually composed primarily of opportunists, whereas in areas where disturbance is less frequent, more equilibrium species are found. This is because equilibrium species are generally better competitors than opportunistic

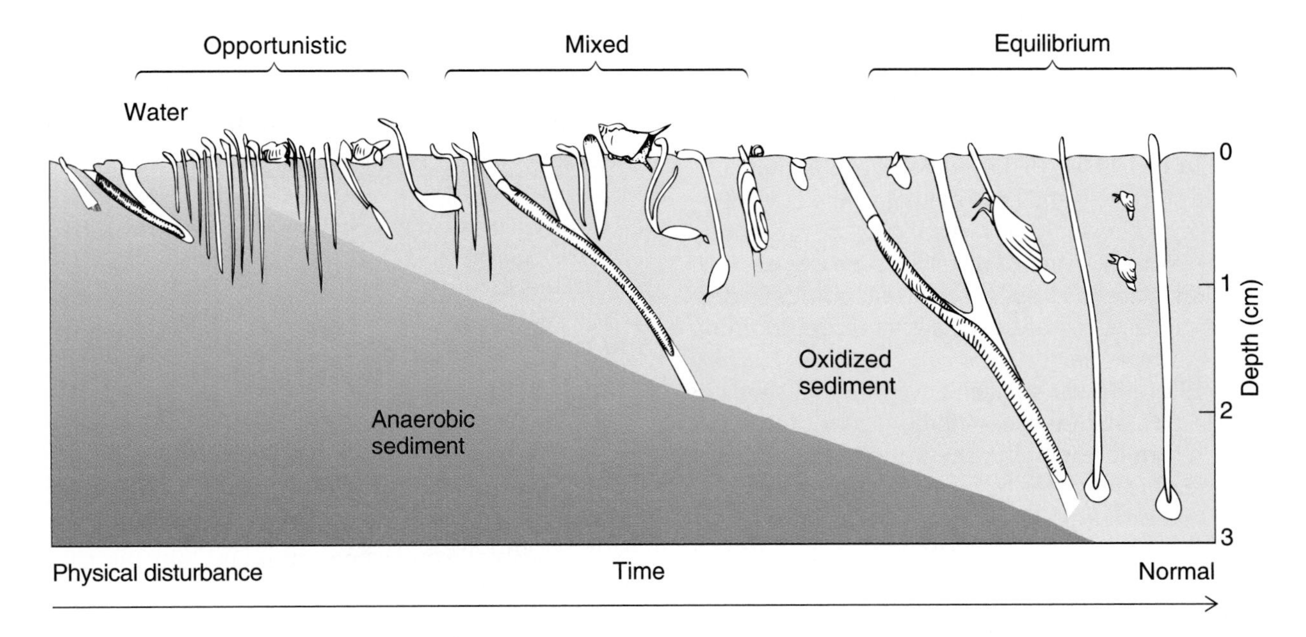

FIGURE 1.20 Changes in a soft bottom community with time after a physical disturbance. (From "Disturbance and Production on the Estuarine Sea Floor," D. C. Rhoads et al., September–October, 1978, *American Scientist*. Reprinted by permission.)

species and, given sufficient time, will outcompete the opportunists.

Because of their different characteristics, there are differences in abundances between these two groups. Communities in which the number of opportunists is high show great fluctuations in numbers of individuals over time, whereas communities composed of equilibrium species are more constant in abundances.

If this situation prevails, it would appear that communities would be composed of either opportunists or equilibrium species, depending on frequency of disturbance. In reality, samples of the bottom usually contain representatives of both groups. How does this happen? Apparently, the equilibrium species are slower to colonize disturbed areas than are opportunists, but they are better competitors once they are there. Once there, they push out (outcompete) the opportunists. Over time, there is a sequence in which the area is settled by opportunists immediately following a disturbance. As time goes on, the equilibrium species settle in and finally outcompete the opportunists. When sampling, then, differing numbers of both groups would result, depending on the time since the last disturbance. Varying numbers of both groups, therefore, can be seen in bottom samples (Figure 1.20).

Part of the reason for different numbers of individuals and species from different areas in the same bottom is because disturbances do not affect the whole bottom equally, but act selectively. The result is a patchy array of communities, each in varying degrees of recovery from disturbance and with differing numbers of opportunists and equilibrium species.

Since wave-induced disturbance decreases with depth, it is not surprising that deeper communities tend to be inhabited by more equilibrium species than shallow water, unless another type of disturbance creates unstable conditions. Such a disturbance could be biological, for example, the reworking of sediment by fishes or burrowing deposit feeders.

COMPARISON OF TERRESTRIAL AND MARINE ECOSYSTEMS

It is quite true that ecological principles apply equally whether one is dealing with terrestrial, freshwater, or marine ecosystems. However, the special conditions prevailing in salt water that result from the physical and chemical properties of water have, directly or indirectly, channeled evolution and adaptation of marine organisms. The result has been some striking differences in the organization of marine communities when compared with terrestrial communities. These differences will be considered next.

Physical and Chemical Differences

Seawater has several physical features that have a profound effect on the organization of marine communities. They are its greater density with respect to air (more than 800 times as dense as air), its greater viscosity (about 60 times that of air), its ability to transmit sound (4 times faster than air), its low electrical resistivity (16 orders of magnitude less than that of air), and its ability to absorb light. Of these factors, the density and viscosity differences have the most widespread effects on marine organisms.

The greater density of seawater means that relatively large organisms and particles can remain suspended in it. This is not possible in air. One significant result of this is that the marine ecosystem has evolved a whole community of small organisms that are perpetually afloat, the plankton. No such comparable community exists terrestrially floating permanently in the air. As a result, all communities in the sea are bathed in a medium that itself contains a community! The presence of this suspended community has, in turn, led to the evolution of animals that are adapted to filter organisms and particles out of seawater. These **filter feeders** are unique to aquatic systems. There is nothing really comparable on land. The closest analog would be spiders, which capture flying insects in their webs. Furthermore, since the waters of the oceans are in perpetual motion, many of these filter feeders are sessile, extending their filtering nets up into the moving water mass. Sessile filtering animals are not possible in the terrestrial environment because the bathing medium flowing by contains no perpetually suspended food organisms as particles. Finally, marine organisms, especially the sessile species, have evolved motile larval forms. Such motile larvae, when put into the plankton, permit dispersal of the species.

As a result of the suspension of a whole community in the bathing medium, it is more difficult to have isolated communities and isolated species distributions in the sea.

An interesting sidelight of the viscosity difference between air and water is that water becomes less viscous as it heats up, whereas air becomes more viscous. Tropical marine animals, therefore, encounter less resistance to movement than do temperate or polar animals, whereas tropical birds flying experience more drag than do temperate or polar birds.

Another physical factor indirectly responsible for differences between marine and terrestrial organisms is **gravity**. Because marine plants and animals are buoyed up by the greater density and viscosity of water, they do not need to have a significant amount of their biomass invested in structural material, such as skeletons or cellulose, in order to hold themselves erect against the force of gravity. Similarly, where movement is concerned, terrestrial animals must raise their mass against the force of gravity for each step. Such movement requires significantly more energy than swimming movements do in aquatic organisms; hence, there is more energy storage.

This absence of large amounts of structural material and energy storage in the bodies of marine organisms is reflected in the differences in the dominant biochemical compounds found in terrestrial and marine organisms. Among terrestrial plants, the dominant structural and storage compounds are carbohydrates; among marine organisms, the predominant body biochemical is protein with smaller amounts of carbohydrates and calcium carbonate for structure. Living organisms composed primarily of carbohydrates tend to be long-lived, slow growing, and rich in stored energy. Protein-dominated organisms, on the other hand, tend to be rapid growing and with lower energy storage.

In contrast to air, water strongly absorbs light. As a result, light entering water can only penetrate to a certain depth before it is completely absorbed. Although this depth varies, it is sufficiently shallow (200–400 m) that the majority of the volume of water in the oceans is without light. This means that autotrophic life and primary productivity are limited to an extremely narrow band near the surface. Whereas terrestrial communities, with the exception of caves, all possess sufficient light for photosynthesis, many marine communities exist without the benefit of an autotrophic component.

Not only does water absorb light, it does so differentially, depending on the particular wavelength. The result is that certain wavelengths penetrate deeper than others. This is not true on land, where nearly the same spectral composition of light impinges at all levels in any community. The importance of this to marine communities is discussed on pages 59–62.

The speed of sound at 20°C in air is 346 m/sec, whereas at the same temperature in water it is 1,518 m/sec, a fourfold increase. In addition, the sound

TABLE 1.4

Number of Higher Animal Taxa in the Three Main World Ecosystems

Taxon	Marine	Terrestrial	Freshwater
Phyla	31	14	16
Classes	85	28	35

wavelength in water increases by a factor of four. The biological consequences of this concern the use of sound by animals to monitor their environment, find food, and avoid predators. The faster movement of sound in water coupled with the more limited range of vision due to the absorption of light by water means that sound is more useful and has a greater range for orienting an animal to its surroundings in water than does vision. Both terrestrial and aquatic animals emit sounds to orient to and detect prey. However, for marine animals to detect prey with sound, the sound waves must be several wavelengths smaller than in air to detect the same size of potential prey. Since in water the wavelengths of sound increase fourfold, marine animals must use shorter wavelengths (higher frequency) than terrestrial animals to detect the same size prey (see Chapter 3 for further discussion).

The physical factor with the greatest difference between air and water, as Denny (1990) points out, is electrical resistivity. Water has a resistivity 10^{16} times less than that of air. The consequences of this are that some aquatic animals have developed an ability to detect electrical activity in other animals at a distance. This sense has not been found in terrestrial animals.

Because of its greater density, water experiences greater changes in pressure with depth than does air. This, as we are now beginning to learn, has profound effects on the metabolism of marine organisms and their distribution. This is discussed more fully in Chapter 4.

It is also worth pointing out that, unlike air, seawater contains energy-rich organic compounds in the form of dissolved organic matter (DOM), which can itself be a potential food source.

A final physical difference between terrestrial and marine systems concerns oxygen. In air, oxygen constitutes a nearly constant 21% of the volume throughout the world. Water holds less oxygen, and its concentration can vary with the temperature and salinity (pp. 5–6). Finally, oxygen diffuses through water at a rate 1/10,000 of the rate in air—once reduced, it regenerates slowly.

These facts concerning oxygen in seawater create a situation for marine organisms in which not only can the oxygen level be reduced in the external environment due to limited solubility or biological activity, but also in the immediate boundary layers around the organism due to the slowness of diffusion. These conditions have led marine organisms to develop rather extensive morphological, physiological, and behavioral adaptations to provide adequate oxygen in the face of variable availability. Some of the more important adaptations include extensive respiratory surfaces, particularly in the form of integumental respiration, and small size, giving a greater surface-to-volume ratio.

Biodiversity Differences

Although the oceans are by far the largest repository of life on the planet, this is not reflected in all the patterns of the biodiversity in the oceans, which are poorly known. At the highest taxonomic level, namely the phylum, the oceans have more diversity than the land (Table 1.4). Of the 34 animal phyla thus far recognized, only 11 occur on land and only one is endemic there, whereas 29 occur in the oceans and 14 are endemic there. At the species level, however, there is considerable disagreement. The estimated number of described marine species is 250,000, with perhaps another 150,000 to 200,000 yet to be described (excluding microorganisms). However, Grassle and Maciolek (1992) have estimated the number of marine species yet undescribed at 10 million, whereas May (1992) considers that the total number of marine species is unlikely to exceed 500,000. The number of terrestrial species is not known within an order of magnitude, but perhaps 1.5 million have been named and estimates of the total number range from about 4 or 5 million to more than 50 million. This large number of terrestrial species is due primarily to the huge numbers of insect species coupled with the large number of vascular plant species (about 350,000 species), which serve as habitats for many insect species. By contrast, the oceans have only 50 species of flowering plants and perhaps 12,000 to 13,000 species of algae.

The oceans and the land both have similar latitudinal gradients in biodiversity, namely an increase in species diversity in many taxa moving from polar regions to the tropics. However, in the oceans there is also a longitudinal gradient in diversity, at least in shallow water. The Indo-Pacific, the area encompassing the Philippines, Indonesia, northern Australia, and New Guinea, has the highest diversity, which falls off both east and west from that center. In the Atlantic, the highest diversity is found in the west (Caribbean Sea) and declines to the east. The highest species diversity in the oceans occurs among the benthic organisms, not among those living in the water column. Among habitats, the highest diversity is found in coral reefs and deep sea benthos.

Life History Differences

There are several life history traits common to many marine organisms that are not seen in terrestrial organisms. In contrast to terrestrial organisms, aquatic photosynthetic organisms and animals may shed both male and female gametes into the water and have external fertilization. Terrestrial plants may shed male gametes (pollen) to be moved by air currents, but female gametes are never shed. Furthermore, there are no marine equivalents of the many terrestrial pollinators, animals that transfer male gametes in return for a reward. Finally, for animals, parental care (brood protection) and energy investment in offspring are much lower in aquatic systems than in terrestrial systems. Strathmann (1990) suggests that this is probably because the abundance of food in water permits feeding at a very small size, because desiccation is not a factor, and perhaps because the lower diffusion constant of oxygen in water and the higher viscosity of water may simply preclude furnishing enough oxygen to a brooded clutch of young.

Structural and Functional Differences

One striking difference between terrestrial and marine communities is the insignificance in the latter of large macroscopic plants. Terrestrial communities are universally dominated by large flowering plants that are persistent and long-lived. With the exception of certain large kelp plants, the communities of the sea have no large photosynthetic organisms and the dominant autotrophs are single-celled microscopic organisms of various groups of algae. In turn, this means that the dominant herbivores of the sea are also small, often microscopic, animals, in contrast to the large-bodied herbivores common to terrestrial communities. There are, for example, few herbivores in the sea to compare in size with the antelopes of the African plains or the bison of the North American grasslands. The dominant marine herbivores are microscopic crustaceans called **copepods** or various protistans (see Figure 2.5).

Another characteristic of terrestrial plants is that significant portions are composed of rigid structural materials such as wood and fiber, which themselves are relatively indigestible by most herbivores. As a result, herbivore grazing in terrestrial communities rarely removes significant amounts of the entire plant community. By contrast, the small herbivores of the seas usually consume the entire autotroph, and a major difference between terrestrial and marine ecosystems is the ability of the herbivores to remove the autotrophs completely (see pp. 70–73).

Terrestrial communities are characterized by a matrix of long-lived plants and a fauna that is generally shorter-lived. In the seas, at least among those communities associated with the sea bottom, the autotrophs are short-lived, and the matrix of the community is of relatively large, long-lived animals. One speaks of redwood forests, oak forests, and tall grass prairies on land, but of coral reefs, clam beds, oyster reefs, and mussel beds in the seas.

As a result of the microscopic nature of most marine autotrophs and herbivores, the majority of large animals in the sea are carnivores. Large marine animals are, therefore, on higher trophic levels than large land animals. From the standpoint of human beings, harvestable food in the form of animal protein is available primarily from large herbivores on land (second trophic level), such as cattle and sheep, whereas harvestable food from the sea is usually in the form of carnivores from the third or fourth trophic levels, such as salmon, tuna, and halibut. A corollary of this is that most marine food chains have about five links (steps) to reach the top carnivore, whereas terrestrial food chains tend to be shorter, averaging three links to reach the top carnivore (Figure 1.21).

There is also a question about the efficiency of energy production and transfer between terrestrial and marine ecosystems. Generally, the production of organic material is higher in terrestrial ecosystems than in marine, but the efficiency of transfer from first to second trophic levels is higher in marine ecosystems. Although starting out with less energy fixed, marine systems lose less with the first transfer. Attempts at further assessments of the efficiency of energy transfer in marine food chains are complicated because of the various feedback mechanisms that tend to "blur" the trophic level to which to assign the animal. The primary group responsible for this blurring of the trophic level is the filter feeders. These animals feed upon suspended particles, and what is retained on their filters is based on

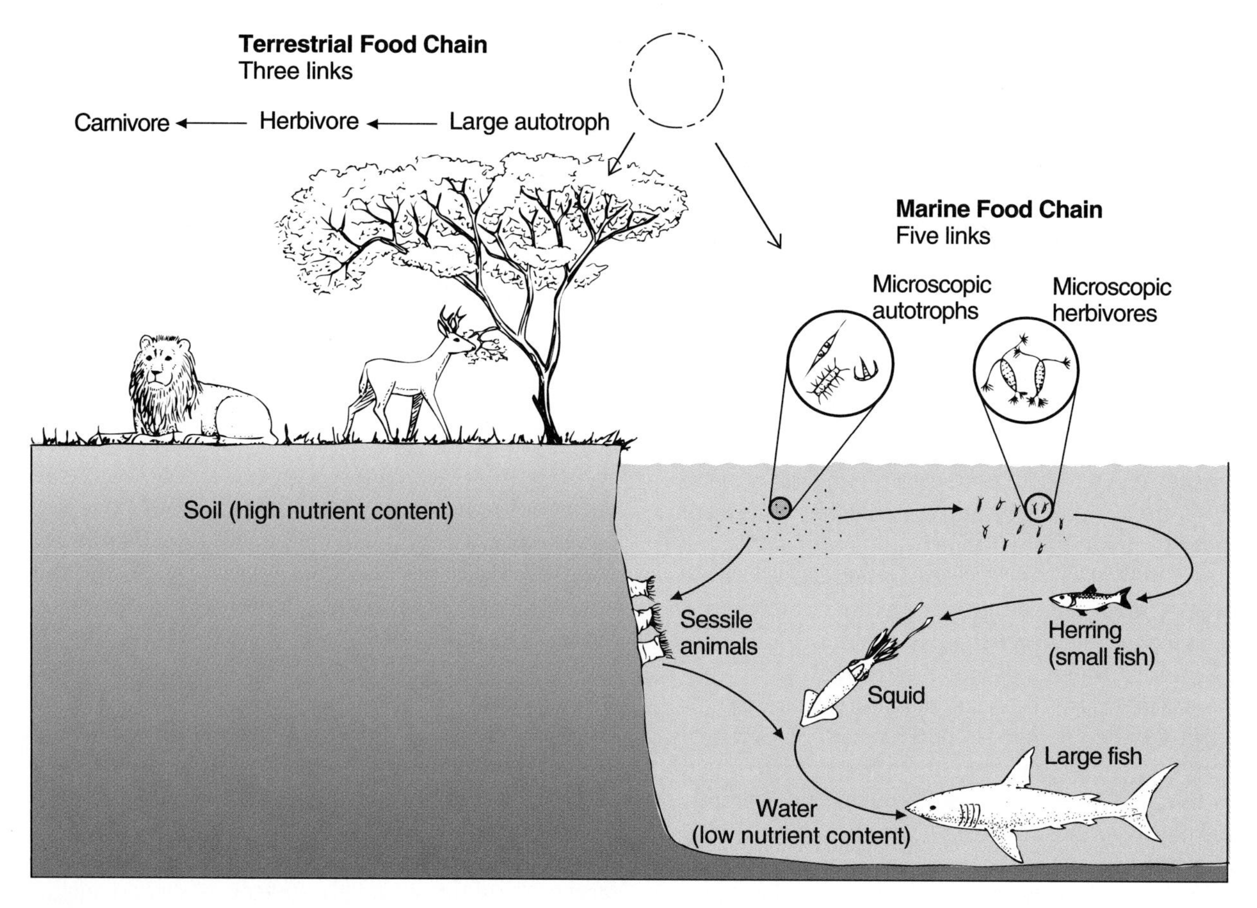

FIGURE 1.21 Diagrammatic representation of the differences between marine and terrestrial food chains.

size, not whether the particle is animal or plant, living or dead. This means they often take in both plant and animal material and feed on several trophic levels. A second reason for blurred trophic levels is the tendency of marine species to switch positions in the trophic hierarchy. Juvenile fishes, for example, may feed on herbivorous copepods, but when they mature they feed on other fishes higher in the trophic spectrum. An additional problem is the ability of various marine organisms to switch trophic levels as adults. Salmon, for example, may feed on squid at one time but switch to lantern fish at another; thus, they shorten the food chain since lantern fishes are themselves the food of squid. Similarly, euphauslid crustaceans (krill) may be herbivores in shallow water and carnivores in the open ocean.

The differences between terrestrial and marine ecosystems should be kept in mind as we explore the various habitats and communities of the sea.

DIVISION OF THE MARINE ENVIRONMENT

The marine ecosystem is the largest aquatic system on the planet. Its size and complexity make it difficult to deal with as a whole. As a result, it is convenient to divide it into more manageable, albeit somewhat arbitrary, subdivisions, each of which can then be discussed in terms of the ecological principles that govern the adaptations of the organisms and the organization of the communities. No universally acceptable scheme of subdivision of the marine environment has yet been proposed. The one followed here is modified from Hedgpeth (1957) and has enjoyed widespread use among biologists for more than 40 years. It probably comes as close as any to a universally acceptable division.

Major Subdivisions of the World Ocean

Beginning with the waters of the open ocean, subdivisions can be made both vertically and horizontally.

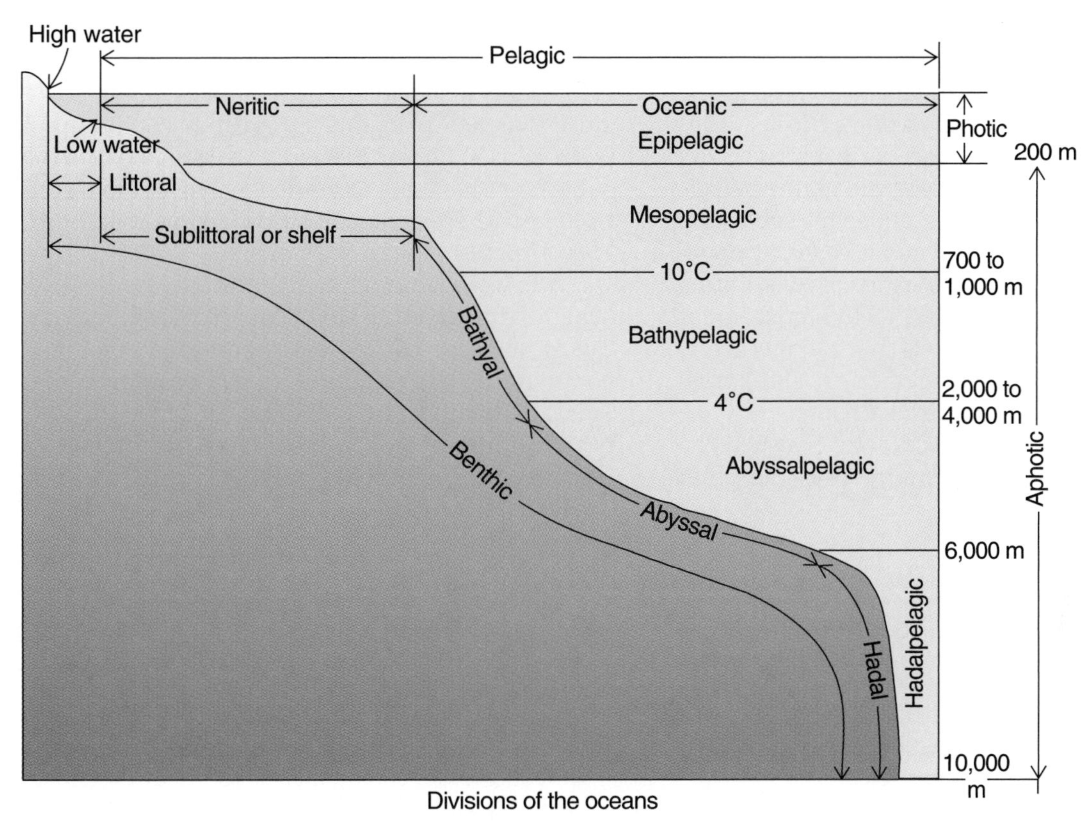

FIGURE 1.22 Divisions of the oceans (not to scale). (Modified from "The Treatise on Marine Ecology and Paleontology", J. Hedgpeth, *Ecology Memoir #67*. Copyright Geological Society of America.)

The entire area of the open water is the **pelagic** realm; pelagic organisms are those that live in the open sea away from the bottom. This is in contrast to the **benthic** realm, which is a general term referring to organisms and zones of the sea bottom. Horizontally, the pelagic realm can be divided into two zones. The **neritic** zone encompasses the water mass that overlies the continental shelves. The **oceanic** zone includes all other open waters (Figure 1.22). Progressing vertically, the pelagic realm can be further subdivided. Two schemes are possible. The first is based on light penetration. The **photic** or **euphotic** zone is that part of the pelagic realm that is lighted. Its lower boundary is the limit of light penetration and varies in depth with clarity of the water. Generally, the lower boundary is between 100 and 200 m. A synonym for this zone is the **epipelagic** zone. Because it is the zone of primary production in the ocean, it is of major importance and the subject of Chapters 2 and 3. The permanently dark water mass below the photic zone is the **aphotic** zone. Some scientists prefer to recognize a transition zone between the photic and aphotic called the **disphotic** zone. This transition area has enough light for vision but not enough for photosynthesis; it extends down to about 1,000 m. Disphotic is synonymous here with mesopelagic.

The pelagic part of the aphotic zone can be subdivided into zones that succeed each other vertically. The **mesopelagic** is the uppermost of the aphotic areas. Its lower boundary in the tropics is the 10°C isotherm, which ranges from 700 to 1,000 m, depending on the area. Next is the **bathypelagic**, lying between 10 and 4°C, or in depth between 700 and 1,000 m and between 2,000 and 4,000 m. Overlying the plains of the major ocean basins is the **abyssalpelagic**, which has its lower boundary at about 6,000 m. The open water of the deep oceanic trenches between 6,000 and 10,000 m is called the **hadalpelagic** (Figure 1.22).

Corresponding to the last three pelagic zones are three bottom or benthic zones. The **bathyal** zone is that area of bottom encompassing the continental slope and down to about 4,000 m. The **abyssal** zone includes the broad abyssal plains of the ocean basins between 4,000 and 6,000 m. The **hadal** is the benthic zone of the trenches between 6,000 and 10,000 m. This entire aphotic area, although encompassing the largest volume and area of the world's oceans, is little known and forms the subject of Chapter 4.

The benthic zone underlying the neritic pelagic zone on the continental shelf is termed the **sublittoral** or **shelf** zone. It is illuminated and is generally populated with an abundance of organisms constituting several different communities, including seagrass beds, kelp forests, and coral reefs. The majority of this area is covered in Chapter 5, but the specialized coral reef communities are discussed in Chapter 9.

Two transitional areas exist, one between the marine environment and the terrestrial, the other between marine and fresh water. The **intertidal** zone or **littoral** zone is that shore area lying between the extremes of high and low tide; it represents the transitional area from marine to terrestrial conditions. It is a zone of abundant life and is well studied. It is the subject of Chapter 6. **Estuaries** represent the transition area where fresh and salt water meet and mix. They form the subject of Chapter 8.

It is well to remember that the classification scheme discussed in this chapter is not absolute but is used for convenience, and organisms and communities may extend into two or more zones. This is particularly true for organisms that are powerful swimmers and divers. These organisms are discussed in Chapter 3.

SUMMARY OF KEY CONCEPTS

- Water is a simple chemical compound consisting of one atom of oxygen covalently joined to two atoms of hydrogen. It is a polar compound bonding by weak hydrogen bonds to other water molecules. This hydrogen bonding gives rise to many of water's unique chemical and physical properties that make life possible.
- The unique chemical and physical properties of water include the highest surface tension of all common liquids, a high viscosity relative to air, a high latent heat of vaporization and fusion, high heat capacity, a peculiar density-temperature relationship, and the ability to act as an almost universal solvent.
- Seawater is pure water in which are dissolved a variety of gases and solids. Most of these dissolved substances have ratios among themselves that are constant, thus allowing the determination of the concentrations of all substances by measuring one. Thus the salinity of seawater is usually measured by determining chlorine concentration.
- The two major gases in seawater are oxygen and carbon dioxide, and their solubility is a function of temperature. Oxygen concentration also depends upon usage and production by organisms.
- Seawater is a buffer employing a carbonic acid–bicarbonate system that maintains a pH around 8.0.
- All major oceans are connected to each other. They are connected to the continents by shallow extensions of the continents called continental shelves. Most of the ocean basins are flat abyssal plains, but these plains are cut by deep trenches in some places and have submarine ridges, volcanoes, and seamounts in other areas.
- Temperature in the oceans varies latitudinally and vertically.
- As a result of different temperatures and salinities, the oceans have water masses of different densities. The uppermost water masses are moved around by action of the prevailing wind belts of the earth to form surface global currents and local waves. Under certain conditions of current movement, coupled with the Coriolis effect, these currents may diverge causing cooler water to rise from deeper depths, a phenomenon called upwelling. Also, some water masses may cool when transported into cold climates, or when formed in polar areas, and sink to form deep-water masses that make up the bottom water in the world's ocean basins.
- Ecology is the science that treats the spectrum of relationships that exist between organisms and their environments and among groups of organisms. The various organisms and the environmental factors may be further organized into various levels such as species, populations, communities, and ecosystems.
- Ecosystems are functional units of variable size composed of living and nonliving parts that interact. Ecosystems have a trophic structure, usually consisting of an autotrophic base where energy is fixed followed by one or more layers of heterotrophs through which the fixed energy is passed. The number of trophic levels is limited by the loss of energy during the transfer between levels.
- Biogeochemical cycling of carbon, nitrogen, and phosphorus between marine organisms and the physical environment is vital to the continued maintenance of life in the oceans.
- Communities and ecosystems have the same trophic structure throughout the world's oceans, but the species composing the structure differ in different geographical areas.
- A niche in ecology is defined as the role of an organism in a community. It may be broad or narrow and can be established for virtually any environmental or biological factor. By contrast, a habitat is the place where an organism is found.

- Most communities have a characteristic species structure of a few species that are abundant or dominant, and define the community, and many less abundant species. Species structure may be measured by species richness or species diversity. There are two schools of thought as to how high species diversity is established and maintained, the equilibrium model and the intermediate disturbance model.
- Communities are not static units; they change through time. Some change in an orderly fashion through a predictable sequence, whereas others progress in an unpredictable fashion. Such ecological succession is explained by the facilitation, inhibition, and tolerance models.
- Populations, communities, and ecosystems are regulated by various chemical and physical factors of the environment. Populations are further regulated by the biological factors of competition, predation, parasitism, and disease.
- Large numbers of marine animals have larvae in their life cycle. Larvae are morphologically different stages from the adult, which usually undergo a profound change in morphology when they become adult.
- There are three patterns of development in marine organisms: planktotrophic larvae, lecithotrophic larvae, and direct development, forming juveniles or nonpelagic larvae. There are several hypotheses to explain under what conditions each of the patterns would prevail, but none account for all the nuances seen in the sea, suggesting that additional factors must be operating.
- Many marine benthic communities are established by larvae settling from the water column. To ensure settlement into the proper environment, larvae respond to various environmental stimuli of which certain chemicals appear to be the most influential.
- Most organisms can be categorized into one of two life history patterns, opportunistic or equilibrium.
- There are some striking differences between terrestrial and marine ecosystems, which are due primarily to the physical and chemical differences in the bathing media, air and water.
- The greater density and viscosity of water compared with air means that organisms can live permanently suspended in water, which is not the case for air. Therefore, we have plankton communities in the water. These in turn allow the existence of sessile animals that filter organisms out of the water column. Water also buoys up organisms against gravity, so that marine organisms do not have significant amounts of biomass invested in structural material to hold them up.
- Water absorbs light and does so differentially for the wavelengths of visible light. This means that light only penetrates to a few hundred meters in the ocean; therefore, the greatest volume of the oceans is without light. Many marine communities are therefore without an autotrophic component.
- Because of its greater density, water also experiences greater changes in pressure with depth, which has profound effects on the metabolism of organisms.
- Water holds less oxygen than air, and oxygen concentration is not constant as in air but varies with temperature and salinity. Some parts of the ocean may approach anoxic conditions.
- There are also differences in biodiversity between the marine and terrestrial environments. The marine environment has the greatest number of phyla, but the terrestrial environment greatly exceeds the marine in numbers of species.
- Certain life history traits seen in marine organisms are not found in terrestrial organisms. Marine organisms shed both male and female gametes into the water, but terrestrial organisms never shed female gametes. There are no marine equivalents of terrestrial pollinators.
- Large plants form the matrix of most terrestrial communities, but the marine autotrophs are dominated by single-celled microscopic autotrophs. This means that the herbivores in the sea are also tiny animals, in contrast to the large herbivores on land. As a result, most marine food chains have five links to reach the top carnivore as opposed to three links on land.
- The marine ecosystem is the largest aquatic system on the planet, and its size and complexity make it difficult to deal with as a whole. It is therefore convenient to divide it into more manageable units that can then be taken up separately.

REVIEW QUESTIONS

ESSAY: Develop complete answers to these questions.

1. What is a thermocline? How does it become established? Why is it that once the thermocline has become established it will not be disrupted by a summer storm? What is a pycnocline?
2. Is there a marine equivalent to the terrestrial concept of ecological succession? Is there a marine climax community(ies)? Compare and contrast the classical or facilitation model of succession with the inhibition and the tolerance models.
3. Energy is said to flow through an ecosystem, but chemicals cycle (i.e., biochemical cycles). Why is this distinction an important one to keep in mind when studying ecosystems?
4. Enumerate and discuss three life history differences between marine and terrestrial organisms.

BIBLIOGRAPHY

Bascom, Willard. 1980. *Waves and beaches, the dynamics of the ocean surface.* New York: Anchor Books.

Briggs, J. C. 1974. *Marine zoogeography.* New York: McGraw-Hill.

Butman, Cheryl. 1987. Larval settlement of soft sediment invertebrates: The spatial scales of pattern explained by active habitat selection and the emerging role of hydrodynamical processes. *Oceanog. Mar. Biol. Ann. Rev.* 25:113–165.

Caswell, H. 1981. The evolution of "mixed" life histories in marine invertebrates and elsewhere. *Amer. Nat.* 117:529–536.

Chia, Fu-shang. 1988. Differential larval settlement of benthic invertebrates. In *Reproduction, genetics and distributions of marine organisms,* edited by J. S. Ryland and P. A. Tyler. Fredensborg, Denmark: Olsen and Olsen, 3–12.

Clements, F. E. 1936. Nature and structure of the climax. *J. Ecol.* 24:252–284.

Connell, J. H. 1975. Some mechanisms producing structure in natural communities: A model and evidence from field experiments. In *Ecology and evolution of communities,* edited by M. L. Cody and J. M. Diamond. Cambridge, MA: Belknap Press, 460–490.

Connell, J. H. 1978. Diversity in tropical rain forests and coral reefs. *Science* 199(4335):1302–1310.

Connell, J. H., and R. O. Slayter. 1977. Mechanisms of succession in natural communities and their role in community stability and organization. *Amer. Nat.* 111:1119–1144.

Crisp, D. J., and P. S. Meadows. 1962. The chemical bases of gregariousness in cirripedes. *Proc. Roy. Soc. London (Biol.)* 150:500–520.

Denny, Mark. 1990. Terrestrial versus aquatic biology: The medium and the message. *Amer. Zoo.* 30(1):111–122.

Dietrich, G., K. Kalle, W. Krauss, and G. Siedler. 1980. *General oceanography: An introduction.* 2d ed. New York: Wiley.

Egler, F. E. 1954. Vegetation science concepts. I. Initial floristic composition, a factor in old field vegetation development. *Vegetation* 14:412–417.

Grassle, J. F., and J. P. Grassle. 1974. Opportunistic life histories and genetic systems in marine benthic polychaetes. *J. Mar. Res.* 32:253–284.

Grassle, J. F., and N. Maciolek. 1992. Deep sea species richness: Regional and local diversity estimates from quantitative bottom samples. *Amer. Nat.* 139: 313–341.

Gross, M. G. 1982. *Oceanography: A view of the earth.* 3d ed. Englewood Cliffs, NJ: Prentice-Hall.

Hedgpeth, J. 1957. Classification of marine environments and concepts of marine ecology. In *The treatise on marine ecology and paleoecology.* Vol. 1, *Ecology.* Memoir No. 67 of The Geological Society of America, edited by J. E. Hedgpeth. New York: Geological Society of America.

Ingmanson, D. E., and W. J. Wallace. 1973. *Oceanology: An introduction.* Belmont, CA: Wadsworth.

Krebs, C. J. 1985. *Ecology, the experimental analysis of distribution and abundance.* New York: Harper and Row.

MacArthur, R. 1960. On the relative abundance of species. *Amer. Nat.* 94 (874):25–34.

Martin, J., *et al.* 1994. Testing the iron hypothesis in ecosystems of the equatorial Pacific Ocean. *Nature* 371:123–129.

May, R. M. 1992. Bottoms up for the oceans. *Nature* 357:278–279.

McEdward, L. R., ed. 1995. *Ecology of marine invertebrate larvae.* Boca Raton, FL: CRC Press.

Obrebski, S. 1979. Larval colonizing strategies in marine benthic invertebrates. *Mar. Ecol. Progr. Ser.* 1:293–300.

Ormond, R. F. G., J. D. Gage, and M. V. Angel, eds. 1997. *Marine biodiversity, patterns and processes.* Cambridge, England: Cambridge University Press.

Palmer, A. R., and R. R. Strathman. 1981. Scale of dispersal in varying environments and its implication for life histories of marine invertebrates. *Oecologia* 48:308–318.

Smith, R. L. 1980. *Ecology and field biology.* 3d ed. New York: Harper and Row.

Smith, R. L. 1986. *Elements of ecology.* New York: Harper and Row.

Steele, J. H. 1985. Comparison of marine and terrestrial ecosystems. *Nature* 313:355–358.

Steele, J. H. 1991. Ecological explanations in marine and terrestrial systems. In *Marine biology, its accomplishment and future prospect,* edited by J. Mauchline and T. Nemoto. New York: Elsevier, 101–106.

Strahler, Arthur. 1971. *The earth sciences.* 2d ed. New York: Harper and Row.

Strathmann, R. R. 1974. The spread of sibling larvae of sedentary marine invertebrates. *Amer. Nat.* 108:29–44.

Strathmann, R. R. 1977. Egg size, larval development, and juvenile size in benthic marine invertebrates. *Amer. Nat.* 111:373–376.

Strathmann, R. R. 1978. The evolution and loss of feeding larval stages of marine invertebrates. *Evolution* 32:899–906.

Strathmann, R. R. 1985. Feeding and nonfeeding larval development and life history evolution in marine invertebrates. *Ann. Rev. Ecol. Syst.* 16:339–361.

Strathmann, R. R. 1990. Why life histories evolve differently in the sea. *Amer. Zoo.* 30(1):197–208.

Strathmann, R. R., and M. F. Strathmann. 1982. The relation between adult size and brooding in marine invertebrates. *Amer. Nat.* 119:91–101.

Thorson, G. 1950. Reproductive and larval ecology of marine bottom invertebrates. *Biol. Rev.* 25:1–45.

Thorson, G. 1966. Some factors influencing the recruitment and establishment of marine benthic communities. *Neth. J. Sea Res.* 3(2):267–293.

Underwood, A. J. 1974. On models for reproductive strategy in marine benthic invertebrates. *Amer. Nat.* 108:874–878.

Vance, R. 1973. Reproductive strategies in marine benthic invertebrates. *Amer. Nat.* 107(955): 339–352.

Whittaker, R. H. 1975. *Communities and ecosystems.* 2d ed. New York: Macmillan.

Wilson, D. P. 1952. The influence of the nature of the substratum on the metamorphosis of the larvae of marine animals, especially the larvae of *Ophelia bicornis* Savigny. *Anna. l'Insti. Oceanogr. Monaco* 27:49–156.

PLANKTON AND PLANKTON COMMUNITIES

As we have seen in the previous chapter, there are certain fundamental physical differences between terrestrial and marine environments, which contribute to the differences we observe in the organization of the communities in the two areas. Perhaps nowhere are these differences more dramatic than in the free-floating and weakly swimming associations of organisms that we collectively call the plankton. It is appropriate to begin a consideration of the ecology of the oceans with the plankton, because the chlorophyll-bearing organisms of the plankton contribute the greatest amount of photosynthesis in the oceans. In fact, some scientists believe that oceanic productivity rivals that of the terrestrial environment and may even be greater. The plankton trap most of the sun's energy that enters the oceans, which is then transferred to the many other communities of the oceans. This vital role of initial fixation of energy makes plankton important in the economy of the oceans. Just as there could be no life on land without the energy-fixing grasses, trees, and shrubs, there could be no life in the oceans without the energy-fixing, minute planktonic organisms.

Definitions

The term **plankton** is a general term. Planktonic organisms have such limited powers of locomotion that they are at the mercy of the prevailing water movement. This is in contrast to the **nekton**, strong-swimming animals of the open sea, capable of exercising movement against the prevailing water flow. Plankton may be further subdivided. **Phytoplankton** are free-floating organisms of the sea that are capable of photosynthesis. **Zooplankton** are the various free-floating animals. **Bacterioplankton** include the various minute heterotrophic and autotrophic bacteria. Recently, we have become aware of various **virioplankton** organisms in the oceans.

Because planktonic organisms are traditionally captured with nets of various mesh sizes, they have also been classified by size. This size classification does not distinguish between photosynthetic and nonphotosynthetic organisms. The most recent and probably most acceptable, though arbitrary, scheme of classification for plankton has seven size classes (Table 2.1). **Megaplankton** are all organisms above 20 cm. **Macroplankton** are organisms 2–20 cm in size. **Mesoplankton** fall between 0.2 and 20 mm in size. **Microplankton** are plankton 20–200 μm in size. These first groups are those usually captured in standard plankton nets and are called **net plankton**. The **nanoplankton** are very small organisms ranging 2–20 μm in size. The **picoplankton** are minute organisms 0.2–2.0 μm in size and consist primarily of bacteria and cyanobacteria. The smallest class is the **femtoplankton**, 0.02–0.2 μm in size. Little is known of this last

TABLE 2.1

Size Classification for Plankton Organisms

	Size Class						
	Femtoplankton	*Picoplankton*	*Nanoplankton*	*Microplankton*	*Mesoplankton*	*Macroplankton*	*Megaplankton*
Class Name	(0.02–.2μm)	(0.2–2.0μm)	(2.0–20μm)	(20–200μm)	(0.2–20*mm*)	(2–20*cm*)	(20-200*cm*)
Virioplankton	—						
Bacterioplankton		—					
Mycoplankton			—				
Phytoplankton			—	—	—	—	
Protozooplankton			—	—	—	—	
Metazooplankton			—	—	—	—	—
Size (m)	10^{-7}	10^{-6}	10^{-5}	10^{-4}	10^{-3} 10^{-2}	10^{-1}	10^{-0}

Source: From *Sea Microbes*, by J. M. Sieburth, © 1979 Oxford University Press.

group. The nanoplankton, picoplankton, and femtoplankton cannot be captured in plankton nets because, in order to filter them out of the water column, the mesh would have to be so fine that when the nets were pulled behind a boat, the water would not pass through. As a result, these organisms can be obtained only by centrifuging samples of seawater, filtering water samples on fine filters such as millipore filters, or allowing them to settle out from water samples.

A final set of terms concerns the life history characteristics of the zooplankton organisms. **Holoplankton** are those organisms that spend their entire lives in the plankton. **Meroplankton**, on the other hand, are those species that spend but a part of their lives in the plankton. Meroplanktonic organisms include a large number of larvae of animals that as adults either live on the bottom or swim as nekton.

THE PHYTOPLANKTON

Although this text is designed to cover the principles of organization of life in the sea, not to detail the various taxonomic groups, it is worthwhile to include here a brief description of the common phytoplankton. The phytoplankton include a wide range of photosynthetic organisms. The larger phytoplankton— those normally captured in nets—consist of only two groups that dominate the net phytoplankton throughout the world: diatoms and dinoflagellates. The smaller phytoplankton, those in the nanoplankton and picoplankton size ranges that pass through nets, are more variable. It is the picoplankton group that now appears to be more important in productivity (see pp. 82–85).

Diatoms

The **diatoms** (class Bacillariophyceae) are easily differentiated from the dinoflagellates because they are enclosed within a unique glass "pillbox" and have no visible means of locomotion. Each box is composed of two valves or frustules, one valve fitting over another (Figure 2.1). The living part of the diatom is within the box. The box is constructed of silicon dioxide, the same material that is a major constituent of glass. Each box is highly ornamented with species-specific designs, pits, and perforations, and this feature has made these organisms very popular with light microscopists and, more recently, with scanning electron microscopists (Figure 2.2).

Diatoms may occur singly, each individual occupying a single box, or they may occur in chains of various kinds, which themselves add to the ornamentation seen in the individual boxes (Figure 2.2; see Figure 2.4).

In reproduction, each diatom divides into two halves; one half then occupies the top valve of the original box, while the other takes the bottom. Each then secretes a valve so that the typical box is re-created. Each of these new valves is secreted *within* the old valve; hence, as the process continues through several generations, the average size of the diatom population decreases. As a result, individuals of species of diatom vary in size. Obviously, there must be a limit to the process of reduction. This occurs after a certain number of generations, when the diatom casts off both valves and becomes a structure called an **auxospore**. Within this spore, new valves are secreted that reestablish the original size of the diatom species (see Figure 2.16). Although this is the general mechanism of diatom reproduction and reestablishment of size, there are some diatoms that undergo replication without valve decrease and do not require auxospore formation to reestablish size.

Diatoms are abundant as species and individuals. Although it is not possible to illustrate all here,

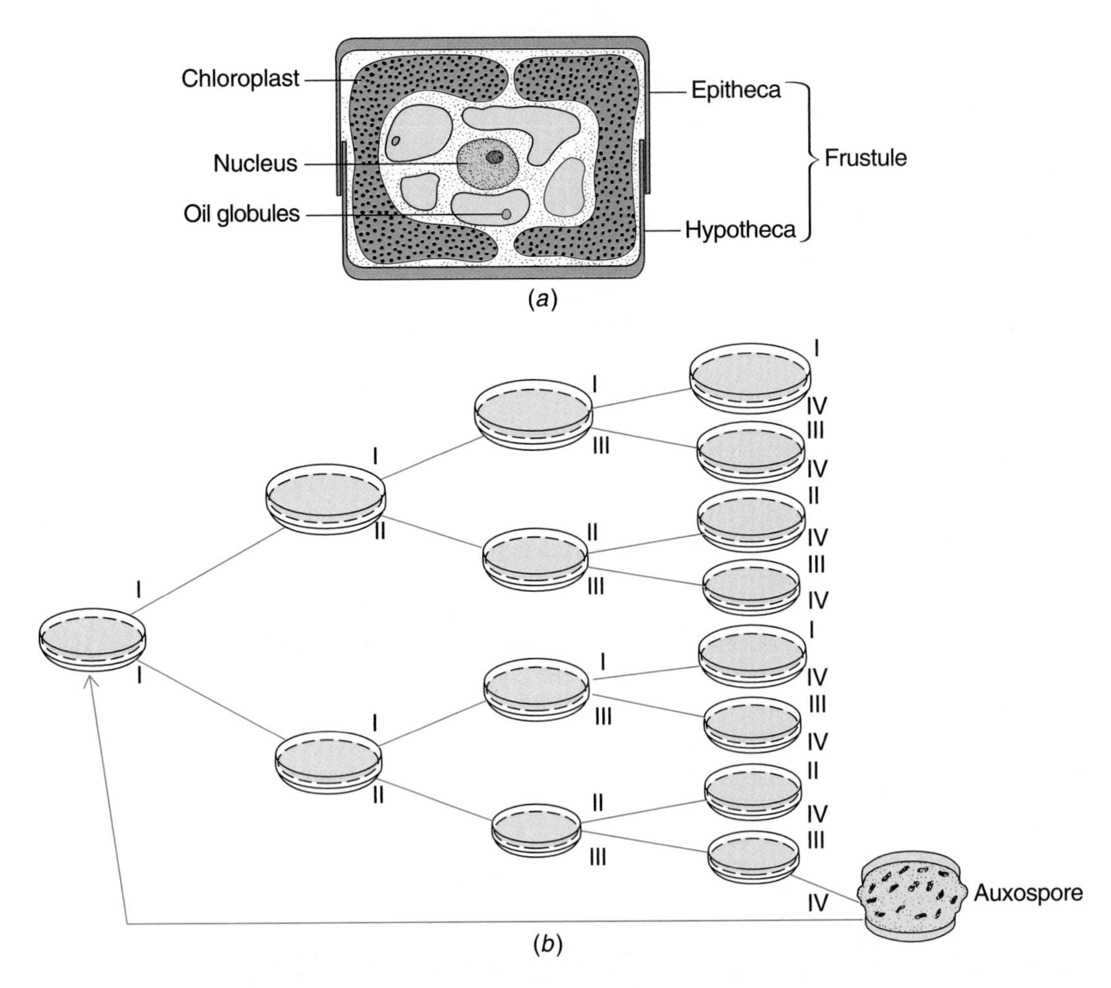

FIGURE 2.1 Diatom features. (a) Diagrammatic representation of the structure of a diatom. (b) Diagrammatic representation of the diminution of size in successive divisions and the restoration of size following auxospore formation.

some of the more common planktonic genera are shown in Figures 2.2 and 2.4.

In recent years it has been discovered that diatoms produce a toxin called **domoic acid**, which may cause mortality in other organisms.

Some diatoms are not planktonic but benthic. These diatoms will be discussed in Chapter 6.

Dinoflagellates

The second major group, the **dinoflagellates** (class Dinophyceae), have two flagella, which they use to move themselves through the water. They lack an external skeleton of silicon but are often armored with plates of the carbohydrate cellulose (Figures 2.3 and 2.4). They are also characterized by having chromosomes that are condensed throughout the cell cycle. Dinoflagellates are generally small organisms and usually solitary, rarely forming chains. They reproduce by simple fission, as do the diatoms. In this case, each daughter cell retains half the original cellulose armor and forms a new part to replace the missing half without any diminution in size; hence, successive generations do not change in size. Some dinoflagellates also are capable of producing toxins that are released into seawater. If dinoflagellates become extremely abundant (2 to 8 million cells per liter), the cumulative effect of all the toxins released may affect other organisms, causing mass mortality. Such extreme concentrations, or blooms, of dinofla-

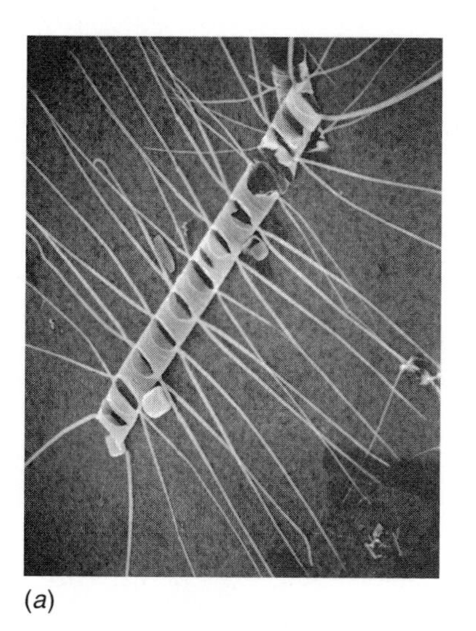

(a)

(b)

(c)

FIGURE 2.2 Scanning electron micrographs of diatoms. (a) *Chaetoceros* sp. (700×). (b) *Thalassiosira decipiens* (1,000×). (c) *Pseudoeunotia doliolus* (1,400×).

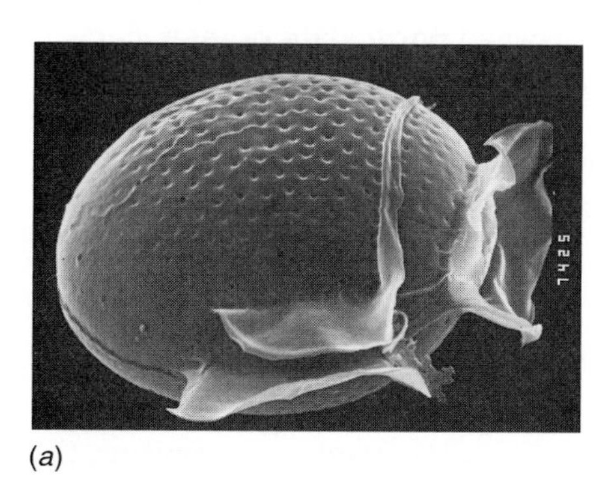

(a)

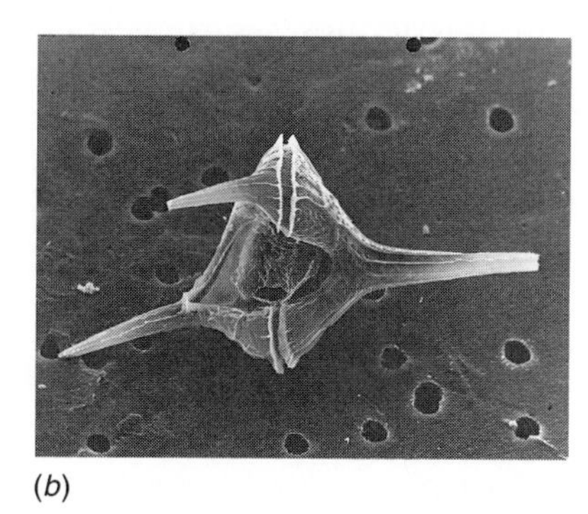

(b)

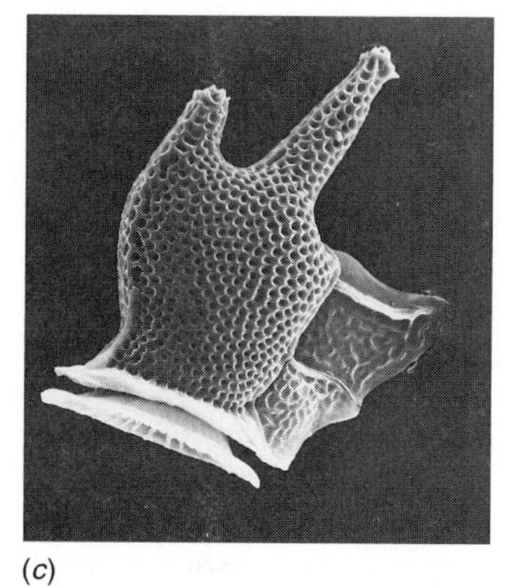

(c)

FIGURE 2.3 Scanning electron micrographs of dinoflagellates. (a) *Dinophysis parva* (21 × 16 μm). (b) *Ceratium candelabra* (140 × 70 μm). (c) *Dinophysis caudata* (160 × 80 μm). (Biophoto Associates/Photo Researchers Inc.)

gellates are called **red tides** and are responsible for massive localized mortality in fish and invertebrates in various places.

Some dinoflagellates have nonmotile stages called **zooxanthellae**, which are symbionts in the tissues of many invertebrates such as corals, sea anemones, and giant clams (see discussion in Chapter 10). Other dinoflagellates, such as the common *Noctiluca* (see Figure 2.24a), are also highly bioluminescent (see Chapter 5 for discussion) and, when present in large numbers, can actually light the wakes of boats and the breaking waves on a beach. Many dinoflagellates, such as *Noctiluca*, are not photosynthetic, but feed by engulfing food particles (phagocytosis), thus playing the role of a grazer.

The Smaller Phytoplankton

Constituents of the nanoplankton and picoplankton size classes (sometimes collectively called nanoplankton) include a number of photosynthetic organisms. These organisms are important in primary productivity and in oceanic food webs, which has only been realized in recent years (see discussion on pp. 82–85). Perhaps most important are the newly discovered, very tiny (0.8–0.6μm) **prochlorophytes**, which are probably the most numerically abundant phytoplankters in the open ocean (10^6 cells/ml). Prochlorophytes are thought to contribute about one-third of all chlorophyll *a* in the open ocean and are estimated to contribute between one-third and one-half of the total ocean productivity. They resemble

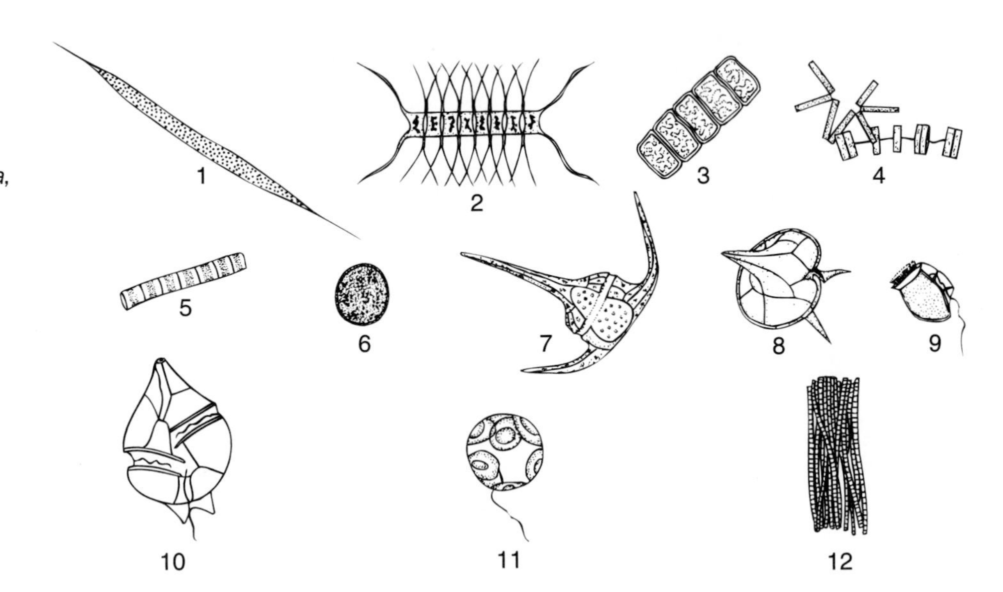

FIGURE 2.4 Diagrammatic representation of some characteristic genera of marine phytoplankton. Diatoms: (1) *Rhizosolenia*, (2) *Chaetoceros*, (3) *Navicula*, (4) *Thalassiosira*, (5) *Skeletonema*, (6) *Coscinodiscus*. Dinoflagellates: (7) *Ceratium*, (8) *Peridinium*, (9) *Dinophysis*, (10) *Gonyaulax*. Coccolithophores: (11) *Coccolithus*. Cyanobacteria: (12) *Trichodesmium*. (Not to scale.)

cyanobacteria but are smaller and have a different combination of pigments. Other important groups are the **haptophytes** (Coccolithophoridae, Haptophyceae) and the **cyanobacteria** (Cyanophyceae, formerly called blue-green algae) (Figure 2.4). The cyanobacteria are prokaryotic cells that possess chlorophyll *a*, but this is not in plastids and occurs in single cells, filaments, or chains. Cyanobacteria are abundant in the tropics, where they occasionally form dense mats of filaments and color the water. The Red Sea was named from the red color of the cyanobacterium *Trichodesmium* (*Oscillatoria*) *erythraeum*.

Haptophytes are tiny eukaryotic flagellates (less than 20 μm) with one or two chromatophores; a unique threadlike appendage, the haptonema; and distinctively marked external calcareous or organic plates. They are abundant in all oceanic waters. The most widespread and abundant haptophytes are the **coccolithophores**, easily distinguished by the tiny calcareous plates (coccoliths) on their outer surface. They have complex life histories, with several morphologically different cells present in the same species, and several modes of reproduction. Coccolithophores are now recognized as a major source of primary production in many ocean areas. Other, less abundant, microalgae include the silicoflagellates (Chrysophyceae), the cryptomonads (Cryptophyceae), and certain motile green algae (Chlorophyceae).

Pelagic bacteria, or bacterioplankton, are also found in all oceans. They are most abundant near the sea surface and are now thought to equal or exceed the total biomass of phytoplankton. They are usually found in association with organic particles in the water column, collectively called **particulate organic carbon (POC)**, or on various gelatinous zooplankton pieces known as marine snow (see p. 87). They are usually motile and gram-negative and occur in coccoid, rod, and spiral forms. They decrease markedly with depth, and their role in the microbial loop of the oceanic food web is now clearly recognized (see pp. 82–85).

THE ZOOPLANKTON

In contrast to the phytoplankton, which consist of a relatively small variety of photosynthetic organisms, the zooplankton are extremely diverse, consisting of a host of larval and adult forms representing most of the animal and many of the protistan phyla. As with the phytoplankton, two groups can be identified: the larger net zooplankton and the smaller microzooplankton (nanozooplankton and picozooplankton). We know much more about the larger net zooplankton because of their size and ease of collection. Marine biologists have long considered these larger net zooplankton to be the most important grazers of the phytoplankton. In recent years, however, the importance of the smaller microzooplankton as grazers has been established (see pp. 82–85). Among the net zooplankton, one group is more important than the others. The subclass Copepoda (subphylum Crustacea, phylum Arthropoda) includes small holoplanktonic crustaceans that dominate the net zooplankton throughout the world's oceans (Figure 2.5). These small animals are of vital importance in the economy of the ecosystems in the oceans because they are one of the primary herbivo-

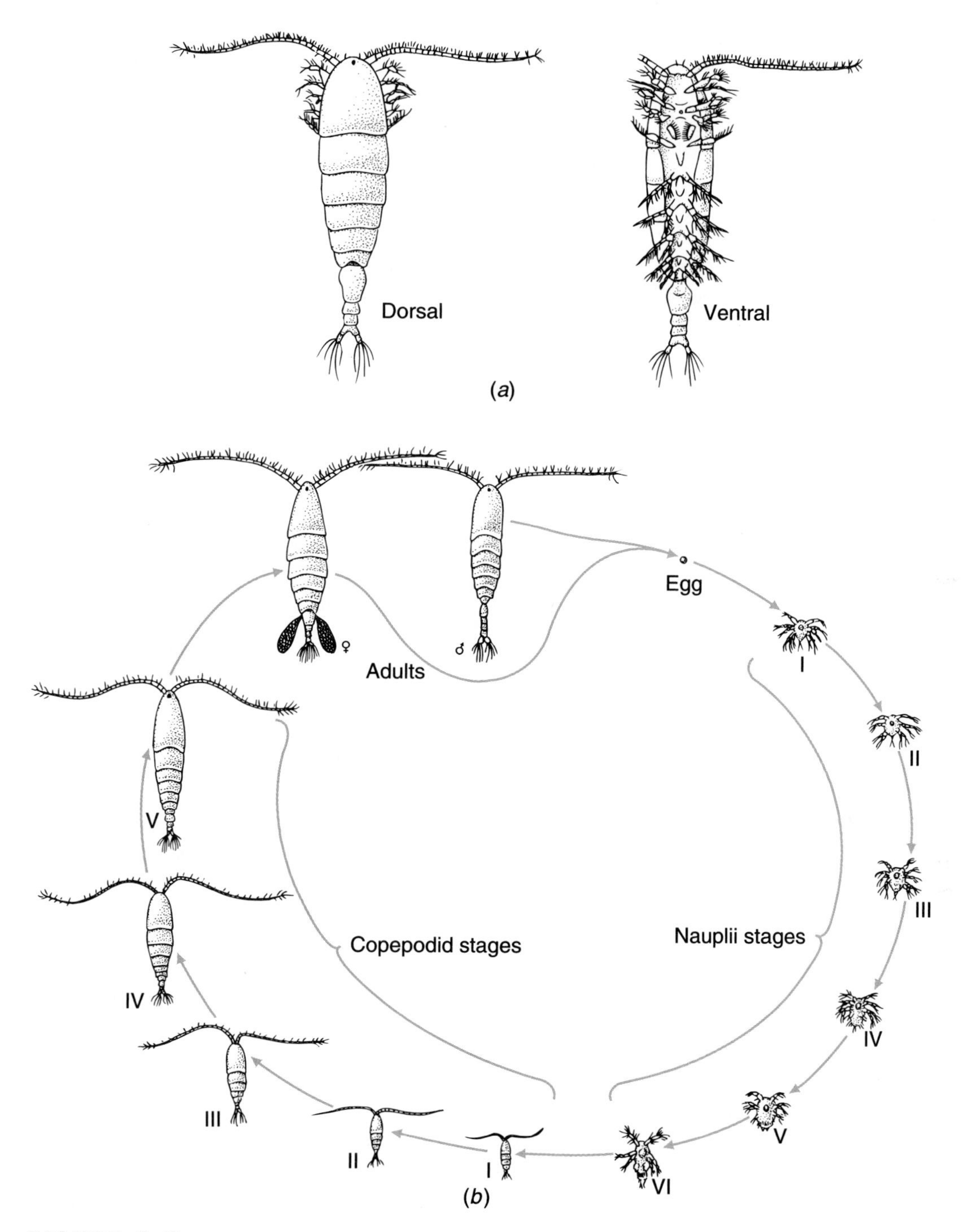

FIGURE 2.5 Copepods. (a) Typical copepod showing major anatomical features. (b) Outline of the typical life cycle of a copepod.

rous animals in the sea. It is they who graze upon the aquatic pastures constituted by the larger phytoplankton, especially in coastal waters, and provide one of the links between the primary production of the autotrophs and the numerous large and small carnivores.

Copepods

Free-living planktonic copepods are generally small, between one and several millimeters in length. They swim weakly, using their jointed thoracic limbs, and have a characteristic jerky movement. They employ their large antennae to slow their rate of sinking. Most of the free-living planktonic copepods have a characteristic body shape and hence are readily recognizable (Figure 2.5a).

Copepods graze on phytoplankton either by using a filtering mechanism that works by removing algal cells from the water or by using the anterior appendages. As Koehl and Strickler (1981) have demonstrated, copepods produce a stream of water past their bodies by flapping several anterior appendages (2nd antennae, mandibular palps, 1st maxillae and maxillipeds). When an algal cell enters this stream of water, the anterior appendages respond by beating asymmetrically, redirecting the water so the entrained algal cell moves toward the animal. The copepod then rapidly flings apart its 2nd maxillae, which then close over the water and the algal cell, forcing water out through the setae while retaining the algal cell (Figure 2.6). Some copepods are carnivorous, seizing prey with their appendages.

In copepods, the sexes are separate, and sperm are transferred to the female packaged in spermatophores. After fertilization, the eggs are enclosed in a sac carried by the female, attached to her body. They hatch as **nauplius** larvae and progress through several naupliar stages and then several more **copepodid** stages before becoming adult (see Figure 2.5). As we shall see, the life history of the adult and larval forms affects the phytoplankton cycles observed in the sea.

Other Zooplankton

Because of the great diversity of zooplankton organisms, it is not practical to cover all in detail. Only the larger and more significant taxonomic groups will be considered. Within each of these taxa are often many different species.

Among the holoplankton, the kingdom Protista is the most important, approaching or even exceeding the copepods as grazers, but is usually overlooked because of the small size of the individuals and their difficulty of capture (Figure 2.7). The dominant groups of the net zooplankton are in the phylum Sarcomastigophora, particularly the order Foraminiferida (forams) and the class Radiolaria (Figure 2.7a, b). Both radiolarians and foraminiferans are single-celled organisms that produce skeletons, calcium carbonate in the case of forams and glass (SiO_2) in the case of radiolarians. Radiolarians are exclusively marine, forams primarily so. Members of these two taxa are so abundant and widespread that their skeletons have formed thick layers of foraminiferan and radiolarian ooze extending over vast areas of the deep-sea floor. Among the nanoplankton, many small protistans are the most important primary consumers. The dominant groups are various ciliates of the phylum Ciliophora and small nonphotosynthetic flagellates of the phylum Sarcomastigophora (Figure 2.7). Both of these groups are exceedingly abundant and are the major grazers of the nanophytoplankton (Figure 2.8).

Holoplanktonic members of the phylum Cnidaria include the various jellyfishes of the classes Hydrozoa (Figure 2.9) and Scyphozoa (Figures 2.10 and 2.11) and the curious, complex hydrozoan colonies known as siphonophores (Figures 2.10–2.12; see Figure 2.25b). Scyphozoan jellyfishes are among the largest plankton organisms, and may occasionally be found in large numbers (see Figure 2.10).

Closely related to the Cnidaria is the phylum Ctenophora (see Figure 2.11; Figure 2.13). With few exceptions, this phylum is entirely planktonic (see Figures 2.11c, d; 2.13, 2.24c, and 2.26a). All are voracious carnivores, capturing food with sticky tentacles or engulfing it with an oversized mouth. Locomotion is via rows of fused, large cilia called ctenes.

The phyla Nemertea and Annelida are represented in the holoplankton by a few highly specialized forms, which are not abundant (Figure 2.14). The annelids include polychaete worms of the families Tomopteridae and Alciopidae, the latter with the largest and best-developed eyes in the Annelida. Most planktonic nemerteans live in deep water.

The phylum Mollusca is the second largest phylum in the animal kingdom. It is usually considered to be composed of slow-moving benthic animals. However, it includes a considerable variety of specially adapted holoplanktonic forms. Perhaps the most highly modified planktonic mollusks are the **pteropods** and **heteropods** (see Figures 2.11e, f and 2.24d). Both groups are probably derived from benthic snails and slugs and are classified in the class Gastropoda. Pteropods are of two types, shelled (order Thecosomata) and naked (order Gymnosomata) (Figure 2.15). Shelled pteropods have fragile shells and swim using their winglike foot. They are particle feeders employing mucous nets and are preyed upon by the faster-swimming naked pteropods. Heteropods are large carnivores with transparent, jellylike bodies (Figures 2.16 and 2.17). A special case is the pelagic shelled snail, *Janthina*, which maintains itself at the surface by clinging to the underside of its own bubble raft (see Figure 2.18; Figure 2.25c). A final group of planktonic mollusks are the squids. Many, perhaps most, squids are powerful, fast

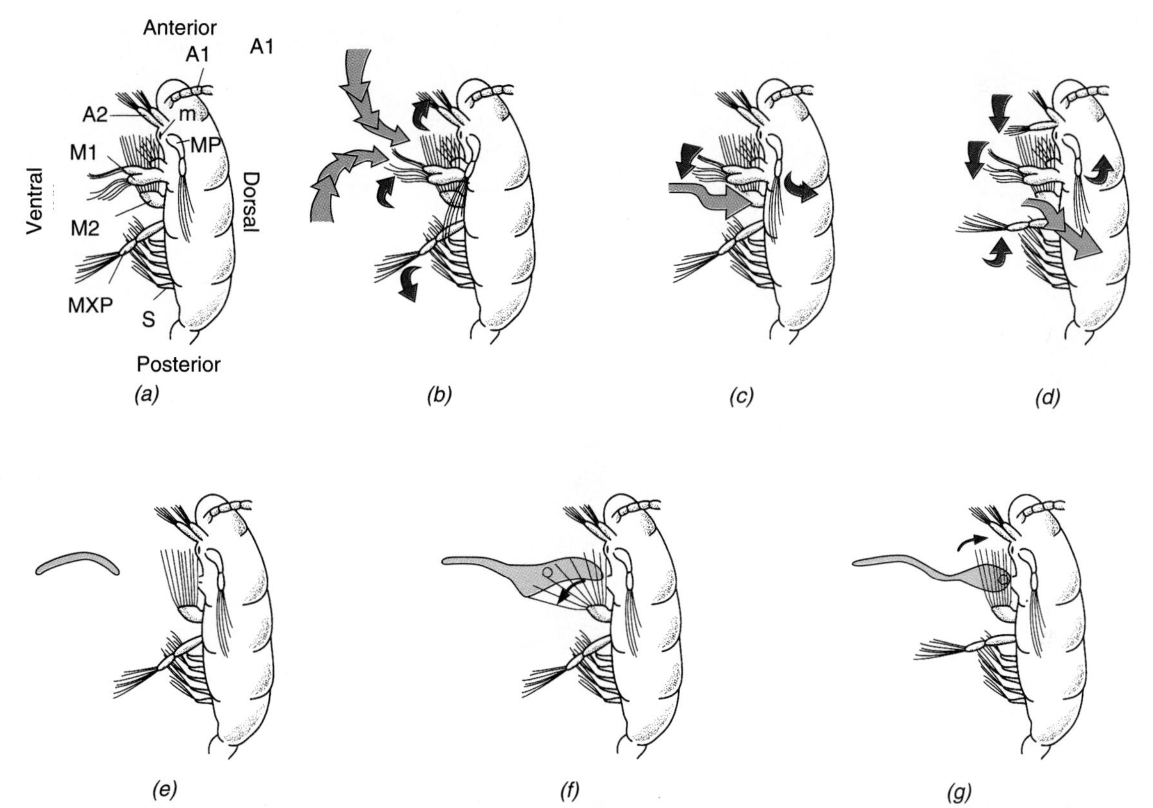

FIGURE 2.6 (a) Diagram of a copepod in feeding position as viewed from the left side. Only the left appendage of each pair is shown. Abbreviations are A1 = 1st antennae; A2 = 2nd antennae; m = mouth; MP = mandibular palp; M1 = 1st maxillae; M2 = 2nd maxillae; MXP = maxilliped; S = swimming appendage. (b–d) Diagrams of feeding appendage movement (darker blue arrows) and water current movement (light blue arrows). Arrows with narrow shaft and wide head indicate movement laterally toward the reader; arrows with a wide shaft and narrow head indicate movement away from the reader. (e–g) Diagrams indicating mechanism of food capture in calanoids. (e) shows movement of water bypassing 2nd maxillae when no cells are present; (f) shows the 2nd maxillae flung open when an algal cell is present (small arrow); (g) shows final capture of algal cell by an inward sweep of 2nd maxillae. (From "Copepod Feeding Currents" M. A. R. Koehl & J. R. Strickler, *Limnology & Oceanography,* Vol. 26, pp. 1061–1072, 1981. Reprinted by permission of the American Society of Limnology and Oceanography.)

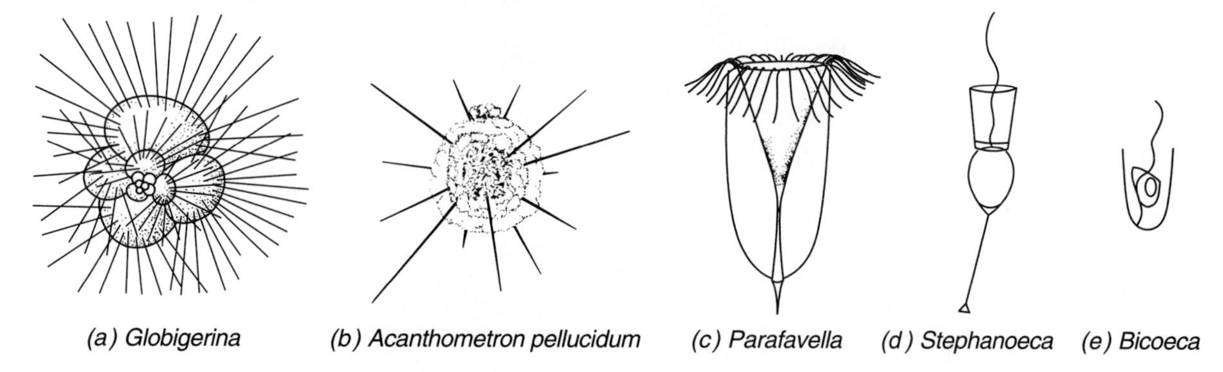

FIGURE 2.7 Holoplankton. Examples of some typical plankton members of the kingdom Protista. (a) Phylum Sarcomastigophora, order Foraminiferida. (b) Phylum Sarcomastigophora, order Radiolaria. (c) Phylum Ciliophora. (d) Phylum Sarcomastigophora. (e) Phylum Sarcomastigophora. (Not to scale.) (a–c after *A Guide to Marine Coastal Plankton and Marine Invertebrate Larvae* by DeBoyd L. Smith. Copyright © 1977 by Kendall/Hunt Publishing Company. Used with permission.)

FIGURE 2.8 Epifluorescence photomicrograph of a DAPI stained flagellate consuming a red fluorescing phytoplankter. (Photo courtesy of Nick Welschmeyer.)

FIGURE 2.9 A Hydrozoan jellyfish *Crassota alba*. (Approximate size is 15 mm.) (Photo courtesy of Dr. George Mastsumoto.)

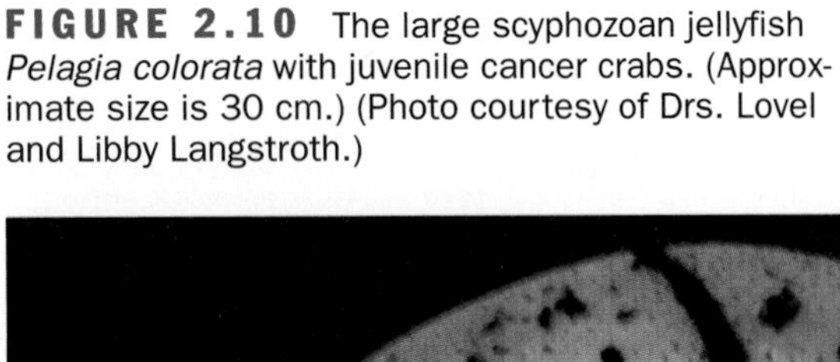

FIGURE 2.10 The large scyphozoan jellyfish *Pelagia colorata* with juvenile cancer crabs. (Approximate size is 30 cm.) (Photo courtesy of Drs. Lovel and Libby Langstroth.)

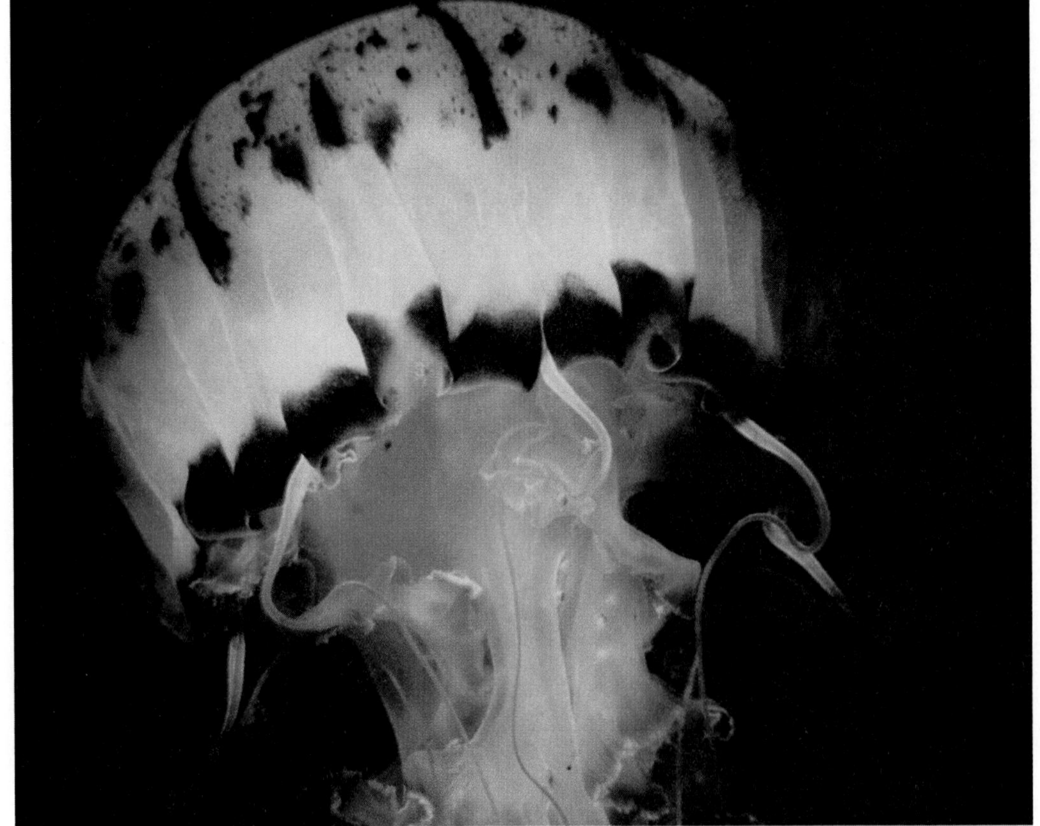

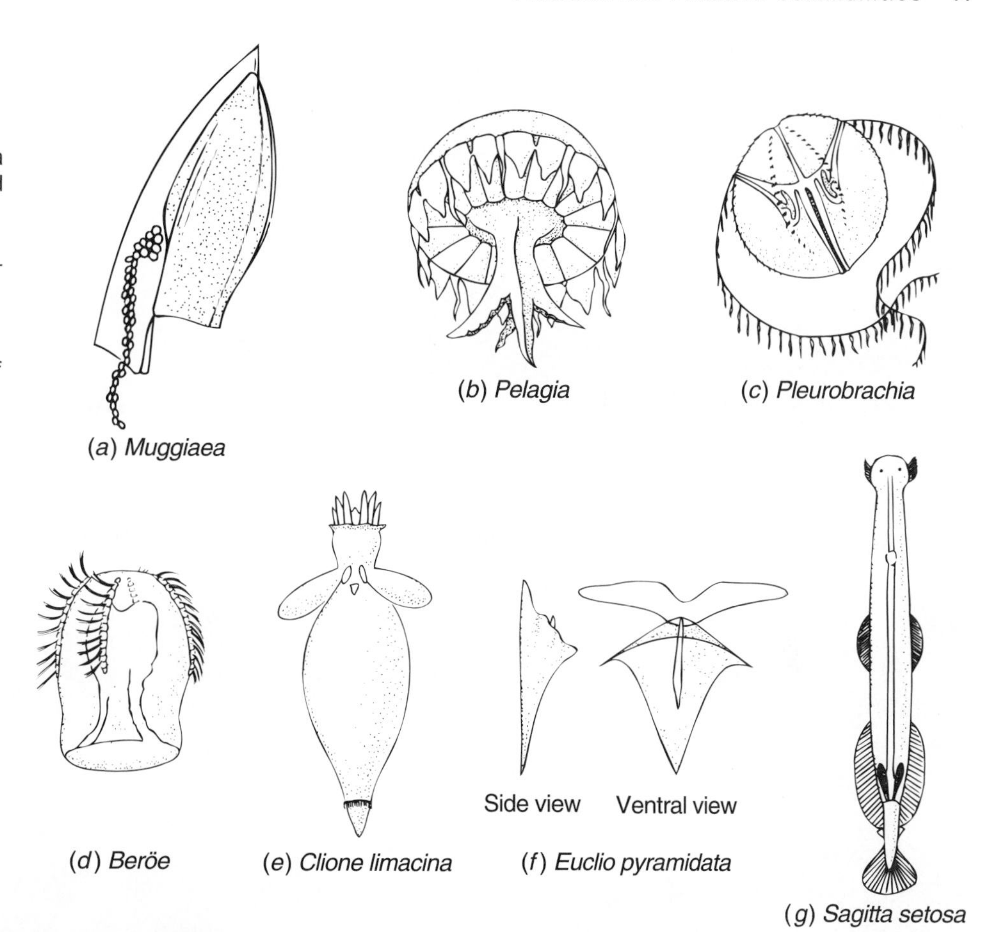

FIGURE 2.11 Holoplankton. Examples of some typical plankton members of the phyla Cnidaria (a, b); Ctenophora (c, d); Mollusca (e, f); and Chaetognatha (g). (Not to scale.) (a, e, f, g from *Marine Plankton: A Practical Guide* by G. E. Newell & R. C. Newell, © 1973 Century Hutchinson. Reprinted by permision of the author. b, c, d from *A Guide to Marine Coastal Plankton and Marine Invertebrate Larvae* by DeBoyd L. Smith. Copyright © 1977 by Kendall/Hunt Publishing Company. Used with Permission.)

FIGURE 2.12 The siphonophore *Nanomia bijuga.* Only the upper portion of the colony is visible in this photo. (Approximate size is 15 cm.) (Photo courtesy of Dr. George Matsumoto.)

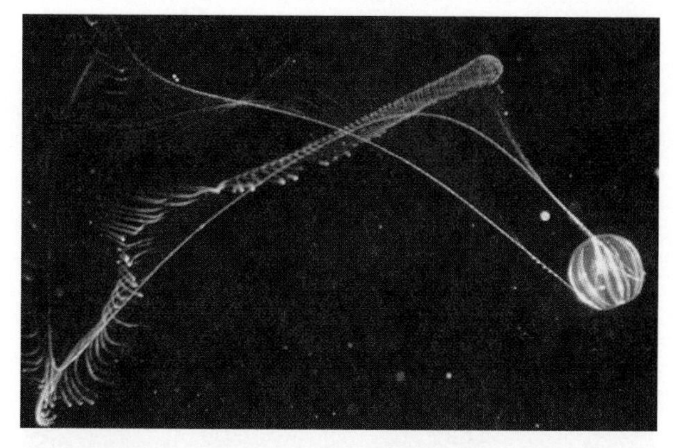

FIGURE 2.13 A ctenophore of the genus *Pleurobrachia.* Note the long tentacles used to capture prey. (Approximate size of body is 2 cm.) (Photo courtesy of Dr. George Matsumoto.)

FIGURE 2.14 A holopelagic polychaete of the genus *Tomopteris*. (Approximate size is 8 cm.) (Photo courtesy of Dr. George Matsumoto.)

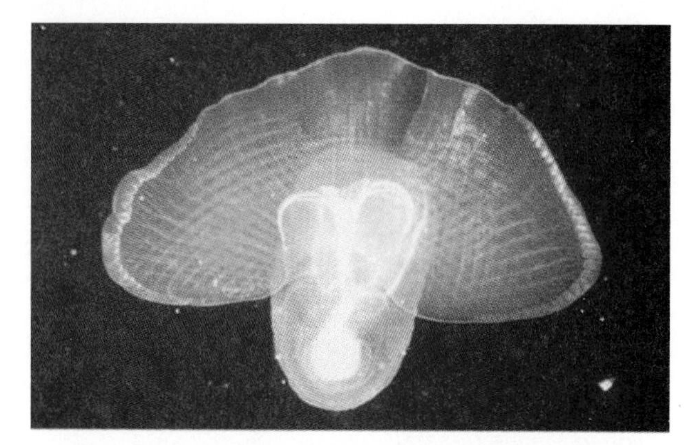

FIGURE 2.15 The largest pelagic pteropod mollusk genus, *Corolla*. These animals feed using a large mucous net, not shown here, that may exceed 2 m in diameter. (Approximate size is 8 cm.) (Photo courtesy of Dr. George Matsumoto.)

FIGURE 2.16 A small heteropod mollusk of the genus *Oxygyrus* with reduced transparent shell. (Approximate size is 20 mm.) (Photo courtesy of Dr. Roger Seapy.)

FIGURE 2.17 The heteropod *Pterosoma planum*, a visual planktonic predator. (Approximate size is 70 mm.) (Photo courtesy of Dr. Roger Seapy.)

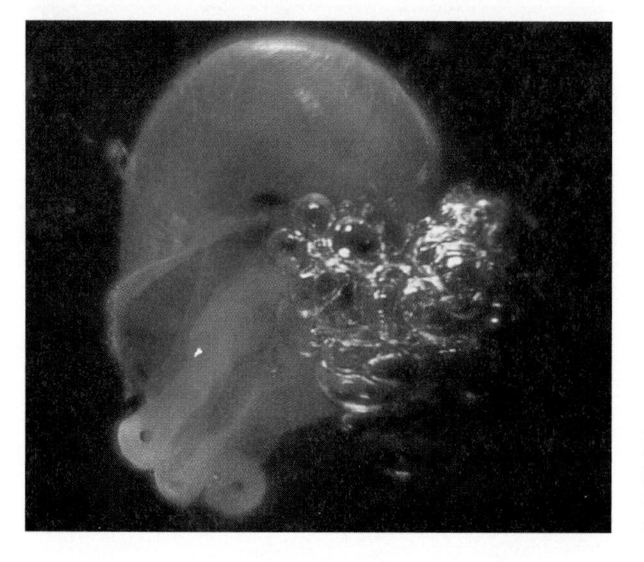

FIGURE 2.18 *Janthina prolongata*, a shelled gastropod living at the air-water interface suspended from a bubble raft. (Approximate size is 3 cm.) (Photo courtesy of Dr. Roger Seapy.)

swimmers that must be considered nekton. There are, however, a significant number of very small squids that are not strong swimmers and must be considered planktonic (see Figure 2.24b; Figure 2.19).

It is among the phylum Arthropoda that the greatest number of plankton organisms occur. In the seas of the world, virtually all belong to the subphylum Crustacea. In addition to the dominant copepods, the holoplanktonic crustaceans include members from the order Cladocera (Figure 2.20a, b), class Ostracoda (Figure 2.20d), order Mysidacea, order Amphipoda, order Euphausiacea (Figures 2.20c and 2.21), and order Decapoda. Most of these taxa are small filter feeders straining algae and/or small animals out of the water.

The phylum Chaetognatha is very small, consisting of about 100 species, which, with the exception of one genus, are planktonic. Chaetognaths, or "arrow worms," are abundant members of the plankton throughout the world (see Figures 2.11g and 2.26d). All are voracious predators of copepods and other planktonic organisms.

The final phylum represented in the holoplankton is the Chordata. Planktonic chordates belong to the classes Thaliacea (salps) and Larvacea (see Figure 2.20e, f; Figure 2.22). These gelatinous-bodied animals are filter feeders. Larvaceans construct a "house" around themselves and pump water through a screen in the house to filter out their food. "Houses" are continually built and shed.

Compared with the holoplankton, meroplankton as a group are even more diverse. A bewildering array of larval forms constitute the meroplankton. These larvae are derived from virtually all animal phyla and from all different marine habitats. The number of larval forms is greater than the number of species in the sea that produce them, because many species go through a series of larval stages before becoming adult. Each of these species may be represented in the plankton by several different stages. This is particularly true in the Crustacea, where some decapods may have as many as 18 different larval stages, and most have more than one. Phyla that are not represented in the holoplankton, such as Bryozoa, Phoronida, Echinodermata, and Porifera, have larval forms (Figure 2.23a–f). Other abundant phyla, such as Nemertea, Mollusca, and Annelida, which have but a few highly adapted holoplankton groups, have planktonic larvae (Figure 2.23g, h, and l). The Crustacea have by far the greatest number of larval types, and these also include larvae of the abundant holoplankton organisms. Even the nekton may be represented here, since many fishes have eggs and larvae in the plankton. Although space does not allow treatment of all the types, next we will discuss some adaptations of the plankton.

FLOTATION MECHANISMS

Plankton tend to have a density (mass per unit volume) that is somewhat greater than that of seawater. This means that any plankton organism will eventually sink in the water column. This can be deleterious for both phytoplankton and zooplankton; phytoplankton would sink below the lighted areas of the sea and be unable to photosynthesize, and zooplankton would sink out of the area where their phytoplankton food occurs. On the other hand, there may be some advantage to a selective sinking in the water column for phytoplankton because there are more nutrients deeper in the water (see pp. 62–65). In general, however, uncontrolled or continued sinking is undesirable for both groups. Since they are weak swimmers unable to cope with winds and currents and since living flesh tends to sink, how do these organisms cope with the problem of staying in the upper layers of the sea?

Principles

To answer the above question, we must consider some physical and chemical characteristics that will bear on this problem. The density of seawater is a function of two factors, its temperature and its salinity—seawater becomes more dense as its salinity increases and less dense as its temperature increases. As a result, wherever there are temperature changes over the course of a year, such as in temperate seas, the density also varies. Another physical characteristic is the viscosity of seawater. Viscosity is related to temperature and salinity in that the more saline water is and the lower its temperature, the more viscous it is and the less rapidly things sink in it.

A second set of principles has to do with the effect of shapes on the rate of sinking in dense liquids. Objects of different shape but similar weights fall at differing rates in both gases, such as air, and liquids, such as water. The rate of fall is proportional to the amount of resistance the body offers to the gas or liquid through which it moves. Objects with a great amount of surface for a given weight tend to fall at slower rates than objects of similar weight with less surface area. Included here is the physical law that surface area increases as the square of the linear dimensions of the object, but the volume increases as the cube of the same dimensions. This means that for a given or fixed shape the smaller the body, the greater the surface area relative to the mass.

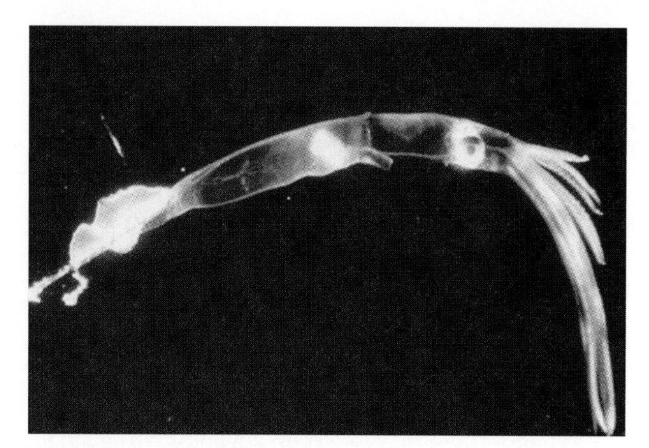

FIGURE 2.19 A small planktonic squid of the genus *Chiroteuthis*. (Approximate size is 15 cm.) (Photo courtesy of Dr. George Matsumoto.)

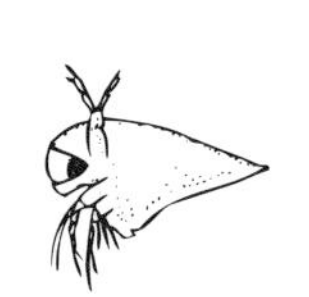

(*a*) *Evadne nordmanni* (♂)

(*b*) *Podon leucarti* (♀)

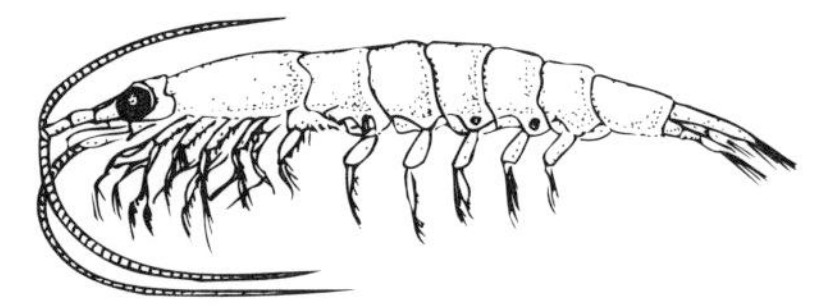

(*c*) *Meganyctiphanes norvegica*

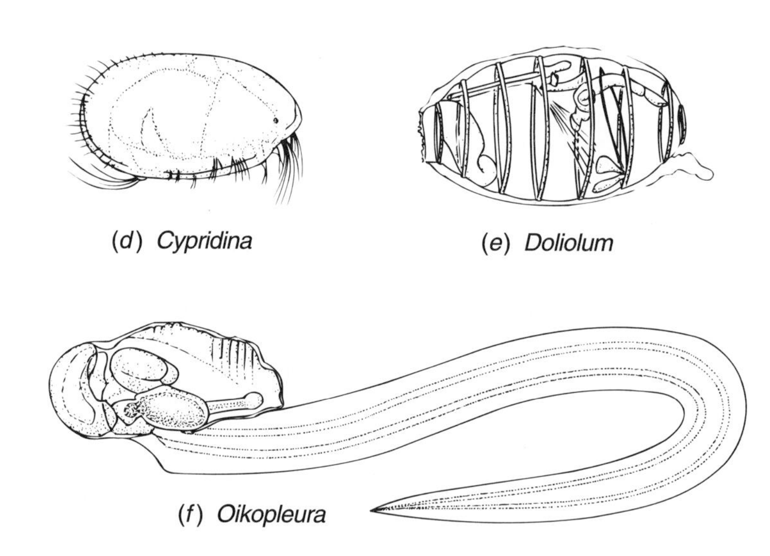

(*d*) *Cypridina*

(*e*) *Doliolum*

(*f*) *Oikopleura*

FIGURE 2.20 Holoplankton. Typical planktonic forms of the phyla Arthropoda (a, b, c, d) and Chordata (e, f). (a, b) Order Cladocera, subphylum Crustacea. (c) Order Euphausiacea, subphylum Crustacea.(d) Class Ostracoda, subphylum Crustacea. e) Salp of the class Thaliacea. (f) Class Larvacea. (Not to scale.) (a–c from *Marine Plankton: A Practical Guide* by G. E. Newell & R. C. Newell, © 1973 Century Hutchinson. Reprinted by permision of the author. d–f from *A Guide to Marine Coastal Plankton and Marine Invertebrate Larvae* by DeBoyd L. Smith. Copyright © 1977 by Kendall/Hunt Publishing Company. Used with Permission.)

FIGURE 2.21 *Euphausia*, an individual of one of the most abundant groups of zooplankton in the oceans of the world. (Approximate size is 2 cm.) (Photo courtesy of Dr. Roger Seapy.)

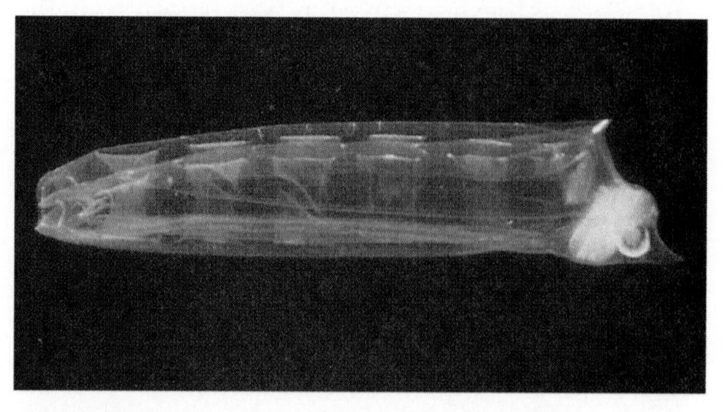

FIGURE 2.22 A salp of the genus *Iasis*. (Approximate size is 5 cm.) (Photo courtesy of Dr. Roger Seapy.)

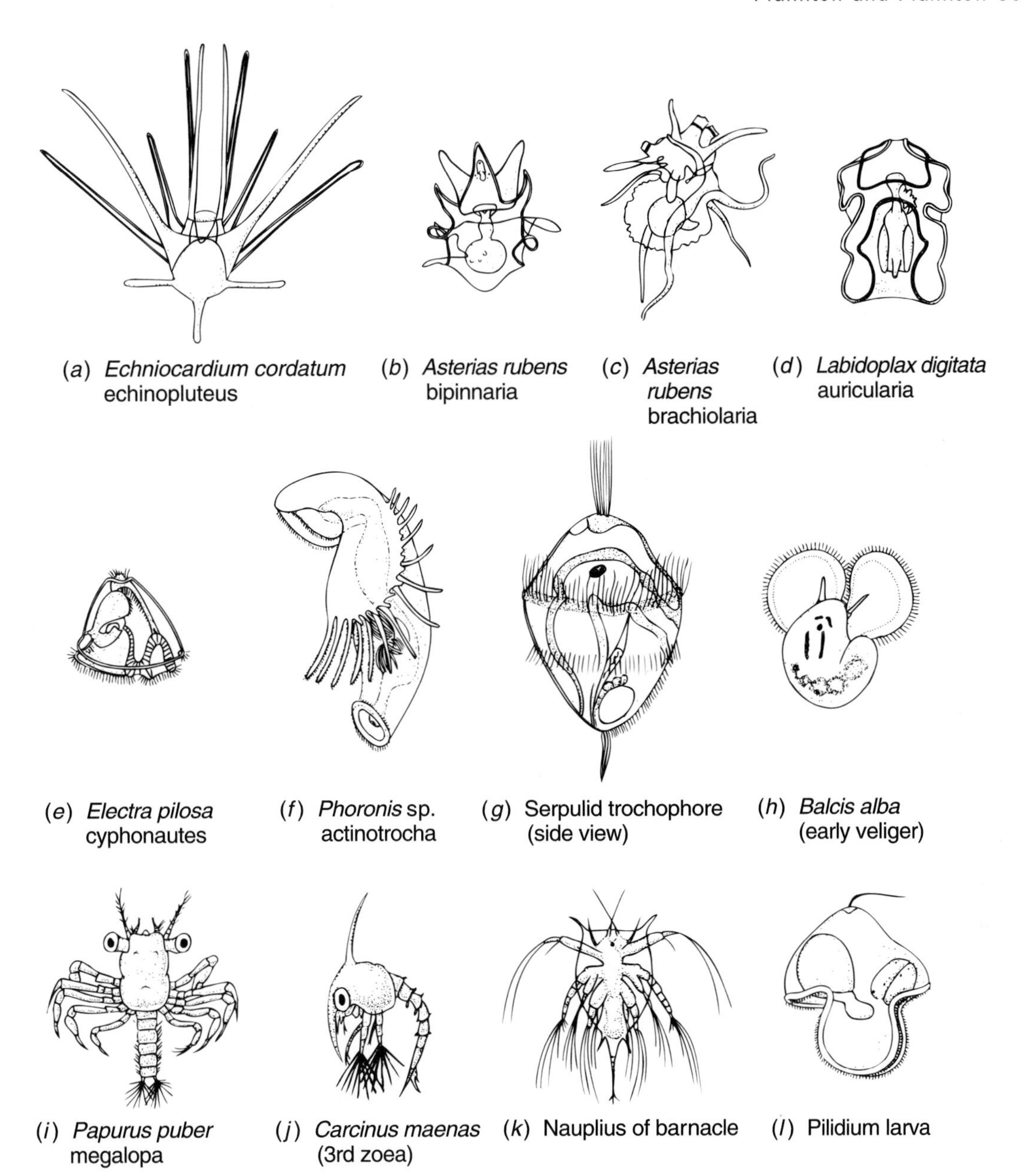

FIGURE 2.23 Meroplankton. Examples from several phyla. (a) Echinodermata, sea urchin. (b, c) Echinodermata, starfish. (d) Echinodermata, sea cucumber. (e) Bryozoa. (f) Phoronida. (g) Annelids, Polychaeta. (h) Mollusca, gastropod. (i, j) Arthropoda, Crustacea, Decapoda. (k) Crustacea, barnacle. (l) Nemertea. (a–j from *Marine Plankton: A Practical Guide* by G. E. Newell & R. C. Newell, © 1973 Century Hutchinson. Reprinted by permision of the author. k, l from *A Guide to Marine Coastal Plankton and Marine Invertebrate Larvae* by DeBoyd L. Smith. Copyright © 1977 by Kendall/Hunt Publishing Company. Used with Permission.)

From the previously mentioned physical laws governing changes in density and viscosity of water and those governing resistance, it is possible to derive a very simple equation that will relate sinking rates of organisms to these parameters. The equation is

$$SR = \frac{W_1 - W_2}{(R)(V_W)}$$

where SR = sinking rate; W_1 = density of the organism; W_2 = density of seawater; $W_1 - W_2$ = amount of overweight (overweight = the amount by which the flesh of the organism exceeds the weight of a similar volume of water); R = surface of resistance; V_W = viscosity of water.

The organisms can do nothing about the density or the viscosity of water. All adaptations of the organism to reduce the sinking rate must be concerned with either reducing the amount of overweight or increasing the surface of resistance.

Let us consider first those adaptations designed to reduce the amount of overweight.

Reduction of Overweight

One mechanism that may be employed to reduce overweight is simply to alter the composition of the body fluids so they are less dense than an equal volume of seawater. It is necessary, however, to maintain the same number of solute particles in the body fluid as in the surrounding seawater in order to avoid osmotic problems. **Osmosis** is the passage of water across semipermeable membranes, such as cell membranes, to equalize the water content on both sides of the membrane. How is this change in composition accomplished? Perhaps the most common way is to replace heavy chemical ions in the body fluids with lighter ones. This allows the animal to maintain the same osmotic condition (same number of solute particles) while becoming lighter with respect to seawater. An example is the dinoflagellate *Noctiluca* (Figure 2.24a), whose internal fluid contains ammonium chloride (NH_4Cl) and which is iso-osmotic with seawater but is less dense, having a specific gravity of 1.01 versus the 1.025 of seawater. Similarly, the cranchid squids, which have very fat, bulbous bodies for squid, are filled with NH_4Cl, making them less dense than seawater (Figure 2.24b).

In a similar fashion, certain planktonic forms such as salps, ctenophores, and heteropods (Figure 2.24c, d) actively exclude heavy ions such as SO_4^{2-} from their bodies and replace them with osmotically similar but lighter chloride ions.

Another flotation mechanism is the development of special gas-filled floats. Since gas is much less dense than a similar volume of water, buoyancy is assured. Perhaps the most familiar examples of this approach are the floats of the Portuguese man-of-war (*Physalia*)(Figure 2.25a, b) and the swim bladders of fishes. If the animal is able to regulate the gas pressure in the float or bladder, it can move up or down in the water column at will. This is what some fishes do.

Similar to the use of gas-filled floats is the employment of liquids that are less dense than water. Prominent among these liquids are oils and fats. These can serve a dual purpose, as fats and oils also act as food reserves for the organisms. Copepods, for example, often store excess food in the form of oil droplets under the carapace, and these droplets aid in buoying the animal. Diatoms also store food as oils.

Whereas some planktonic organisms have employed the various flotation mechanisms to alter their density and remain afloat, many others have not. If we return to our original equation, we can see that, if overweight cannot be reduced, the only other option is to increase the surface of resistance to the water and at least slow the rate of sinking.

Changes in Surface of Resistance

Surface of resistance may be increased in a number of ways, and all may be observed in planktonic organisms.

One of the most common observations made about plankton is that the general body size is small. This is particularly true for tropical plankton. As we noted earlier, surface area increases as the square of the linear dimension, whereas volume increases as the cube. This means that the smaller the organism, the greater the surface area relative to volume. By remaining small, plankton organisms offer far more surface area of resistance to sinking per unit volume of living material than if they were large. Small size is particularly important for tropical plankton, because they live in warmer and thus less viscous water, where sinking would be faster.

The other way of increasing the surface of resistance is to change the shape of the body. If one were to drop a round ball and a flat coin of equal weight into water, the ball would sink more quickly because it has less surface of resistance to the fluid. In a like manner, planktonic organisms have evolved various flattened body shapes or appendages. Such examples (Figure 2.26) include a host of different species from many different phyla.

Even more common than changes in body shape is the development of various spines and body projections. These structures add considerable resistance, but little to the weight. They may also deter predators by preventing the predator from ingesting or swallowing the organism. Such adaptations are common in the various diatoms (see Figure 2.2a), radiolarians (see Figure 2.25b), foraminiferans, and crustaceans.

Water Movements

A final mechanism of buoyancy has to do not with the organisms, but with the nature of water movements

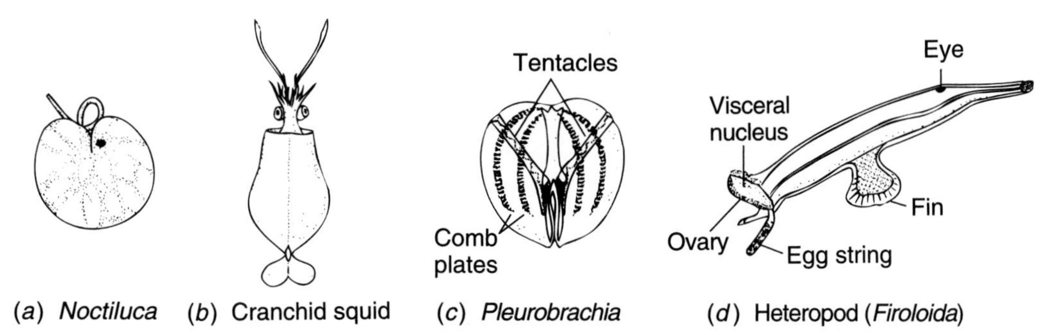

FIGURE 2.24 Flotation mechanisms. Examples of organisms that exclude heavy ions, forming body fluids less dense than seawater. (Not to scale.) (From *An Introduction to the Study of Tropical Plankton*, J. H. Wickstead, 1985.)

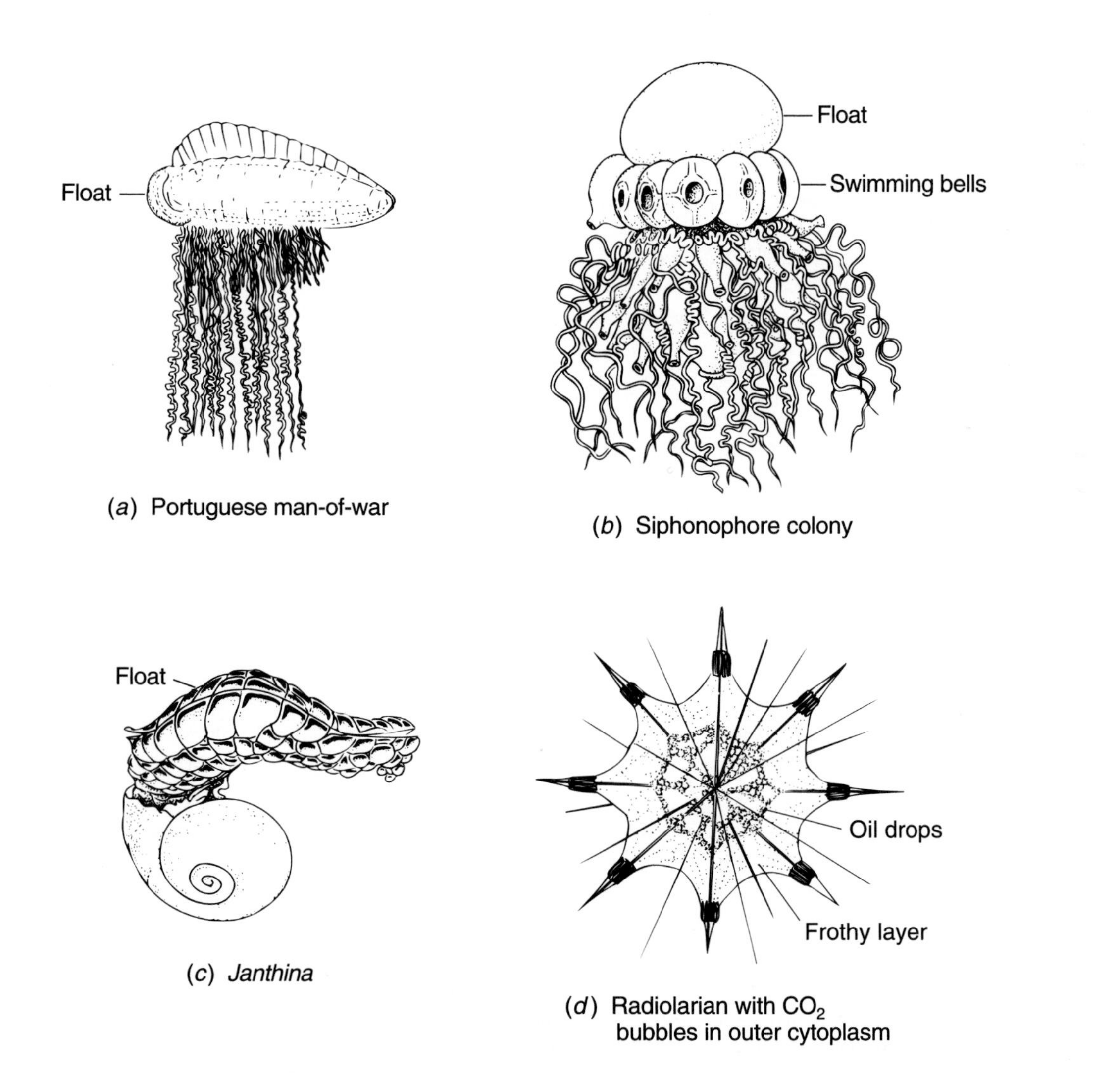

FIGURE 2.25 Flotation mechanisms. Examples of organisms that use gas- or fluid-filled floats. (a) and (b) are both siphonophore colonies. (a, from *Marine Plankton: A Practical Guide*, G. E. Newell & R. C. Newell, © 1973 Century Hutchinson. Reprinted by permission of the author. b, d, from L. Hyman, 1940, *The invertebrates*, Vol. 1, *Protozoa to Ctenophora*, McGraw-Hill. By permission.)

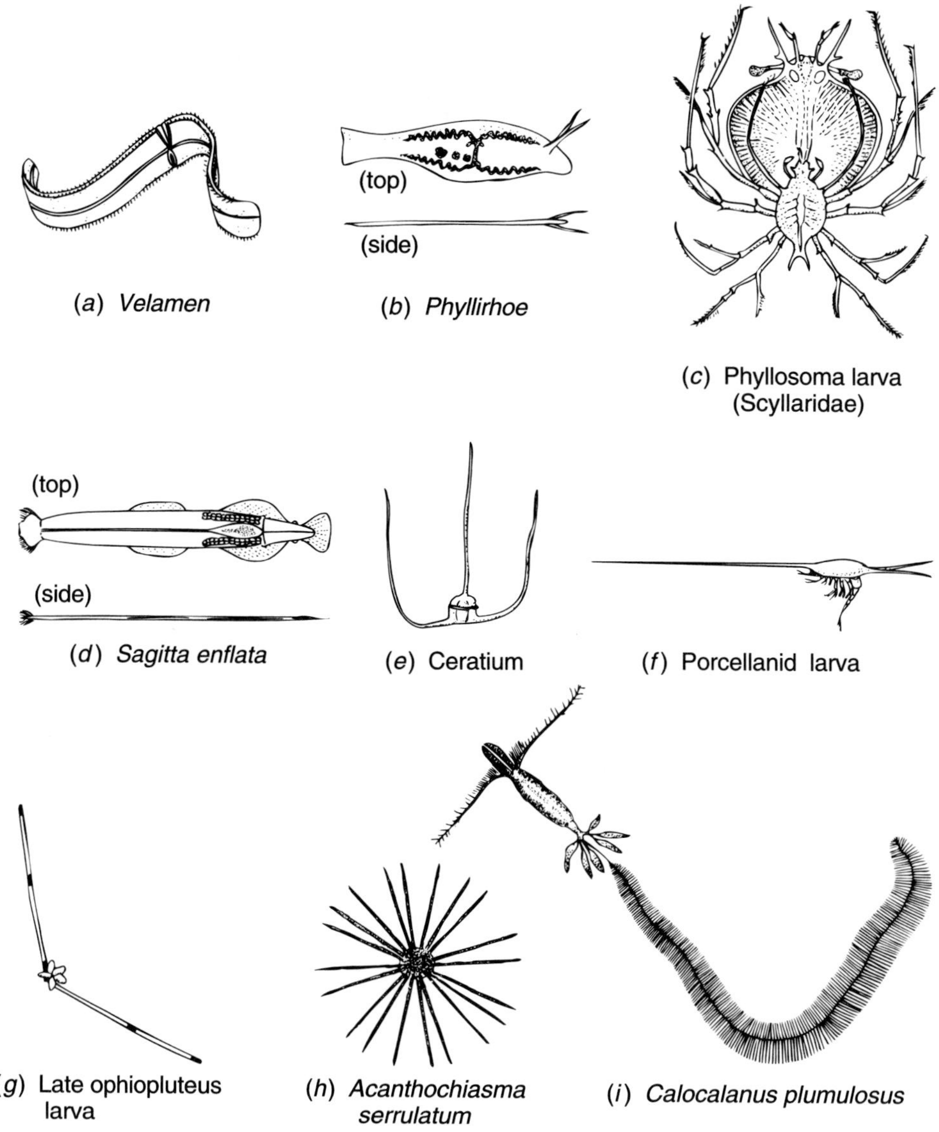

FIGURE 2.26 Flotation mechanisms. Examples of organisms that are flat or that have spines: (a) Phylum Ctenophora. (b) Phylum Mollusca, class Gastropoda. (c) Phylum Arthropoda, class Crustacea. (d) Phylum Chaetognatha. (e) Order Dinoflagellata. (f) Phylum Arthropoda, order Decapoda. (g) Phylum Echinodermata, class Ophiuroidea. (h) Phylum Protozoa, class Sarcodina. (i) Phylum Arthropoda, subphylum Crustacea. (Not to scale.) (a-g from *An Introduction to The Study of Tropical Plankton*, J. H. Wickstead, 1985, Century Hutchinson. h from *Marine Plankton: A Practical Guide*, G. E. Newell & R. C. Newell, © 1973 Century Hutchinson. Reprinted by permission of the author. i, from *The Cambridge Natural History*, Vol. 4, *Crustacea*, reprinted in 1968 by Wheldon and Wesley.)

(*a*) *Velamen*

(top)
(side)
(*b*) *Phyllirhoe*

(*c*) Phyllosoma larva (Scyllaridae)

(top)
(side)
(*d*) *Sagitta enflata*

(*e*) Ceratium

(*f*) Porcellanid larva

(*g*) Late ophiopluteus larva

(*h*) *Acanthochiasma serrulatum*

(*i*) *Calocalanus plumulosus*

in the ocean. In the ocean, the surface water heats up during the day and cools at night. This alternating heating and cooling changes the density and leads to the creation of convection cells, which are small units of water that are either sinking or rising according to their density. These are gentle movements, and plankton may be moved by them.

On a small scale, the action of the wind blowing over the water can produce localized vertical movement of plankton due to the production of **Langmuir convection cells**. Langmuir cells are produced when wind speeds are above 3 meters per second. Each Langmuir cell is a few meters wide and hundreds of meters long. In each cell, wind causes the water to move away from the center (divergence) toward the outside, where it meets the water from the adjacent cell. These two convection currents converge and downwell along the line of convergence for a short distance. They then move horizontally, where they again meet water moving in a similar direction from the adjacent cells, whereupon they rise to the surface to complete the cycle (Figure 2.27). Plankton organisms in such cells will tend to be carried downward at the margins of the cell (convergence) and upward in the center (divergence). On a larger scale, upwelling (pp. 64–65) can carry plankton organisms toward the water surface.

PRIMARY PRODUCTION

Almost all life on the earth depends on chlorophyll bearing organisms being able to use the energy of sunlight to synthesize energy-rich organic molecules from inorganic materials. This is the process of photosynthesis. The general equation for this process is

$$6CO_2 + 6H_2O \xrightarrow[\text{nutrients}]{\text{sunlight}} \underset{\text{energy-rich organic compounds}}{C_6H_{12}O_6} + 6O_2$$

The basis for nearly all life in the sea is the photosynthetic activity of the aquatic autotrophs. There are, however, significant differences in the form of the organisms and in the location and extent of maximum photosynthetic activity imposed by the particular physical and chemical conditions of the oceans (see pp. 59–66). These will be covered subsequently. Photosynthetic organisms are diverse and widespread in the world's oceans. They include the pelagic phytoplankton and cyanobacteria, benthic microalgae, benthic macroalgae (kelps, seaweeds), seed plants (sea grasses, mangroves, salt marsh plants), and symbiotic algal cells in corals and other animals. However, the pelagic phytoplankton (including prochlorophytes and cyanobacteria) are responsible for perhaps 95% of all marine primary productivity. It is for this reason we consider primary productivity in detail here.

Primary Productivity

Primary productivity is defined as the rate of formation of energy-rich organic compounds from inorganic materials. Primary productivity is usually considered synonymous with photosynthesis, but this is not quite correct, since a minor amount of primary productivity may be produced by chemosynthetic bacteria (see Chapter 4). Here, however, primary productivity is confined to photosynthesis.

The total amount of organic material fixed in the primary productivity process is termed the **gross primary production**. Since some of this total production must be used by the plants themselves to operate their life processes, collectively called respiration, a lesser amount is available for transfer or use by the other organisms of the sea. **Net production** is the term given to the amount of total production left after losses from plant respiration and available to support other trophic levels.

For either gross or net primary production, the rates are usually expressed in terms of grams of carbon fixed per unit area or volume of seawater per time interval. For example, production may be reported in grams of carbon per square meter per year (g C/m^2/yr), which represents the production integrated over all depths from the surface to the base of the euphotic zone.

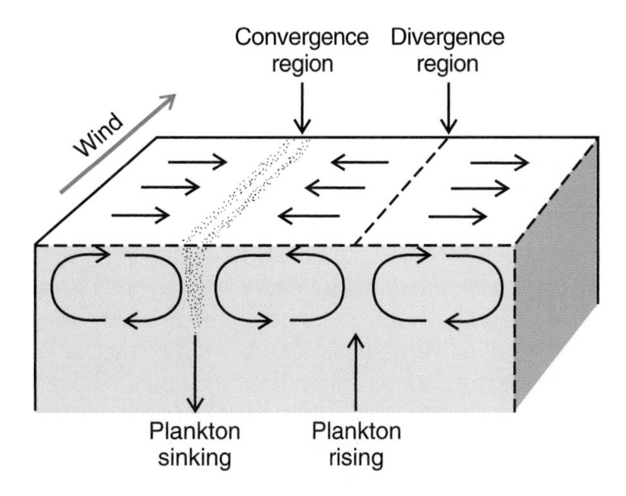

FIGURE 2.27 Langmuir convection cells induced by wind.

A few other terms require definition at this point. The **standing crop**, as applied to autotrophs, is the total amount of the organism's biomass present in a given volume of water at a given time. Primary productivity and standing crop can vary considerably on a time scale of a few days to a year, and this variation is the result of a large number of factors that act both directly and indirectly on the photosynthetic process within autotrophs and on the autotrophs themselves. These factors will be discussed later in this chapter.

As we saw in Chapter 1, water has a profound effect on the light that penetrates it. Because water absorbs light, there is less light energy available as one goes deeper, and eventually, light disappears. For photosynthetic organisms, this means that at some depth, the light energy is just sufficient for the organism to fix energy at a rate equal to the rate at which it uses energy in its own metabolic processes. In other words, there is a depth at which the respiratory usage of energy by the photosynthetic organism equals the ability of the photosynthetic mechanism to produce energy. If the organism goes deeper, the respiratory needs continue at the same rate, but the decreasing light is insufficient to allow the photosynthetic process to keep up and so there is a net loss of energy. The depth at which the rate of respiration of a photosynthetic organism is just equaled by the rate of photosynthesis is called the **compensation depth**. Above this depth, the rate of photosynthesis exceeds that of respiration, and there is a net production of carbon, or a net primary production.

The compensation depth varies over the world's oceans because it depends on the clarity of the water. The clearer the water, the greater the light penetration and the deeper the compensation depth. In general, the compensation depth is deeper in clear, open ocean waters and shallower in inshore waters where large amounts of particulate matter are found in the water.

Since the compensation depth varies with depth of light penetration, it should also be possible to define the compensation depth in terms of light intensity. In fact, this is true. The **compensation intensity** is the light intensity at which photosynthesis equals respiration. Although this intensity is somewhat different for different phytoplankton species, it is approximately that depth at which the intensity has been reduced to 1% of the surface light intensity. The compensation depth can then be approximated as the depth to which 1% of the incident light penetrates.

It is well to remember that the compensation depth varies, not only in different geographic areas of the world's oceans as noted previously, but also from day to day and season to season in any one geographical area, as a result of changes in light with season (see pp. 59–62), changes in water clarity, and availability of nutrients.

Measurement of Primary Productivity

If we return to the photosynthetic equation on page 55, we can see that it should theoretically be possible to measure the rate of production of organic compounds by measuring the rate of appearance or disappearance of some component of that equation. If one could measure the rate of disappearance of CO_2 or the appearance of O_2, this would be a measure of the rate of photosynthesis. In fact, this is done, but for various reasons it is practical only for the two components mentioned.

The classical method used for many years to measure primary productivity is called the **light-dark bottle method**. In this method, two identical bottles are employed. One bottle is completely transparent, while the other is made completely opaque by painting it black or covering it with some opaque substance like aluminum foil. Into each is placed the same volume of seawater taken from the body of water for which you wish to estimate productivity. The water contains the naturally occurring phytoplankton and zooplankton organisms. The oxygen content of the water to be added to the two bottles is determined on a separate sample to establish the initial concentration. The bottles are then stoppered, attached to a line, and returned to the appropriate depth in the sea. In the bottle that is dark, no photosynthesis can occur, but the phytoplankton and zooplankton within continue to respire and use up oxygen. In this bottle, then, the original oxygen content will decrease as the respiring phytoplankton and zooplankton use up the dissolved oxygen. In the light bottle, on the other hand, photosynthesis continues in excess of respiration; hence, oxygen builds up over the initial values, since it cannot escape. After the two bottles have incubated for a period of time determined by the investigator, the bottles are brought up, unstoppered, and the water is analyzed for oxygen content. Usually, the oxygen content is determined by the Winkler method.

Once the oxygen content is known for each bottle after incubation, it is possible to calculate the rate of photosynthesis from these results and the initial concentration of oxygen of the water when put into the bottles. To understand how this is done, it is necessary to consider what has happened in each bottle. When the original seawater was added to each bottle, it contained a certain amount of oxygen already present (initial oxygen concentration $= O_1$). In the light bottle, this oxygen was used by the organisms for respiration. At the same time, however, the phytoplankton organisms were producing more oxygen through photosynthesis. The final oxygen concentration in the light bottle, then, is the result of oxygen added by photosynthesis plus initial oxygen concentration, less the oxygen used in respiration. Since we know the initial amount of O_2, in order to estimate the total amount of oxygen produced by photosynthesis, we need to know how much is consumed in respiration. If there were no respiration, only photosynthesis, in the light bottle, the increase in oxygen would be greater than that actually observed. This is where the dark bottle comes in. In the dark bottle, only respiration occurred and the oxygen concentration decreased. Since both bottles have the same volume and were incubated the same length of time, the decrease in oxygen in the dark bottle is a measure of the total respiration in the light bottle. To find the total amount of oxygen produced in photosynthesis, then, it is necessary to make a few simple additions and subtractions. First, the initial oxygen concentration of the water (before incubation) is subtracted from the final oxygen concentration in the light bottle, and in the dark bottle the final oxygen concentration is subtracted from the initial concentration. The value of the subtraction in the case of the light bottle is the **net community photosynthesis**, or the amount of photosynthesis in excess of respiration by both autotrophic and heterotrophic organisms. This is called **new production**. The value from the dark bottle is simply the respiration of the organisms. **Gross photosynthesis** is then obtained by adding the amount of oxygen respired to the net community photosynthesis, or subtracting the dark bottle from the light bottle. It is a measure of the total amount of photosynthesis that occurred during incubation. It is possible to convert the rates of oxygen production to rates of carbon dioxide assimilated by using a conversion factor. These factors are different depending on the organic compound produced (carbohydrate, lipid, or protein), but an average conversion factor would be 1.2 to 1.4.

Of course, the values for both gross and net community photosynthesis will differ with different incubation depths, because light values change with depth and affect the photosynthetic process. This traditional method was originally only applicable in waters where there was considerable photosynthetic activity, because the sensitivity of the Winkler method is too low to pick up oxygen changes under sparse phytoplankton populations and short 2- to 4-hour incubation times. Longer incubation times would permit use of the Winkler method but cannot be done because various "bottle effect" errors, such as the increased growth of oxygen-consuming bacteria on the walls of the container, become too large. More recently, the Winkler method has been electronically coupled to optical titration, which gives greater sensitivity.

Another method for estimating primary productivity is the **^{14}C method**. This has been the preferred technique for calculating primary productivity for the past 30 to 35 years. Here, radioactive ^{14}C is added to a bottle containing a volume of seawater populated by phytoplankton. For experimental and technical reasons the ^{14}C is usually introduced as $H^{14}CO_3^-$ (bicarbonate). A known quantity of $H^{14}CO_3^-$ is added, and the bottle is incubated for a period at the appropriate depth in the sea (or in special racks on deck). At the end of the incubation period, the bottles are brought up and the water filtered onto fine membrane filters or glass fibers, which catch all phytoplankton organisms. The amount of radioactivity on the filters is measured with a counter. The amount of ^{14}C that appears on the membrane (e.g., in the particles) is proportional to the rate of production. To correct for possible nonphotosynthetic uptake of ^{14}C, a dark bottle is often run along with the light bottle. One calculates the carbon uptake from the following equation:

$$\text{C uptake} = \frac{{}^{14}\text{C in particles on the filter} \times \text{available inorganic carbon} \times 1.05}{\text{Total } {}^{14}\text{C added}}$$

The available inorganic carbon in the sample is usually measured by alkalinity titration. In most marine waters with a salinity greater than 30 psu, the total available inorganic carbon is about 25 mg C per liter. The 1.05 factor is included in the equation to correct for the fact that phytoplankton discriminate between ^{14}C and ^{12}C, absorbing ^{14}C more slowly.

Both of the methods discussed have errors and problems associated with them, and both rest upon certain assumptions. In the case of the light-dark bottle method, the assumption is that conditions in both bottles are similar except for light. This may not be the case and will introduce error. Bacterial growth tends to be accelerated by the introduction of surfaces; hence, the increased bacterial numbers may affect the oxygen levels in the bottles. This problem is common to both ^{14}C and light-dark bottle methods. Another problem common to both methods is that phytoplankton organisms enclosed in a bottle do not act normally, and this may be reflected in their photosynthetic rate. For the light-dark bottle technique, a serious problem is comparative insensitivity when concentrations of phytoplankton are low or when waters are polluted or contain high bacterial concentrations. The chief additional problems with the ^{14}C method are that there may be cell breakage on the filters, allowing some ^{14}C to leak out, and that the method does not account for ^{14}C fixed in photosynthesis and then respired out as $^{14}CO_2$, which then is redissolved in seawater. Since we cannot account for this latter problem, it is not always clear whether the method measures gross or net photosynthesis or something in between.

Recently, various independent methods used to assess carbon dioxide uptake or oxygen release have suggested that the ^{14}C method underestimates primary productivity, particularly in waters with low nutrient contents and low phytoplankton densities. Part of this problem may lie in the technique (reviewed by Peterson, 1980). Part of the problem also may result from inadvertent contamination from metals associated with bottle samples and ^{14}C stock solutions.

Because of these shortcomings, newer methods of assessing primary productivity have been and are being attempted. Included here are the use of

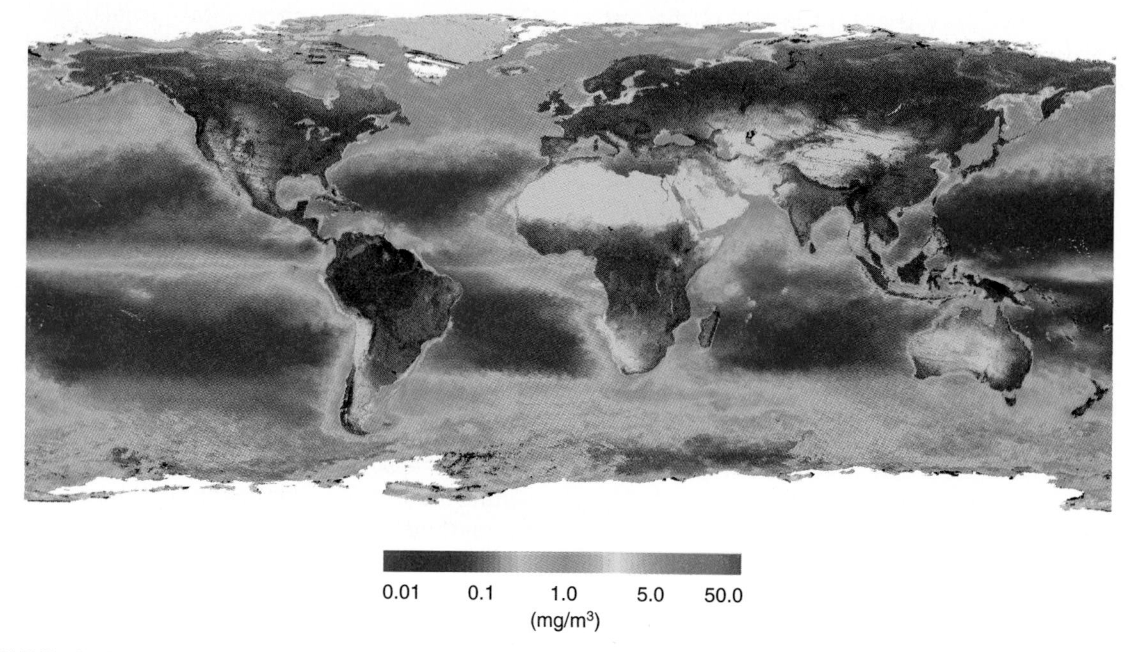

FIGURE 2.28 Global net primary productivity in grams of carbon per square meter per year for the biosphere calculated from satellite-derived measurements of surface chlorophyll and absorbed photosynthetically active solar radiation. (Courtesy of NASA/Orbimage Seawifs maps.)

large-scale optical assessment via satellite imaging and bio-optical methods of measuring photosynthesis, specifically from specially designed fluorometers. In some ways these instruments measure "physiological health" of the phytoplankton. The type of instrument used in these studies is called a pulse amplitude modulated (PAM) fluorometer. Yet another variant, the fast repetition rate (FRR) fluorometer, attempts to measure the instantaneous photosynthetic rate.

Net primary productivity (NPP) for both terrestrial and oceanic biomes has now been calculated from the absorbed photosynthetically active solar radiation (APAR) and *e*, which is an average light utilization efficiency, in the following equation:

$$\text{NPP} = \text{APAR} \times e$$

For both the oceans and the terrestrial environment, the APAR can be derived from satellite measurements, in the case of the oceans the measurement of chlorophyll and in the terrestrial environment from estimates of greenness. However, *e* cannot be accurately measured from space and must be determined with field measurements. Figure 2.28 shows the net primary productivity for the biosphere as determined by this method.

Standing Crop

Standing crop may be autotrophs, animals, or both. The **standing crop** (at any point in time) is the result of the difference between the factors tending to increase the numbers of individuals—reproduction and growth—and those factors tending to decrease biomass or numbers—death and sinking or lateral transport out of the area. If reproduction and growth rates are, or have been, high and death or removal low, then standing crop will be high, and vice versa.

Standing crop of phytoplankton in the oceans is a difficult factor to measure accurately. This partly is due to the patchy distribution of plankton organisms, the problems of sampling (see pp. 85–87 for further discussion of this), and problems inherent in the methods. The usual method of measuring standing crop is to measure some component common to all phytoplankton, usually the chlorophyll *a* content of a given volume of seawater. Since all phytoplankton must have chlorophyll to photosynthesize, the total amount of chlorophyll in a given volume of seawater should be a direct measure of the total biomass of phytoplankton present. Chlorophyll can be measured by taking advantage of its ability to fluoresce when excited by an appropriate wavelength of light or by directly extracting the chlorophyll from the plants with a chemical such as acetone and then measuring the intensity of color in a colorimeter. The problems with this method are numerous. The method assumes, for example, that cellular chlorophyll content is constant, but it is not; it varies among the phytoplankton species, as well as among cells of the same species, depending on their condition. It also varies with time of day and light intensity. These problems are not insignificant and constitute a substantial problem in assessing standing crop.

Other methods of assessing biomass have been used. One is the measurement of **adenosine triphosphate (ATP)**. ATP is found only in living cells and disappears rapidly on death. It is a useful indicator of standing crop. The method devised by Holme-Hansen and Booth (1966) depends on the use of the protein luciferin and the enzyme luciferase. These two compounds in the presence of ATP produce light (bioluminescence), which is then measured quantitatively by a photocell. Drawbacks of the method are that ATP content is variable, depending on several biological and physical factors, making extrapolation of ATP levels to units of living cells difficult; interference by microzooplankton is a problem; and biologists cannot agree on the percentage of ATP measured in a unit volume of water that should be attributable to animals or bacteria. Although not used extensively today, a potential advantage of the ATP method is that once sufficient data are available to assign values accurately to all components of the plankton, this method will give biomass values for zooplankton, phytoplankton, and bacterioplankton collectively. Today ATP is not widely used for measuring standing crop, but fluorescence microscopy is used for enumerating picoplankton, bacteria, heterotrophic flagellates, and viruses.

FACTORS AFFECTING PRIMARY PRODUCTIVITY

Since the entire ecosystem of the world's oceans depends for energy almost exclusively on the photosynthetic activity of the small phytoplankton confined to the thin layer of lighted surface waters of the oceans, it is important to understand the conditions under which their productivity is either enhanced or inhibited. This is the first step toward understanding how our oceans function.

Physical and Chemical Factors

If we consider terrestrial plants, there are a series of physical and chemical factors that affect their growth, survival, and productivity. Critically important factors would include light, temperature, nutrient concentration, soil, and water. Considering the freely floating phytoplankton, we find that the number of these factors that are important is considerably reduced. For autotrophs already suspended in water, water is a given and soil is of no importance. Temperature, which can vary enough in the terrestrial biosphere to influence productivity, has a considerably narrower range in the marine environment. Because this range almost never encompasses lethal limits for life and because the changes are always gradual due to the inherent physical properties of water, temperature is of lesser importance to productivity in the ocean. Of all the physicochemical factors affecting terrestrial plant production, only two, **light** and **nutrient supply**, are significant in limiting productivity in the ocean. Since, however, the phytoplankton are suspended in water and affected by forces that move water, and because both light and nutrients are also affected by the water masses, a new, very important factor not seen on land enters into consideration. This factor is a composite one we may call **hydrography**, and comprises all those factors that act to move water masses around in the oceans, such as currents, upwelling, and diffusion. It is the interplay of these three factors—light, nutrients, and hydrography—that provides the limits to phytoplankton productivity in the oceans and the geographical differences we observe. It is now necessary to consider how each factor acts separately to limit or enhance production and then to consider how all three act in concert to produce the observed productivity patterns of the oceans.

Light Photosynthesis is possible only when the light reaching the autotrophic cell is above a certain intensity. This means that the phytoplankton are limited to the uppermost layers of the ocean where light intensity is sufficient for photosynthesis to occur. The depth to which light will penetrate into the ocean, and hence the depth at which production can occur, depends on a number of factors. These factors include absorption of light by the water, the wavelength of light, transparency of the water, reflection from the surface of the water, reflection from suspended particles, latitude, and season of the year.

A number of meteorological features influence light before it even touches the surface of the water (Figure 2.29). Factors such as clouds and dust interfere with light in such a way that lesser amounts survive the passage through the atmosphere to reach the water surface. This reduces the available light initially, without reference to water conditions.

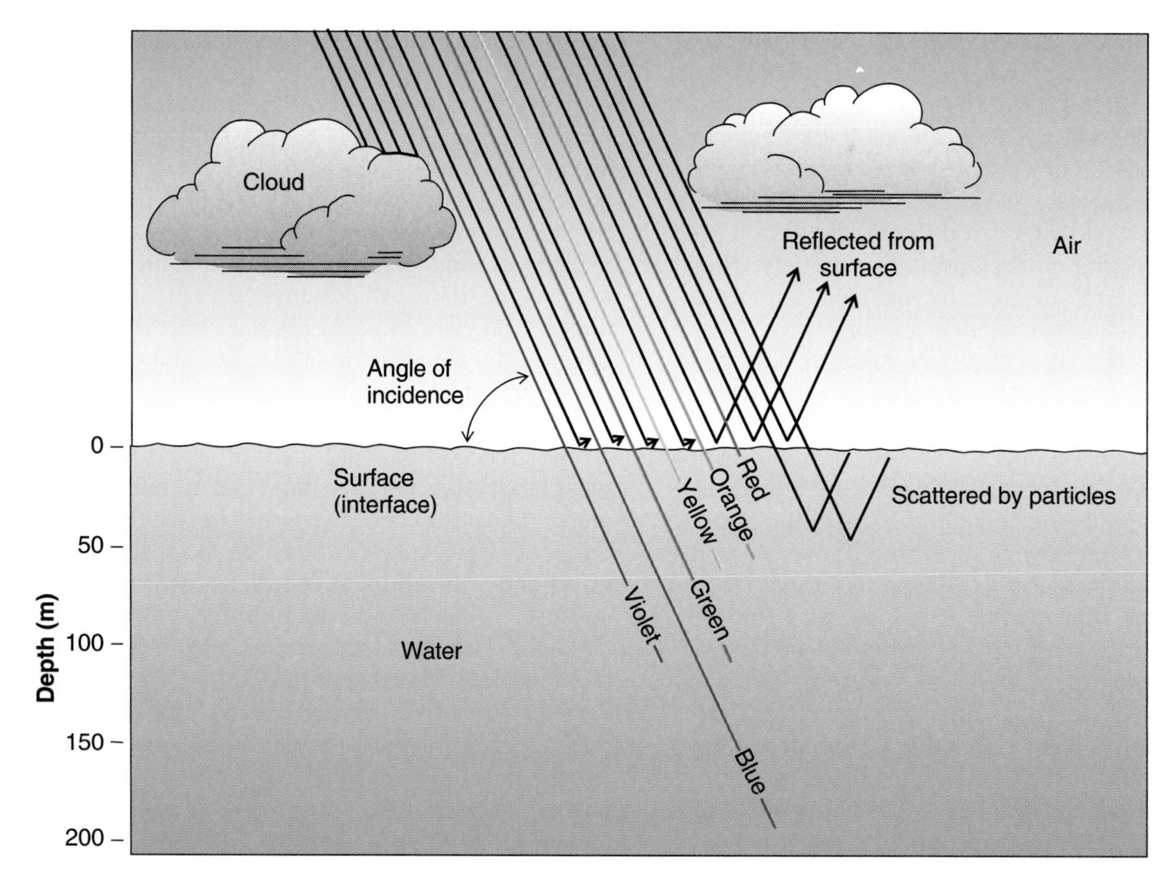

FIGURE 2.29 The fate of incident light in the ocean.

When light strikes the surface of the water, a certain amount of light is reflected; the amount depends on the angle at which the light strikes the surface of the water. If the angle from the horizontal is low, a large amount will be reflected. Conversely, the nearer the angle is to 90° (that is, perpendicular to the horizontal surface of the water), the greater the penetration and the less the reflection. Reflected light is lost to the system, so from the standpoint of phytoplankton productivity, maximum penetration is the most desirable.

The angle at which light strikes the surface of the water is directly related to the maximum height of the sun above the horizon. In the tropical regions of the earth, the sun is directly overhead at midday, or virtually perpendicular to the sea surface, giving an angle for maximum penetration of light into the water column. This position changes little with seasons, so light conditions are maximal all year long. As one progresses toward the poles from the equator, the sun may be directly overhead during the summer months, but it may be far from this position at other times of the year. The closer one approaches the poles, the greater becomes the difference in the height of the sun above the horizon among the seasons of the year. This reaches its maximum in the Arctic and Antarctic regions, where the sun is absent during the winter or is so low to the horizon that no light can penetrate the water. The presence of ice in these areas also reduces light penetrating into the water. This means that as one moves away from the equator to the north or south, the amount of light penetrating the surface of the ocean, and available for plant use, changes significantly with season. As a result, the amount of photosynthesis that can occur also changes and is maximal in the summer and minimal in the winter. Only in the tropics is the light optimal all year (Figure 2.30).

The portion of light that enters the water column is subject to further reduction from two additional processes acting on it within the water. The first is scattering from various suspended particles in the water column. Suspended living or dead particles intercept the light and either absorb it or scatter it. Scattered light experiences a longer path length, which in turn results in further attenuation. Second, water itself absorbs light, making it unavailable for the phytoplankton. The amount absorbed is a function of

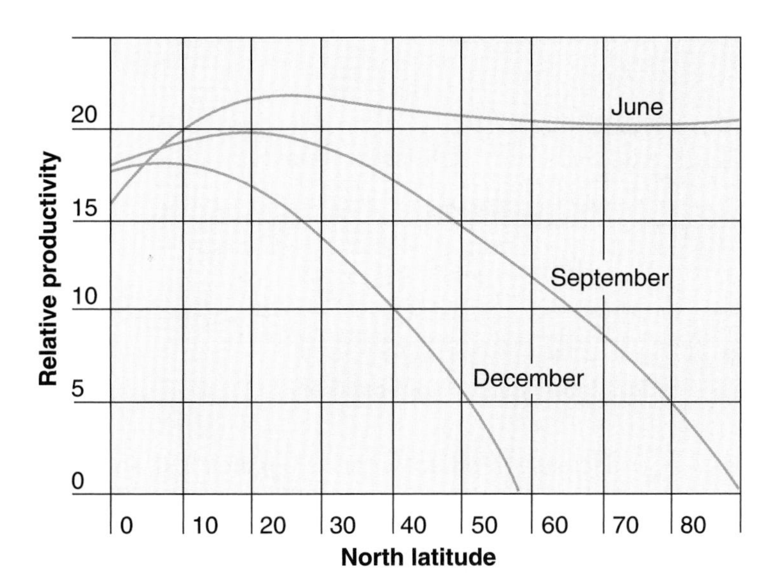

FIGURE 2.30 Relative amount of photosynthesis as a function of latitude for three seasons of the year. (From "Geographical Variations in Productivity" from *The Sea*, Vol. 2, ed. M. N. Hill, © 1963 John Wiley & Sons.)

wavelength and the path length or water depth in perfectly clear, clean water. This absorption of light by water is the reason most water masses of the ocean are dark below a certain level. Because of this absorption, photosynthesis is automatically restricted to the thin, uppermost lighted layer. In the clearest ocean water, this may correspond to the upper 150–200 m. Water, however, does not absorb all wavelengths of light equally. Sunlight, as it arrives at the surface of the ocean, is composed of radiation in a spectrum of wavelengths measured in nanometers (1 nanometer = 10^{-9} m). This spectrum includes all the visible colors ranging from violet to red (see Figure 2.29), or wavelengths from about 400 to 700 nm. As these wavelengths enter seawater, the violet and red components are quickly absorbed by the water. The green and blue components are absorbed less rapidly, so they penetrate most deeply. Eventually, they too are absorbed by the water. All the red and violet light is absorbed within the first few meters of even the clearest seawater, whereas 10% of the blue light may penetrate to more than 100 m under similar circumstances. It is important to realize that, for any given wavelength of light, a certain fraction of its remaining intensity is lost by absorption with each additional increment of water depth. Even though blue and green light penetrate deeper into the water column, the intensity decreases with depth, and it is intensity that is needed by plants. The intensity, relative to the surface, can be estimated from the **extinction coefficient**: $I_z = I_0 e^{-kz}$, where I_0 and I_z are the intensities at the surface and at depth; k is the extinction coefficient (for pure water, $k = 0.035$); and z is the depth or path length in meters.

The depth to which a given intensity of light penetrates is a function of the transparency of the water and the differential absorption by the water. Since differential absorption by water is constant, the changes in depth of effective light penetration are due primarily to particle concentration. Where there are large numbers of particles in the water, such as in coastal waters, the depth of light penetration may be severely reduced and the amount of light insufficient for photosynthesis below a few meters. On the other hand, in the clearest tropical water, where few interfering particles exist, light intensity may be sufficient for photosynthesis down to 100–150 m.

Phytoplankton photosynthesis is light dependent. The rate of photosynthesis is high at intermediate light levels and decreases as the light intensity either decreases or increases (Figure 2.31). On the other hand, the rate of respiration of the phytoplankton cells does not change much with depth. This means that, as the photosynthetic cells go deeper in the water column, the rate of photosynthesis declines as the light intensity decreases, until at some point the photosynthetic rate equals the respiration rate. At this point, there is no net production of organic material. As defined earlier, this depth is called the compensation depth, and it is approximated by the depth to which 1% of the incident radiation penetrates. The compensation depth marks the lower limit of the photic zone and varies geographically from a few meters in very turbid inshore waters to depths of 120–150 m or more in the open waters of tropical oceans. It also varies seasonally in temperate areas where high turbidity during certain seasons (the plankton bloom) reduces it to a few meters

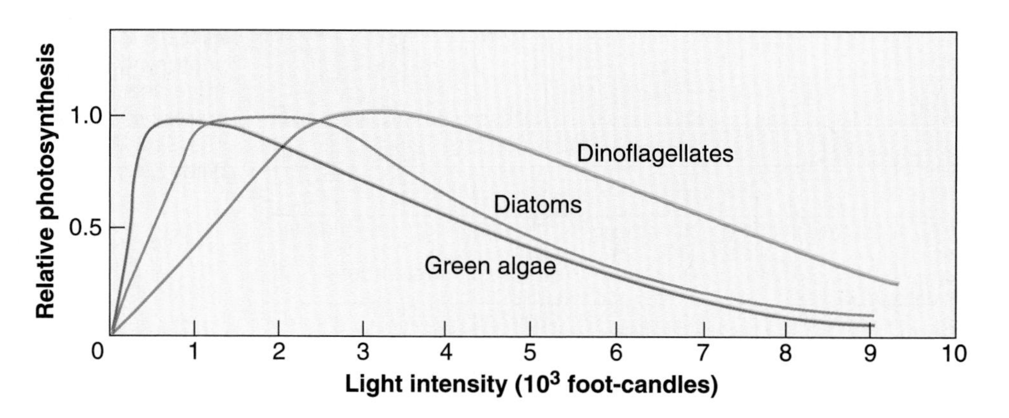

FIGURE 2.31 Photosynthesis versus light intensity curves for three groups of phytoplankton. (From *Biological Oceanographic Processes* 3/e T. R. Parson et al. Copyright © 1984 Butterworth Heinemann. Reprinted by Permission.)

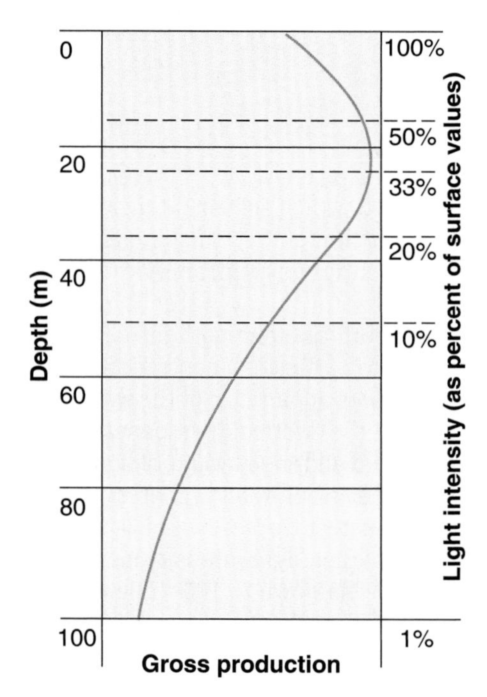

FIGURE 2.32 Inhibition of photosynthesis in the surface layers of water as measured by gross production at different levels in the water column. (After "Productivity, Definition & Measurement" by E. Steeman-Nielsen from *The Sea*, Vol. 2, ed. M. N. Hill, © 1963 John Wiley & Sons.)

(see p. 63), whereas at other times the sparse populations of organisms increase it.

The compensation depth also changes with season, due to the change in the position of the sun, and it may be virtually absent during the winter months in high latitudes.

For most phytoplankton, the photosynthetic rate varies with light intensity. Near the surface of the water column, where the light intensities are the highest, most species show a leveling off or a decrease in photosynthesis (Figures 2.31 and 2.32). This is due either to an inhibition at high light levels or saturation of the photosynthetic apparatus such that it is not possible to increase rates. Different species have different curves of photosynthetic rate when plotted against light intensity, giving different optimal light intensities for maximum photosynthesis (see Figure 2.31). This may be of considerable significance in seasonal succession (see p. 80).

Nutrients The major inorganic nutrients required by phytoplankton for growth and reproduction are nitrogen (as nitrate, NO_3^-; nitrite, NO_2^{2-}; or ammonium, NH_4^+) and phosphorus (as phosphate, PO_4^{3-}). Diatoms and silicoflagellates also require silicate (SiO_2) in significant amounts. Other inorganic and organic nutrients may be required in small amounts. All of these nutrients are of great importance, partly because they occur in such small amounts in seawater. They are the limiting factors for phytoplankton productivity under most conditions, and the oceans of the world can be considered nutrient-poor deserts when compared with terrestrial counterparts. For example, fairly rich agricultural land contains about 0.5% nitrogen in the upper meter of soil. In one cubic meter of soil, this amount of nitrogen is sufficient to permit the production of 50 kg of dry organic matter. Terrestrial plants, under ideal conditions, can produce several kilograms of dry organic matter in excess of their own need per year per square meter; thus, the reservoir of nitrogen in that one cubic meter is sufficient to allow a plant to grow for many years. As a result, it is possible in terrestrial systems to have long-lived plants that continue to grow for many years. Forests are a good example, for they represent the accumulation of years of organic production. Of course, eventually, unless the nitrogen is renewed in the soil, nutrient exhaustion will limit growth.

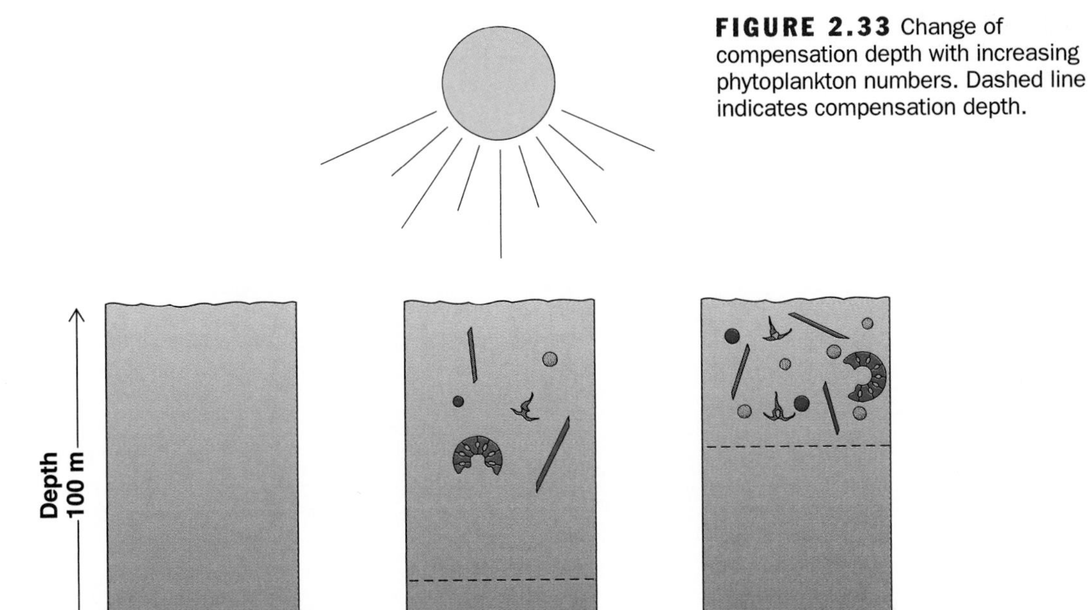

FIGURE 2.33 Change of compensation depth with increasing phytoplankton numbers. Dashed line indicates compensation depth.

On the other hand, the richest ocean water contains only about 0.00005% nitrogen, or 1/10,000 of the amount in soil. This means that a cubic meter of such water could permit a production of only 5 g of dry organic matter. In contrast to soil, however, where plant roots may penetrate only 1 m, oceanic phytoplankton should, theoretically, have access to the nitrogen in a column of water that extends as deep as the plants can exist. What is this depth and does it compensate for the greatly reduced nitrogen concentration? As we learned in the previous section, phytoplankton are limited in depth by light. If we assume the most ideal light conditions, we might have plant production as deep as 100–120 m. This would mean the plants would potentially have access to a volume of 100 m^3 (1 $m^2 \times 100$ m deep) with a potential production of 500 g of dry weight (5×100), or about 1/100 of the amount on dry land. In fact, production levels of that magnitude never occur in the ocean, and the maximum amount of organic production that can accrue under a square meter of seawater under ideal conditions is only about 25 g. Why the discrepancy?

Several factors act in concert to reduce the biomass yield. First, the previous example assumed that the nitrogen content of seawater was constant throughout the water column. Such is usually not the case. The upper layers of water usually have a reduced concentration compared with lower waters, as will be shown. Also, due only to light absorption by water, the production at 100 m would be less than at 10 m. More important, the increasing numbers of phytoplankton cells have a profound effect. As the phytoplankton population grows in the upper 100 m of water, the plants absorb more and more of the light. As a result, there is less and less light penetrating to the deeper levels. Less light means that the compensation depth begins to move upward and becomes shallower (Figure 2.33). The original 100-m reservoir of nutrients is reduced, and as plants increase in numbers, more and more of the water column and nutrient supply become inaccessible to the phytoplankton; thus, the total potential productivity is reduced. Finally, as the phytoplankton grow, they absorb the nutrients, and these nutrients are not available to other plants.

The result of all these factors acting together is to reduce the theoretical primary production. The reduction in the compensation depth and the absorption of nutrients is great. It has been estimated that by the time a phytoplankton population reaches a density of 2 g/m^3, an original 100-m compensation depth has decreased to as little as 3.5 m, and all nitrogen has been transferred into phytoplankton bodies. In terms of production and of standing crop of autotrophic material, the oceans appear to be biological deserts

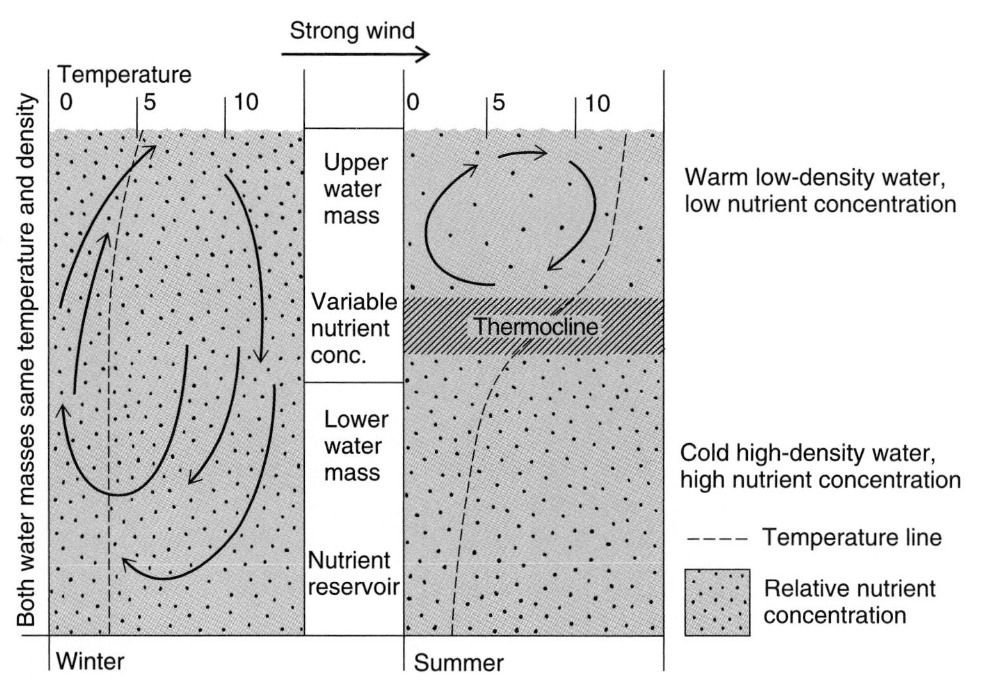

FIGURE 2.34 Thermal structure of temperate ocean water masses in two seasons, winter and summer, showing the relationships between nutrient concentration, temperature, density, and wind-induced mixing.

when compared with fertile land. The rate of production between land and sea, on a global scale, may not be significantly different, but the differences in nutrient concentration preclude that rate from continuing in the sea for long periods of time.

If, however, the nutrients are used up in the upper, lighted zone, an untapped reservoir remains in the water mass below the photic zone. Since this mass is several orders of magnitude larger in volume than that of the lighted (photic) zone, it represents a considerable reservoir of nutrients that could enhance production greatly. If this vast reservoir could be tapped, high rates of production and large standing crops could be sustained on a long-term basis.

Unfortunately, certain physical factors prevent general access to this reservoir over most of the world's oceans. Water, as we learned in Chapter 1, has different densities, depending on its temperature and salinity. Cold, saline water is more dense than warm, less saline water. Below the photic zone throughout the world's oceans, the water is cold and, therefore, dense. In the photic zone, the water varies in temperature. In the tropics it is warm, and therefore less dense, all year. In the temperate zone, this water is warm in summer and cold in winter. In polar regions, it is cold and dense year-round. In the tropics, the difference in density between the warmer upper and colder lower layers is of such magnitude that the two do not mix; hence, the nutrients cannot reach the photic zone. In the temperate seas the same situation prevails in the summer, but in the winter the temperature of the two water masses becomes similar and mixing can occur. In polar regions, there are no significant differences in temperature, and mixing can occur year-round.

Another component is required to tap the nutrient reservoir—a mechanism to accomplish mixing. Even though the water masses approach the same density, some force must be exerted to mix the masses. The force available over most of the world's oceans is the wind. Strong winds blowing over the surface of the water create the mixing force (Figure 2.34). The wind is of sufficient strength to cause mixing of the water masses and to transport nutrient-rich water into the euphotic zone, when the two masses are similar in density and temperature. In polar seas, this is possible at all times (when not covered by ice), and in temperate seas, only in the winter. The great differences in density between the photic and aphotic zones all year in the tropics and in the temperate zones in the summer are of such magnitude, however, that no wind (even from storms such as hurricanes) is strong enough to mix the layers. This means that the upper layer remains nutrient-poor in the tropics at all times and in the temperate zone in summer, unless specific hydrographic conditions overcome the stability of the density differences.

The special hydrographic conditions that bring nutrient-rich deep water up into the photic zone are upwelling, divergence of currents, and the slow upward trickle caused by eddy diffusion at the thermocline.

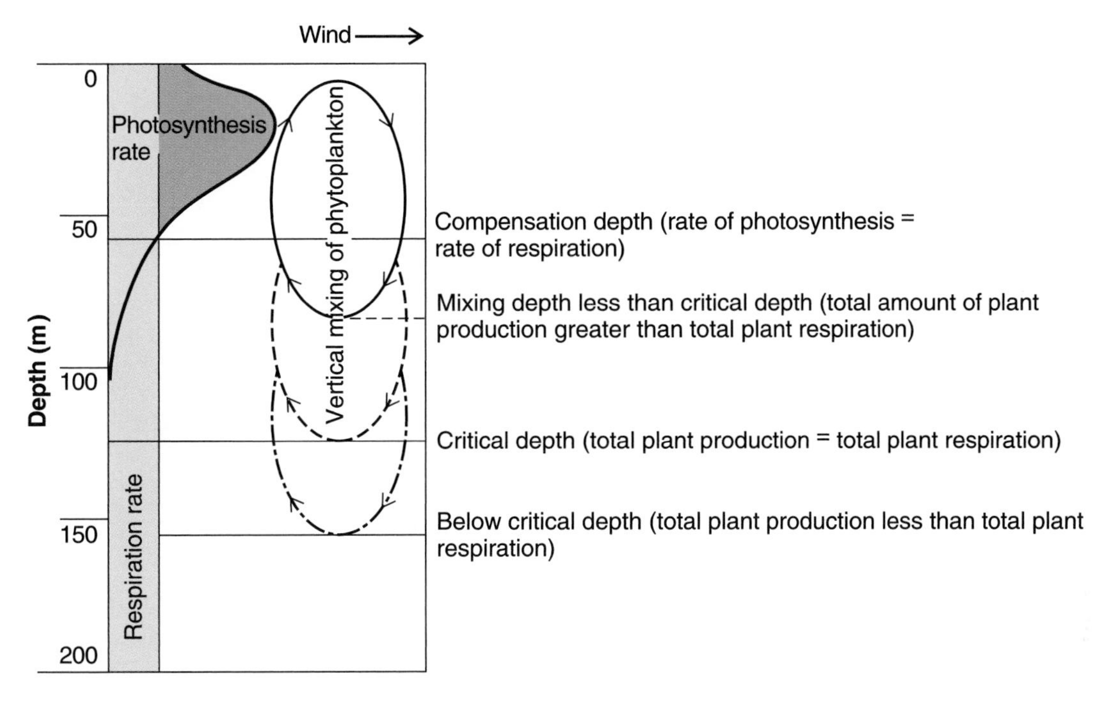

FIGURE 2.35 The relationship between the compensation depth and the critical depth. Critical depth is the depth to which the total phytoplankton biomass may be circulated and still spend enough time above the compensation depth to have a total production equal to its total respiration during the same time period. (Modified from *Productivity Of The Seas* D. H. Cushing, Oxford biology Reader, #78, 1975. Copyright 1975 Butterworth Heinemann. Reprinted by permission.)

Upwelling occurs where the surface water moves away from shore and is replaced by nutrient-rich deeper water brought to the surface from a few hundred meters. Persistent upwelling along the west coast of North America is responsible for high productivity in that region. Divergences are the result of transverse ocean currents flowing away from each other, bringing up deeper nutrient-rich water. Such a divergence occurs between the North Equatorial Current and the Equatorial Countercurrent in the Pacific; this is one area in the open tropical ocean where productivity can be high, due to upwelled nutrients (see Figure 1.14). It should be noted here that it now seems that some fraction of the nutrients, particularly nitrogen and perhaps also some trace metal nutrients, come from dust settling out of the atmosphere. Perhaps the best example of the importance of trace metal nutrients comes from the experiments by Martin et al. (1994) in adding iron to an open ocean area near the Galapagos Islands. This area had sufficient nitrates and phosphates to sustain high productivity because of its position in the aforementioned tropical divergence zone but had naturally low productivity. Addition of iron resulted in a doubling of plant biomass and a fourfold increase in plant production, indicating that iron, indeed, can limit productivity under certain conditions.

Turbulence and Critical Depth Not only does vertical mixing bring up nutrients, it also carries phytoplankton cells into the depths. As long as vertical mixing is confined only to the upper illuminated zone, the phytoplankton cells can be carried downward only a short distance and will remain where there is sufficient light for photosynthesis. When mixing includes the lower water mass, however, it is possible for the plant cells to be carried well below the compensation depth. If the mixing is especially vigorous, the phytoplankton may spend most of their time below the compensation depth, and there will be no net production, because the time spent in the lighted zone in active photosynthesis is insufficient to fix as much organic matter as is used when they are below the compensation depth. This effect of turbulence has led to the development of the critical-depth concept (Figure 2.35). The **critical depth** is that depth at which total gross photosynthesis of the phytoplankton in the water column equals total respiration. It is

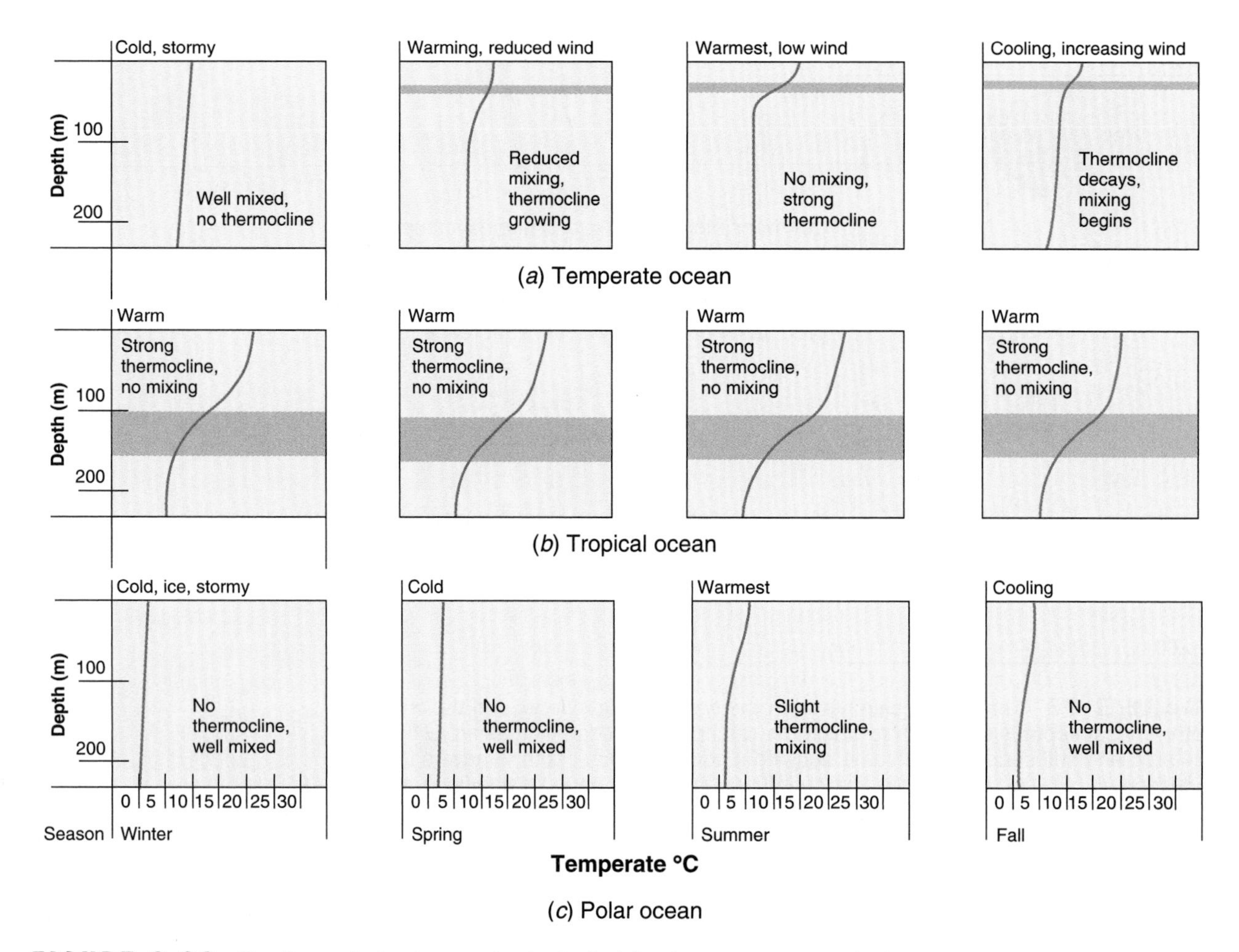

FIGURE 2.36 The thermal structure and extent of mixing in temperate, tropical, and polar seas during the four seasons of the year.

calculated over a 24-hour period and is different from the compensation depth, which is the depth at which the rate of photosynthesis equals the respiration rate. The critical depth is always greater than the compensation depth, because we are dealing with a vertical mixing process in which the phytoplankton population is circulated between lighted and unlighted areas. When it is in the light, the phytoplankton population can photosynthesize in excess of respiration and build up organic matter. The decisive point is the length of time spent in each area; in turn, this depends on how deep the mixing occurs. Deeper mixing means more time spent below the photic zone, where respiration uses up organic material more rapidly than the short time in the lighted zone can compensate for. Whenever the wind-driven vertical mixing is less than the critical depth, photosynthesis is greater than respiration and a net production occurs.

Geographical Variations in Productivity

Having considered the role of light, transparency, and nutrients separately, it is possible to consider how these factors interact with each other to produce the geographical or latitudinal variations in productivity we observe in the oceans.

Temperate Seas In the temperate zone seas, the amount of light varies seasonally. As a result, the amount of solar energy entering the water varies, which alters the temperature in the upper water layers. The thermal structure of the water column, therefore, changes seasonally (Figure 2.36). In the summer months, the sun is high, days are long, and the upper layers heat up and become less dense than underlying layers. In other words, the water column is thermally stratified and no mixing occurs. In the fall, the amount of solar energy entering the water column decreases, days become shorter, upper layers cool,

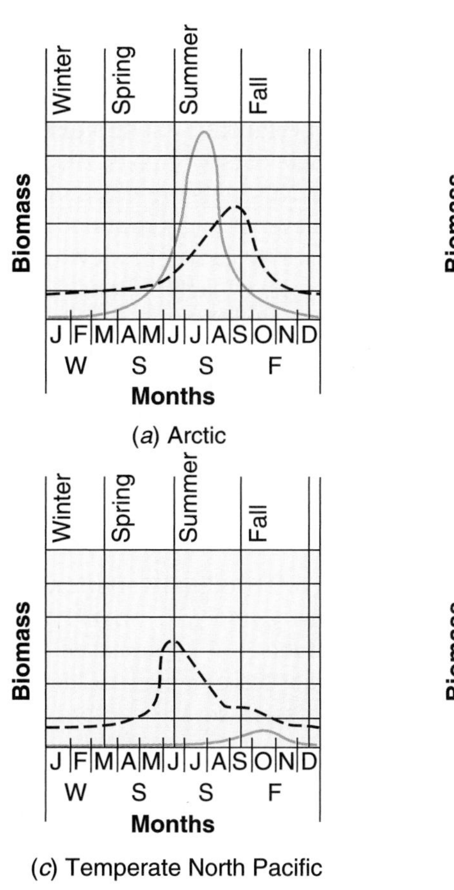

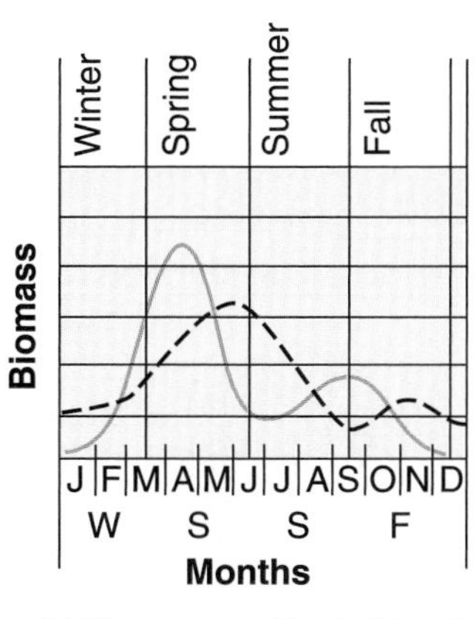

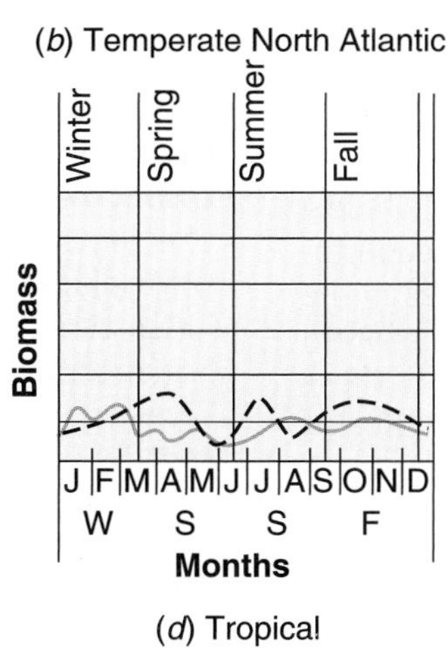

FIGURE 2.37
Summary of the seasonal cycles of net phytoplankton and zooplankton in four different geographical areas. The solid line represents phytoplankton; the dashed line represents zooplankton. (Modified from *Biological Oceanographic Processes* 3/e T. R. Parson et al. Copyright © 1984 Butterworth Heinemann. Reprinted by Permission.)

and thermal stratification decreases. Finally, a point is reached where the temperature of the surface layers has been reduced to such an extent that the density of the layer is little different from that of the underlying mass. At this point, mixing can occur whenever sufficient wind is available. In winter, usually the storm season in the temperate zone, the sun is lowest on the horizon, solar energy input to the water is at a minimum, thermal stratification is at a minimum or absent, and mixing occurs. With the onset of spring, the days become longer, the solar energy increases, the upper layers begin to rise in temperature, and the system moves toward reestablishment of thermal stratification.

In contrast to the tropics, all the major factors that affect productivity change seasonally in temperate seas. This is reflected in the change in production over the year (Figure 2.37b), with a major peak in spring, a lesser peak in the fall, and low productivity in winter and summer. We may explain this as follows: The low winter productivity is the result of low light levels due to the low position of the sun on the horizon and because the winter storms mix the isothermal water column and carry phytoplankton cells below the critical depth. In the spring, the increased light and solar energy increase the temperature of the upper layers. With increasing temperature come increasing differences in density between upper and lower layers. Under such conditions the wind cannot mix the water to as great a depth as in winter; at some point, phytoplankton cells are no longer carried below the critical depth. Since nutrients in upper layers have been replenished during the winter mixing, conditions are good for phytoplankton growth, and we observe the **spring bloom** (Figure 2.37b). As spring passes into summer, the water column becomes more thermally stratified, mixing with lower levels ceases, and light conditions reach optimal levels. Because mixing ceases due to stratification, nutrient replenishment ceases and production falls, even though light levels are optimal. With the advent of fall, the thermal stratification begins to break up and nutrients are returned to upper levels. If, in the fall, the mixing alternates with calm weather such that the plants spend more of their time in the upper layers and are not carried below the critical depth, a small

TABLE 2.2

Production for Several Different Geographical Areas

Location	*Productivity in C/m2/year*
Long Island Sound (temperate inshore)	380
Continental shelf	100–160
Tropical oceans	18–50
Temperate oceans	70–120
Antarctic oceans	100
Arctic Ocean	<1

Source: After Raymont, 1980.

bloom will occur because of the increased nutrients. This bloom declines in late fall, due to decreasing light and increased mixing. In the winter, low light levels and deep mixing of the water column keep productivity low. Note that there are differences in the seasonal cycle curves between the temperate North Atlantic and North Pacific in Figure 2.37c and d. These differences are the result of somewhat different hydrographic conditions, coupled with different nutrient concentrations and availability (see pp. 75–77).

Tropical Seas In the tropical seas, the upper waters are well lighted throughout the year because the sun does not show marked changes in height above the horizon. Light conditions are, therefore, optimal for phytoplankton production. At the same time, the continual input of energy from the sun maintains the surface layers of water at temperatures much higher than those in deeper waters. This means there is a great difference in density between surface and deep waters; hence, mixing does not occur. We say that such waters are thermally stratified. This **thermal stratification** extends throughout the year (see Figure 2.36). In the tropical seas, the light conditions are optimal for high productivity. Because the sun's energy creates a thermal stratification in the water column that prevents mixing and the upward transport of nutrients, however, the productivity is low but constant all year (Figure 2.37d). Tropical seas are very clear and have the deepest compensation depths, but they are that way because there are few phytoplankton in the water column due to the low nutrient content.

Polar Seas In polar areas, productivity is restricted to a single short period in the polar summer, usually July or August in the Arctic (Figure 2.37a). At this time, the snow cover on the ice has disappeared, allowing sufficient light to enter the water through the ice to permit phytoplankton growth. In areas outside of the permanent ice pack, breakup of the ice at this time opens the leads, allowing sufficient light to enter the water and permitting phytoplankton growth. Following this single burst of growth, the production quickly declines. Nutrients are not limiting and the water column is never strongly stratified (see Figure 2.36c). The reason for the lack of production at other times is due primarily to lack of light. Light intensity is insufficient for a fall bloom, and during the long winter, light is either absent or prevented from reaching the water column by a layer of snow over the ice pack. (See a more detailed analysis of polar seas in Chapter 5.)

Productivity in Inshore and Coastal Waters

The previously discussed latitudinal variations in phytoplankton productivity apply to open ocean areas away from the influence of landmasses. The situation in the water masses adjacent to land is somewhat different. There are several factors that contribute to this difference.

First, inshore waters tend to receive a considerable input of the critical nutrients, PO_4^{3-} and NO_3^-, due to runoff from the adjacent land (where, as we have seen, the nutrients are far more abundant). Because of this input, inshore waters usually do not show nutrient depletion. A second factor contributing to the difference is the water depth. Most inshore waters are shallower than the critical depth; thus, the phytoplankton cannot be carried below this depth in any kind of weather. Given sufficient light, production can occur at any time, even in the winter. A third factor is that shallow inshore waters rarely have a persistent thermocline; hence, no nutrients are locked up in bottom waters. A final influencing factor is the presence of large amounts of terrigenous debris in the water, which may act to restrict depth of the photic zone and counteract the high nutrient concentration and shallow depth.

Interaction of these factors on a latitudinal basis produces changes, both in the cycle of productivity and in the total production when compared with offshore areas (Table 2.2). In temperate regions, instead of a bimodal production cycle, as seen offshore, production remains high all through the summer. Nutrients are not limiting, due to runoff from land and lack of a permanent thermocline. Yearly average production in inshore temperate waters is higher than in offshore waters due to the greater nutrient concentrations and lack of

TABLE 2.3

Annual and Seasonal Net Primary Productivity (NPP) of the Major units of the Biosphere.

	Ocean NPP		*Land* NPP
Seasonal			
April to June	10.9		15.7
July to September	13.0		18.0
October to December	12.3		11.5
January to March	11.3		11.2
Biogeographic			
Oligotrophic	11.0	Tropical rainforests	17.8
Mesotrophic	27.4	Broadleaf decidious forests	1.5
Eutrophic	9.1	Broadleaf and needleleaf forests	3.1
Macrophytes	1.0	Needleleaf evergreen forests	3.1
		Needleleaf deciduous forest	1.4
		Savannas	16.8
		Perennial grasslands	2.4
		Broadleaf shrubs with bare soil	1.0
		Tundra	0.8
		Desert	0.5
		Cultivation	8.0
Total	48.5		56.4

[a] Ocean color data are averages from 1978 to 1983. The land vegetation index is from 1982 to 1990. All values are in petagrams of carbon (1 Pg = 10^{15} g). Ocean NPP estimates are binned into three biogeographic categories on the basis of annual average C_{sat} for each satellite pixel, such that oligotrophic = $C_{sat} < 0.1$mg m^{-3}, mesotrophic = $C_{sat} > 1$mg m^{-3}, and eutrophic = $C_{sat} > 1$mg m^{-3}.

critical depth problems. That production is not even higher inshore probably is due to the presence of large amounts of light-absorbing debris in shallow water, and the fact that in offshore water, production can occur to a greater depth. In other words, in shallow waters production is limited to the upper 5–10 m, whereas offshore it may go as deep as 50 m. In tropical waters, the difference between inshore and offshore waters is particularly dramatic. Inshore tropical waters have a productivity as much as ten times that of offshore waters. This must be attributed in large part to the increased nutrient concentration inshore compared with offshore areas.

Inshore production is further enhanced through contributions of benthic autotrophs, a component not present in offshore areas. The production of benthic algae and seagrasses is considered in Chapter 5 but is mentioned here for completeness.

Whereas we may have higher productivity in inshore or neritic waters, we also have greater variability over time and space due to local geography, river and stream discharge, storms, and tides. As a result, it is more difficult to predict the yearly productivity.

PRIMARY PRODUCTIVITY OF THE BIOSPHERE

Employing primarily data obtained from satellite measurements of absorbed solar radiation and using two computer models, Field et al. (1998) have estimated the total net primary productivity of the entire biosphere and compared the percentages of that productivity produced by the oceans and the terrestrial environment. They have further subdivided these estimates into seasonal and biogeographic components (Table 2.3). Strikingly, these estimates from satellite data give an ocean productivity that is twice calculations made before the availability of satellite data. Thus, the annual net primary productivity of the oceans is 48.5 petagrams (1 petagram = 10^{15} grams) of carbon, while that of the terrestrial environment (exclusive of areas of permanent ice and snow) is 56.4 petagrams. How do we reconcile this roughly equal productivity with the aforementioned discussion of the reasons for low productivity in the oceans? First of all, there is the discrepancy of the amount of land and ocean on the planet as well as the amount of the terrestrial realm that is under permanent ice and snow. Also, the terrestrial environment has extensive areas of extreme desert, which Field et al.

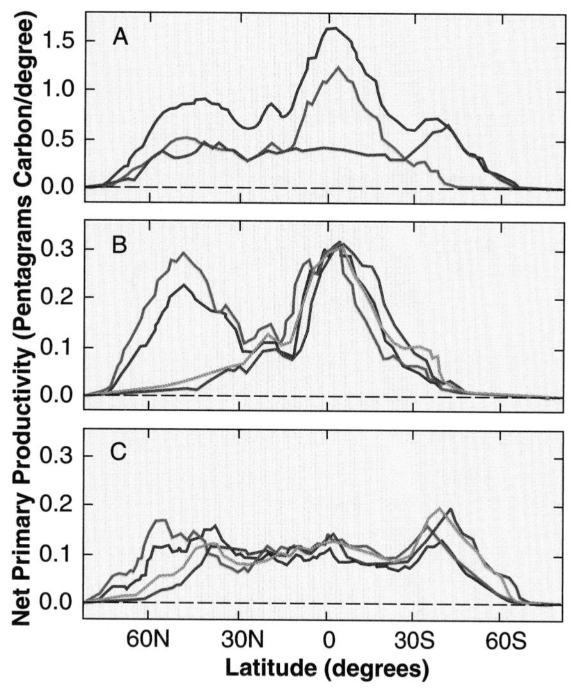

FIGURE 2.38 Latitudinal distribution of global net primary productivity. (A) Global total (blue line), land total (green line), ocean total (red line). (B) Land net primary productivity April to June (blue line), July to September (green line), October to December (orange line), and January to March (red line). (C) Ocean net primary productivity, four seasons as in B. (Reprinted with permission from Field et al., *Science*, Vol. 281, p. 239, 1998. Copyright © 1998 American Association for the Advancement of Science.)

indicate have less productivity than the vast tropical and subtropical oceanic gyres (see pp. 87–88). Given these caveats, if we calculate the net primary productivity on an areal basis, the terrestrial environment has an average annual productivity of 426 g C/m^2 while the figure for the oceans is 140 g C/m^2. This lower figure stems from the aforementioned light absorption by the water and particles, the lack of abundant nutrients, and the competition between the phytoplankton and the medium for the existing light. In fact, according to Field et al., on average only 7% of the photosynthetically available radiation (PAR) incident on the ocean is actually absorbed by the phytoplankton, while in the terrestrial environment the leaves absorb about 31% of the PAR. In addition, the phytoplankton of the oceans of the world represent only 0.2% of the global photosynthetic biomass but manage to account for almost half of the biosphere's net primary productivity. This is primarily because the turnover time of plant organic matter in the oceans is only about 2–6 days, whereas on land it averages 19 years.

Both the oceans and land are patchy, with areas of high and low productivity. In high-productivity areas in both oceans and land, the maximal rate of production of carbon is similar at about 1,000 to 1,500 g C/m^2/yr, but the oceanic areas that have such rates are much more spatially restricted. For example, on land 25% of the area can yield a net primary productivity of 500 g C/m^2/yr, while the comparable areal proportion in the oceans is 1.7%.

On a latitudinal basis, the net primary productivity of the biosphere reaches a maximum in three places (Figure 2.38). The largest centers on the equator and is primarily due to the high terrestrial productivity in the tropical forests. The next largest is in the temperate latitudes of the Northern Hemisphere and is again primarily due to the seasonal high terrestrial productivity. Finally, the smallest peak occurs in the temperate latitudes of the Southern Hemisphere and is mainly due to enhanced nutrient availability in the ocean due to the southern subtropical convergence and the reduced land area. Figure 2.39 summarizes the global oceanic annual net primary productivity.

THE OCEAN ECOSYSTEM: THE CLASSIC MODEL

We now begin a discussion of the interactions among the various planktonic organisms, beginning with the larger or net plankton. The following sections will detail first the classical, or earlier, concepts that describe the interactions of the larger phytoplankton, diatoms and dinoflagellates, and the larger zooplankton, primarily copepods. These ecological interactions still dominate coastal, upwelling, and polar seas. This classical model is followed by a discussion of the importance of the microplankton.

Grazing

As on land, plant production is transferred into the food chain of the pelagic community through the grazing activity of herbivores. Because the phytoplankton are small, the herbivores are also small. Many protistan and invertebrate planktonic species are herbivores, but by far the dominant larger herbivores, those that graze on the larger diatoms and dinoflagellates, in all oceans are various species of copepods (see Figure 2.5). The dominance of the herbivorous larger or net zooplankton by copepods is so great (70–90% of biomass) that it is possible to consider only the copepods in discussing the effects of grazing on the larger or net phytoplankton populations. Furthermore, in most

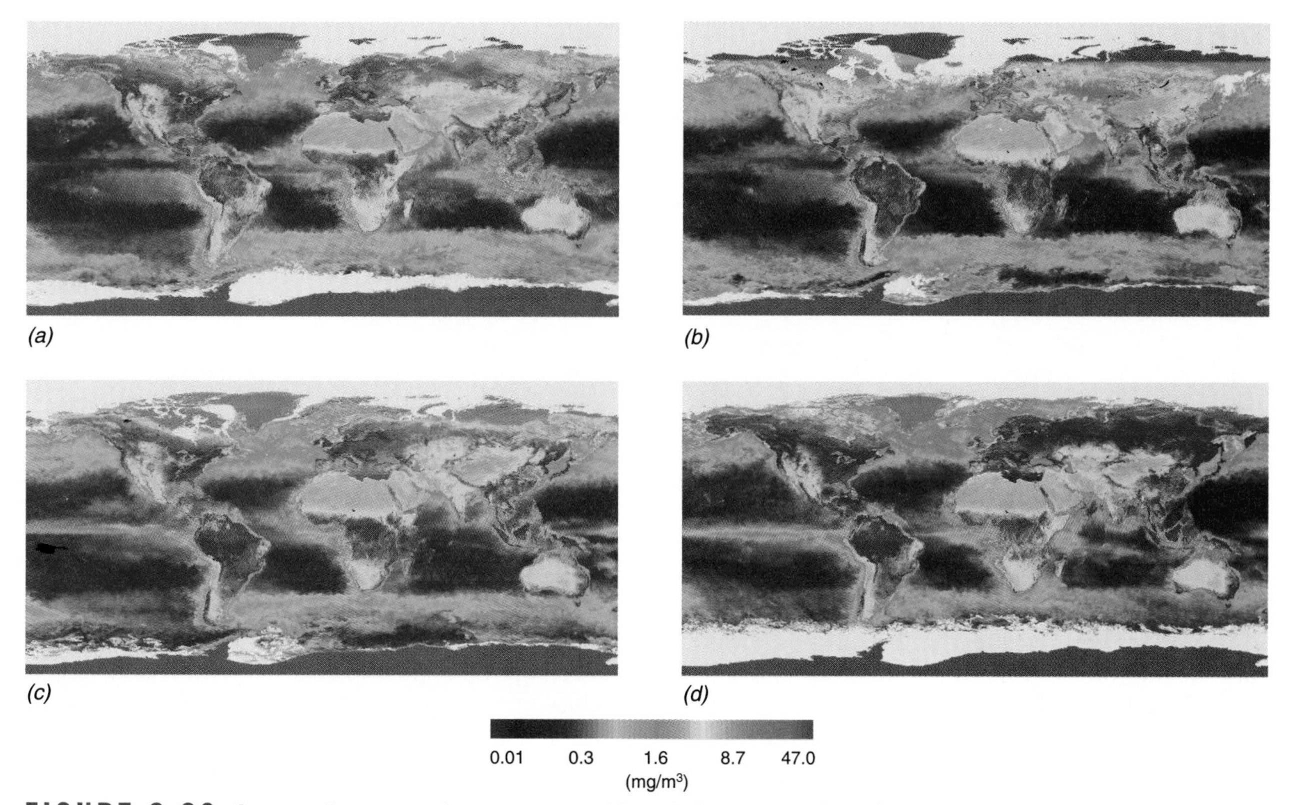

FIGURE 2.39 Seasonally averaged upper ocean chlorophyll concentrations for the world's oceans derived from satellite coastal zone scanner data. (a) Northern hemisphere Spring, Mar. 98–May 98; (b) Northern hemisphere, Summer, June 98–Aug. 98; (c) Northern hemisphere, Fall, Sept. 97–Nov. 97; (d) Northern hemisphere, Winter, Dec. 92–Feb. 98.(Courtesy of NASA/Orbimage Seawifs maps.)

temperate seas, only from one to three species of copepod are numerically dominant.

Thus far, we have considered the changes in net phytoplankton production as a function only of interacting physical factors. In the oceans, however, phytoplankton exist together with various herbivores. What effect do these herbivores have on the phytoplankton populations? Evidence suggests a great effect.

Careful inspection of the numbers of net phytoplankton cells per unit volume of water has shown that for the temperate North Atlantic Ocean the rate of decline of phytoplankton numbers following the spring bloom is steeper than would be predicted based on the nutrient decrease. Furthermore, nutrients never disappear completely from the upper waters, as would be expected if they were limiting. It can also be shown that the curve of zooplankton abundance in the temperate North Atlantic is similar in form to that of the phytoplankton, only delayed slightly in time (see Figure 2.37b). All of these data have suggested to some workers (Riley, 1946; Cushing, 1959; Harvey et al., 1935) that the copepods are responsible for the regulation of the larger phytoplankton populations. This is reinforced by data from the North Pacific Ocean where there is no spring bloom of phytoplankton (see Figure 2.37c). This is presumably the result of the presence of feeding copepods in the North Pacific at that time as opposed to the North Atlantic.

The rapidity with which copepods can remove phytoplankton cells is remarkable (Table 2.4). In experimental conditions, Fleming (1939) established a diatom population that initially had 1 million cells per liter and divided once each day. It was grazed by a copepod population adjusted to remove 1 million cells per day. Under such conditions, the diatom population would remain in a steady state of 1 million cells per liter. When the copepod density was doubled (double the grazing rate), the effect was to reduce the diatom population to 27,000 cells per liter in five days. If the copepod numbers and their grazing rate were increased five times, the diatoms were eliminated in five days. Since the rate of diatom reproduction remained unchanged in the above scenarios, any increase in copepod numbers in the

TABLE 2.4

Changes in the Numbers of Individuals of a Phytoplankton Population with a Constant Rate of Reproduction When Subjected to Two Different Grazing Intensities

	Phytoplankton Population Density (cells/liter)		
Time in Days	*Initial Grazing Intensity*	*Grazing Intensity Doubled*	*Grazing Intensity Increased Fivefold*
0	1,000,000	1,000,000	1,000,000
1	1,000,000	500,000	62,000
2	1,000,000	250,000	3,900
3	1,000,000	125,000	240
4	1,000,000	62,500	15
5	1,000,000	31,250	1

Source: Adapted from Fleming, 1939

plankton could have a potentially significant effect on the numbers of diatom individuals. It is important to understand that this decrease in numbers is independent of any change in the photosynthetic rate, or the rate of primary production. It is quite possible to have a diatom population decreasing in numbers due to grazing, while increasing the rate of primary production. Of course, with fewer algal cells, the total amount of carbon fixed would decline with declining algal cell populations, but it would still be possible for the *rate* of fixation of carbon in each cell to increase. It is for this reason that the population density or standing crop of phytoplankton may be a poor or misleading indicator of primary productivity.

It is thus possible for the standing crop of phytoplankton to decline due to grazing, while the rate of primary production increases or remains steady. Under such circumstances, the diatom population divides rapidly because the high photosynthetic rate allows the buildup of protoplasm for cell division. As fast as the diatoms reproduce, however, they are consumed by the copepods and other herbivores. In this case, the major portion of the carbon fixed in photosynthesis appears not in the standing crop of phytoplankton, but in the standing crop of zooplankton. Does such a situation occur in the oceans, and if so, what is the magnitude of the carbon fixed that appears in the zooplankton versus phytoplankton? Plankton ecologists have used various mathematical equations employing the concentrations of nutrients, water transparency, light availability, and photosynthetic rates to calculate what the productivity of a given body of water should be and then comparing it with what is observed. The results for many areas have been remarkable. For example, Hart (1942) reported that, in the Antarctic seas around South Georgia Island and in the English Channel, the standing crop of phytoplankton was 2% of calculated production. In oceanic areas of the Antarctic seas where grazing was more intense, Mare (1940) found that the standing crop of phytoplankton was only 0.5% of calculated production. In other words, the majority of primary productivity resides not in the bodies of the algal cells but in the zooplankton. These data also show that copepods are extremely efficient grazing animals and are capable of reducing the populations of phytoplankton drastically in patches of the ocean. This contrasts with the situation in natural terrestrial communities, where the grazers rarely remove all the vegetation and the standing crop of plants is high.

The ability of copepods to decrease the larger, net phytoplankton populations to very low levels means that the sudden decrease in the phytoplankton curve after the spring bloom in temperate waters could be caused by the grazing of increased numbers of copepods rather than being a result of nutrient depletion. This concept is supported by the finding that in the Antarctic seas, where nutrients are not limiting due to upwelling, the phytoplankton population also drops precipitously after the spring bloom. Additional evidence comes from the fact that grazing, and hence breakup of phytoplankton cells during passage through the copepod guts, releases the nutrients, which are then excreted back into the water. Under these circumstances, nutrients would be constantly regenerated in the upper layers and productivity never halted. Additional support for grazing as a control mechanism for phytoplankton is furnished by the patchiness of plankton. In the oceans, there are dense patches of phytoplankton and clear patches with copepods. One explanation is that copepods overgraze the phytoplankton.

If the copepods are so efficient at reducing the phytoplankton population, how does the spring bloom in the North Atlantic ever get started? The answer to this is that the copepods reproduce more slowly than the phytoplankton; thus, in early spring, following the winter low period, the phytoplankton can respond more rapidly to the improving light conditions and lowered turbulence. They reproduce rapidly and build up to high levels in absence of significant grazing. The copepods, also at low levels, begin to breed as well, but their cycle is longer (weeks versus days for phytoplankton). They cannot build up large enough populations until after the phytoplankton have bloomed (see subsequent section on copepod cycles). After that delay, however, when their populations are high, they quickly graze down the plant population to the low summer level. Copepod levels also decline because of less food and more predation, and because copepod egg production depends on food available to the breeding adult.

The previous situation prevails in the temperate and subpolar North Atlantic. In the tropics, where nutrients limit population levels of phytoplankton to low levels at all times, the situation is different. In these areas, there are no real pulses in either the phytoplankton or copepod cycles. Instead, there is a steady but inconspicuous consumption by copepods of the small phytoplankton crop.

Copepod Cycles Because the copepods dominate the larger zooplankton in all world oceans as the major grazers of the larger, bloom-forming diatoms and dinoflagellates, and because they can determine the form of the phytoplankton population curve, it is worthwhile to review the major aspects of their life cycle. This contributes to our understanding of their grazing patterns. This section will concentrate on copepod cycles in temperate and polar waters.

Relatively few of the numerous copepod species have been thoroughly studied, but in most temperate and polar waters, only a few species predominate. For some of these species, notably *Calanus finmarchicus* of the Atlantic Ocean, the life cycle is well known. Since other species similar to *Calanus* also dominate Arctic and North Pacific waters and have somewhat similar life cycles, a general scheme may be established.

All copepods have a similar pattern of development. The eggs hatch as naupliar larvae and pass through a series of six naupliar stages. They then enter another larval stage called the copepodid, passing through five of these stages before becoming adult. For the large, dominant Atlantic copepods of the species *Calanus finmarchicus*, the early winter months are spent in the last or fifth copepodid stage in the deeper water of the open ocean. Here they persist without feeding, living on fat or oil reserves in their bodies. They do not migrate at this time (see p. 76). Those copepodids that survive the winter begin their final molt to the adult stage in late winter or early spring. At this time, the total numbers of individuals are at a minimum for the year. After molting to the adult condition, the copepods migrate to the surface waters and begin feeding. They do not begin breeding, however, until after the phytoplankton numbers have begun to increase. This delay in the onset of breeding permits the phytoplankton to escalate their numbers to produce the high population numbers of the spring bloom. This first brood of copepod eggs then goes through the naupliar and copepodid stages and itself matures and reproduces. The resulting increase in numbers of copepods then begins to reduce the phytoplankton numbers, creating the drop in the phytoplankton curve and the increase in the zooplankton curve seen in Figure 2.37b. This cycle of generations continues through the spring and summer, and the number of generations per year depends on the environmental conditions. Off Scotland and Maine, for example, there may be three complete generations. Further north, the number is reduced to two off Norway, and in Arctic seas, such as off Greenland, to one. Further south, off the Middle Atlantic states and English Channel, four generations may be produced. Each cycle, from egg production through adult to egg production again, takes about two months under favorable conditions. The life span of these large calanoid copepods is about two months for those individuals born at this time (Figure 2.40a), which means adults do not live long after breeding.

The last brood produced in late summer does not complete the developmental cycle. Instead, when they reach Stage V copepodids, these animals migrate into deeper water and remain there through the winter, subsisting on stored food reserves. They return to the surface waters in the spring to feed before being able to lay eggs. For this generation, the life span is much longer, perhaps five to seven months, depending on the area. Egg production is influenced by the abundance of phytoplankton available as food. Successive generations of copepods build up large numbers in the spring, feeding on the spring bloom of phytoplankton. Similarly, the decrease in copepod numbers in the summer reflects the decrease in phytoplankton due to excessive grazing by the large numbers built up in the spring.

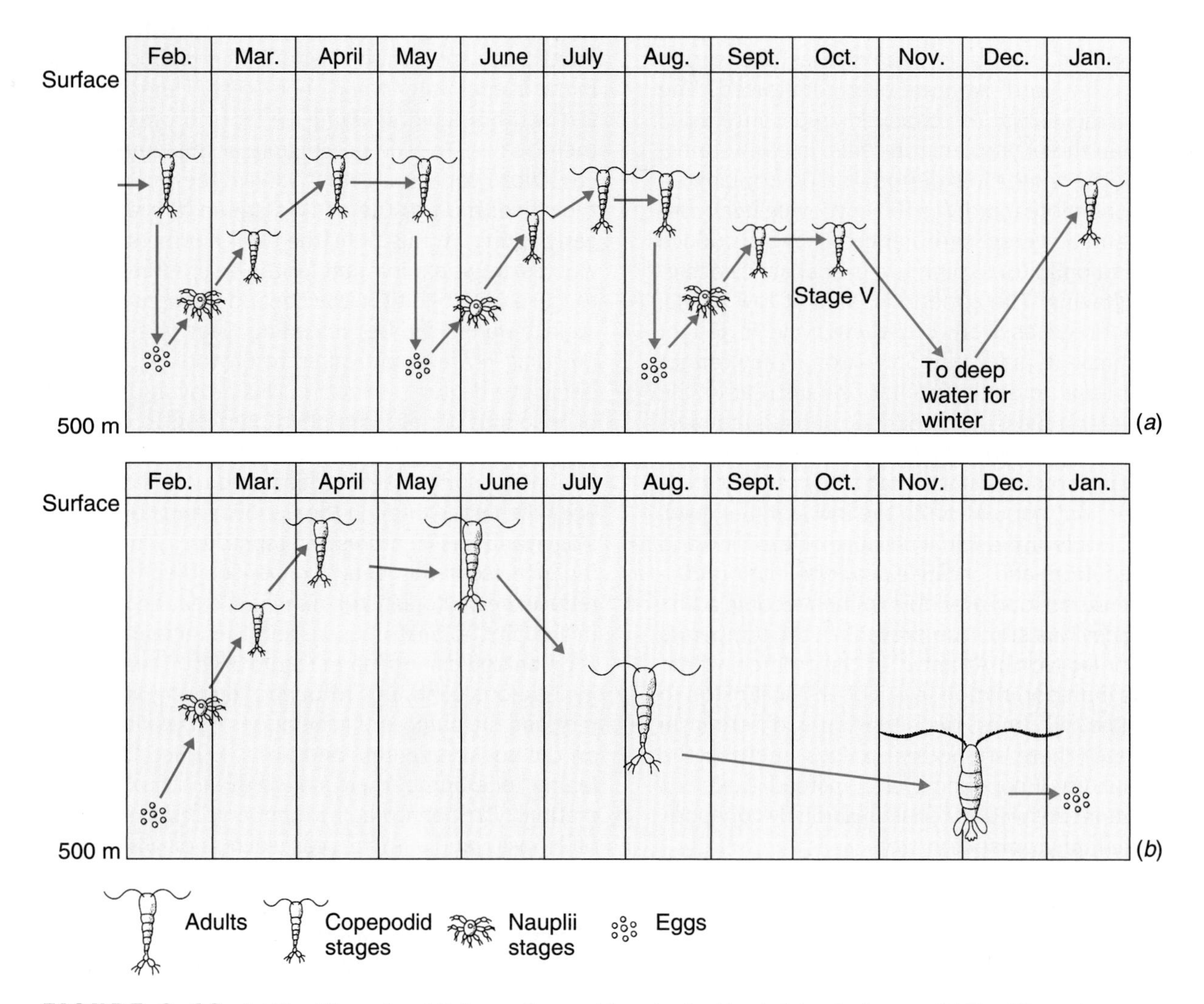

FIGURE 2.40 (a) The life cycle of *Calanus finmarchicus* in the North Atlantic Ocean. (b) The life cycle of *Neocalanus plumchrus* in the North Pacific Ocean.

In summary, the life cycle of *Calanus finmarchicus* is about two months from egg through larval stages to adult, and successive generations are produced through the spring and summer in temperate and polar waters. The number of generations decreases as one moves poleward because of the increasing shortness of the season. Copepod numbers reflect phytoplankton densities.

The dominant calanoid copepods of the North Pacific Ocean, *Neocalanus plumchrus* and *Neocalanus cristatus*, have a somewhat different life cycle than *C. finmarchicus* (Figure 2.40b). This, coupled with certain environmental differences between the two oceans, leads to the different form of the seasonal curves for zooplankton and phytoplankton seen in Figure 2.37c. To begin with, the two *Neocalanus* species are larger than *C. finmarchicus* and feed on both phytoplankton and zooplankton. *Calanus finmarchicus*, on the other hand, is primarily a herbivore. The *Neocalanus* species are also less efficient in their grazing activity. Finally, the seasonal cycles are different. Both *N. plumchrus* and *N. cristatus* have only one generation per year. Both species migrate to depth (300 m) in the summer and overwinter there as adults with sufficient energy reserves to mate and lay eggs. The eggs hatch and the larval copepods are back in the surface waters in the late winter ready to graze before the onset of the spring phytoplankton bloom.

While the above is the situation for the dominant, large *Calanus*-type copepods, there are other smaller copepods in both the Atlantic and Pacific oceans, such as *Acartia*, *Pseudocalanus*, and *Microcalanus*, that occur with *Neocalanus* and *Calanus* and also differ slightly in life cycle. In these forms, the generation time is shorter, and they tend to have more generations per year. Although they build up Stage V copepodids in the fall, there is no conclusive evidence of overwintering and it may be that these small forms breed throughout the year, albeit at a reduced rate in the winter and at other periods of low phytoplankton availability. This latter type of life cycle with continuous breeding year-round is found in many tropical copepods. Because of their small size, the constancy of the conditions, and the continuous phytoplankton growth, development is rapid and many broods are produced per year. This spreads out the grazing pressure more or less equally over the year to coincide with steady phytoplankton populations.

In general, then, the copepod cycles are maintained for maximum use of almost all larger or net phytoplankton production. However, a certain, but as yet undetermined, amount of energy is lost through fecal pellets and exoskeletons sinking out of the photic zone.

Copepod Influence on Atlantic and Pacific Plankton Community Ecology At this juncture, it is useful to consider the differences in the life cycles of the dominant copepods in the North Atlantic and North Pacific alongside what we know about the slight differences in environmental factors between the two oceans to demonstrate how these interactions can lead to the very different seasonal cycles of the larger phytoplankton and zooplankton in the two areas.

The environmental differences between the two oceans are slight but significant. At the same latitude the Atlantic is somewhat warmer, 8.5°C at depth in the winter compared with the Pacific, which is 3.8°C. The winter mixing extends to beyond 200 m in the Atlantic, but only to 100 m in the Pacific. Finally, the nutrient levels in the Pacific are always above 5 μm/liter, whereas in the Atlantic they are reduced below 1 μm/liter in the summer. Other physical and chemical factors are similar between the two oceans. There are, however, some differences in organisms. Among the photosynthetic organisms, the Atlantic has larger numbers of diatoms and dinoflagellates, whereas the Pacific has larger numbers of small flagellates. The growing season is also shorter in the Atlantic than in the Pacific.

Putting all these factors together with the life history differences among the copepod species produces surprisingly different results, as Parsons and Lalli (1988) have noted. Beginning in the winter, we find that in both oceans the dominant copepods overwinter at depth, but the difference in temperature means that the Atlantic copepods have a higher metabolic rate, so by early spring, they have insufficient reserves left to lay eggs. In contrast, the larger Pacific copepods that overwinter at a lower temperature, hence lower metabolic rate, retain sufficient reserves to lay eggs without resuming feeding (Figure 2.41).

In early spring, the phytoplankton growth begins early in the Pacific but proceeds slowly because of reduced light and temperature. In the Atlantic, phytoplankton growth is delayed due to the depth of the mixed layer. Once it begins, however, the light is better and it proceeds more rapidly, aided by the higher temperatures. In the Pacific, as a result of the slow phytoplankton growth and the continual presence of smaller herbivores in the surface waters, combined with the entrance into the surface waters of the recently hatched copepodids of *Neocalanus*, the grazers keep pace with the plants, preventing any sudden bloom. In the Atlantic, on the other hand, the zooplankton cannot keep up with the plant growth and reproduction aided by the high nutrient content and the lack of a dominant copepod grazer. A grazer is absent because *Calanus finmarchicus* must feed first to build up enough reserves to produce eggs, which in turn must have a finite development time before hatching. As a result, there is a delay in the grazer population increase, and a phytoplankton bloom ensues. In other words, there is an efficient coupling of the phytoplankton and zooplankton populations in the Pacific but not in the Atlantic. By late spring, the phytoplankton providing the bulk of the primary productivity in the Pacific are small in size and low in density. This means the larger *Neocalanus* cannot efficiently graze on them, and they turn to carnivorous feeding on the smaller zooplankton, thus suppressing their numbers. In the Atlantic, the majority of the phytoplankton consists of large diatoms in high density, and they are grazed efficiently by *C. finmarchicus*. This large food supply leads to rapid growth and maturity of *Calanus*, and they begin to produce the next generation.

In early summer, the *Neocalanus* have reached maturity and migrate to depth to avoid higher surface temperatures. Their large size and large food reserves ensure that they will survive and be able to produce eggs. Meanwhile, the grazing in the surface layers is taken over by smaller zooplankton now released from the feeding pressure of the *Neocalanus*. These smaller zooplankton keep the grazing pressure on the phytoplankton throughout the fall and winter when *Neocalanus* are at depth. The higher level of nutrients permits some phytoplankton growth even during the low light levels of the winter. In the Atlantic, by summer the phytoplankton have exhausted the nutrients

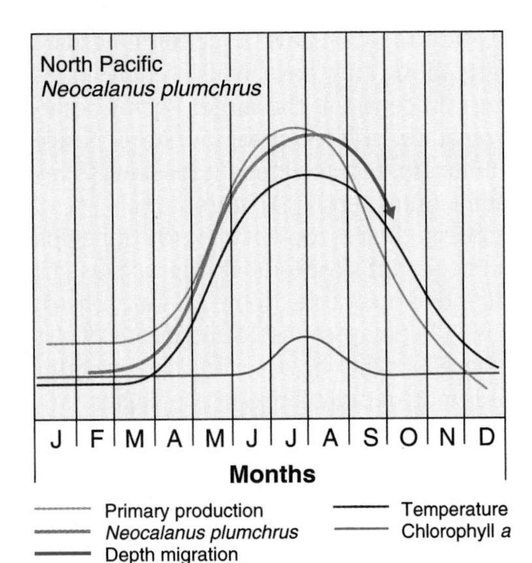

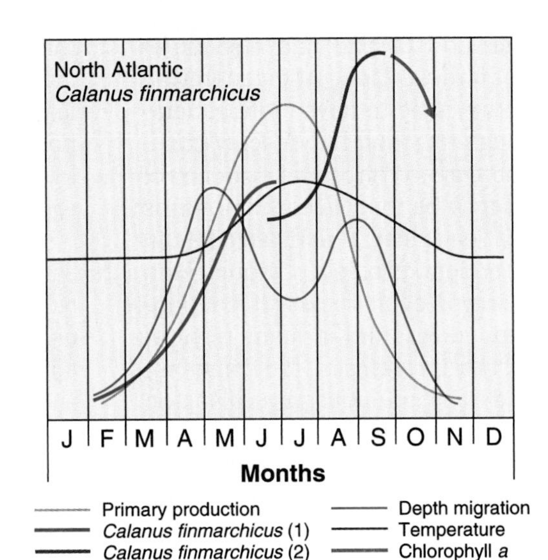

Neocalanus plumchrus		*Calanus finmarchicus*
Overwinters with low metabolic loss at 3.8°C. *Result*: Copepod has sufficient reserves to lay eggs without feeding in the spring.	*Winter*	Overwinters with high metabolic loss at 8.5°C. Copepod has insufficient reserves to lay eggs without feeding in the spring.
Phytoplankton growth starts slowly but is limited by light and temperature. *Result*: Zooplankton can track the spring bloom and suppress any increase in chlorophyll *a*; efficient phasing.	*Early Spring*	Phytoplankton growth cannot start in mixed layer > 200 m. When growth starts, it is rapid due to higher temperatures. Zooplankton cannot keep up with phytoplankton growth and chlorophyll *a* increases; inefficient phasing.
Phytoplankton providing most of the primary production are all small (< 200 mμ). *Result*: Low phytoplankton cell density and small size favors alternate strategy of carnivorous feeding on microzooplankton resulting in loss of food chain efficiency. Relatively slow growth of zooplankton and slow development rate (i.e., stages) leads to large size.	*Late Spring*	A much larger proportion of the phytoplankton is made up of large diatoms. High phytoplankton stocks and large phytoplankton size favor herbivory with rapid zooplankton growth rate and development rate to next generation, resulting in small size of adults.
N. plumchrus gradually reaches Stage V and migrates to depth to avoid high surface temperatures of summer. *Result*: Large size achieved of aestivating *N. plumchrus* assures winter survival at low temperature.	*Early Summer*	*C. finmarchicus* grows first new generation (1). Growth rate is rapid, but phytoplankton exhaust winter nutrient supply. *C. finmarchicus* is ready to produce a second generation (2), which begins by growing faster than the first due to higher temperatures; phytoplankton production must now be based on efficient recycling of nutrients.
Annual cycle of *N. plumchrus* in surface layers entirely replaced by late summer opportunistic copepods and carnivores. *Result*: Phytoplankton shows small increase after *N. plumchrus* departure but this is again suppressed by summer herbivores.	*Late Summer*	Second generation of *C. finmarchicus* grows to provide immature copepodites, which must overwinter. Other herbivores together with carnivorous zooplankton develop. Second generation is more capable of surviving winter than first generation since cooling trend favors larger body size of late development stages.

FIGURE 2.41 A comparison of the differences in the community ecology of the larger plankton organisms of the North Pacific and North Atlantic oceans. (Modified from "Comparative Oceanic Ecology of the Plankton Communities of Subarctic Atlantic & Pacific Oceans" Parsons & Lalli, *Oceanographic Marine Biology Annual Review*, Vol. 26, pp. 317-359,)

at about the time *C. finmarchicus* begin to produce the second generation. This second generation of *Calanus* begins growth at a faster rate than the parent generation because of high temperatures and large food supply. The phytoplankton numbers, however, now begin to decrease due to nutrient exhaustion and grazing by the high numbers of *Calanus*. Phytoplankton production now becomes dependent on nutrient regeneration in the euphotic zone. Finally, the second generation of *C. finmarchicus* grows to the last copepodid stage but with little reserve energy, thanks to the declining phytoplankton population, and migrates to depth to overwinter.

A probable, but as yet unproven, result of the poor coupling of the phytoplankton and zooplankton cycles in the Atlantic is that much of the carbon fixed in the spring bloom ends up as detritus that sinks and probably sustains large numbers of benthic fishes there. By the same token, the closer coupling of the phytoplankton and zooplankton cycles in the Pacific may be responsible for the greater numbers of pelagic fishes in the Pacific.

More recently, work by Miller et al. (1991) has provided additional information that has increased our understanding of how the North Pacific pelagic ecosystem functions. These workers were also interested in the problem of why, in contrast to the North Atlantic, the North Pacific waters always had substantial nutrient levels and no substantial spring phytoplankton blooms. Their results showed that the phytoplankton in the North Pacific are always dominated by nanoplankton-sized organisms and that the larger phytoplankton, such as diatoms, are less common. Furthermore, the *Neocalanus* were not grazing these smaller phytoplankters, which indicated there had to be another set of smaller grazers. These microzooplankton grazers were mostly protozoans and probably the main food of the *Neocalanus*. More interestingly, they also discovered that these protozoans had very rapid growth rates and could always overtake and suppress any incipient blooms of the small phytoplankton. Because the mixed layer is so shallow in the North Pacific, the phytoplankton can overwinter in the lighted zone, albeit at a low level due to reduced light. In addition, this provides the microzooplanktonic protozoans a continued food source. These protozoans also provide the answer to the persistent higher nutrient levels. The micrograzers produce very small fecal matter that does not sink out of the lighted zone but is recycled in the system, mainly as NH_4^+, which itself seems to be favored over nitrate by the phytoplankton. This explanation of the situation in the North Pacific has been called the **mixing and micrograzer hypothesis**.

One last part of the North Pacific dilemma needs to be explained and that is what controls the stocks of the larger phytoplankton. As noted, the diatoms and dinoflagellates are persistently present. Why do they not bloom as in the Atlantic? There is no satisfactory answer to this question at present. It could be that the *Neocalanus* species are capable of preventing this, but another suggestion by Martin and Fitzwater (1988) is that this area is lacking in iron, which facilitates growth of the smaller phytoplankton at the expense of the larger ones. The diatoms and dinoflagellates, therefore, cannot attain bloom levels without this necessary trace element (see also the section on iron limitation in the intertropical convergence).

Vertical Migration One of the more puzzling phenomena in the sea is the vertical migration of zooplankton. Vertical migration is the name given to the daily migration of certain zooplankton organisms down into the depths during daylight hours and up into the surface waters at night. It is entirely separate from the seasonal migration noted above, where copepods descended in the fall to deeper waters to overwinter. Vertical migration may be just a special case of the more general phenomenon of the regular vertical migration of many open ocean pelagic animals from one depth zone to another and back. Movements of these other so-called "deep scattering layers" are discussed in Chapter 4. We cover here the special case of the vertical migration of epipelagic plankton.

Daily or **diel** vertical migrations have been known for more than 100 years. What makes this phenomenon so puzzling is that it is difficult to offer a satisfactory explanation as to why such small organisms should expend significant amounts of their limited energy resources for this purpose. The distances traversed in these daily movements may be 100–400 m. Since we are considering animals of only a few millimeters in size, this is the equivalent of humans walking 25 miles to work and then 25 miles back at night.

Vertical migration has been observed in all zooplankton taxa, but not all zooplankton migrate. With respect to taxonomic group, size, or feeding habits, there are few generalities that can be drawn from a comparison between those that do migrate and those that do not. This inability to find correlations makes its occurrence even more difficult to understand.

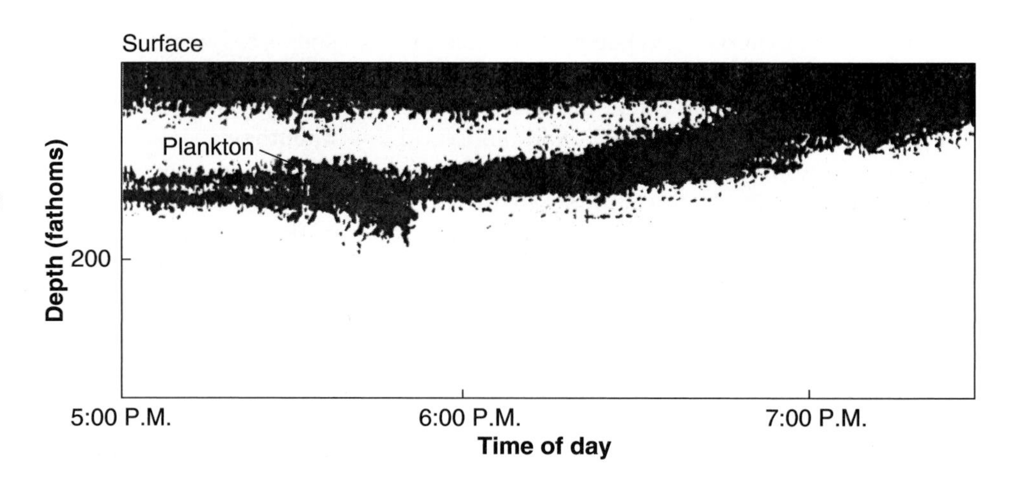

FIGURE 2.42 Sonogram record of movement of vertical migrating plankton. (After G. B. Farquhar, ed., *Proceedings of an international symposium on biological sound scattering in the ocean,* 1970, U.S. Government MC Report 005.)

It is now generally agreed that the major stimulus that initiates and controls vertical diel migrations is light. Vertical migrators respond negatively to light, moving themselves deeper as the surface light intensity increases. Conversely, they move toward the surface as the surface light intensity decreases. In a typical pattern, the migrators are found at the surface during the night; with the approach of dawn, they begin their descent. As the light intensity increases through the morning, the animals move deeper, usually keeping themselves at a certain light intensity. At noon or whenever the intensity is maximal, they will be at the deepest position. As the sun sinks through the afternoon and light intensity decreases, they begin their journey toward the surface again, arriving there around sunset and remaining until just before dawn. As might be expected, reverse migrators proceed in exactly the opposite way, descending as the light decreases and rising as the light increases. Since the advent of sonar or echo-sounding gear in World War II, it has been possible actually to observe this phenomenon by watching the trace formed on the recorder by sound reflected from the bodies of these animals as they move up and down. Their bodies form a layer in the water column (Figure 2.42).

Since the ocean is not uniform in its physical parameters in a vertical direction, other factors may modify the light-regulated scheme. For example, temperature may also influence migration. Some migrators are limited by temperature change in the depth to which they go. In such cases, the maximum depth to which they descend is set by temperature, which overrides the stimulus of light. Another constraint is the presence of the bottom. In open ocean areas where submarine ridges or mountains rise to within daytime depths of the vertical migrators, the migrators may congregate at the bottom. Such concentrations of animals are a prime food source and may be why both submarine banks and seamounts are often densely populated with predators.

Although light is the main stimulus to which the animals respond and the guide to their positioning in the water column, it does not explain why the animals make the journey. What advantage can there be for such energy-demanding movements? Over the years, a number of explanations have been offered by marine ecologists. At present, four have been presented, but there is no universal agreement among ecologists as to which, if any, of the explanations is correct.

Perhaps the most popular explanation, which cannot be attributed to any single person, but has been reviewed by McLaren (1963) and Haney (1988), is that the zooplankton move from the lighted waters into dimly lit, deeper waters to avoid predation by visual, diurnal predators, such as various fishes, cephalopods, and birds that forage at the surface. Similarly, reverse diurnal migrations are a response to avoid nocturnal surface-dwelling predators or diurnally migrating nonvisual predators. The following are problems with this explanation:

1. The daytime depths of residence for many zooplankters still have sufficient light for visual predators to operate.
2. Other zooplankton migrate deeper than would be required by this hypothesis.

3. Many of the migrators possess and use bioluminescent organs that seem to counteract the effect of the darkness.
4. Many potential predators are also migratory.

Support for this hypothesis has come mainly from investigations in freshwater lakes. These studies have demonstrated that in systems with few or no visual predators, vertical migration is lacking; whereas in those systems with such predators, vertical migration is common. Ohman (1990), however, has demonstrated that the marine copepod *Pseudocalanus newmani* may undergo vertical migration, reverse migration, or no migration depending on whether the primary predators are visually hunting planktivorous fish, nocturnally feeding nonvisual zooplankton, or are absent. This study also suggests that diel vertical migration is not a fixed trait in zooplankton species but a dynamic behavioral trait that may be modified by prevailing conditions.

A second hypothesis, somewhat related to the first, is light-damage avoidance. This hypothesis suggests that zooplankton organisms migrate to avoid damage from solar radiation. Since damaging ultraviolet light does penetrate several meters deep in clear oceans, this seems a reasonable hypothesis. It is also known that certain pigments, primarily carotenoids, protect organisms from damage by ultraviolet light. We have, therefore, a dilemma: The presence of pigment may protect the zooplankton from light, enabling them to inhabit lighted waters, but it also makes the animal more visible to visual predators. It seems we must assume that the greater risk of predation due to the high visibility of pigmented animals is offset by either the escape response of the zooplankton (vertical migration) or the absence of visual predators. At present, no experimental or field studies have shown a direct relationship between light damage and vertical migration.

A third hypothesis for the occurrence of vertical migration, suggested by Hardy (1953), is that it allows the weakly swimming zooplankton to change their horizontal position. To understand how this is possible, it is necessary to return to the previously described principle known as the Ekman spiral (Chapter 1). As one moves downward from the surface waters, the speed and direction of a surface current also change. The speed decreases and the direction of its movement changes to the right of the surface-most current (see Figure 1.13). This means that a descending zooplankton organism will encounter currents that move not only more slowly but in slightly different directions. Once the zooplankter has descended into the deeper, slower-moving water, the surface currents will bring different water across and above the animal. If the zooplankter ascends again, it will find itself in a different water mass. What is the value of this? Since zooplankton organisms lack the ability to swim against currents, they would automatically be confined to a given water mass in the ocean. Given their great ability to graze out the phytoplankton from an area, they would soon be without food and starve. By simply sinking into the depths and then rising again, however, they would come up in a different surface water mass, potentially ungrazed. This explanation is supported by the universal observation that phytoplankton and zooplankton are very patchy in occurrence. This mechanism might also be employed by zooplankton to avoid or leave very dense phytoplankton patches. Hardy and Gunther (1935) have suggested that a toxic substance is produced by the phytoplankton, which has deleterious effects on zooplankton.

A final reason for diel migration, suggested by McLaren (1963) and McAllister (1969) and refined by Enright (1977) and McLaren (1974), concerns production and energetics. This two-part hypothesis suggests (1) the rate of phytoplankton production is greater when subjected to discontinuous nocturnal grazing as opposed to a population subjected to continuous grazing; and (2) the zooplankton obtain an energetic advantage by spending a portion of their time in cold, deep waters rather than maintaining themselves constantly in the warm, upper waters. The second part of this hypothesis rests on the ability to demonstrate that more energy is saved by remaining in deeper, cold waters during the day than is expended in making the migration. It has been demonstrated that a neutrally buoyant zooplankter requires remarkably small amounts of energy to migrate, but it remains to be proved that the amount saved by remaining at depth is in excess of this amount. Implicit in this energetics theory is that feeding in warm water increases the efficiency of feeding and assimilating food, while residence at depth improves efficiency of growth.

This hypothesis is supported by the observation that in polar regions, where the temperature differential between deeper and surface waters is lowest, vertical migration is least often observed. Conversely, the strongest vertical migration patterns are observed in the tropics, where the greatest differential exists between surface water and depth. However, attempts to verify this hypothesis both in the laboratory and field have failed so far.

This final argument may also be used to explain the overwintering of copepods at depths where phytoplankton is scarce. Such overwintering at low temperatures conserves energy.

In summary, it can be stated that vertical migration is a worldwide phenomenon of certain zooplankton. The stimulus is primarily light, although it may be modified by other factors, such as temperature. Diel migration occurs most commonly in strongly thermally stratified seas and becomes suppressed or disappears where seas approach isothermal conditions. It is also absent in temperate seas during winter months. The hypotheses for its occurrence include avoidance of predation, avoidance of damaging ultraviolet light, change of position in the water column, and as a mechanism to increase production and conserve energy. It is probable, as Haney (1988) suggests, no single explanation for occurrence will be found true in all systems, and there are multiple causes for vertical migration in different systems, as well as within a single system.

Seasonal Succession in Phytoplankton

In addition to the marked changes in abundance observed in the phytoplankton over the course of a year, there is also a marked change in species composition. This change in the dominant species from season to season is called **seasonal succession**. Under seasonal succession, one or more species of diatom, dinoflagellate, or coccolithophore dominate the plankton for a shorter or longer period of time and then are replaced by another set of species. This pattern is repeated yearly. This "succession" is different from the typical terrestrial ecological succession in which various plant associations replace each other until finally a so-called "climax" community develops, which persists through time.

Figure 2.43 displays the dominant phytoplankton species at several geographic locations to emphasize the widespread occurrence of this successional phenomenon. Some successional changes occur even in polar (East Greenland) and tropical (Bermuda) areas.

What are the factors causing this phenomenon? Margalef (1963) and Smayda (1980) have suggested several. Considering that the seasonal succession is most often and clearly seen in temperate seas, which have a marked change in temperature during a year, temperature has been suggested as a cause. This may be one of the factors, but it is unlikely to be solely responsible because certain dominant species recur at different temperatures. Furthermore, temperature changes rather slowly in seawater, and the replacement of dominant species often is much more rapid.

Another suggested reason is the change in nutrient level over the year, with differing concentrations favoring different phytoplankton species. While this factor may also contribute, observations suggest population changes are not closely correlated, that phytoplankton populations rise and fall much more quickly than nutrient concentrations change.

Yet another explanation for succession is that it is the result of **biological conditioning** of the seawater by the organisms in it. By this we mean that one group of organisms, when dominant in the water column, secretes or excretes very small quantities of organic compounds or **metabolites** that have an effect on other organisms, either inhibiting or promoting their growth. At the same time, these organic compounds could affect the very organisms producing them, making the water either more or less suitable for their own existence.

These organic metabolites could, and probably do, include a number of different classes of organic compounds. Some are likely toxins, such as those released by the dinoflagellates during red tides, which inhibit growth of other photosynthetic organisms. In such cases, the population explosion of dinoflagellates is so great that the water becomes brownish red in color from the billions of dinoflagellate cells. Although each cell secretes a minute amount of toxin, the massive dinoflagellate numbers cause the toxin to reach concentrations that kill many creatures. This toxin can be concentrated in such filter-feeding organisms as clams and mussels, rendering them toxic to humans at certain times and places.

Another class of metabolite is the **vitamins**. It is now known that certain phytoplankton species have requirements for certain vitamins and that there are considerable differences between species as to requirements. The B vitamins, especially vitamin B_{12}, thiamine, and biotin, seem to be the most generally required. Some species may be unable to thrive until a particular vitamin, or group of vitamins, is present in the water. These are produced only by another species; hence, a succession of species could occur whereby first the vitamin-producing phytoplankter is present and then the vitamin-requiring species follows.

Other organic compounds that may inhibit or promote various species include amino acids, carbo-

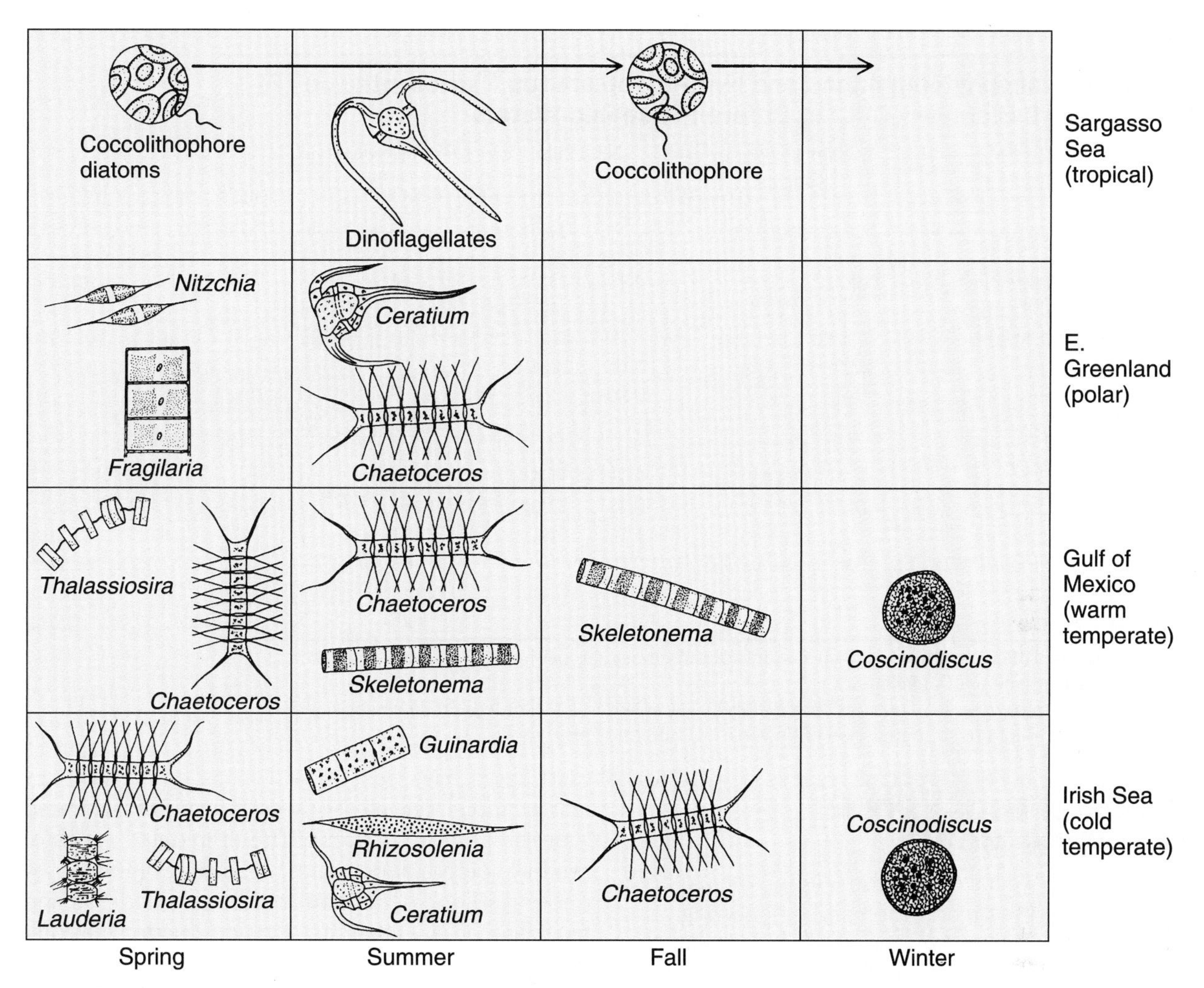

FIGURE 2.43 Seasonal succession of the dominant phytoplankton in four different geographical areas. With the exceptions of coccolithophores and *Ceratium*, which is a dinoflagellate, all genera are diatoms.

hydrates, and fatty acids. Although it is suspected that these organic metabolites may have an important role in species succession and it has been demonstrated in the laboratory that phytoplankton species vary both in their ability to produce necessary vitamins and in their requirements for such in order to grow, evidence is still sketchy as to their real role in the sea.

There is also evidence to suggest that grazers, particularly selective grazers, can influence the phytoplankton species composition. Many copepods and invertebrate larvae pick out selected phytoplankton species from mixed groups, changing the species composition.

A growing body of evidence now suggests that all of the factors considered here are operating simultaneously to produce species succession. The importance of any factor will vary with the particular phytoplankton species and the environmental conditions.

To summarize, there is a general seasonal pattern of succession, progressing from diatom to coccolithophorid- or dinoflagellate-dominated systems (Figure 2.43). In temperate waters, small-celled, rapidly dividing diatom species initiate the spring bloom. They are succeeded first by larger diatoms in the late spring and then by a summer warm-water dinoflagellate community. In tropical and subtropical waters, coccolithophores and dinoflagellates are generally dominant all year, although they may be interrupted by periodic diatom blooms.

TABLE 2.5

Summary of Nanophytoplanktons and Net Phytoplankton Photosynthesis (P) and Chlorophyll Biomass (CHL) in Different Oceanic Waters[a]

	Nanophyto-plankton		*Net Phyto-plankton*		*Percentage net Plankton*	
Region	P	*chl*	P	*chl*	P	*chl*
A. Tropical						
Eastern Equatorial Pacific	.06	20.7	0.1	3.0	14	13
North Equatorial Pacific		21.1		1.0		4
North Equatorial Atlantic					10	22
Sargasso Sea	22.5	4.3	3.6	0.4	14	9
Caribbean	0.3	12.2	0.03	1.2	9	9
Indian Ocean					7	10
B. Upwelling						
Peru Current	2.7	18.2	0.2	1.6	7	8
California Current	0.7	15.6	0.1	2.6	12	14
Costa Rica Dome	1.8	—	1.2	—	41	—
Equatorial Upwelling (91°W)	0.4	—	2.0	—	83	—
C. Temperate						
Northeast Pacific					<25	<25

[a] Numbers are means. Photosynthesis is in micrograms of carbon per liter per hour. Chlorophyll is given in milligrams per square meter of surface.

Source: Modified from Malone, 1980 in *Physiological Ecology Of Phytoplankton*, I. Morris, editor. Copyright © 1980 Blackwell Scientific. Reprinted by permission of Universit of California Press.

THE OCEAN ECOSYSTEM: A CHANGING MODEL

Up to this point, we have been considering the usual explanations for the organization and functioning of the marine plankton system. Much of this classical description, while still valid as a basic framework and still true in polar seas, coastal seas, and upwelling areas, has been considerably modified in the past 20 years, primarily due to the advent of new technology. This new technology includes such techniques as satellite oceanography, remote sensing, epifluorescence and differential interference microscopy, scanning and transmission electron microscopy, ATP assays, nucleopore filters, and new methods for measuring minute amounts of various particles and chemicals. We now take up these new developments to build the final picture of the plankton ecosystems of the world as we understand them today.

Microorganisms and Productivity

The classical methods for studying phytoplankton productivity from net samples employing the ^{14}C method had the larger net phytoplankton, diatoms and dinoflagellates, as the dominant primary producers. In an attempt to understand the relative significance of nanophytoplankton and picophytoplankton to the total phytoplankton productivity, a number of workers using various new methods have now demonstrated that the phytoplankton of the epipelagic zones of both temperate and tropical seas are dominated by photosynthetic organisms in the nanoplankton and picoplankton size fraction. A similar situation may prevail in polar seas, but there are currently few data to support it. The domination is not only by numbers of individuals, but by the amount of photosynthesis as well. In most cases, Malone (1980) suggests that the nanoplankton account for 80% or more of all photosynthetic activity and 75% of the phytoplankton biomass in open ocean waters, especially in the tropics and subtropics (Table 2.5). Pomeroy (1974) even suggests that much nanophytoplankton photosynthesis is by organisms that are less than 30 μm in size. In the past, the contribution of these tiny photosynthetic organisms was ignored because they passed through nets, were easily destroyed on filters, were sparsely distributed, and only could be studied alive.

Nanoplankton and picoplankton seem to be less dominant in coastal (neritic) and upwelling areas, where net phytoplankton play a proportionately greater role but where productivity and biomass are also more variable.

Not only are the small size classes of phytoplankton abundant, they also show less seasonal variability in biomass than the larger net phytoplankton.

They do not show the characteristic curves shown in Figure 2.37. This lack of variability is particularly noticeable in tropical and warm temperate zones, such as the central gyre of the North Pacific and the Sargasso Sea area in the Atlantic. These areas seem to be steady state systems with very little seasonal or geographic change, according to McGowan (1974). Marine biologists attribute this steady state situation to the stable hydrographic regime, which gives rise to the continued low nutrient concentration. Low nutrient concentration favors smaller photosynthetic organisms that have a proportionately greater surface area to absorb the nutrients but a lower relative need. Small organisms also sink more slowly.

The dominant groups of nanophytoplankton and picophytoplankton include coccolithophores, very small flagellates of various kinds, and several kinds of prokaryotic and eukaryotic cells including the newly discovered prochlorophytes. According to Chisholm et al. (1988), members of this latter group were discovered using flow cytometry, are smaller than coccoid cyanobacteria, and reach densities of 10^5 cells per ml of water in certain deep areas of the photosynthetic zone.

A final consideration, neglected in analyses of plankton communities until now, is that of symbiosis. Many of the larger planktonic protozoa, particularly the Foraminiferida and Radiolaria, contain intra- and extracellular photosynthetic symbionts of different taxa. These symbionts are probably responsible for a significant, but unknown, fraction of primary productivity, according to Fenchel (1988). Even more interesting is the finding by Stoecker et al. (1987) that many planktonic ciliates retain the functioning chloroplasts of their algal food in a short-term "chloroplast symbiosis" that has previously been described only for some marine mollusks (see Chapter 10). The contribution of this type of symbiosis to the pelagic planktonic productivity is unknown.

Nanoplankton, Respiration, and Grazing

If the majority of photosynthetic organisms are minute, they would be consumed by small heterotrophic organisms. Just as our phytoplankton nets are too large to capture nanophytoplankton and picophytoplankton, so, it appears, are the filtering systems of the dominant adult copepods. What, then, are the primary consumers of these minute phytoplankters? We are not certain, but two groups of the kingdom Protista appear most significant. They are the nonphotosynthetic flagellates (phylum Sarcomastigophora) and various ciliates (phylum Ciliophora; see Figure 2.7). Both of these groups can ingest the tiny phytoplankton cells. The flagellates act like amoebae and engulf the individual cells, while the ciliates filter the cells from the water column.

How significant are these heterotrophs? Their importance can be estimated by measuring the respiration rate in the epipelagic zone of the ocean. The respiration rate is more difficult to measure than the photosynthetic rate because the size spectrum of organisms is larger than producers, because the total respiratory rate is very low, requiring methods of concentration of organisms, and because of inherent problems in techniques. Despite these problems, the respiration rate of net plankton versus nanoplankton has been estimated by such investigators as Pomeroy and Johannes (1968) and Hobbie et al. (1972), resulting in a close agreement among them that the respiration rate of nanoplankton exceeds that of net plankton by a factor of 10. What this, in turn, suggests is that nanozooplankton are consuming most of the energy fixed by the autotrophs. Contrary to our classical model of diatoms and dinoflagellates being grazed by copepods as the first step in the epipelagic food chains, we now believe the initial link is from nanophytoplankton and picophytoplankton to ciliates and small flagellates. This is not to say that the classical chain is not valid, only that it is probably of less importance, at least in the open ocean.

Before we can conclude our discussion of this new concept of how the food web is organized in the open ocean, we must consider a few final components.

Bacteria, Particulate, and Dissolved Organic Matter

Early studies on the epipelagic zone determined that the water had a rather stable and static amount of dissolved organic matter, which was generally considered too low to support any significant number of bacteria except on particles where organic matter was concentrated. Recently, more sophisticated analyses for dissolved and particulate carbon and detection of living material have established a different picture.

The amount of dissolved organic carbon (DOC) in the waters of the world's oceans constitutes one of the largest reserves of organic carbon on our planet. The total amount has long been considered stable, with residence times of thousands of years. At the same time, we know that perhaps one-fourth of the carbon fixed by photosynthesis is lost to seawater as dissolved organic matter "leaked" from cells. How can we reconcile this continual production with generally stable amounts measured in the water? Where does it go and why does it not build up? The answer appears to be that microorganisms, primarily bacteria, may account for this discrepancy through direct uptake of the dissolved organic matter. From field studies, Azam (1998) has stated that such direct uptake may account for up to 50% of the total oceanic primary productivity and that bacteria not only take up dissolved organic matter but attack both living and dead organic particles, thus freeing more dissolved organic matter. Fuhrman (1992) has suggested that bacterial biomass exceeds that of the phytoplankton in oligotrophic oceans. Clearly the microbial loop is a major biological player in the total carbon flux in the oceans.

The previous findings strongly suggest that bacteria are abundant in the water column, more abundant than we had thought. The introduction of direct plate counts with fluorescent microscopy has now demonstrated population numbers of from 10^5 cells/ml in oligotrophic waters to 5×10^6 cells/ml in coastal waters. According to Hobbie and Williams (1984) most are free living as **bacterioplankton**, are heterotrophic, and are about 0.4/μm in diameter. Larger bacteria are associated with planktonic particulate matter. These bacteria and their importance were missed by earlier scientists because of their very small size, which permitted them to pass through most filters then in use. Only the new nucleopore filters retain them. According to Sieburth (1982), bacteria passing a 1-μm filter may account for 90% of the uptake of dissolved organic carbon in the water column.

Bacterioplankton also have a very high assimilation efficiency according to Pomeroy (1974), up to 60–65% as compared with the classic 10–20% efficiency for larger organisms. Added to their large numbers, this means that these bacteria are converting a substantial amount of DOC into particulate organic carbon (POC) in the form of bacterial cells. It also suggests how what was formerly considered "lost" production (i.e., the carbon lost by leakage from the phytoplankton) is not really lost but is recycled back into the food chains. The bacteria, in turn, are consumed by different nanozooplankton organisms and can channel a significant amount of energy into planktonic food webs. This has been termed the **microbial loop** by Pomeroy (1974) and is illustrated in Figure 2.44.

The bacteria are also partially responsible for the regeneration of nutrients in the photic zone, permitting the continued productivity of the phytoplankton even in the absence of an influx of nutrient-rich water. Such regeneration is particularly important in the highly stratified waters of the tropics and subtropics.

In addition to these heterotrophic bacteria, Johnson and Sieburth (1979) showed that photosynthetic bacteria are ubiquitous in seawater. They are responsible for a significant fraction of the photosynthetic activity, particularly in oligotrophic waters. Together with the heterotrophic bacteria, these two groups are the major components of the picoplankton-size class.

Finally, before constructing a modern view of a planktonic food web, we need to mention one last group of potentially significant organisms that may affect our construct. These are the **viruses**. The existence of viruses in the marine environment has been known for more than 40 years, but it is only in the last few years that their potential role in plankton systems has been investigated. Viruses have no metabolic machinery of their own and must rely on a host organism to exist and reproduce. Our knowledge of the significance of viruses in planktonic systems is in its infancy. Fuhrman (1999) noted that viruses are the most common biological units in the world's oceans and may number as many as 10 billion per liter. He further suggests that viruses probably infect all marine organisms and may significantly influence many ecological processes including nutrient cycling, bacterial and algal diversity, and systems respiration and algal bloom control. At our current level of knowledge, that is about all that can be said. All of these findings mean that we have had to change our view of the planktonic food web.

The Marine Planktonic Food Web

In the classical marine planktonic food web, we had the net phytoplankton—the diatoms and larger dinoflagellates—as the dominant producers. They were consumed, in turn, by copepods and other crustaceans. Copepods were preyed upon by various carnivores, both fishes and other invertebrates. A final trophic level in this progression included the top carnivores, such as large fishes, cephalopods, and marine birds and mammals. With our new evidence we must conclude that the dominant primary producers are various nanophytoplankton, such as coccolithophores, and such picoplankton as cyanobacteria and prochlorophytes, supplemented by the bacterioplankton, which make available and recover potentially "lost" DOC. These, in turn, are consumed by various small protistans, mainly tiny flagellates and ciliates, which then are fed upon by larger net zooplankton. These interactions are summarized in Figure 2.44.

Where does the classical model fit into this scheme? According to Kiorboe (1993), all strongly stratified oligotrophic water masses are characterized by the dominance of a microbial type of food web, while those areas of the oceans that are weakly stratified or are mixed or turbulent are dominated by the net phyto-

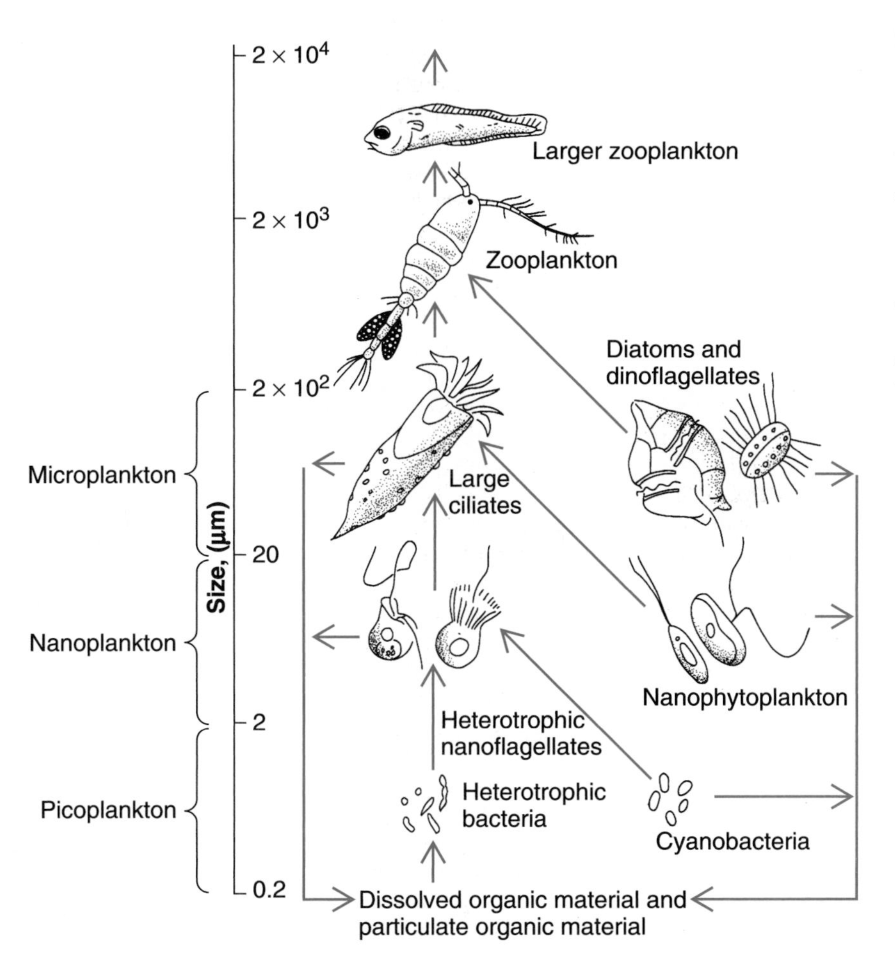

FIGURE 2.44 Diagrammatic representation of the microbial loop and its relationship to the "classical" plankton food web. Autotrophs are on the right, heterotrophs on the left. (Modified from "Marine Plankton Food Chains" T. Fenchel, *Ann. Review Ecology & Systematics,* Vol. 19, pp. 19–38, 1988. With permission from the Annual Review of Ecology and Systematics, Volume 19, © 1988 by Annual Reviews, www.AnnualReviews.org.)

plankton and a classic-type food chain. A comprehensive view of both the classical planktonic food web and the microbial loop is given in Figure 2.45.

Spatial Distribution of Plankton

A characteristic feature of plankton populations is that they tend to occur in **patches**. These patches vary spatially on a scale of a few meters to hundreds of kilometers. They also vary in time—daily, seasonally, and annually—as well as vertically in the water column. It is this patchiness and its constant change in time and space that has made it so difficult for plankton biologists to learn about the ecology of plankton. At its most extreme, because the water in which the plankton is suspended is constantly moving, each sample taken by a plankton biologist removes a different volume of water, so each sample is unique and replicates do not exist. This makes it difficult to generate applicable principles and is one reason for our unsatisfactory understanding of plankton ecology.

Patches may be generated by both physical and biological effects. Physical effects are particularly important. Large-scale patchiness can be due to changes in the stability of the water column, resulting in changes in nutrient availability (see pp. 62–65). Another cause of patchiness is the large-scale **advection**, or movement of water masses in which the plankton are embedded. On the largest scales, the various ocean current systems produce convergent **gyres** in some areas, such as the North Pacific and Sargasso Sea, and divergent systems in others, such as the upwelling areas at the equator. Such areas cover large parts of the globe and define major plankton biomes.

On a smaller scale of tens to hundreds of kilometers, another source of patchiness comes from ocean **eddies** or **rings**. Rings are masses of water rotating either clockwise (anticyclonic) or counterclockwise (cyclonic) that have broken off from larger ocean current systems and have moved into different water masses. Because they have water of different physical and chemical characteristics from the waters into which they move and carry plankton characteristic of the system in which they originated, they are biologically different from the surrounding

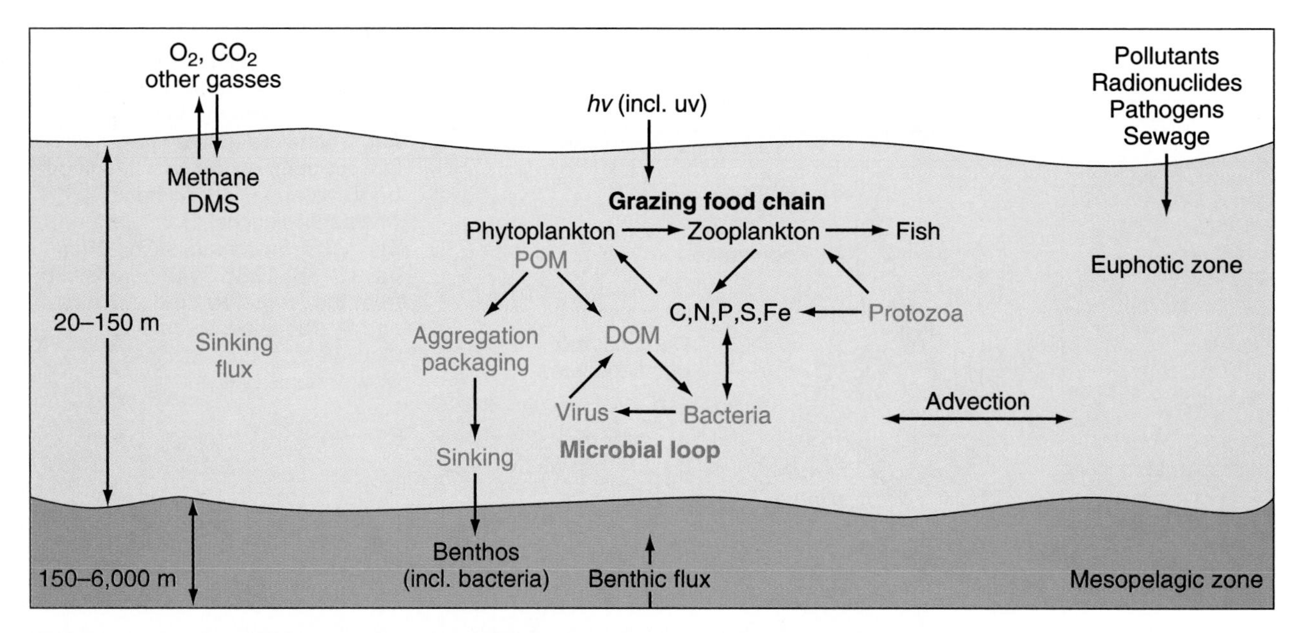

FIGURE 2.45 Modern view of the pelagic food web, emphasizing the microbial loop as a major path for organic matter and indicating its interaction with the classic grazing food chain and the sinking flux. (Reprinted with permission from Azam, *Science,* Vol. 280, p. 694, 1988. Copyright © 1998 American Association for the Advancement of Science.)

water. Such rings are best known from the North Atlantic, where they are spawned from the Gulf Stream (Figure 2.46).

Another source of large-scale patchiness is the presence of **coastal fronts** in neritic waters. As water flows over the continental shelf, or other regions where the water is shallow enough, the friction with the bottom causes vertical turbulence in the water column. This turbulence usually transfers nutrients and changes the water chemistry; thus, the plankton populations are also changed.

Physical processes responsible for small-scale patchiness include Langmuir cells (see p. 54). These cells usually give rise to patches several hundreds of meters long because the plankton organisms accumulate along the lines of convergence (see Figure 2.27). At the smallest scale, simple local turbulence is responsible for plankton patches.

Plankton may also exhibit patchiness vertically. Physical factors contributing to this type of patchiness include light intensity, nutrients, and density gradients in the water column. Phytoplankton in particular tend to be unevenly distributed vertically, leading to the existence of a **chlorophyll maximum**. In the chlorophyll maximum area, the chlorophyll concentration increases 2–20 times what it is in other areas of the photic zone. Such chlorophyll maxima are found in all temperate and tropical seas. They occur at greater depths (100–150 m) in the tropics but are a persistent feature of these waters. In temperate waters, the maximum usually occurs only during the summer. The chlorophyll maxima usually occur at depths where the light intensity is 1% of what it is at the surface (compensation depth). Although many reasons have been suggested for the occurrence of such maxima, there is no universally acceptable explanation for its occurrence.

Biological factors causing patchiness usually create small-scale anomalies. A host of biological phenomena may be involved, but a few stand out. The ability of zooplankton to migrate vertically and graze out phytoplankton cells at a rapid rate (see pp. 70–73) can create patches where phytoplankton organisms are rare and zooplankton abundant. Similarly, active swimming by certain zooplankton organisms can cause them to aggregate in dense groups. One particular biological feature that needs to be emphasized is the phenomenon of marine snow. **Marine snow** is the name given to the various kinds of amorphous particulate material derived from living organisms found in the water column. Marine snow consists of cast-off mucous nets produced by planktonic filter feeders, such as pteropod mollusks (see Figure 2.15) and tunicates (see Figure 2.22), and the remains of dead gelatinous larger zooplankton, such as salps (see Figure

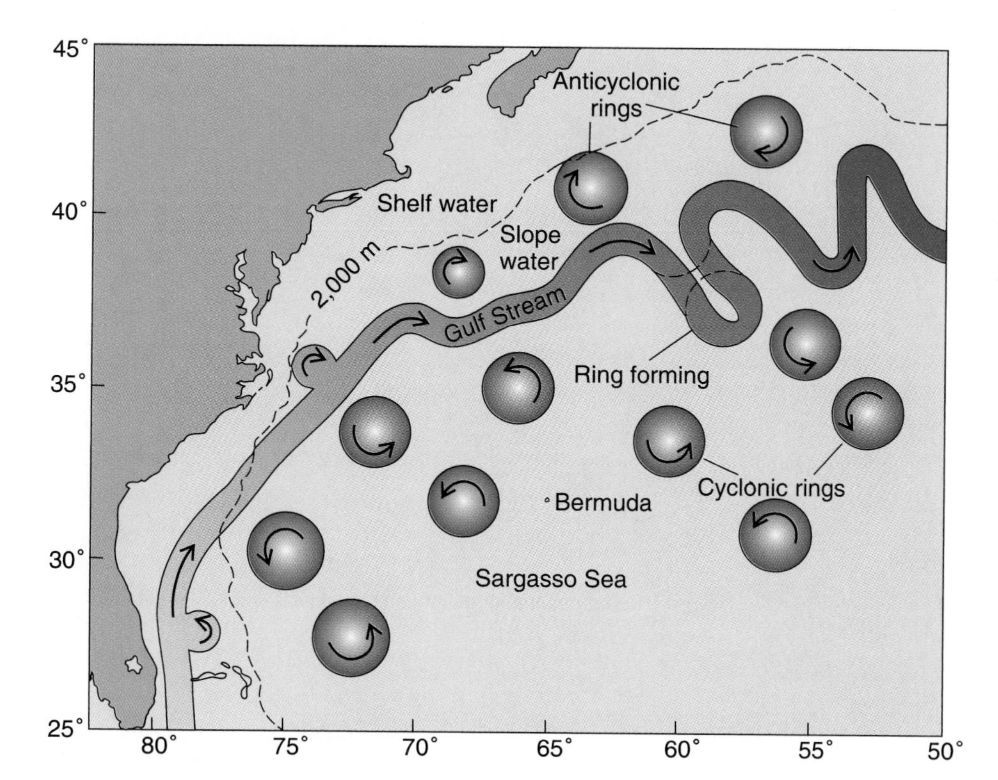

FIGURE 2.46
A schematic representation of the path of the Gulf Stream and the distribution and movement of rings. (From *Oceanus*, Vol. 17. No. 3. Copyright © 1973 Woods Hole Oceanographic Institution.)

2.20) and jellyfishes (see Figure 2.10). The size of marine snow particles is usually 0.5 cm or less. These floating particles attract or accumulate living organisms as well as other nonliving floating and sinking particles, such as diatom frustules and fecal pellets. They become floating microcosms with populations of organisms not found in the surrounding open water. Marine snow produces a solid substrate for organisms that must attach to a surface to live. Since these gelatinous particles are also a locus for bacteria and small flagellates, there may be the regeneration of a certain amount of nutrients from the entrapped fecal pellets and other particulate matter. This, in turn, may lead to the enhancement of the numbers of photosynthetic organisms, such as dinoflagellates. This microcosm is qualitatively and quantitatively different with respect to its component organisms than the organisms living free in the surrounding water. Each piece of marine snow is, in fact, a small patch.

Marine snow is widespread in the world's oceans and is one means of creating heterogeneous or patchy conditions in an otherwise homogeneous environment. Because marine snow is destroyed by plankton nets, the existence of these particles and their associated fauna and flora was only recognized when divers began looking at plankton in the natural environment.

The Major Plankton Biomes

Given all the uncertainties so far discussed about the study and understanding of plankton, it may be surprising to learn that we now have enough data to recognize certain widespread patterns of the distribution and abundance of plankton. It is possible to establish broad associations, communities, or biotic provinces, and make some reasonable statements concerning the composition of these associations and their change in time. It is fitting to close this chapter with such a summary.

As a result of the oceanic circulation that was outlined in Chapter 1, both the Atlantic and Pacific oceans have the water circulated in several large gyres. The two largest are the central gyres north and south of the equator in both the Atlantic and Pacific oceans. They are largely warm-water tropical or subtropical systems. Between them lies an area of water flowing east to west, the tropical equatorial zone. Poleward from each central gyre lies a temperate subpolar gyre (Figure 2.47). Transition zones exist between the gyre systems, acting much as ecotones between terrestrial communities.

Extensive work by McGowan (1974), Reid et al. (1978), and Hayward and McGowan (1979) has established that each of these gyre systems in the Atlantic

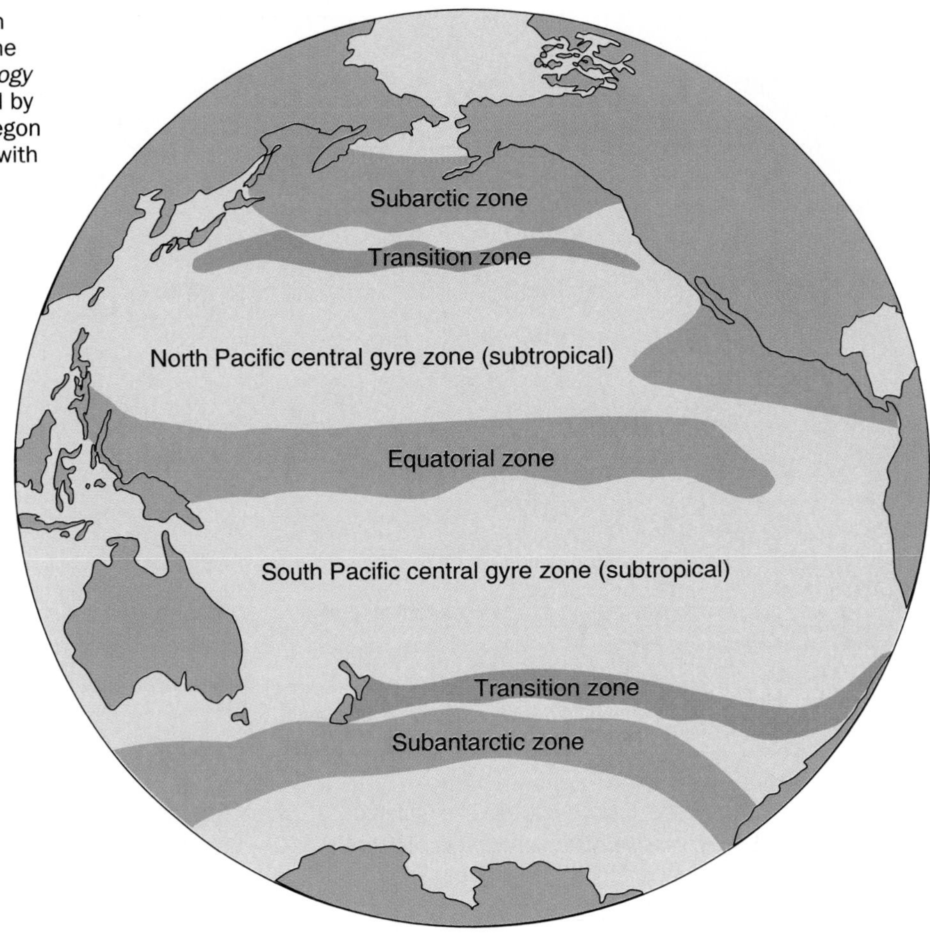

FIGURE 2.47 The main plankton biotic provinces in the Pacific Ocean. (From *The Biology Of The Oceanic Pacific*, edited by Charles B. Miller, © 1974 Oregon State University Press. Used with permission.)

and Pacific have characteristic associations of plankton species distinct enough to qualify as biotic provinces or communities. They are also partially closed systems, showing little mixing with adjacent gyres in the transition zones.

Some of the differences between these biotic provinces are due to the different physical and chemical characteristics of the water masses. The large central North and South Atlantic and Pacific gyres are tropical and subtropical with permanent thermoclines, warm upper waters, and low nutrient concentrations. The subpolar gyres have cold water, high nutrients, and undergo seasonal changes (see pp. 66–68). Productivity is low in the central gyres, a condition termed **oligotrophic**. The subpolar gyres, by contrast, are high in productivity, a condition termed **eutrophic**.

A number of features characterize all these biotic provinces. First, despite what we know about the general patchiness of plankton, the number and kinds of species within each province are very uniform, on spatial scales ranging from a few meters to hundreds of kilometers. As McGowan (1974) has said, they are "monotonous." Furthermore, at least for the very large central gyres, the species composition and relative abundances are also uniform for up to 16 years. They appear to be very stable systems.

These systems are also apparently old, at least in the Pacific Ocean. Evidence there suggests these sets of species or similar ones have been in existence for about 10 million years. With respect to the features of the plankton in the central subtropical gyres, the individual zooplankton organisms tend to be small and often transparent or blue, whereas subpolar zooplankton organisms are larger. The biomass is also smallest in the central gyres and greatest in subpolar and equatorial provinces. The central subtropical gyres and equatorial zones have the greatest numbers of zooplankton species, and the subpolar zones have the fewest. Carnivores are also relatively more abundant in the subtropical gyres and equatorial regions.

Phytoplankton organisms are more broadly distributed throughout the various provinces. No phytoplankton organisms appear to be endemic to any given province.

Perhaps the most puzzling question still to be answered for these large systems is what ecological mechanisms regulate them and keep them so uniform and stable. One of the best ways to answer this question would be to set up controlled experiments. This is not possible in the open ocean, so biologists have had to use more indirect methods. At present, it appears that the North Pacific Gyre, at least, is food limited due to the low nutrient levels, but there is no evidence of competition among the zooplankton filter feeders for the limited phytoplankton. Neither is there any evidence for niche partitioning and food specialization among more than 200 species of copepods. One might expect such partitioning in a system as stable as this. What does regulate these systems? We simply do not know.

SUMMARY OF KEY CONCEPTS

- Plankton is a general term referring to organisms that are at the mercy of the currents.
- Plankton may be subdivided by size into seven different groups; by photosynthetic capacity into three groups; or by life history characteristics into two groups.
- Phytoplankton are photosynthetic organisms of which the larger ones are dominated by the diatoms and dinoflagellates.
- Nanophytoplankton are the most important photosynthetic organisms in the oceans and include the prochlorophytes, coccolithophores, and cyanobacteria.
- The zooplankton are very diverse, including animals from most phyla, but are dominated in the larger size classes by the copepods.
- Larval stages, meroplankton, are more diverse than holoplankton because many species have multiple larval stages.
- Plankton have a density slightly greater than that of seawater but are buoyed up by flotation mechanisms.
- Sinking rate is a function of the overweight of the organism, the surface of resistance, and the viscosity of the water.
- To reduce sinking, plankton lower the overweight by altering the chemical composition of the body, by employing gas-filled floats, and by using compounds less dense than water.
- Plankton may reduce their sinking rate by increasing the surface of resistance of the body through different shapes and use of spines or other projections.
- Langmuir convection cells also serve to keep plankton afloat.
- Primary production in the oceans is about 95% based in the phytoplankton.
- Primary production is the rate of formation of energy-rich organic compounds from inorganic materials.
- Because water absorbs light, less energy is available for photosynthesis as you go deeper in the ocean. At a depth called the compensation depth, the rate of photosynthesis equals the rate of respiration. The compensation intensity is the light intensity at which photosynthesis equals respiration.
- The classic method for measuring primary productivity is the light-dark bottle, which measures the change in oxygen content. This has been mostly superceded today by the ^{14}C methods, which use the radioactive isotope of carbon.
- Standing crop is the amount of biomass of a particular group of plants or animals present in the water at a given time.
- The factors affecting primary productivity in the ocean are light, nutrient content, and a composite factor called hydrography that comprises all those factors that tend to move water masses around.
- The depth to which light will penetrate with enough intensity to permit photosynthesis depends on a number of factors including wavelength, absorption by water, transparency of the water, reflection by particles, latitude, and season of the year.
- The major nutrients needed by phytoplankton for growth and reproduction are nitrates and phosphates, which together act to limit productivity because they are in short supply in most lighted areas of the ocean or are only seasonally available from the deep reservoir of the ocean depths.
- Vertical mixing of the water column replenishes the nutrients to the lighted areas of the temperate oceans in

winter, but the permanent, large difference in density between the lighted surface waters of the tropics and the deep waters does not permit mixing; this leaves most of the tropics with low nutrient levels and productivity.

- Vertical mixing also carries the phytoplankton cells below the compensation depth, circulating them between lighted and unlighted areas; this gives rise to the critical depth concept.
- In temperate seas, the water column is well mixed in winter and nutrients are abundant, but lack of light reduces productivity. In spring, a thermocline develops that reduces mixing while light and photosynthesis are increasing. In the summer, the thermocline is well established and no mixing occurs to replenish the nutrients, so that while light is abundant the nutrients are limiting. In fall, the thermocline decays and mixing begins to replenish the nutrients, but the diminishing light reduces photosynthesis and productivity.
- In tropical seas, the thermal stratification extends throughout the year but light is optimal. Thermal stratification prevents upward transport of nutrients, so tropical seas have low productivity and hence clear water.
- In polar seas, productivity is restricted to a single short period in the polar summer when the light is sufficient for productivity. Thermal stratification is usually not present, but production is strongly reduced by insufficient light through most of the year.
- In inshore waters, productivity is enhanced through runoff of the critical nutrients from the land, the lack of a permanent thermocline, and the fact that the water column is shallower than the critical depth.
- The total net primary productivity of the biosphere is almost equally split between oceans and land, and the ocean productivity as estimated from satellite data is two times greater than calculations before satellite data.
- On an areal basis, the net primary productivity of the oceans is about a quarter of that in the terrestrial environment. This reduced productivity stems from light absorption and competition between the phytoplankton and water for light as well as lack of nutrients.
- Primary producers in the ocean, while responsible for nearly half the net primary productivity, make up only 0.2% of the global primary producer biomass. They have a turnover time averaging 2–6 days, whereas terrestrial primary producers have an average turnover time of 19 years.
- Globally, net primary productivity reaches three maxima at three latitudes. The largest centers on the equator, the next largest is in the temperate latitudes of the Northern Hemisphere, and the smallest peak occurs in the temperate latitudes of the Southern Hemisphere.
- The classic model of the interaction between the phytoplankton and zooplankton grazers has the diatoms and dinoflagellates as the dominant autotrophs and the copepods as the dominant grazers of them.
- Copepods are capable of grazing the diatom and dinoflagellate populations down to very low levels and can control the numbers of phytoplankton. Grazing may also release nutrients to the upper water mass, permitting continued productivity even in the presence of a strong or permanent thermocline.
- Differences in the life histories of the dominant copepods, the presence or absence of additional grazers, and the differing hydrographic conditions between the North Atlantic and North Pacific oceans explain the differences in community ecology of the plankton communities in these two areas.
- Vertical migration is the diel movement of certain zooplankton down to depth during daylight and up to the surface at night, or vice versa. The hypotheses to explain its occurrence include avoidance of predation, avoidance of damaging ultraviolet light, change of position in the water column, and as a mechanism to increase production and conserve energy.
- Seasonal succession is the change in species composition of the phytoplankton with season of the year. Reasons for this succession include temperature changes, changes in nutrient level, and biological conditioning of the water.
- With the use of new techniques, we now have a different understanding of the structure of the plankton communities of the open oceans. Under the microbial loop concept, the dominant photosynthetic organisms are nanophytoplankton and picophytoplankton organisms, which are grazed by ciliate protozoa and heterotrophic nanoflagellates, and these in turn are consumed by the larger net zooplankton. Part of the carbon fixed by these tiny cells "leaks" out as dissolved organic matter, which in turn is taken up by bacterioplankton, and they in turn are consumed by the heterotrophic nanoflagellates.
- Plankton tend to be distributed in patches. Patchiness may be caused by physical factors such as advection, gyres, fronts, and eddies or by biological factors such as grazing, vertical migration, and marine snow.
- Major plankton biomes exist in the Pacific Ocean, and they are associated with the major gyre systems.

REVIEW QUESTIONS

ESSAY: Develop complete answers to these questions.

1. Are phytoplanktonic plants the only photosynthetic organisms present in the plankton? What noneukaryotic organisms may be important in primary production in the ocean? Discuss.
2. What is primary production? How does it differ from photosynthesis? How does one go about measuring primary production? How do gross, net, and total primary production differ? How is standing crop used to determine production? Discuss.
3. Name two factors that limit primary production in marine environments, and discuss their importance. (Instructors may wish to suggest other limiting factors.)
4. Is phytoplankton production nutrient limited or limited by grazing? Are phytoplankton populations limited by "bottom-up" or by "top-down" processes? Explain these terms and the processes involved.
5. Why is there less seasonal variability in biomass in the nanoplankton of tropical and warm temperate waters, such as the central gyre of the North Pacific and Sargasso Sea? Discuss.
6. Discuss the ecological importance of the plankton populations occurring in patches.

BIBLIOGRAPHY

Anonymous. 1982. Marine pelagic protozoa and microzooplankton ecology. *Ann. Inst. Oceanog. Suppl.* Tome 58.

Azam, F. 1998. Microbial control of oceanic carbon flux: The plot thickens. *Science* 280:694–696.

Azam, F., T. Fenchel, J. G. Field, J. S. Gray, L. A. Meyer-Reil, and F. Thingstad. 1983. The ecological role of water column microbes in the sea. *Mar. Ecol. Prog. Ser.* 10:257–263.

Bougis, P. 1976. *Marine plankton ecology*. New York: Elsevier.

Capriulo, G. M., and D. V. Nirivaggi. 1982. A comparison of the feeding activities of field collected tintinnids and copepods fed identical natural particle assemblages. In *Marine pelagic protozoa and microzooplankton ecology. Ann. Inst. Oceanog. Suppl.* Tome 58:325–334.

Chisholm, S. W., R. J. Olsen, E. R. Zettler, R. Goericke, J. B. Waterbury, and N. A. Welschmeyer. 1988. A novel free-living prochlorophyte abundant in the oceanic euphotic zone. *Nature* 334:340–343.

Cushing, D. H. 1959. On the nature of production in the sea. *Fish. Invest. London, 2d ser.*, 22 (6):1–40.

Cushing, D. H. 1975. The productivity of the sea. *Oxford Biology Reader*, no. 78. London: Oxford University Press.

Cushing, D. H., and J. J. Welsh, eds. 1976. *The ecology of the seas*. Philadelphia: Saunders.

Denton, E. J. 1974. Buoyancy in marine animals. *Oxford Biology Reader*, no. 54. London: Oxford University Press.

Enright, J. T. 1977. Diurnal vertical migration: Adaptive significance and timing. Part 1, Selective advantage: A metabolic model. *Limn. Oceanog.* 22:856–872.

Falkowski, P., R. T. Barber, and V. Smetacek. 1998. Biogeochemical control and feedbacks on ocean primary production. *Science* 281:200–206.

Fenchel, T. 1988. Marine planktonic food chains. *Ann. Rev. Ecol. Sys.* 19:19–38.

Field, C. B., M. J. Behrenfeld, J. T. Randerson, and P. Falkowski. 1998. Primary production of the biosphere: Integrating terrestrial and oceanic components. *Science* 281:237–240.

Fleming, R. H. 1939. The control of diatom populations by grazing. *Journal du Conseil Permanent International pour l'Exploration de la Mer.* 14:210–227.

Fraser, J. 1962. *Nature adrift, the story of marine plankton*. London: G. T. Foulis and Co., Ltd.

Fuhrman, J. 1992. Bacterioplankton roles in cycling of organic matter: The microbial food web. In *Primary productivity and biogeochemical cycles in the sea*, edited by P. G. Falkowski and A. D. Woodhead. New York: Plenum Press, 361–383.

Fuhrman, J. 1999. Marine viruses and their biogeochemical and ecological effects. *Nature* 399:541–548.

Fuhrman, J., and C. A. Suttle. 1993. Viruses in marine planktonic systems. *Oceanography* 6(2):52–63.

Haney, J. F. 1988. Diel patterns of zooplankton behavior. *Bull. Mar. Sci.* 43(3):583–603.

Hardy, A. C. 1953. Some problems of pelagic life. In *Essays in marine biology* (Richard Elmhurst Memorial Lectures). Edinburgh: Oliver and Boyd, 101–121.

Hardy, A. C. 1956. *The open sea: Its natural history*. Vol. 1, *The world of plankton*. Boston: Houghton Mifflin.

Hardy, A. C., and E. R. Gunther. 1935. The plankton of the south Georgia whaling grounds and adjacent waters, 1926–27. *Discov. Rep.* 11:511–538.

Hart, T. J. 1942. Phytoplankton periodicity in Antarctic surface waters. *Discov. Rep.* 21:263–348.

Harvey, H. W., L. H. N. Cooper, M. V. Lebour, and F. S. Russell. 1935. Plankton production and its control. *J. Mar. Biol. Assoc.* 20:407–441.

Hayward, T. L., and J. A. McGowan. 1979. Pattern and structure in an oceanic zooplankton community. *Amer. Zool.* 19:1045–1055.

Hill, M. N., ed. 1963. *The sea*. Vol. 2, *The composition of sea water, comparative and descriptive oceanography*, Section IV, Biological oceanography. New York: Wiley, 347–485.

Hobbie, J. E., and P. J. LeB. Williams, eds. 1984. *Heterotrophic activity in the sea*. New York: Plenum Press.

Hobbie, J. E., O. Holme-Hansen, T. T. Packard, L. R. Pomeroy, R. W. Sheldon, P. J. Thomas, and W. J. Wiebe. 1972. A study of the distribution and activity of microorganisms in ocean water. *Limn. Oceanog.* 17:544–555.

Holme-Hansen, O., and C. R. Booth. 1966. The measurement of adenosine triphosphate in the ocean and its ecological significance. *Limn. Oceanog.* 11:510–519.

Johnson, P. W., and J. McN. Sieburth. 1979. Chroococcid cyanobacteria in the sea: A ubiquitous and diverse phototropic mass. *Limn. Oceanog.* 24:928–935.

Kiorboe, T. 1993. Turbulence, phytoplankton cell size and the structure of pelagic food webs. In *Advances in marine biology*, Vol. 29, edited by J. H. S. Blaxter and A. J. Southward. New York: Academic Press, 1–72.

Kirchman, D. L. (editor) 2000. *Microbial ecology of the oceans*. New York: Wiley-Liss. 542 pp.

Koehl, M. A. R., and J. R. Strickler. 1981. Copepod feeding currents: Food capture at low Reynolds numbers. *Limn. Oceanog.* 26:1062–1073.

Malone, T. C. 1980. Algal size. In *The physiological ecology of phytoplankton*, edited by I. Morris. Berkeley: University of California Press, 433–463.

Mare, M. F. 1940. Plankton production off Plymouth and the mouth of the English Channel in 1939. *J. Mar. Biol. Assoc.* 24:461–482.

Margalef, R. 1963. Succession in marine populations. In *Advancing frontiers of plant sciences*, Vol. 2, edited by R. Vira. 137–188.

Martin, J. H., and S. Fitzwater. 1988. Iron deficiency limits phytoplankton growth in the northeast Pacific subarctic. *Nature* 331:341–343.

Martin, J. H., et al. 1994. Testing the iron hypothesis in ecosystems of the equatorial Pacific. *Nature* 371:123–129.

McAllister, C. D. 1969. Aspects of estimating zooplankton production from phytoplankton production. *J. Fish. Res. Bd. Can.* 26:199–220.

McGowan, J. A. 1974. The nature of oceanic ecosystems. In *The biology of the Pacific Ocean*, edited by C. B. Miller. Corvallis: Oregon State University Press, 9–28.

McGowan, J. A., and P. W. Walker. 1985. Dominance and diversity maintenance in an oceanic ecosystem. *Ecol. Monogr.* 55:103–118.

McLaren, I. A. 1963. Effects of temperature on growth of zooplankton and the adaptive value of vertical migration. *J. Fish. Res. Bd. Can.* 20:685–727.

McLaren, I. A. 1974. Demographic strategy of vertical migration by a marine copepod. *Amer. Nat.* 108:91–102.

Miller, C. B., ed. 1974. *The biology of the Pacific Ocean*. Corvallis: Oregon State University Press.

Miller, C. B., B. W. Frost, P. A. Wheeler, M. R. Landry, N. Welschmeyer, and T. M. Powell. 1991. Ecological dynamics in the subarctic Pacific, a possibly iron-limited ecosystem. *Limn. Oceanog.* 36(8):1600–1615.

Morris, I., ed. 1980. *Studies in ecology*. Vol. 7, *The physiological ecology of phytoplankton*. Berkeley: University of California Press.

Newell, G. E., and R. C. Newell. 1973. *Marine plankton, a practical guide*. London: Hutchinson.

Ohman, M. D. 1990. The demographic benefits of diel vertical migration by zooplankton. *Ecol. Monogr.* 60(3):257–281.

Parsons, T. R., and C. M. Lalli. 1988. Comparative oceanic ecology of the plankton communities of the subarctic Atlantic and Pacific oceans. *Oceanog. Mar. Biol. Ann. Rev.* 26:317–359.

Parsons, T. R., M. Takahashi, and B. Hargrave. 1984. *Biological oceanographic processes*. 3d ed. New York: Pergamon.

Peterson, B. J. 1980. Aquatic primary productivity and the ^{14}C-CO_2 method: A history of the productivity problem. *Ann. Rev. Ecol. Sys.* 11:359–385.

Platt, T., ed. 1981. Physiological bases of phytoplankton ecology. *Can. Bull. Fish. Aquatic Sci.* 210:346.

Pomeroy, L. R. 1974. The ocean's food web, a changing paradigm. *Bioscience* 24:499–504.

Pomeroy, L. R., and R. E. Johannes. 1968. Respiration of ultraplankton in the upper 500 meters of the ocean. *Deep-Sea Res.* 15:381–391.

Raymont, J. E. G. 1980. *Plankton and productivity in the oceans*, 2d ed. Vol. 1, *Phytoplankton*, Vol. 2, *Zooplankton*. New York: Macmillan.

Reid, J. L., E. R. Brinton, A. Fleminger, E. L. Venrick, and J. A. McGowan. 1978. Ocean circulation and marine life.

In *Advances in oceanography*, edited by H. Charnock and G. Deacon. New York: Plenum Press, 65–130.

Riley, G. A. 1946. Factors controlling phytoplankton populations on Georges Bank. *J. Mar. Res.* 6:54–73.

Russell-Hunter, W. D. 1970. *Aquatic productivity*. New York: Macmillan.

Sherr, E. B., and B. F. Scherr. 1991. Planktonic microbes: Tiny cells at the base of the ocean's food webs. *Trends Ecol. Evol.* 6(2):50–54.

Sieburth, J. M. 1979. *Sea microbes*. New York: Oxford University Press.

Sieburth, J. M. 1982. The role of heterotrophic nanoplankton in the grazing and nurturing of planktonic bacteria in the Sargasso and Caribbean Seas. *Ann. Inst. Ocean. Suppl.* Tome 58:285–296.

Smayda, T. J. 1980. Phytoplankton species succession. In *The physiological ecology of phytoplankton*, edited by I. Morris. Berkeley: University of California Press, 493–570.

Smith, DeB. L., and K. B. Johnson. 1996. *A guide to marine coastal plankton and marine invertebrate larvae*. 2d ed. Dubuque, IA: Kendall/Hunt.

Stoecker, D. K., A. E. Michaels, and L. H. Davis. 1987. A large proportion of marine planktonic ciliates contain functional chloroplasts. *Nature* 326:790–792.

Waterbury, J. B., S. W. Watson, R. R. L. Guillard, and L. E. Brand. 1979. Widespread occurrence of a unicellular, marine planktonic cyanobacterium. *Nature* 277:293–294.

Wickstead, J. H. 1965. *An introduction to the study of tropical plankton*. London: Hutchinson.

Wimpenny, R. S. 1966. *The plankton of the sea*. New York: Elsevier.

OCEANIC NEKTON

In contrast to the plankton, the nekton comprises those organisms that have developed powers of locomotion so that they are not at the mercy of the prevailing ocean currents or wind-induced water motion. They can move at will through the water. Most are large animals and include the largest and fastest-moving organisms in the sea. Plankton are dominated by a host of invertebrate animals, but nekton are predominantly vertebrates. The fishes are the most numerous nekton, both in species and individuals, but representatives of every vertebrate class except amphibians are found here.

***Nekton** in the larger sense includes all organisms capable of sustained locomotion against the water motion without consideration of habitat. In this chapter, we will consider only nektonic animals distributed in the epipelagic zone of the open ocean, the oceanic nekton. The ecology and adaptations of nekton associated with inshore waters, such as kelp beds and coral reefs, and with the deep sea will be considered in subsequent chapters. The special adaptations of the oceanic nekton are different from those of nekton living in deeper waters and inshore waters; hence, we justify this division of the nekton.*

Because these animals are fast swimming and often wide ranging in the immense open seas of the world, they are difficult to study at sea and are virtually impossible to keep captive under natural conditions. We know very little about most aspects of the ecology or life history of these forms. In the dearth of such field data and virtual absence of laboratory information, we have been forced to infer much of their ecology by studying anatomical and physiological characteristics of captured specimens.

Most students will have little opportunity for direct contact with living oceanic nekton except for fishes and occasional seal, sea lion, or porpoise sightings or strandings. However, the immensity of the ocean area inhabited by this fauna and its significant role in the ecology of the upper layers of the seas of the world make it important to study. Some nektonic fishes, such as the tunas, sustain a major world fishery. This fauna also includes the whales, perhaps the most popular group of marine organisms in the public eye, over which controversy has raged in recent years with respect to their continued survival. Helping whales survive requires some knowledge of the habitat in which they live.

COMPOSITION OF THE OCEANIC NEKTON

The oceanic nekton is composed of a wide variety of bony fishes, sharks, and rays, as well as lesser numbers of marine mammals, reptiles, and birds. The only invertebrates that can be considered oceanic nekton are certain of the cephalopod mollusks.

Several different groups of fishes may be recognized in the nekton (Figure 3.1). First, there are fishes that spend their entire lives in the epipelagic. These fishes may be termed **holoepipelagic**. Included here are certain sharks (thresher shark, mackerel shark, blue shark), most flying fishes, tunas, marlins, swordfish, saury, oarfish, and others (Figure 3.2). These fishes often lay floating eggs and have epipelagic larvae. They are most abundant in the surface waters of the tropics and subtropics.

A second group of oceanic fishes is termed **meroepipelagic**. These fishes spend only part of their life cycle in the epipelagic. This is a more diverse group and includes fishes that spend their adult lives in the epipelagic but spawn in inshore waters (herring, whale shark, dolphin, halfbeaks) or in fresh water (salmon) (Figure 3.3). Still other fishes enter the epipelagic only at certain times. At night certain of the deeper-water fishes, such as lantern fish, migrate to the surface layers to feed. This latter group, with all other deeper-living fishes (collectively called midwater fishes), is considered in the next chapter.

FIGURE 3.1 A school of pelagic fishes. (Photo courtesy of Drs. Lovell and Libby Langstroth.)

Most fishes spend their early lives in the epipelagic but their adult lives in other areas. These

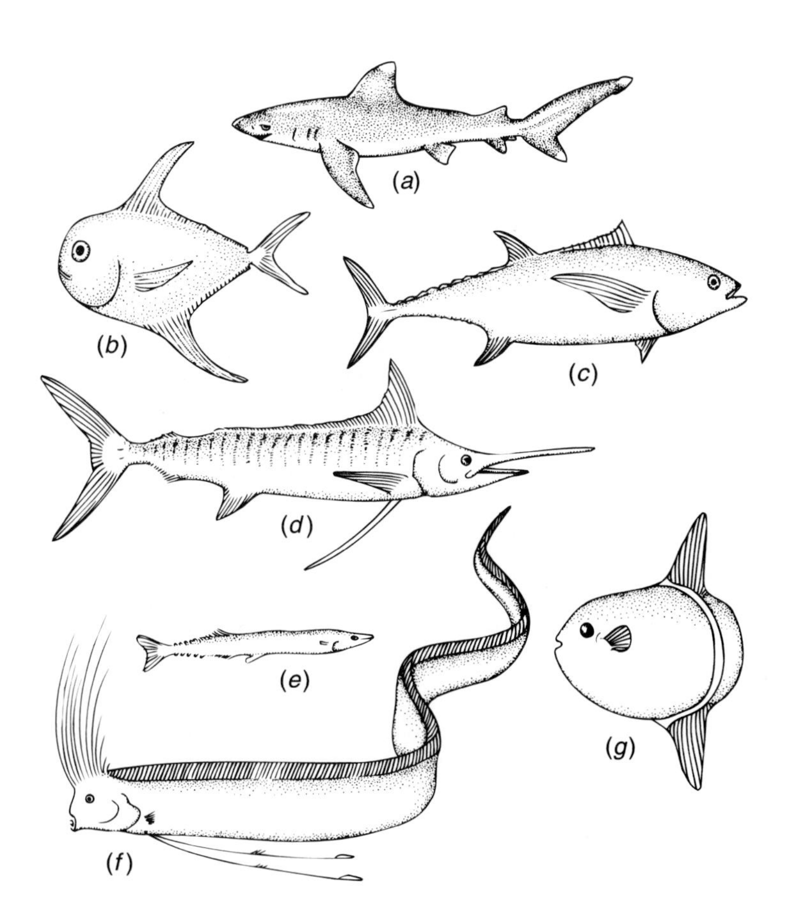

FIGURE 3.2 Some characteristic holoepipelagic fishes. (a) White tip shark (*Pterolamiops longimanus*). (b) Spiny marine bream (*Taractichthys longipinnus*). (c) Yellowfin tuna (*Thunnus albacares*). (d) Striped marlin (*Tetrapterus audax*). (e) Saury (*Cololabis saira*). (f) Oarfish (*Regalecus glesne*). (g) Ocean sunfish (*Mola mola*). (Redrawn from N. V. Parin, 1970, *Ichthyofauna of the epipelagic zone*, trans. from the Russian by the Israel Program for Scientific Translations.) (Not to scale.)

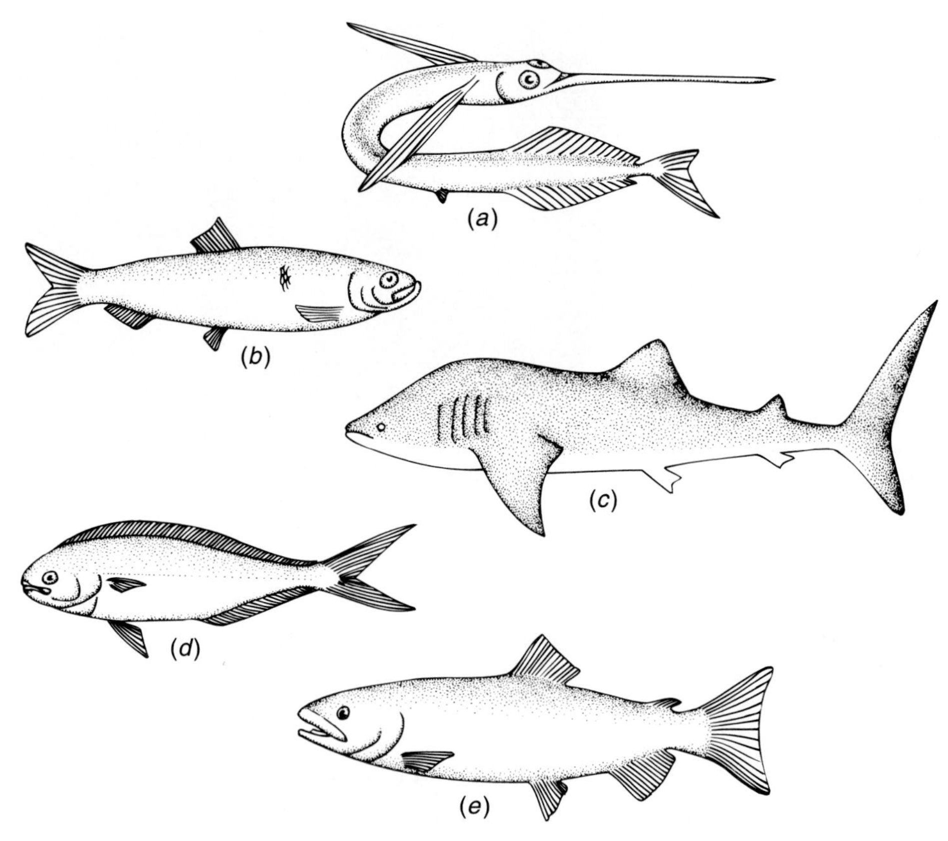

FIGURE 3.3 Some characteristic meroepipelagic fishes. (a) Ribbon halfbeak (*Euleptorhamphus viridis*). (b) Herring (*Clupea harengus*). (c) Whale shark (*Rhincodon typus*). (d) Dolphin (*Coryphaena hippurus*). (e) Salmon (*Oncorhynchus keta*). (Redrawn from N. V. Parin, 1970, *Ichthyofauna of the epipelagic zone*, trans. from the Russian by the Israel Program for Scientific Translations.) (Not to scale.)

juveniles form a persistent part of the epipelagic fauna but can best be considered meroplankton since they have a restricted ability to move. They will not be considered further here.

The second component of the oceanic nekton is the marine mammals. Nektonic marine mammals include the whales (order Cetacea) (Figure 3.4) and the seals and sea lions (order Carnivora, suborder Pinnipedia) (Figure 3.5). Other marine mammals exist, such as manatees and dugongs (order Sirenia) and sea otters (order Carnivora), but these animals are not pelagic because they occupy inshore waters at all times. Little is known of the role of manatees and dugongs in any community. The role of sea otters in the economy of inshore waters is discussed in Chapter 5, but their new connection to the open ocean is considered at the end of this chapter.

Nektonic reptiles are almost exclusively turtles and sea snakes. Marine iguanas exist in the Galápagos Islands, and saltwater crocodiles inhabit many island areas of the Indonesian and Australian region, but these again are littoral animals that rarely venture out of sight of land. The fossil record indicates that, during the Cretaceous period some 60 million years ago, marine reptiles were much more common and varied than today. At that time, large plesiosaurs, ichthyosaurs, and mosasaurs roamed the warm seas.

Technically, most seabirds are not nektonic, since they fly over the open ocean or swim on the surface. They, however, do feed in the upper layers and some may dive to depths near 100 m to feed. They also occur in large numbers in some areas where it is likely they have a significant effect on marine prey populations. They, therefore, enter the economy of these waters and are considered here. Perhaps the only group of truly nektonic birds are the flightless penguins of the Southern Hemisphere, but cormorants and other seabirds do dive for food and spend much time swimming.

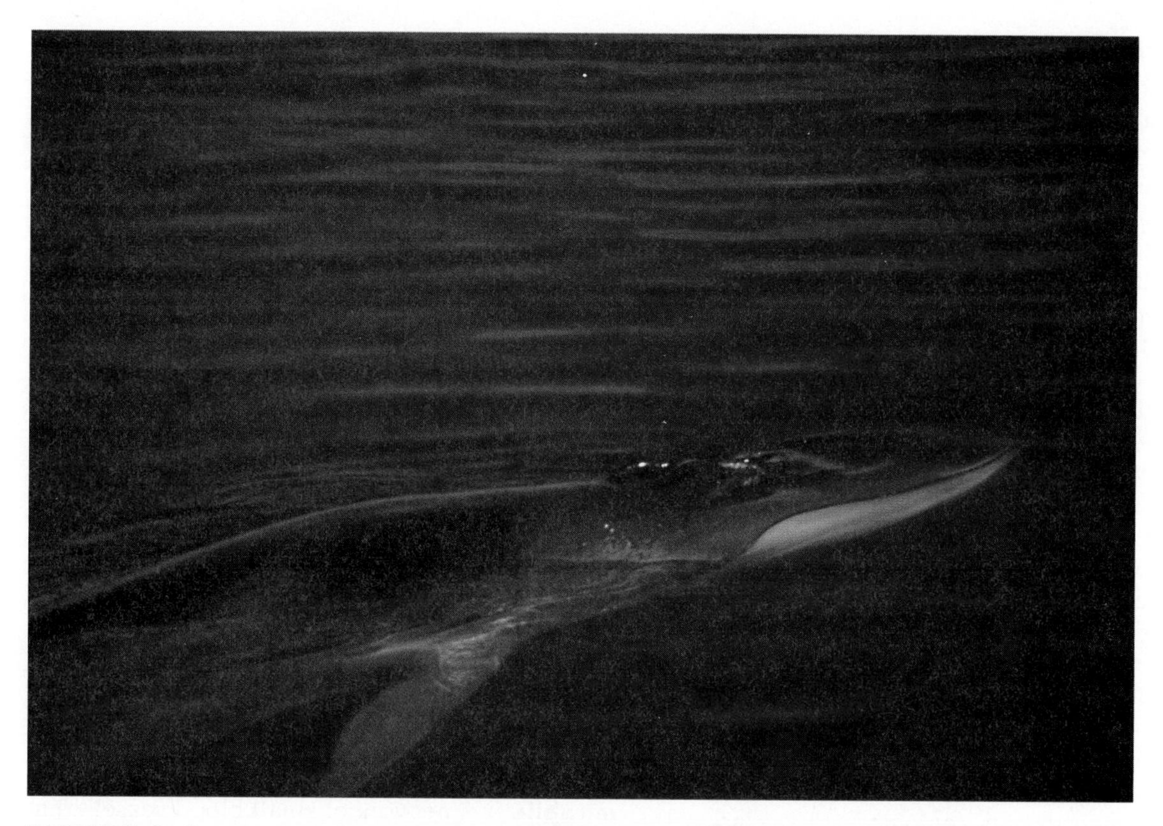

FIGURE 3.4 A fin whale (*Balaenopterus physalus*). (Photo courtesy of Dr. Jim Harvey.)

FIGURE 3.5 Antarctic fur seal (*Arctocephalus australis*). (Photo courtesy of John Heine.)

ENVIRONMENTAL CONDITIONS

The environmental factors acting in the epipelagic zone inhabited by the nekton are, of course, the same as those reported in the preceding chapter for the plankton (see pp. 59–66), and include light, temperature, density, and currents. However, the relative importance of different factors in selecting adaptations and life strategies of the nekton may be different. It is important to note that the perception of this environment is very different for a large, fast-swimming fish or mammal than it is for a small copepod.

Several environmental conditions should be emphasized as being important to the nekton and for which adaptations have evolved. First, this is an area with a profound three-dimensionality. Second, there is no large solid substrate anywhere; thus, the animals are always suspended in a transparent medium with no shelter from predators. Because of the lack of structures in this area, there is nothing visible for animals to take bearings on when moving from place to place horizontally. Finally, the lack of a substrate means there is no firm support for the animals, most of which have flesh somewhat more dense than the surrounding seawater.

Three-dimensionality, combined with the lack of any obstacles, facilitates the evolution of adaptations for great mobility. Great mobility and ability to cover large distances, in turn, selects for the development of nervous and sensory systems that will provide and process the information necessary to navigate in the area, find and capture food, and avoid predators.

Similarly, the lack of any shelter, coupled with the large size of most nekton forms, leads to development of faster swimming speeds as one of the few options left to escape predators and, in turn, to enable them to capture food. This lack of shelter also leads to development of camouflage as another protective option. Continuous suspension of the more dense body of nektonic animals in the less dense water leads to the progressive development of various adaptations toward keeping afloat.

ADAPTATIONS OF OCEANIC NEKTON

Buoyancy

Perhaps the most significant adaptations of nektonic animals are those that keep the animals suspended in the water and propel them through the water swiftly. It is these adaptations that have caused the strong convergence that we observe in morphology of otherwise very different animals.

The first order of business in living in the epipelagic is to stay afloat. This, as we saw in the last chapter, is also the primary concern of the plankton. Most nektonic animals have densities very close to that of seawater. Since living tissues are generally heavier than seawater, the fact that many of these large animals are nearly neutrally buoyant means there must be lower-density areas in their bodies that counteract the higher density of most tissues.

Most fishes have a **gas** or **swim bladder** in their bodies. This structure counteracts the denser flesh of the fish and gives the fish a neutral buoyancy (Figure 3.6). Most fishes can regulate the amount of gas in the gas bladder and change their buoyancy state. Two types of gas bladder systems are known: the **physostome**, in which there is an open duct between the gas bladder and the esophagus, and the **physoclist**, in which there is no duct. Physostomous fishes can move gases in and out of the bladder via the duct by gulping air at the surface, but they usually fill the swim bladder through the gas gland and a rete mirabile system. The **rete mirabile** is a network of small blood vessels that branch off a large vessel. Physoclistous fishes also secrete gas into the swim bladder through the gas gland and the rete mirabile, but they must void the gas through a special gas absorptive section of the gas bladder called the **oval**.

In very fast moving fishes, such as bonito (*Sarda*) and mackerel (*Scomber*), which also move vertically in the water column, the gas bladder cannot adjust quickly enough to compensate for pressure changes and maintain neutral buoyancy; thus, it becomes more of a liability. Denton and Marshall (1958) noted the fastest fishes lack gas bladders because they get enough lift from the body surface and/or pectoral fins during rapid locomotion.

Gas-filled cavities in the form of lungs also help sustain neutral buoyancy of all the air breathing nektonic animals. Some marine mammals have accessory air sacs (Figure 3.6). These animals can regulate their buoyancy through the amount of air they hold in their lungs. Birds also have additional air sacs. They have bones with air channels that are mostly an adaptation for flight but can also provide buoyancy. In most diving seabirds (excluding penguins), air trapped under the feathers provides the greatest amount of buoyancy. Among marine mammals, the

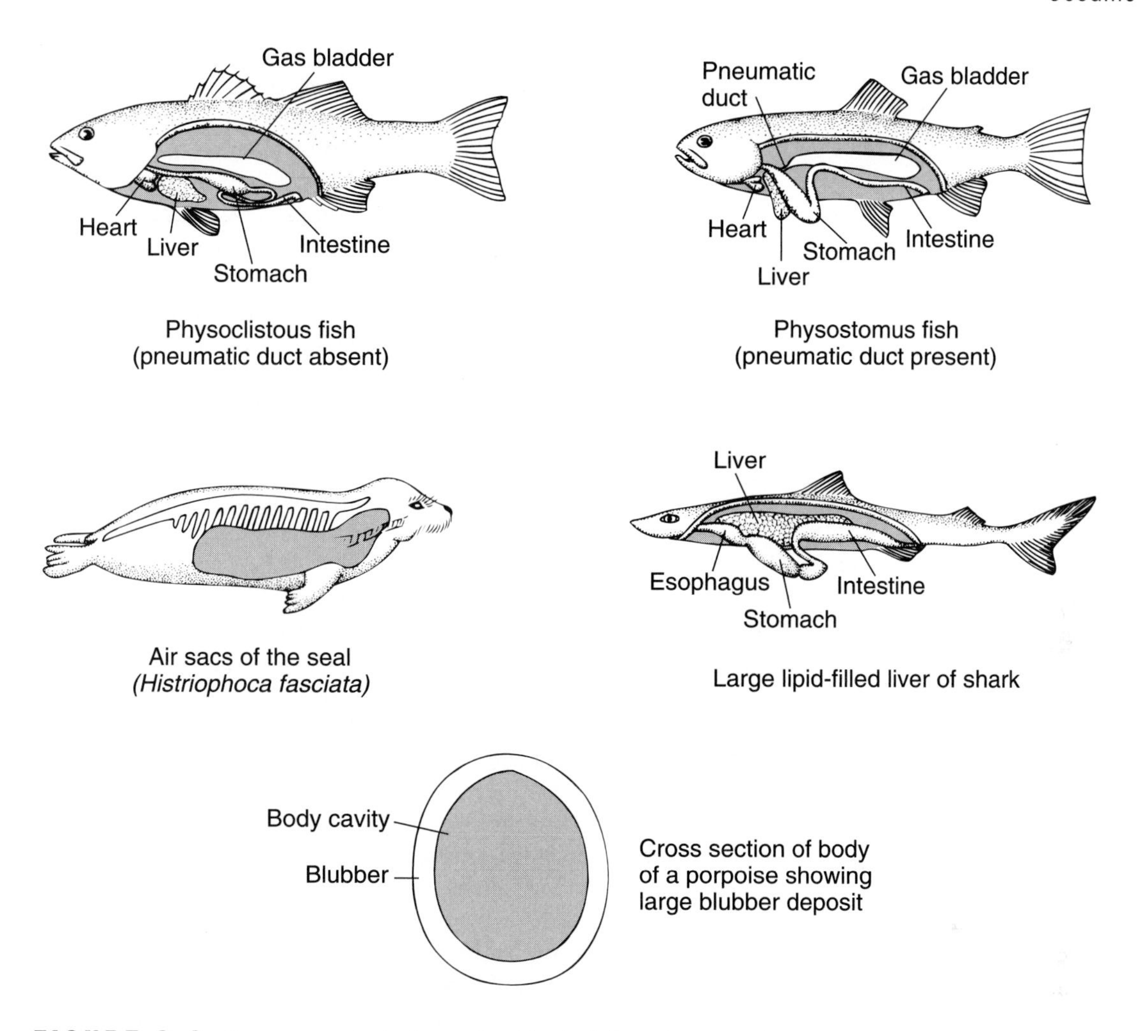

FIGURE 3.6 Buoyancy adaptations of nektonic fish and mammals.

sea otter and fur seals also use air trapped in their dense wool undercoat for buoyancy.

Another mechanism of ensuring neutral buoyancy is replacement of the heavy chemical ions in the body fluids with lighter ones. We observed this in plankton as well (see pp. 51–52). The only nektonic animals in which this occurs are a few of the squids. Some squids have body cavities and tissues in which heavy sodium ions are replaced with lighter ammonium. As a result, an equal volume of body fluid is less dense than the same volume of seawater. This is a common mechanism among plankton, but it is rare among nekton because for it to be effective the amount of ammonium-dominated fluid must be large. Large, fluid-filled spaces give the animal a rotund appearance, markedly decreasing its ability to move rapidly.

Increasing buoyancy by reducing bone or other hard parts is not an option, since strong, rigid skeletons are needed for the muscle system to propel the animals through the water. This is, of course, a marked contrast to the plankton.

Still another mechanism to increase buoyancy is to lay down lipid (fat or oil) in the body. Lipid is less dense than seawater and can be important in contributing to buoyancy. Large amounts of lipid are present in many nektonic fishes, primarily those that lack swim bladders, such as sharks, mackerels, bluefish (*Pomatomus*), and bonito. Presumably, the lipid, at least partially, substitutes for the swim bladder. Lipid may

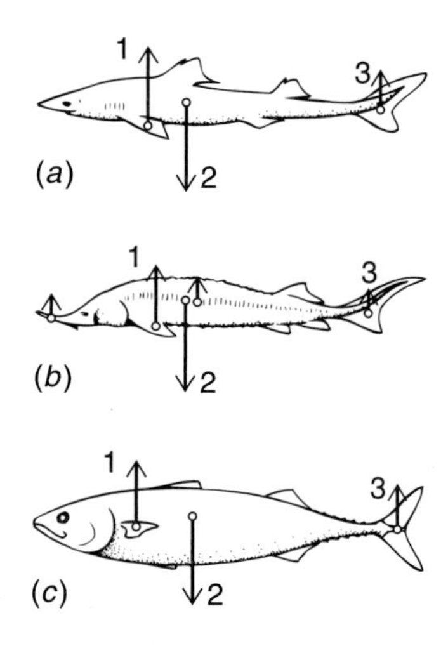

FIGURE 3.7 Various tail and fin shapes in fishes showing the lift provided. (a) Shark (*Squalus acanthias*) with slightly heterocercal tail. (b) Sturgeon (*Acipenser stellatus*) with heterocercal tail. (c) Mackerel (*Scomber scombrus*) with homocercal tail. (1) Force created by the pectoral fins or analogs; (2) force due to residual weight; (3) vertical component of propulsive force furnished by the caudal fin. (Modified from Y. G. Aleyev, 1977, *Nekton*, Dr. W. Junk BV, with kind permission from Kluwer Academic Publishers.)

be deposited in various body parts, such as muscles, bone, internal organs, and the body cavity, or may be localized in one organ. In pelagic sharks, for example, the lipid is localized in the enlarged liver, and in many shark species fat deposition in the liver is part of their development; thus, some young sharks are negatively buoyant and gradually become neutrally or positively buoyant as they mature and fat stores build up in the liver. In marine mammals, the lipid usually is deposited as a layer of fat just below the skin (blubber), where it not only aids buoyancy but insulates against heat loss, stores energy, and adds to streamlining.

In addition to these static means of maintaining or increasing buoyancy, certain nektonic animals also show some hydrodynamic mechanisms for producing additional buoyancy during movement. Perhaps the most common are the formation of lifting surfaces in the anterior region, usually represented by pectoral fins or flippers, and the presence of a heterocercal tail. A heterocercal tail has an upper lobe that is larger and better developed. In this system, the fins or flippers act as movable ailerons, just as on an airplane, and when canted at the appropriate positive angle, cause the individual to rise in the water column as the tail provides the propulsive thrust (Figure 3.7). If the tail is heterocercal, its motion also provides a portion of the upward thrust. In some forms, the lift provided by the fins or flippers is aided by the whole anterior portion of the body, which can also be inclined at an angle to provide lift. The best development of these dynamic buoyancy forces occurs in negatively buoyant forms.

In general, the more primitive fishes tend to have hydrodynamic (water movement) adaptations to create lift, whereas the more advanced forms seem to evolve static or passive means to achieve neutral buoyancy. This is because less energy is expended to obtain neutral buoyancy than to have to move constantly to achieve the lift necessary to keep a body suspended in the water. Because of air-filled lungs, air-breathing mammals tend to be nearly neutrally buoyant.

Locomotion

A second, and related, group of adaptations in nektonic animals are those that enable the animals to move through the water. These adaptations can be divided into two groups: those necessary to create the propulsive force and those that reduce the resistance of the body to passage through the water.

The force necessary to propel a nektonic animal through the dense water is created by some part or parts of the animal's body. The most common way of moving forward is by undulating the body or the fins. Virtually all nektonic fishes show these types of motion. In the undulatory mechanism, the animal moves itself forward by sweeping the posterior part of the body and fins from side to side. This throws the body into a series of short curves that start at the head and move down the body. This side-to-side motion is created by alternate contractions of the body musculature, first on one side and then on the other, in interaction with forces from the reacting water. Elastic properties of body and fins and the lateral shape of the fish determine the phase relationships between the muscle contraction pattern and the undulating movements of the body. In most fishes, the wave of muscle activity along the body on each side is much faster than the wave of curvature. In the saithe (*Gadus virens*), for example, the lateral muscles behind the head and near the tail are instantaneously active.

The amplitude of lateral movements of fishes with a slender body increases gradually toward the tail. Propulsive forces are generated because the lateral movements accelerate a mass of water around the fish. The head enters undisturbed standing water. Just behind the head acceleration of water begins and continues along the body until the end of the tail

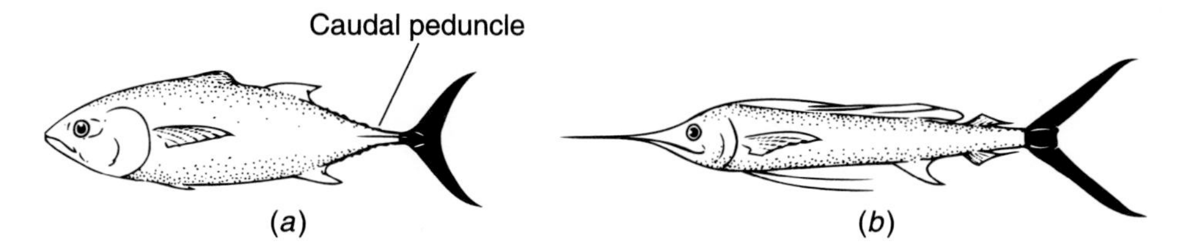

FIGURE 3.8 Fast-swimming fishes with the characteristic lunate tail and narrow caudal peduncle. (a) Tuna (*Thunnus thynnus*). (b) Sailfish (*Istiophorus platypterus*).

blade releases the jet of water in a direction opposite to the swimming direction.

In short, stubby fishes the propulsive wave is less apparent because the amplitude on the body is low, but it shows a sharp increase near the caudal peduncle. This peduncle is usually narrow or even dorsoventrally flattened. The prime characteristic of this group is their lunate tail blade (Figure 3.8). Propulsive forces are generated exclusively by the tail blade acting as an undulating hydrofoil. A similar propulsive system is used by the whales, but the movements of the tail are up and down rather than side to side. The speed of a fish is related to the speed with which the muscle contraction waves pass down the body; it also depends on other aspects of the body shape. In general, relatively short, stubby fishes (such as tuna) are faster than very long, narrow fishes (such as eels) (Figure 3.9).

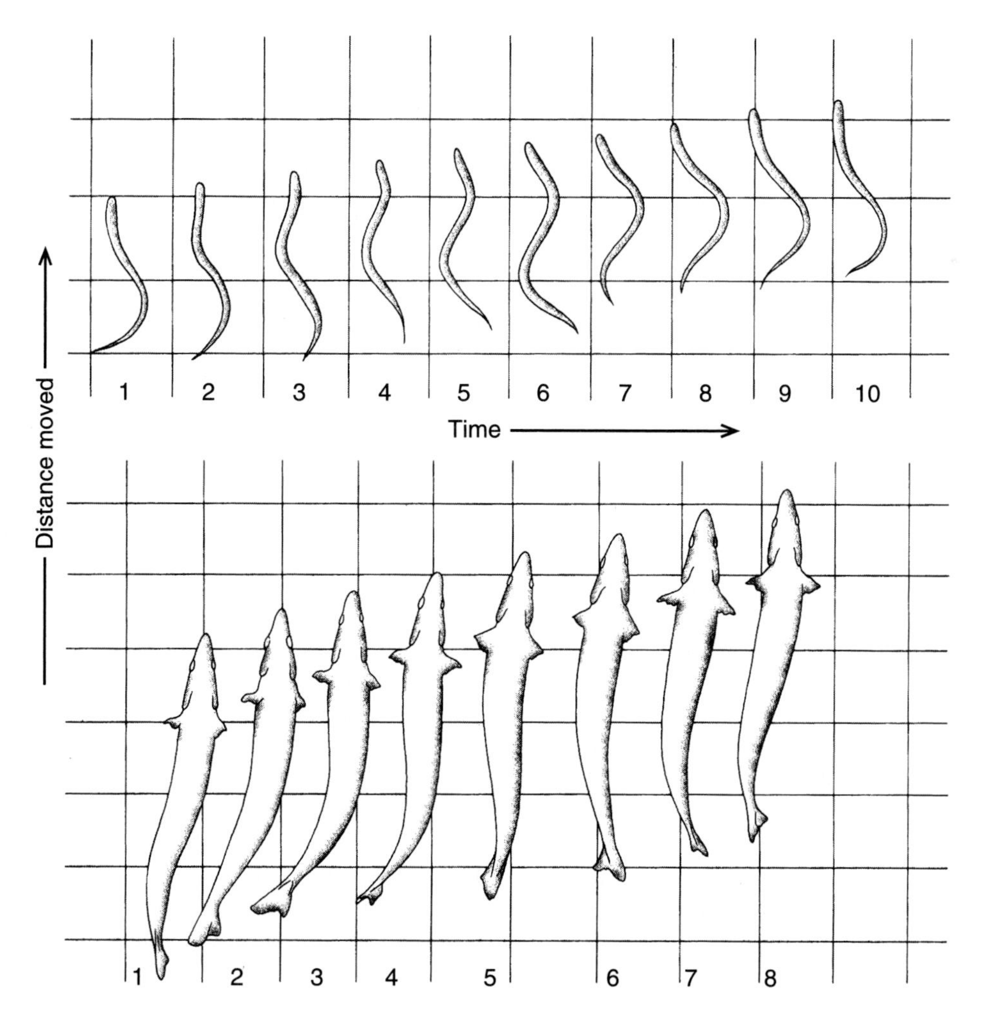

FIGURE 3.9 Propulsion in an elongated fish (top; length-to-width ratio ≅ 0.05) and a stubby fish (bottom; length-to-width ratio ≅ 0.1). (Adapted from Y. G. Aleyev, 1977, *Nekton*, Dr. W. Junk BV, with kind permission from Kluwer Academic Publishers.)

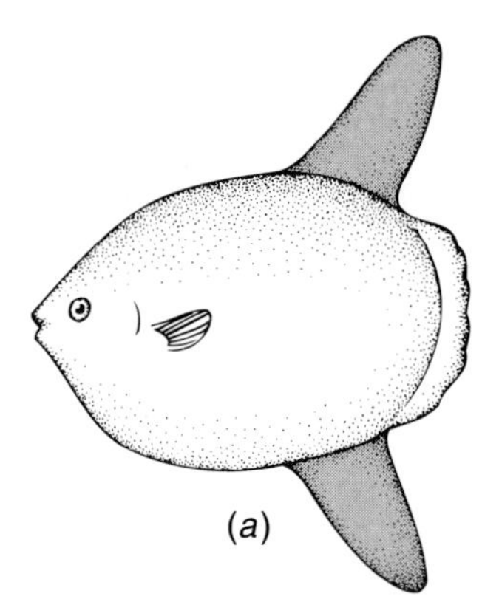

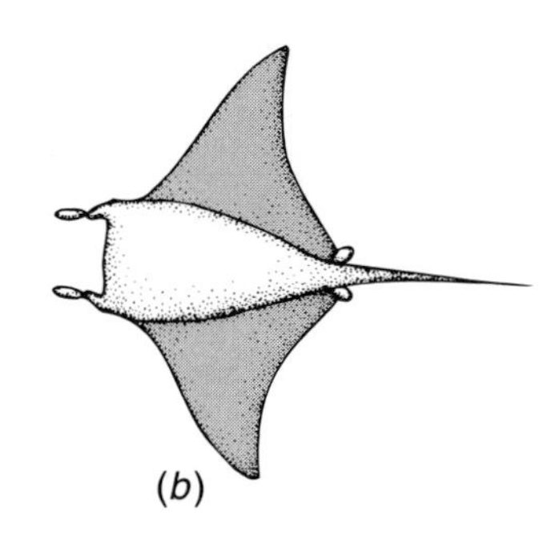

FIGURE 3.10 Fishes using fins for locomotion. (a) Ocean sunfish (*Mola mola*). (b) Manta ray (*Manta hamiltoni*). (Not to scale.) (From Y. G. Aleyev, 1977, *Nekton*, Dr. W. Junk BV. Reproduced by permission of Dr. W. Junk BV, with kind permission from Kluwer Academic Publishers.)

Another mode of propulsion is to employ undulatory movements of fins. In this mode of locomotion, the body remains stationary and the fins are moved in one of several ways to cause forward motion. This form of locomotion is slower than the previous ones. Examples of this type of movement are found in the rays, such as the manta ray (*Manta hamiltoni*), certain squids (*Todarodes*), and ocean sunfish (*Mola mola*; Figure 3.10). Some elongated fishes, such as oarfish and ribbonfish, undulate the dorsal fin while holding the body rigid. In most fishes, however, the lateral fins are used for maneuvering or lift, as in sharks with a heterocercal tail.

Except for cetaceans and sea snakes, the common form of propulsion among marine air-breathing vertebrates is through paddling movements of either the forelimbs or hindlimbs, or both. The limbs of marine turtles, seals and sea lions, and penguins are modified into flat, paddle-shaped appendages that the animals use to move themselves through the water, as we would employ oars (Figure 3.11). The speed of movement through the water using paddles depends on the frequency of the stroke. In organisms that employ few strokes, such as turtles, speed is low. In others, such as penguins—where up to 200 strokes per minute have been recorded by Brooks (1917) in *Pygoscelis papua*—speed may be very fast (10 m per second for the 200 strokes per minute).

Seals and sea lions, when swimming, cannot reach the speeds seen in whales. Eared seals (sea lions) swim using their front flippers as paddles, but earless seals use their webbed hind feet, spread vertically like the double caudal fin of fishes. The final type of propulsion is "jet propulsion" using water. This form of propulsion is the domain of the squids, who can produce very rapid movement, up to 16 m per second.

It has recently been discovered that swimming efficiency is improved in cetaceans by wrapping the body in an elastic sheath. This sheath acts like a rubber band to snap the tail flukes back once they have been raised, so that swimming is less energetically demanding than originally proposed.

Surface of Resistance and Body Shape

Because water is a very dense medium, it is difficult to move an object through it, and it is even more difficult to move an object rapidly. To propel a body through water requires more energy than to move the same body at the same speed through air. It requires less energy to move an object through water if it has a shape that reduces the surface of resistance to the water to a minimum. Since nektonic animals must move and since they have finite energy resources, a major set of adaptations must be made to reduce the surface of resistance and the drag of the body.

There are several types of drag or resistance to movement that must be overcome. **Frictional resistance** is proportional to the amount of surface area in contact with the water (Figure 3.12). Minimum frictional resistance is produced by a spherical object, which has the minimum surface area for a given volume of all geometric shapes. If, however, one wishes to move a nektonic object through water, another type of resistance becomes of great importance. This is **form resistance**, where the drag is proportional to the cross-sectional area of the object in contact with the water. In such a case, the spherical object presents a very large area; hence, it is an unsuitable shape for nektonic animals. To reduce form resistance to a minimum, the shape should be relatively long and thin, such as a thin cylinder or wire. The final type of resistance that must

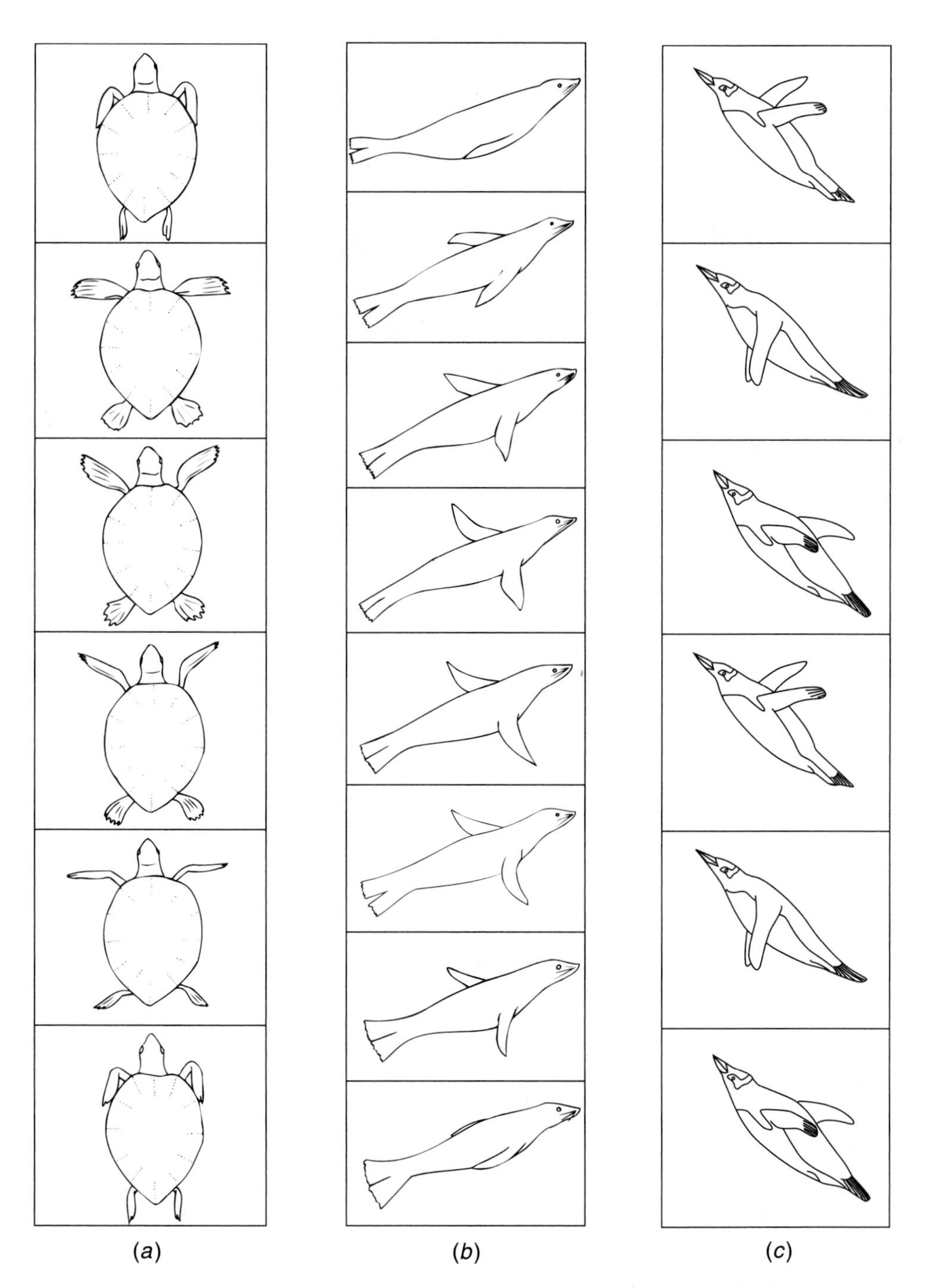

FIGURE 3.11 Swimming by paddle motions in three classes of marine vertebrates. (a) The green turtle (*Chelonia mydas*). (b) The fur seal (*Arctocephalus pusillus*). (c) The penguin (*Pygoscelis papua*). (Redrawn from Y. G. Aleyev, 1977, *Nekton*, Dr. W. Junk BV, with kind permission from Kluwer Academic Publishers.)

be overcome is the so-called **induced drag** or *turbulence* due to changes in speed and direction of the flow. The flow around small animals or slow, large ones is usually smooth and called *laminar*. With increasing speed or size, the laminar flow is disrupted and thrown into vortices or eddies, greatly increasing the drag. The flow pattern along large, fast animals is not laminar but turbulent. The water particles vibrate and have less

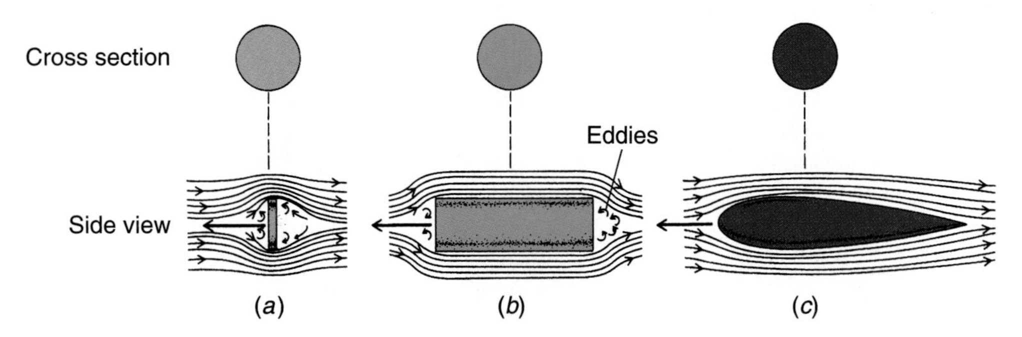

FIGURE 3.12 Drag forces on variously shaped objects moving through water.(a) A flat disk. (b) A cylinder. (c) A teardrop-shaped object. All have the same cross-sectional shape and area, but they differ in their three-dimensional form. The teardrop shape is best for a fast-moving fish and has about 1/14 the resistance of that on the disk in a.

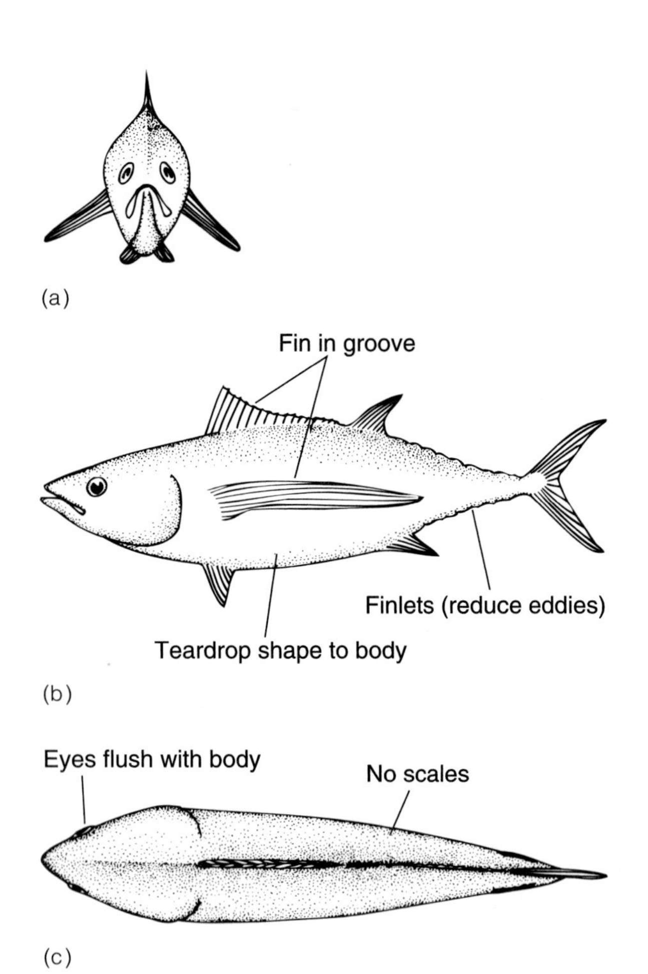

FIGURE 3.13 Three views of a tuna, showing the adaptations necessary for fast movement. (a) Front view. (b) Side view.(c) Top view.

tendency to be thrown into eddies. The turbulent flow follows the contour of a streamlined body, even at high speeds where a laminar flow would have been disrupted. Swordfish generate the turbulent flow with the rough sandpaper-like skin on their sword. This greatly reduces drag and makes high speeds possible. Tunas have an area of scales behind the head called the corselet, which acts as a microturbulence generator. The best streamlined bodies have a teardrop shape, somewhat blunt in front and tapered to the posterior end. Such a shape gives the lowest resistance for the largest volume if the ratio between the largest diameter and the length is about 0.22. Whales, dolphins, and tunas closely approach this ideal shape (Figure 3.13).

In addition to body shape, nektonic animals have other adaptations to reduce drag. These adaptations generally fall under the category of streamlining the external surface of the body so that there are no protuberances that may break the smooth flow of water over the body and increase drag. To achieve this, species that move rapidly, such as tunas, are structured so that almost all normally protruding body structures are recessed into depressions or grooves from which they may be elevated only when needed. In fast-moving fishes, eyes, though large, do not protrude beyond the sides of the body. Pectoral and pelvic fins fit into grooves except when in use, and body scales are reduced or absent (Figure 3.13). Similarly, in marine mammals, the hair is lost or reduced in length because it produces more drag than bare skin. Mammary glands are flattened, and genitalia of the males and nipples of the females do not protrude beyond the skin, except when in use.

Defense and Camouflage

The strongest adaptations seen in nektonic animals are those that relate to achieving rapid forward movement in the water column. These adaptations are of such importance that they take precedence over, or preclude, any adaptations for defense against predators, if such adaptations would mean decreasing rapid movement. This immediately limits the number of possible defensive adaptations. At the same time, most of these nektonic animals are large as adults, which means most have few predators; the largest (whales) often have few predators other than humans and killer whales (Figure 3.14). The need for elaborate defensive mechanisms is, therefore, somewhat reduced.

Certain defensive mechanisms remain possible within the framework of fast locomotion. Of these, the most common is that of *camouflage*. Since we are now dealing with an environment that has no physical hiding places—organisms are visible in all three dimensions—camouflage mechanisms could take one of three directions: transparency of the body, cryptic coloring, and alteration of body shape (cryptic body shape).

If the body of an organism is transparent and suspended in the transparent surface water of the ocean, the animal will become invisible in the water. As we saw in the plankton, such transparency is a common defensive adaptation of many of those species. It is, however, not found among the nekton. The reason is that as the size and thickness of nektonic animals increases, it becomes much more difficult to keep a body transparent, particularly when the body is highly muscular for propulsion. This is why camouflage by transparency is not an option for nekton.

Camouflage through alteration of body shape is possible so long as the shape does not interfere with fast locomotion. Among nektonic vertebrates, the most common manifestation is to develop a ventral keel to the body to eliminate a conspicuous shadow on the belly of the animal when viewed from below. What is the origin of this shadow and how does body shape help reduce it? To answer this, we must know something about how light appears in the surface waters of the ocean. When light enters the water, the rays penetrate in a cone-

FIGURE 3.14 A killer whale (*Orcinus orca*) surfacing through the pack ice in Antarctica. (Photo courtesy of Dr. John Oliver.)

like path. At the same time, some of the light is reflected or scattered back in all directions by the particles in the water. This scattered light illuminates objects in the water from different angles other than the surface, but its intensity is much less than the downwelling surface light. If an animal is suspended in this water column, it is illuminated most intensely from above, while the scattered light illuminates from the sides and below. Because light intensity from the side and below is so low, a shadow appears under the animal where its body has cut out the strongly downwelling surface light (Figure 3.15). If the body is now extended ventrally as a keel to form a sharp ventral edge rather than a rounded edge, the shadow will be eliminated when viewed from below. This is because all surfaces on the body now are oriented so that none is illuminated exclusively by the diffuse scattered light;

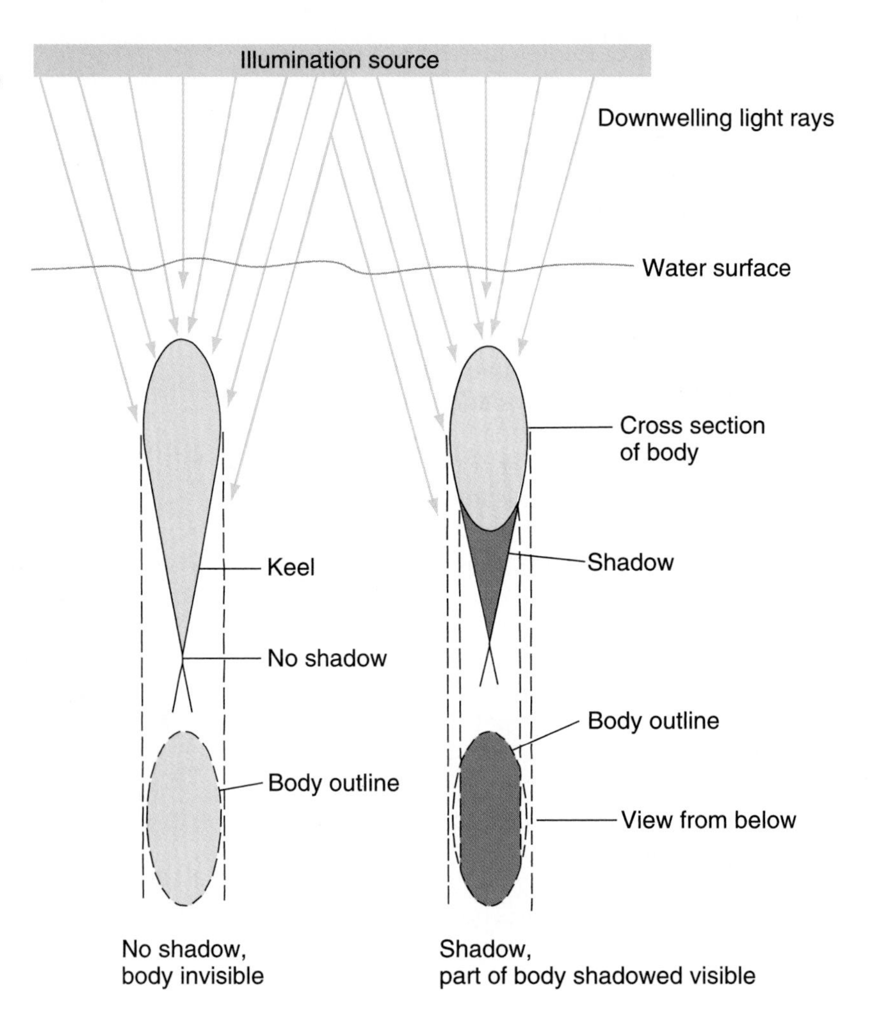

FIGURE 3.15 Diagram showing how a keel on the ventral surface of an animal eliminates the dark shadow normally cast downward by an unkeeled animal. The presence of the shadow means that an animal living deeper and looking up would see the unkeeled nektonic animal due to the shadow, but would not see the keeled animal, which would blend into the lighted background. (Modified from Y. G. Aleyev, 1977, *Nekton*, Dr. W. Junk BV, with kind permission from Kluwer Academic Publishers.)

there is also at least some component of the intense surface light (Figure 3.15). Without this conspicuous shadow, an animal viewed from below becomes virtually invisible in the downwelling light; hence, it is camouflaged from deeper-dwelling predators. The elimination of the shadow is enhanced if the keel has reflecting surfaces on it in the form of white pigment or scales.

Cryptic coloration is also characteristic of most nektonic animals. In the lighted upper waters of the ocean, the dominant spectral colors are blues and greens. When water is viewed from above at the surface, it appears greenish or bluish looking down into the depths. It should not be surprising then that many nektonic forms have dark blue or green on their dorsal surfaces so that potential visual predators looking down would be hard-pressed to see their form against the general blue-green background. At the same time, when viewed from below, the water looks brighter or whiter toward the surface. Any dark organism swimming in this area would be conspicuous if viewed from below, even if keeled to prevent a shadow. If, however, the ventral side is colored white or silver to maximize light reflectance or to blend with the downwelled light, the animal will become invisible. As a result, we observe that many nekton are bicolored, or *countershaded*, dark green or blue above and white or silver below (Figure 3.16). For certain of the highly surface-dwelling vertebrates, such as porpoises, the color pattern is more complex, with irregular bands of light and dark that actually mimic the pattern of wave-roughened surface waters themselves (Figure 3.17). Some dolphins that live within tuna schools are countershaded gray above and white beneath, with white spots and speckles in the gray and black spots and speckles in the white. Colored in this manner, these animals are difficult to perceive from the side, underwater, amid a school of tuna.

FIGURE 3.16 Contrasting color patterns as seen on Dall's porpoise (*Phocoenoides dalli*). (Photo courtesy of Dr. Jim Harvey.)

FIGURE 3.17 Cryptic coloring on the sides of a Pacific white-sided porpoise (*Lagenorhynchus obliquidens*), mimicking the wave-roughened surface of the water. (Photo courtesy of Nancy Black.)

Among the abundant flying fishes there are large pectoral large fins. These fishes escape predators by propelling themselves out of the water and gliding for long distances on these winglike fins.

Other than the adaptations already discussed, there are few specialized morphological structures developed as defense against predators. Most likely this is because the development of various spines and armor would interfere with the ability to move quickly by increasing resistance. It also may be partly because the larger nekton are the top predators in the system and have few or no other nekton consuming them.

Sensory Systems

Since nektonic animals are large, fast moving, and primarily predaceous, they might be expected to have well-developed sense organs. This is generally the case, and with few exceptions, such as lateral lines in fishes, the senses are no different from those possessed by other vertebrates in different habitats. Lateral lines are rows of small tubes, open to the water and containing sensory pits sensitive to pressure changes in the water.

Sharks and rays possess special sense organs called ampullae of Lorenzini, which are sensitive to minute electric currents in the water. Kalmijin (1971) has demonstrated that predaceous cartilaginous fishes can use this electroreception to find prey. A geomagnetic sensory system in marine mammals seems to play an important role in long-distance navigation and may also explain the various mass strandings of some toothed whales.

Most sensory information received by nekton comes via light (vision), sound (hearing), or chemicals (olfaction). Eyes tend to be well developed and complex in all forms, but the size of the eye relative to size of body varies greatly. The eyes are usually set laterally on the head in such a way that the fields of view for each do not overlap, yet encompass large areas on either side. The lack of overlap of fields of view means that binocular vision is small or absent among most nekton except for pinnipeds.

Echolocation

Among the mammals of the nekton, it is the sense of hearing that has generated the greatest number of specialized adaptations. These adaptations, in turn, suggests the great importance of hearing to these animals. The importance of sound to nektonic mammals resides in the fact that sound travels more than four times faster in water than in air and has a much greater communication range than does sight. As a result, most nektonic animals show strong development of sound-receiving structures.

In the terrestrial environment, enhanced sound reception is usually indicated in external morphology through the enlarged external ears or pinna of mammals. Such structures, however, would create an excessive drag on aquatic vertebrates; hence, they are reduced or absent. To make up for their absence, aquatic mammals tend to develop other structures of the head to receive sound waves.

Sound reception and production are most highly developed in those cetaceans that use them for **echolocation** in much the same way we use sonar to determine depth. In echolocation or **sonar**, sound waves are sent out from a source in a particular direction. These sound waves pass uninterruptedly through the water until they encounter an object with a different refractive index than water. When they strike such an object, they are reflected and return to the source. The time interval between the production of the sound and its movement to a target and back after reflection is a measure of the distance between the source and the object (Figure 3.18). As the distance changes, so will the time necessary for the sound "echo" to return. Continual production of sound waves and sensory evaluation of the reflected waves during swimming give a nektonic animal a constant check on all objects in its path. Knowing the distance to objects makes it possible for the echolocating animal either to avoid them (predators) or to close in on them (food source).

Low-frequency sound is used by some animals to communicate among themselves in the water column, such as with the "singing" of humpback whales and low frequencies (20–80 Hz) produced by blue whales. Low-frequency sound, however, does not produce information as to the fine structure of objects. To obtain that information, high-frequency sound waves must be produced and reflected from the object for the animal to discriminate detail. Most nektonic animals that have highly developed echolocation also can vary the frequency of the sound produced.

It is in the toothed whales that echolocation reaches its zenith. These animals possess elaborate morphological modifications of the head and respiratory systems that permit them to send and receive sound waves varying over a wide range of

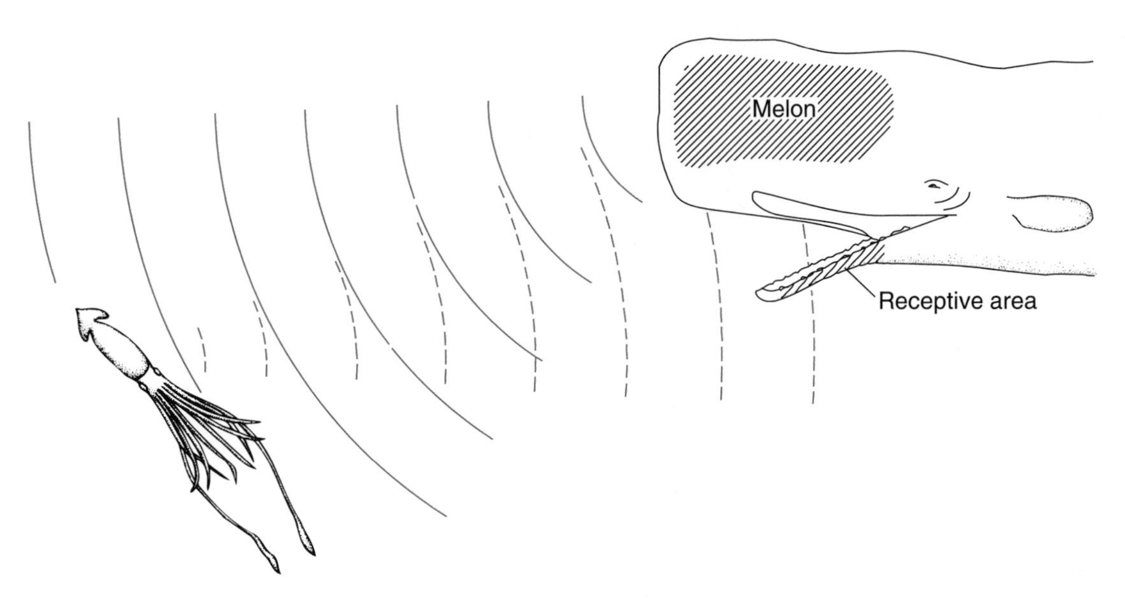

FIGURE 3.18 Echolocation in a sperm whale. Sounds are made in the nasal passages, focused by the melon, and sent out (solid lines). The returning echoes (broken lines) are received by the lower jaw and transmitted to the ear by bone conduction.

frequencies. Discriminatory ability of high-frequency sound is quite remarkable. Dolphins, for example, are reported by Kellogg (1958) to be able to distinguish between two fish species of similar size and shape. Norris et al. (1961) report porpoise able to distinguish even between objects differing only in thickness.

Toothed whales have a peculiarly bulging, rounded forehead and, associated with it, a dorsally placed external nasal opening or blowhole. Internally, a complex series of air sacs is associated with the nasal passages leading from the blowhole to the lungs. The rounded forehead is caused by the presence underneath of a large, fat-filled structure called the **melon**. This fatty organ reaches its greatest development in the sperm whales, where it is called the spermaceti organ; it extends as much as 40% of the entire animal's length. The relationship of these structures is diagrammed in Figure 3.19.

Although we do not yet completely understand how the elaborate echolocation system operates in sound production and reception, enough is understood to permit a description of how we believe this apparatus functions. Sounds are produced by the toothed whales by movement of air through the nasal passage and associated air sacs. Sound is produced by recirculating internal air during diving. Special muscles in and around the nasal passages and air sacs allow these channels to change shape and volume and change the frequency of the sound produced. The fatty melon is apparently used as an acoustical lens to focus and direct the train of outgoing sound waves, and this directionality increases with increasing frequency of the sound, so these animals can pinpoint objects with high-frequency sound. Directionality is further enhanced by the bones of the peculiarly shaped skull of these toothed whales. Reception of the reflected waves is centered in the bone and fat deposits of the lower jaw and in the middle ear. In contrast to most mammals, where the middle ear resides in bone attached to the skull, the toothed whale's middle ear is loosely attached to the skull by ligaments and provided with special air- and fat-filled spaces.

The elaborate morphological modifications of the head region of toothed whales are primarily directed to producing and receiving a wide range of sound frequencies, which allow the animals both to navigate without bumping into objects and to pinpoint potential food organisms for capture. Toothed whales also possess very large brains relative to their body size, brains that are second only to those of humans in development of the cerebral hemispheres. It is possible that these large brains are necessary to allow them to process rapidly the acoustical information received.

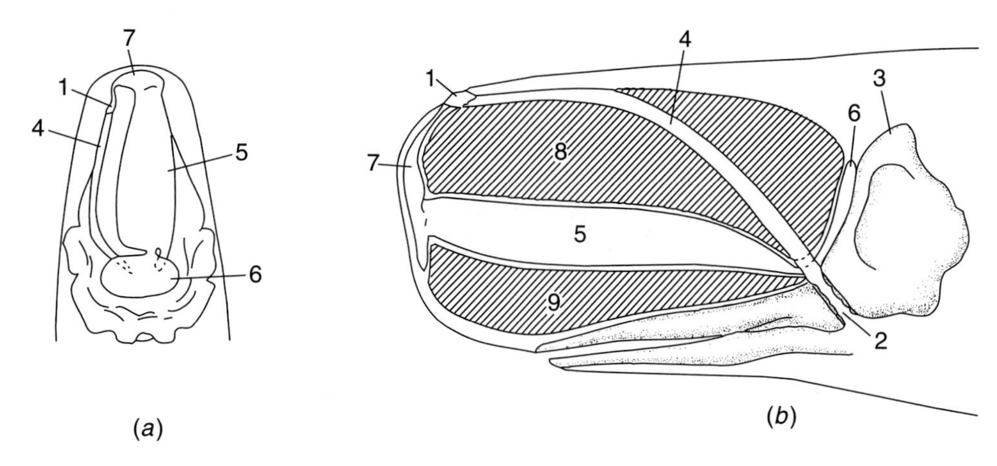

FIGURE 3.19 The anatomy of the head of a sperm whale (sound production, sound reception, and buoyancy control). (a) Dorsal view of the head of a sperm whale (*Physeter macrocephalus*). (b) Side view of the head of a sperm whale. (1) Nasal opening or nostril; (2) bony nasal duct; (3) sound-focusing surface of cranium; (4) left nasal passage; (5) right nasal passage; (6) frontal air sac; (7) distal air cavity; (8) and (9) upper and lower spermaceti sac (acoustic lens). (From Y. G. Aleyev, 1977, *Nekton*, Dr. W. Junk BV. Reproduced by permission of Dr. W. Junk BV, with kind permission from Kluwer Academic Publishers.)

Reproduction and Life Cycle

Among epipelagic fishes, no special reproduction mechanisms are apparent that would sharply set them apart from their benthic or shallow-water relatives. Characteristically, however, holonektonic bony fishes, such as tuna and marlin, spawn eggs that float and undergo development in the open ocean waters. Some even have threadlike structures that permit them to adhere to various kinds of floating debris. Because these floating eggs are planktonic, they are subject to tremendous losses due to predation. As a result, the fishes produce huge numbers of eggs to offset the equally large losses. Parin (1970) reports that skipjack and albacore tuna, for example, produce 2 and 2.6 million eggs, respectively, while striped marlin spawn over 13 million, and ocean sunfish produce an incredible 300 million eggs. Spawning is often intermittent and is often extended over a period of many months.

Among the pelagic sharks, a different reproductive strategy is observed. These fishes produce only a few large eggs or embryos. Parin (1970) notes that the thresher shark (*Alopias*), for example, has but two embryos and the blue shark (*Prionace glauca*) up to 54. If so few eggs were to start out as small as those of bony fishes and have to undergo development for a substantial time in the plankton, the chances any would survive are very poor. Sharks enhance the chances for survival of their few offspring by producing very large eggs and by retaining the eggs in the female for a much longer period of time so that, at birth or hatching, the young are large in size and immune to most potential predators.

We know relatively little about the growth of pelagic fishes, but available knowledge suggests that the growth rate is very rapid. Tunas, for example, seem to increase their weights by 2–6 kg per year and length by 20–40 cm (Table 3.1). Correlated with this rapid growth, most nektonic fish seem to be short-lived; even the large tunas appear to live only 5–10 years. By contrast, the pelagic sharks may live 20–30 years.

Marine birds and turtles retain the reproductive characteristics of their terrestrial relatives. All produce shelled eggs, which are laid on land. Marine birds often congregate in large groups to nest on islands or cliffs inaccessible to terrestrial predators (Figure 3.20). This ensures that the usually helpless young (**altricial**) will survive until old enough to fly. It also makes such birds extremely vulnerable to human predation or pollution, since a large fraction of the existing population of a species may be present in one small area and can be wiped out easily. For example, as a result of the massive *Exxon Valdez* oil spill in Alaska, about 300,000 murres were killed, representing probably 70% of the breeding population in that area. Most marine birds have definite breeding seasons and may migrate thousands of miles from their feeding grounds to their breeding grounds.

TABLE 3.1

Age and Length in Three Species of Tuna[a]

	Age, Years							
Species	1	2	3	4	5	6	7	8
	Length, cm							
T. alalunga	18–66	32–84	54–89	73–94	70–95	82–100	98–105	—
T. obesus	45–50	62–70	74–94	85–116	97–138	115–155	—	—
T. albacares	54–103	70–136	85–155	99–154	109–134	110–130	127–145	135–160

[a]Albacore (*Thunnus alalunga*), big-eye tuna (*Thunnus obesus*), and yellow fin tuna (*Thunnus albacares*).
Source: After Parin, 1970.

Marine turtles all lay eggs in holes that they excavate in the sand above the high-tide level on beaches in various parts of the tropics. This is the only time these animals normally return to land. Immediately upon hatching, the young turtles instinctively head for the ocean, where their continued development occurs, but about which we know virtually nothing. As with birds, turtles migrate thousands of miles and congregate off certain beaches for breeding. Females haul out to lay eggs only on certain beaches, and since both eggs and adults are considered excellent food by humans, most marine turtles have been drastically reduced in numbers in recent years in all areas of the world (Figure 3.21). Some sea snakes give birth to living young in the water; others lay eggs on beaches.

With respect to reproduction in marine mammals, there are two groups: those that give birth on land and those that give birth in the water. We know considerably more about reproduction in the group that breeds on land because they may be observed by humans at that time. Knowledge of the reproductive patterns of

FIGURE 3.20 Sea birds (black-legged kittiwakes) nesting on a cliff in the Pribilof Islands. (Photo by the author.)

FIGURE 3.21 Sea turtle (*Lepidochelys olivacea*) laying eggs on a beach. (Photo courtesy of Gregory Dimijian.)

those giving birth in the water is generally limited to observations made on captive animals in aquaria.

Seals, sea lions, and walruses all give birth to live young on land or on floating ice (Figure 3.22). Their young often cannot swim at birth and may require some time before they are capable of venturing into the water. Harbor seal (*Phoca vitulina*) pups, however, can swim immediately after birth. During this terrestrial period, the pups grow very rapidly on their mother's rich milk and gain the strength and insulating layers of fat and fur they need to survive in the cold, open water. Many sea lions and seals, such as fur seals, Steller sea lions, and elephant seals, are polygamous and territorial on the breeding grounds (Figure 3.23). The largest and most aggressive males (territorial bulls) collect varying numbers of females into a harem and occupy a small area of beach, which they protect from other bulls (Figure 3.24). Should another bull attempt to steal a female, take over a harem, or encroach upon the territory occupied by the bull and his females, the resident territorial bull will fight. Although such fights may be mostly bluff and noise, serious fights occur in which one or the other bull may be seriously injured or killed. In the seals and sea lions where such territorial behavior occurs, the males are usually much larger in size than the females.

This territorial behavior and aggressiveness on the breeding grounds do not extend to their life away from there. This behavior suggests that breeding space is limiting, whereas pelagic food and space are not. This method of breeding also means that relatively few males, the territorial bulls, do the breeding, and most others are excluded (Figure 3.24). As with the marine birds, many of these pinnipeds migrate long distances to their breeding grounds; for example, the northern fur seal (*Callorhinus ursinus*) is pelagic all over the northern Pacific Ocean, but most migrate back each summer to breed on two small islands making up the Pribilofs in the Bering Sea.

In contrast to the pinnipeds, the cetaceans give birth to their offspring in the water. The young of whales, therefore, must be able to swim at birth and instinctively know how to surface for air. They also remain closely associated with the mother (Figure 3.25). Whereas young pups of sea lions may be left on the breeding grounds for days while the female forages for food in the open ocean, the juvenile whales always remain close to their mothers, where they are protected from predators. As with the pinnipeds, certain cetaceans may also undergo migration for breeding purposes. Often, this is a migration of thousands of miles from the feeding areas in cold water to the calving grounds in warmer waters (Figure 3.26). The animals move to warm waters for calving because the newborn do not have the insulating blubber of the adults. They have a better chance of survival in warm water until the additional insulating layers can be added.

The young of both pinnipeds and whales grow very rapidly, adding many kilograms per day. Blue

FIGURE 3.22 Females (brown) and pups (black) of the northern fur seal (*Callorhinus ursinus*) on the breeding grounds in the Pribilof Islands. (Photo by the author.)

FIGURE 3.23 A Steller's sea lion rookery on Año Nuevo Island in California. (Photo by the author.)

FIGURE 3.24 A breeding bull and females, northern fur seal (*Callorhinus ursinus*). (Photo by the author.)

FIGURE 3.25 A humpback whale (*Megatera novaeangeliae*) with calf. (Photo courtesy of Tony Stone Images.)

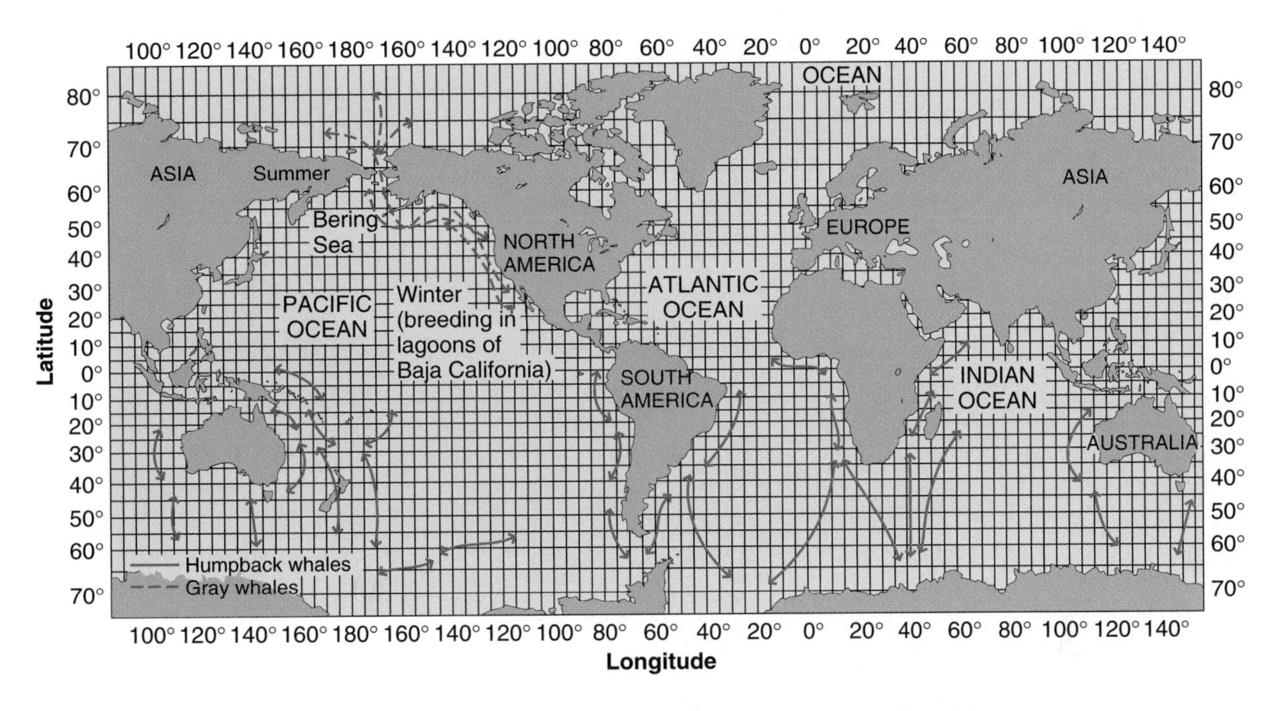

FIGURE 3.26 Migratory routes of whales. Migratory routes of humpback whales (*Megaptera novaeangliae*) in the Southern Hemisphere and migratory route of the gray whale (*Eschrichtius robustus*) in the North Pacific. (a from *Whales and Dolphins*, E. J. Slijper. Copyright © 1976 Springer-Verlag. Reprinted by permission. b, Reprinted with the permission of Atheneum Books for Young Readers, an imprint of Simon & Schuster Children's Publishing Division from *A Natural History Of Marine Mammals* by Victor B. Scheffer. Copyright © 1976 Victor B. Scheffer.)

whales, for example, can grow from 3 tons at birth to 23 tons at weaning, seven months later. The reason for such rapid growth is that the milk of pinnipeds and cetaceans is extraordinarily rich in fats (as much as ten times as much fat as in cow's milk) and is produced in large quantities.

Pinniped pups are nursed throughout the time they are on the rookeries. At the end of the season, in some species, they are abandoned by the females and must fend for themselves in the open ocean. Cetacean calves are nursed for varying times depending on the species. Nursing lasts 15 months in sperm whales but may last several years in pilot whales.

Because of the enormous amount of energy put into milk production to sustain a single offspring, usually one young at a time is produced, and young are produced annually (most pinnipeds) or at even longer intervals (walrus, certain whales; Figure 3.27). Stocks of these animals can be easily reduced, and it may take a long time to recover their numbers.

Most marine mammals are long-lived. Gray seals have been recorded to live 46 years and harbor seals 36 years, while small whales, such as the bottlenose porpoise, may live 32 years and large whales, such as the sperm whale and fin whale, live up to 77 and 80 years, respectively. Correlated with the long life span is the delayed onset of sexual maturity and reproduction. Male fur seals, for example, do not become harem masters until 9 or 10 years of age, and sperm whales apparently do not breed until about age 20.

Migrations

As we noted in the previous section, a large number of marine mammals, birds, and reptiles undertake extensive migrations for breeding or other purposes. This migration for breeding is a common characteristic of air-breathing marine vertebrates.

Nektonic fishes also undertake extensive horizontal migrations equivalent in distance to those taken by the air breathers. These migrations are of great importance but, unfortunately, have been little analyzed. Migrating holonektonic fishes include the various tunas and the saury.

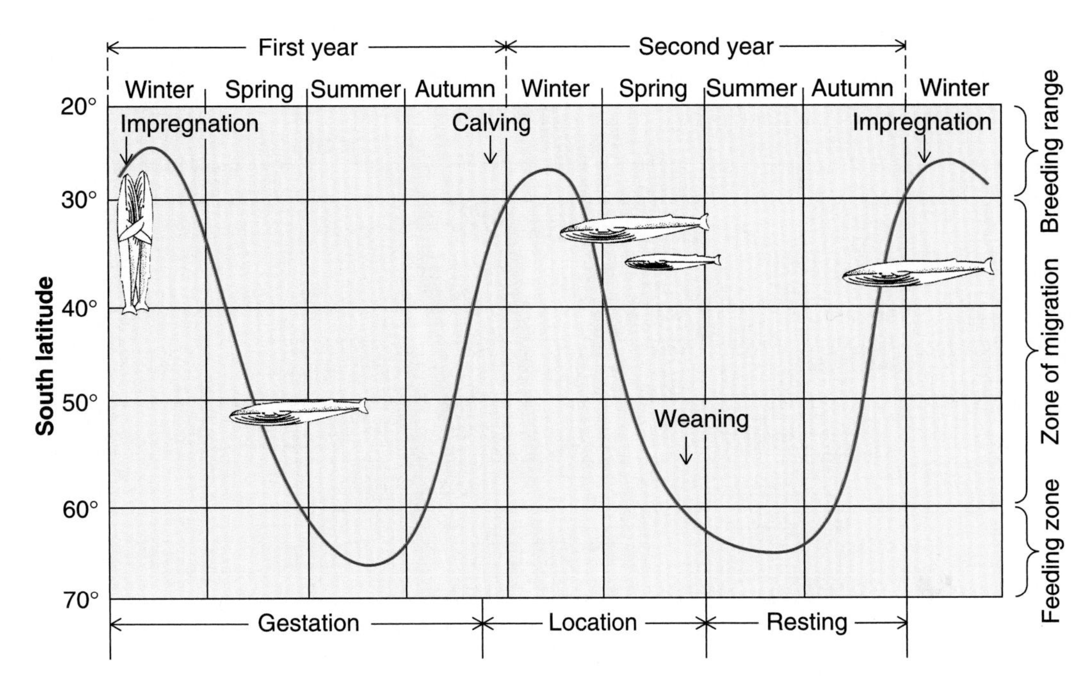

FIGURE 3.27 The reproductive cycle of the fin whale (*Balaenoptera physalis*) in the Southern Hemisphere. (From *The Stock Of Whales*, N. A. Mackintosh, 1965, Fishing News Books. Reprinted by permission of Blackwell Science.)

Salmon, which spend most of their lives dispersed in the open ocean, migrate back to freshwater streams in which they were spawned to reproduce. In this respect, they are similar to the marine air-breathing vertebrates. Salmon have the ability to return to breed in the same stream where they hatched. This requires considerable navigational abilities on the part of the fish. From recent studies, it appears that part of the key to this navigational ability resides in the fishes' sense of "smell"; that is, these fish navigate to find the coast but then use olfaction to follow various waterborne "scents" to locate their home streams. The suggested mechanism is that, when the young salmon migrate down the streams to enter the sea, they are imprinted with the "odors" of the various streams in succession. This sequential memory is then reversed on their return, allowing them to find their way back to the same stream.

Equally remarkable migrations are made by the various species of sea turtles. As with the salmon, these animals migrate from distant feeding grounds to congregate on one or a few beaches to lay their eggs. How do they locate these beaches? In the case of beaches along the shores of continents, this is not as difficult to understand. Here, a turtle could simply follow along the shoreline until the beach with the right "smell" came along. Many turtles along the Atlantic coast of South America, however, regularly nest on tiny Ascension Island, some 1,400 miles out in the middle of the Atlantic Ocean. Others migrate equally long distances to specific beaches on islands or mainland coasts. How do they find such a small target? Since the turtles must navigate most of the distance to a mere pinpoint of an island or beach out of sight of land and without any landmarks in the ocean, they must have some ability first to know where they are in order to set a course to find the nesting ground. This is an ability that Lohman (1992) has called "map sense."

Although we still do not know exactly what cues are used by the turtles to fix their positions and navigate, recent studies have suggested some likely possibilities. Perhaps the most significant is the ability to detect magnetic fields. This is theoretically a good choice since several geomagnetic parameters vary in a predictable manner according to latitude. This ability to detect and follow magnetic fields is used by certain marine mammals, and there is now evidence that turtles may also have it. A second mechanism is to orient to wave propagation or direction. Many areas of the open ocean have consistent wave and swell direction that would permit use in navigation. A final mechanism would be to

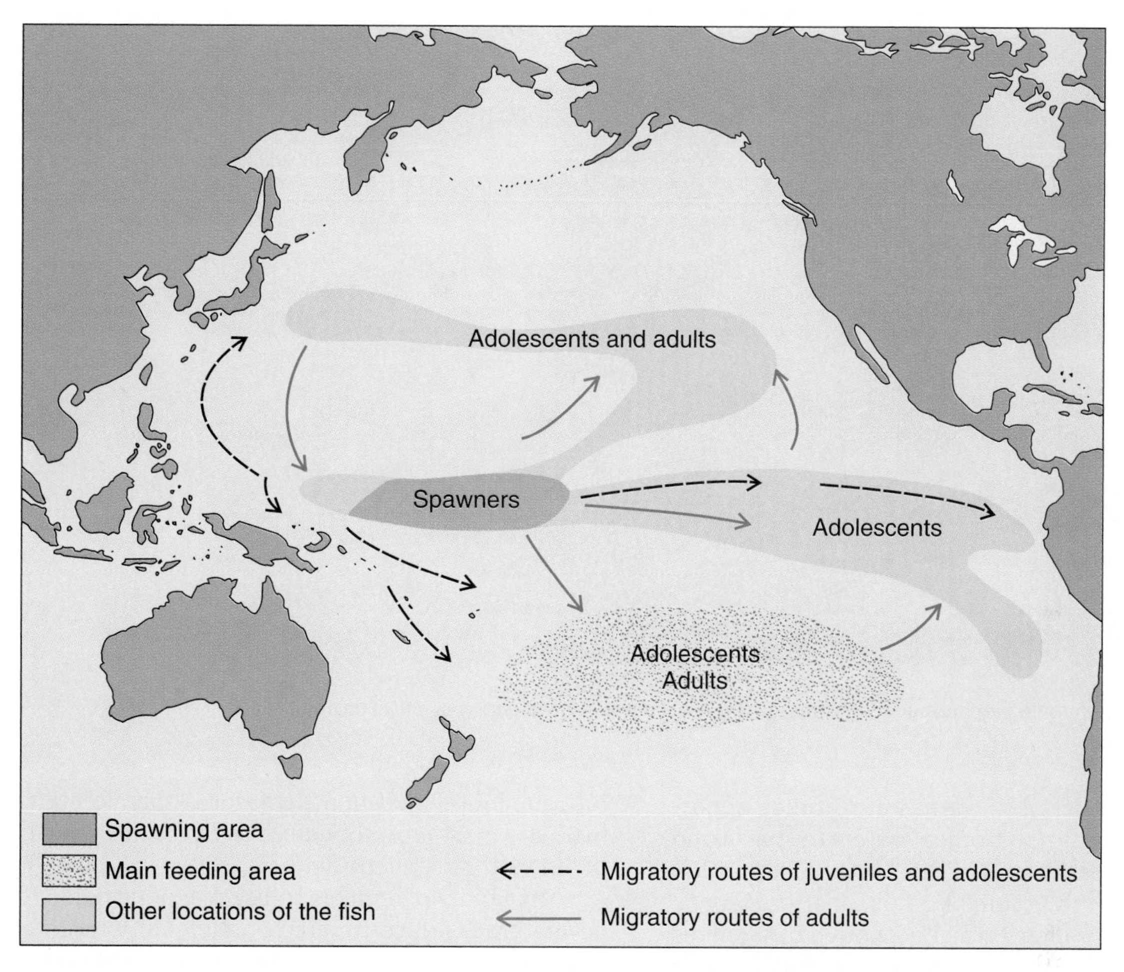

FIGURE 3.28 Centers of distribution and migratory routes of the big-eye tuna (*Thunnus obesus*) in the Pacific Ocean. (Modified from *The Rand McNally atlas of the oceans*, Rand McNally.)

follow some chemical cue, such as that used by salmon. Initial experiments with hatchling sea turtles by Lohman (1992) suggest that some combination of cues from the earth's magnetic field and a seasonal consistent pattern of wave or swell may be the most likely mechanisms.

Other extensive and complex migrations are those undertaken by tunas and their relatives. Tunas are primarily tropical fishes that make extensive migrations across oceans within the tropical zone and also move into the temperate waters during the warmer seasons. Bluefin tunas tagged in Florida have been recovered across the Atlantic in the Bay of Biscay, and Pacific albacore tagged off California have been captured in Japanese waters. It is not entirely clear why these fishes should make such monumental journeys, but it appears such migrations permit the fishes to exploit their food resources more completely and reduce the possibility of destroying or eating out their food in any one area. Such migrations allow these fast-moving fishes to take advantage of the rich food areas of the temperate zone. One of the most important factors that guides these migrations is water temperature. As primarily tropical fishes, tunas venture into the temperate waters when the water temperature rises above 20°C. Off California, the cold upwelled water is very rich in food organisms, but the tunas usually enter these waters in the summer when surface temperatures reach 20–21°C. If the year is cold and temperatures of the surface waters do not rise to this point, the tunas are absent. The Atlantic blue fin tuna, however, regularly summer off Newfoundland in water below 20°C, so not all

FIGURE 3.29 Two male elephant seals (*Mirounga angustirostris*). (Photo courtesy of Dr. Jim Harvey.)

tunas are restricted to warm water. Tunas apparently always return to tropical waters to spawn and to spend the early parts of their lives (Figure 3.28).

More recently, information on albatross tagged so that their movements can be followed by satellites has shown that birds nesting in Hawaii regularly make foraging trips as far east as the west coast of North America to procure food for their chicks. These trips entail a round trip distance of about 6,000 miles!

There may also be migratory differences between the sexes. In northern elephant seals (*Mirounga angustirostris*) in the North Pacific Ocean, Stewart and DeLong (1995) found that the males migrate to the Gulf of Alaska whereas the females migrate offshore (Figure 3.29).

The ability of migratory animals to make accurate journeys across what appear to be featureless seas has generated a profound interest in marine scientists about how these animals manage to navigate accurately. Over the years, a number of hypotheses have been suggested to explain migration that have implicated a wide variety of sensory systems. Suggested mechanisms have included use of some sort of celestial compass such as humans use, odor or chemical imprinting as suggested for salmon, wave direction as noted for turtles, cueing on underwater topography, echolocation, detection of thermal structure, and most recently, detection of the flux density of the earth's magnetic field.

The ability of animals to use magnetic fields to navigate was not considered possible until recently because of the lack of this sense in humans and the apparent lack of any known biological mechanism capable of transmitting the weak geomagnetic field into an animal's nervous system. The discovery of magnetotactic bacteria with linear chains of membrane-bound crystals of magnetite (Fe_3O_4) provided the biological mechanism for transducing the geomagnetic field to the nervous system. Subsequently, the finding of membrane-bound magnetite with abundant nervous connections in tuna by Walker et al. (1984) demonstrated the existence of an anatomical basis for magnetotaxis in a migrating nektonic animal.

Magnetic reception has now been suggested as the system used for long-distance navigation by turtles and marine mammals, but experimental proof has been difficult to come by due to the lack of good small-scale geographic maps of magnetic anomalies and the difficulty of setting up laboratory tests.

The most interesting data that seem to confirm the presence of this magnetic sense in whales come from the well-known phenomenon of mass strandings.

If cetaceans are using the flux density of the earth's magnetic field to navigate by, they can use it in two ways. First, the topography of the local field can be used like a road map, with the animals following certain magnetic contours. Second, a timer based on the regular changes or fluctuations of the magnetic field permits the animals to know their position and progress. Problems arise in this system whenever the geomagnetic contour lines cross land or when the pattern of the timer is disrupted by irregular fluctuations in the magnetic field. Whenever these two conditions occur, magnetically navigating whales may beach themselves. Analysis of mass strandings by Kirschvink (1990) for the east coast of the United States and Klinowska (1990) for England have demonstrated that all sites of mass stranding had only one physical feature in common, the geomagnetic contour lines crossed from the sea onto land. If the whales were following a strategy of traveling parallel to the geomagnetic lines, which would work well in the open ocean, but which gives no clue as to where land and water meet, they would strand themselves. Furthermore, the dates of such strandings were consistent with times when there were irregular fluctuations in the normal patterns of geomagnetic fields.

There are also short-term movement patterns related to food availability and predator avoidance. Many dolphins, such as dusky dolphins, Hawaiian spinners, and Bay of Fundy harbor porpoises, for example, show offshore nighttime and onshore daytime movements.

Although many migrations are undertaken for breeding purposes, such as those of certain of the large baleen whales and the northern fur seal, there may be other reasons for migration as well. Predation is one possibility, and it may be that some baleen whales migrate to warmer locations to calve because killer whales are less abundant at low latitudes; thus, such a ploy would protect the young.

Special Adaptations of Marine Birds and Mammals

The warm-blooded marine mammals and birds require some special adaptations to permit them to live in ocean waters. These special adaptations are primarily concerned with maintaining temperature, diving, and osmotic regulation.

Water has a higher thermal conductivity than air, which means it is quick to extract heat from a warm body. Humans experience this when they become chilled after a short time in water of even 80°F; in air of the same temperature, they would be comfortable indefinitely. Marine mammals, which maintain elevated body temperatures with respect to the surrounding water, must adapt to prevent their body heat from being drained away.

One means to slow the rate of heat loss is to have a large body. As you may recall from the plankton section (see p. 52), the ratio of surface area to volume of any body is lower for a large organism than for a small organism. The larger the body, the smaller the surface area relative to the volume of the body. Since heat production is a function of the volume of a body while heat loss is a function of wetted surface area, this means that large animals, which have a low surface area to volume ratio, can thermoregulate better than small animals. Thermoregulation is aided by the smaller surface area in contact with the environment through which heat may escape. All nektonic marine mammals are large, and it may be that the reason there are no mouse-sized marine mammals is that they would simply die from chilling. There are small marine birds (petrels, auklets), but these animals are only completely immersed in the water for a short time during intermittent dives. Others, such as gulls, rarely dive, and only a fraction of their body is in contact with the water at any time.

A second adaptation that prevents or reduces heat loss is a thick, insulating layer of blubber or fat just beneath the skin. This layer reaches its greatest thickness in whales, where it may be 2 feet thick. In such pinnipeds as the walrus and elephant seal, subcutaneous fat may constitute as much as 33% of the weight. Blubber and fat are poor conductors of heat; hence, they protect the animal from losing internal heat. The thicker the blubber or fat layer, the less the heat loss. Marine mammals inhabiting polar waters, therefore, have thicker layers of fat than temperate and tropical species.

A final adaptation concerns the circulatory system. The areas of a marine mammal that offer the greatest surface area to the water and the greatest heat loss, and also lack much of the protective layer of fat, are the fins, flippers, and flukes. What adaptations prevent massive heat loss through these extremities? In cetaceans, the answer is that the arteries that bring the warm blood out to these extremities are surrounded by a number of smaller veins that bring blood back to the central core of the mammal. Because of this arrangement, the heat of

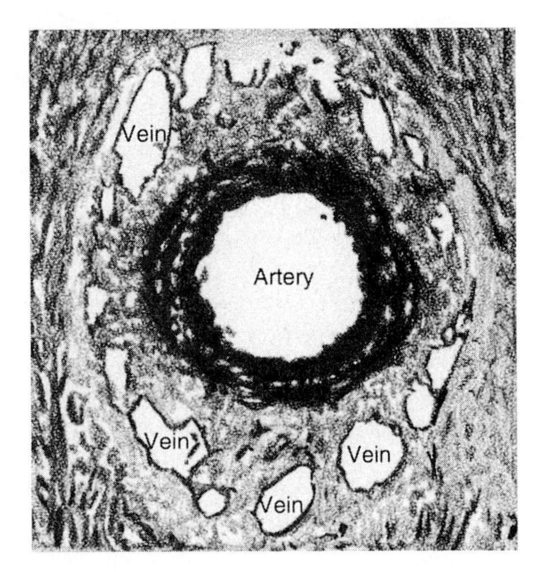

FIGURE 3.30 Section of the tail fluke of a bottle-nosed dolphin (*Tursiops truncatus*), showing the arrangement of veins surrounding an artery (25×). (From P. F. Scholander and W. E. Schevill, *Journal of Applied Physiology*, Vol. 8, p. 281, 1955. reprinted by permission.)

the blood in the arteries can be absorbed by the cooler blood returning in the veins before it is lost to the external water through the thin flesh of the outer extremities (Figure 3.30). This is a countercurrent system of circulation designed to save heat.

Because most of their adaptations are designed to conserve body heat, marine mammals (especially pinnipeds) may, on occasion, become too warm. Hot, still days may mean considerable stress from overheating. On these uncommon occasions, the animals must act to dissipate heat. They do this by waving their flippers in the air while increasing the blood flow to the extremities and restricting the flow back to the core through the veins. The result is greater heat loss and subsequent cooling. Seals and sea lions may also open their mouths and pant like a dog.

Most nektonic marine mammals (particularly the pinnipeds and the whales) regularly dive, or are capable of diving, to depths far greater than that of humans. Whereas exceptional humans may be capable of free dives as deep as 60 m with a breath-hold duration of 6 minutes, various seals and porpoises are known to dive to depths of 160–1,800 m and hold their breaths for 6–48 minutes. The champion is the sperm whale, which can dive as deep as 2,250 m and hold its breath for 80 minutes. How is it possible to do this? The answer rests with a number of anatomical and physiological adaptations.

One of the first questions to be asked regarding these prolonged deep dives is how do these animals avoid the bends? The **bends** is a serious affliction of human divers in which the nitrogen breathed under pressure in deep water as part of their air supply bubbles out of solution in the blood when a diver ascends too quickly. These bubbles then lodge in the joints, brain, and lungs, causing paralysis and even death. The key here is that humans get the bends due to breathing gas *under pressure*, as from a scuba tank. Since marine mammals do not inhale pressurized gas, but depend on gas inhaled at the surface at regular atmospheric pressure, this condition has less chance of occurring. The chance of its occurrence is further reduced by certain other adaptations, the most important of which is the collapse of the lungs. In a deep dive, the pressure of the outside water causes the gas exchange tissues of the lungs of the diving mammal to collapse. This collapse forces the residual gas in the absorptive area of the lung into the nonabsorptive cartilaginous air passageways; thus, diffusion of nitrogen into the blood stops. With no nitrogen entering the blood, there can be no nitrogen to bubble out when the animal surfaces. This collapse of the lungs is aided by the fact that diving marine mammals have few ribs that are attached to the sternum, and the sternum is shortened. The rib cage, therefore, is easily pushed in. Finally, it has also been suggested that the peculiar "foam" of emulsified fat droplets and mucus that many whales have in their respiratory passages absorbs the nitrogen gas, so it cannot enter the bloodstream where it could cause the bends. Recently Williams et al. (2000) have documented that the collapse of the lungs during a dive reduces the animals' volume and hence buoyancy making them negatively buoyant and allowing them to descend rapidly without expending any energy.

A second problem with deep diving is to explain how the animals can survive so long without access to a supply of oxygen. A consideration of this problem will suggest that the only mechanisms for surviving are for the animals to store more oxygen than nondiving forms do and to conserve carefully the amount they have. Since the major organ system involved in oxygen transport is the circulatory system, it is there we may look first for adaptations.

Many diving birds and mammals have a larger blood volume than their terrestrial relatives. The elephant seal, according to Elsner (1969), has a blood volume of 12% of its body weight, whereas humans have a blood volume of only 7% of body weight and domestic dogs only 9%. This larger volume allows more oxygen to be held in the body. This oxygen capacity is increased further, as Scholander (1940) notes, because marine mammals also possess a higher oxygen capacity per unit volume of blood than terrestrial mammals (40 ml O_2/100 ml blood in elephant seals versus 16–24 ml O_2/100 ml in humans), but it is not as great as might be expected. Neither the increased blood volume nor the increased O capacity is sufficient to account for the ability of pinnipeds and cetaceans to remain underwater for extended periods. For example, harbor seals have about the same weight as humans and possess about twice the amount of oxygen in the form of increased blood volume and oxygen capacity. However, rather than being able to submerge twice as long as humans, they can stay under five to ten times as long (Figure 3.31). Other adaptations must also be at work.

Marked slowing of the heartbeat during the period of submersion is one of these additional adaptations. This **bradycardia** is common to all diving, air-breathing vertebrates. The decrease in heartbeat is quite dramatic. For example, in the Pacific bottle-nosed dolphin, *Tursiops truncatus*, the heartbeat under experimental conditions drops from about 90 beats per minute at the surface to about 20 beats per minute during a 5-minute dive (Figure 3.32).

Of more importance than slowing the heartbeat, however, are two other adaptations. First, the circulatory system cuts off the blood supply to various organs and organ systems during a dive, including the muscles, digestive system, and kidneys. This cutoff conserves the limited oxygen supply in the blood for the more sensitive and vital tissues, such as the brain and central nervous system. Thus, a limited supply of oxygen lasts longer, as it is used only by selected organs. In this way the animal may remain submerged much longer than if it had to supply oxygen to all its tissues. The second adaptation is related to the first: The muscular system and other organs are extremely tolerant of anaerobic conditions and continue to function when the blood flow is cut off. This results in the buildup of large amounts of lactic acid in the muscles during a dive. Lactic acid is the end product of anaerobic metabolism. The muscular system can continue to function in the absence of a blood supply not only because it is very tolerant of lactic acid and anaerobic conditions, but because of yet another adaptation. The muscular system of marine mammals is rich in an oxygen-containing compound called **myoglobin**. This compound has a structure very similar to hemoglobin, but it is better at storing oxygen. When the animal dives, the muscle blood supply is shut off. The muscles initially have a large supply of oxygen in the myoglobin, but as the dive continues, the oxygen of the myoglobin is depleted, and the muscles continue to operate anaerobically, building up lactic acid. It now appears, however, that few dives are of a duration that exceeds the aerobic capacity of the animals.

FIGURE 3.31 A harbor seal (*Phoca vitulina*) resting underwater on a rock. (Photo courtesy of Drs. Lovell and Libby Langstroth.)

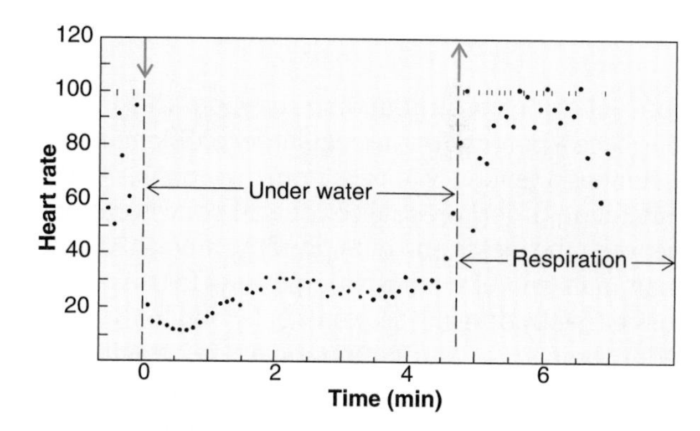

FIGURE 3.32 Change in heart rate in the bottle-nosed dolphin (*Tursiops truncatus*) during diving. Arrows indicate the beginning and end of the dive. (Reprinted by permission of *Nature*, Vol. 212, No. 407, p. 23. Copyright © 1966 Macmillan Magazines, Ltd.)

When a dive has been completed and the animal returns to the surface and begins breathing again, the blood is recharged with oxygen, the blood supply is restored to the muscles and other organs, and oxygen replenishes the myoglobin. The blood picks up the accumulated lactic acid, where it is eliminated by oxidation in the liver. At this point, the animal is ready to dive again.

Marine mammals and birds have salt concentrations in their blood and body fluids lower than

FIGURE 3.33 Albatross (*Dionedea exulsans*) showing the tubular nostrils. (Photo courtesy of Johnny Johnson.)

concentrations of surrounding seawater. This means they face a potential osmoregulatory problem in that water has a tendency to pass from their bodies to the outside in order to equalize the salt concentrations between the inside and outside of the animal. Marine mammals and birds must compensate for this water loss either by drinking seawater or by obtaining water from their food. Marine mammals derive most of their water from metabolic breakdown of their prey. If they drink seawater, they gain an unwanted quantity of salt, which they then must get rid of in some way. Marine birds and turtles eliminate this excess salt through special salt-secreting glands in the orbit of the eye. Marine mammals, however, do not have such glands, nor do they have sweat glands as do terrestrial mammals. The only organ left to them for elimination of salt is the kidney. Excess salt must be removed via the kidneys and washed out of the body with large amounts of water. The urine is very concentrated, and the large multilobular kidneys are very efficient at removing salts without losing much water.

As noted previously, we now know from satellite tracking of certain albatross that these birds regularly travel immense distances to forage for food for their chicks (see p. 118). This has raised the question of how they, and related procellariiform seabirds such as petrels and shearwaters, find their patchily distributed food sources, mainly krill, squid, and fish, in a vast and featureless ocean. The first clue to the answer is that these birds possess an anatomically elaborate sense of smell, which in contrast is poorly developed in most birds. This includes a very elaborate olfactory epithelium housed within a tubular-shaped nose as well as large olfactory bulbs in the brain that relay information to higher centers. Because of the tubular nature of the nostrils, these birds are also called "tube-nosed" seabirds (Figure 3.33).

Nevitt (1999) has demonstrated that at least some of these birds respond quickly to the presence of the gas dimethyl sulfide (DMS) in the air. It is known that when krill feed upon phytoplankton, their prey release dimethyl sulfide, which can persist for hours or even days. The presence of krill in turn tends to attract their predators, which may in turn be preyed upon by the birds and other predators. Since krill occur in patches that are unpredictable in the vast reaches of the oceans, a mechanism that would allow them to be detected from large distances would make foraging more certain and less wasteful of energy. Nevitt (1999) suggests that some of the tube-nosed birds find these patches by detecting the presence of DMS, or perhaps other specific olfactory clues, in the air.

ECOLOGY OF NEKTON

As we noted in the introduction, because of the difficulties of observation and experiment with natural populations of nekton, we know very little about the ecology of these animals. Most of our information concerns the feeding relationships and probable trophic links.

Feeding Ecology and Food Webs

Basically, all adult nekton are carnivores preying on either the smaller plankton or other nekton. The largest number of nekton are predators on other nekton. The plankton feeders consume the larger zooplankton and include such fishes as flying fish and sardines (Clupeidae), but the best-known plankton feeders are the baleen whales. Baleen whales include all of the larger whales, such as the blue whale, the largest animal that has existed on this planet. These tremendous animals have large mouths devoid of teeth; they have instead sheets of **baleen** or whalebone, which hang down like a curtain from the roof of the mouth cavity. Baleen consists of a series of closely spaced, parallel plates fringed on the free edge. These plates form a sieve mechanism. In right and bowhead whales, the mouth is very large with long, finely fringed baleen that permits them to sieve out small copepods and euphausiids (krill) from large amounts of water while slowly moving forward. In the more streamlined rorqual whales, which range in size from the 9-m minke to the 25-m blue, the

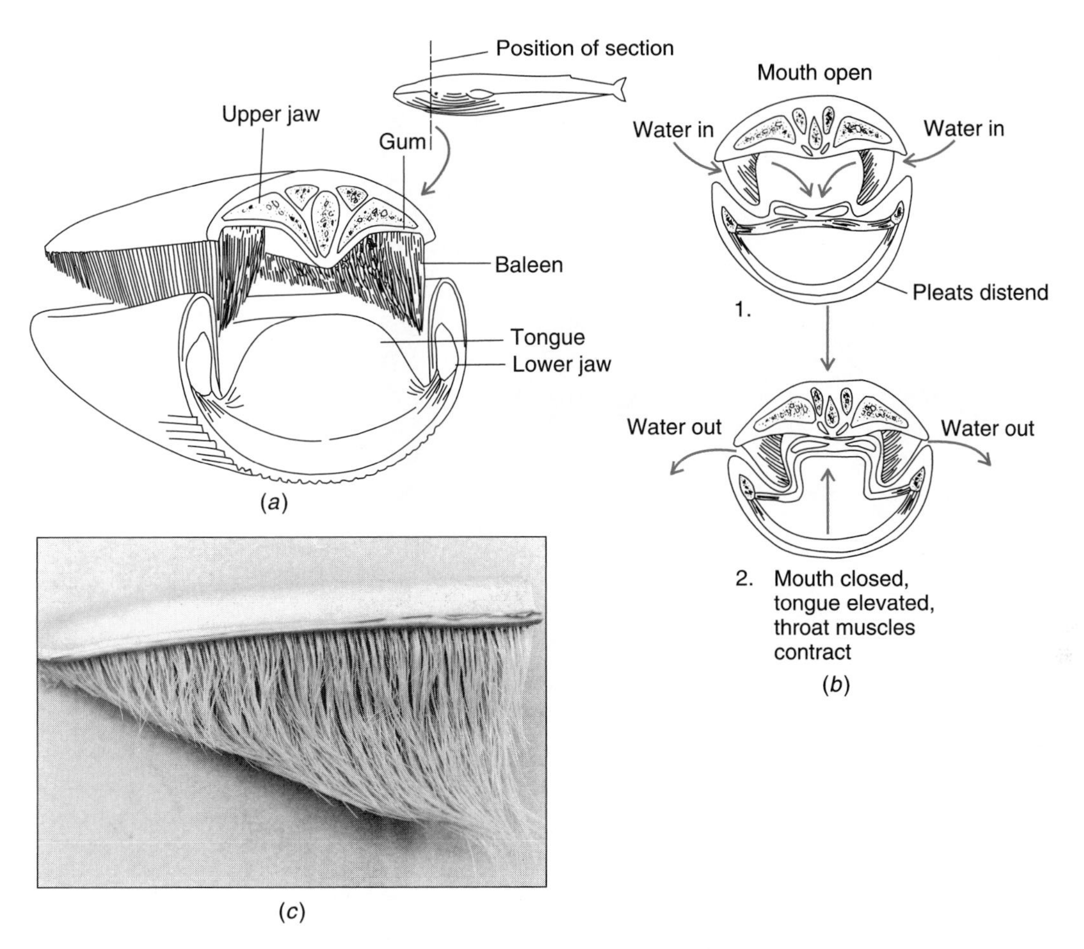

FIGURE 3.34 (a) Cross section of a baleen whale head, indicating the functional anatomy of the feeding apparatus. (b) The feeding process in a baleen whale: (1) mouth open and tongue depressed, water and plankton flow in; (2) mouth closed, tongue raised, water passes out through the baleen, which strains out the plankton. (c) Photograph of the baleen from a gray whale (*Eschrichtins robustus*). (a, b modified from *Whales*, E. J. Slijper, © 1979 Hutchinson & Company, Random House, Ltd.)

feeding is somewhat different. These animals have relatively smaller mouths with short baleen. They can, however, open their mouths to a greater than 90° angle between the upper and lower jaw. These animals feed by lunging into masses of krill or schools of small fishes with wide open mouths, taking in large volumes (up to 70 tons) of water and prey. The large volume distends the accordion-like furrows or pleats in the throat. The animals then contract the throat muscles and push the water out through the baleen while keeping the mouth partially closed. Prey are trapped against the baleen (Figure 3.34). Baleen whales are not the only marine mammals to feed on krill; the crab-eater seal of the Antarctic does so also, but it filters out the krill on the cusps of its peculiarly shaped teeth (Figure 3.35). Penguins are also important consumers of krill.

One exception to this plankton feeding among baleen whales is the gray whale (*Eschrichtius robustus*), which is a bottom feeder. Gray whales dive to the bottom and take in large quantities of sediment. They filter this sediment through their baleen and retain small invertebrates, such as amphipods. Gray whale feeding leaves large, irregular depressions on the shallow seafloor, and the whales are important disturbance forces of these areas.

Plankton-feeding fishes are varied and include such common forms as flying fish, salmon, and the largest fishes, the whale sharks. As with the baleen whales, the dominant food organisms are various

FIGURE 3.35 The teeth of the Antarctic crab-eater seal (*Lobodon carcinophagus*) used to filter euphausiids. (Photo courtesy of AIBS from September 1964 cover of *BioScience*. Used by permission of Dr. V. B. Scheffer.)

crustaceans of the zooplankton, including euphausiids, copepods, and amphipods. The type of zooplankton taken varies among different areas of the ocean, as well as with different seasons. In the case of the Pacific Ocean salmon, the diet varies considerably, not only with season, but also among different years. There is some evidence that this change in diet with years is related to the competition for food among the several salmon species. When there are years of high abundance of certain salmon, the intensified competition leads to an expanded range of food items taken and to changes in the composition of the diet. For example, Parin (1970) notes that in 1956, when pink salmon (*Oncorhynchus gorbuscha*) were low in numbers, 73.2% of the diet of chum salmon (*O. keta*) consisted of euphausiids. In 1957, however, when pink salmon were abundant, chum consumed ctenophores and jellyfish (72.6% of their diet). It does not appear that planktivores are specialized on any one kind of plankton; rather, they are size selective.

Whereas few of the vertebrates of the epipelagic zone of the oceans are plankton feeders, there are a large number of small fishes living in the mesopelagic zone (see Chapter 4) that are planktivores and that migrate into the epipelagic at night to feed. Thus, the competition for plankton may be greater than first anticipated, but our understanding of these relationships is poor at present. These same small fishes also form a food source for some of the carnivorous fishes of the epipelagic.

The nekton-feeding fishes, birds, and mammals dominate the open-ocean nekton. Their diet generally includes fishes, squids, or large crustaceans. The size of prey taken generally depends on the size of the predator, with the larger species taking progressively larger prey species. The largest carnivore in the oceans is the sperm whale (*Physeter macrocephalus*), which preys upon the largest squid, diving deeply to obtain them. Apparently, the sperm whales are the only animals preying upon the giant squid, *Architeuthis*. (The stomach of one sperm whale yielded a giant squid 10.5 m long and weighing 184 kg.) The second largest carnivore of the nekton is the killer whale (*Orcinus orca*), which preys upon fishes, penguins, porpoises, seals, and sea lions (see Figure 3.14). These animals have also been known to attack the much larger baleen whales, which they do in packs of 3–40. In such attacks on gray whales (*Eschrichtius robustus*), the killer whales usually go after the tongue of the whale. The capacity of these animals is truly enormous; one specimen, reported by Slijper (1976), was captured with parts of 13 porpoises and 14 seals in its stomach.

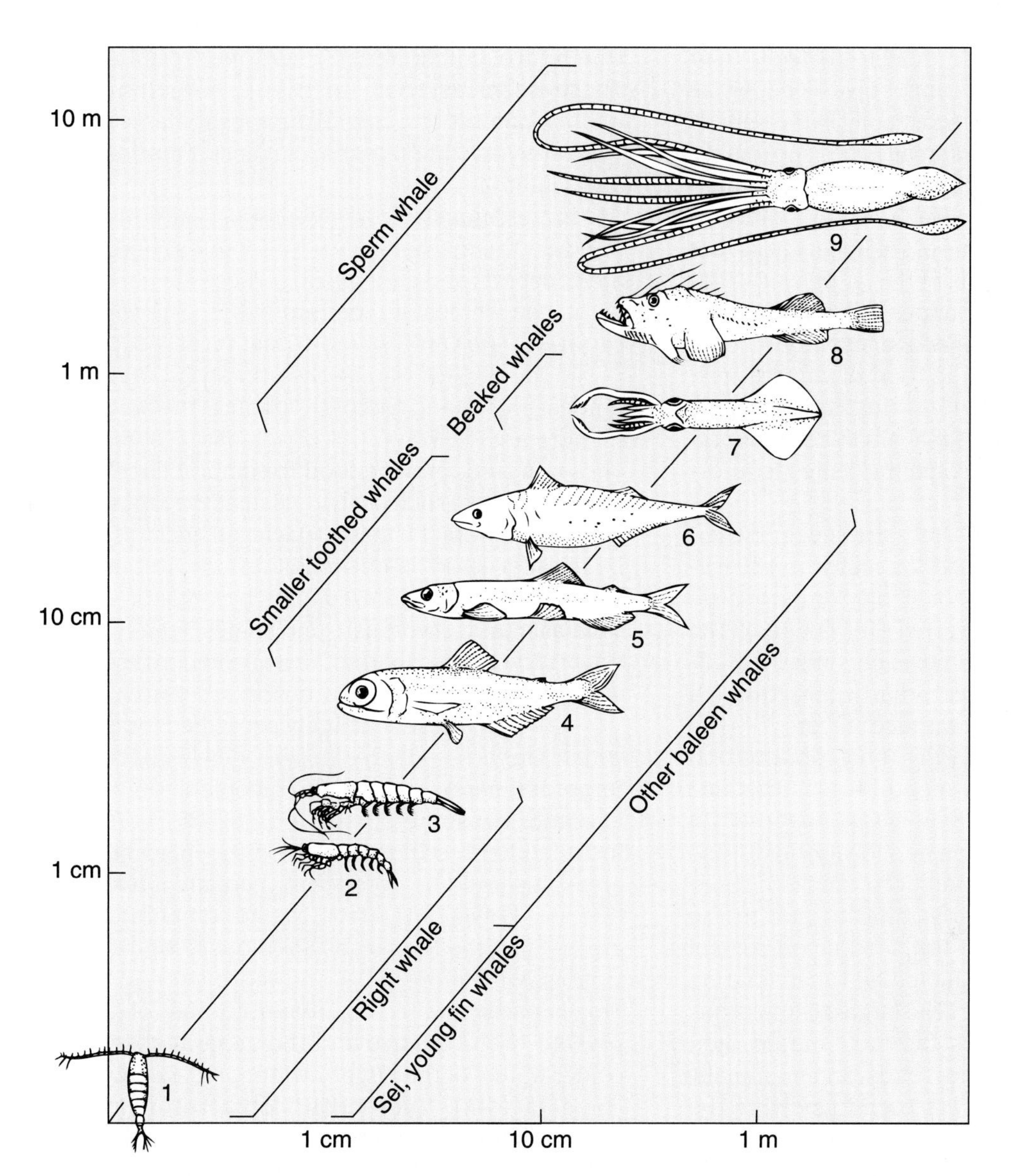

FIGURE 3.36 Size ranges of preferred food items of cetacean groups on a logarithmic scale: (1) calanoid copepods; (2, 3) various euphausiids; (4) myctophid (lantern) fish; (5) capelin; (6) mackerel; (7) onychoteuthid squid; (8) bathypelagic angler fish; (9) architeuthid squid, sporadically eaten by sperm whales. (From *The Ecology of Whales and Dolphins*, D. Gaskin. Reprinted by permission of Heinemann Educational Publishers, a division of Reed Educational & Professinal Publishing Ltd.)

The smaller-toothed whales, seals, and sea lions tend to feed on various squids and fishes, which they capture and swallow whole. Both groups have been attacked by humans because of the belief that they feed heavily on such commercially important fishes as salmon and, therefore, compete with humans. Although they take commercially important fishes, it remains to be proven that such fishes constitute a significant part of their diet. The truly fish-eating whales include dolphins of the genera *Delphinus*, *Stenella*, *Tursiops*, and *Lagenorhynchus*. Examination of the stomachs of five species of delphinids revealed a diet composed of 90% of mesopelagic lantern fish. These dolphins have a short, slender rostrum with many teeth. Those porpoises feeding on squid tend to have reduced numbers of teeth, but increased ridges in the palate to hold slippery squid. Included with squid eaters are the rare beaked whales (Ziphiidae) and the pilot whales. The types of food and size ranges for most whales are summarized in Figure 3.36.

Pelagic birds are generally more restricted in their feeding areas, because they cannot dive as deeply as the marine mammals. They feed mainly in the surface waters. They consume various small fishes and squids. Since certain species often occur in tremendous numbers, it may be that they have a significant effect on the fish and squid populations of the upper layers, but we are ignorant of the details of this at present. They may also be competing with various fishes and smaller marine mammals for the same food resource.

We know most about the food of various nektonic fishes. Of these fishes, the commercially important species, such as tuna, albacore, and marlin, are best known. They consume fishes, squids, and crustaceans.

Tunas have been intensively studied, and their food is extremely varied, with as many as 180 different food items reported for skipjack tuna (*Katsuwonus pelamis*). Part of this varied diet is due to the great areas over which they move, encountering different arrays of food items. Small tunas tend to feed on species in the surface layers of the epipelagic zone, whereas larger tunas obtain food from depths where they capture mesopelagic organisms. As with planktivores, the composition of the diet varies with season and location. Marlins and swordfish have diets similar to tunas.

Nektonic sharks have a significant role in the feeding relationships of nekton, feeding on several different species of fishes, including tunas, marine birds, mammals, and squids. The largest carnivorous shark, the white shark, has the most varied diet, which includes not only fishes, birds, and mammals, but also other sharks. One was even found with the remains of a basking shark in its stomach.

One of the most consistent features of feeding in nektonic fishes is the general lack of selectivity or specialization. Most fishes feed on any food present of the appropriate size.

With the limited information now available, it is possible to outline the food webs in the open epipelagic zones in the oceans of the world, but only in a very simplified and general way, as much more study is needed to define all of the links and their relative importance. These food webs are different among the tropical and polar oceans (Figure 3.37). In such cold temperate oceans as the North Pacific, the main net phytoplankton are diatoms, which are consumed by various zooplankton. These serve as food for plankton-feeding fish, mainly salmon and migrating mesopelagic forms. Top carnivores include the sharks and marine birds and mammals.

In the Antarctic, primary production is also centered on the diatoms, and the foremost herbivores are copepods and euphausiids. These are consumed by baleen whales and certain fishes. The top-level carnivores are various marine mammals.

In the tropical waters, the primary producers are dinoflagellates and coccolithophores. The herbivores include a wide variety of zooplankton organisms. These, in turn, provide food for a number of planktivorous fishes, such as flying fish at the surface and lantern fish from the mesopelagic. All are food for the larger, first-level predatory fishes and squids. This level of predators is preyed upon by larger predators, such as marlin, swordfish, and sharks. Finally, at the top are the largest sharks, the white and mako, which attack tuna, swordfish, and marlin.

Although these are highly simplified food webs, which do not include all the newly discovered important nanoplankton and picoplankton producers and consumers (see pp. 82–85), they are included here to illustrate the differences between areas. It should be noted that there are marked differences between the tropics and temperate-polar areas. First, the food webs are more complex in the tropics, with more links and more trophic levels. This is partly due to the greater number of species present in the tropics and the general absence in colder waters of the larger, swifter predaceous fishes (tuna, marlin). The second observation is that marine mammals and birds play a larger role in the food webs of the polar regions than the tropics. Indeed, larger numbers of marine mammals occur in colder seas than in warmer seas.

Ecological Significance of Marine Mammals

Although there is considerable interest by people and the media in the plight of the world's whales and dolphins and their past and future decimation by humans, there is little scientific information about their ecological importance in either pelagic or inshore ecosystems. As indicated earlier in this chapter, the importance of many, if not most, nektonic

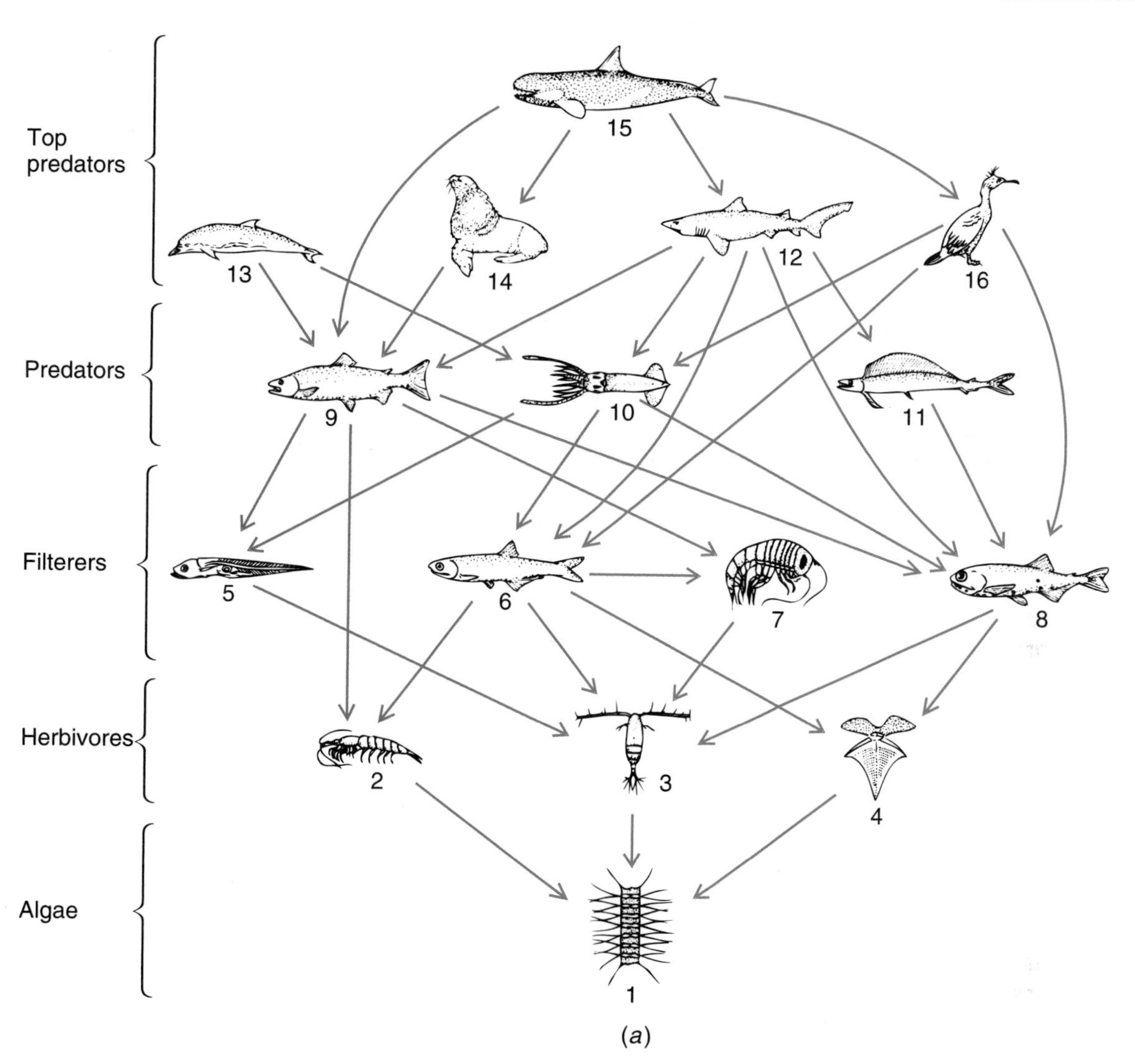

FIGURE 3.37 Food webs and trophic structure of the pelagic community in three different geographical areas. (a) Cold temperature waters. Algae: (1) diatoms. Herbivores: (2) euphausiids; (3) copepods; (4) pteropods. Filterers: (5) juvenile fish; (6) anchovy; (7) hyperiid amphipods; (8) lantern fish. Predators: (9) salmon; (10) squid; (11) lancet fish. Top predators: (12) mackerel shark; (13) porpoise; (14) seals and sea lions; (15) killer whale; (16) marine birds. **(CONTINUES ON PAGES 128 AND 129)**

organisms in the dynamics of the open ocean is hard to determine. Preliminary investigations, as reviewed by Katona and Whitehead (1988), suggest that although the productivity of cetaceans is low, the biomass is large enough to be significant on an ocean-wide basis. This significance can be appreciated by noting that it has been postulated by Kanwisher and Ridgway (1983) that cetaceans probably consume more prey than the entire world's fisheries. If sustained, this suggests that the energy flow through the cetacean component of the open ocean food webs may be extremely important. This is particularly true for such areas as the Antarctic seas, where Kanwisher and Ridgway (1983) estimated that the unexploited

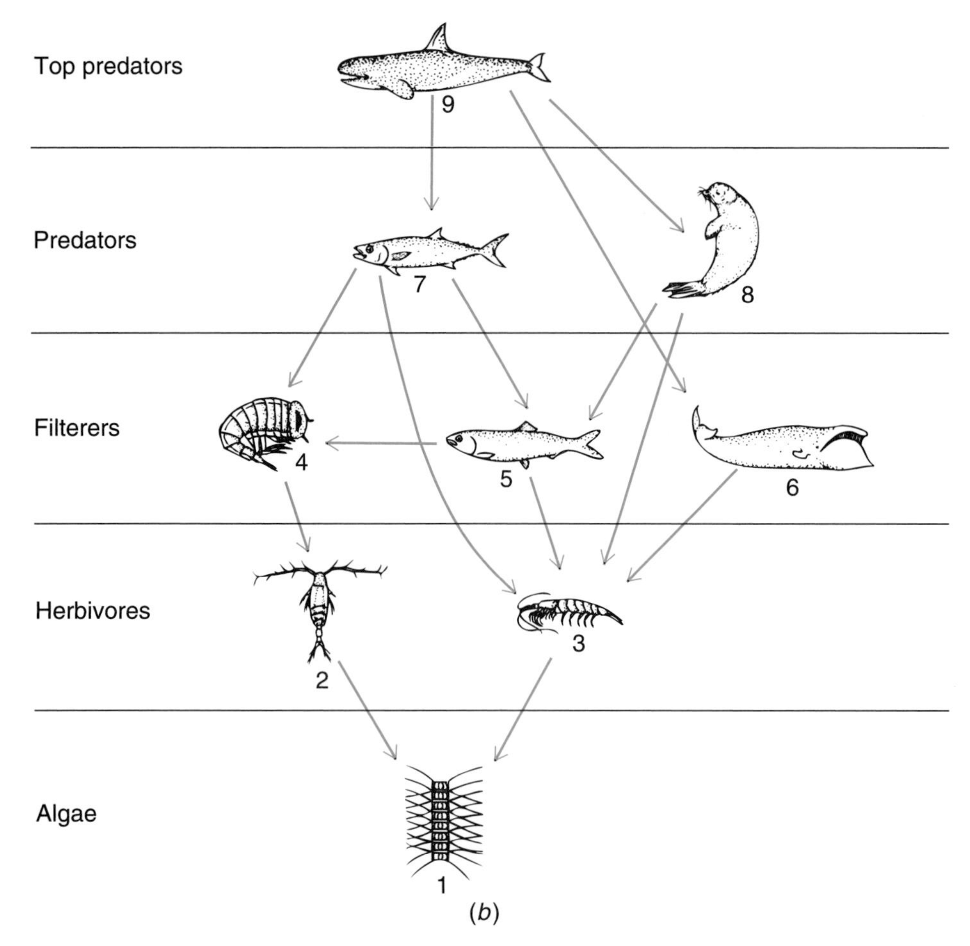

FIGURE 3.37 (CONTINUED) (b) Antarctic seas. Algae: (1) diatoms. Herbivores: (2) copepods; (3) euphausiids. Filterers: (4) hyperiid amphipods; (5) planktivorous fish; (6) baleen whales. Predators: (7) predatory fish; (8) seals and sea lions. Top predators: (9) killer whale.

baleen whale population there consumed 190 million metric tons of krill (*Euphausia superba*) per year, about two times the current world fisheries catch (see Chapter 11). Now that the numbers of these Antarctic whales have been severely reduced (by perhaps 90%) by hunting, evidence to sustain their importance has come by looking at the results of this unparalleled overexploitation. Populations of Antarctic birds and pinnipeds tripled following the destruction of the whales. The reason for this population explosion is simply suggested to be the availability of krill not consumed by the whales.

Whales are important also in that their carcasses sink rapidly and provide a significant food resource for deep-sea benthic creatures (see following chapter). Gray whales disturb large areas of the subarctic Bering and Chukchi seas on a scale equivalent to large geological or meteorological forces (see Chapter 5). Finally, whales produce a considerable amount of the sound in the world's oceans, and it is possible that such signals may be used as cues by other species. Clearly, while whale removal from the ocean ecosystem would probably have little effect on the primary productivity or the total biomass of organisms, it would certainly alter some systems, perhaps more than we realize. Certainly, the richness of our human lives would be much decreased.

The ecological significance of other marine mammals in various communities is less well understood, but suggestive. Preen (1995) has hypothesized that

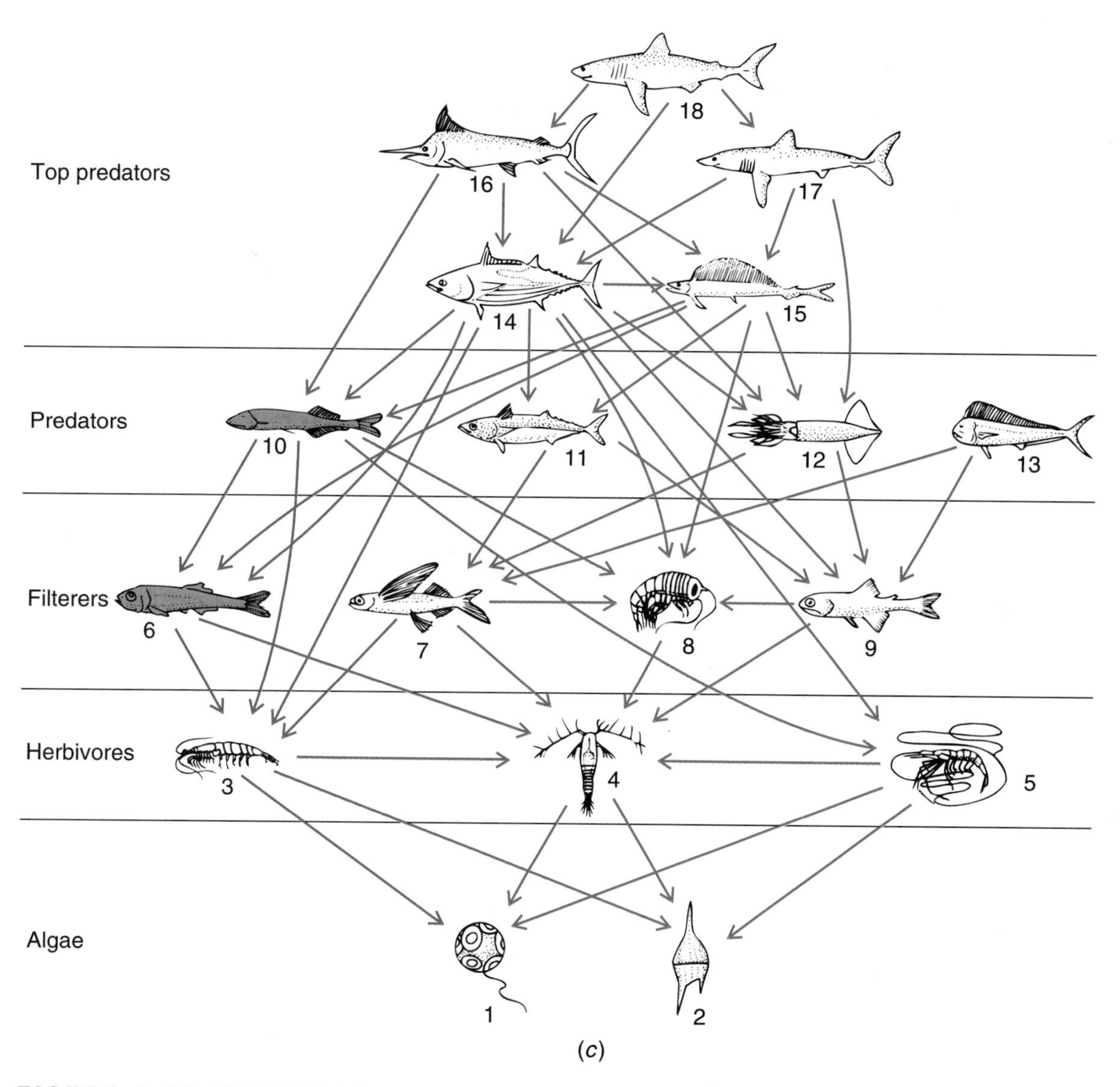

FIGURE 3.37 (CONTINUED) (c) Tropical seas. Algae: (1) coccolithophores; (2) dinoflagellates. Herbivores: (3) euphausiids; (4) copepods; (5) shrimp. Filterers: (6) vertically migrating mesopelagic fishes; (7) flying fishes; (8) hyperiid amphipods; (9) lantern fish. Predators: (10) mesopelagic fish (11) snake mackerel; (12) squid; (13) dolphin (*Coryphaena*). Top predators: (14) tuna; (15) lancetfish; (16) marlin; (17) medium-sized sharks; (18) large sharks. (a–c, from *Icthyofauna of the Epipelagic Zone*, N. V. Pari, translated from Russian by the Israel Program for Scientific Translations.)

dugong grazing on seagrass beds in Australia reduced seagrass biomass by up to 96% and prevented the expansion of the dominant species. It has also been suggested that the severe reduction of whales and several species of fishes in the Bering Sea that were major predators of krill and other zooplankton led to the release of krill for other species. This, coupled with other environmental changes, may have led to the fish assemblage in the area becoming dominated by the less nutritious pollock (*Theragra chalcogramma*). It has thus been hypothesized that because of the decline of the more nutritious forage species of fishes, the populations of harbor seals (*Phoca vitulina*) and Steller's sea lions (*Eumetopias jubatus*) have suffered

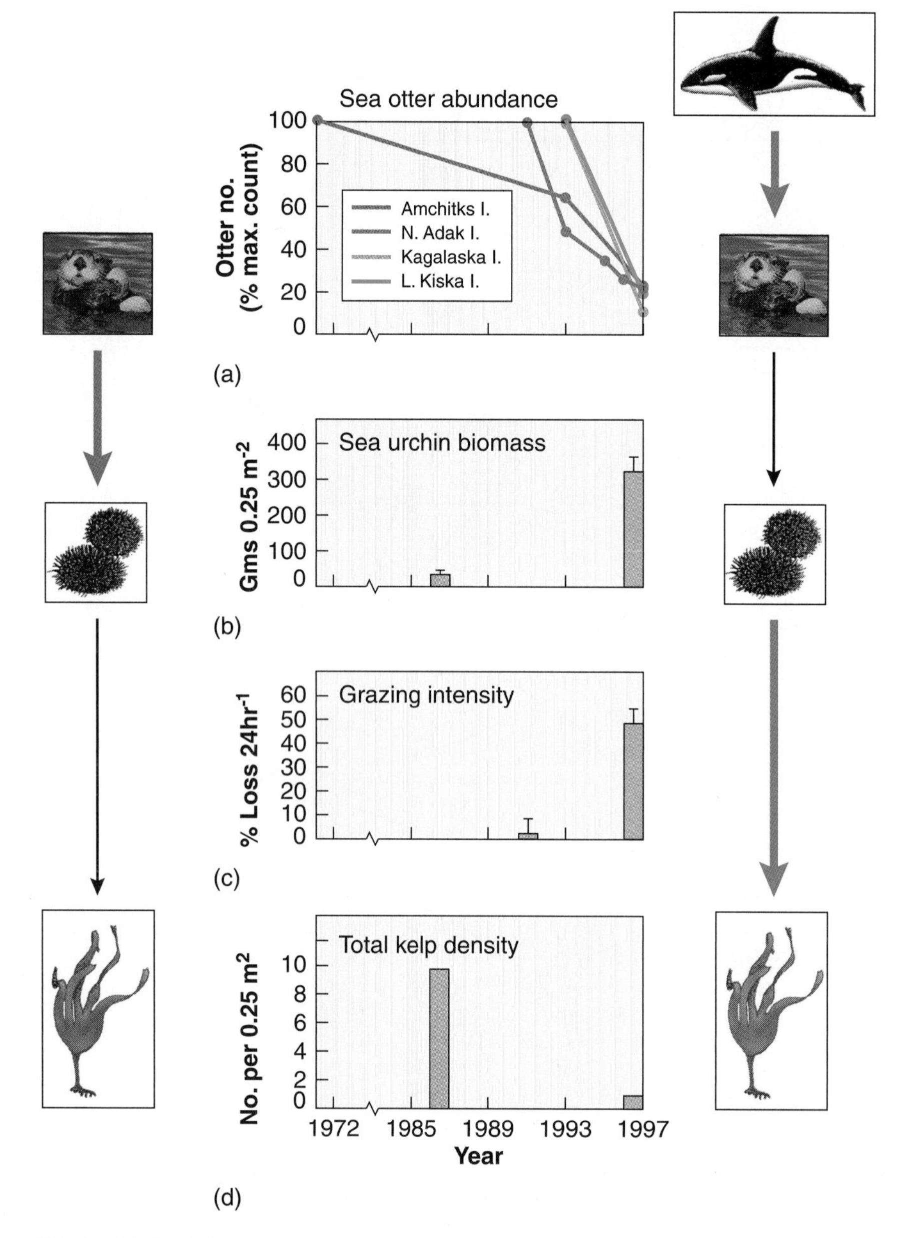

FIGURE 3.38 (A) Changes in sea otter abundance over time in the Aleutian Islands with concurrent changes in (B) sea urchin biomass, (C) grazing intensity, and (D) kelp density. The marginal diagrammatic representations show the proposed mechanisms of change. The one on the left shows the organization of the kelp forest system before the decline of the sea otter, and the one on the right indicates how the system changed with the advent of the killer whale as the terminal predator. Heavy red arrows represent strong and light black arrows weak interactions. (Reprinted with permission from "Killer Whale Predation on Sea Otters Linking Oceanic and Nearshore Systems" J. A. Estes et al., *Science,* Vol. 282, pp. 473–473, 1998. Copyright © 1998 American Association for the Advancement of Science.)

sharp declines recently from lack of food. Estes et al. (1998) have extended this hypothesis a step further by linking the recent declines in the sea otter (*Enhydra lutris*) populations in the western Aleutian Islands to predation by killer whales (*Orcinus orca*), never before known to feed on otters, because their main food source of seals and sea lions has declined so dramatically (Figure 3.38).

SUMMARY OF KEY CONCEPTS

- Nekton includes all organisms capable of sustained locomotion against water motion.
- Nekton are difficult to study in nature and difficult or impossible to keep in captivity.
- The oceanic nekton comprises a wide range of fishes and a lesser number of mammals, reptiles, birds, and cephalopod mollusks.
- Environmental conditions significant to nekton include the three-dimensionality of the medium, transparency of the medium, lack of shelter from predators, and the absence of physical boundaries or structures.
- Buoyancy adaptations of nekton include the presence of a gas bladder in most fishes, air sacs and blubber in marine mammals, air channels in the bones of marine birds, lipid-filled livers in sharks, and hydrodynamic adaptations to provide lift.
- Adaptations to increase locomotory ability fall into two groups: those serving to create the propulsive force and those reducing the resistance of the body to passage through the water.
- Most fishes, whales, and sea snakes propel themselves using undulating movements of the body, whereas pinnipeds, turtles, and birds paddle using the limbs.
- To reduce drag and therefore increase locomotory speed, most nekton have teardrop-shaped bodies and reduced forms of those projections from the body that would break the smooth flow of water over the body.
- Defense against predators in nekton includes camouflage through color patterning and through alteration of body shape to eliminate or reduce shadows.
- Nektonic animals tend to have well-developed sensory systems, and most information comes to them via light (vision), sound (hearing), or chemicals (olfaction).
- Sound reception has generated the greatest number of specialized adaptations and has been developed most highly in cetaceans, who use echolocation or sonar to navigate and locate and identify food.
- Most pelagic bony fishes produce immense numbers of eggs that float and are subject to equally huge losses; in contrast, pelagic sharks produce few large eggs that hatch into large juveniles that are much less subject to predation.
- Marine birds and reptiles retain the reproductive characteristics of their terrestrial relatives producing shelled eggs laid on land.
- Whales give birth in the water; pinnipeds give birth on land. The young of whales and pinnipeds both grow very rapidly, sustained by milk that is very high in fat content. All marine mammals appear to be long-lived.
- Many nekton undertake extensive migrations for breeding or feeding. The means of navigation in marine mammals and turtles appears to be linked to an ability to detect the earth's magnetic field.
- Marine birds and mammals have special adaptation to survive in cold water. These include fat layers to reduce heat loss and circulatory systems that conserve heat.
- Many marine mammals are deep divers. The adaptations to permit such diving include collapse of the lung, large amounts of myoglobin, large blood volume, bradycardia, cut off of blood to certain tissues and organs, and great tolerance of anaerobic conditions.
- Most nekton are carnivores preying upon either the smaller plankton or, in most cases, on other nekton.
- The food webs and trophic structure of the pelagic community differ in different geographical areas of the oceans. In cold temperate waters, the top carnivores are sharks and marine birds and mammals. In Antarctic waters, the top carnivores are various marine mammals. In the tropics, the top carnivores are various fishes.
- The ecological significance of marine mammals in the open-ocean systems is hard to determine, but the demise of the baleen whales in the Antarctic has led to large increases in Antarctic birds and pinnipeds due to the availability of krill not consumed by the whales, which suggests that the whales were important in this ecosystem. Other significant mammals are dugongs and various pinnipeds.

REVIEW QUESTIONS

ESSAY: Develop complete answers to these questions.

1. Compare and contrast feeding in the blue whale with that of the gray.
2. Discuss the ecological significance of the Cetacea in the world's oceans.
3. Why do marine mammals usually produce only one offspring at a time? How can the species survive with such a low fecundity?
4. Discuss why nektonic animals, although possessing very different evolutionary histories, show a strong convergence in morphology.
5. Compare and contrast nektonic organisms with planktonic ones. Which of these groups possesses greater diversity? Discuss.
6. Nektonic organisms live in a profoundly three-dimensional world, but it is not uniform. Discuss how this world is structured vertically.

BIBLIOGRAPHY

Aleyev, Y. G. 1977. *Nekton*. The Hague: Dr. W. Junk BV.

Andersen, H. T., ed. 1969. *The biology of marine mammals*. New York: Academic Press.

Brooks, W. S. 1917. Notes on some Falkland Island birds. *Bull. Mus. Comp. Zool.* 61:135–160.

Carey, F. G. 1973. Fishes with warm bodies. *Sci. Amer.* 228(2):36–44.

Carr, A. 1965. The navigation of the green turtle. *Sci. Amer.* 212(5):79–86.

Clarke, M. R. 1979. The head of the sperm whale. *Sci. Amer.* 240(1):128–141.

Denton, E., and N. Marshall. 1958. The buoyancy of bathypelagic fishes without a gas-filled swim bladder. *J. Mar. Biol. Assoc.* 37(3):753–767.

Dunson, W. A., ed. 1975. *The biology of sea snakes*. Baltimore: University Park Press.

Elsner, R. 1969. Cardiovascular adjustments to diving. In *The biology of marine mammals*, edited by H. T. Andersen. New York: Academic Press, 117–145.

Estes, J. A. , M. T. Tinker, T. M. Williams, and D. F. Doak. 1998. Killer whale predation of sea otters linking oceanic and nearshore ecosystems. *Science* 282:473–476.

Gaskin, D. E. 1982. *The ecology of whales and dolphins*. London: Heinemann.

Hardy, A. 1965. *The open sea: Its natural history*. Vol. II, *Fish and fisheries*. Boston: Houghton Mifflin.

Kalmijin, A. J. 1971. The electric sense of sharks and rays. *J. Exp. Biol.* 55:371–383.

Kanwisher, J. W., and S. H. Ridgway. 1983. The physiological ecology of whales and porpoises. *Sci. Amer.* 248(6):110–121.

Katona, S., and H. Whitehead. 1988. Are cetaceans ecologically important? *Oceanog. Mar. Biol. Ann. Rev.* 26:553–568.

Kellogg, W. N. 1958. Echo ranging in the porpoise. *Science* 128:982–988.

Kirschvink, J. L. 1990. Geomagnetic sensitivity in cetaceans: An update with live stranding records in the United States. In *Sensory abilities of cetaceans, laboratory and field evidence*, edited by J. A. Thomas and R. A. Kastelein. New York: Plenum Press, 639–649.

Klinowska, M. 1990. Geomagnetic orientation in cetaceans: Behavioural evidence. In *Sensory abilities of cetaceans, laboratory and field evidence*, edited by J. A. Thomas and R. A. Kastelein. New York: Plenum Press, 651–663.

Kooyman, G. L., and R. J. Harrison. 1971. Diving in marine mammals. *Oxford Biology Reader*, no. 6.

Lohman, K. J. 1992. How sea turtles navigate. *Sci. Amer.* 266(1):100–106.

Nevitt, G. 1999. Foraging by seabirds on an olfactory landscape. *Amer. Sci.* 87:46–53.

Norris, K. S. 1969. The echolocation of marine mammals. In *The biology of marine mammals*, edited by H. T. Andersen. New York: Academic Press, 391–423.

Norris, K. S., J. H. Prescott, P. V. Asa-Dorian, and P. Perkins. 1961. An experimental demonstration of echolocating behavior in the porpoise, *Tursiops truncatus* (Montagu). *Biol. Bull.* 120:163–176.

Parin, N. V. 1970. *Ichthyofauna of the epipelagic zone*. Jerusalem: Israel Program for Scientific Translations.

Preen, A. 1995. Impacts of dugong foraging on seagrass habitats: Observational and experimental evidence for cultivation grazing. *Mar. Ecol. Prog. Ser.* 124:201–213.

Scheffer, V. B. 1976. *A natural history of marine mammals*. New York: Scribner.

Scholander, P. F. 1940. Experimental investigations on the respiratory function in diving mammals and birds. *Hvalradets Skrifter Norske Videnskaps Akad., Oslo* 22:1.

Seachrist, L. 1994. Sea turtles master migration with magnetic memories. *Science* 264:661–662.

Slijper, E. J. 1976. *Whales and dolphins*. Ann Arbor: University of Michigan Press.

Stewart, B. S., and R. L. DeLong. 1995. Double migrations of the northern elephant seal *Mirounga angustirostris*. *J. Mammalogy* 76(1):196–205.

Walker, M. M., J. L. Kirschvink, S-B. R. Chang, and A. E. Dizon. 1984. A candidate magnetic sense organ in the yellowfin tuna, *Thunnus albacares*. *Science* 224:751–753.

Williams, T. M. et al:, 2000. Sink or swim: Strategies for cost–efficient diving by marine mammals. *Science* 288:133–136.

Würsig, B. 1979. Dolphins. *Sci. Amer.* 240(3):136–148.

Würsig, B. 1989. Cetaceans. *Science* 244:1550–1557.

DEEP-SEA BIOLOGY

The most extensive habitat on the planet inhabited by living organisms consists of the permanently cold, dark waters and bottom of the deep oceans. The shallow waters fringing the continents and islands of the world make up less than 10% of the total area of the world's oceans. The lighted upper waters of all oceans are an even smaller fraction of the total volume of ocean living space. Of the 70% of the planet's surface covered with water, perhaps 85% of the area and 90% of the volume constitute the dark, cold environment we call the deep sea. Considering the entire volume of the biosphere, 79% of it is composed of marine water with depths greater than 1,000 m.

Although this area is the largest habitat on earth, its biology is the least known and explored. This is primarily due to the difficulty of sampling this remote area and to the fact that, until recently, it was virtually inaccessible to humans. Although the recent advent of deep manned and remote-operated submersible vehicles would now allow much of this vast area to be observed by scientists, very little of it has been visited by these vehicles because of the high cost of operating them.

It is important to consider the deep sea, even though this huge area is unlikely to be sampled or observed by most users of this book, because it comprises such a huge amount of the biosphere and because it is a frontier where new and exciting discoveries are being made. It is also vital to study the deep sea because it may become increasingly important as a source of materials and as a dumping site for human wastes.

This chapter outlines our present understanding of the basic principles of adaptations and organization of life forms in this vast and intriguing area.

ZONATION

When we refer to the **deep sea** in this chapter, we mean that part of the marine environment that lies below the level of effective light penetration for phytoplankton photosynthesis in the open ocean and deeper than the depth of the continental shelves (>200 m). The upper part of the deep sea receives some light and has been called the *dysphotic zone* (Figure 4.1). The dysphotic zone is a transition area. The entire area below it is permanently dark. One term for this area is the **aphotic** zone as opposed to the lighted, or **photic**, zone, where all primary production occurs. In tropical waters, the start of the aphotic zone is deeper (~600 m), whereas in temperate waters, it is shallower (~100 m).

Several schemes of zonation for the deep sea have been proposed over the years and reviewed by Menzies, George, and Rowe (1973), but none has been universally acceptable to all scientists working in the

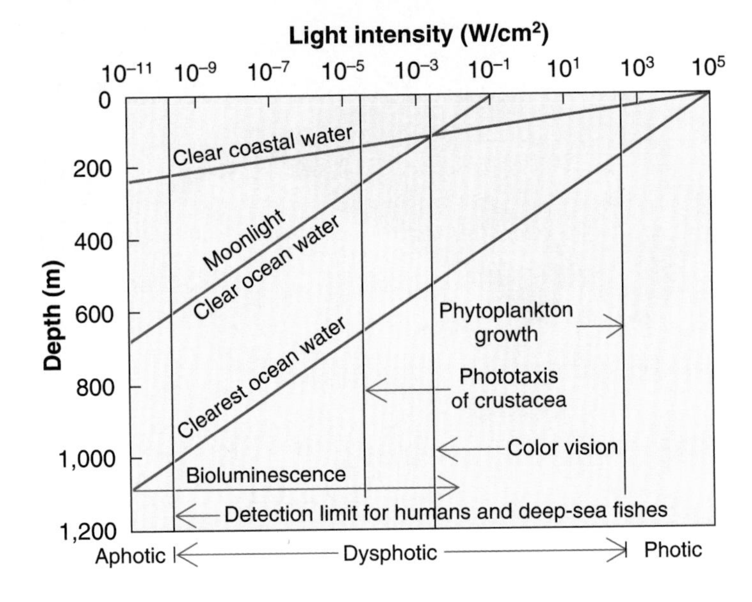

FIGURE 4.1 Schematic diagram showing the penetration of sunlight into the clearest ocean water and into clear coastal water in relation to minimum intensity values for some biological light receptors. (From *Biological Oceanographic Processes*. 3/e T. R. Prsons et al 1984. Copyright © 1984 Butterworth Heinemann. Reprinted by permission.)

area. There are a number of reasons for this, but the primary reason is the lack of sufficient ecological information. Most schemes of zonation have been produced by simply dividing the water column into discrete zones based on depth, or temperature changes, or both. Another approach has been to divide the deep ocean on the basis of species abundance, distribution, or associations. A problem here is that certain species are distributed at different depths in different ocean basins, depending on special hydrological or ecological characteristics of the region. Similarly, the changes in temperature occur at different depths in different areas. On the other hand, kinds of species may be better for indicating zones than the associated environmental parameters. We shall follow the zonation scheme outlined in Chapter 1 and illustrated in Figure 1.22. The relevant deep-sea zones are listed in Table 4.1 and briefly reviewed here.

The deep sea can be divided into two primary areas: benthic (associated with the bottom) and pelagic (associated with the open water) zones. Since the physical environment differs between them, these two areas are inhabited by different associations of organisms. We know more about benthic deep-sea fauna than about pelagic fauna. The benthic fauna, as noted in Chapter 1, can be subdivided into those that occupy a **bathyal** zone on the continental slope and those living in a large **abyssal** zone constituting most of the deep-sea floor. The animals of the trenches live in a separate **hadal** (ultra-abyssal) zone, but little is known of their biology.

In the pelagic division, there is an upper zone immediately below the photic zone, which contains animals, many of which migrate vertically into the photic zone at night. This is usually called the **mesopelagic** zone (= dysphotic, twilight). This zone is inhabited by many species of animals with well-developed eyes and a variety of light organs. The dominant fishes here are black, and the crustaceans are red. Because this is the most accessible of the

TABLE 4.1

Deep-Sea Faunal Zones

Light	*Pelagic Zones*	*Depth Range*	*Benthic Zones*	*Depth Range*
Present (photic)	Epipelagic or photic	0–200 m	Continental shelf or sublittoral	0–200 m
Present (disphotic)	Mesopelagic	200–1,000 m (?)	Bathyal	200–4,000 m (?)
Absent (aphotic)	Bathypelagic	(?) 1,000–4,000 m (?)		
	Abyssalpelagic	(?) 4,000–6,000 m (?)	Abyssal	4,000–6,000 m (?)
	Hadalpelagic	6,000–10,000 m (?)	Hadal	6,000–10,000 m

Note: (?) Boundary uncertain.
Source: After Hedgpeth, 1957.

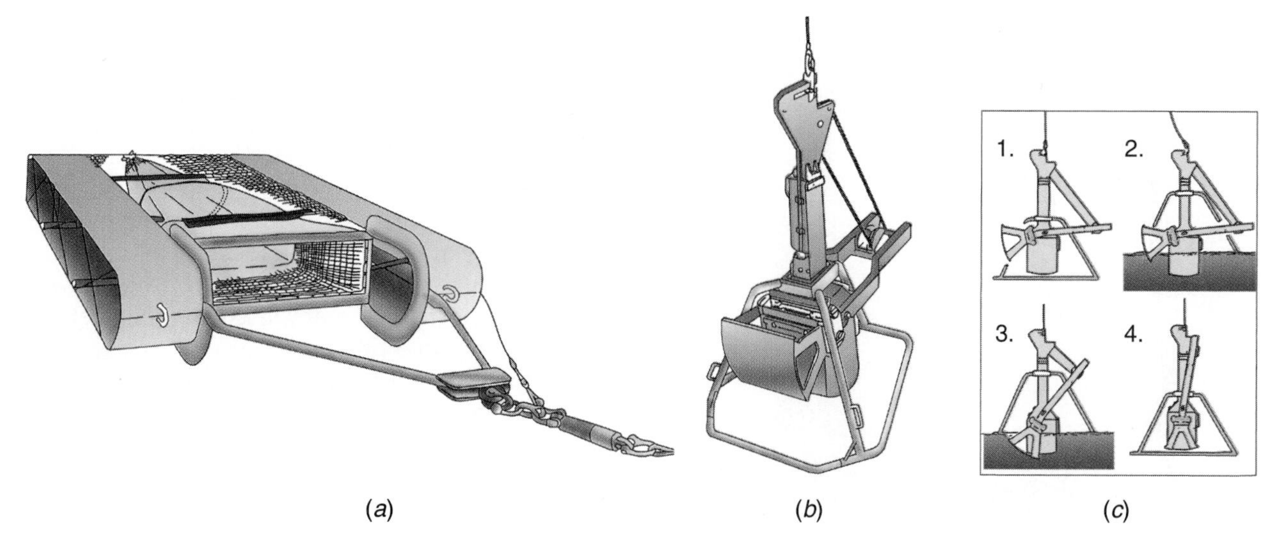

FIGURE 4.2 Two types of deep-sea sampling gear. (a) The epibenthic sled. (b) Box core. (c) Sequence showing the box core taking a sample. (a, from *The Abyssal Environment and Ecology of The World's Oceans*, R. J. Menzies et al., © 1973 John Wiley & Sons. b, c from *Deep Sea Biology*, John Gage and Paul Tyler. Copyright © 1991 Cambridge University Press. Reprinted by permission.)

deep-sea pelagic zones, we know the most about it. The numbers of organisms appear to be the highest of all the deep-sea pelagic zones. This zone extends down to about 700–1,000 m, the depth varying with location, clarity of water, and other factors.

Below the mesopelagic, it is much more difficult to establish any pelagic zones that are universally acceptable. The region between the lower limit of the mesopelagic and the upper limit of the deep trenches (6,000 m) has been divided into two zones by Hedgpeth (1957): an upper **bathypelagic** and a lower **abyssalpelagic**, a division continued here. The boundaries of these zones are uncertain and vary from one authority to another. These areas are characterized by low numbers of species and individuals, considerably fewer than observed in the mesopelagic. The organisms tend to be white or colorless, with reduced eyes and bioluminescent organs.

The open water of the trenches is often called the **hadalpelagic** zone. It appears set off from the other areas, and we are so ignorant of it that we do not know whether a separate pelagic community exists in these areas.

SAMPLING THE DEEP SEA

In the preceding section, we alluded to the problems of sampling and obtaining information about the deep sea. Before continuing this discussion, it is useful to indicate briefly how our information is obtained from the deep sea and why it is difficult to obtain and often biased. An analogy has been made that characterizing the deep sea and its communities using the gear we have traditionally used is like characterizing the terrestrial communities using samples taken with a butterfly net towed behind an airplane! This may be exaggerated, but it is close enough to have a ring of truth.

The traditional types of gear used to sample the deep-sea benthos include various types of nets, grabs, and dredges of similar design to those used in shallow water. Representative types are illustrated in Figure 4.2. Sampling of pelagic animals is done with several types of midwater trawls (Figure 4.3). Because the density of organisms is so low in pelagic areas below the mesopelagic, these midwater nets must have a large gape and sieve a lot of water to obtain any number of animals and to reduce the problem of animals simply avoiding the nets.

The main problem in sampling the deep sea, pelagic or benthic, is that the deeper the sampling, the more cable necessary to get to depth and the longer the time needed to make a single haul. Furthermore, as more and more cable is played out in a deep haul, the greater the weight and the greater the chance for snarling the cable or otherwise fouling the gear. In order for most nets towed on the bottom to fish correctly, an amount of line equaling one and a half to three times the depth must be

FIGURE 4.3 Rigging a large midwater trawl for fishing. (Courtesy of Dr. Greg Cailliet.)

played out. If one were trawling at only medium depths of 4,000 m, this would be about 5,000–6,000 m, or 5–6 km (3–4 miles) of cable. Not only do few vessels have the capability of carrying such amounts of cable, but it takes a long time to feed that amount out and then retrieve it. Twenty-four hours may be required to make one trawl in the trenches. One can thus appreciate why the number of samples from deep water is not great.

Still other problems plague the deep-sea biologist. In the shallow waters of the continental shelf, it is possible to tell whether the trawl or dredge is fishing on the bottom by "feeling" the vibrations of the cable. In the deep sea, so much cable is out that the weight of it far exceeds the drag of dredge or trawl, and vibration in the cable does not reflect the action of the sampling device. Failure of the gear to contact the bottom or to fish correctly was suggested by Menzies (1964) as responsible for up to 50% of deep-sea sample failures. Finally, with so much cable played out, the chance for snarling, knotting, or tangling it in the gear is increased. Most deep-sea sampling today avoids such gear problems by using depth-indicating devices, such as an acoustical "pinger," on the gear with shipboard readout so that the gear can be better controlled.

A major problem in sampling the deep-sea benthos is the basic qualitative nature of most dredging devices. They sample an unknown area of the bottom. Because densities of species and individuals are difficult or impossible to estimate from these samples, comparative studies of deep and shallow water have been difficult. Fortunately, recent investigations in the deep sea have employed large box corers or multicorers, which overcome this problem by sampling a known bottom area (Figure 4.2b).

Finally, in sampling the midwater animals, there is the problem of net avoidance. Trawling in deep water usually means the ship must travel relatively slowly because of the great weight of the cable it drags. This means that many fast-swimming pelagic animals may avoid the net by swimming out of its path. If this happens to any great extent, characterizing the communities from such net hauls gives a misleading idea of what is in fact there.

Since 1970, new instrumentation has permitted deep-sea biologists to consider aspects of the ecology of deep-sea animals that were not possible in earlier years. Deep-sea cameras of various designs have done much to increase our understanding of the real nature of deep-sea communities, particularly bottom communities. The advent of deep manned and remotely operated submersibles with various devices for capturing organisms, including pressurized retrieval systems, has further advanced our understanding of these areas by permitting observation and retrieval of living organisms for experimental studies (Figures 4.4, 4.5, and 4.6).

ENVIRONMENTAL CHARACTERISTICS

Before considering the life in the deep ocean, it is necessary to establish the physical and chemical conditions under which the organisms must function. The major point to emphasize in considering these factors is that, at any given level or position in the deep ocean, these factors remain constant over long periods of time.

Light

In the deep ocean, the entire area is without sunlight, except at the very upper parts of the mesopelagic, where some light penetrates. Since this area is either dark at all times or has such extremely low light levels, photosynthesis is not possible, precluding any photosynthetically based primary productivity. What light is present is usually produced by the animals themselves (see pp. 152–154). This lack of light means the organisms have to rely on senses other than vision to find food and mates and to maintain various interspecific and intraspecific associations. The lack of sunlight may also have had a selective effect on the locomotory habits of the animals and on their propulsive systems.

FIGURE 4.4 (a) The deep submersible *Alvin*. (b) The deep submersible *Alvin* in the water. (Photos courtesy of a, Norbert Wu b, Dale Stokes/Mo Yung Productions/Norbert Wu.)

FIGURE 4.5 A remote-operated vehicle being deployed. (Photo by the author.)

Pressure

Of all the environmental factors acting on deep-sea organisms, pressure shows the greatest range. Pressure increases 1 atmosphere (atm) (14.7 lb/in.^{2} or 1 kg/cm^2) for each 10 m in depth. Since depths, including trenches, vary from a few hundred meters to more than 10,000 m, the range of pressure is from 20 to more than 1,000 atm. Most of the deep sea is under pressures between 200 and 600 atm. In no other marine environment does pressure exhibit such a great range or play such a potential role in the distribution of organisms.

Until recently, we lacked detailed information on the direct effects of pressure on most deep-sea organisms. This was primarily because virtually all organisms trawled from the deep sea arrived at the surface dead or dying. Only in recent years, with the advent of traps

FIGURE 4.6 Some deep-water benthic animals from the bathyal zone of the eastern Pacific. (a) *Tromikosoma panamense* (echinoid) (b) *Colessendeis* sp. (pycnogonid) (c) *Umbellula lindahli* (pennatulid). (d) *Freyella microplax* (brisingid sea star) (e) *Granelodone pacifica* (octopod) (f) *Scotoplanes globosa* (sea cucumber).

that incorporate a special pressure-maintaining chamber, have undamaged larger metazoan animals been retrieved from the deep sea in good condition. Some of these have been maintained for experimental purposes, and we are obtaining more knowledge of the biological effects of pressure.

That pressure must have a significant effect on deep-sea organisms can be extrapolated from experiments done on one group of organisms first successfully retrieved alive from the deep sea. Bacteria have been retrieved alive and cultured from the deepest areas of the ocean. Figure 4.7 shows the results of culturing a deep-sea bacterium under different pressure regimes. The fact that bacteria from the deeper areas virtually cease growth and reproduction at lower pressures, but grow and reproduce actively at higher pressures corresponding to their natural environment, suggests there are special adaptations to pressure that can profoundly affect organisms. Since metazoan animals are anatomically and physiologically more complex than bacteria, it is probable that multicellular animals will evince an even stricter pressure-dependent physiology.

There is now considerable evidence that both proteins and biological membranes are strongly affected by pressure and must be modified to work in animals living in the deep sea. For proteins, Siebenaller and Somero (1989) have shown the direction of adaptation has been to reduce the pressure sensitivity of the structure at the cost of other aspects of the enzyme systems. For example, the enzyme systems of deep-sea fishes have higher activation energies but lower catalytic efficiencies than their relatives in shallower water.

Studies were done by Siebenaller and Somero (1978) on the enzyme systems of two closely related fishes living at different depths. These studies demonstrated that differences in hydrostatic pressure as small as 100 atm or less are sufficient to alter markedly the functional properties of enzymes, namely, their ability to bind to the appropriate substrate, and the rate at which the reaction proceeds. Such changes are comparable to the changes in enzyme activity caused by temperature differences. These results strongly suggest that hydrostatic pressure plays a major role in organisms' adaptation to the deep-sea environment.

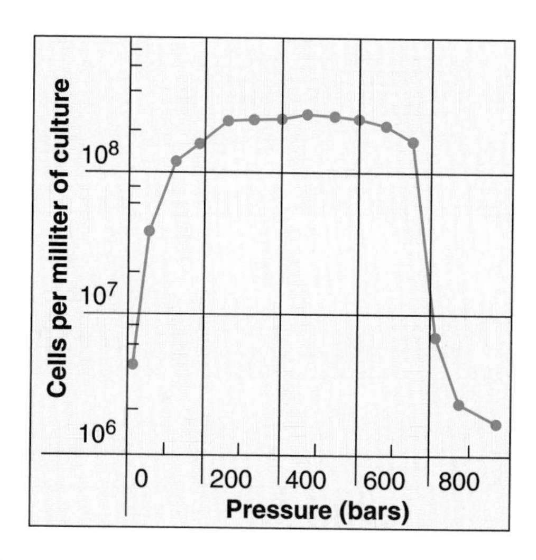

FIGURE 4.7 Graph of the amount of growth of a deep-sea bacterium at different pressures. (Reprinted with permission from "Isolation of a Deep Sea Barophilic Bacterium and Some of its Growth Characteristics" A. A. Yayanos et al., *Science,* Vol. 205 (4408), pp. 808–809, 1979. Copyright © 1979 American Association for the Advancement of Science.)

Somero (1982), in a review of physiological and biochemical adaptations of deep-sea fishes, has pointed out other differences in muscle enzymes between shallow and deep-dwelling fishes. Deep-dwelling fishes tend to have muscle enzymes that are less efficient and are present in the cells in low concentrations. The combination of these two characteristics probably results in the low metabolic rates, as well as the sluggish movements, of some deep-sea fishes. Most of these fishes are "float-and-wait" predators. Unlike the muscle enzymes, however, the enzymes of the nervous system seem not to be inefficient or reduced.

With respect to biological membranes, the major effect of pressure is on the lipids that make up the membranes. Because lipids are more compressible than water and can undergo phase changes under pressure, there is potential for pressure effects on membranes. Higher pressures reduce the fluidity of membranes. To adapt to high pressures, bacteria incorporate more fluid lipids into their membranes. This adaptation is termed the **homeoviscous adaptation**. Cossins and Marshall (1989) have demonstrated for mitochondrial membranes in the livers of

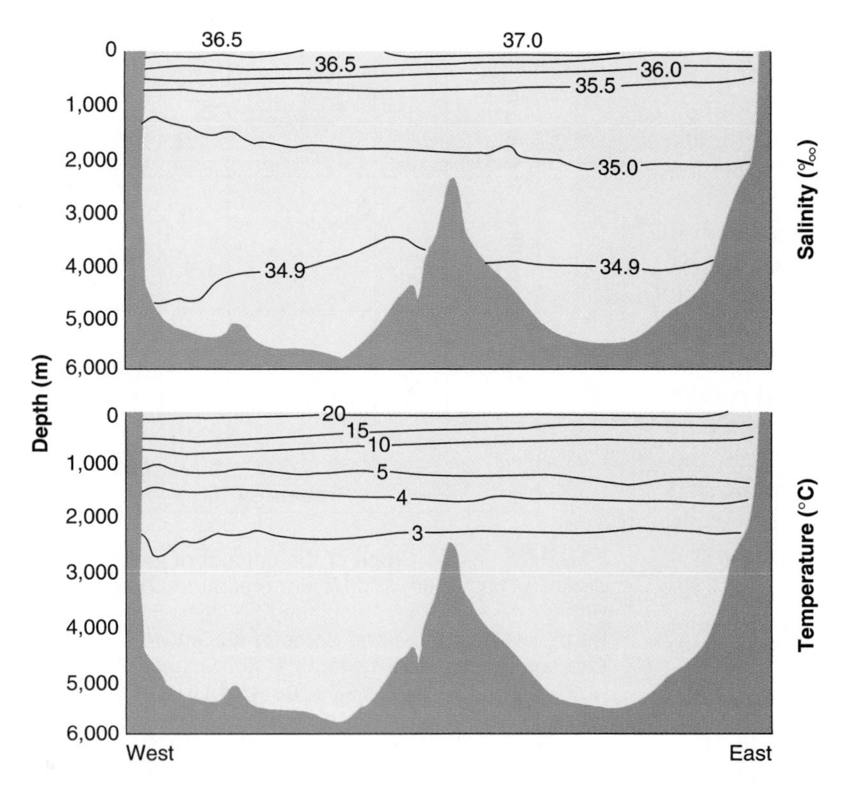

FIGURE 4.8 Salinity and temperature changes across the North Atlantic Ocean with respect to depth. (After *Introduction to Oceanography*, D. A. Ross, p. 305, © 1982 Prentice Hall, Upper Saddle River, NJ.)

deep-sea fishes that the ratio of saturated fatty acids to unsaturated ones declined with increasing depth, suggesting a similar adaptation.

Many laboratory studies on various kinds of cells from protozoan to mammalian have demonstrated that pressure affects the morphology of the cells, including the ability to form mitotic spindles and to undergo mitosis. Zimmerman and Zimmerman (1972) reported that amoebas lose their pseudopods and ball up, and *Tetrahymena* loses its ability to locomote with cilia. Pressure also has a significant effect on certain physiological and biochemical processes, such as typical muscle physiology. Perhaps the most dramatic effect of pressure is its effect on such macromolecules as proteins. Protein synthesis and function are markedly affected by pressure in laboratory situations, usually adversely. Since proteins are vital in living systems as enzymes and as major building blocks, it would seem that deep-water animals must have some special adaptation with respect to protein structure and function in order to survive.

The body of laboratory studies of the observed effects of pressure on shallow-water animals, coupled with the known effects of pressure on cells and macromolecules (protein) and the recent experimental evidence from living deep-sea animals, are strong indications that the physiology of deep-sea animals is profoundly affected by pressure and that pressure may be of considerable significance in zonal patterns of distribution.

There is another effect of pressure that affects organisms that secrete calcium carbonate shells. As pressure increases and temperature decreases, the solubility of calcium carbonate increases, as does its tendency to dissociate. This means that as one goes deeper into the ocean, it becomes more and more difficult for organisms to lay down shells. In the deep sea, the number of shell-forming species decreases, and the remaining ones tend to have thin shells.

Salinity

Below the first few hundred meters in the world's oceans, the salinity is remarkably constant throughout the depths (Figure 4.8). There are some minor differences in salinity, but none that are ecologically significant, except perhaps in the Mediterranean Sea.

Temperature

The two areas of greatest and most rapid temperature change in the oceans are the transition zone between the surface waters and the deep waters, the area known as the **thermocline**, and the transition between the deep-sea floor and the hot water flows at the hydrothermal vents. Thermoclines vary in thickness from a few hundred meters to nearly a thousand meters (Figure 4.9). Below the thermocline, the water mass of the deep ocean is cold and far more homogeneous. Thermoclines are strongest in the tropics, where the temperature of the epipelagic zone is usually above 20°C. From the base of the epipelagic, the temperature drops over several hundred meters to 5 or 6°C at 1,000 meters. It continues to decrease toward the bottom, but the rate of change is much slower (Figure 4.9). Below 3,000–4,000 m, the water is isothermal. What is ecologically significant is that, at any given depth, the temperature is practically unvarying over long periods of time. There are no seasonal temperature changes, nor are there any annual changes. Perhaps nowhere on earth is there another habitat with such a constant temperature. This, however, is in direct contrast with the hydrothermal vents. In these systems the temperatures of the water as it emerges from the "black smoker" chimneys may be as high as 400°C (it is kept from boiling by the high hydrostatic pressure) while within a few meters it may be back down to 2–4°C (see pp. 161–170).

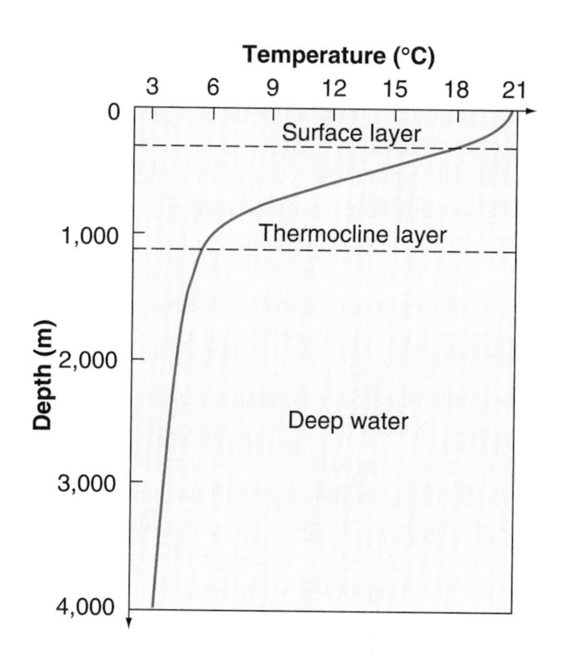

FIGURE 4.9 The thermocline layer in the ocean. (From *Oceanology: An Introduction*, D. E. Ingmanson & W. J. Wallace, © 1973 Wadsworth Publishing Company.)

Oxygen

Even though the deep ocean waters are far from a source of oxygen replenishment, either by interaction with the atmosphere or through production by autotrophs, there are essentially no abyssal or hadal areas that lack sufficient oxygen to support life. The few exceptions include the Carioca trench off Venezuela in the Caribbean and the Santa Barbara basin off California. They are anaerobic at the bottom. The oxygen present in the deep water masses entered the water when this now deep water mass was at the surface. Virtually all the water of the deep sea has its origin at the surface in the Arctic or Antarctic areas (see Chapter 1, pp. 10–15). Here, the oxygen-rich cold water sinks and flows north or south to make up the deep water of the world's oceans. Since it no longer gains oxygen after sinking, the reason these deep masses are not depleted of oxygen from respiration of deep-water organisms is that the density of deep-water organisms is so low and they have such low metabolic rates that together they are insufficient to deplete the oxygen supply. That this situation is true is supported by the fact that the oxygen concentration declines in the 20 m or so just above the bottom in the deep sea. It is also near the bottom that the most dense concentrations of organisms occur in the deep sea.

Another peculiar feature of oxygen concentration in the deep sea is the presence of the **oxygen minimum zone** at 500–1,000 m (see Figure 1.3). Oxygen values are higher both above and below this zone. In the oxygen minimum zone, oxygen values may fall to less than 0.5 ml/liter. The occurrence of this zone is mainly the result of respiration of organisms, coupled with the lack of interchange of the water mass with more oxygen-rich water. The reason that an oxygen minimum zone occurs at 500–1,000 m and not deeper is that the numbers of animals in deeper waters are so low that they never deplete the available oxygen, whereas at 500–1,000 m, organisms are abundant. Above 500 m, oxygen depletion does not occur, even though the animal biomass is high, because there is constant replenishment from air and by autotrophs.

Animals living in the oxygen minimum zone often have special adaptations to extract oxygen from the low-oxygen environment and/or can subsist with minimal amounts of oxygen.

Food

The deep sea is removed from the photosynthetic zone and has no primary production except for certain areas where chemosynthetic bacteria are found (see pp. 161–170). All organisms living in the deep sea depend on food that ultimately is produced elsewhere, where photosynthesis is possible, and subsequently transported into the deep sea. The deep sea, along with caves in the terrestrial environment, are, therefore, unique among the world ecosystems in that they have virtually no indigenous primary productivity.

Potential food must come into the deep sea by sinking from the surface waters. Since the populations of organisms are greatest in the upper layers of the oceans, the chance that any food particle will be able to sink through all these voracious animals without being consumed or decaying is small. As a result, relatively few food particles are available to the animals of the deep sea as opposed to those in the surface waters. Exceptions occur, however, such as the eastern equatorial Pacific where within 2° of the equator, high surface productivity provides a high carbon input in the form of phyoplankton detritus. This paucity of food is probably the reason the density of deep-sea animals is very low and the size of most is small. Without sufficient energy in the form of food, large numbers of organisms cannot be sustained.

The probability that a particle of food will decay or be consumed during sinking increases with increasing depth, since there is more time and there are more organisms to consume it. Thus, the deeper the organisms, the less the food available to them.

A portion of the particles sinking from the photic zone are particles of material not directly suitable as food. An example would be fecal pellets and shed chitinous exoskeletons of crustaceans. Most organisms cannot digest chitin. These materials, however, are acted upon by bacteria during descent into the deep sea and, after reaching the bottom, are converted into suitable food in the form of bacterial protoplasm. Because the residence time for a food particle is longer on the bottom than in the water column, there is an increase in bacteria in the bottom oozes of deep sea; thus, this allows increased numbers of larger organisms to survive. Densities of the benthic infaunal organisms may exceed that of pelagic organisms at the same depth. This also explains the reduction in oxygen in the near-bottom water.

The amount of food available to the deep sea is correlated with the amount of primary productivity in the surface waters above or else with the proximity to a secondary source, such as organic debris from terrestrial habitats. Deep-sea areas under productive surface waters and near islands or continents have more food than deep waters under low surface productivity or far from land.

Several types of food sources are available in the deep sea. Directly available foods include those deep-sea organisms that spend their early or larval stages in the lighted upper waters and then migrate into the depths, where they furnish food to the predators. An example is the group of deep-sea angler fishes whose young live in the upper waters of the ocean, while the adults are all restricted to the deep sea. Another directly available food source is the large, dead bodies of marine mammals and fishes that may sink quickly before being completely consumed (Figure 4.10). This would seem an unpredictable type of food, but Isaacs and Hessler of Scripps Institution of Oceanography have shown, by placing baited cans with cameras attached into the deep sea, that food is quickly detected and brings in consumers from great distances (Figure 4.11). That this unlikely food source must be reliable is also supported by the fact that several groups of deep-sea gastropod mollusks have also evolved that live on whale and shark remains (see p. 164).

Indirect food sources include indigestible animal and plant remains, which first must be acted on by bacteria. Examples are chitin and wood and cellulose of land plants. Other potential sources are colloidal or dissolved organic material in the water and gelatinous plankton matter (marine snow). At present, the relative importance of these sources cannot be evaluated.

■ We can summarize by saying that food is very scarce in the deep sea, when compared with other areas of the oceans, and that it decreases with depth and distance from land. This food scarcity is responsible for the low densities and small sizes of deep-sea organisms. ■

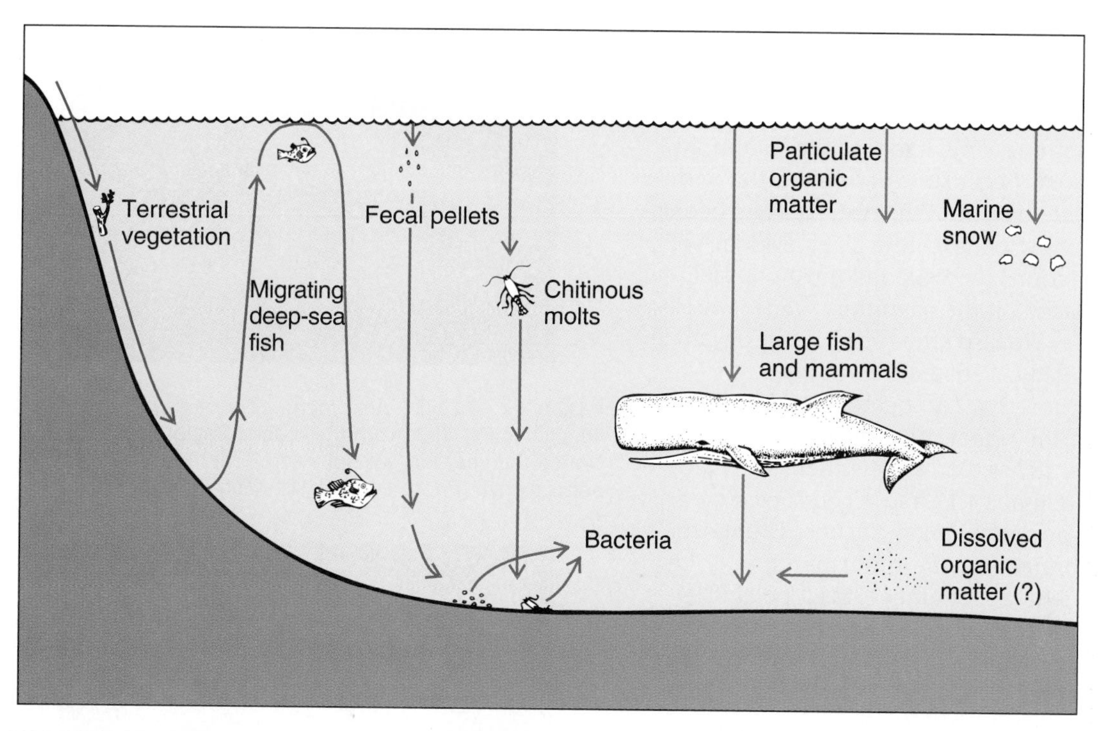

FIGURE 4.10 Food sources to the deep sea.

FIGURE 4.11 Two photographs showing the rapid attraction of deep-sea animals to bait and its subsequent consumption. (a) At the time of reaching the bottom. (b) Six hours and 40 minutes after reaching the bottom, showing consumption of bait by the amphipod *Hirondella gigas* (at 9,605 m in the Philippine trench). (Photos courtesy of Dr. Robert Hessler, Scripps Institution of Oceanography.)

ADAPTATIONS OF DEEP-SEA ORGANISMS

The organisms that live in the deep sea possess a number of adaptations correlated with the conditions of the environment. Because experimentation with living metazoan animals from the deep sea has been limited until recently, we have little direct evidence of how many adaptations function and must offer educated guesses based on the conditions under which we know they live. We also know little about physiological and biochemical adaptations at present.

One adaptation that can be observed, even in organisms from the upper mesopelagic, is that of *color*. Mesopelagic fishes tend to be either silvery gray or a deep black (Figures 4.12 and 4.13). They are not countershaded, as are epipelagic fishes. The mesopelagic invertebrates, on the other hand, may be either purple, bright red, or orange. Mesopelagic jellyfish are often a dark purple, whereas the crustaceans, such as copepods, mysids, and shrimps, are brilliant red (Figures 4.14–4.19). Since these organisms live in virtually unlighted waters, black organisms will be invisible, but why the red color? Young (1983) has suggested that since most bioluminescent light is blue, the red pigmentation would protect these animals from the revealing rays of the bioluminescent flashes used by predators.

Organisms living even deeper, in the abyssal and bathyal zones, are often colorless or a dirty white; they lack pigment. This is particularly true for animals dwelling on the bottom, although some taxa, such as anemones, may be colorful. Fishes, however, may be black at all depths.

Another adaptation seen particularly in mesopelagic and upper bathypelagic fishes is the presence of large eyes. The eyes in these fishes are much larger relative to body size than those in fishes of the lighted epipelagic zone (Figure 4.20). Often, the presence of large eyes is correlated with the presence of light organs (see pp. 152–154). These fishes dwell in the uppermost parts of the deep sea, where some small amounts of light may penetrate. Many also migrate vertically into the epipelagic zone at night. The large eyes give them maximum light-collecting abilities in low-light areas. The large eyes are also presumably needed to detect the low light intensities of the light organs. Increased surface area is only one visual adaptation. These fishes also have enhanced "twilight vision" derived from the pigment rhodopsin and an increase in the density of rods in the retinal area.

FIGURE 4.12 *Anoplogaster cornuta*, a "daggertooth" fish from the mesopelagic zone. Notice the very large mouth and large teeth for seizing prey. (Photo courtesy of Michael Kelly.)

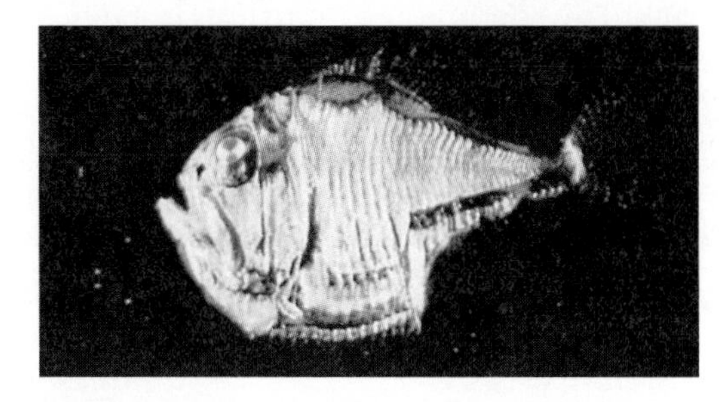

FIGURE 4.13 *Argyropelecus lychnus*, a mesopelagic hatchetfish. Note the lines of black photophores along the ventral portion of the body, the tubular eyes and silvery reflecting scales (Photo courtesy of Michael Kelly.)

FIGURE 4.14 A mysid crustacean of the genus *Boreomysis*. (Photo courtesy of Gary McDonald.)

Fishes dwelling in the deepest parts of the ocean (abyssalpelagic, hadalpelagic) show a different trend. Many of these animals have small eyes or lack eyes entirely (Figure 4.21). Because they dwell in permanent

FIGURE 4.15 A deep-sea prawn of the genus *Notostomus*. Note the deep orange-red color, characteristic of many deep-sea crustaceans. (Photo courtesy of Dr. Roger Seapy.)

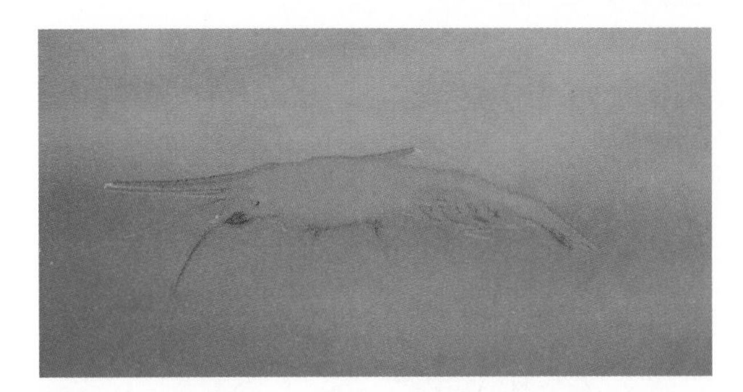

FIGURE 4.16 The giant deep-sea mysid *Gnathophausia*, an example of deep-sea gigantism. (Photo courtesy of Dr. Roger Seapy.)

darkness, eyes are not necessary. Generally, fishes in depths of 2,000 m and above have eyes, often large; below 2,000 m, eyes are small, degenerate, or lost. Often, bottom dwellers have no eyes.

Yet another adaptation in fishes is **tubular eyes** (see Figure 4.13). Fishes with peculiar tubular eyes are found in several families. They give the fishes an

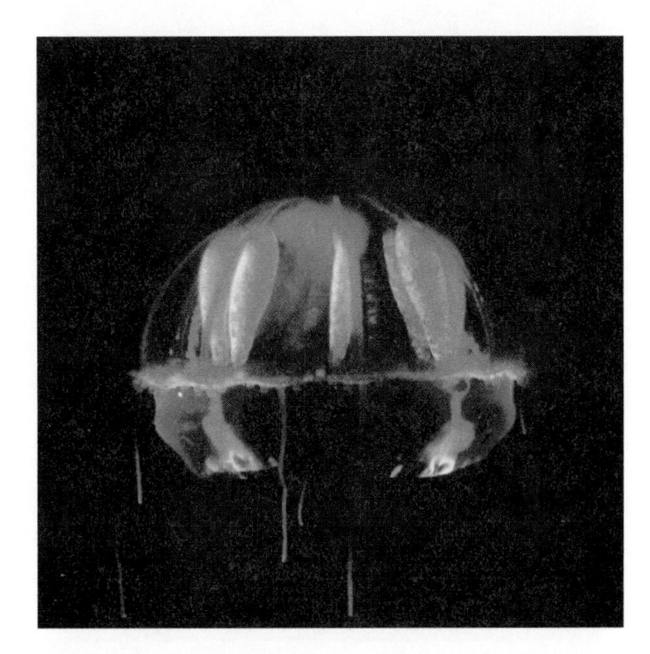

FIGURE 4.17 A deep-sea jellyfish, *Benthocodon hyalinus*. Note the characteristic deep purple color. (Photo courtesy of Dr. George Matsumoto.)

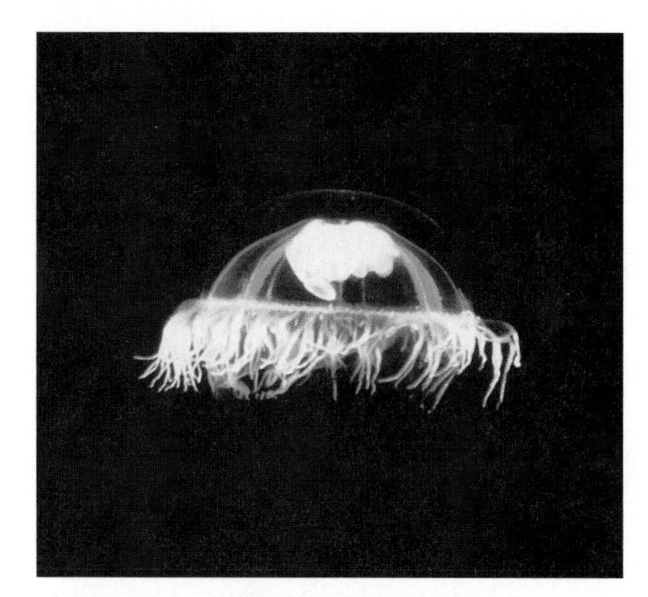

FIGURE 4.18 The deep-water jellyfish *Arctopodema amplum.* (Photo courtesy of Dr. George Matsumoto.)

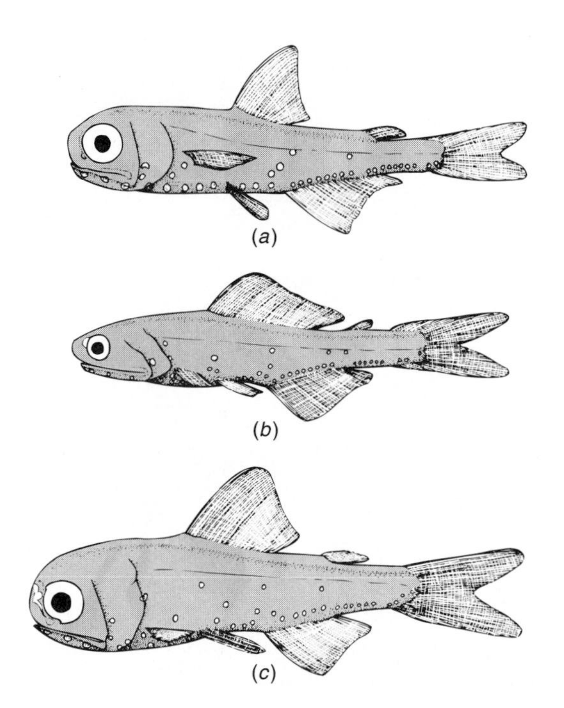

FIGURE 4.20 Large eyes in three mesopelagic fishes of the family Myctophidae. (a) *Myctophum punctayum.* (b) *Lampanyctus elongatus.* (c) *Diaphus metopoclampus.* (After *Aspects of Deep Sea Biology*, N. B. Marshall. Copyright © 1954 Philosophical Library, New York. Reprinted by Permission.)

FIGURE 4.19 A deep-sea ctenophore, *Lampoctena sanguineventer.* (Photo courtesy of Dr. George Matsumoto.)

FIGURE 4.21 Fish with reduced eyes, a gulper eel of the genus *Saccopharynx.* (Photo courtesy of Michael Kelly.)

FIGURE 4.22 The dimorphic eyes of the squid *Histioteuthis.* (Photo courtesy Dr. Roger Seapy.)

extremely bizarre appearance. In these fishes the eye is a short black cylinder topped with a hemispherical, translucent lens. Each eye has two retinas, one at the base of the cylinder and one on the wall of the cylinder. These tubular eyes tend to give wide binocular vision so that a light flash will be seen by both eyes, making it more likely that the brain will receive adequate stimulation. Therefore, the tubular eye appears to be an adaptation to increase sensitivity. Why this type of eye should be developed in the deep sea is not clear.

Among the invertebrates, certain squids of the family Histioteuthidae exhibit another peculiar adaptation. They have one eye much larger than the other (Figure 4.22). Such squids also display numerous light-producing organs called photophores. Work by Young (1975) has demonstrated that these squids, which live in the mesopelagic at depths between 500 and 700 m, usually orient with the arms and small eye downward and with the large eye upward. In this position, the large eye gathers the faint light downwelling from the surface, while the smaller eye responds to light from photophores. It has been further demonstrated with another midwater squid, *Abraliopsis*, that the squid responds to downwelling light by turning on its photophores just enough to match the downwelling light; hence, it becomes invisible when viewed from below (i.e., prevents silhouetting).

Food scarcity in the deep sea seems to be the reason for another series of adaptations. Most deep-sea fishes have large mouths, larger relative to their size than fishes of other areas (Figures 4.23 and 4.24; also see Figure 4.12). Furthermore, the mouth is often filled with long teeth recurved toward the throat, an adaptation to ensure that whatever is caught does not escape. Even more bizarre is that, in some fishes, the mouth and skull are so hinged that

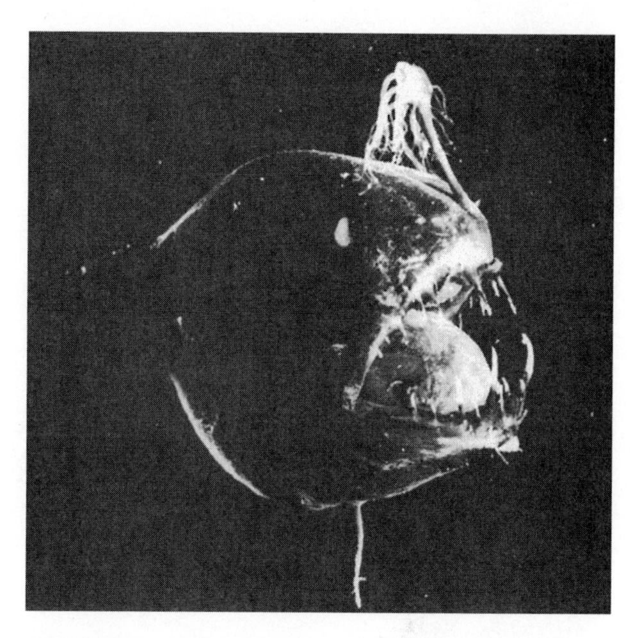

FIGURE 4.23 *Linophryne*, a genus of fish with a very large mouth. (Photo courtesy of Michael Kelly.)

FIGURE 4.24 The deep-sea fish *Aristostomias scintillans.* Note the greatly expanded mouth cavity that allows the animal to swallow very large prey. The long barbel from the jaw region has a luminescent organ at the end. (Photo courtesy of Dr. Roger Seapy.)

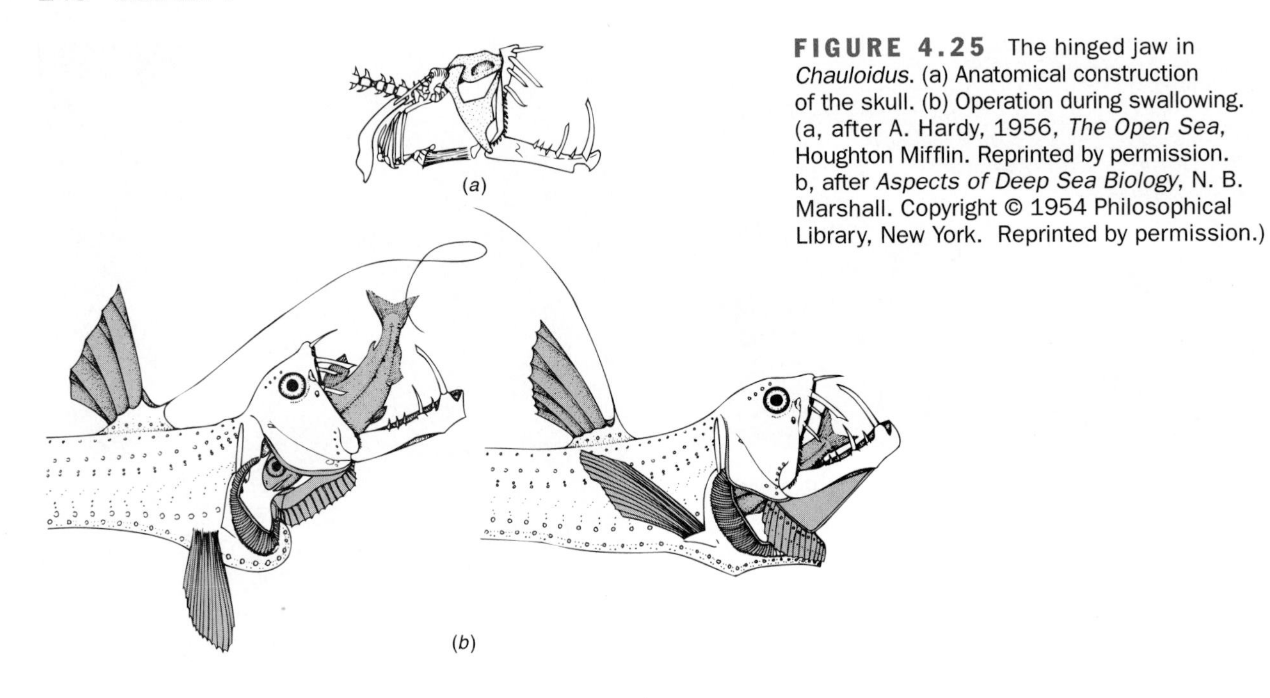

FIGURE 4.25 The hinged jaw in *Chauloidus*. (a) Anatomical construction of the skull. (b) Operation during swallowing. (a, after A. Hardy, 1956, *The Open Sea*, Houghton Mifflin. Reprinted by permission. b, after *Aspects of Deep Sea Biology*, N. B. Marshall. Copyright © 1954 Philosophical Library, New York. Reprinted by permission.)

FIGURE 4.26 Deep-water angler fish, *Melanocetus johnsoni*, before and after swallowing a larger fish. (From *The Open Sea*, A. Hardy, © 1956 Houghton Mifflin.)

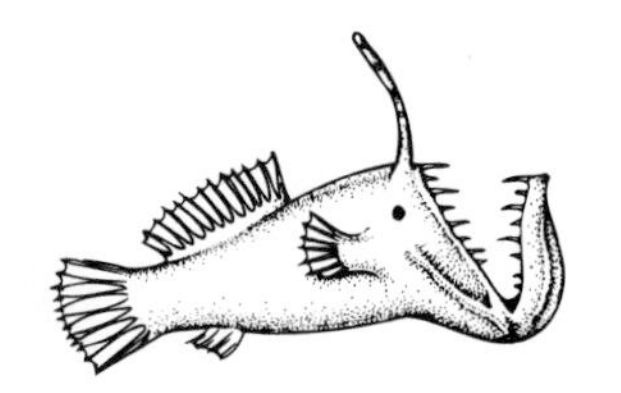

FIGURE 4.27 A deep-sea angler fish, *Melanocetus johnsoni*. Note the extremely large mouth and the luminescent "lure" above the head. (Photo courtesy of Dr. Roger Seapy.)

the animals can open their mouths much wider than their own bodies, enabling them to swallow prey larger than themselves (Figures 4.24–4.27). Food is not passed up simply because of size! Still other fishes, such as the angler fish (Ceratoidea), have responded to the scarce food by establishing themselves as traps, using a highly modified part of the dorsal fin as a lure (Figure 4.27). In this case, the lure is a luminescent organ. Other types of lures are found in many other fishes, such as the stomiatoids, and occur on barbels attached to the chin (Figure 4.28). One curious adaptation reported by McAllister (1961) is that many carnivorous fishes have black linings to their coeloms, so bioluminescence from ingested prey will not shine out and attract predators to them!

Because food is scarce and the density of organisms is very low, there is a potential problem of finding a mate in this vast, dark area. One adaptation to this problem is found in angler fish, *Ceratias*, in which the large individuals are all female. The males are very tiny and are parasitic, attached to the body of the female (Figure 4.29). Since the males are always present to provide the sperm, a search for a mate is not necessary. Of course, the male has first to find the female, and this is presumably accomplished via olfaction.

Body size of deep-sea organisms is a peculiar paradox. Since food is scarce, it might be anticipated that most organisms would be small. Indeed, when considering fishes, this appears to be the case. Most fishes captured in the deep sea are smaller than corresponding relatives in shallow water. As far as we now know, the deep sea contains few large fishes, though a few large ones have been photographed by the aforementioned baited "monster camera." It is also possible that the larger animals are present but are able to avoid our capture methods. For example, no specimen of the giant squid genus *Architeuthis* has ever been captured in a net or observed by a camera or video in situ, but we know they exist because moribund specimens have washed ashore and parts of the squid have been found in the guts of sperm whales. On the other hand, certain species of some invertebrate groups, particularly amphipods, isopods, ostracods, mysids, and copepods, attain a much greater size in deep waters than their relatives in shallow water. This phenomenon of larger size with increasing depth is called **abyssal gigantism** (see Figure 4.16). The size that certain of these abyssal giants attain is remarkable. The isopod *Bathynornus giganteus* reaches 15 in. (42 cm), and the amphipod *Alicella gigantea* reaches 6 in. (15 cm). *Gigantocypris*, the giant mesopelagic ostracod, has a diameter of several centimeters. The copepod *Gausia princeps* reaches a size of 10 mm, nearly ten times the size of most calanoid copepods.

FIGURE 4.28 Luminescent barbel of the stomiatoid fish *Idiacanthus* sp. (Photo courtesy of Norbert Wu/Peter Arnold Inc.)

Scientists currently disagree about the reasons for abyssal gigantism, but there are two main theories. The first attributes it to the peculiarities of metabolism under conditions of high pressure. The second suggests that the combination of low temperature and scarce food reduces the growth rates in these crustaceans and increases their longevity and the time it takes them to reach sexual maturity, such that larger size is obtained. The extreme expression of this combination would be the abyssal giants.

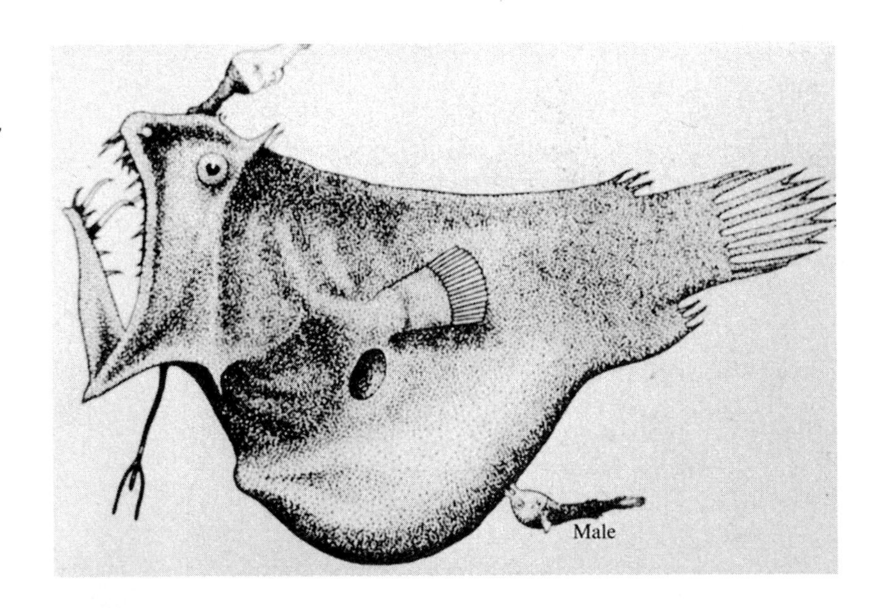

FIGURE 4.29 Large female angler fish, *Linophryne argyresca*, with attached small parasitic male. (From "Deep Sea Angler-Fishes (Ceratpodea)" C. T. Regan & E. Trewavas, *Dana Report*, No. 2, 1932. Reprinted by permission of Oxford University Press, Ltd., and The Carlsberg Foundation.)

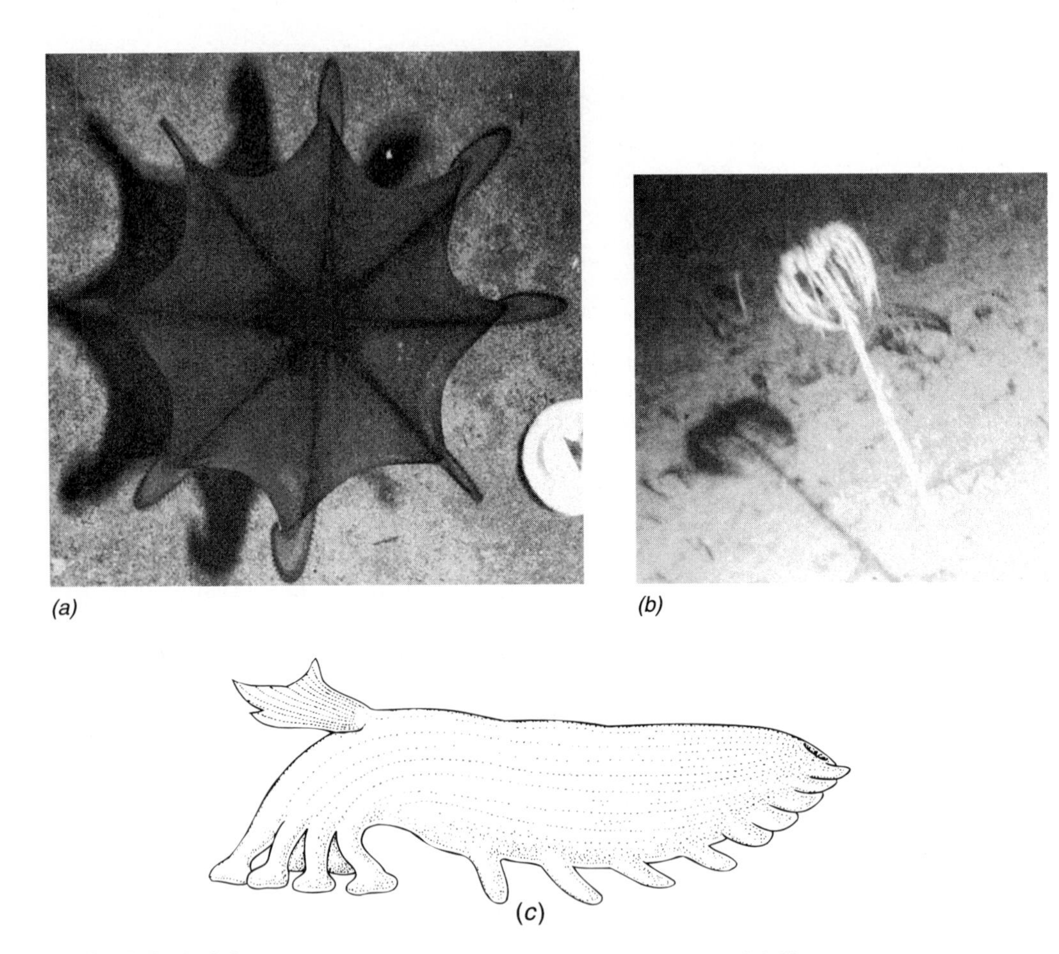

FIGURE 4.30 Adaptations to raise animals above the ooze. (a) Cirrate octopod. (b) A stalked crinoid. (c) A sea cucumber of the genus *Peniagone*. (a, courtesy of Dr. Clyde Roper. b, courtesy of Dr. H. Mullins. c, redrawn from Hansen, Galathea Report, Vol. 2.)

TABLE 4.2

Anatomical Differences in Mesopelagic and Bathypelagic Fishes

Features	*Mesopelagic, Plankton-Consuming Species*	*Bathypelagic Species*
Color	Many with silvery sides	Black
Photophores	Numerous and well developed in most species	Small or regressed in most species; a single luminous lure on the female of most angler fish
Jaws	Relatively short	Relatively long
Eyes	Fairly large to very large with sensitive pure-rod retinas	Small or regressed, except in the males of some angler fishes
Olfactory organs	Moderately developed in both sexes of most species	Regressed in females, but large in males of *Cyclothone* species and angler fish
Central nervous system	Well developed in all parts	Usually weakly developed system
Myotomes	Well developed	Weakly developed
Skeleton	Well ossified, including scales	Weakly ossified; scales usually absent
Swim bladder	Usually present, highly developed	Absent or regressed
Gill system	Gill filaments numerous, bearing very many lamellae	Gill filaments relatively few, with a reduced lamellar surface
Kidneys	Relatively large, with numerous tubules	Relatively small, with few tubules
Heart	Large	Small

Source: From *Deep Sea Biology: Developments and Perspectives* by N. B. Marshall, © 1980 Garland STPM Press.

The large size may also be the result of natural selection. Large size, long life, and delayed sexual maturity confer certain advantages to organisms in the deep sea. These include production of larger eggs and subsequently larger young, which can then feed on a wider range of food sizes. This precludes special food in this food-short area. Large animals are also more mobile and can cover more area in search of both food and potential mates. Increased longevity means a longer period of sexual maturity; hence, these animals have greater time to find mates.

Although these abyssal giants have captured much attention, they are rare. Most of the deep-sea benthic infauna (polychaete worms, crustaceans, and mollusks) are much smaller than their shallow-water counterparts. In fact, the small size of benthic invertebrates is the major characteristic of deep-sea fauna, not gigantism. Given the vast area of the deep sea, some giants should be expected, and similar examples of giant taxa can also be found in the less extensive shallow-water areas, for example, the giant clams found on tropical reefs.

The bottom substrate of most deep-ocean areas is a soft ooze. Benthic organisms inhabiting this soft material tend to have delicate bodies, long legs, or in the case of sessile animals, long stalks to raise them above the ooze (Figure 4.30). Some fishes have long, narrow fins that serve the same purpose.

In general, as one goes deeper in the oceans, the anatomical organization of the animals becomes more simplified. This is best seen in a comparison of the features of mesopelagic and bathypelagic fishes as outlined in Table 4.2. It is also reflected in the generally slow mode of life of most deep-sea creatures. Measurements of oxygen uptake on the deep-sea floor at several localities by Smith and Hinga (1983) gave results from 0.02 to 0.1 ml of oxygen per m^2 per hour, or more than 100 times less than that measured in shallow water. Part of this is due to the reduced density of organisms, but part also reflects the low metabolic activity of the organisms present.

Biochemical adaptations to the deep sea have been difficult to ascertain in the past because of the lack of live animals to study. However, newer data on chemical composition suggest that profound changes may occur. In fishes, Childress and Nygaard (1973) report that the water content of the body tissues increases with increasing depth, while the lipid and protein concentrations decrease. In other words, they become more like jellyfish. In crustaceans, Childress and Nygaard (1974) found that the protein content also decreased with depth. Presumably, this decrease in both groups is related to the scarcity of food from which to produce the protein. Caloric content also decreases with depth.

A final set of adaptations in deep-sea organisms has to do with bioluminescent organs and will be discussed separately.

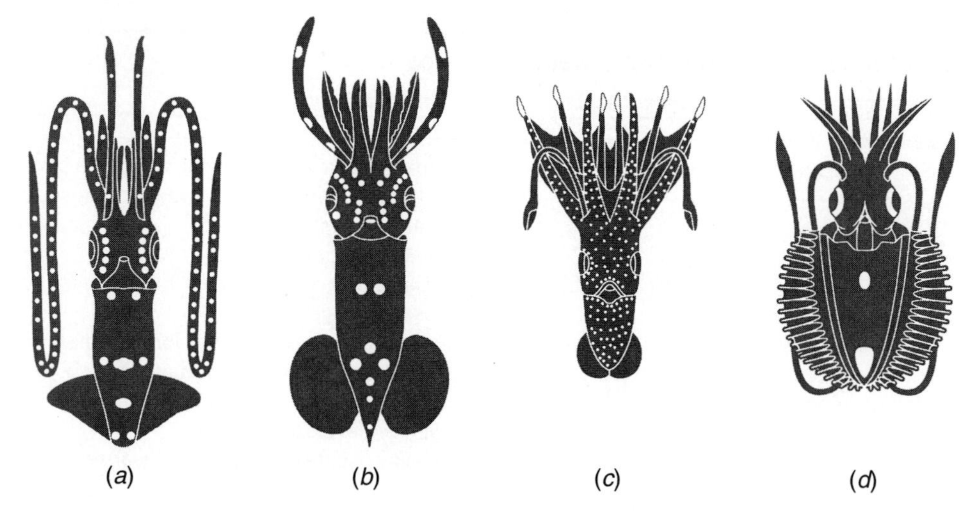

FIGURE 4.31 Light organ patterns of some deep-sea squids. (a) *Nematolampus.* (b) *Pterygioteuthis.* (c) *Histioteuthis.* (d) *Ctenopteryx* male. (From *The Biology of Cephalopods*, eds. M. Nixon & J. B. Messenger, © 1977 Academic Press, Ltd.)

Bioluminescence in the Deep Sea

Marine bioluminescence is not confined to deep-sea organisms. Indeed, it is a widespread phenomenon in the sea. Many people have observed it in the form of "phosphorescent seas," which result from light production in the surface waters by untold millions of dinoflagellates. Whereas bioluminescence is widespread, it is in the deep-sea mesopelagic organisms that it reaches its highest and most complex development. In the mesopelagic zone, as many as 70% of the organisms have the ability to produce light. The prevalence of bioluminescence in the deep sea and the large numbers of organisms that can produce light have been documented by photos and videos taken by remote-operated vehicles (ROV) and manned submersibles; hence, it is appropriate to discuss the phenomenon at this time.

Bioluminescence is the production of light by living organisms. The mechanism of light production is well known from studies of such terrestrial organisms as fireflies, and the same mechanism is used by aquatic organisms. In the reaction, a substrate called luciferin is oxidized in the presence of the enzyme luciferase to produce a molecule that emits light. Luciferin and luciferase may differ from organism to organism, but the reaction is similar. The spectrum of color produced varies from species to species, but in general in the mesopelagic, the wavelength ranges between 460 and 480 nanometers.

Not only are large numbers of deep-sea organisms capable of producing light, they also have developed the most elaborate organs for producing this light. These light-producing organs are called **photophores** and are particularly abundant in fishes and squids (Figure 4.31), but are present in other invertebrates as well (Figure 4.32; see also Figures 4.13, 4.24, 4.27, and 4.28).

The largest number of animals with photophores is found in the upper areas of the deep sea, the mesopelagic and upper bathypelagic. The incidence of organisms with bioluminescence decreases in the deepest parts of the sea.

Photophores in deep-sea organisms are of several types and range in structure from simple to complex. The simplest photophores consist of a series of glandular-like cells that produce the light or a simple glandular cup holding a bacteria culture that produces the light (see also p. 454). In either case, the cells or cup are surrounded with a screen of black pigment cells. The more elaborate photophores have in addition one or more of the following: lenses to focus the light, a color filter, or an adjustable diaphragm of pigment cells. Still others are able to move the photophores by muscular action. Some squids combine their photophores with overlying layers of skin that contain chromatophores, allowing them to alter the color or intensity of the light produced.

That the production of light is of considerable adaptive significance to its possessor is demonstrated by (1) its prevalence in the deep sea, and (2) the complex anatomical organs present to produce it. It is also apparent that bioluminescence serves different functions in different animals. Several different hypotheses have been suggested for the use of bioluminescence in the deep sea. These have been reviewed by Young (1983), who suggests that most are used primarily as a means of food capture or defense against predators.

Many organisms, particularly fishes and squids, have their photophores placed ventrally. For those animals living in the mesopelagic zone, where some light downwells from the surface, a visual predator looking up at them could see a silhouette. The photophores, when producing a similar intensity as the downwelled light, would cause the silhouette to disappear and make the animal less vulnerable to predation. This is, in effect, camouflage. Since some organisms can adjust their downward photophores to the incident radiation level from above, this is a viable function.

Photophores might also be used to produce a "blinding" flash of light, which would momentarily startle a potential predator and allow disengagement and escape. Since such light production may also have the opposite effect of attracting more distant predators, however, this use of bioluminescence contains an element of risk. Some deep-sea squids (*Histioteuthis dispar*) distract potential predators by shooting out a luminescent cloud rather than a black "ink" cloud as is done by squids in lighted waters. This cloud masks the escape of the squid.

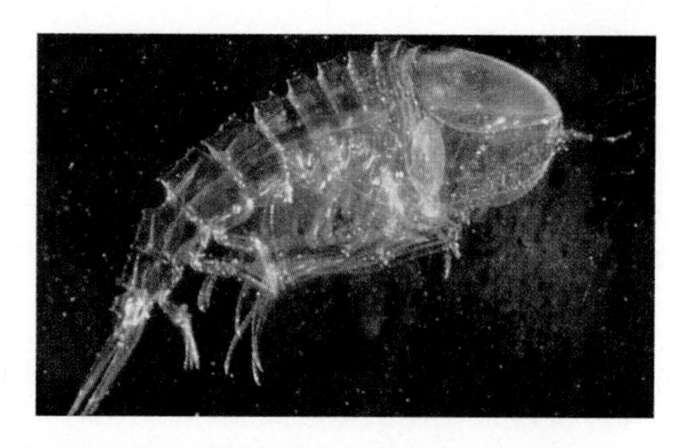

FIGURE 4.32 A mesopelagic amphipod of the genus *Cystisoma* showing the transparent body and huge eyes. (Photo courtesy of Dr. Roger Seapy.)

As noted before, photophores might also be employed as lures to attract potential prey within range of the predator. This is undoubtedly the function of the elaborate light organ (esca) on the modified dorsal fin (illicium) of the angler fish and the "lures" of other midwater fishes (see Figures 4.24, 4.27, and 4.28). Photophores may also be used for simple illumination of an area so a predator may pick out its prey.

A final set of uses involves recognition. In a perpetually dark area, such as the deep sea, the particular pattern of distribution of photophores over the body of the animal would be visible at a distance. Elaborate patterns of photophores occur on fishes in particular. The pattern is different for different species; thus, these patterns would allow individuals to recognize their own species. This may be important in maintaining schools and in finding potential mates (Figure 4.33). In the latter case, certain fishes also have photophore patterns that differ between males and

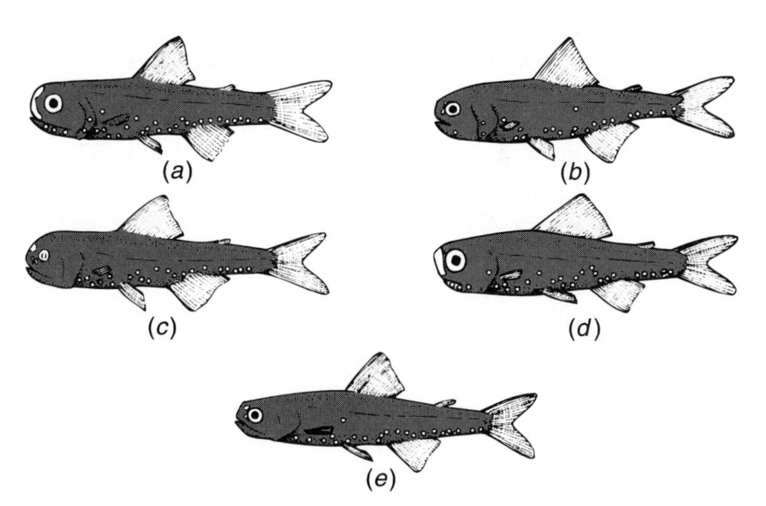

FIGURE 4.33 Species-specific light organ patterns (open circles) in the lantern fish genus *Diaphus*. (a) *D. macrophus*. (b) *D. lucidus*. (c) *D. splendidus*. (d) *D. garmani*. (e) *D. effnlgeus*. (From *Aspects of Deep Sea Biology*, N. B. Marshall. Copyright © 1954 Philosophical Library, New York. Reprinted by permission.)

females to ensure mate selection. It is interesting to note here that one of the first scientists to observe living mesopelagic fishes in their natural habitat, William Beebe, noted while observing from a bathyscape that he could tell the different myctophid (lantern fish) species in the dark from their photophore pattern. If he could, it seems likely that the fish can also.

Other uses of photophores may occur and more may become recognized with increased knowledge of this area, but the previously presented hypotheses represent the current most plausible hypotheses.

■ We can say in summary, paraphrasing Young (1983), that in this homogeneous twilight area without physical barriers or structures for concealment and with abundant predators, life becomes a peculiar battle in which the judicious use of bioluminescence is of major importance in concealing one from predators, finding and attracting prey, and ensuring survival. No other area on earth is quite like it. ■

COMMUNITY ECOLOGY OF THE BENTHOS

Although we have samples of organisms from the pelagic environment of the deep sea, the problems of capture and net avoidance make the available information on the community composition of the pelagic fauna incomplete at best and suspect at worst. The exception to this is the mesopelagic, where many samples and observations are available. Considerably more is known about the deep-sea benthic communities because the sedentary and sessile nature of the animals makes net avoidance less important. As a result, the samples taken and the observations of this area by deep submersibles and cameras have given us a reasonable idea of its composition. Ecologically, all benthic organisms throughout the world's oceans can be divided into two groups on the basis of position. The **epifauna** are benthic organisms that live on or are otherwise associated with the surface of the bottom. **Infauna** are organisms that live within the substrate.

Faunal Composition

Virtually all the major animal groups have representatives in the deep sea, but the relative abundance of groups varies. Crustaceans, particularly isopods, amphipods, tanaids, and cumaceans, are common in the deep sea. In the Atlantic abyssal area, they constitute 30–50% of the fauna. Polychaete worms are also abundant, making up from 40–80% of the fauna in the Atlantic. Particularly common in abyssal areas are sea cucumbers (Holothuroidea), usually of large size, and brittle stars (Ophiuroidea). Holothurians are often the most common organisms in deep-sea photographs (Figure 4.34). This, plus the fact that they form 30–80% of the biomass of living organisms taken in some trawls, suggests they are a major component of the abyssal community. Since holothurians are deposit feeders, the abyssal oozes are likely an excellent food source, hence their abundance.

Starfish (Asteroidea), sea lilies (Crinoidea), and sea urchins (Echinoidea) are also present, but not in abundance (Figures 4.35 and 4.36). Among the sponges, the deep sea is populated by the glass sponges (Hexactinellida), a group rarely found in shallow water. Sea anemones (Anthozoa), sea pens (Pennatulacea), sea fans (Gorgonacea), hydroids (Hydrozoa), and zoanthids (Anthozoa) are the common representatives of the phylum Cnidaria (Figures 4.37 and 4.38). A number of fishes are also present, but their mobility makes it difficult to estimate relative abundance. Figure 4.39 diagrammatically summarizes zonation changes and typical abyssal animals.

In the trenches (hadal zone) the available data suggest that there is a higher percentage of peracaridean crustaceans, polychaetes, and holothurians and a lower percentage of sea stars, sea urchins, and brittle stars in comparison with the abyssal.

The major fishes of the bottom waters of the deep sea are rat tails (Macrouridae), cusk eels, bythidids (brotulas), liparids (snailfishes), and certain eels (*Synaphobranchus*, *Cyema*) (Figure 4.40). A marked characteristic of the deep sea is the relative lack of the more derived acanthopterygian fishes and the domination by the evolutionarily older or primitive fishes. In pelagic zones below the bathypelagic, fishes tend to be rare, while bottom-dwelling fishes are generally slow moving and sedentary.

One of the dominant features of the deep-sea benthos is the very small size and fragility of most individuals. For example, in exploration of the shallow waters of the continental shelf, a 1-mm sieve is often used to screen out infaunal organisms. Such a screen used in the deep sea would retain practically no organisms. Furthermore, the deep-sea animals would fall apart unless carefully handled. The fragility of these animals is due to several circumstances, including the fact that

FIGURE 4.34 A group of deep-sea holothurians of the genus *Scotoplanes* on the bottom in 3,000 m. Holothurians are one of the major groups of large megafauna organisms characteristic of sedimentary areas of the deep sea. (Photo courtesy of Dr. Barbara Hecker.)

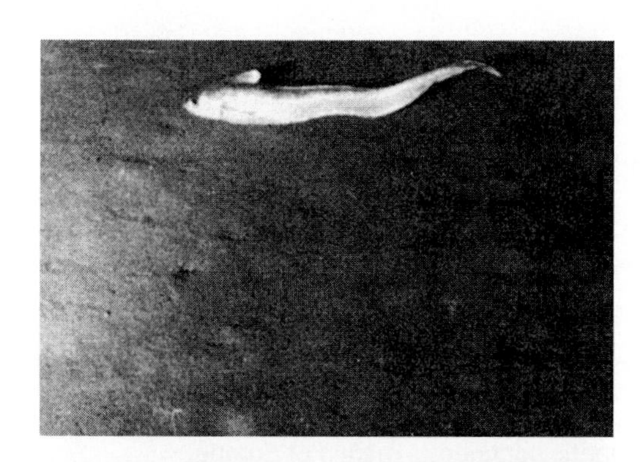

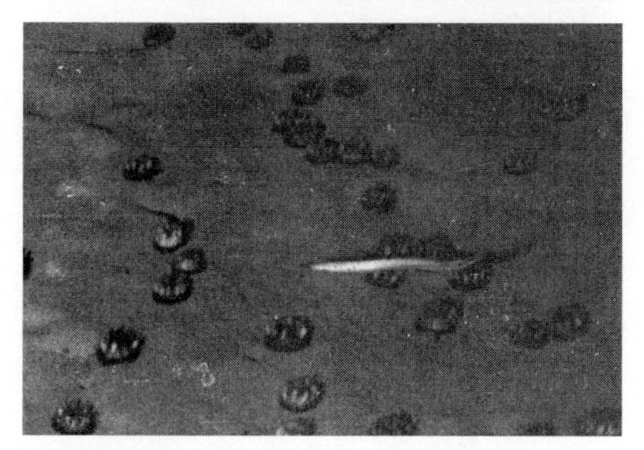

FIGURE 4.35 Two views of the bottom at 1,300 m on the Gay Head–Bermuda transect showing variation in abundance of organisms. (a) Area barren of large benthic animals. The fish is a rattail. (b) Concentration of the echinoid *Phormosoma placenta.* (Photos courtesy of Dr. Fred Grassle; reprinted with permission from J. F Grassle *et al.*, *Deep-Sea Res.*, Vol. 22., Pergamon Journals Ltd., 1975.)

FIGURE 4.36 A typical deep-sea brisingid starfish. (Photo courtesy of Dr. Barbara Hecker.)

FIGURE 4.37 Sea pens of the genus *Kophoblemnon* in 3,000 m of water. (Photo courtesy of Dr. Barbara Hecker.)

FIGURE 4.38 A sea anemone, *Liponema brevicornis.* (Photo courtesy of Dr. Barbara Hecker.)

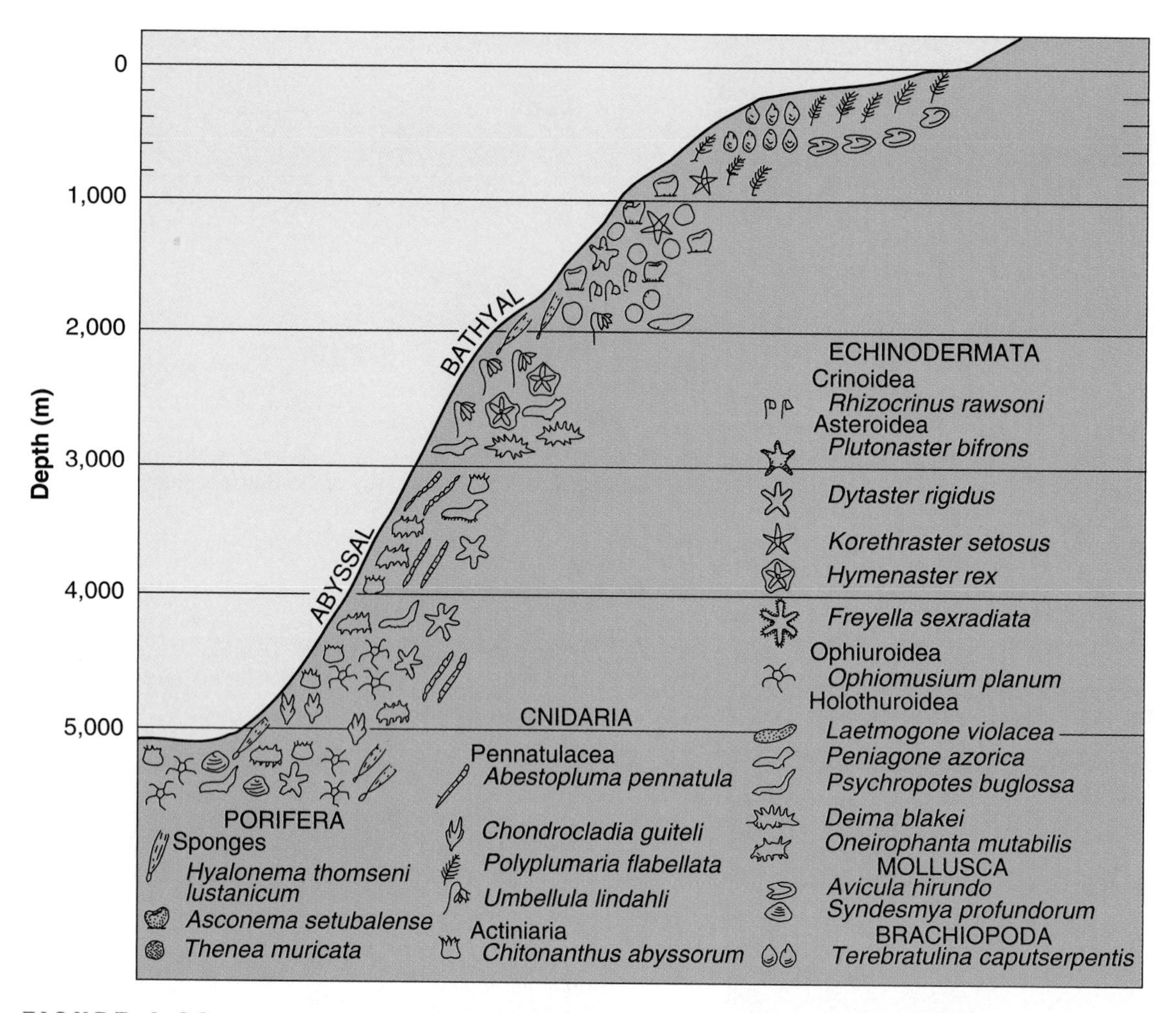

FIGURE 4.39 Zonation of benthic animals with depth on the Atlantic continental margin of Europe. (From *The Face of the Deep.* B. Heezen and C. Hollister, © 1971 Woods Hold Oceanographic Institution.)

FIGURE 4.40 A rattail fish (*Coryphaenoides filifer*) at 3,000 m. A brisingid starfish is in the foreground. (Photo courtesy of Dr. Barbara Hecker.)

the environment is virtually motionless and that calcium carbonate is difficult to deposit.

Community Structure

In contrast with the shallow-water benthos or the epipelagic, suspension feeders are scarce in the deep-sea benthos. According to Jumars and Gallagher (1982), no more than 7% of the species are suspension feeders because of very slow water movement and few food particles in the water. As a result, there is relatively little food to take out of the water and a filter feeder would have to expend more energy creating a water current than it would receive in food—a losing situation.

Carnivores also are less abundant among the invertebrates, although this is uncertain due to the difficulty of separating omnivores and scavengers from carnivores. Herbivores are present in low numbers, feeding mainly on plant remains, such as wood. The majority of animals, however, some 80% by number, are deposit feeders ingesting the organic-rich sediment.

Most attempts at analyses of the age structure of these benthic organisms through size-frequency graphs have given the same results no matter what time of the year they were sampled. Whereas such monotonous similarity in these graphs suggests that these animals reproduce throughout the year and have no seasonal breeding time, we do know that there is seasonal breeding for some deep-sea animals. Life spans for most of the deep-sea animals are not known.

In an attempt to understand colonization and recruitment rates in the deep-sea benthos, Grassle and Morse-Porteous (1987) set down trays of defaunated sediments at two sites in the North Atlantic (1,800 and 3,600 m). Results indicated similar rates of colonization at the two depths but at much slower rates than observed in shallow water. Furthermore, even after five years, the species composition and density of the experimental trays were different from those of the surrounding natural area. The most common colonizers were polychaete worms of the families Spionidae, Capitellidae, and Sigalionidae. Experiments with organically enriched but azoic sediments demonstrated that organisms would preferentially settle in large numbers in the organically enriched areas or patches. Screening out predators resulted in high densities of colonizers. It appears that normally heavy predation coupled with occasional escapes from predation may explain the patchiness of the fauna.

Diversity

Because of the low density of benthic populations, the small size of most invertebrates, and the paucity of adequate samples, the deep sea was considered a biological desert for many years. In the early 1960s, this notion was abandoned as sampling efforts intensified and methods of dredging and sample processing improved. As a result, the desert idea was replaced by the opposite belief, first postulated by Sanders (1968), that the deep sea is a highly diverse community. The concept of species diversity is founded on the intuition that large groups of co-occurring species are characterized by a greater complexity of biological interactions. As noted in Chapter 1, species diversity has two important components: the number of species and their patterns of relative abundance. For example, if two areas have the same number of species, one area has a greater diversity if it has a more even distribution of individuals among the resident species. The idea of a highly diverse deep-sea fauna is based entirely on their relative abundance patterns. While the total number of individuals is sparse, each deep-sea species is commonly represented by only a few individuals. In other words, most deep-sea species are rare. This pattern had led to the popular contention that deep-sea diversity is as high as that of coral reefs, or among the highest in the world. In contrast, shallow-water assemblages are often dominated by one or several very abundant animals, but there are also many rare species.

The suggestion that the deep sea is a highly diverse area had the salutory effect of generating even more interest in and sampling of the deep sea by other scientists, the more critical evaluation of the Sanders work, and the generation of alternative hypotheses.

If one accepts the idea of a highly diverse deep sea, there are several conflicting hypotheses attempting to explain this diversity. The first is termed the **stability-time hypothesis**, first suggested by Sanders (1968). This says that high diversity occurs because highly stable environmental conditions have persisted over long periods of time and have allowed species to evolve that are highly specialized for a particular microhabitat or food source. Since most of the benthic animals are detritus feeders in the deep sea, this means they have become very specialized for a particular narrow range of particle sizes. Such specialization is virtually unknown among deposit feeders elsewhere in the oceans and has not been demonstrated in deep-sea benthos. Moreover, the stability-time

TABLE 4.3

Summary of General Biological Characteristics of Deep-Sea Organisms

Reproduction and Development
Few eggs, large, yolk-rich
Slow gametogenesis
Late reproductive maturity
Reduced gonadal volume
Slow embryological development
Breed usually once (semelparous)
Physiological
Low metabolic rate
Low activity level
Low enzyme concentration
High water content
Low protein content
Small size
Ecological
Slow, indeterminate growth
High longevity
Slow colonization rate
Low population densities
Low mortality due to low predation pressure

hypothesis has been falsified as an explanation of diversity patterns among the insects found in sugarcane fields and among the marine crustaceans inhabiting coral heads.

A second hypothesis is the **cropper** or **disturbance theory** suggested by Dayton and Hessler (1972). The basic premise is that organisms generally increase in numbers until they reach the limit of some resource that is in least abundance. At that point, competition occurs, and in the ensuing competition, species are eliminated (principle of competitive exclusion). Since in the deep sea, animals are scarce compared with the total amount of space available to inhabit, competition for space does not occur (except, perhaps, for animals living on hard substrates, which are rarer). The only other resource likely to be competed for is food. As noted, food is in very short supply. The cropper theory suggests that none of the deep-sea animals are food specialists and that they are generalists feeding on anything they can engulf, not distinguishing whether it is living or dead; hence, the term cropper, as one that feeds indiscriminately on anything smaller than itself. High diversity is maintained because the intense "cropping" by all levels and sizes of animals prevents any species from building up its population to the point where it would be competing with another for food. High diversity is due, then, to intense predation at all levels, which allows a large number of species to persist, eating the same food, but never becoming abundant enough to compete. A central argument against this theory is that heavy predation should result in the evolution of definite life history features. These include production of large numbers of young, a short life span, and fast growth. As far as is known (Table 4.3), these are the exact opposite of the life history characteristics found in the deep-sea organisms, which grow slowly, live long, and produce few offspring. However, if cropping results in differential mortality of age groups, either set of life history traits can, in theory, evolve.

A final hypothesis, which does adequately explain the original reports that diversity increases with increasing depth, is the **area hypothesis**. This is a very simple idea. Since species number is positively correlated with area (i.e., more area equals more species), diversity is highest in the deep sea because it covers the greatest area. Both diversity and area of sea bottom decrease with decreasing depth. This correlation is significant. Although the actual numbers of species per unit bottom area have not been established, trends in relative abundance patterns can be related to the size of large biogeographical areas. This correlation, however, applies only to samples taken in the Atlantic Ocean. Workers in the Pacific Ocean have not found a general increase in diversity with increasing depth. On the contrary, both components of diversity (species density and relative abundance) are highest at intermediate bathyal depths rather than in the abyss.

Additional data and analyses of various data sets have produced a picture that, though still not completely resolving the diversity issue, has helped to clarify some aspects of it and put the issue more in perspective. Rex (1981) has demonstrated for some benthic organisms in certain areas that diversity does not continue to increase as one goes deeper into the ocean. In fact, he has shown that diversity is higher at intermediate depths and decreases both downward and upward to the continental shelf, giving a curve of parabolic shape (Figure 4.41).

The reasons for this parabolic diversity pattern and deep-sea diversity in general still remain unanswered. Data now available, however, suggest there

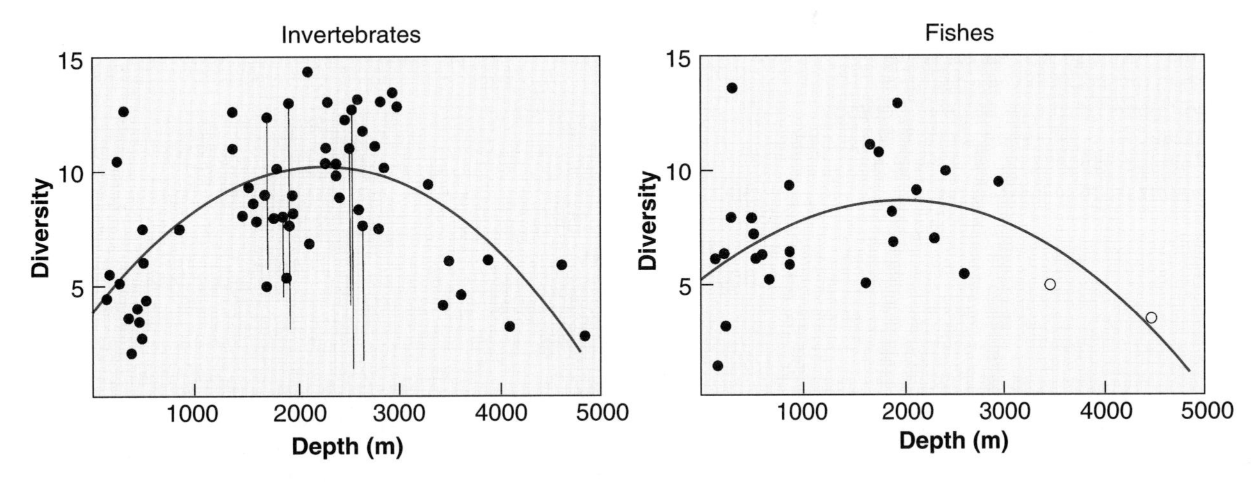

FIGURE 4.41 Patterns of species diversity in the deep-sea benthos based on samples from the northwest Atlantic Ocean. (With permission from "Community Structure in the Deep Sea Benthos" M. Rex. *Annual Review of Ecology and Systematics*, Vol. 12, © 1981 by Annual Reviews. www.AnnualReviews.org.)

are various patterns of distribution of deep-sea benthos that change with depth, and at different depths, the patterns result from differing sets of interacting factors. For example, it is now apparent from photographs and observations that the deep ocean floor is not a homogeneous, featureless plain but has various mounds, holes, fecal pellets, and tracks. Furthermore, these features, because of the stability of the water, persist for long periods of time and are, therefore, suitable "microhabitats" for various organisms. This can lead to increased diversity just as it does in the topographically diverse rocky intertidal areas (Chapter 6) and coral reefs (Chapter 9). In other words, the deep sea is "patchy," and organisms are distributed patchily. If one samples several patches, one naturally gets higher diversity. Moreover, the patches exist on different scales of size and so do the organisms. This means that diversity also changes on different size scales; thus, Rex (1981) showed that for small animals maximum diversities occurred at greater depths. This perhaps is due to longer persistence of microhabitat features or to the absence of competition with larger organisms. Predation and competition may also act differentially. In the deeper waters, the density of animals is low, due to decreased energy input from above. Predators are rare because their prey is rare or absent. In such cases, it may be that competition or microhabitat specialization is the dominant structuring force. In shallower depths, predators are more abundant due to increased prey density and may act strongly to structure the community, relegating competition or microhabitat specialization to a minor role.

Grassle and Morse-Porteous (1987) have suggested that high diversity in the deep sea is maintained by a combination of three factors: (1) the patchy distribution of organically enriched areas in a background of low productivity; (2) the occurrence of discrete small-scale disturbance events (mainly from biogenic agents) in an area of otherwise great constancy; and (3) the lack of barriers to dispersal among species distributed over a very large area.

■ In summary, it appears none of the early hypotheses are by themselves adequate to explain all deep-sea diversity patterns, and the observed patterns are the result of many factors interacting simultaneously with some factors having more influence in certain areas, at certain depths, or at certain times. These factors include microhabitat diversification due to a heterogeneous environment, energy input, competition, predation, and rate of disturbance. When we have more quantitative samples from more areas of the deep sea, perhaps this whole issue of diversity and the forces that structure the organization of communities will become clearer. ■

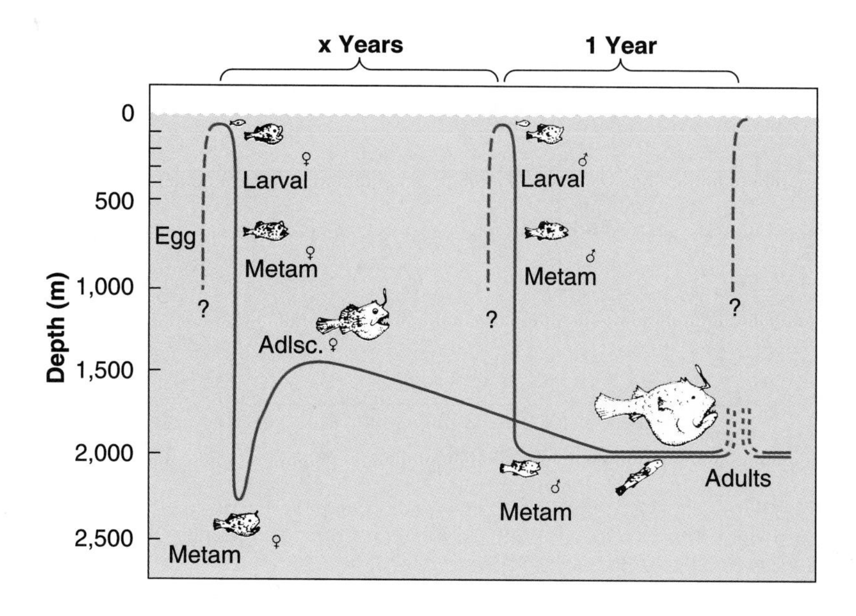

FIGURE 4.42 Life cycle of an angler fish. The fish spends its larval life at the surface and descends upon metamorphosis. (From *Aspects of Deep Sea Biology*, N. B. Marshall. Copyright ©1954 philosophical Library, New York. Reprinted by permission.)

Life History Patterns

General life history patterns with respect to reproduction and development are more difficult to generalize, primarily because we still lack data about these aspects of the life history of most deep-sea animals. In concordance with the low metabolic rate and low activity rate, however, we find that perhaps most deep-sea animals have slow or delayed sexual maturity, smaller gonads, and slow embryological development. Some, in fact, may reproduce only once during their life span (semelparous). Meade et al. (1964) report hermaphroditism to be common among fishes.

In shallow-water areas and terrestrial areas, there are usually cyclic changes in some environmental parameter that initiate and control reproduction. As a result, reproduction is usually confined to a particular time of year. In the deep sea, where light is absent and nearly all environmental parameters are monotonously constant, it would appear that a cyclic pattern of reproductive activity should be absent. What little evidence we have so far, however, is mixed. Reproductive data from benthic organisms in the deep sea off San Diego, studied by Rokop (1974, 1979), showed that reproduction is constant and continuous throughout the year with no seasonal peaks in four bivalve mollusks. However, Lightfoot et al. (1979) found a distinctly seasonal cycle of reproduction in two other bivalve species from Rockall Trough in the Atlantic. Similar findings of both seasonality and lack of it exist for various Crustacea, while continuous reproduction appears to be the pattern in deep-sea echinoderms. In some cases, seasonal reproduction has been linked to a pulse of particulate organic matter sinking from the euphotic zone after a bloom in the surface waters.

Since food is scarce, it might be suspected that most deep-sea organisms would produce few offspring, which would not spend time as larvae in the open water (see Chapter 1, pp. 23–25, on larval strategies). However, what information we have seems to suggest a mixture of strategies. In gastropod mollusks where planktotrophy, lecithotrophy, and direct development can be readily inferred from the shape, sculpture, and number of whorls in the larval shell, the data suggest that planktotrophic development increases with depth. Bouchet and Waren (1979) have reported for the Atlantic that at depths to 1,000 m less than 25% of the gastropods had planktotrophic development, while below 4,000 m over 50% had planktotrophic development. Most of the remainder had direct development. Most fishes also have pelagic eggs and larvae.

Two general reproductive patterns exist in the deep sea. In the first case, the early stages of the life history are spent in the lighted surface waters, and the juveniles migrate down to adult depths as they mature or metamorphose. Such is the pattern exhibited by certain deep-sea angler fish (Figure 4.42),

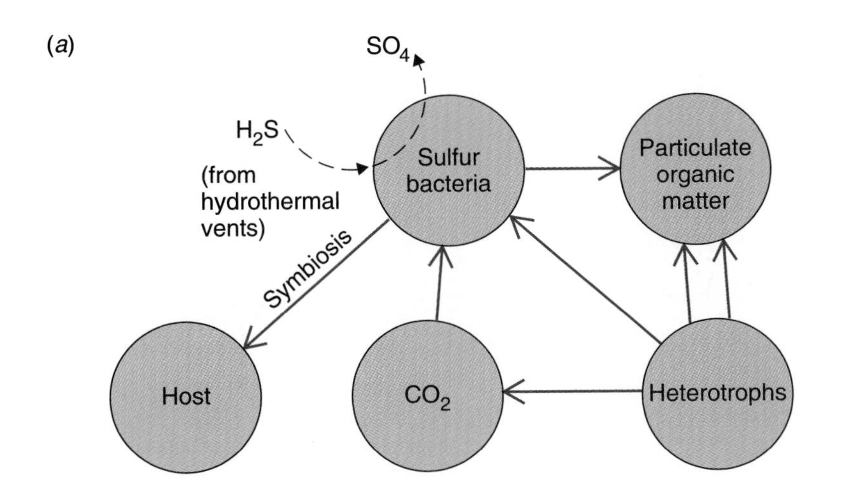

(b)

Light + H_2O

$CO_2 \rightarrow [CH_2O] + O_2$

Green plant photosynthesis

O_2 + $4H_2S$

$CO_2 \rightarrow [CH_2O] + 4S + 3H_2O$

Bacterial chemosynthesis

FIGURE 4.43 Chemosynthesis in hydrothermal vent communities. (a) Sulfur bacteria in the vents use the energy in hydrogen sulfide gas to build complex organic molecules and grow. The source of carbon is carbon dioxide gas, which is also dissolved in the water. The bacteria are either eaten by heterotrophs, live symbiotically with the larger animals (to which they furnish organic material), or die and form particulate organic matter (detritus), which in turn is used as a source of food by heterotrophs. (b) Comparison of photosynthesis and chemosynthesis. (a, after "Analysis of Marine Ecosystems" A. R. Longhurst, © 1981 Academic Press, Ltd.)

myctophids, stomiatids, many gastropod mollusks, and certain squids. In the other, no migration occurs and the young stages are spent in the same area as the adult. Such is the case with the unique cephalopod *Vampyroteuthis infernalis* and certain prawns. Table 4.3 summarizes the general biological features of deep-sea animals.

Hydrothermal Vent and Cold Seep Communities

Perhaps one of the most significant and exciting discoveries in deep-sea biology, indeed in marine biology in general, occurred in 1976. At that time, a series of dives with the deep submersible *Alvin* was made in the Galápagos rift zone some 200 miles northeast of the islands. As reported by Ballard (1977), these dives to 2,700 m revealed a spectacular abundance of hitherto unknown marine animals living in and around four hot-water geysers in an otherwise barren ocean floor. The water temperature in these vent areas was 8–16°C, considerably above the normal temperature prevalent at the bottom at these depths (about 2°C). This warm water was very high in reduced sulfur compounds, primarily H_2S. The hydrogen sulfide was found to be used as an energy source by bacteria (Figure 4.43; see also Chapter 6 on sulfide bacteria in mud shores).

Since that time, further exploration with submersibles and with towed camera systems has revealed similar vent communities, at depths ranging from 1,500 to 3,200 m, at several additional locations in the eastern Pacific Ocean (Figure 4.44). In addition, other hydrothermal vent communities have been reported from the western Pacific, in the Marianas Back Arc Basin (3,590–3,660 m), in the North Fiji and Lau basins (1,990 m), and in a subduction zone off eastern Japan. Tunnicliffe et al. (1998) report that about 30 hydrothermal sites have now been the subject of study by biologists (Figure 4.44). Similar communities of organisms have been found that at least in part depend upon chemoautotrophic production associated with the emission of reducing chemicals that are not always associated with the edges of tectonic plates. These are called **cold seep** communities and have been found in 24 deep-sea locations in the Atlantic, the eastern and western Pacific, and the

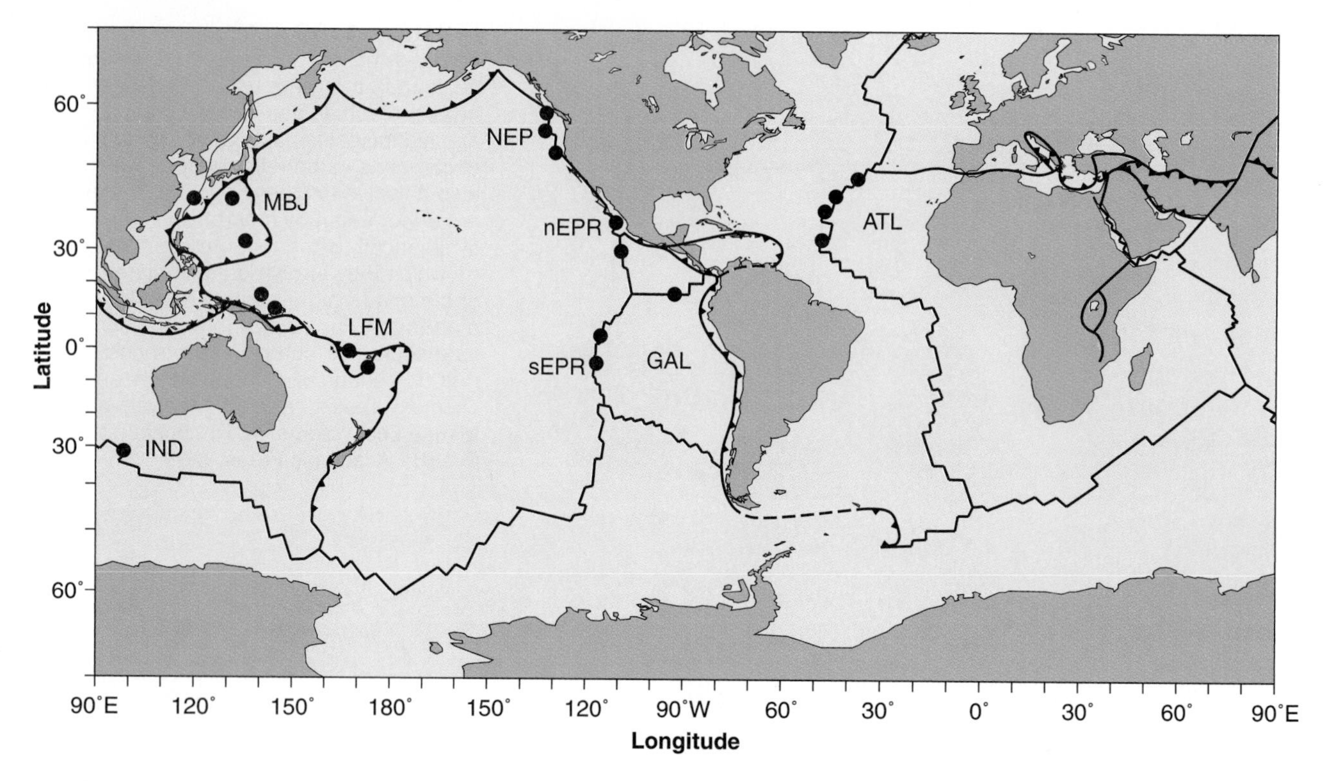

FIGURE 4.44 Distribution of major vent fields around the world. Each field may include several sites. IND = southeast Indian Ridge, LFM = Lau, Fiji, Manus, Woodlark, and Lihir sites; MBJ = Marianas, Bonin, and Okinawa sites; NEP = Explorer, Juan de Fuca, and Gorda ridges; nEPR + 9° through 21°N East Pacific Rise; GAL = Galápagos Rift; sEPR = 17° through 25°S; ATL = mid-Atlantic Ridge sites. (From "A Biogeographical Perspective of Deep-Sea Hydrothermal Vent Fauna" Tunnicliffe et al., *Adv. In Marine Biology*, vol. 34, p. 360, 1998, AP London.)

Mediterranean Sea at depths between 400 and 6,000 m (Figure 4.45). The largest number of sites, 13, are found off North America. According to Sibuet and Olu (1998), seeps are found on both active and passive plate margins. Known American seep sites include the Gulf of Mexico at the base of the Florida Escarpment at 3,200 m, on the continental slope off Louisiana at 600–700 m, in Monterey Bay at 900 to 3,500 m, and two sites in the Atlantic, one off of eastern Canada and the other off North Carolina. Undoubtedly, more will come to light as deep-sea exploration continues. Some of these cold seeps are associated with "cold" hypersaline brines or other hydrocarbon seeps. The locations of the presently known hydrothermal vents and hydrocarbon seeps are shown in Figures 4.44 and 4.45.

All these vent areas and cold seeps display an astonishing assemblage of large (megafauna) animals forming a unique association for the deep sea. Truly they are "oases" in an otherwise low-density, low-productivity area. Each hydrothermal vent system seems to have an ecologically similar set of dominant species, though not always a similar set taxonomically. Tunnicliffe (1992) has reported a total of one new class, three new orders, and 22 new families of animals from the vent systems. Eighty-two percent of the 443 species recovered from the vents since their discovery in 1976 appear to be endemic to the vents, and most were previously unknown (Figure 4.46). The dominant animals, by size, at these vents usually include one or more of the following: large clams, mussels, and vestimentiferan worms (Figures 4.47, 4.48, and 4.49). The latter belong to a special deep-sea taxon formerly known as the Pogonophora and are characterized by the complete lack of a digestive tract. (Pogonophora used to be classified as a phylum but has now been reduced to a family of polychaete worms.) Vestimentiferans appear absent

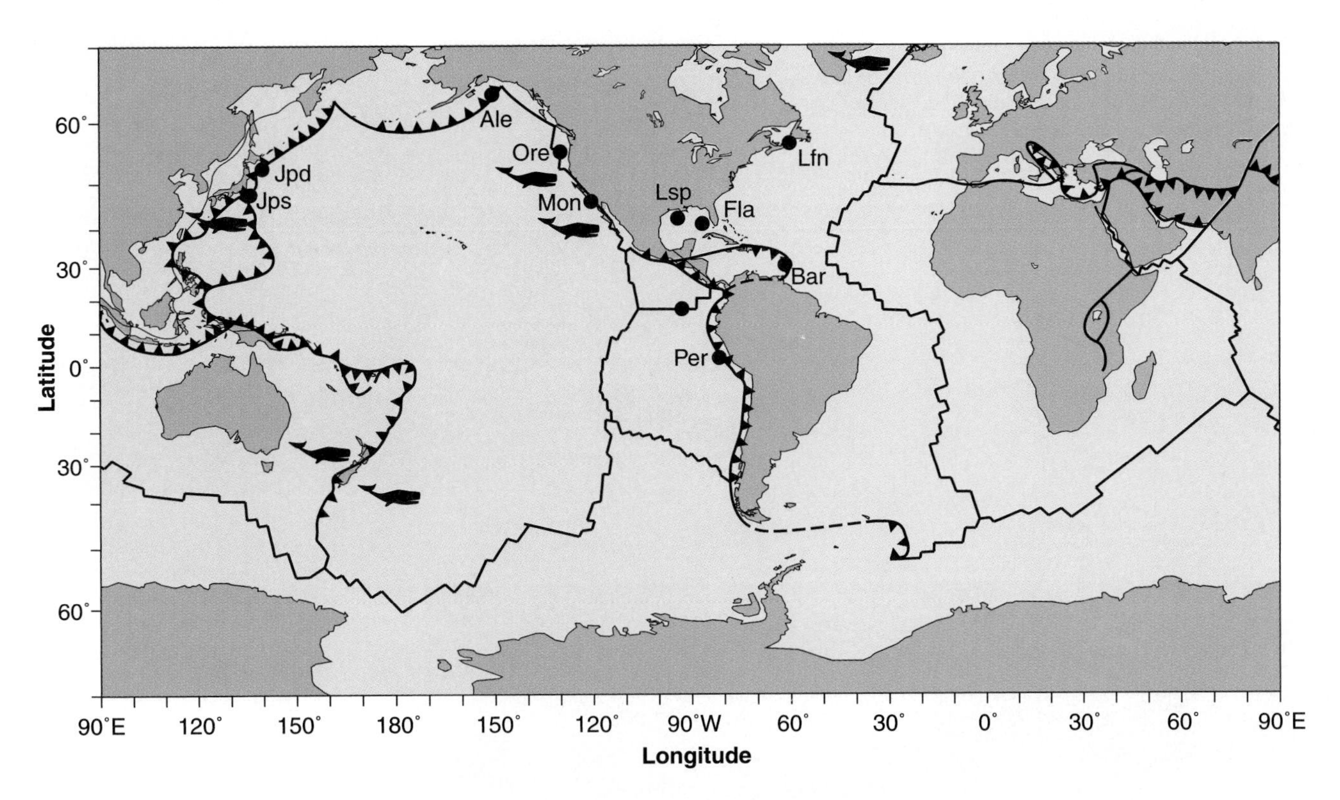

FIGURE 4.45 Locations of cold seeps (black circles) and whale carcasses (black whale symbols) in the world's oceans. Only the better known ones are illustrated. Jps = Shallow Japan; Jpd = Japan Trench; Ale = Aleutian Trench; Ore = Oregon Margin; Mon = Monterey Canyon; Per = Peru Margin; Lsp = Louisiana Slope; Fla = Florida Margin; Bar = Barbados; Lfn = Laurentian Fan. (From "A Biogeographical Perspective of Deep-Sea Hydrothermal Vent Fauna" Tunnicliffe et al., *Adv. In Marine Biology*, vol. 34, p. 383, 1998, AP London.)

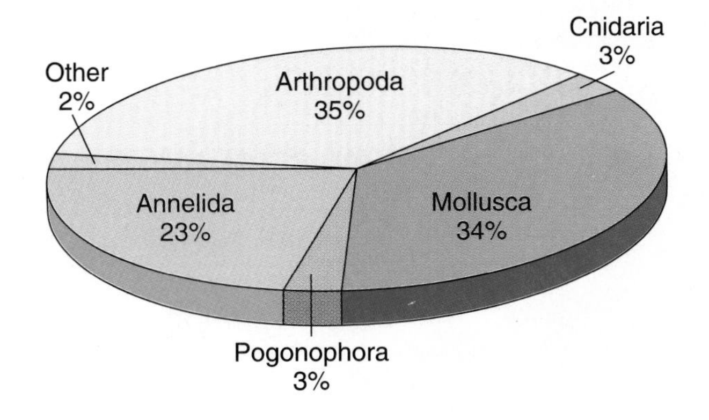

FIGURE 4.46 Overall composition of the total invertebrate fauna as recorded from the vents. (From "A Biogeographical Perspective of Deep-Sea Hydrothermal Vent Fauna" Tunnicliffe et al., *Adv. In Marine Biology*, vol. 34, p. 381, 1998, AP London.)

from the West Pacific vent systems where the dominant symbiont-containing animal is a new species of gastropod mollusk, *Alvinoconcha hessleri*. Other large animals associated with vents include crabs, a large number of snails, a variety of polychaete worms, an odd benthic siphonophore, enteropneust worms (called "spaghetti worms" from their appearance in photos; Figure 4.50), and five new species of fishes. The faunal components of the vent systems are primarily ancient and highly adapted to live in these

FIGURE 4.47 A vent clam field dominated by the giant white clam *Calyptogena magnifica.* (Photo courtesy of Woods Hole Oceanographic Institution. Photo by George Somero.)

FIGURE 4.48 Large mussels (*Bathymodiolus*) and galatheid crabs of the genus *Munidopsis* around a hydrothermal vent. (Photo courtesy of Woods Hole Oceanographic Institution. Photo by Robert Hessler.

FIGURE 4.49 Giant vestimentiferan tube worms, *Riftia pachyptila,* near a Galápagos thermal vent. (Photo courtesy of Woods Hole Oceanographic Institution. Photo by John Donnelly.)

FIGURE 4.50 Mats of enteropneust worms, *Saxipendium coronaturn*, called "spaghetti worms," draped over rocks around hydrothermal vents. (Photo courtesy of Woods Hole Oceanographic Institution. Photo by James Childress.

areas with highly toxic chemicals. This suggests that the suite of adaptations necessary for exploiting the vents represents a considerable barrier to evolving other groups to colonize these habitats.

The cold seeps and hypersaline brine communities are less well known. According to Sibuet and Olu (1998), the dominant seep species are bivalve mollusks of several families, pogonophoran worms, and certain sponges. Of the 211 species known from seeps at present, a large majority are endemic and only 13 species are shared between vents and seeps. Diversity at some of the seep sites also appears to be higher than at vents. Whereas the vent systems seem to depend primarily on the utilization of sulfides for energy, the cold seep areas may depend on sulfide and/or methane (Figure 4.51).

The final, and probably the least known, of these deep-sea sulfide communities was discovered in 1987 when the submersible *Alvin* chanced upon a 21-m whale skeleton at 1,240 m in the Santa Catalina Basin off southern California. Subsequent study of this skeleton revealed chemolithoautotrophic bacteria and certain other animals associated with vent systems on the outside of the bones, and vesicomyid clams (like the *Calyptogena* at the vents) in the sediments around the bone. In this case, the sulfide source was produced by anaerobic bacteria inside the whale bones decomposing the large lipid content of the bone (Figure 4.52). As the sulfide diffused out of the bone, it provided the energy source for the bacteria. Although whales have been much reduced throughout the world's oceans, Smith (1992) estimates that such sulfide-rich whale falls may have an average spacing of one per 25 km in the North Pacific and may give credence to the hypothesis that such falls may be the stepping stones that permit the sulfide-based communities to disperse over the vast distances between the vent systems.

All of the organisms of these sulfide or methane communities are confined to an area within only a few meters of the sulfide or methane source. Both vents and seeps may be separated from other such associations by hundreds or thousands of kilometers. The latter has aroused the interest of marine biologists curious about how such communities not only persist, but how they disperse to find these vents; they are virtually looking for a needle in a haystack. Although we still have much to learn about these communities, we have made much progress since their discovery.

All vent systems and cold seeps depend on the primary productivity of **chemolithoautotrophic bacteria** to survive. Potential energy in the form of reduced sulfur compounds, such as hydrogen sulfide, is spewed from the hot vents. Cold seeps ooze either sulfides or, in some cases, methane or other hydrocarbon (see Figure 4.51). These reduced compounds are oxidized by the chemolithoautotrophic bacteria to release energy, which they use to form organic matter from carbon dioxide (see Figure 4.43). Since the vent water is anoxic, the bacteria depend on the surrounding cold ambient water for the oxygen necessary to produce the organic material. Because sulfide is rapidly oxidized by oxygen in the ambient water and vent waters are rapidly diluted, sulfide does not persist in the water column for any distance from the vents. This is why the organisms that require both oxygen and sulfide must live in a very narrow volume where vent water and ambient water begin to

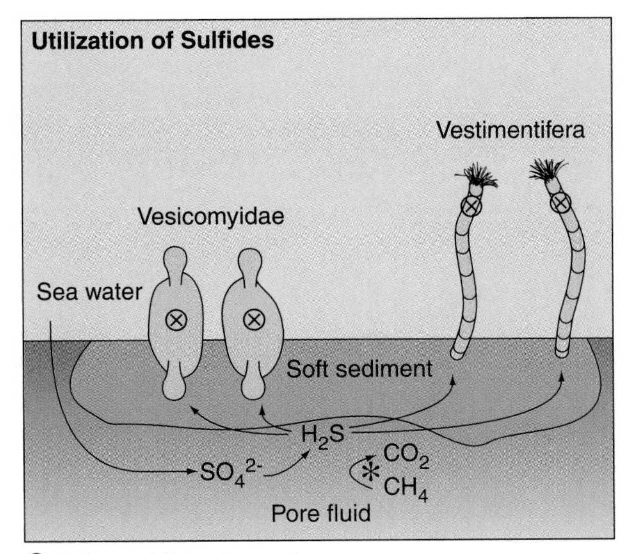

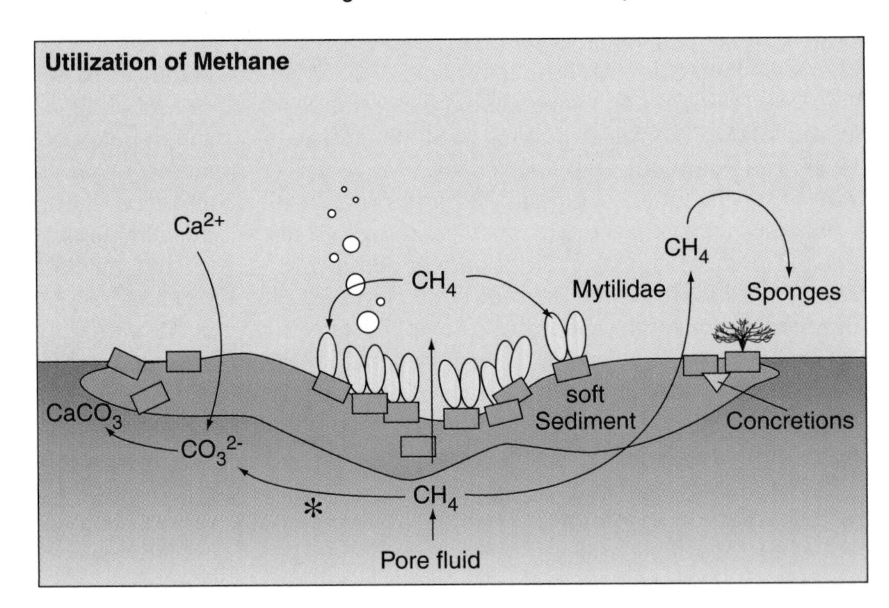

FIGURE 4.51 Diagrammatic representation of the utilization of sulfides by clams and tube worms and of methane by mussels and sponges. The activity of both methanotrophic bacteria and sulfate-reducing bacteria in the sediments results respectively in the oxidation of methane from pore fluid to carbon dioxide and in the reduction of seawater sulfate to sulfide. (From *Deep Sea Research II,* Vol 45, p. 532, 1998. Copyright 1998 with permission from Elsevier Science.)

interact. Productivity is very high. Karl et al. (1984) measured production rates at one vent of 19 μg C/g/h. The primary consumers are the various animals that filter the bacteria from the water, graze the bacterial film from the rocks, or are locked in a symbiotic association with the bacteria (see Figure 4.51).

The animals that form a symbiotic association with the chemolithoautotrophic bacteria in the vent systems are of particular interest. The large vestimentiferan worm, *Riftia pachyptila*, and the largest clam, *Calyptogena magnifica*, both contain symbiotic chemosynthetic bacteria in their tissues. *Riftia* has no digestive tract as an adult, and there is evidence that it obtains its nutrition from the bacteria. Although the clam retains its ability to filter feed, it too apparently derives most of its nutrition from the symbiotic bacteria. What is curious about this system is that for the bacteria to function they must have a reduced sulfide source. According to Arp and Childress (1983), this is brought to them by the blood of *Riftia*

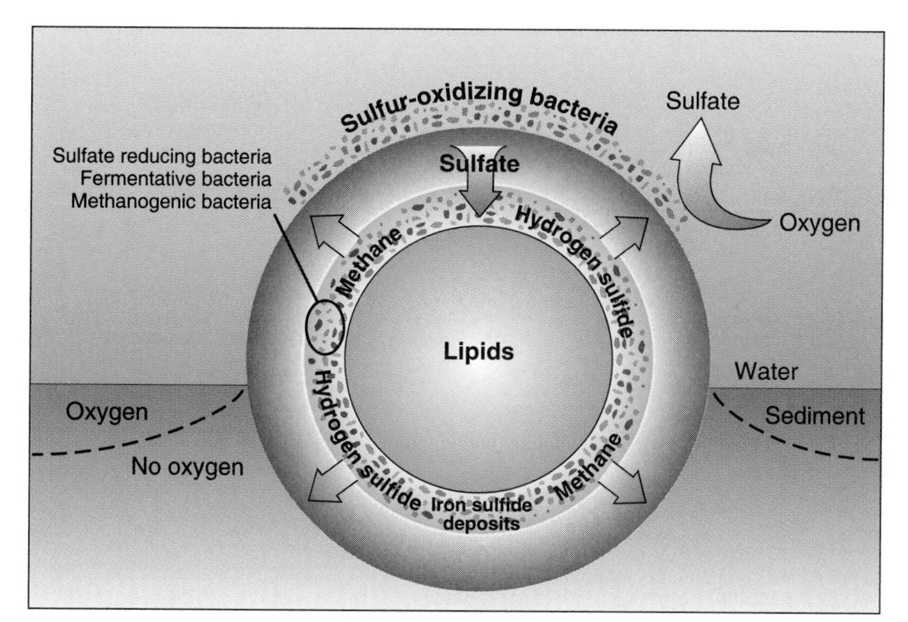

FIGURE 4.52 Diagrammatic representation of the chemical changes occurring in whale bone. (After Whale Falls, C. Smith, 1992, *Oceanus*, Vol. 35, No. 3, p. 77. Copyright © 1992 Woods Hole Oceanographic Institution.)

and *Calyptogena* (see Figures 4.47 and 4.49). Under normal circumstances this would poison or kill a typical aerobic animal, since sulfides bind to the cytochrome system, interfering with oxygen uptake and, therefore, respiration. The bloods of *Riftia* and *Calyptogena*, however, have in them a protein factor that binds to the sulfide to transport it, freeing it again in the presence of the bacteria that take it up. This is assumed since there is no inhibition of oxygen uptake in the presence of sulfide. How other vent animals deal with sulfide is not as yet known. In the cold seeps and brine pools, by contrast, the dominant mussels (*Bathymodiolus*) have a methane-based symbiosis. The seeps and brine pools are rich in methane, which is taken up by the gills of the mussels, where intracellular bacteria oxidize the methane and provide energy for the mussels and the bacteria (see Figure 4.51).

Hessler and Smithey (1983) have noted that vent animals are distributed in the vent systems with respect to water temperature (Figure 4.53). Water temperature in the vent systems is quite variable. Water that rises directly to the surface following contact with the hot subsurface rock can exit at 350°C, forming the "black smoker" chimneys. No animals are associated with these extreme temperatures. Other vent water may mix with cooler water on its way to the surface, giving rise to temperatures that vary from near ambient to 50°C. Most of the vent animals are associated with water that is below 30°C and with hydrogen sulfide levels of 400 μm or less. It is curious that rather than being exposed to water that is the average of the mixing of vent and ambient, most vent animals appear to live where they encounter discrete parcels of the two water types and must be adapted to considerable temperature change that occurs rapidly. As noted above, most vent animals appear to be restricted to the vent areas, and we do not know what keeps them from invading the adjacent deep-sea area, nor do we know what prevents the normal deep-sea fauna from invading the vent systems.

Childress and Fisher (1992) suggest that the vent systems are very unstable environments that can vary in structure and water flow over time periods as short as weeks or months. There are also long-term changes brought on by a reduction in the flow of vent water and eventual cessation of water flow. Vent communities may also be destroyed by the outpouring of fresh lava. Just such an event was reported by Tunnicliffe (1992) for a site on the East Pacific Rise.

According to Grassle (1985), the life span of individual vents is only in the tens of years, and scientists have observed inactive vents with the remains of vent animals. This suggests that these vent species have life spans also measured in decades. Recent estimates of the life span of *Calyptogena* by Grassle (1985) gave the ages of the oldest at 30 years. However, the growth rate and size reached by *Calyptogena*

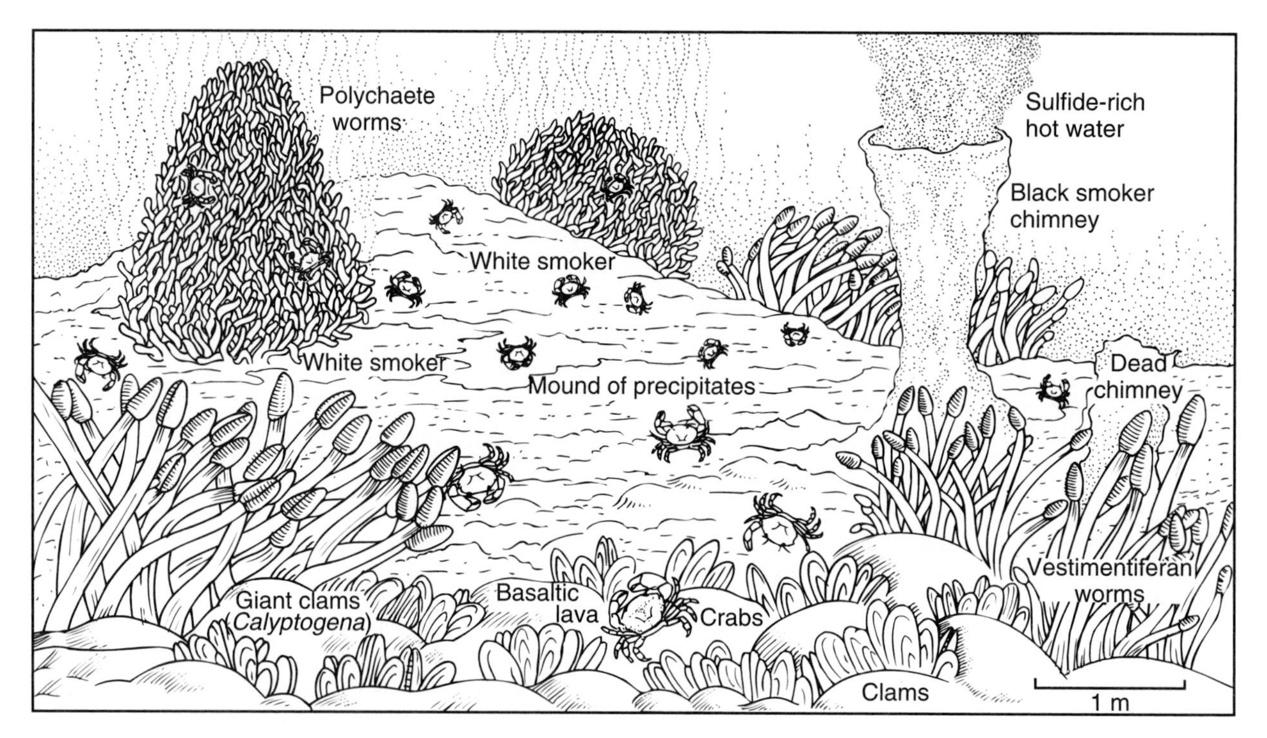

FIGURE 4.53 A diagrammatic representation of a typical vent community, showing the presence of the larger macrofaunal organisms. (Modified from "The Geology of Deep Sea Hot Springs" K. M. Haymon and K. C. McDonald, *American Scientist*, Vol. 73, No. 5, pp. 441–449, 1985. Copyright © 1985 American Scientist. Reprinted by permission.)

varies depending on the vent site, suggesting that the growth rates among vent animals may be more variable than we think. Given these circumstances, it would seem that the vent animals would experience strong selective pressure to grow rapidly to sexual maturity, produce many young, and have an efficient means of dispersal. What we know now suggests that the dominant animals do grow much faster than surrounding deep-sea animals away from the vents, but we are uninformed about reproductive potential for most. In the case of dispersal, we are similarly without much data except for the Mollusca. Among the gastropod mollusks at the vents, a strange situation seems to prevail. Most species seem to lack planktotrophic larvae and have nonplanktotrophic development; hence, these animals have low dispersal. It is an enigma how such development could facilitate the location of widely separated vent systems, but somehow it must, because Grassle (1985) reports that close relatives of present-day vent animals have existed for 200 million years.

Certain areas of the hydrothermal vents in the Galápagos have been the object of follow-up studies, reported by Childress (1988), that have demonstrated how quickly these areas can change. The particular area called "the Rose Garden," when discovered in 1979, was dominated by large numbers of vestimentiferan worms and contained significant, but fewer, numbers of large clams and mussels. When revisited in 1985, a little over five years later, the worms had markedly declined, while the mussels had dramatically increased and dominated the area. The clams were also abundant. Among the associated fauna, filter feeders had generally declined, siphonophores and enteropneusts were virtually absent, and mobile scavengers and carnivores, such as galatheid crabs and gastropod mollusks, were more abundant. According to Hessler et al. (1988), these striking community changes seem to be a result of two factors. The first was a change in the physical conditions that was manifested mainly by a decline in the vent water flux, which led to a decline in the vestimentiferans, and probably also the filter feeders, but favored the mussels. The second was continuing recruitment into the area, which favored the crabs, mussels, and clams. It is also possible, however, that the vesti-

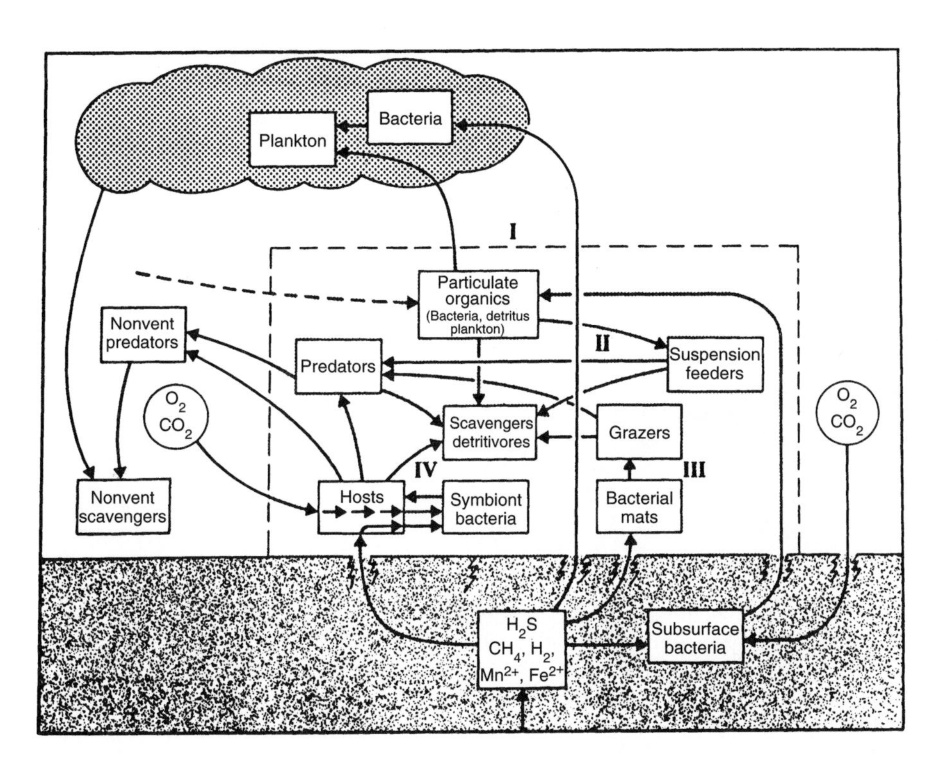

FIGURE 4.54 The food web of a hydrothermal vent community. The four distinct carbon-energy pathways are indicated by the roman numerals. I, Export production of bacteria and particulate organic material in the vent plume. II, Suspension-feeding pathway involving indigenous vent filter feeders. III, Grazer-scavenger pathway involving bacterial mats. IV, Symbiotic chemolithoautotrophic bacterial pathway involving large invertebrate hosts. (After "A Biogeographical Perspective of Deep-Sea Hydrothermal Vent Fauna" Tunnicliffe et al., *Adv. In Marine Biology*, vol. 34, p. 341, 1998, AP London)

mentiferans declined as the mussels grew and increased, effectively cutting off the vestimentiferans from their sulfide source by absorbing the sulfide before it could get to the more distant absorbing crowns of the worms. This latter scenario does not require any change in the water circulation.

It would appear that mortality, at least among the vestimentiferan worms, is very high. Tunnicliffe et al. (1990) have shown that, for the vestimentiferan *Ridgeia pescesae* on the Juan de Fuca vent system, 44% of the animals were removed over a 26-day period due to collapsing sulfide/sulfate chimneys on which they lived or to predation by fish.

As a result of observations at the Rose Garden, a general pattern of community change over time has been suggested by Hessler et al. (1988) for the Galápagos hydrothermal vents. Beginning with an established vent system dominated by vestimentiferans, clams, and mussels, one sees the vestimentiferans gradually diminish, while the clam and mussel populations grow. This may be the result of a decrease in water flux or sulfide concentration or the result of competition with the more efficient mussels. As flow diminishes in the vents, suspension feeders will decline as will the clams, leaving the mussels as the most persistent. Since the same suite of dominant organisms is not present at all hydrothermal vents, the applicability of this successional scheme is unknown.

Although most attention has been directed at the vestimentiferan worms and the large clams and mussels because of their size and symbiotic relationship with the chemolithoautotrophic bacteria, it should be noted that most vent species do not have symbiotic bacteria and feed by grazing on bacteria or filtering. Although at our current state of knowledge it is not possible to quantify a food web for the vents, our knowledge is sufficient to outline a trophic structure (Figure 4.54). Such a structure is characterized by the simplicity of the relationships and limited by the supply of chemicals required for the chemolithoautotrophic reactions that supply most of the energy that fuels the system. There are four pathways for carbon and energy flow within the food web. The first is the export of production in the form of bacteria and organic particulate matter through the vent plume that is found at levels of 100–500 m above the benthic vent community and is carried away by currents. This production is fed upon by various plankton organisms, which occur in abundance in the plume. The second path is a suspension feeding loop in which the bacteria, particulate organic matter, and perhaps some

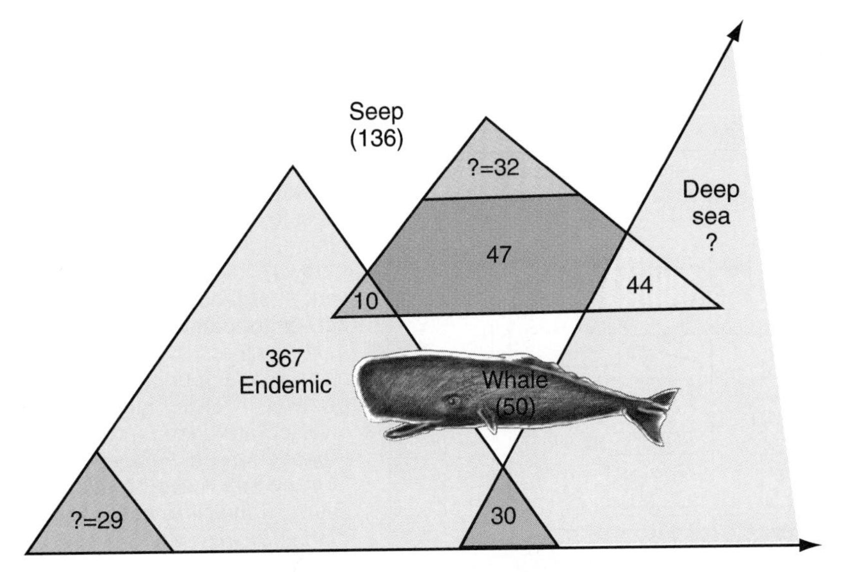

FIGURE 4.55 Overlap of species among sulfide-rich habitats and the deep sea. Vent and seep habitats are represented by triangles overlapping with each other and with the deep sea. Species numbers in the triangles represent the number of endemics; the numbers of overlap species are also indicated (? means species not yet identified). The whale represents the species found on decaying whale bones. Of the total whale bone species, two are known from whale bones, seeps, and vents; one is found at bone, vent, and deep sea; four are known from vents and bones; while one is known only from bone and seeps; the rest are endemic to bones. (From "A Biogeographical Perspective of Deep-Sea Hydrothermal Vent Fauna" Tunnicliffe et al., *Adv. In Marine Biology*, vol. 34, p. 393, 1998, AP London)

indigenous plankton are removed from the water by resident vent filter feeders. A third trophic pathway is by grazing. The chemolithoautotrophic bacteria and other bacteria form mats over the bottom and are grazed upon by numerous benthic animals among which the gastropod mollusks are the most abundant. The final pathway is through the symbiotic chemolithoautotrophic bacteria that reside in the tissues of the dominant vestimentiferan worms, clams, and mussels. This pathway has the greatest biomass in the vent systems. In turn, these dominant animals are eaten by various predators both in and out of the vent community. Scavengers receive input from all the pathways, and some of the vent animals may be involved in more than a single pathway.

Biogeographically, the three chemosynthetic-based communities display a very high degree of endemism and very low overlap with each other or the surrounding deep-sea benthos (Figure 4.55). The seeps have a greater depth distribution than the vents and also show a decrease in species richness with depth. Single-site species richness is higher at seeps then at vents and may be due to the longer-term stability of the seeps.

MIDWATER COMMUNITY ECOLOGY

Because of the increased difficulty in sampling pelagic deep-sea organisms due to net avoidance and migration, the lack of observations on these animals from deep submersibles, their avoidance of light, and their small size, we know less about the ecology of midwater animals than we do about the deeper but sedentary and sessile benthos.

Deep Scattering Layers

Early in World War II, when the use of echo sounders was being experimented within the oceans to detect submarines and the bottom, the devices (now known as sonar) regularly began to pick up echoes. They

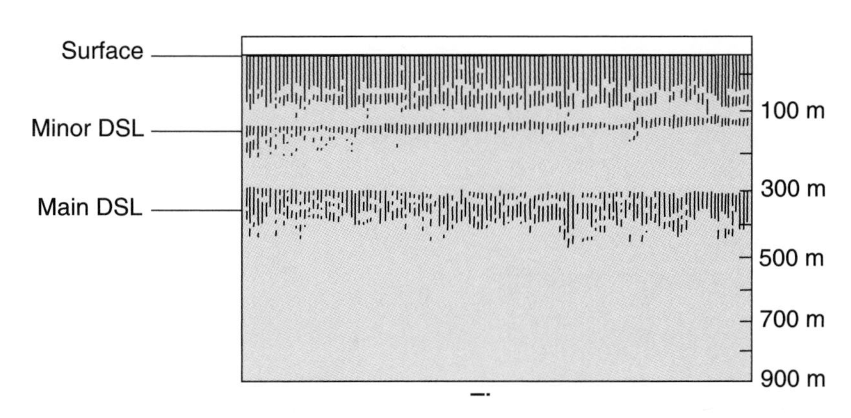

FIGURE 4.56 A sonogram recording showing the presence of the deep scattering layers in the Indian Ocean. (After G. B. Farquhar, ed., 1970, Proceedings of an international symposium on biological sound scattering in the ocean. U.S. Government MC Report 005.)

came, not from the bottom, but from sources in the water column itself, hundreds or even thousands of meters above the known bottom. These areas of sound reflection in midwater were termed **deep scattering layers (DSLs)**. The depths of the DSLs varied. During the day, there were often two or three layers varying in depth from 200 to 700 m (Figure 4.56). At night, these different layers migrated toward the surface, where they often merged into a single broad band. At dawn, the layers returned to depth. Although first interpreted as the result of some physical discontinuity in the water, the DSLs are now known to be concentrations of midwater animals, and the movement of the layers toward the surface at night is the result of a vertical migration similar to that undertaken by surface zooplankton (see pp. 77–80). Presumably, these midwater animals are migrating into the surface waters at night to feed on the abundant plankton.

The major animals forming the DSLs include the lantern fish (Myctophidae) and other similar mesopelagic fishes (see Figure 4.20), shrimplike crustaceans such as euphausiids (krill), sergestid and pasaphaeid shrimp, various squids, and certain deep-floating hydrozoan cnidarians called siphonophores (see Figures 4.14, 4.15, 4.16, and 4.19). The composition of the scattering layer varies with geographical area.

Deep scattering layers have been found in all oceans except the Arctic. They are best developed in areas with high surface productivity and are faint in areas of low productivity.

Species Composition and Distribution

The pelagic zones of the deep sea are dominated by small fishes of the families Myctophidae, Gonostomatidae, and Sternoptychidae, and euphausiid and decapod crustaceans (Figure 4.57). Blackburn (1977) has noted that these combined groups constitute more than 80% of the biomass, at least in the mesopelagic zone. One estimate for the Pacific Ocean has fishes making up 20–45% of the biomass, shrimps 15–25%, and euphausiids 35–50% in the upper 1,000 m of water. Minor components include other crustaceans, such as mysids, amphipods, and copepods, chaetognaths, scyphozoan jellyfish, and cephalopods. This dominance of the pelagic deep-sea fauna by small fishes, euphausiids, and shrimps seems to be widespread in all oceans, but as Omori (1974) has noted, the relative abundance of each group changes with latitude and depth (Figure 4.58). Seasonal changes in the composition of the assemblage appear minor but have received little study.

There are two components of the pelagic species assemblages in the mesopelagic zone. One includes those animals that do not migrate vertically—certain fishes (*Cyclothone*, *Sternoptyx*), mysids, some shrimps (sergestids, penaeids, carideans), and a few euphausiids, amphipods, cephalopods, and cnidarians. The other component is called the **interzonal fauna** and undergoes a diurnal vertical migration. These animals live below 450 m by day and migrate into surface waters by night. This component includes most of the mesopelagic fishes (Myctophidae, Gonostomatidae) and most euphausiids and decapods; hence, the composition of the community changes depending on whether it is sampled at night or during the day (Figure 4.58).

Food and Feeding

Information on the food of midwater organisms is sparse, but is most adequate for two dominant

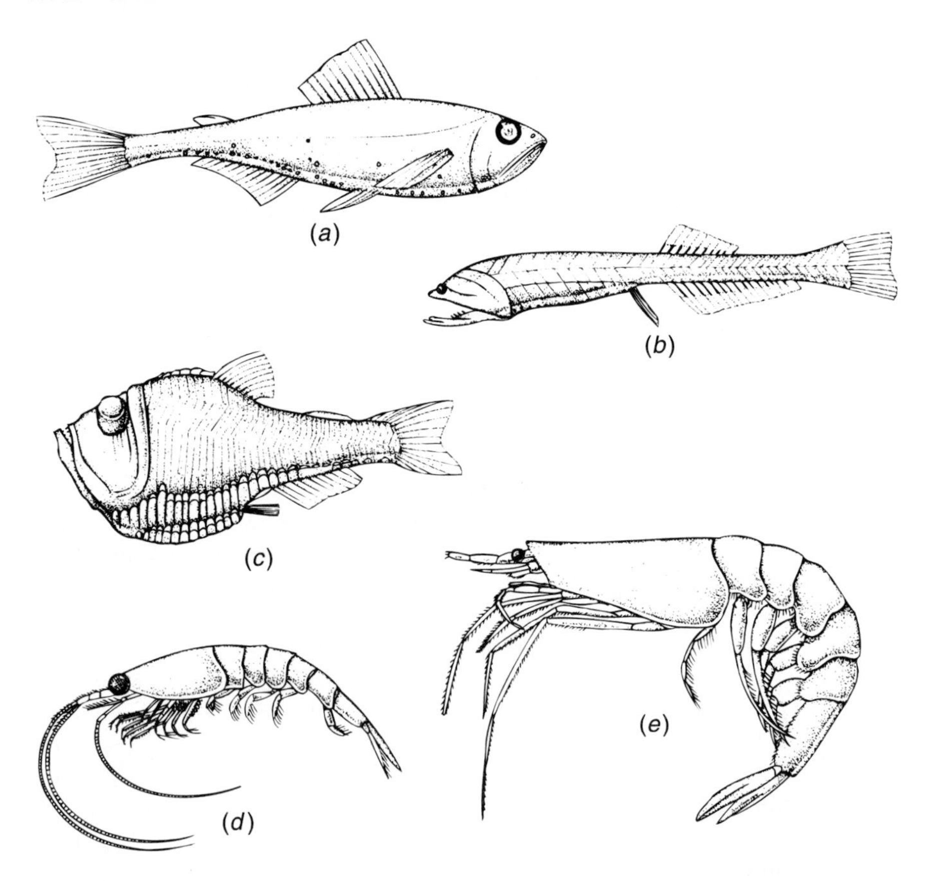

FIGURE 4.57 Representatives of the major groups of organisms of the pelagic zones of the deep sea. (a) Lantern fish, *Tarletonbeania crenularis* (Myctophidae). (b) Gonostomatid fish, *Cyclothone pallida* (Gonostomatidae). (c) Hatchet fish, *Argyropelecus affinis* (Sternoptychidae). (d) Euphausiid, *Euphausia pacifica*. (e) Sergestid shrimp, *Sergestes similis*.

groups, the fishes and decapod crustaceans. Studies of the diet of midwater fishes have been concentrated on the numerically abundant families Myctophidae, Gonostomatidae, and Sternoptychidae. Whereas there is often considerable variation in species composition in the diets of these fishes, the principal components in all cases, as noted by Hopkins and Baird (1977), are smaller crustaceans, such as copepods, euphausiids, ostracods, amphipods, and small decapod shrimps (Table 4.4). Of these groups, Gartner et al. (1997) indicate copepods are consumed the most, but there may be considerable selectivity.

Changes in the diet of these fishes occur with age. In general, the larger and older individuals take a greater proportion of the larger crustaceans than do the smaller fishes (Table 4.4). Seasonal changes in diet have also been observed for some fishes, and they may reflect availability of prey. Geographical variation in diet also occurs in widely distributed species. Although it is generally thought that fishes in these energy-poor deep-water environments should be opportunists and generalists, Gartner et al. (1997) report that true generalists are rare.

Omori (1974) reports most decapod crustaceans are also predators or omnivores using various available food items. Analyses of stomachs of midwater shrimps suggest that copepods, euphausiids, and other small crustaceans constitute the majority of the food. There is also evidence of a certain amount of cannibalism. The shrimps are not specialists, but opportunists, taking what is available within a certain size range.

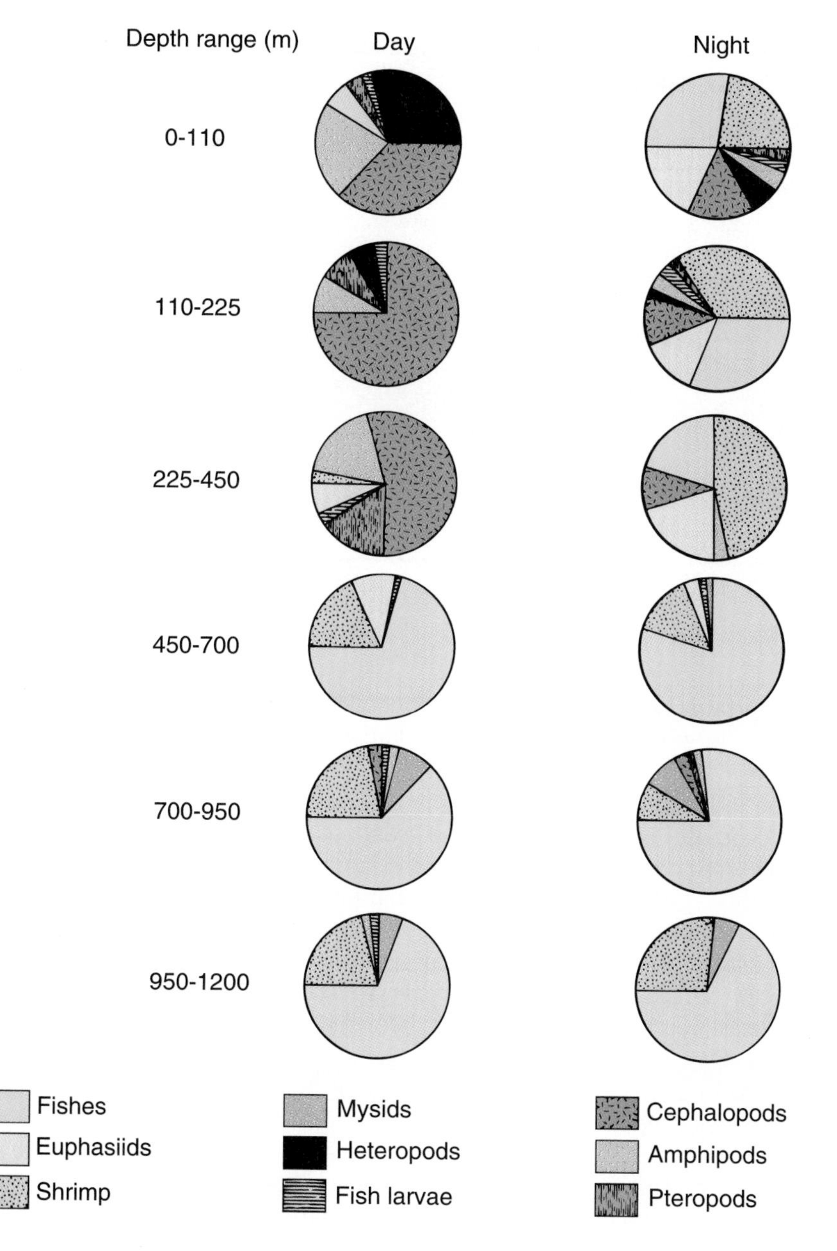

FIGURE 4.58 Differences in biomass composition of the midwater community in the tropical Pacific with depth and between day and night. (Based on *Oceanic Sound-Scattering Prediction*, N. R. Andersen & B. J. Zahuranec eds., 1977 Plenum Publishers.)

In spite of the paucity of food in the pelagic areas of the deep sea—a condition that causes the organisms to have the varied diets observed among fishes and crustaceans—there is a certain amount of selectivity and food resource partitioning among animals. The most common form of food resource partitioning occurs when species feed in different depth zones in the water column. Different prey items or different prey abundances occur in different depth zones; thus, predators feeding in these different zones encounter and consume prey populations of differing composition. Even within a given depth zone, the prey composition of the resident predators varies, indicating a certain degree of resource partitioning. This is probably due to anatomical differences in the prey capture mechanisms of the predator. Such structural

TABLE 4.4

Changes in Diet with Size and Age in Six Midwater Fishes

	Size Range (mm)	Percent of Food <5 mm	Percent of Food >5 mm	Trend in Diet Composition (Increasing Percent)
Argyropelecus aculeatus (WN central Atlantic)	10–20	99	1	Ostracods, copepods → fish
	50–60	12	88	
Argyropelecus affinis (off southern California)	21–34	94	6	Copepods, ostracods → euphausiids, salps, chaetognaths
	52–68	68	33	
Argyropelecus sladeni (off southern California)	10–20	66	34	Ostracods, copepods, chaetognaths → euphausiids
	30–50	24	76	
Sternoptyx diaphana (Pacific subantarctic)	10–24	100	0	Copepods, ostracods → euphausiids
	40–80	60	40	
Sternoptyx obscura (off southern California)	10–20	86	14	Copepods → ostracods, amphipods, chaetognaths
	30–50	63	37	
Valenciennellus tripunctulatus (E. Gulf of Mexico)	30–55	85	15	Small copepods → larger copepods
	30–35	58	42	

Source: Oceanic Sound-Scattering Prediction, N. R. Andersen and B. J. Zahuranec, eds. 1977

modifications allow them to prey only on certain prey sizes or shapes. Since many organisms migrate through depth zones on a diurnal basis, the available prey composition varies over a 24-hour period at any given depth. Therefore, a predator that feeds during the day at a given depth may encounter different prey than one feeding at the same depth at night.

In a study of the potential competition among mesopelagic fishes and shrimps in the eastern Gulf of Mexico, Hopkins and Sutton (1998) found considerable niche overlap among the species if single niche parameters were considered. However, if the spectrum of niches, including food size, food composition, and vertical distribution, were considered together, then they found considerably less niche overlap and resource partitioning existed at the species level. They attributed the high species diversity of the fishes and shrimps in the Gulf of Mexico to such partitioning.

Life History Patterns of Midwater Organisms

Most fishes and decapod crustaceans of the mesopelagic zone are short-lived creatures. According to Mauchline (1977), myctopid and gonostomatid fishes tend to become sexually mature at from one to three years and probably live no more than two to four years.

The reproductive cycle among midwater fishes is not as yet well known. Data from Clarke (1973, 1974) on species of myctophids and stomiatoid fishes from the central North Pacific Ocean suggest that the fishes have a seasonal reproductive cycle and spawn in the spring and summer. Juveniles do not appear to undergo vertical migration as do adults, and they are distributed in shallower water than are the adults.

Decapod shrimps of the dominant mesopelagic genera *Sergestes*, *Pasiphaea*, and *Gennadas* mature in their first or second year and have life spans of one to three years. Euphausiids of midwater zones generally mature in their first year and live perhaps two years. On the other hand, the bathypelagic and abyssalpelagic crustaceans are estimated to live as much as two to seven times longer.

The lower temperatures and the greater scarcity of food in the bathypelagic and abyssalpelagic zones probably decrease the rate of growth, delay the onset of sexual maturity, and therefore, increase the longevity of these animals. The extreme expression of this trend is abyssal gigantism (pp. 149–151).

Mesopelagic decapod shrimps also show seasonal reproductive activity, usually spawning in late winter or spring in temperate zone seas. In bathypelagic shrimps, it appears that spawning is prolonged over more of the year. Little is known about the movements and distribution of larval shrimps.

SUMMARY OF KEY CONCEPTS

- The deep sea is that part of the marine environment that lies below the level of light penetration and beyond the continental shelf.
- Deep-sea pelagic faunal zones include mesopelagic (dysphotic), bathypelagic, abyssalpelagic, and hadalpelagic.
- Deep-sea benthic faunal zones include bathyal, abyssal, and hadal.
- It is difficult to sample the deep sea because of the need for large amounts of cable, the uncertainty that the gear is sampling correctly, the qualitative nature of most gear, and the time involved in obtaining a sample.
- Newer instruments to sample the deep sea include manned and unmanned submersibles.
- In the deep sea, the major environmental factors that affect animals and their interactions are temperature, pressure, absence of light, food, and oxygen content.
- The range of pressure is greatest in the deep sea and has a profound effect on the metabolism of the animals, including enzyme systems, membranes, and the ability to secrete calcium carbonate.
- Oxygen is adequate for life in most of the deep sea except in certain areas, called oxygen minimum zones, where very low levels may preclude some organisms.
- Food is scarce in the deep sea and decreases with depth and distance from land, leading to low densities and small size of deep-sea organisms.
- Adaptations observed in mesopelagic animals include red color in crustaceans, black color in many fishes, tubular eyes or large eyes in fishes, large mouths with recurved teeth to enable the animal to feed on large prey, and lures to attract prey.
- Most deep-sea animals are small, but some crustacean groups show abyssal gigantism.
- As one goes deeper, the anatomical organization of the animals becomes simplified and they show increased water content.
- Bioluminescence is widespread, particularly in the mesopelagic zone; there, many animals produce light, often in elaborate photophores.
- Bioluminescence is of major importance in concealing animals from predators, finding and attacking prey, finding mates, and ensuring survival.
- Faunal composition of the deep-sea benthos includes both epifaunal and infaunal organisms, and virtually all major animal phyla have representatives.
- Community structure of the deep-sea benthos shows a paucity of suspension feeders due to the low density of food in the water, low density of carnivores and herbivores, and dominance by detritus feeders.
- Age structure, life span, colonization times, and recruitment patterns are poorly known.
- Deep-sea diversity, particularly in the benthos, is very high; there are several theories—stability time, cropper, and areal—that have been suggested to explain it, but none appear to be individually adequate to explain all the patterns. The patterns are probably the result of many factors acting simultaneously. These factors include microhabitat diversification, energy input, competition, predation, and rate of disturbance.
- Two general reproductive patterns exist in the deep sea. Either the young stages are spent in the surface waters and then the animals migrate to the deep sea as they mature or metamorphose, or no migration occurs and the young stages are spent in the same area as the adults.
- One of the most significant discoveries in deep-sea biology was the finding of a previously unknown set of organisms living around hot water springs called hydrothermal vents. These vents have subsequently been discovered at plate boundaries in a number of areas in the Pacific and Atlantic oceans.
- Similar communities, called cold seeps, have been discovered where reducing chemicals are seeping out from the seafloor.
- Both seeps and vents are areas where reducing chemicals, sulfides and methane, are produced and the existence of the communities depends on the primary productivity of the chemolithoautotrophic bacteria that oxidize the sulfides and methane and use the energy to fix carbon dioxide into organic compounds.
- The communities at the vents and seeps are unique, and they have a high diversity and biomass, with the dominant animals displaying a special symbiotic relationship with the sulfide-oxidizing bacteria. The animals are confined to an area within a few meters of the sulfide source, and the seeps and vents may be separated from each other by hundreds or thousands of kilometers.
- Individual vents have a life span only in the tens of years, whereas the seep systems may have longer life spans.
- Observations of the vents at the Galápagos over time suggest that initially they are dominated by vestimentiferan worms, clams, and mussels, but that over time the worms diminish while the clams and mussels grow. As the flow of the vents diminishes over time, many animals disappear and the mussels appear to be the most persistent.
- More recently, a similar sulfide-driven community has been discovered on whale bone.
- Midwater community ecology is less well known because lack of observation, net avoidance, light avoidance, and small size of the animals.

- Deep scattering layers are concentrations of sound-reflecting midwater animals found in all oceans except the Arctic.
- Midwater organisms seem to be dominated by various small fishes and crustaceans, which seem to have niche partitioning that permits them to avoid competition for food; this allows very diverse numbers of species to coexist.
- Most mesopelagic animals are short-lived, a few years, whereas the bathypelagic and abyssal pelagic animals are longer-lived.

REVIEW QUESTIONS

ESSAY: Develop complete answers to these questions.

1. Choose one of the following questions to discuss. How are deep-sea fishes like the Venus flytrap? Why are most deep-sea fishes float-and-wait predators? Compare and contrast sit-and-wait predators in terrestrial systems and float-and-wait predators in the marine systems.
2. Why do we believe that bioluminescence in the deep sea is of great importance to the organisms possessing it? Discuss.
3. At what oxygen concentrations are the deep-sea waters? Why are they not zero? Where is the oxygen added to the deep-sea waters? Why are the deep-sea waters not depleted by organisms found in these waters?
4. Consider the three hypotheses that attempt to explain the relatively high species diversity of the deep-sea benthos. Discuss the relative merits of each. (Instructors may wish to ask students for ways to develop tests of these hypotheses.)
5. Why are planktonic larval forms not as advantageous for benthic animals of the abyss as they are for animals of other benthic environments?

BIBLIOGRAPHY

Andersen, N. R., and B. J. Zahuranec, eds. 1977. *Oceanic sound scattering prediction*. New York: Plenum Press.

Angel, M. V., and M. Sibuet, eds. 1990. Deep sea biology. *Prog. Oceanog.* 24 (1–4):1–343.

Anonymous. 1984. Deep-sea hot springs and cold seeps. *Oceanus* 27:3–78.

Arp, A. J., and J. J. Childress. 1983. Sulfide binding by the blood of the hydrothermal vent tube worm *Riftia pachyptila*. *Science* 219:295–297.

Ballard, R. D. 1977. Notes on a major oceanographic find. *Oceanus* 20(3):35–44.

Blackburn, M. 1977. Studies on pelagic animal biomasses. In *Oceanic sound scattering prediction*, edited by N. R. Andersen and B. J. Zahuranec. New York: Plenum Press, 283–299.

Bouchet, P., and A. Waren. 1979. Planktotrophic larval development in deep water gastropods. *Sarsia* 64:37–40.

Brauer, R., ed. 1972. *Barobiology and the experimental biology of the deep sea*. North Carolina Sea Grant Program, University of North Carolina, Chapel Hill.

Bruun, A. 1957. Deep-sea and abyssal depths. In *The treatise on marine ecology and paleoecology*, edited by J. E. Hedgpeth. Vol. I, *Ecology*. Memoir 67, Geol. Soc. Amer., 641–672.

Childress, J., ed. 1988. Hydrothermal vents: A case study of the biology and chemistry of a deep sea hydrothermal vent of the Galápagos Rift. The Rose Garden in 1985. *Deep-Sea Res.* 35 (10–11A):1677–1849.

Childress, J. J., and C. R. Fisher. 1992. The biology of thermal vent animals: Physiology, biochemistry and autotrophic symbioses. *Oceanog. Mar. Biol. Ann. Rev.* 30: 337–441.

Childress, J., and M. Nygaard. 1973. The chemical composition of midwater fishes as a function of depth of occurrence off southern California. *Deep-Sea Res.* 20:1093–1111.

Childress, J., and M. Nygaard. 1974. Chemical composition and buoyancy of midwater crustaceans as a function of depth of occurrence off southern California. *Mar. Biol.* 27:225–238.

Clarke, T. S. 1973. Some aspects of the ecology of lantern fishes (Myctophidae) in the Pacific Ocean near Hawaii. *Fish. Bull.* U.S. 71:401–434.

Clarke, T. S. 1974. Some aspects of the ecology of stomiatoid fishes in the Pacific Ocean near Hawaii. *Fish. Bull.* U.S. 72:337–351.

Cossins, A. R., and A. G. Marshall. 1989. The adaptation of biological membranes to temperature and pressure: Fish from deep and cold. *J. Bioenerg. Biomembr.* 21:115–135.

Dayton, P., and R. H. Hessler. 1972. The role of biological disturbance in maintaining diversity in the deep sea. *Deep-Sea Res.* 19:199–208.

Dietz, R. S. 1962. The sea's deep scattering layers. *Sci. Amer.* 207(2):44–50.

Ernst, W. G., and J. G. Morin, eds. 1982. *The environment of the deep sea*. Englewood Cliffs, NJ: Prentice-Hall.

Gage, J. D., and P. A. Tyler. 1991. *Deep sea biology*. New York: Cambridge University Press.

Gartner, J. V., R. E. Crabtree, and K. J. Sulak. 1997. Feeding at depth. In *Deep sea fishes*, edited by D. J. Randall and A. P. Farrell. San Diego, Academic Press, 115–182.

Grassle, J. F. 1985. Hydrothermal vent animals: Distribution and biology. *Science* 229:713–717.

Grassle, J. F., and L. Morse-Porteous. 1987. Macrofaunal colonization of disturbed deep sea environments and the structure of deep sea benthic communities. *Deep-Sea Res.* 34:1911–1950.

Hardy, A. 1956. *The open sea*. Vol. 1, *The world of plankton*. New York: Houghton Mifflin.

Haymon, K. M., and K. C. McDonald. 1985. The geology of deep sea hot springs. *Amer. Sci.* 73(5):441–449.

Hedgpeth, J. 1957. Classification of marine environments. In *The treatise on marine ecology and paleoecology*. Vol. 1, *Ecology*. Geol. Soc. of Amer. Memoir 67, 18–27.

Heezen, B. C., and C. D. Hollister. 1971. *The face of the deep*. New York: Oxford University Press.

Hessler, R. R., and W. M. Smithey, Jr. 1983. The community structure of megafauna. In *Hydrothermal processes at seafloor spreading centers*, edited by P. A. Rona et al. New York: Plenum Press, 735–770.

Hessler, R. R., W. M. Smithey, M. A. Boudrias, C. H. Keller, R. A. Lutz, and J. Childress. 1988. Temporal changes in the megafauna at the Rose Garden hydrothermal vent (Galápagos rift; eastern tropical Pacific). *Deep-Sea Res.* 35 (10–11):1681–1709.

Hopkins, T. L., and R. C. Baird. 1977. Aspects of the feeding ecology of oceanic midwater fishes. In *Oceanic sound scattering prediction*, edited by N. R. Andersen and B. J. Zahuranec. New York: Plenum Press.

Hopkins, T. L. and T. T. Sutton. 1998. Midwater fishes and shrimps as competitors and resource partitioning in low latitude oligotrophic ecosystems. *Mar. Ecol. Prog. Ser.* 164:37–45.

Idyll, C. P. 1976. *Abyss*. New York: Crowell.

Ingmanson, D. E., and W. J. Wallace. 1973. *Oceanology: An introduction*. Belmont, CA: Wadsworth.

Jones, M. E., ed. 1985. Hydrothermal vents of the eastern Pacific, An overview. *Bull. Biol. Soc. Washington*, no. 6: 1–547.

Jumars, P. A., and E. D. Gallagher. 1982. Deep-sea community structure: Three plays on the benthic proscenium. In *The environment of the deep sea*, edited by W. G Ernst and J. G. Morin. New York: Prentice-Hall.

Karl, D. M., D. J. Burns, K. Orrett, and H. W. Jannasch. 1984. Thermophilic microbial activity in samples from deep sea hydrothermal vents. *Mar. Biol. Lett.* 5:227–231.

Lightfoot, R., P. A. Tyler, and J. D. Gage. 1979. Seasonal reproduction in deep sea bivalves and brittlestars. *Deep-Sea Res.* 26A:967–973.

Longhurst, A. R. 1981. *Analysis of marine ecosystems*. New York: Academic Press.

MacDonald, A. G. 1975. *Physiological aspects of deep sea biology*. New York: Cambridge University Press.

Marshall, N. B. 1954. *Aspects of deep-sea biology*. New York: Philosophical Library.

Marshall, N. B. 1980. *Deep sea biology: Developments and perspectives*. New York: Garland STPM Press.

Mauchline, J. 1972. The biology of bathypelagic organisms, especially crustacea. *Deep-Sea Res.* 19:753–780.

Mauchline, J. 1977. Estimating production of midwater organisms. In *Oceanic sound scattering prediction*, edited by N. R. Andersen and B. J. Zahuranec. New York: Plenum Press.

McAllister, D. E. 1961. A collection of oceanic fishes from off British Columbia with a discussion of the evolution of the black peritoneum. *Bull. Nat. Mus. Canada* 172:39–43.

Meade, G. W., E. Bertelson, and D. M. Cohen. 1964. Reproduction among deep sea fishes. *Deep-Sea Res.* 11:569–596.

Menzies, R. J. 1964. Improved techniques for benthic trawling at depths greater than 2000 meters. *Biol. Antarctic Seas. Antarctic Res. Set., Amer. Geophys. Union* 1:93–109.

Menzies, R. J., R. Y. George, and G. T. Rowe. 1973. *The abyssal environment and ecology of the world oceans*. New York: Wiley.

Nixon, M., and J. B. Messenger, eds. 1977. *The biology of cephalopods*. New York: Academic Press.

Omori, M. 1974. The biology of pelagic shrimp in the ocean. *Adv. Mar. Biol.* 12:233–324.

Parsons, T. R., M. Takahashi, and B. Hargrave. 1984. *Biological oceanographic processes*. 3d ed. New York: Pergamon.

Regan, C. T., and E. Trewavas. 1932. *Deep sea angler-fishes [Ceratoidea]*. Dana report no. 2. The Carlsberg Foundation and Oxford University Press.

Rex, M. A. 1981. Community structure in the deep sea benthos. *Ann. Rev. Ecol. Sys.* 12:331–353.

Robison, B. H. 1995. Light in the oceans midwaters. *Sci. Amer.* 273(1):60–64.

Rokop, F. 1974. Reproductive patterns in the deep-sea benthos. *Science* 186:743–745.

Rokop, F. 1979. Year around reproduction in deep sea bivalve molluscs. In *Reproductive ecology of marine invertebrates*, edited by S. E. Stancyk. Columbia: University of South Carolina Press, 189–198.

Rona, R. P., K. Bostrum, L. Laubier, and K. L. Smith, eds. 1983. *Hydrothermal processes at seafloor spreading centers*. New York: Plenum Press.

Ross, D. A. 1982. *Introduction to oceanography*. Englewood Cliffs, NJ: Prentice-Hall.

Rowe, G. T., ed. 1983. *The sea*. Vol. 8, *Deep sea biology*. New York: Wiley.

Sanders, H. L. 1968. Marine benthic diversity, a comparative study. *Amer. Nat.* 102:243–282.

Sanders, H. L., and R. R. Hessler. 1969. Ecology of the deep-sea benthos. *Science* 163:1419–1424.

Siebenaller, J., and G. N. Somero. 1978. Pressure adaptive differences in lactate dehydrogenases of congeneric fishes living at different depths. *Science* 201(4352):255–257.

Siebenaller, J., and G. N. Somero. 1989. Biochemical adaptation to the deep sea. *Crit. Rev. Aquat. Sci.* 1:1–25.

Sibuet, M., and K. Olu. 1998. Biogeography, biodiversity and fluid dependence of deep cold-seep communities at active and passive margins. *Deep-Sea Res.* II 45:517–567.

Smith, C. R. 1992. Whale falls. *Oceanus* 35(3):74–78.

Smith, K. L., and K. R. Hinga. 1983. Sediment community respiration in the deep sea. In *The sea*, Vol. 8, *Deep sea biology*, edited by G. T. Rowe. New York: John Wiley and Sons, 331–370.

Somero, G. N. 1982. Physiological and biochemical adaptations of deep sea fishes: Adaptative responses to the physical and biological characteristics of the abyss. In *The environment of the deep sea*, edited by W. G. Ernst and J. G. Morin. Englewood Cliffs, NJ: Prentice Hall, 256–278.

Tunnicliffe, V. J. 1991. The biology of hydrothermal vents: Ecology and evolution. *Oceanog. Mar. Biol. Ann. Rev.* 29:319–407.

Tunnicliffe, V. J. 1992. Hydrothermal vent communities of the deep sea. *Amer. Sci.* 80(4):336–349.

Tunnicliffe, V. J., F. Garrett, and H. P. Johnson. 1990. Physical and biological factors affecting the behaviour and mortality of hydrothermal vent tubeworms (vestimentiferans). *Deep-Sea Res.* 37(1):103–125.

Tunnicliffe, V. J., A. G. McArthur, and D. McHugh. 1998. A biogeographical perspective of the deep-sea hydrothermal vent fauna. *Adv. Mar. Biol.* 34:353–442.

Yayanos, A. A., A. S. Dietz, and R. van Boxtel. 1979. Isolation of a deep sea barophilic bacterium and some of its growth characteristics. *Science* 205(4408):808–809.

Young, R. E. 1975. Function of the dimorphic eyes in the midwater squid, *Histioteuthis dofleini*. *Pac. Sci.* 29(2):211–218.

Young, R. E. 1983. Oceanic bioluminescence: An overview of general functions. *Bull. Mar. Sci.* 33:829–845.

Zimmerman, A. M., and S. B. Zimmerman. 1972. Commentary on high pressure effects on cellular systems. In *Barobiology and the experimental biology of the deep sea*, edited by R. W. Brauer. North Carolina Sea Grant Program, University of North Carolina.

SHALLOW-WATER SUBTIDAL BENTHIC ASSOCIATIONS

In this chapter we deal with the bottom communities and conditions for life in the shallow waters fringing the landmasses of the world: the continental shelf region of the world's oceans. We have considerably more knowledge about this region than we have about the deep sea. It is more accessible and more studied because the continental shelf areas are the major fishing grounds of the world. This chapter covers major subtidal associations except for coral reefs. Because of their many unique properties, coral reefs are covered separately in Chapter 9. The shallow subtidal area is contiguous at its upper end with the intertidal regions of the world's oceans, and at least some of the principles we discuss in this chapter also will be applicable to the intertidal zone.

Coverage and Definitions

This chapter covers that area of the oceans that lies between the level of lowest low water on the shore to the edge of the continental shelf at a depth of about 200 m. In our scheme of classification, this area is known as the **sublittoral**. Overlying it are the waters of the **neritic** zone (see Figure 1.22 on the scheme of zonation). Most of this zone is composed of soft sediments, sand and mud, and a much lesser area of hard substrate. On an areal and numerical basis, the bottom is dominated by infaunal organisms.

Infaunal organisms are usually divided into categories based on size. **Macrofauna** are organisms that are greater than 0.5 mm in size. The term **meiofauna** is applied to organisms that lie within the size range of 0.5–0.062 mm. A final size class is **microfauna**, which includes organisms below 0.062 mm in size. This group is primarily composed of protozoa and bacteria. This chapter will deal primarily with macrofauna. Meiofauna and microfauna are covered in Chapter 7.

ENVIRONMENTAL CONDITIONS

The continental shelf waters show more variability in environmental conditions than do either the epipelagic of the open ocean or the deep sea. Perhaps the most important physical factor that acts on the bottom communities is turbulence or wave action. In these shallower waters, the interaction of waves, currents, and upwelling creates turbulence. This turbulence keeps inshore waters from becoming thermally stratified except for brief periods, at least in the temperate zone. As a result, nutrients are rarely limiting or locked in a bottom reservoir. Productivity is generally higher than in similar waters offshore because of nutrient abundance, from both runoff from land and recycling (see Table 2.2). This

FIGURE 5.1 A shallow-water soft bottom. Note the ripple marks. (Photo courtesy of Tony Stone Images.)

high productivity sustains large populations of both zooplankton and benthic organisms.

Wave action is an important factor in this area. Long-period ocean swell and storm waves have an effect that can extend to the bottom in these shallow waters. In soft bottoms, the passage of such waves may cause large surging motions in the bottom water, which greatly affects the stability of the substrate. The substrate particles may be moved around and resuspended. This has a profound effect on the infaunal animals in the substrate. Wave action also determines the type of particles present. Heavy wave action removes fine particles by keeping them in suspension, leaving mainly sand; thus, fine silt sediments can occur only in areas that have low wave action or are too deep to be affected by wave action.

Salinity in this region is more variable than in the open ocean or deep ocean. Except for areas where large rivers discharge massive amounts of fresh water, however, the salinity does not change enough to be of ecological significance.

Temperature is also more variable in inshore waters and shows seasonal change in the temperate zone. These temperature changes may be used by organisms as cues to begin or end various activities, such as reproduction.

Light penetration in these turbulent waters is reduced when contrasted with open ocean areas. The combination of large amounts of debris, both from land and from breakup of kelp and seagrasses, plus the high plankton densities due to abundant nutrients, reduce light penetration from 10 to 20 m to just a few meters.

The food supply is abundant in this area, partly because of the increased productivity by the plankton but also because of the production of attached plants, such as kelp and seagrasses. This is one of the few areas in the sea where macroscopic plants have any significance in production. A final food source is runoff from land. Although there are large autotrophs in the sublittoral shelf, there are relatively few large grazing animals. The major use of kelps and seagrasses as food occurs only after they have been broken up into detrital particles.

The soft bottoms in the sublittoral are essentially without large topographic feature diversity, and the vast expanses extend monotonously for long distances. Small feature diversity exists in the form of ripple marks, worm tubes, fecal mounds, and so on (Figure 5.1). Lacking large topographic relief, the only apparent difference from one place to another is that of substrate grain size and composition (silica vs. calcium carbonate). Subtidal hard substrates, on the other hand, may have considerable relief with many potential habitats. The lack of major topographic relief in the infaunal areas means fewer habitats for animals to occupy and fewer potential ways for making a living. The number of infaunal species is generally less than the number of epifaunal species, because there are fewer niches available. The most abundant infaunal

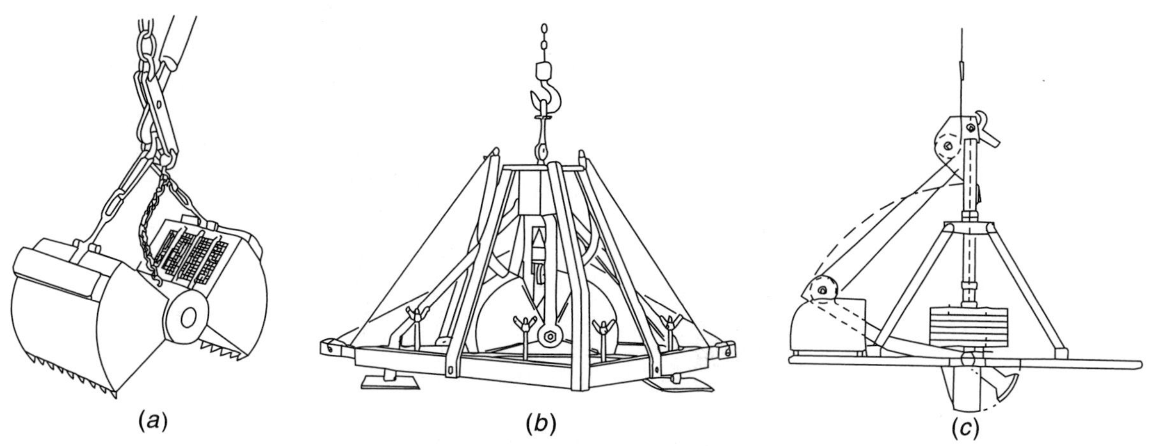

FIGURE 5.2 Various types of grab samplers. (a) Peterson grab. The jaws close as the cable is pulled upward, thus securing the sample. (b) Smith-McIntyre grab. This is a spring-loaded grab, in open position here, which will close when it hits the bottom. (c) Reineck box corer. The "box" is closed by a knife edge actuated by pulling up on the cable. (a and c from *Methods for the Study of Marine Benthos*, 2d ed., N. A. Holmes & A. D. McIntyre eds., 1984 Blackwell Science. b from Hardy, A. C., 1959 *The Open Sea: It's Natural History, Part II, Fish and Fisheries*, New Naturalist Series, 37. Collins, London.)

animals are deposit feeders, ingesting the abundant detritus that rains down, or are suspension feeders, filtering out of the water column the abundant plankton or floating detritus. A lesser number, mainly worms, crustaceans, mollusks, and echinoderms, are predators. Bottom fishes, on the other hand, are predominantly carnivores.

It is convenient to divide the subtidal continental shelf region into four major habitats, which are readily recognizable and topographically or physiognomically different, and discuss each of these separately. The four are (1) open unvegetated sedimentary environments (the most common areally), (2) hard substrates dominated by low-encrusting plants and animals, (3) kelp beds and forests, and (4) seagrass beds.

UNVEGETATED SEDIMENTARY ENVIRONMENTS

Sampling

From the times of Petersen in the early 1900s (see subsequent section) until about the 1960s, the common method for sampling the organisms of the soft sediments of the continental shelf was to drop a **grab** (Figure 5.2). Various types of benthic grabs have been developed over the years, and each has certain advantages and disadvantages. Most have the advantage of being quantitative; that is, they take a sample of known areal extent from the bottom. Standard treatment of these grab samples has been to screen out the organisms on one or a series of screens. Mesh sizes of 1.0 mm and 0.5 mm have been the most commonly employed.

Although such sampling was quantitative and gave a good indication of the kinds of organisms present and their relative abundance, it was a remote method in which the benthic biologist never actually saw the community. It was also not possible to do any manipulative experiments with the community. This was changed in the 1960s with the advent of scuba (self-contained underwater breathing apparatus), which allowed the biologist to investigate an underwater community firsthand and to undertake manipulative experiments, such as using cages to keep various organisms in or out of the community. Thus, changes in time could be followed. Experimental ecology arrived with scuba, but because it was not used until later in sedimentary environments as opposed to the hard substrates, we have fewer data for the soft sediments.

Grabs are used today, particularly in areas below safe diving depths (about 30 m). Within safe diving depths, firsthand work by divers is more popular and productive.

Other sources of data include remote, towed cameras or video systems and side-scan sonar. The latter is capable of recording, very accurately, small-scale detail of the bottom topography.

Infaunal Associations

There are four taxonomic groups of dominant macrofauna present in sublittoral soft bottoms: class Polychaeta, subphylum Crustacea, phylum

Echinodermata, and phylum Mollusca. Polychaete worms are the most abundant and are represented by numerous tube-building and burrowing species. The dominant crustaceans are the larger ostracods, amphipods, isopods, tanaids, mysids, and a few of the smaller decapods. They mainly inhabit the surface of the sand and mud. Mollusks are represented primarily by various burrowing bivalve species and some scaphopods, with a few gastropods at the surface. Echinoderms particularly common in the subtidal benthos include brittle stars, heart urchins, sand dollars, sea cucumbers, and a few predatory sea stars.

The first quantitative work in marine ecology was done on the sublittoral benthic infauna. In the first years of the twentieth century, C. G. Joh. Petersen, a Danish biologist, began his investigations on the benthos of the Danish seas. He was interested in evaluating the role of bottom organisms in supporting the fish populations on which the Danish fishing industry depended. To do this on a quantitative basis, he invented a grab, the now-famous Petersen grab, which picked up a definite area of the bottom, usually 0.5 or 0.1 m^2 (Figure 5.2a). Since he had a sample that covered a known area of bottom, he could then count the numbers of organisms in those samples and extrapolate the numbers in the whole bottom area. Once this was done, he would have some estimate of the amount of food available to the fishes and could guess the biomass of fish that could be supported.

Analyzing many of these grab samples over the years, Petersen (1918, 1924) observed that extensive bottom areas were occupied by recurrent groups of species and that other areas were inhabited by different associations of species. In all cases, he observed that relatively few species made up most of the individuals and biomass. These findings contrasted with nonquantitative dredge samples taken in the same waters. In the latter case, the faunal lists might be nearly identical in the two areas. The value of the quantitative study was that for the first time it gave a way to evaluate differences between communities based on relative abundance, not just presence or absence of species.

After several years of evaluating these quantitative samples, Petersen observed that the different areas and their dominant organisms remained relatively constant and uniform over time. He then proposed these associations as communities and named them on the basis of the dominant animals. For example, he had a *Macoma balthica* community from inner Danish waters in 8–10 m of water, which was dominated by the bivalve *M. balthica* but was also characterized by the polychaete *Arenicola marina*, and the bivalves *Mya arenaria* and *Cardium edule* (Figure 5.3). In this same manner, he named and characterized a whole series of communities from the sublittoral in Scandinavian waters.

Following Petersen's work, various benthic biologists began to investigate other shallow-water benthic areas around the world. The leader of this school was another Dane, G. Thorson. What Thorson (1955) discovered, particularly for the temperate zone waters, was that communities of organisms similar to those found by Petersen were found in similar habitats around the world. The concept of **parallel bottom communities** arose from this discovery. In this ecological concept, similar sediment types at the same depths around the world contain similar communities. Granted, the species are not the same, but they are closely similar ecologically and taxonomically (Figure 5.3). Additional work in tropical waters by Thorson (1966) revealed that this concept did not extend to such areas, at least not in the same way.

That certain recurrent groups occupy the same substrate types over large areas of the world's ocean seems established. This suggests that such associations are not random but represent real interacting systems in which some combination of factors ensures the persistence of the community.

Petersen described the communities, and others later validated their continued existence in various parts of the world. What was lacking was an explanation of how such communities persisted and what kinds of interactions ensured continued success. Since Petersen's time, answers to these questions have come to light and will be discussed in the following sections.

Community Pattern and Structure

Most marine organisms have patchy distributions. We noted this with respect to the plankton in Chapter 2 (see pp. 85–87). It is also the situation one encounters when analyzing benthic organisms. Even in the early days of Petersen, not all quantitative samples had either the same species or the same relative abundances. They also varied with season. Petersen was able to characterize his communities primarily on a statistical basis dependent on the presence of a large number of samples. Even in these seemingly monotonous open sedimentary

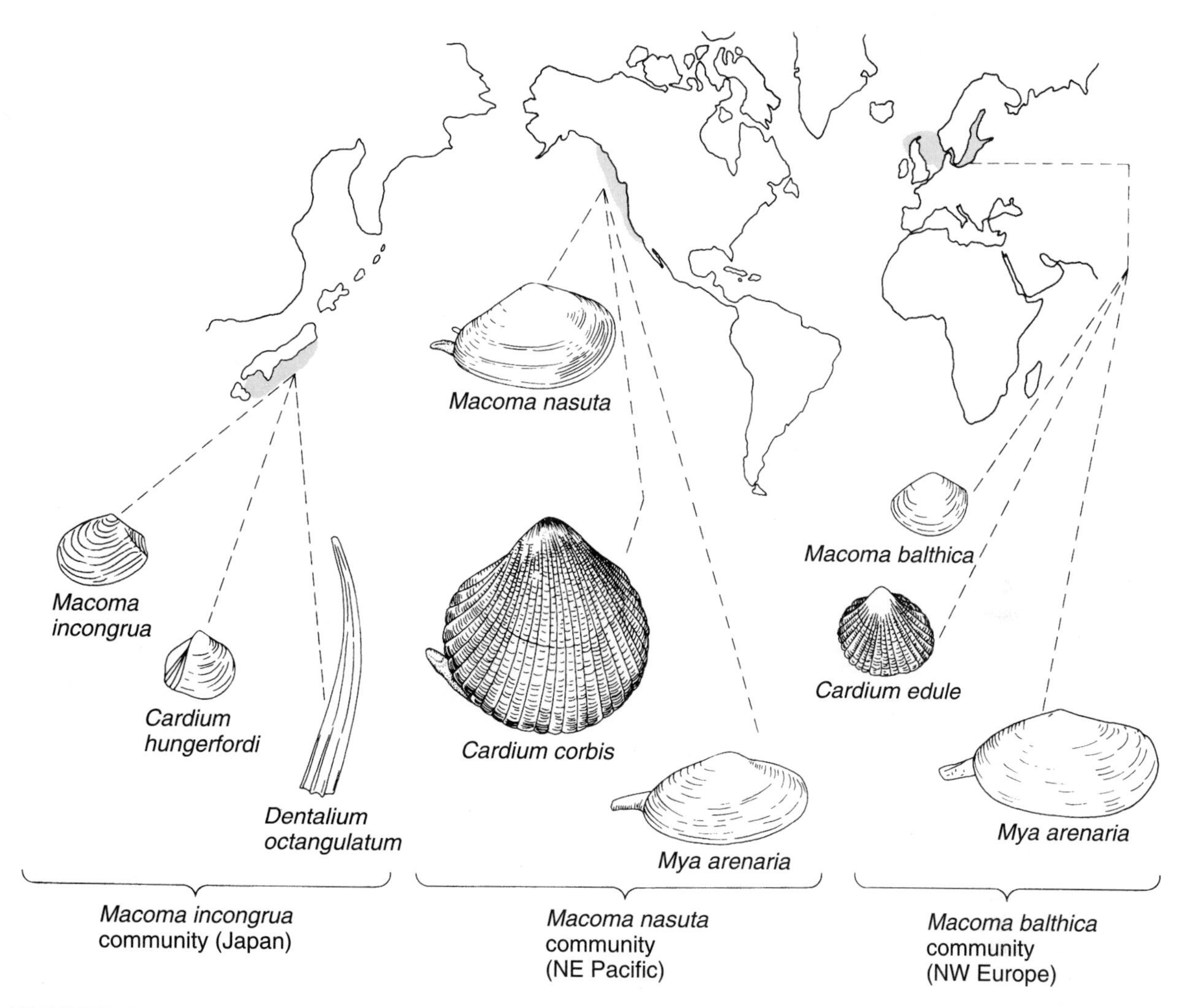

FIGURE 5.3 Diagram illustrating the parallel bottom communities dominated by the bivalve *Macoma* in three different areas in the North Temperate Zone. (Modified from "Bottom Communities" in *The Treatise on Ecology and Paleontology*, Vol. 1, G. Thorson, 1957, eds. J. Hedgpeth, © 1957 Geological Society of America.)

environments removed from the more physically variable surface or intertidal areas, patchiness of organisms exists—both in time and space. It is the result of the various physical factors and interactions of organisms and gives rise to the diverse patterns we observe in the distribution of benthic associations. Certain cycles that may occur on a daily, seasonal, or annual basis or over a period of years also originate in the interactions that produce patchiness. A further complicating factor is that variability in benthic associations occurs not only on a horizontal scale, but on a vertical scale as well, since many of these infaunal organisms burrow to different depths in the substrate.

Within any sedimentary benthic community there are usually hundreds of macrofaunal (greater than 0.5 mm in size) organisms, such that it is difficult to work with individual species in any attempt to understand the various ecological relationships. Fortunately, as Woodin (1983) has pointed out, the organisms in these environments can be classified into a limited number of ecological categories. One can then use the functional groups to compare assemblages of species and to attempt to arrive at generalizations regarding how the communities operate. A functional group includes all the species of different animal taxa that use and affect the

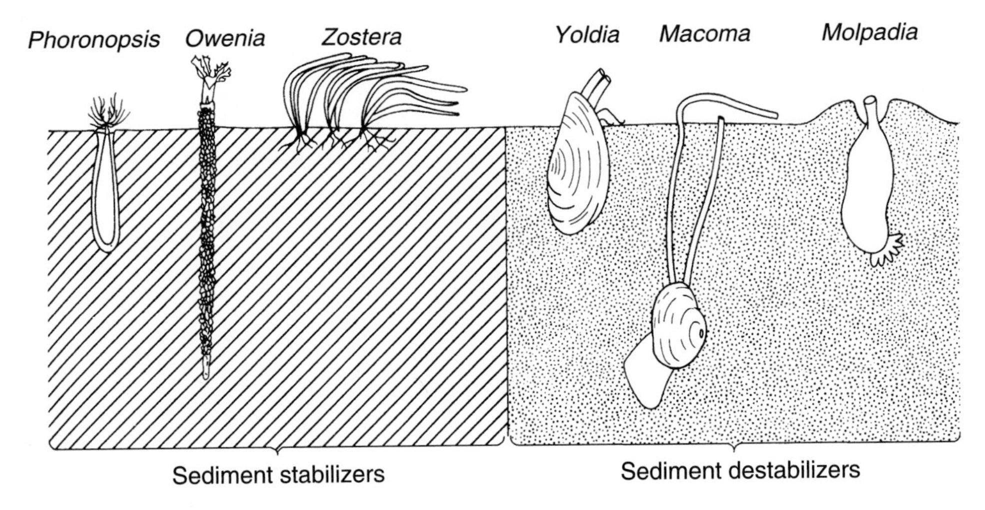

FIGURE 5.4 Representatives of the two functional groups of sedimentary organisms: sediment stabilizers and sediment destabilizers or bioturbators.

environment in similar ways. For example, in Chapter 1 we noted the division of animals into opportunistic or equilibrium species. That is one functional classification.

Another approach would be to classify various species as **sediment destabilizers** and **sediment stabilizers**. Destabilizers or **bioturbators** include both motile and sedentary organisms whose activities, either movement in the case of the motile ones or biological activities in the case of the sedentary ones, cause the sediment to move, become resuspended, erode, or otherwise change. Examples of this kind of organism are mobile deposit-feeding clams, such as *Nucula*, *Yoldia*, and *Macoma*, studied by Rhoads and Young (1970) and the sedentary deposit-feeding sea cucumber *Molpadia oolitica*, which Rhoads and Young (1971) found destabilized the sediment by its feeding and defecation (Figure 5.4). Sediment stabilizers include various seagrasses, whose roots bind sediments and whose upper parts change the local hydrographic regime (see pp. 210–212). They also include various invertebrate tube builders, which, when in high density, form mats or areas in which sediment is stable. Examples of tube dwellers that stabilize sediments include amphipod crustaceans, phoronid worms, tube-building anemones and polychaetes (Figures 5.4 and 5.5). This classification of sediment stabilizers and destabilizers is an improvement on Woodin's (1976) earlier classification of burrowing deposit feeders, tube builders, and suspension feeders.

Another functional division could be the more traditional trophic divisions of herbivore, omnivore, and carnivore. Whereas it is usually easy to classify most infaunal organisms into such functional divisions, some are intermediate. For example, maldanid polychaetes are tube builders, but they are also deposit feeders. To place such an organism in a stabilizing or destabilizing category will require more information regarding their effect on the substrate. Note also that any functional group includes a number of different taxa of invertebrates and may further include different nutritional types. For example, we may have mobile burrowers that are predators, deposit feeders, or suspension feeders, and they may also be divided into those that are opportunistic and those that have reached equilibrium in their life history features. In other words, categorization by functional groups can be done several ways and does not necessarily exclude certain other features of the life history of the organism. The only trophic mode absent from open sedimentary communities is macroautotrophy. There are no large autotrophs, and any herbivores in this system must feed on plant material that is either in the plankton (suspension feeders) or benthic microalgae (diatoms) or that is brought in from outside as debris (deposit feeders or scavengers).

Community Structure and Change

Basically, changes in sedimentary community structure occur through either physical factors or biologi-

FIGURE 5.5 Subtidal tube-dwelling anemone-like cnidarian *Pachyceriathus fimbriatus*. (Photo courtesy of Drs. Lovell and Libby Langstroth.)

cal factors, or some combination of the two. The simplest cases involve the dominance of physical factors. For example, along the Pacific coast of California, Oliver et al. (1979) have demonstrated that subtidal benthic communities are zoned along a gradient of increasing wave disturbance to the bottom. In shallow water (less than 14 m), the bottom sediments are regularly disrupted by waves, and the dominant inhabitants are various small, mobile crustaceans. In the deeper zones (greater than 14 m), the fauna is dominated by tube-building polychaete worms. Evidence showed that wave action prevented the establishment of worm tubes in the shallow water. Wherever wave action was diminished, the polychaete zone became shallower, and wherever wave action was stronger, the worm zone was driven deeper.

A more complicated example in which both biological and physical factors act is seen in an East Coast example. Mills (1969) studied a sandy area in Barnstable Harbor, Massachusetts, that had for some time been dominated by the snail, *Ilyanassa obsoleta*. This community had low biomass and low numbers of individuals of associated organisms probably due to the mechanical disturbance of the sediments by the snails. Because of the low density of infaunal animals, this sandy area was colonized in spring by a tube-building amphipod, *Ampelisca abdita*. The *Ampelisca* quickly built a dense mat of tubes that excluded *Ilyanassa*. Selective deposit feeding by the *Ampelisca* allowed fine sediment particles to accumulate, and the increased topographic diversity created by the tubes also changed the physical environment of the area. This led to a change in the biological environment and the colonization of the mats and sediments by an abundance of polychaete worms favoring fine sediments. The *Ampelisca* tube mats, however, are not stable, and stormy weather, particularly in winter, causes large areas to wash out. This permits the return of *Ilyanassa*, and the cycle can start all over again.

The role of biological factors in deciding community structure is more complex. The dominant biological factors that alter community structure are competition, predation, and recruitment. We shall consider each separately and then consider some test cases where two or more interact.

Competitive interactions among organisms can be direct or indirect. Direct competition involves either contact between organisms or the direct interference in some aspect of the other organism's life cycle. Indirect competition usually involves one organism interrupting or preventing resource use by another.

Direct interactions, including those often seen between organisms on hard substrates, such as overgrowth, pushing off other organisms, and so forth (see Chapter 6), are not as common in sedimentary environments. This is probably because such activity involves having a solid substrate to push against. It also may be inherent in colonial organisms, conspicuously absent in sedimentary substrates. Direct competition, involving aggressive activities leading to exclusion or spacing out, has been reported by several authors for various polychaetes and small crustaceans, but most studies were done in the laboratory. The regular distribution patterns found by Levin (1981), Reise (1979), Wilson (1981), and others suggests that these kinds of actions occur. Some direct interactions that displace other species have been reported by Levinton

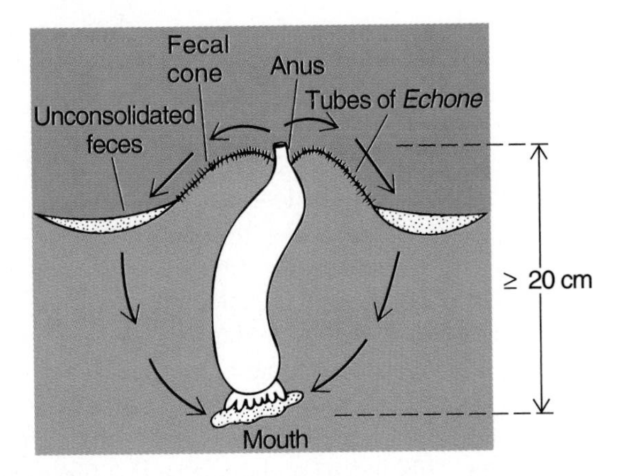

FIGURE 5.6 *Molpadia oolitica*. Diagrammatic representation of feeding position, showing surface cone and depression and subsurface excavation produced by intensive feeding. (From "Animal-sediment relations in Cape Cod Bay, Massachusetts, II." D. C. Rhoads & D. K. Young, 1971, vol 3. Copyright © 1971 Springer-Verlag.)

(1977), who showed that movement of the mobile bivalve *Yoldia limatula* within the substrate disrupts the burrows of the more sedentary bivalve *Solemya velum*, leading to its replacement. Myers (1977) reported a similar situation in which a burrowing sea cucumber does the same thing to small bivalves.

Indirect interactions leading to changes in community structure and organization appear to be more common in sedimentary communities. These indirect actions usually involve some sort of change in the sediment condition.

Burrowing deposit feeders are most abundant in soft, muddy sediments, areas with a high concentration of organic matter. It has been observed in several areas that, where burrowing deposit feeders are prevalent, suspension feeders are rare or absent. The deposit-feeding organisms burrow through the top few centimeters of bottom and create a very loose, unstable layer of fine particles including fecal pellets. This layer is easily resuspended by the slightest water motion. The resuspended sediment clogs the fine filtering structures of the suspension feeders, making feeding impossible. Furthermore, the constant reworking and settlement of resuspended particles leads to burial of the newly settled larvae of suspension feeders, suffocating them. It does not kill the larvae of the deposit feeders because they burrow into the more compact substrate underneath. This exclusion of one trophic group by modification of the environment by another is called by Rhoads and Young (1970) **trophic group amensalism**. In this way, deposit feeders maintain their own communities and exclude suspension feeders. An example of this is seen in Rhoads and Young's (1970) work in Buzzard's Bay, Massachusetts, where they showed that reworking of the sediments by three deposit-feeding clams, *Nucula proxima*, *Macoma tenta*, and *Yoldia limatula*, effectively excluded the suspension feeders.

Suspension-feeding organisms seem more abundant where substrate is more sandy, where the organic matter is less, and where deposit feeders would find less food and more difficult burrowing. In the absence of deposit feeders, the suspension feeders can establish themselves. Once established, they may also exclude any potential deposit feeders by filtering larvae out of the water.

A more complicated situation occurs over large areas of Cape Cod Bay, Massachusetts, where a large sea cucumber, *Molpadia oolitica*, occurs in high densities (2–6 per square meter) in silt-clay muds. This animal is a sedentary deposit feeder that lives head down vertically in the substrate. It ingests sediment at depth and deposits loose fecal matter at the surface, forming a mound around the anal opening. The reworking of the sediment produces a loose, high-water-content, easily suspended sediment layer on the surface of the substrate except in the immediate vicinity of the fecal cone. The areas between the cones are unstable and exclude suspension feeders (amensalism). On the other hand, the stable cone areas attract the tube-building, suspension-feeding polychaete *Euchone incolor* and other tube-dwelling polychaetes. These tubes, in turn, help further stabilize the fecal cone area and attract other suspension feeders (Figure 5.6). Note in this example that *Molpadia* is a relatively large animal. To effect changes that will influence other animals, an animal must be large enough or abundant enough to change large areas of the sediment.

Tube-building organisms can stabilize the substrate with their tubes (see Figure 5.5). They also extend down into the substrate. By stabilizing the substrate, they prevent resuspension of fine particles. At the same time, the presence of the tubes in the substrate in sufficiently high densities restricts the available space for burrowing animals. The burrowers cannot burrow through the rigid tubes; thus, the tube builders indirectly restrict burrowing animals. Such exclusion by interference with normal activities is an example of **competitive interference**. Although burrowing deposit feeders rarely co-occur

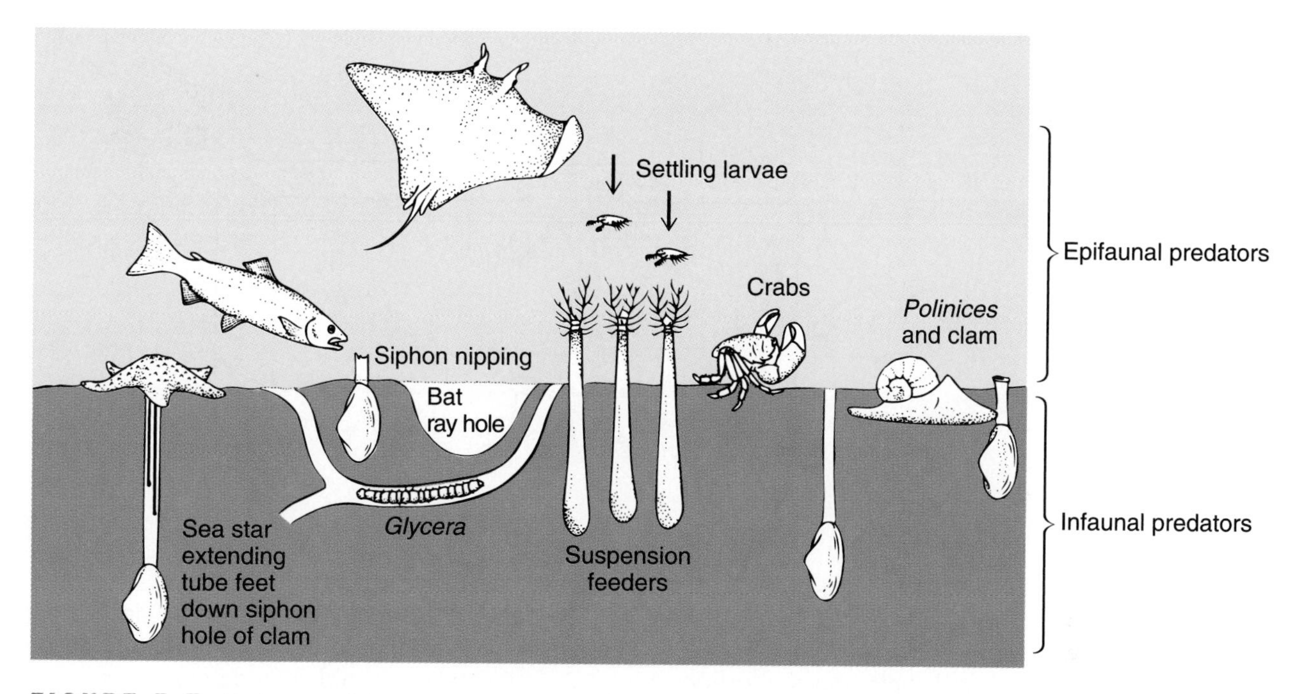

FIGURE 5.7 Various predator types and their effects on shallow-water sedimentary communities.

with suspension feeders or with tube dwellers, and vice versa, it is possible to find these different functional groups in the same geographical area, usually with sharp boundaries between them. These boundaries usually do not coincide with any physical discontinuity and are presumed to result from either predation or some other physical or biological disturbance that has removed the resident organisms. Since most animals dwelling in sedimentary environments are solitary and do not reproduce asexually, any patch created by disturbance must be reoccupied either by immigration of adults or settlement of larvae (see pp. 22–27 in Chapter 1).

The second biological factor important in determining the species structure of infaunal communities is predation. Both invertebrate and vertebrate predators take a toll on the infaunal organisms. Predator activity may be responsible for clearing certain small areas of macrofauna and creating a disturbance that will be followed by recolonization. This is one way that patchy distributions occur on the bottom. Such clearing of areas is known to result from feeding activities of certain flatfish, rays, whales, and walruses (see Chapter 3).

Sediments, as we have noted, are three-dimensional substrates, and it may be possible for some prey organisms to use this three-dimensionality as a refuge from predators. If, for example, the predator cannot dig into the substrate after its prey and the prey can burrow deeply, then the depth becomes a refuge for the prey. If a predator can dig as far down into the substrate as the prey can, however, then there is no refuge. Moreover, the predator in its digging activities may have a widespread effect on other members of the infaunal community by destroying or disrupting the surrounding sediments. For example, VanBlaricom (1982) showed, in a subtidal area off southern California, that large rays created pits in the sand in the process of removing resident clams. In digging these large pits, most other infauna were destroyed, and these areas of substrate became places where resettlement could take place.

Predators, depending on their position and effect on the sediment, have been classified into three different types by Woodin (1983). **Surface predators** are exposed at the surface and take organisms at or near the surface without disrupting the sediment structure (Figure 5.7). Most fishes and various crabs are surface predators, as are certain invertebrates that may be infaunal themselves but capture and ingest other infaunal adults, juveniles, or larvae. A peculiar group that needs to be included here consists of browsers that take off only parts of the prey. An example would be the nipping of siphons from clams by various fish species, as reported by deVlas

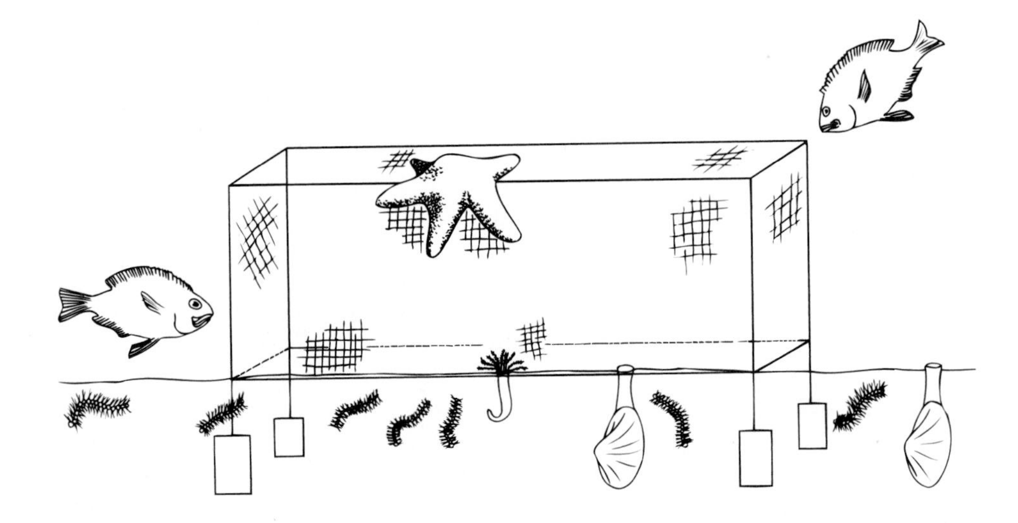

FIGURE 5.8
Effects of a cage that excludes predaceous starfish and fish from the infauna; thus, the infauna density inside the cage is increased compared with that outside.

(1979) and Peterson and Quammen (1982). An interesting "cannibalistic" relationship exists between various suspension feeders and their own and other larvae. These larvae may be captured and ingested as they settle to the substrate and are intercepted by the feeding nets of the suspension feeders. A special case includes such predators as worms of the genus *Glycera*, which form burrows in the substrate but feed mainly at the surface, emerging from their burrows to seize prey (Ockelmann and Vahl, 1970).

Burrowing predators include those that Woodin (1983) terms "weasel" predators, meaning they move down various tubes or channels provided by the deeper-dwelling prey and attack them. An example would be various nemertean worms and perhaps some gastropods that have a long proboscis that can be extended down into the tubes. An interesting case is that of the sea star *Pisaster brevispinus*, which feeds on deep-dwelling clams on the Pacific coast of North America. According to Van Veldhuizen and Phillips (1978), *Pisaster* somehow senses the buried clams as it moves across the surface. It then extends the tube feet from around the mouth area deep into the sediment (it can extend them to a length of 20 cm), attaches them to the clam, and pulls the clam to the surface without greatly disturbing the sediment. It can also extend its stomach down to the clam to digest it (Figure 5.7).

Digging predators excavate holes to get to their food and include such different animals as the blue crab *Callinectes sapidus*, the gastropod *Buccinum undatum*, and the rays. Digging predators have a greater effect on the general infaunal organisms than do weasel predators, because they also disturb the sediment.

A final category includes predators that burrow through the substrate and live within it at all times. These are the **infaunal predators** and include such organisms as nemertine worms and a number of carnivorous polychaete worms.

Different predatory types have different effects on their environment and create diverse patterns in sedimentary communities. In general, the digging predators have the most widespread and recognizable effects because they create a sediment change. Predation by surface predators most likely affects the size classes of prey, particularly if there is a deep refuge for the adults, which leaves the juveniles most vulnerable. The siphon nippers would have their greatest effect on the growth of the prey. The effects of other predators are difficult to predict a priori without knowledge of prey preferences, relative predator densities, and other factors.

It is difficult to determine the relative roles of predator-induced and wave-induced disturbances on the structure of soft-bottom communities. This is mainly because it has been difficult in the past to devise good experiments that will exclude predators from areas of the bottom while allowing wave action to proceed. Virnstein (1977) and other ecologists have devised various "cages" that, when placed over areas of the bottom, exclude or include predaceous fishes and invertebrates. When carefully done, such studies have produced increased numbers of infaunal organisms within the cage as compared with outside (Figure 5.8). This suggests that predation is important in determining at least the numbers of infaunal organisms present and perhaps also the species composition.

It has been particularly difficult to estimate the importance of infaunal predators in determining benthic community structure, because we cannot observe them and have difficulty devising ways of excluding them in manipulative experiments. Indeed, most of the ecological studies of the effects of predation on sedimentary communities have been based on studies of epifaunal predators. In a review, Ambrose (1991) has noted that infaunal predators are often abundant and with few exceptions seem able to significantly reduce the infaunal prey populations. Interestingly enough in 63% of the cases, the prey was another predator, suggesting multiple trophic levels may be a characteristic of these soft-bottom communities. Another characteristic of these infaunal predators is that they confine their activities mainly to the surface-most layers, affecting shallow-dwelling taxa more than deep-dwelling ones.

Although studies quantifying the ability of the infaunal predators to reduce prey numbers are rare, data from long-term studies in the Wadden Sea in northern Europe have shown that the nemertine and polychaete predators can consume from 20 to 50% of the standing stock of prey. These figures also suggest that predation by these infaunal predators may help explain the lack of an established competitive dominant in manipulative experiments where the epifaunal predators have been excluded while the infaunal predators remained. Perhaps the exclusion of epifaunal predators does not always result in increased infaunal prey densities because the infaunal predators take the place of the excluded epifaunal predators.

■ We can summarize our understanding of the relative roles of competition and predation on the structure of sedimentary communities by paraphrasing from a review by Wilson (1991). Wilson concludes there is no evidence for both predation and competition affecting community structure, and there is also no evidence of any competitive dominant in any soft-sediment system. Although we have made much progress in understanding the organization of infaunal communities in the last 25 years, we still lack the ability to predict the effects of either competition or predation. We still lack knowledge of growth rates, life spans, and the nature of the food and population dynamics of the dominant soft-bodied invertebrates, particularly deposit feeders. Without these data, it is still too early to generate any unified theory. ■

Most benthic infaunal and epifaunal invertebrates are recruited from larvae settling out from the plankton. Therefore, a primary process that could produce changes in infaunal adult abundances and community structure is recruitment. Despite this fundamental relationship, there are few studies analyzing the role of recruitment in determining adult abundances and community structure. This is probably because it is difficult to isolate the effect of recruitment from various postrecruitment mortality sources. In a review of the role of recruitment, Menge and Farrell (1989) suggest recruitment may be an influence in structuring adult communities when one or more of three conditions prevail. First, the recruiting organisms are generally free of other sources of mortality, such as predation and competition. Second, the recruiting organisms settle in such numbers that they either swamp the predators with sheer numbers or keep pace with the predation rate. Finally, the recruitment rates of all species in the community are low and none of the organisms settling are competitive dominants. It is probably safe to conclude that recruitment has the potential to influence community structure in these sedimentary communities, or in any benthic community, but the lack of definitive studies permits us only to speculate.

The role of recruitment in determining the structure of soft-sediment communities was more recently reviewed by Olafsson et al. (1994), who concluded that despite the now-extensive studies on the population dynamics of soft-bottom marine invertebrate communities, there is not yet compelling evidence demonstrating that recruitment shapes the adult communities. However, they also point out the difficulties in separating settlement intensity, mortality, and variability from various postsettlement processes that may also affect mortality, making any conclusions tentative at present. They also note that several postsettlement processes—namely predation by large epifaunal predators, inhibitory adult-juvenile interactions, sediment disturbance, and food limitation—could also regulate adult community structure, particularly if several operate jointly.

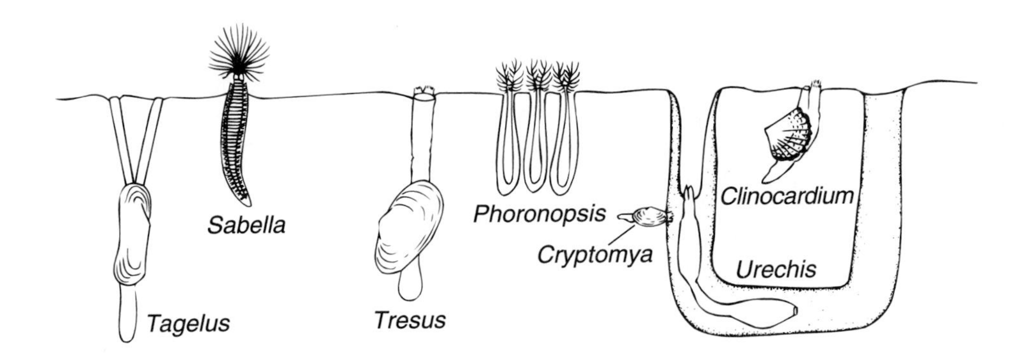

FIGURE 5.9
Various infaunal suspension-feeding types distributed at different depths in the substrate.

Vertical Distribution and Competition

Competition among benthic marine invertebrates is generally restricted to food and space. In the subtidal soft-bottom areas of the oceans, where the dominant species are suspension and deposit feeders, competition is usually for space. Food, in the form of either detritus or plankton, is abundant. Suspension feeders, of course, must have their feeding structures in the water column. It is possible, however, to get more suspension feeders in a given area of bottom if, instead of placing them all on the surface, they are distributed vertically in the substrate with only the feeding organ at the surface (Figure 5.9). Depending on the orientation of the animal in the substrate, the feeding structure may take up less room than the whole animal. Deposit feeders may feed not only at the surface but at any level in the substrate that contains organic material. Such animals do, however, require a connection to the surface to obtain oxygen from the overlying water. In a similar manner, more deposit feeders could be accommodated and direct space competition reduced if the different species were distributed vertically in the substrate.

Such vertical distribution of species is a well-known feature of soft bottoms. It occurs in both mud and sand (Figure 5.10) and in both suspension- and deposit-feeding organisms. Experiments by Peterson (1977), in which certain species were removed from given depth strata, resulted in an increase in abundance of other species occupying the same depth level. Still other experiments, in which the density of given species at given levels was increased above natural levels, resulted in rapid emigration out of the high-density areas by the species. These experiments suggest there is competition for space among species, and it results not only in vertical spacing of species but also in maintenance of certain densities.

Competition for space may also be the reason for the occurrence of **commensalism** among certain species (see Chapter 10 for discussion). Competition is so severe that it selects for organisms that share space. An example is the clam *Cryptomya californica*, which lives with its siphons in burrows of the shrimp *Callianassa californiensis* or the worm *Urechis caupo* (see Figure 5.9).

Community Patterns, Stability, and Variability

Community structure in open sedimentary environments changes over differing time periods and produces certain patterns or patches that recur on differing areal and time scales. We shall now take what we know of physical and biological factors and their roles in structuring communities and see how these act together over time to create varying community patterns. It is our purpose here to document the types of patterns, explain how they might have come about, and why they change, using some examples from the scientific literature. We note, as Woodin (1983) does, that there is a definite geographical bias to our fund of knowledge. Most well-studied cases come either from the temperate Pacific coast of North

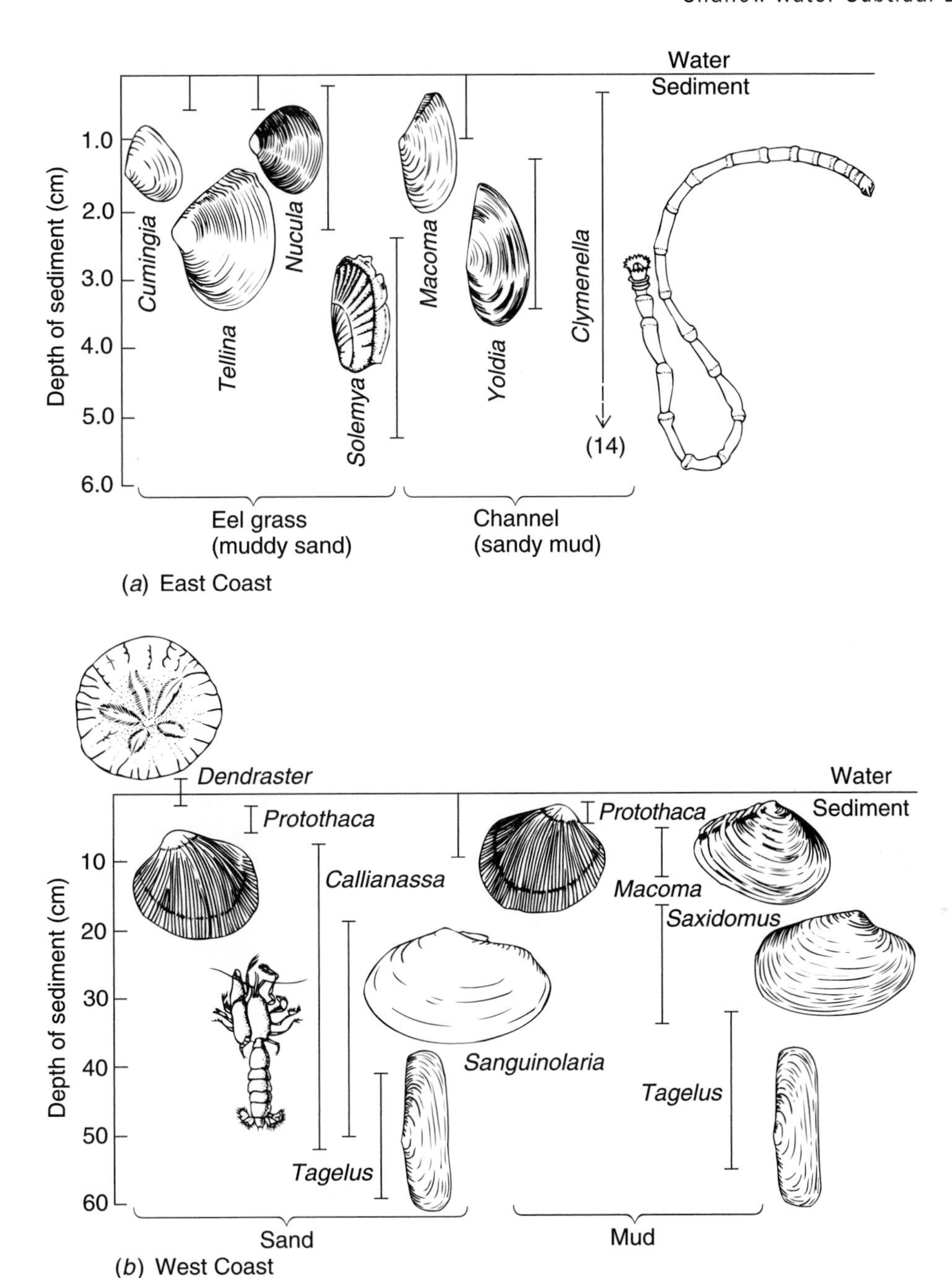

FIGURE 5.10 Vertical distribution of benthic infauna in (a) an Atlantic coast bay, Quisset Harbor, Massachusetts, and (b) a Pacific coast lagoon, Mugu Lagoon, California. (a from "Ecology of Shallow Water Deposit Feeding Communities" in *Ecology of Marine Benthos,* edited by B. Coull. Copyright © 1977 University of South Carolina Press. b from "Competitive Organizations of the Soft-Bottom Macrobenthic Communities of Southern California Lagoons" C. H. Peterson, 1977, *Marine Biology* Vol. 43, pp. 343–359.)

America or from the temperate Atlantic coast of North America and Europe. These two areas have definite differences in infaunal and predator abundances, the Pacific coast being most dense and, in general, relatively greater in abundance of large predators than the Atlantic coast. As a result, the relative importance of competition and predation in structuring these communities differs between the two areas.

In some instances, communities are very stable over periods of several years. This stability seems to result from the dominance of one or a few physical factors, or a dominant animal that controls its own environment and is long-lived. A good example is the previously discussed sea cucumber *Molpadia oolitica*. As we noted earlier, *Molpadia*, by its feeding activities, creates a bottom topography of stable cones or mounds interspersed with unstable depressions. By creating these unstable areas, *Molpadia* prevents other organisms, mainly suspension feeders, from settling in the area. The exception is the more stable area of the fecal cones, where the suspension feeders can and do settle and add to the stability of the cones. Such areas, once established, remain for years, perhaps because of the life span of *Molpadia*. On the southern California coast, Fager (1968) found that a shallow subtidal sand community dominated by nine large epifaunal invertebrates (two sea anemones, one sea pansy, one brittle star, three snails, one starfish, and one hermit crab) remained constant for a six-year period. He also found that this community lacked substantial interspecific interaction, suggesting that the community was structured by physical factors alone.

There are also situations in which the communities show regular cyclical oscillations between alternate states. For example, in the aforementioned sedimentary system in Barnstable Harbor, Massachusetts, Mills (1969) demonstrated that the area was dominated at certain times by the gastropod *Ilyanassa obsoleta* but at other times by the tube-dwelling amphipod *Ampelisca abdita*. In this case, both physical and biological factors act, but alternately. *Ampelisca* is apparently a dominant space competitor. It also interferes with *Ilyanassa* feeding, and so it outcompetes *Ilyanassa* in seasons with low storm activity, leading to the mat development. In stormy winter weather, however, the tube mats are rolled up by the heavy wave action, freeing the substrate for *Ilyanassa*; thus, this community alternates between two states. Such regular patterns may also result from pulses of reproductive activity and subsequent settlement of young. Typically, the pattern is to have high numbers of juveniles in spring and summer. This is followed by more or less massive mortality produced by competitive interactions or predation; hence, there are smaller numbers in winter.

The importance of predation in sedimentary environments is demonstrated by a study by Virnstein (1977) in subtidal sand communities in Chesapeake Bay. Virnstein excluded by cages several large predators, including the blue crab (*Callinectes sapidus*) and two bottom-feeding fish, the spot (*Leiostomus xanthurus*) and the hogchoker (*Trinectes maculatus*). He found that inside the cages, where predators were absent (at least the epifaunal predators), the infaunal invertebrates increased in density and diversity. The species that had the largest population increases were the shallow-living opportunistic ones, those most likely to be preyed upon by epifaunal predators. The deep-dwelling species showed little change, primarily because they probably avoid epifaunal predation by their habits.

Although we lack much data for multiyear changes in sedimentary communities, we do know that in some cases there are changes that occur at long intervals. One way of obtaining such a pattern would be through the variable recruitment success of the dominants. For example, if a dominant, long-lived species produces many juveniles each year, but in most years all or nearly all die, and then in one year all or most live to grow up, the resulting adult population and the community will show oscillations spaced in years. Such a situation seems to prevail for the amphipod *Pontoporeia affinis* in the Baltic Sea (Figure 5.11). Such a long-term cycle may affect other dominants also. For example, in the Baltic Sea, *Macoma balthica* oscillates in dominance with *Pontoporeia*, because at high densities of *Pontoporeia*, the juveniles of *Macoma* are outcompeted. The oscillatory period is about six or seven years.

The previous discussion centered on the role of a single factor—or of two or more factors acting alternately—to impose a pattern of stability. In most benthic sedimentary environments, however, there is a mosaic of patches, each with a different structure of species. These are usually the result of several factors acting simultaneously or in some sequence. They are the most complicated situations and can be understood best by again resorting to some well-studied examples.

One of the simplest situations was that studied by Kastendiek (1976) at depths of 2–13 m on sand at Zuma Beach in southern California. Here the dominant species are the sea pansy *Renilla kollikeri*; the sand dollar *Dendraster excentricus*; the nudibranch

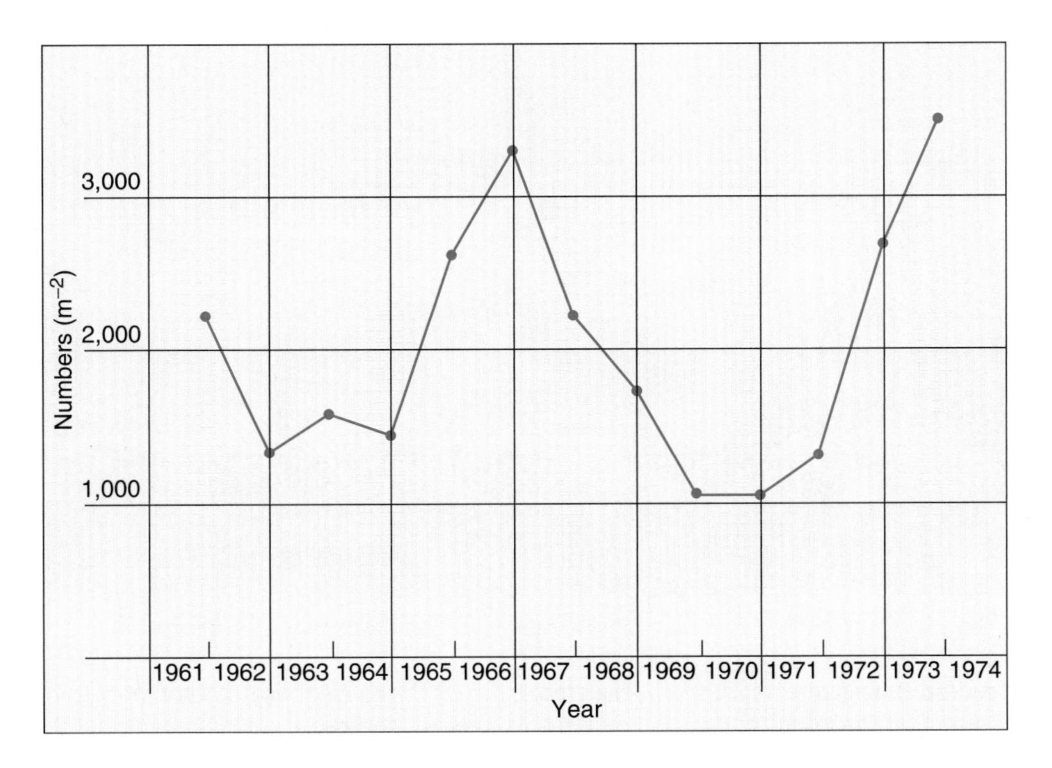

FIGURE 5.11 Graph of the population changes in the density of the amphipod *Pontoporeia affinis* in the Baltic Sea. (From *The Ecology of Marine Sediments*, John Gray. Copyright © 1981 Cambridge University Press. Reprinted by permission.)

Armina californica; and the starfish *Astropecten armatus*. Off Zuma Beach, *D. excentricus* forms a dense band (up to 1,500 individuals per square meter) that lies parallel to the shore and runs the entire length of the beach (Figure 5.12). The band moves into shallower water during calm periods of the year and offshore during stormy times. *Renilla kollikeri* occupies an area inshore of the sand dollar bed. *Armina californica* is a specialist predator of *R. kollikeri*, and *A. armatus* feeds on both *R. kollikeri* and *D. excentricus* (Figure 5.13). In this system, *Renilla kollikeri* is prevented from moving further shoreward because the increased wave motion in shallower water prevents the sea pansy from anchoring itself. It cannot extend further seaward because of the sand dollar bed. When *R. kollikeri* enter the sand dollar bed, the *Dendraster excentricus* uproot and eliminate them. During that time of year when the sand dollars move shoreward, that portion of the *R. kollikeri* bed it occupies is destroyed.

FIGURE 5.12 Sand dollars, *Dendraster excentricus*, in a characteristic upright feeding position in a subtidal bed in California. (Photo courtesy of Drs. Lovell and Libby Langstroth.)

Armina californica and *Astropecten armatus* have a synergistic effect. *Astropecten armatus* is repelled by the expanded autozooids of the *R. kollikeri*. *Renilla kollikeri*, however, has a characteristic behavior when attacked by *A. californica*. This is to close up all polyps and raise itself up off the substrate. The colony thus acts like a sail. The turbulent water then uproots the *R. kollikeri* and tumbles it away from the *A. californica*; thus, *A. californica* gets only one or two bites before the escape

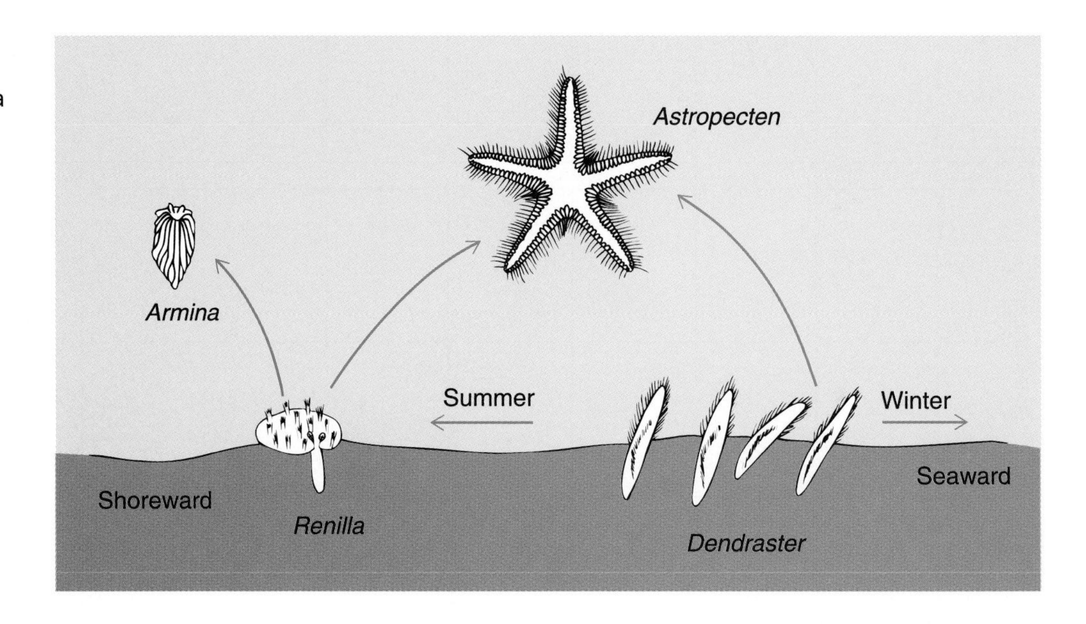

FIGURE 5.13
The food web of a southern California subtidal sand flat.

response. However, a closed-up R. *kollikeri* is readily eaten by A. *armatus*, so escape from one predator is not escape from the other. *Astropecten armatus*, however, is capable of ingesting only R. *kollikeri* up to 40 mm in rachis diameter. *Armina californica*, in turn, is able to consume R. *kollikeri* colonies if they become lodged against objects, such as sand dollars.

The sand dollar bed forms a barrier to *Astropecten armatus*, which normally is distributed seaward of the bed and is not able to penetrate shoreward of it. *Astropecten armatus* also consumes sand dollars. What controls the edges of the sand dollar beds is not known.

In this system, there are both physical and biological controls on the community. *Renilla kollikeri* is controlled seaward by *Dendraster excentricus* and shoreward by waves, and the population is further regulated by predators, of which one is unable to attack it due to the sand dollar bed. Should the sand dollar bed disappear, Kastendiek predicts that *Astropecten armatus* and *Armina californica* would destroy the R. *kollikeri* population.

A final example is that of VanBlaricom (1982), which demonstrates how a temporal mosaic of patches of different groups of infauna may come to exist in sedimentary environments. VanBlaricom investigated the community structure of an open sedimentary environment at 17 m off La Jolla in southern California. In this area, the infaunal community is disturbed by two large rays—the round sting ray *Urolophus halleri* and the bat ray *Myliobatis californica*—that dig pits to expose their prey. In so doing, they also destroy most of the other infauna in the area of the pit or expose them to other predators. As a result, the pits represent open areas available for settlement. These pits also accumulate food material in the form of organic detritus. VanBlaricom showed that the pit areas attract first a set of opportunistic invertebrates that are nocturnal swimmers and feed on detritus. Later animals tend to be subsurface deposit feeders that displace the early colonists. Depending on the time since its creation, each pit may have a different set of organisms. This leads to a mosaic on the bottom of differing infaunal associations representing various stages in recolonization. The picture is further complicated by the activities of still other predators. The starfish *Astropecten verrilli* consumes crab larvae soon after they settle. This reduces the abundances of two crabs, *Cancer gracilis* and *Portunus xantusii*. The two crabs, in turn, feed on infaunal organisms, so sea star predation is important to maintenance of the infaunal community. Finally, the speckled sand dab (*Citharichthys stigmaea*), a fish that is a visual predator, ingests invertebrates stirred up by the ray digging, further depleting the infauna in the pits.

■ In summary, marine subtidal infaunal communities of unvegetated sedimentary environments are characterized by patchy distributions of organisms and certain variability in species abundances and composition over time. The communities are maintained by settlement of larvae from the

FIGURE 5.14 Anemones of the genus *Tealia* and the hydrocoral *Stylantheca* (*Allopora*) *porphyra* with surrounding tunicates, sponges, and coralline algae on subtidal rocks of central California. (Photo courtesy of Drs. Lovell and Libby Langstroth.)

plankton (see Chapter 1). The ability of larvae to choose areas for settlement coupled with the ability to delay metamorphosis ensures that settlement is not random. The patchiness and variability result from continual random disturbances created by water motion or biological activity such as predation, which eliminates the population in a small area (patch). This is followed by recolonization and change with time due to a complex series of biological interactions involving different life history strategies, opportunist versus equilibrium; competitive interference; and exclusion and predation. The result is that different areas have different species abundance and composition depending on the time from the last disturbance and the relative stability of the substrate. Competition for space is ameliorated through vertical stratification of the organisms.■

ROCKY SUBTIDAL COMMUNITIES

Rocky subtidal surfaces are probably not as abundant throughout the world as are sedimentary substrates, but they are common in some areas. Included in the category of hard-substrate communities are those with a set of low-growing encrusting plants and animals, as well as kelp forests and coral reefs. We discuss only the first here. Kelp forests are covered in the next section, and coral reefs are dealt with in detail in Chapter 9.

Usually the subtidal, hard substrates are densely covered with low-growing plants and sessile or sedentary animals (Figure 5.14). The prominent motile animals are usually sea urchins, crabs, and gastropod mollusks. There is little, if any, open primary space (Figure 5.15). Most of these surfaces are without significant topographic relief, giving a homogeneous two-dimensional surface. Nevertheless, they have a high species diversity or species richness, leading biologists to ask how so many species can coexist in such a relatively unstructured flat area. In contrast to the sedimentary environment, there is no third dimension (depth) to be occupied by additional species. Furthermore, there does not appear to be a fine partitioning of the ecological niches as in other environments. Most of the animals of this area are either colonial encrusting species able to spread over surfaces by asexual means or single individuals that settle gregariously, forming large, space-occupying clumps. Both groups feed by suspension feeding. Included in the first group as dominants are such animals as sponges, tunicates, bryozoans, and various cnidarians, such as anemones and hydroids. The second group is composed of bivalve mollusks, algae, sea urchins, and gastropod mollusks. These animals do not appear to be food limited.

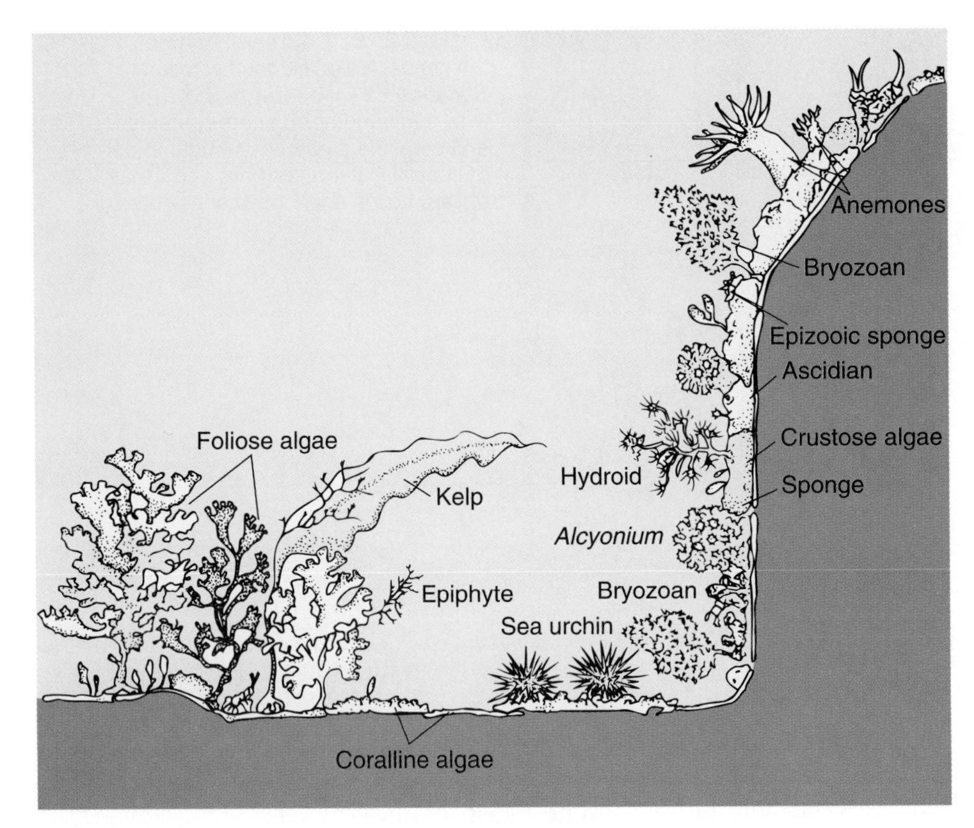

FIGURE 5.15 A diagrammatic representation of the characteristic flora and fauna inhabiting the vertical and horizontal surfaces of some Massachusetts subtidal rocks. Horizontal surfaces are dominated by a canopy of foliose algae and other epiphytic species. If the surfaces are being actively grazed by sea urchins, coralline algae predominate. Vertical surfaces support diverse assemblages of invertebrates and algae that compete for space; some of these species are seasonal, such as the erect bryozoans and hydroids that form a canopy layer during the warm months. (Modified from "The Ecology of the Rocky Subtidal Zone" K. Sebens, *American Scientist*, Vol. 73, No. 5, pp. 548-557, 1985. Copyright © 1985 American Scientist. Reprinted by permission.)

It now appears from studies done primarily in Massachusetts by Sebens (1985, 1986), in Maine by Witman (1987), and in Australia by Fletcher (1987) that the organization and persistence of these areas, and the co-occurrence of many species, are explained by a combination of biological and physical factors that include storm disturbance, competition for space, grazing, recruitment, and mutualism. The disturbances generate patches of free space that are subsequently colonized by larvae or juveniles of nearby adults, or by spread of the adjacent colonial organisms. Competition among these organisms for the limited space, coupled with seasonality of certain organisms and mutualistic interactions among certain species, completes the cycle. This means that in any area at any time there is either a mosaic of patches of different ages with different sets of or different abundances of species, resulting in the high species diversity, or a different zonation pattern. Now let us examine how this comes about using a few examples.

In the Massachusetts subtidal, Sebens (1985, 1986) found the most significant biological disturbance was the sea urchin *Strongylocentrotus droebachiensis*. Where these urchins have been abundant for a period of years on flat or sloping surfaces, they will have grazed off most sessile and sedentary organisms, leaving only a thin layer of encrusting coralline red algae (*Lithothamnion, Phymatolithon, Clathromorphum*). Although other animals also graze the bottom areas, none appear to have the effect of the urchin. Within this generally depauperate grazed area, there is one refuge from the urchin grazing and other predation: the beds of the mussel *Modiolus modiolus*. Experiments by Witman (1985) demonstrated that consumption of several prey items was much greater outside than inside the mussel beds. Two groups of predators patrol these areas, a diurnal set consisting of several fishes, and a nocturnal set of crabs (*Cancer irroratus* and *C. borealis*) and the lobster (*Homarus americanus*).

Where the urchins are rare or absent, the hard bottom consists of a 10- to 20-cm thick canopy of fleshy algae. Below this canopy are encrusting coralline algae and patches of various colonial animals, such as ascidians, plus various small, motile crustaceans and the tubes of polychaete worms and amphipod crustaceans.

Vertical rock walls in this area usually are inaccessible to urchins and have a different association of organisms. On these walls are a variety of encrusting, mostly colonial invertebrates dominated by a colonial compound ascidian, *Aplidium pallidum*, the sea anemone *Metridium senile*, the octocoral *Alcyonium siderium*, and several sponges. This community appears to be structured by competition among the organisms for space. Since most of these organisms are colonial, competition involves direct overgrowth of one by another whenever their edges meet. The species on the vertical faces have been shown by Sebens (1985, 1986) to exist in a competitive hierarchy that ranges from those able to overgrow all others to those that can be overgrown by all others. In general, the colonial species with the larger or more massive growth form are competitively dominant. Thin, flat colonies are competitively inferior.

Experimental addition of urchins to vertical faces results in the removal of all the encrusting forms except the coralline algae. Caging urchins out of horizontal surfaces results in a dense growth of foliose algae.

The previous situation would seem to lead to monotonous landscapes of sea urchins and encrusting coralline algae on horizontal surfaces and to massive aggregations of the dominant space competitor, *Aplidium*, on vertical walls. This, however, does not happen, and there are horizontal areas of foliose algae and vertical walls with various numbers of other inferior space competitors. How does this come about? It has to do with a number of life history and growth features of some of these invertebrates. The sea urchins, for example, are preyed on by fishes, several crabs, and (uncommonly) lobsters. They are also destroyed by an undetermined disease and winter storms. Whenever massive urchin mortality occurs in an area, the community changes and will remain in that alternate state until urchin grazing again reduces the area to crustose corallines. This may take some time, so we get patches. Also, urchins are not able to gain access to vertical walls because they cannot attach themselves to the colonial animals. If some disturbance does clear the vertical surface of *Aplidium*, the urchins may gain access and keep that patch of vertical surface grazed down to corallines for a long period of time. Such an area may exist next to a vertical face with a well-developed association of colonial animals dominated by *Aplidium*.

Inferior colonial competitors may exist with *Aplidium* for several reasons. In the case of the erect octocoral *Alcyonium*, it tends to grow upward as a juvenile. At the same time, *Aplidium* refrains from reproducing for its first two years. On an open surface, therefore, if *Alcyonium* and *Aplidium* settle at the same time, the time lag in *Aplidium* growth will allow *Alcyonium* to grow up high enough to avoid overgrowth. Although *Aplidium* will eventually grow over the surface, it will not displace the *Alcyonium*, which expands above it, maintaining only a narrow attachment to the rock under the *Aplidium*. *Alcyonium* also inhibits urchin grazing either physically, by presenting a vertical obstruction, or through action of its nematocysts on the urchins. This, in turn, may aid the other members of the wall community. *Alcyonium* has crawling larvae, which will not mature unless they settle more than 2 cm away from the parent colony. They also need a certain type of substrate—either rock or coralline algae. Since neither is usually present where the adults are, there is usually no recruitment. Since the dispersal is so space limited, there may be walls that lack any of the species only meters away from adult *Alcyonium*. The anemone *Metridium*, also an inferior space competitor, is found in patches but because of a different set of factors. Once established, *Metridium* holds its space by using its special "sweeper tentacles," heavily armed with nematocysts, for actively deterring any other competitor. Meanwhile, it reproduces asexually by fragmenting and forms a dense clone that can persist for years.

Further north, in the Gulf of Maine, Witman (1987) investigated subtidal rocky surfaces on extremely exposed sites at depths of 4–20 m. Here the deeper surfaces (11–18 m) were dominated by the mussel *Modiolus modiolus*, while the shallower areas of 4–8 m were dominated by the kelps *Laminaria digitata* and *L. saccharina*. Experimental manipulations in these two communities demonstrated the causes of this dramatic zonation pattern. The key species was again the urchin *Strongylocentrotus droebachiensis*. The shallower-dwelling kelps were prevented from moving deeper by grazing by the sea urchins, thus maintaining the integrity of the mussel bed. Storm-generated dislodgment was the most significant factor in the mortality of the mussels. In the absence of urchins, the kelps would attach to the mussels, making them 30 times more subject to dislodgment and mortality by storms. Therefore, the urchins increase the *Modiolus* survivorship by removing the kelp and decreasing the probability of dislodgment. This *Modiolus-Strongylocentrotus* interaction is mutualistic in that it takes significantly more force to dislodge the urchins from the *Modiolus* surfaces than from surrounding bare rock surfaces or kelp beds. The *Modiolus* beds also provided a refuge from predation for the recently recruited small urchins.

Finally, the ability of both kelp and mussels to recover from dislodgment due to storm disturbance indicated that kelp recovery was much faster, the kelp occupying all open patches. This suggested the kelps were competitively superior. In the absence of sea urchin grazing, they would probably dominate the entire depth range. Similar subtidal zonation patterns, with a shallow zone dominated by kelps and a deeper zone dominated by urchins, have been reported for other temperate regions in Nova Scotia and New Zealand and suggest a more universal pattern, perhaps maintained by a similar mechanism.

A somewhat similar mechanism was found by Fletcher (1987) to maintain subtidal rocky areas on the east coast of Australia. In this area, the subtidal rock is covered with encrusting coralline algae on which are found high densities of a suite of grazers that includes a sea urchin, *Centrostephanus rodgersii*; three limpets, *Patelloida alticostata*, *P. mufria*, and *Cellana tramoserica*; and two turbinid gastropods, *Australium tentiforme* and *Turbo torquata*. Manipulative experiments here, however, gave different results depending on which group of grazers was removed. Removal of all grazers resulted in the rapid colonization of the crustose algae by fleshy, foliose algae. If only the turbinid snails were removed, there was no significant change in the community. If limpets and urchins were removed, the fleshy algae covered 80–100% of the area within a year. If only the urchins were removed, the fleshy algae increase was slowed and only approached 100% after two years. Finally, where only limpets were removed, there was an initial algae increase followed by a decline to near conditions found where no grazers were removed.

The differential responses of the community, which depended on which set of grazers was removed, resulted from the more complex interactions among the suite of limpets, turbinid snails, and urchins. Whereas it seemed that the presence of the sea urchins was necessary to maintain the encrusting coralline community, it also appeared that the continued maintenance of at least the populations of the three species of limpets was also dependent on the urchins. However, the mechanism or mechanisms responsible for this are presently unknown.

Although these studies emphasize the importance of urchins as the dominant grazers and the keystone species in determining the type of community formed on subtidal hard substrates, one cannot discount the importance of other grazing animals. This has been demonstrated by Ayling (1981), who investigated three communities each dominated by a different group of animals: (1) encrusting coralline algae, (2) sponges, and (3) ascidians in the subtidal of New Zealand. There were three groups of grazers: an urchin, *Evechinus chloroticus*; a suite of herbivorous gastropods; and a grazing fish, *Parika scaber*. At high densities, the urchin grazing was responsible for the perpetuation of a low-diversity community consisting primarily of crustose coralline algae. However, removal of the suite of grazing gastropods resulted in a change in the community to one with more fleshy ephemeral algae, even in the presence of the urchins. Removal of the grazing fish, which primarily fed on sponges and ascidians, caused a major shift in the community structure from one dominated by crustose coralline algae to one dominated by ascidians and sponges. Finally, disease and parasites also played a role. Ayling reported episodic outbreaks of a bacterial or fungal disease that destroyed large numbers of the otherwise long-lived sponges.

■ In summary, we have on hard substrates a two-dimensional surface. In areas so far studied, all in cold or warm temperate seas, the communities are dominated by an encrusting community of coralline algae, ascidians, or sponges that compete for space among themselves by overgrowth; beds of mussels forming large clumps; or kelps. In all cases, sea urchins appear to be the keystone species, for without them the community changes. Where urchin densities are high, the communities are dominated by crustose coralline algae or mussels, both very resistant to urchin grazing. Where urchins are absent or in fewer numbers, communities of fleshy algae or kelp, or both, dominate. Physical disturbance, primarily by storms but also by disease and predation, causes mortality among the urchins, freeing the more ephemeral algae or kelps to occupy more space. This "typical" picture may be modified by suites of smaller gastropods and fish grazers, both of which may alter the community structure to favor certain encrusting organisms. Finally, the differences in the life histories of various competitively inferior organisms may lead them to persist even in the face of their superior competitors. ■

The existence in the same area of different groups of organisms or patches of organisms is what Sutherland (1974) has called **alternate stable states**. These alternate stable states result when some factor or set of factors disturbs the dominant species, causing a community to switch to a new composition, which in turn remains stable until another significant disturbance occurs.

Although there has been some argument among ecologists as to whether or not alternate stable states is a real phenomenon, the fact remains that very different community structures can and do exist near one another under virtually identical environmental conditions. This suggests some perturbing factor has caused a community to change from one configuration to another, and there is some mechanism that allows the two configurations to persist. One of the most intriguing examples of this phenomenon is that reported by Barkai and McQuaid (1988) for two subtidal rocky communities in South Africa. Off the west coast of South Africa, there are two islands: Malgas and Marcus. They are separated by 4 km, so both experience similar environmental conditions. However, the rocky subtidal communities are quite different in each. At Malgas Island, the rock lobster *Jasus lalandii* and algae dominate the community, and the lobsters make up 70% of the biomass. Few other organisms, save for a few gastropods of the species *Burnupena papyracea*, are found. At Marcus Island, by contrast, rock lobsters are rare, and the community is dominated by extensive beds of the mussel *Choromytilus meridionalis*, large populations of the carnivorous gastropod *B. papyracea*, and numerous holothurians and sea urchins. Why such extreme differences, and how are they maintained? Barkai and McQuaid demonstrated through experimental manipulations that on Malgas Island the rock lobsters prey on the settling mussels, preventing their establishment. They also prey on the gastropod *Burnupena* but are deterred from preying on those gastropods that have a protective bryozoan (*Alcyonidium nodosum*) on their shell. As a result, some gastropods continue to persist in the face of heavy lobster predation. At Marcus Island, when rock lobsters were transferred into the subtidal, all were attacked and killed immediately by the carnivorous gastropod *Burnupena*, which reversed the normal predator-prey relationship established at Malgas Island. Since there are normally no lobsters at Marcus Island, what do the *Burnupena* prey on? The answer seems to be that they scavenge on dead or dying mussels but are unable to successfully attack healthy mussels. Therefore, it appears that on both islands predation maintains the current communities, but instead of differences in predators at each, the change is maintained by a reversal of predator-prey roles (Figure 5.16). These two different communities persisted through the four years of the study as alternate stable states maintained by the very unusual mechanism of predator-prey reversal. Since the local fishermen have reported that the two islands had similar lobster populations 20 years ago, the unanswered question is how the change came about.

The concept of alternate stable states is not limited to subtidal communities. It is applicable to many marine communities, and we shall see it again in our considerations of intertidal and coral reef systems.

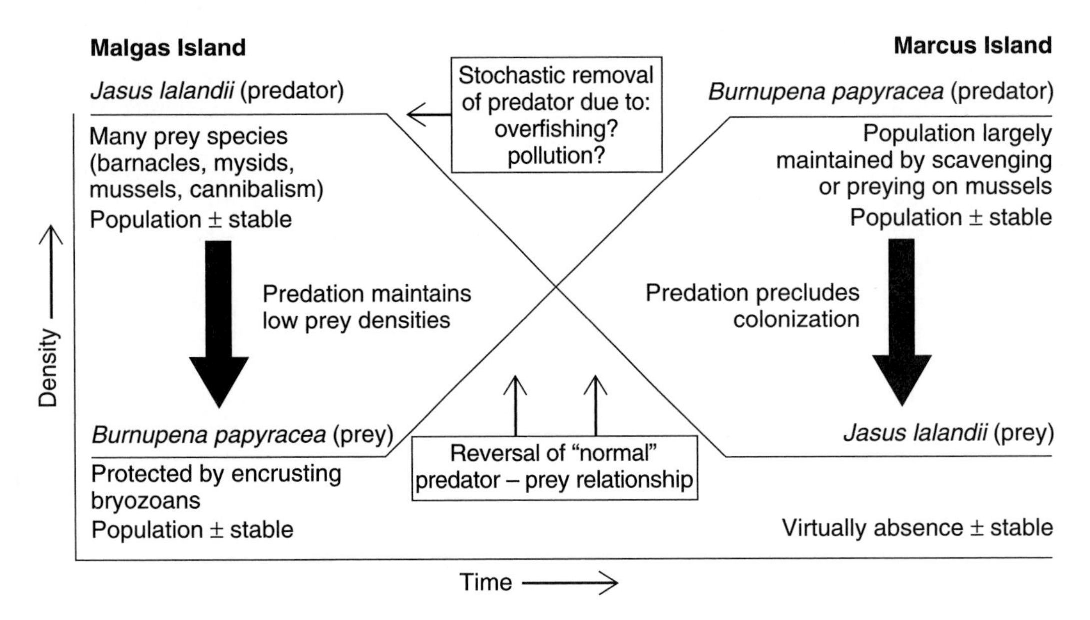

FIGURE 5.16 Schematic summary of rock lobster and whelk predator-prey role reversal at Marcus and Malgas islands. (Reprinted with permission from "Predator-prey Role Reversal in a Marine Benthic Ecoysystem" A. Barkai & C. McQuaid, 1988, *Science*, vol. 242, pp. 62-64. Copyright © 1988 American Association for the Advancement of Science.)

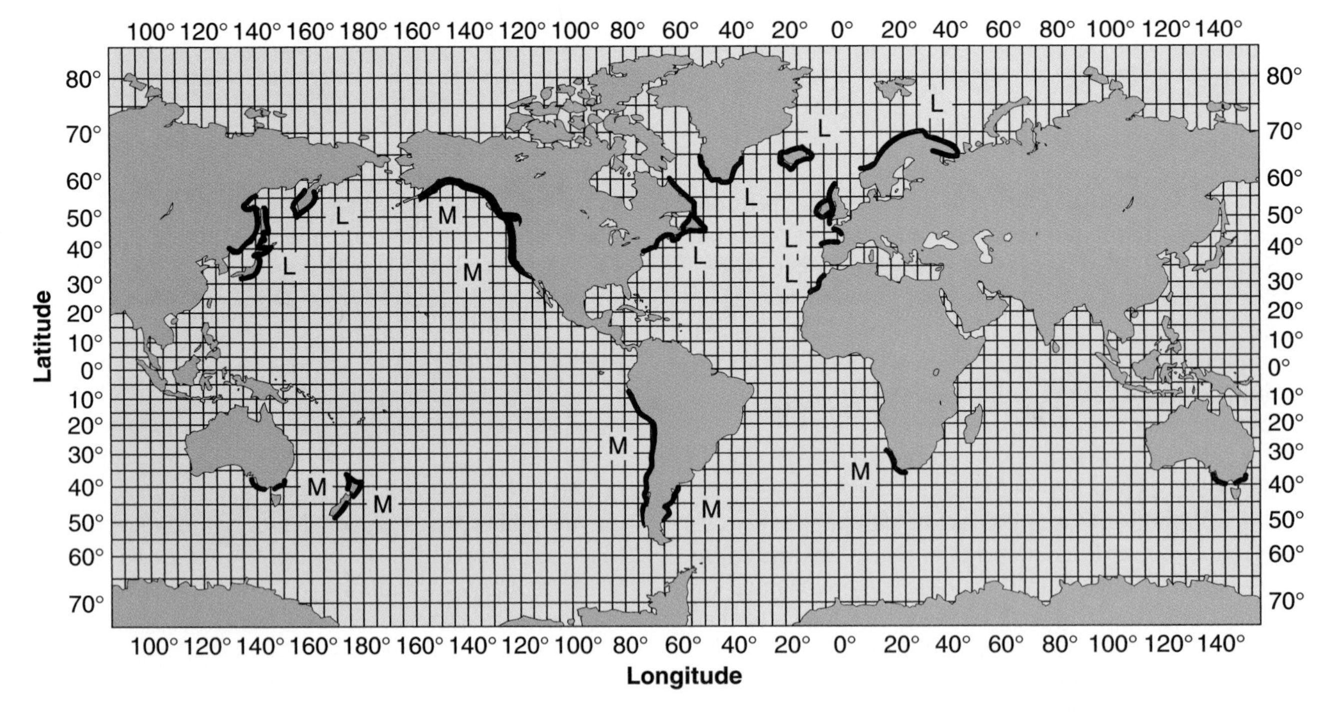

FIGURE 5.17 Distribution of kelp beds and kelp forests in the world, indicating the dominant genera. (L) *Laminaria*; (M) *Macrocystis*. (Reprinted with permission from "Seaweeds: Their Productivity and Strategy for Growth" K. H. Mann, *Science*, Vol. 182, pp. 975-983, 1973. Copyright © 1973 American Association for the Advancement of Science.)

FIGURE 5.18 A kelp bed composed of an overstory of *Laminaria dentigera* and an understory of the surfgrass *Phyllospadix*. (Photo courtesy of Drs. Lovell and Libby Langstroth.)

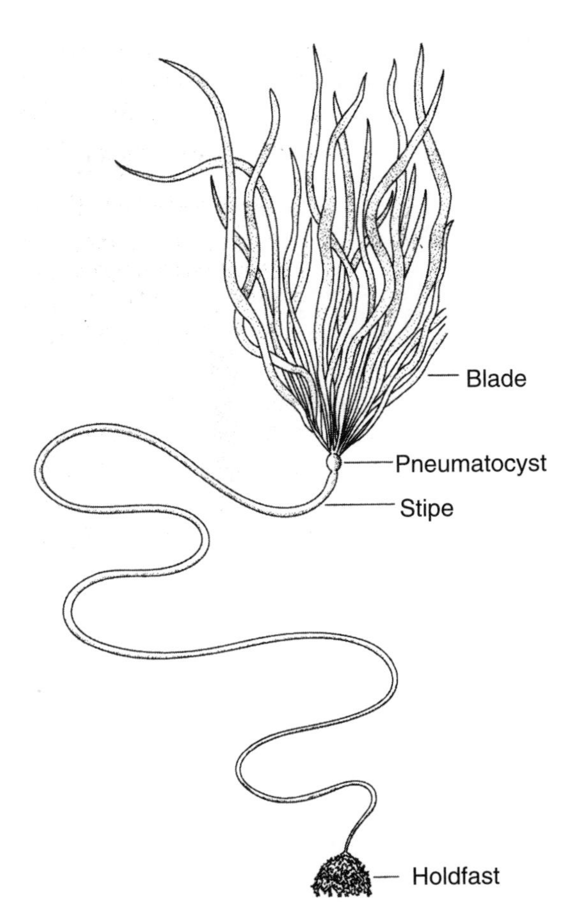

FIGURE 5.20 A kelp plant (*Nereocystis*) and its structure.

FIGURE 5.19 A view inside a giant kelp forest dominated by the kelp genus *Macrocystis* on the California coast. (Photo courtesy of Drs. Lovell and Libby Langstroth.)

KELP BEDS AND FORESTS

Throughout a large part of the cold temperate regions of the world, hard subtidal substrates are inhabited by a community dominated by very large brown algae known collectively as **kelps** (Figure 5.17). These associations are technically known as **kelp beds** if the algae do not form a surface canopy and **kelp forests** where there is a floating surface canopy (Figures 5.18 and 5.19). These terms are, however, often used interchangeably.

Structure and Distribution

Kelps are attached to the substrate by a structure called a **holdfast** rather than by true roots (Figure 5.20). From the holdfast arises a stemlike or trunklike **stipe**, which ends in one or more broad, flat **blades**. At the base of the blade is a **pneumatocyst** or *float*, which keeps the blade at the surface. Kelps obtain their nutrients directly from the seawater as do the

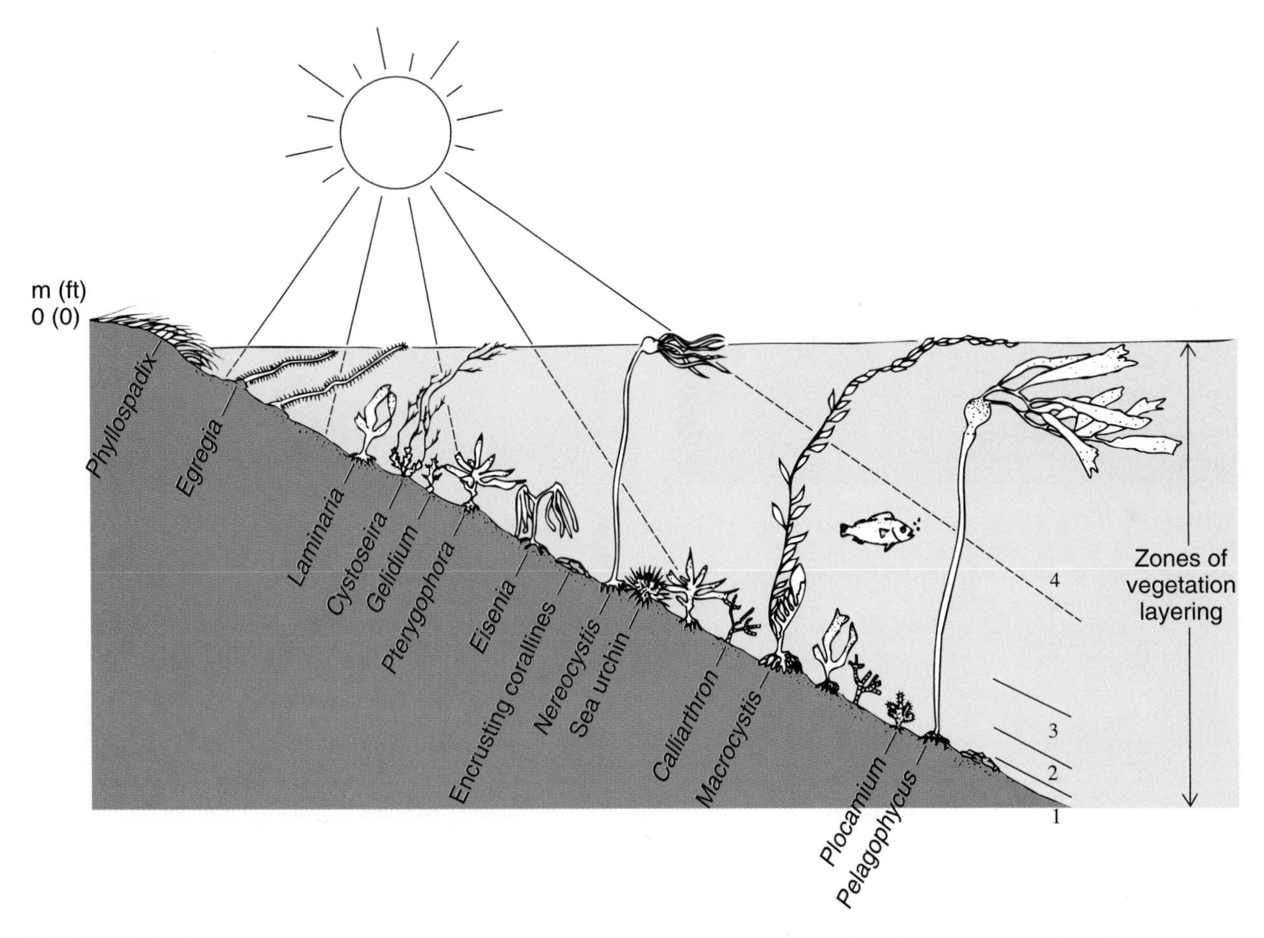

FIGURE 5.21 The distribution of some common seaweeds within giant kelp forests on the Pacific coast of North America. Plants in the four zones of vegetation layering include (1) small filamentous species and encrusting coralline algae; (2) bottom-canopy plants, such as *Gelidium*, *Calliarthron*, and *Plocamium*; (3) understory-canopy kelps, such as *Pterygophora*, *Eisenia*, and *Laminaria*; and (4) midwater and surface-canopy plants, such as *Egregia*, *Macrocystis*, and *Nereocystis*. This is a generalized diagram and some species do not occur at the same site. (Modified from M. S. Foster and D. R. Schiel, 1985, *The ecology of giant kelp forests in California: A community profile*, U.S. Fish and Wildlife Service Biological Report 85 (7.2).)

organisms of the phytoplankton. They depend on the constant movement of water past them to avoid nutrient exhaustion. Since these shallow waters are constantly moved by wave action and currents, and the nutrients are replenished by turbulence, upwelling, and runoff from land, nutrient depletion does not occur over most of the range. However, at the southern end of the range in North America, from San Diego to the southern limit of kelps in Baja California, nutrient depletion can occur in late summer and fall and during El Niño events.

In contrast to the monotonous, level landscape of the soft-bottom or the short-growing colonial organisms of other subtidal rock surfaces, the kelps form an extensive three-dimensional habitat composed of several vertical layers or strata (Figure 5.21). As a result, a large number of potential habitats are available, and the variety of life is greater.

The major kelps that dominate the structure of the kelp forests are the genera *Macrocystis* and *Nereocystis*. Kelp beds are dominated by the genera *Laminaria*, *Pterygophora*, and *Ecklonia*. The Pacific coast of both North and South America is dominated by *Macrocystis*, whereas *Laminaria* is dominant in Atlantic waters and in Japan and the Sea of Okhosk (see Figure 5.17). In contrast to most algae, which are small and generally never exceed a few centimeters or a decimeter in length, the major kelps are giants, with

(*a*)

(*b*)

FIGURE 5.22 (a) The surface of a kelp forest showing the thick canopy of blades. (b) The interior of a California kelp bed. (a, photo courtesy of Peter Arnold; b, photo courtesy of Norbert Wu.)

lengths the equivalent of the height of trees on land. *Macrocystis* and *Nereocystis* on the Pacific coast of North America may reach 20–30 m in length. Such massive autotrophs grow upward from the bottom and spread their blades at the surface of the water, where they obtain the maximum amount of light. These blades form a canopy similar to that of terrestrial forests, cutting off light to the substrate below (Figure 5.22). Below the canopy, extending up from the bottom, is a set of "understory" algae forming another layer (Figure 5.23; see also Figure 5.21).

The extent of the kelp beds and forests on various coasts depends on several factors. First, a hard substrate must be available for attachment. Second, the kelps must have light and can establish themselves only in water depths where the young, small autotrophs receive enough light to grow. The amount of light received at any depth is a function of water clarity, and where clear water prevails, kelp beds and forests extend out from shore to depths of 20–30 m. Where such shallow water is extensive, the beds may extend as much as several kilometers from shore. In more turbulent, murky water, the depth of the beds may be more restricted. Finally, kelps seem limited by high temperature and associated low nutrients, as well as by grazing. Kelp beds and forests occur throughout the world wherever the water is cool but are absent from warm temperate and tropical areas. Their considerable extent along the Pacific coast of North and South America is due to the cold upwelling on the North American coast and the cold water of the Humboldt Current along the South American coast.

Ecology and Life Cycle

Kelps have a phenomenal growth rate. *Nereocystis luetkeana* has been recorded by Scagel (1947) as growing in length by 6 cm/day, and, according to North (1971), *Macrocystis pyrifera* commonly grows 50 cm/day on the California coast. Such growth rates mean plants grow from the bottom to the surface in a short time. Kelp forests and beds are also extremely productive. Mann (1973) reports that the

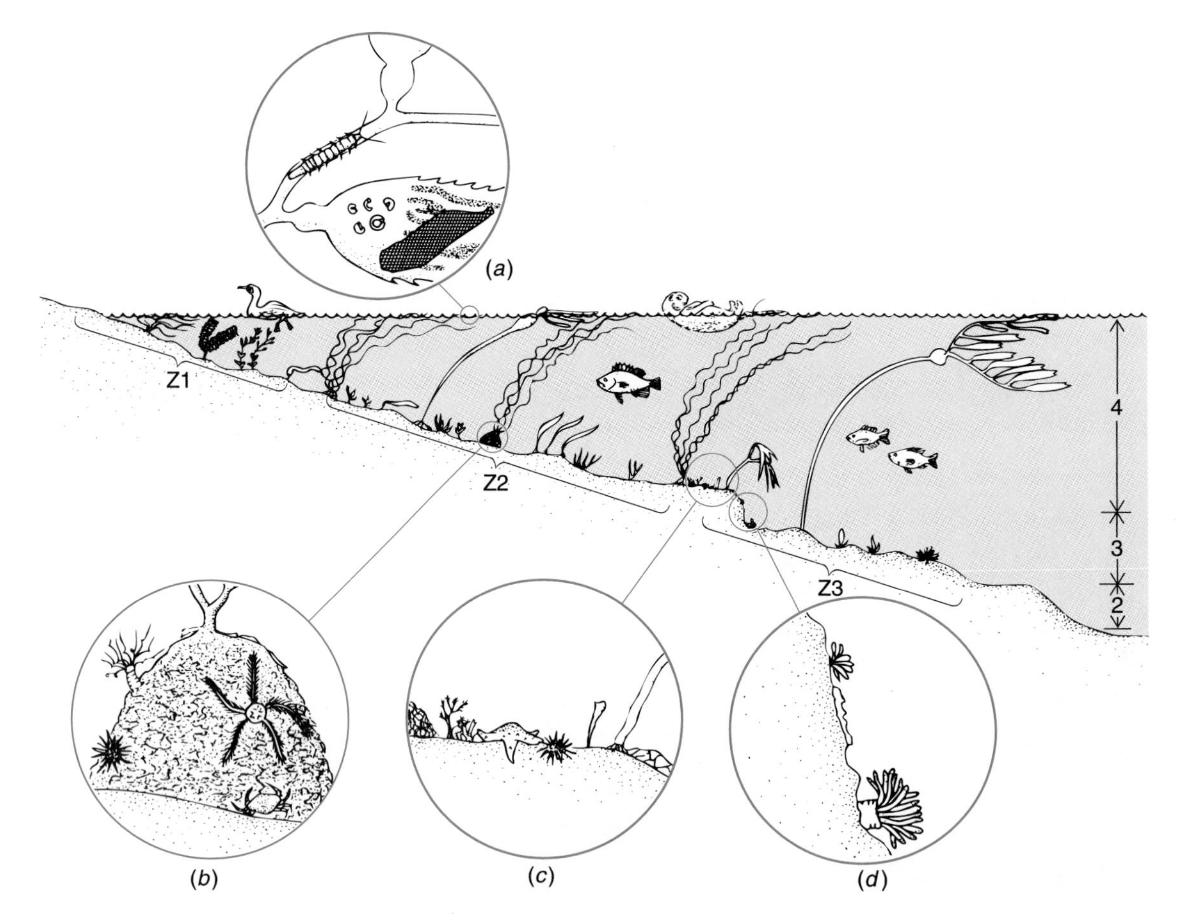

FIGURE 5.23 Cross section showing the inhabitants of a generalized giant kelp forest. The numbers to the right indicate vegetation layers (see legend for Figure 5.21). Three broad zonal associations along the depth gradient are shown: (Z1) inshore of the giant kelp community; (Z2) within the giant kelp community; and (Z3) offshore from the giant kelp community. Various subcommunities or associations are indicated by the circular diagrams: (a) animals associated with the surface of *Macrocystis* and other seaweeds (polychaetes, isopods, bryozoans); (b) animals found in giant kelp holdfasts (small sea urchins, brittle stars, crustaceans, polychaetes) (although shown on the outside, these organisms occupy the spaces within the holdfast); (c) plants and animals characteristic of horizontal surfaces (various sea stars, urchins, benthic fishes, understory algae); (d) organisms most common on vertical surfaces (primarily sessile animals, such as sponges, ascidians, bryozoans, sea anemones). Some of the organisms shown do not necessarily co-occur at any one site. (From M. S. Foster and D. K. Schiel, 1985, *The ecology of giant kelp forests in California: A community profile*, U.S. Fish and Wildlife Service Biological Report 85 (7.2).)

net annual productivity ranges from 800 g C/m^2 in California kelp beds to perhaps 2,000 g C/m^2 in Indian Ocean kelp beds. These figures are several times the production of the phytoplankton on the same per-area basis.

Kelps are perennial plants, often losing stipes and blades but regrowing new ones from the holdfast. The life span for *Macrocystis pyrifera* plants in California ranges from three to seven years maximum. The life cycle of kelp alternates between an asexual macroscopic stage termed the **sporophyte** and a sexual microscopic stage termed the **gametophyte** (Figure 5.24). Only the sporophyte stage is considered here, because almost all of the ecological work has been done with this stage and little is known about the ecology of the gametophyte in the natural environment. However, studies by Reed et al. (1988) and Deysher and Dean (1986) have suggested that patterns of recruitment of sporophytes, at least in southern California, are determined in part by episodic, infrequent occurrences of "recruitment window" conditions that affect the gametophytes.

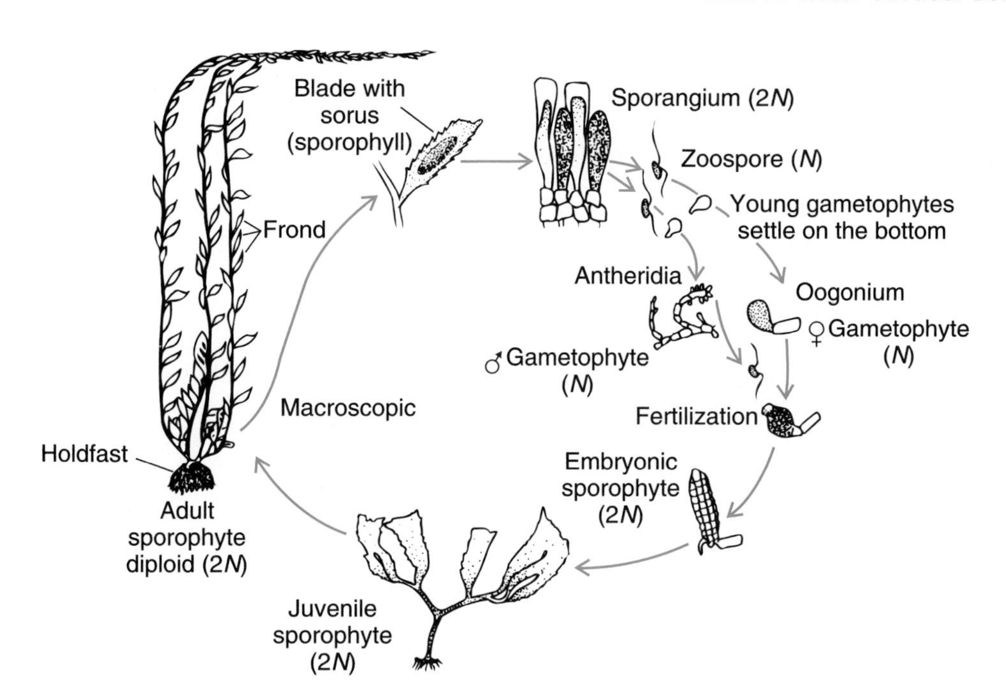

FIGURE 5.24
Life cycle of a typical kelp plant (*Macrocystis*). (From M. S. Foster and D. K. Schiel, 1985, *The ecology of giant kelp forests in California: A community profile*, U.S. Fish and Wildlife Service Biological Report. 85 (7.9).)

The kelps provide the framework for the kelp forest or kelp bed community. Associated with these dominants are many other species of algae, invertebrates, and fishes.

Despite the enormous productivity of these kelps, relatively few herbivores graze directly on the plants. It has been estimated that only 10% of the net production enters the food webs of the kelp bed through grazing. The remaining 90% enters the food chains in the form of detritus or dissolved organic matter.

Since the general economy of kelp beds and forests depends on the continued existence of the giant kelps, it is of interest to know the causes of kelp mortality. Adult plants are only occasionally destroyed by grazing herbivores, but they are vulnerable to destruction by mechanical forces, mainly wave action, and to nutrient depletion, mainly nitrogen. Since they occur in shallow water, often on open coasts, storm waves can have a devastating effect. The waves rip the plants up from the substrate and cast them up on the beach. Furthermore, once a plant has been uprooted, its movement through a bed, as Rosenthal and associates (1974) have demonstrated, often results in entangling among the stipes and blades of other plants. The result is that even more plants are pulled away from the substrate and destroyed. Severe storms can virtually denude kelp beds of the stipes and blades, but if the holdfast and some meristem tissue from the kelps remain, new stipes and blades can be produced.

The large El Niño event of 1982–83 that produced devastating effects on the coral reef communities in the Pacific Ocean (see p. 413) also had catastrophic effects on the extratropical kelp forests. Those of the California coast have been best documented. The storm season of 1982–83 was the most severe in decades, perhaps the worst of this century, according to Tegner and Dayton (1987). These storms caused massive mechanical damage to the kelp forests. In San Diego, the Point Loma kelp forest was reduced from 600 hectares in 1982 to 40 hectares. Off the Palos Verdes Peninsula in Los Angeles, the cover was reduced from 196 hectares to 18. In addition to the loss of canopy, the severity of the waves also ripped up holdfasts, leading to further loss. Finally, recovery in many places, particularly in southern California, was delayed because plants not directly killed by the storms were retarded in growth by the high water temperatures and reduction in nitrogen that accompanied the storms. Devastation from the El Niño was not universal, and some areas seemed to escape damage for unknown reasons. Exceptions included the sandy areas around Santa Barbara, where old holdfast material served as recruitment and growth centers. Most of these forests recovered in a few years.

Whereas storms, especially severe ones, are usually devastating to existing kelp forests, they may have vastly different effects on the subsequent community that develops, which is dependent on the state of the community before the advent of the storms. Ebeling et al. (1985) made regular observations on a southern California kelp forest over a five-year period. During that time, the forest was subjected to two severe storms. The first storm removed the canopy of *Macrocystis* but left intact the understory kelps, such as *Pterygophora californica*. This resulted in a drastic reduction of the drift kelp used as food by the urchins *Strongylocentrotus purpuratus* and *S. franciscanus*. The urchins then emerged from their sheltered cracks and crevices and, in the absence of effective predators, consumed most of the remaining living kelp. This, in turn, weakened the other parts of the food web and led to further declines in kelp forest associates, such as fish. The second storm in the area caught the urchins in the open and destroyed them, as well as clearing surfaces. The newly opened surfaces were then colonized by kelp plants that produced a fully developed kelp forest a year later. Thus two storms led to dramatically different communities, which were dependent on the type of community existing immediately prior to the storm.

Another potential cause of mortality is grazing, particularly by various sea urchins and fishes. Here the evidence is more equivocal. There is little doubt that sea urchins of many species can have significant effects on kelp beds and forests. These effects may range from wholesale removal of plants to selective feeding and even to provision of suitably cleared space in which the settlement of new kelp plants may take place. The creation by urchins of so-called "barren grounds" has been reported in many regions in which kelp occur. In recent years, greatly increased numbers of urchins in certain areas, such as New England and southern California, have coincided with diminished kelp abundance—undoubtedly because of urchin grazing pressure. However, the causes of the increased urchin abundance—a kind of population explosion—are not always clear. In southern California, for example, the urchins do not have to undergo a population explosion. They can simply change their behavior from feeding on drift algae to active foraging on live plants, or vice versa. This has been demonstrated by Harrold and Reed (1985), who studied the grazing activity of red urchins (*Strongylocentrotus franciscanus*) in adjacent barren areas and kelp-dominated areas. In barren areas, they found drift algae sparse; hence, the urchins were poorly nourished and were actively grazing all surfaces. In the kelp areas, the urchins were well nourished, lived mainly in cracks, moved little, and fed on abundant drift kelp.

During the study, a substantial recruitment of kelp occurred on both the barren site and the kelp site. On the barren site the large recruitment transformed the site into a kelp-dominated area. This increased the amount of drift algae, and the urchin population underwent a behavioral transformation, moving into cracks and crevices, feeding on drift algae, and reducing active surface grazing to levels found in kelp-dominated areas. There was a change in the community structure facilitated by a behavioral change in the dominant urchin, which was triggered by a switch in the mode of feeding. The switch was dependent on the availability of drift algae, which was dependent on the abundance of new kelp. In this case, the grazing intensity was independent of urchin density, and the ability of the kelp to colonize a barren area successfully seemed regulated by a favorable set of hydrographic conditions (Figure 5.25).

The previous studies demonstrate that it is not necessary to alter the number of urchins to effect a change in the kelp forest community structure, and that storms can have a dramatic effect on both urchins and kelp. What about predators? Until recently, it had been assumed that urchin populations are kept in check by various predators and that many, if not all, urchin population explosions are the result of dramatic decreases in the major predator. In other words, urchins are capable of destroying kelp beds, but under normal conditions predation checks prevent this destruction from occurring. In most cases, however, the evidence to support this contention is very poor or cannot be separated from other effects. For instance, it was originally thought that increases in urchin densities in eastern Canada and New England wiping out the kelp beds and leading to "urchin barrens" of coralline algae were due to the reduced numbers of lobsters, which were believed to prey on the urchins. Later, as Elner and Vadas (1990) report, it was found that lobsters feed mainly on crabs and rarely on urchins and, further, that the urchins were subject to massive die-offs from disease brought about by high water temperatures with a periodicity of 10–15 years, thus allowing the reestablishment of the kelp beds (Figure 5.25). Similarly, in southern California, the sheephead fish (*Semicossyphus pulcher*) was thought to be the major predator on urchins, but more recent studies have suggested that it is not. It is likely that storms or changes in the inshore waters in the area, due to sewage outfalls and perhaps to hydrographic changes caused by El Niño, have led to changes in numbers of both kelp and urchins or to a change in behavior on the part of the urchins that has manifested itself in changes in community structure.

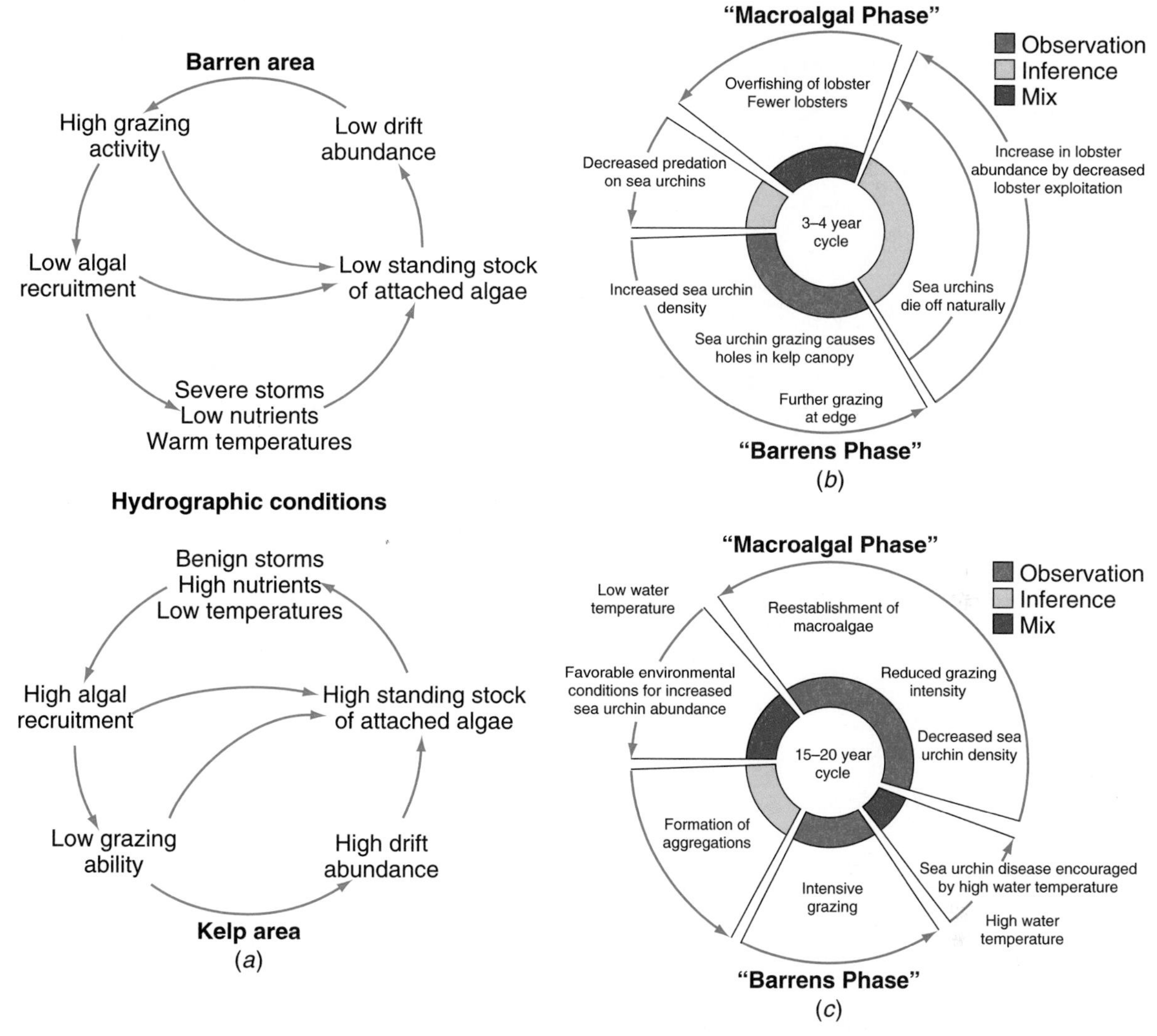

FIGURE 5.25 (a) Diagrammatic summary of how barren areas and kelp areas are maintained at San Nicolas Island, California. (b) Proposed sequence of events causing advent of urchin barrens in the sublittoral off Nova Scotia according to the paradigm of the lobster as the keystone predator of urchins. (c) Proposed sequence of events causing a 10- to 15-year temperature and disease related cycle between urchin barrens and macroalgal dominance. (a, after C. Harrold and D. Reed, 1985, Food availability, sea urchin grazing and kelp forest community structure. *Ecology* 66(4): 1160–1169. b, c, from "Inference in Ecology: The Sea Urchin Phenomenon in the Northwestern Atlantic" R. W. Elner and R. L. Vadas, *American Naturalist*, Vol. 13, pp. 108-125, 1990. Copyright © 1990 University of Chicago Press. Reprinted by permission.)

The only well-documented case where urchins have been shown to be controlled by a predator is in Alaska. Estes and Palmisano (1974) demonstrated that the sea otter (*Enhydra lutris*) is the dominant predator on urchins. In the Aleutian Islands, these investigators found that islands with large otter populations had low urchin densities and a lush growth of kelps. Islands where otters were absent were virtually without kelps and had enormous urchin densities.

In a later study, Estes and Duggins (1995) demonstrated that over periods of 3–15 years at sites where kelp or urchins had been either continuously present or absent, the abundances of each had remained unchanged. However, at sites where sea otters were recolonizing, sea urchin biomass declined dramatically, by 50% in the Aleutian Islands and by almost 100% in southeastern Alaska. At colonizing sites in southeastern Alaska, the increase in kelp was

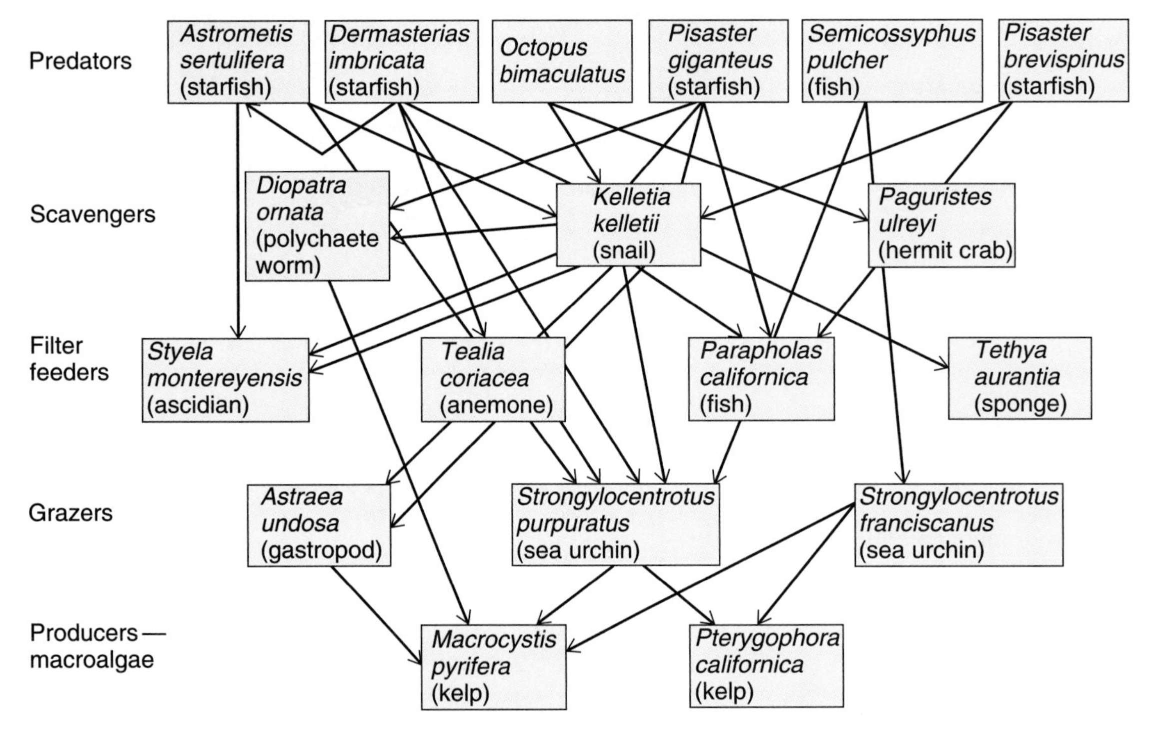

FIGURE 5.26 The food web of a southern California kelp forest. (From R. J. Rosenthal, W. D. Clarke, and P. K. Dayton, 1974, *U.S. National Marine Fisheries Service Fisheries Bulletin*, Vol. 72.)

fast and dramatic. In the Aleutian Islands, however, kelp recovery at the recolonization sites was much slower. This difference was because the urchins had significant annual recruitment and growth in the Aleutians, whereas in southeastern Alaska urchin recruitment was episodic, with years separating recruitment pulses. The rapid return of urchins in the Aleutians tended to slow kelp recovery, whereas the long absence of urchins in southeastern Alaska permitted the quick development of the kelp forest.

The almost complete destruction of the sea otter during the nineteenth century along most of the Pacific coast probably allowed the increase in sea urchin numbers and could have permitted the gradual decrease in the kelp beds. The recovery of the sea otter in California and Alaska has been followed by decreased sea urchin numbers and increased kelp forests in a few areas studied.

The importance of urchins to the integrity of kelp beds in California can be suggested by comparing the recovery of the beds from the two most devastating El Niño events in this century, those of 1957–59 and 1982–83. In the former case, according to Tegner and Dayton (1987), the kelp forests had dense populations of the sea urchins *Strongylocentrotus purpuratus* and *S. franciscanus*, which normally feed on drift kelp. After the 1957–59 event, these urchins survived and, because of lack of drift, moved out to feed on the adult plants, reducing them even more. Persistence of these urchins precluded kelp forest recovery until well into the 1960s. By contrast, the 1982–83 El Niño was a stronger event with a more devastating effect on the kelp forests (see preceding section). Yet most of these forests recovered in a shorter time, probably because, by 1982–83, the fishery for sea urchins had reduced the populations so much that urchin grazing could not prevent the recruitment of new kelp or the regrowth of surviving plants. It is also possible that the anomalous environmental conditions associated with the 1982–83 El Niño may have had deleterious effects on the urchin recruitment.

Other predators of sea urchins exist in kelp beds, notably several starfish species. In some areas, they have been suggested as important in controlling urchin populations (Figure 5.26). However, we now know that sea urchin recruitment is highly variable and that they can also be destroyed by disease or large ocean swells, so that predation is not the whole story of control of the urchins. The role of other organisms in structuring the kelp bed community is not as well known, except for the recent advent of killer whales as predators of sea otters in Alaska (see Chapter 3, pp. 129–130).

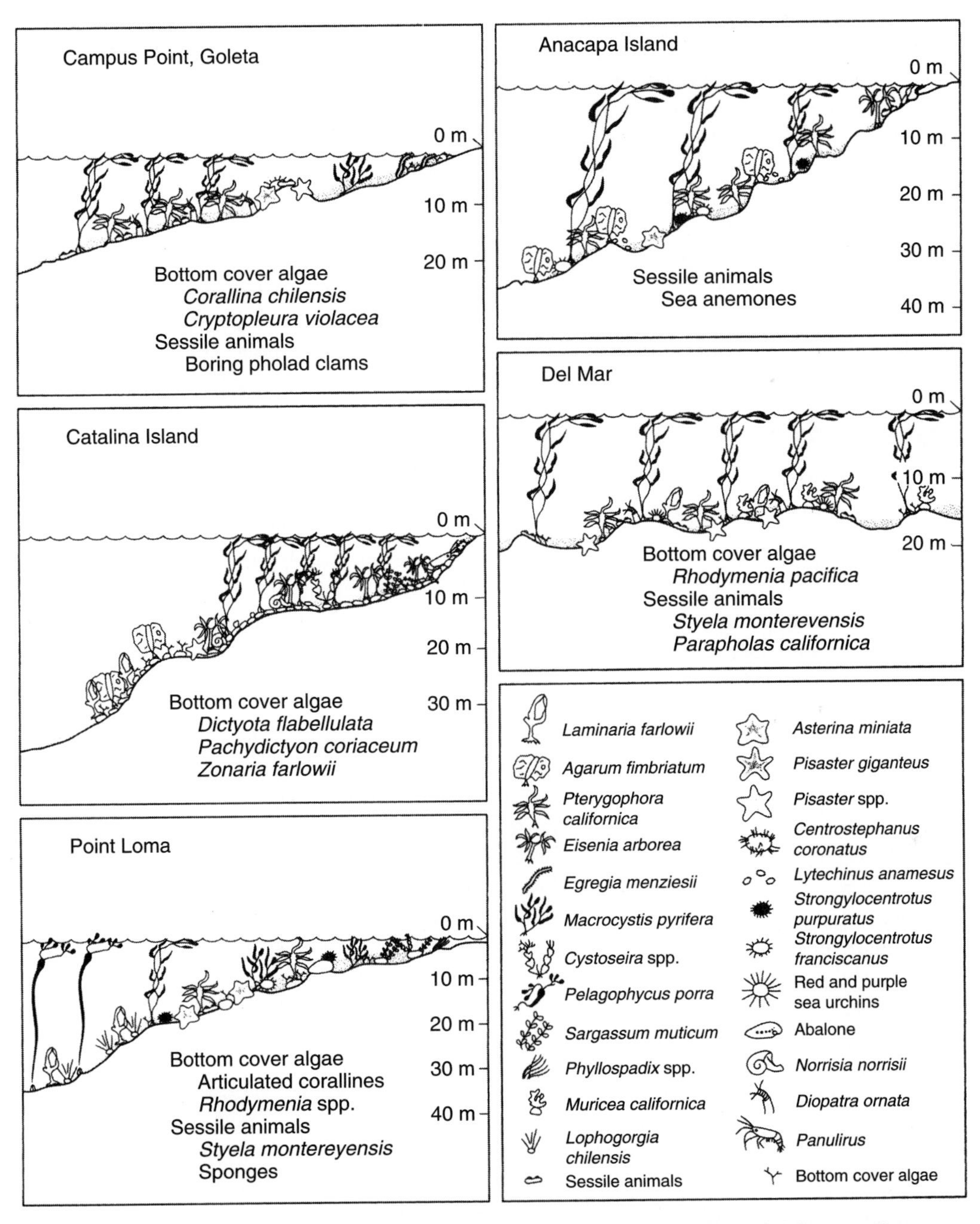

FIGURE 5.27 Different zonation patterns of the dominant organisms of five southern California kelp forests. (From M. S. Foster and D. R. Schiel, 1985, *The ecology of giant kelp forests in California: A community profile*, U.S. Fish and Wildlife Service Biological Report 85 (7.2).)

A generalized food web for invertebrates for a California kelp forest is given in Figure 5.26. In this qualitative food web, note that the primary producers have only the two species of urchins as major grazers. The other grazer, *Astraea undosa*, is a herbivorous gastropod that is unlikely to exert any considerable effect on the kelp. Sea otters are not present in this area, and urchin control is presumably through the activities of the four predaceous starfish, octopus, and one fish coupled with the aforementioned recruitment variability, disease, and wave action.

Kelp forests also show various patterns of zonation, but the zonation is not consistent from one area to another (Figure 5.27). Zonation is determined by changes in physical factors with depth, such as light and wave action; by differences in topographic relief; and partly by competition among species. A general zonation pattern would have a shallow zone nearest the shore that lacks kelp and is dominated by surf grass and small coralline algae. Kelp are absent because the wave surge removes them. At about 5 m, depth conditions are right for kelp, and the forest develops and extends to 20–30 m. As greater depths are approached, the light intensity becomes limiting, and the edge of the kelp forest seems to be set at that point where there is insufficient light for kelp growth.

SEAGRASS COMMUNITIES

Many areas of the shallow sea bottom are covered with a lush growth of aquatic "grasses," collectively called seagrasses. Seagrasses are flowering plants adapted to live submerged in seawater.

Composition and Distribution

Seagrasses worldwide encompass only about 50 species, according to den Hartog (1977), a small number in comparison with their ecological importance. Seagrass beds form dense carpets of as many as 4,000 blades per square meter over extensive areas of the bottom, making them one of the most conspicuous communities of the shallow waters of temperate and tropical seas (Figure 5.28). They may have a standing biomass of 2 kg/m^2. Seagrass beds or meadows are usually well defined, by a visible boundary, from surrounding unvegetated areas. They can vary in size from isolated patches to a continuous carpet that covers many square kilometers. They also vary in the density of the plants.

Most seagrass species are very similar in appearance. They have long, thin, straplike leaves that have air channels and a monopodial growth form. The plants arise from a creeping rhizome (Figure 5.29). Compared with freshwater aquatic plants, the seagrasses have a lesser number of species and are less diverse in morphology.

Seagrasses occur from the midintertidal region to depths of 50 or 60 m. They seem most abundant, however, in the immediate sublittoral area. The number of species is greater in the tropics than in the temperate zones. All types of substrates are inhabited by these grasses, from soupy mud to granitic rock, but the most extensive beds occur on soft substrates. The dense, intertwined rhizomes and roots of the grasses form a strong mat that penetrates the substrate and secures the plant against water motion, thereby stabilizing the bottom sediments.

Environmental Conditions

Seagrass beds are strongly influenced by several physical factors. The most significant is water motion—currents and waves. Since seagrass systems exist in sheltered as well as relatively open areas, they are subject to differing amounts of water motion. For any given seagrass system, however, the water motion is relatively constant. Seagrass meadows under high water motion tend to form a mosaic of individual mounds, whereas meadows in areas of low water movement tend to form flat, extensive carpets. The seagrass beds, in turn, dampen wave action, particularly if the blades reach the water surface. This damping effect can be significant such that 1 m into a seagrass bed the wave motion can be reduced to zero. Currents are also slowed as they move into the bed.

The slowing of wave action and currents means that seagrass beds tend to accumulate sediments. However, this is not universal and depends on the current regime under which the bed exists. Seagrass beds under the influence of strong currents tend to have much of the lighter particles, including seagrass debris, moved out, whereas beds in weak current areas accumulate lighter detrital material. It is interesting that temperate seagrass beds (*Zostera*) accumulate sediments from sources outside the beds, whereas tropical seagrass beds derive most of their sediments from within.

Since most seagrass systems are depositional environments, they eventually accumulate organic material that leads to the creation of fine-grained sediments with a much higher organic content than that of the surrounding unvegetated areas. This, in turn, reduces the interstitial water movement and the oxygen supply. The high rate of metabolism of the microorganisms in the sediments causes the sediments to be anaerobic below the first few millimeters. According to Kenworthy et al. (1982), anaerobic processes of the sediment microorganisms are an important mechanism for regenerating and recycling

FIGURE 5.28 A *Posidonia* seagrass bed in the Mediterranean Sea. (Photo courtesy of John Heine.)

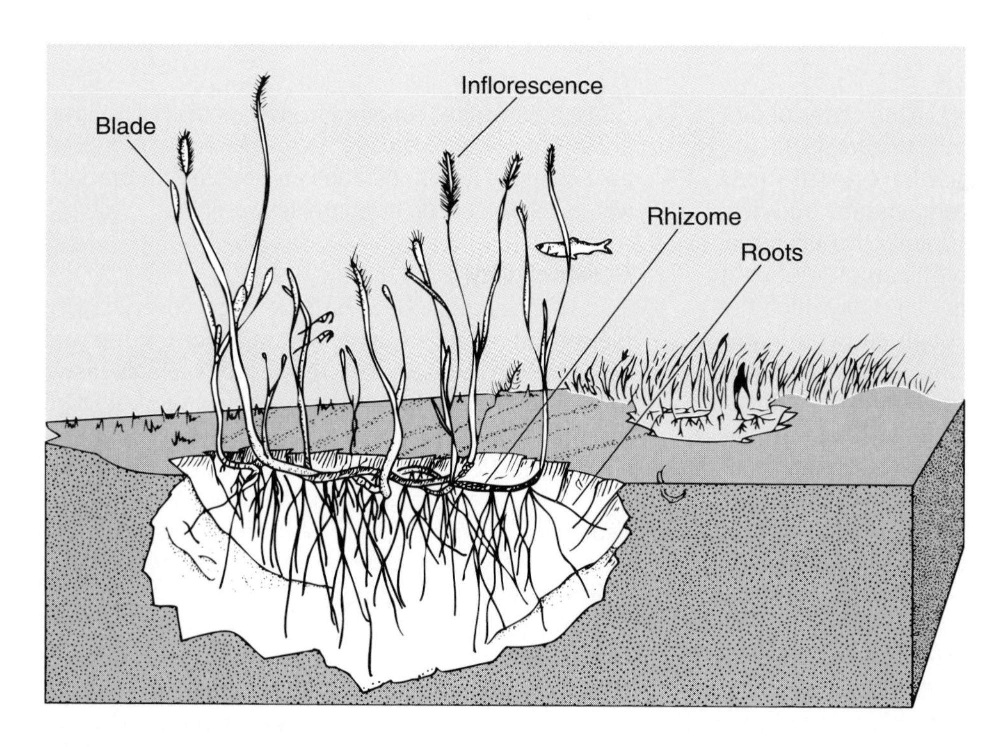

FIGURE 5.29 Diagrammatic three-dimensional illustration of an eelgrass meadow. (Redrawn from G. W. Thayer, W. J. Kenworthy, and M. S. Fonseca, 1984, *The ecology of eelgrass meadows of the Atlantic coast: A community profile*, U.S. Fish and Wildlife Service Program, FWS/OBS-84/02.)

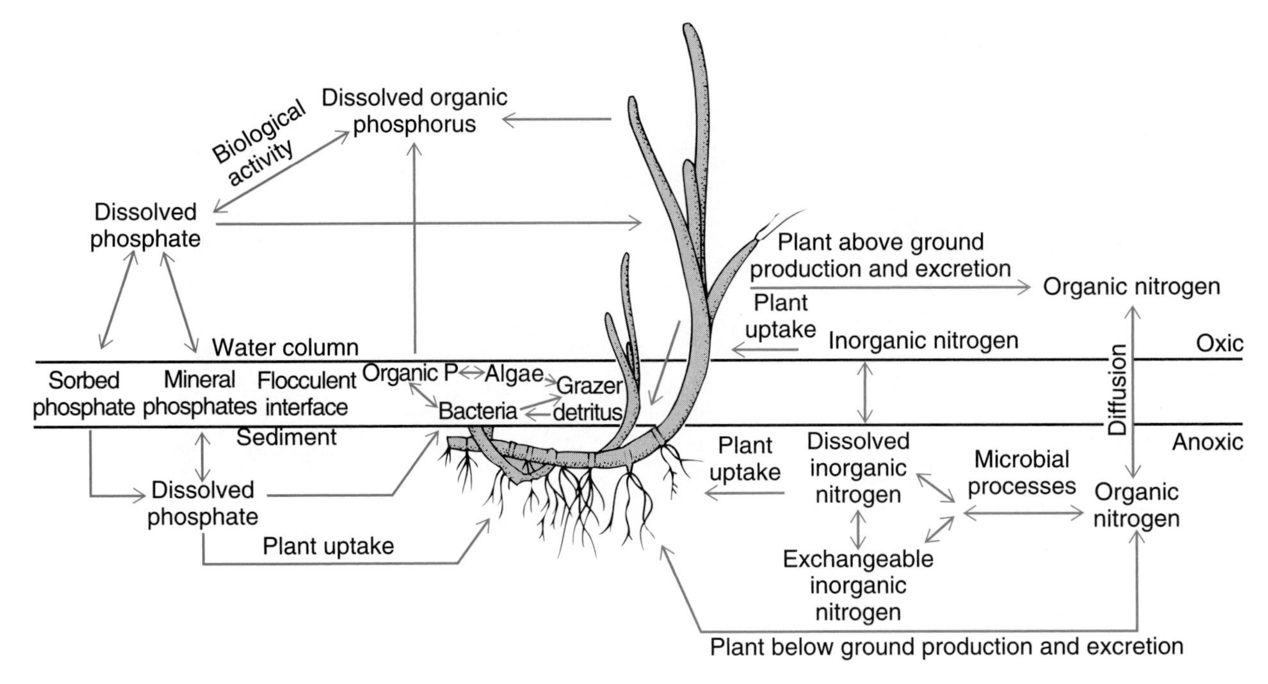

FIGURE 5.30 Diagrammatic representation of the phosphorus and nitrogen cycles in a seagrass bed. (Modified from G. W. Thayer, W. J. Kenworthy, and M. S. Fonseca, 1984, *The ecology of eelgrass meadows of the Atlantic coast: A community profile*, U.S. Fish and Wildlife Service Program, FWS/OBS-84/02.)

nutrients and carbon, assuring the high rates of productivity measured in these beds (Figure 5.30).

Other physical factors that have an effect on seagrass beds include light, temperature, and desiccation. Water depth and turbidity, for example, together or separately control the amount of light available to the plants and the depth to which the seagrasses may extend. Although early on it was suggested by Setchell (1929) that temperature was critical to growth and reproduction of eelgrass, *Zostera marina*, it has since been shown that this particularly widespread seagrass grows and reproduces at temperatures between 2 and 4°C in the Arctic (Hudson's Bay) and at temperatures up to 28°C in Chesapeake Bay. Still, extreme temperatures in combination with other factors may have dramatic deleterious effects. For example, in areas of the subarctic and cold temperate North Atlantic, ice may form in winter. Robertson and Mann (1984) note that when the ice begins to break up, the wind and tides may move the ice around, scouring the bottom and uprooting the eelgrass. In contrast, at the southern end of the eelgrass range, in North Carolina, temperatures over 30°C in summer cause excessive mortality. Seagrass beds also decline if subjected to too much exposure to the air. The effect of desiccation is often difficult to separate from the effect of temperature. Most seagrass beds seem tolerant of considerable changes in salinity and can be found in brackish waters as well as full-strength seawater.

Productivity

Because seagrass beds are densely covered with plants, the result of extensive nutrient cycling and regeneration, and because they cover such extensive areas in continental shelf waters, they have a very high productivity rate and contribute significantly to the total production of inshore waters. McRoy and McMillan (1977) have estimated the production of temperate beds to be 500–1,000 g C/m^2/year, and McRoy and Helfferich (1977) report values of over 4,000 g C/m^2/day for *Thalassia* beds in the tropics.

Seagrass beds are among the most productive areas of the oceans (compare with plankton productivity, Table 2.2; and that of kelp, pp. 203–204). Most of these data on productivity, however, are for the two most studied species: *Zostera marina* (eelgrass) of the North Temperate Zone and *Thalassia testudinum* (turtle grass) of the tropics. Little is known about the other species.

In contrast to other productivity in the ocean, which is confined to various species of algae dependent on nutrient concentrations in the water column, seagrasses are rooted plants that absorb nutrients from the sediment or substrate. They are, therefore, capable of recycling nutrients into the ecosystem that would otherwise be trapped in the bottom and unavailable.

Structure and Biological Interactions

Most seagrass beds generally have only one or at most a few dominant species of seagrasses. Communities in which two or more grasses are present have been little studied and may potentially be different in biological interactions. In general, the number of seagrass species increases from the temperate zone to the tropics. On the Atlantic coast of the United States, the only seagrass present in temperate waters is eelgrass, *Zostera marina*. In the Caribbean Sea, three main species are found: *Thalassia testudinum*, *Halodule wrighti*, and *Syringodinium filiforme*. They occur in various mixed stands.

Seagrass beds expand at different rates depending on the environmental conditions, but expansion can be rapid. Thayer et al. (1975) found in North Carolina that a bed covering 30% of a bay in 1969 had increased to cover 55% by 1973. Unfortunately, we do not have data on how long these beds remain in any given environment.

Seagrass beds are complex systems whose physical structure is dominated by the leaves, roots, and detritus of the seagrasses themselves. Furthermore, they contain large numbers of epiphytic and epizoic organisms, burrowers, and motile animals, all interacting in ways only partly known at present. The roots and leaves provide horizontal and vertical complexity, in addition to abundant food. The diversity of organisms within seagrass beds is much higher than in the unvegetated surrounding areas.

The various trophic levels represented in the seagrass beds undergo shifts in abundance in response to various changes in the seagrasses, as well as to seasonal changes in environmental factors. Therefore, there may be considerable temporal variation in the composition and abundance of the associated organisms.

On the seagrass blades, there is usually a community of epiphytic and epizoic organisms that are so abundant that the biomass of these organisms exceeds that of the seagrass blade on which they reside. This community may also reduce productivity of the seagrass by preventing or reducing light intensity at the surface of the grass blade. Little is known of the importance of this component of the community to the general economy of the beds.

Despite the obvious position of seagrass beds as primary production units in inshore waters (and in direct contrast to the situation in the terrestrial environment, where grasses are heavily grazed by a variety of vertebrate and invertebrate herbivores), surprisingly few animals consume seagrasses directly. It appears that the major grazers are certain birds, a few fishes, sea urchins, and the green sea turtle (*Chelonia mydas*). Despite the small numbers of grazers, their effect may be significant in certain areas. Among the birds, the black brant (*Branta nigricans*) and American brant (*Branta bernicla*) geese are major grazers of seagrasses. McRoy (1966) reported that the black brant consumes 4% of the standing crop of eelgrass in Izembeck Lagoon in Alaska. Moreover, Cottam (1934) indicated that on the East Coast, 80% of the diet of American brant consisted of eelgrass. Feeding by these and other herbivorous birds can have a dramatic effect on seagrass beds. For example, Wilkins (1982) estimated that in one year, Canada geese (*Branta canadensis*) consumed 20% of the eelgrass standing crop in lower Chesapeake Bay, while in Australia black swans (*Cygnus atratus*) uprooted 94% and consumed 82% of net production of *Zostera muelleri*. In many of these cases, the beds are almost cleared of parts of the seagrasses that are above the sediments. Similarly, in the tropics, urchins such as *Lytechinus variegatus* have been reported by Camp et al. (1973) to denude large areas of seagrass. The widespread tropical urchin genus *Diadema* has also been implicated in the destruction of tropical seagrass beds. Green turtles are known to feed on seagrasses, but because of their current low numbers and absence of historical data, we know little of what effect this grazing may have had. Ogden (1980) reports, however, that in the few areas where these turtles are still abundant they create very visible grazing scars in seagrass beds. This suggests that at high population densities in the past they may have had significant effects, especially since in the Caribbean Sea the diet of the turtle seems to be almost exclusively seagrasses. Among fishes, the pinfish (*Lagodon rhomboides*) of the American East Coast appears to feed heavily on eelgrass and can reduce small beds to a short stubble.

A wide variety of predators, both resident and migratory, is associated with seagrass beds, and they have diverse feeding strategies. Despite their abundance, there is little experimental evidence that they

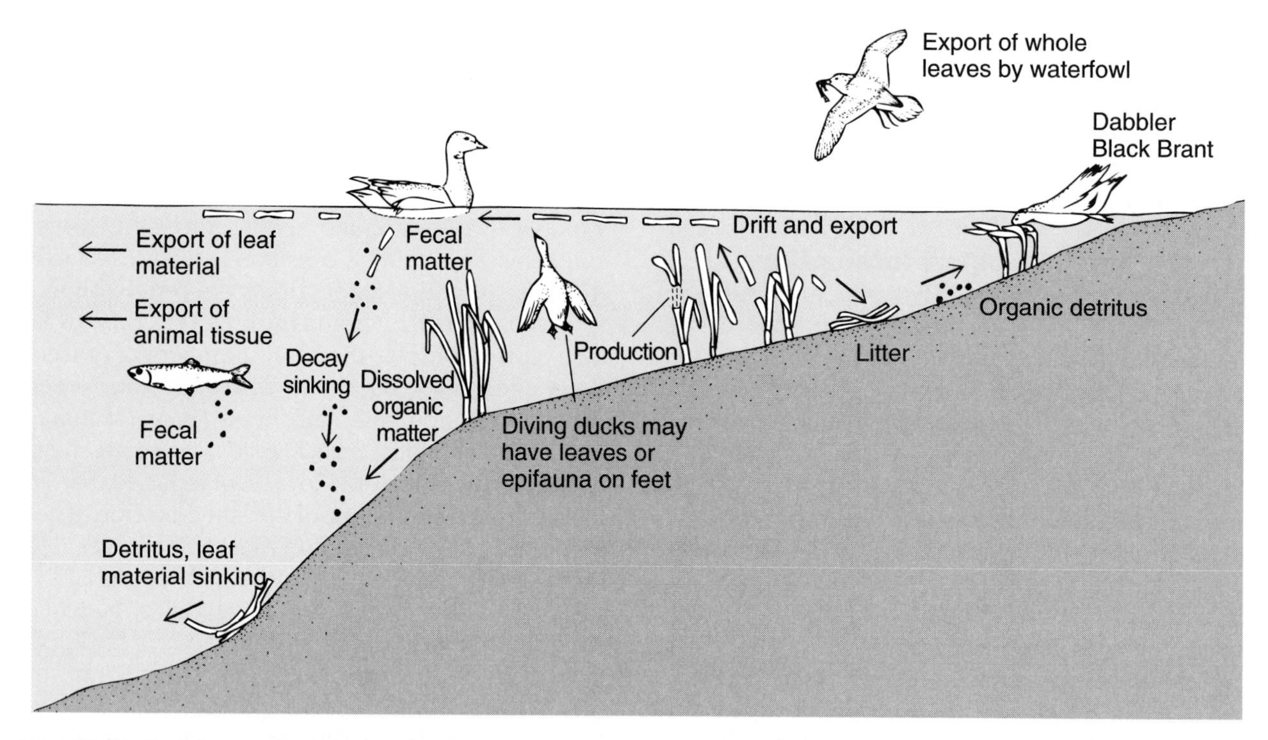

FIGURE 5.31 Diagrammatic representation of the principal pathways of movement of organic material out of a seagrass bed. (Modified from J. C. Zieman, 1982, *The ecology of the seagrasses of South Florida: A community profile*, U.S. Fish and Wildlife Service Program, FWS/OBS-82/25, 124.)

regulate either the structure of the seagrass beds or the abundance of prey. Various experiments where predators have been caged out of areas usually have not demonstrated significant increases of prey inside the cages. There is, however, evidence that seagrass beds may serve as refuges from predation when contrasted with unvegetated areas around them. Peterson (1982), for example, found that densities of two clams, *Mercenaria mercenaria* and *Chione cancellata*, were reduced in areas where seagrasses were removed relative to their densities in control areas where there was seagrass cover. He attributed this to greater predation by three species of whelks of the genus *Busycon*. In this case, the seagrass bed provided a refuge, probably by binding the sediment with its roots, thereby preventing the whelks from gaining access to the clams.

Fishes are abundant in seagrass beds, and food habit studies on the American East Coast by Thayer et al. (1975) indicate that many of them feed within the bed, removing considerable biomass (Figure 5.31). In North Carolina, Thayer et al. (1975) found that the total mass of food in the guts of fishes leaving a meadow was three times that of the fish entering. In the tropics, on the other hand, Ogden (1980) reports that most herbivorous fishes are not resident in the turtle grass beds, but migrate in during the night from surrounding reefs. Grazing by such fishes, as well as by urchins, is responsible for the formation of "halos" of grazed areas around reefs.

Certain larger fishes, such as sharks and rays, are actually or potentially of importance in structuring seagrass communities. For example, Orth (1975) reports that the feeding of the cownose ray (*Rhinoptera bonasus*) destroys large areas of seagrass beds on the East Coast. These fish dig into the bottom to pull out clams and in so doing destroy seagrasses and alter the sediments. Not only do these rays drastically reduce clam densities (from 60 to 100 *Mya arenaria* per square meter to zero) but that of the associated fauna as well.

Relationships to Other Systems

With the exceptions noted, relatively little of the living seagrass tissue is consumed directly by grazers. Most of the seagrass organic production either is transformed into detritus and consumed or further broken down in situ, or else it is transported

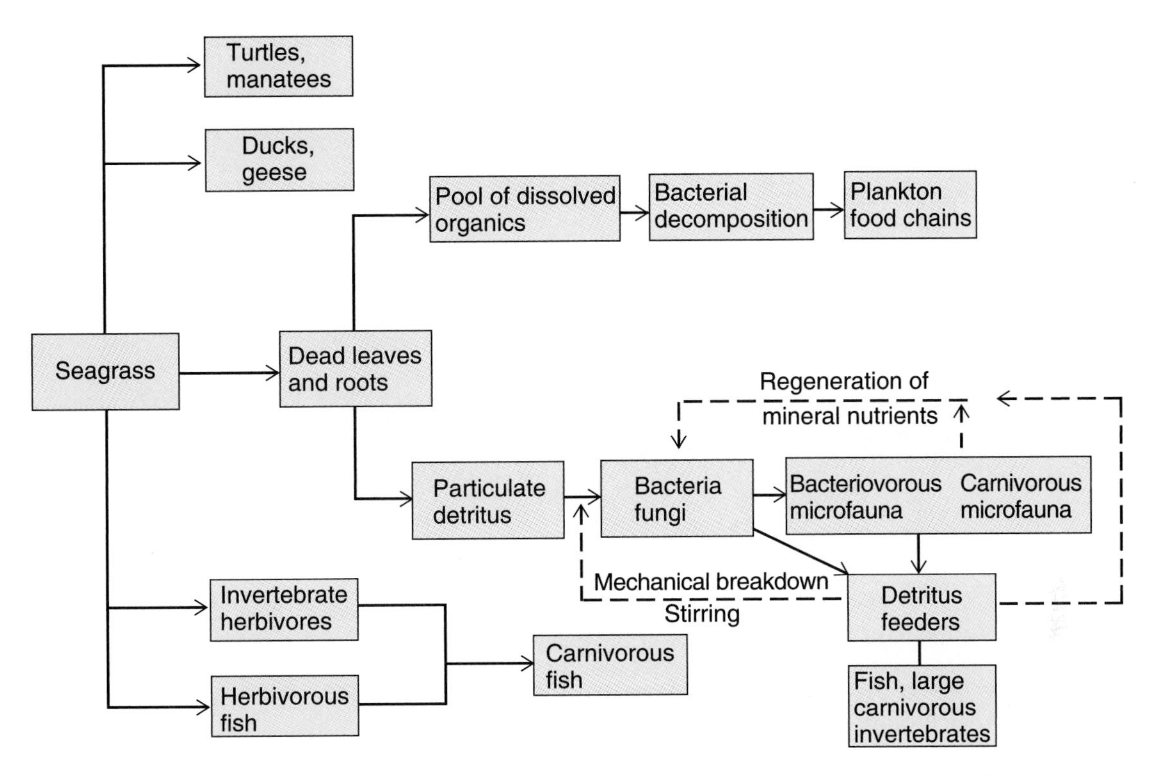

FIGURE 5.32 The pathway for the channeling of eelgrass into the food web. (Modified from *Seagrass Ecosystems*, C. P. McRoy & C. Helfferich, eds., 1977, Marcel Dekker.)

out of the system to be used by other shallow-water communities. How is this large food source channeled to other shallow-water systems? There are at least four ways. Organic material can be transported as whole plant parts with associated organisms, as detritus, as dissolved organic matter, or as living tissue or fecal material of various motile vertebrates (Figures 5.31 and 5.32). The importance of seagrass beds to adjacent communities varies, depending on geography, hydrographic conditions, and depth. Shallow-water beds tend to export more material because they are more often disturbed by water movement and by ice scour, and because they can be reached more easily by various birds. Deep-water beds, on the other hand, are usually inaccessible to birds and also are less disturbed by water motion. In these beds, however, movement by fishes may be more significant. The detrital material produced in seagrass beds may be transported to communities at considerable distances from the seagrass beds, enriching them. It has been suggested as a source of energy for organisms as far removed as the abyssal benthos.

Ecology

Ecologically, seagrass beds serve a number of important functions in inshore areas. They are a major source of primary productivity in shallow waters around the world and thus an important source of food for many organisms (in the form of detritus). In addition, they stabilize the soft bottoms, on which most species grow, primarily through the dense, matted root system. This stabilization of the bottom by roots is sufficiently durable to withstand storms as severe as hurricanes. This system, in turn, shelters many organisms. As a result, many animals are commonly found in seagrass beds that do not have a direct trophic relationship to the grasses. The beds are nursery grounds for many species that spend their adult lives in other areas. Included among these species are several of commercial importance, such as the shrimp *Penaeus duorarum* of southern Florida. These beds may also harbor adults of commercially important species, such as the bay scallop, *Argopecten irradians*, off the East Coast. Indeed, Pohle et al. (1991) have shown through field experiments that the juveniles of the bay scallop have dramatically increased

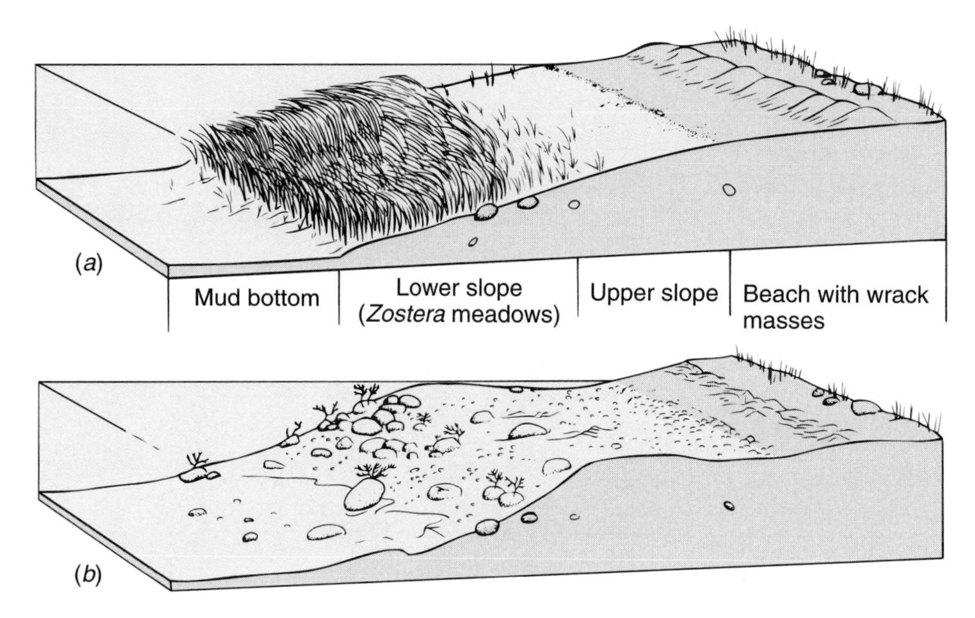

FIGURE 5.33 Representation of (a) the conditions of the eelgrass beds of Scandinavia before the onset of the wasting disease, and (b) the extreme conditions following the complete destruction of the beds. (From *Seagrass Ecosystems*, C. P. McRoy & C. Helfferich, eds., 1977, Marcel Dekker.)

survival in the face of crab predators when they are tethered by their byssal threads to the blades of seagrasses at distances of 20–35 cm above the substrate. Destruction of seagrass beds, therefore, may have a serious effect on certain commercial fisheries.

Seagrass beds may also trap sediment and, therefore, build up the bottom. Such a situation has been recorded for the *Posidonia* beds in the Mediterranean Sea, which build extensive terraces up from bottom. When such a situation prevails and the buildup approaches the surface, the floating leaves break the force of the waves, forming a calm water habitat on the bed.

The leaves of the seagrasses also create a protective canopy, shielding the inhabitants of the bed from the effects of strong sunlight. Where the beds become intertidal, the leaves may cover the bottom substrate at low tide, protecting the inhabitants from desiccation. In Florida, seagrasses are the primary substrate for most attached macroalgae. Very few algae colonize soft sediments, so with limited hard substrate in the sublittoral zone, this is an important function.

The importance of seagrass beds and what might happen when they are destroyed was dramatically emphasized by a natural destructive event that affected the dominant seagrass *Zostera marina* in the North Atlantic Ocean. In the early 1930s, the *Zostera* beds on both sides of the Atlantic Ocean were destroyed by a mysterious "wasting disease." There had been no universal agreement about what caused this destruction until recently. Short et al. (1987) have now confirmed that a pathogenic strain of the protistan *Labyrinthula* causes the symptoms. During the epidemic of the 1930s, the *Zostera* beds in the Pacific Ocean were not affected.

The effects of the *Zostera* disappearance were different in different areas. In Europe, the level of the beach fell; with the loss of sediment, many bared rocks sprouted a new growth of algae in place of *Zostera*; and new sandbars formed (Figure 5.33). The organisms also changed, but less than was anticipated. The eelgrass beds had been dominated by detritus-feeding animals. After the loss, the same bottom area had more suspension feeders and, in general, more species. In Europe, very few annual species associated with *Zostera* disappeared completely. In the United States, on the other hand, destruction of the *Zostera* beds meant that the animals associated with the beds almost completely disappeared. The reason for this difference is probably that, in European waters, the alga *Fucus* quickly replaced *Zostera*, offering the same protected

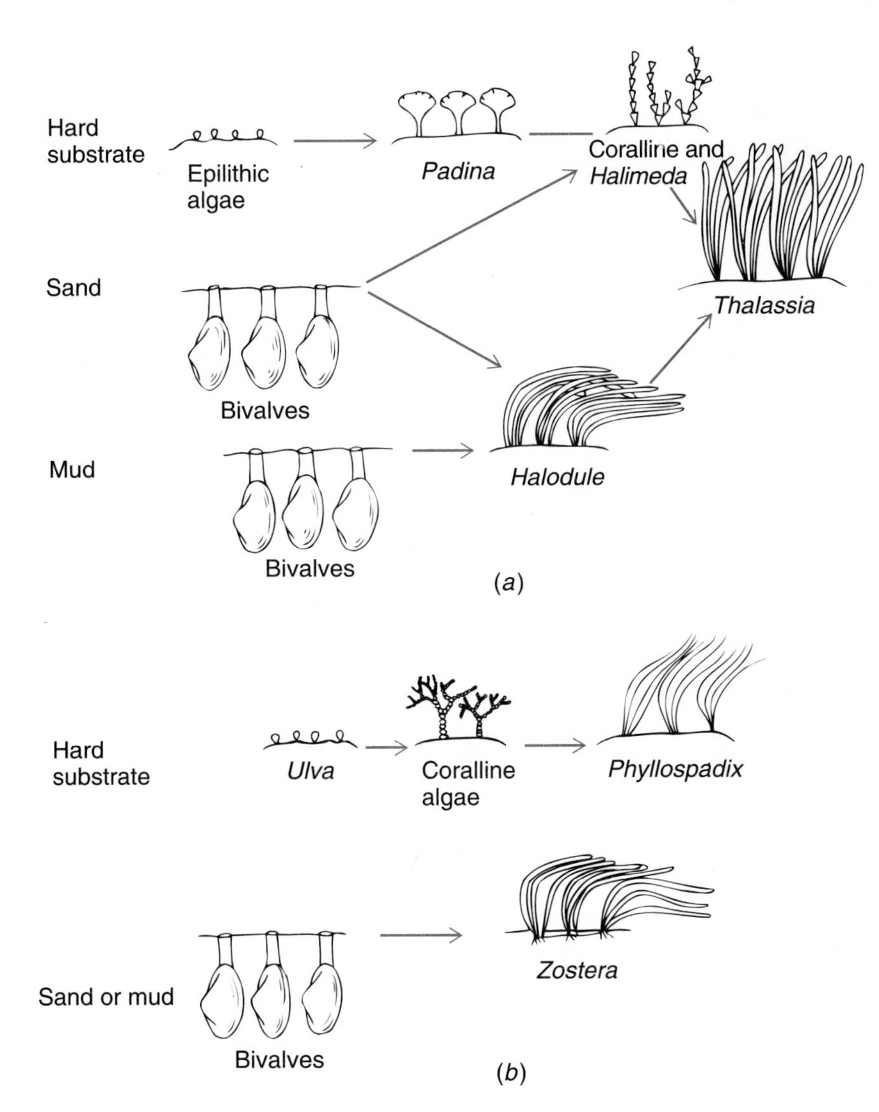

FIGURE 5.34 Successional sequences in seagrass beds. (a) Tropical areas. (b) Temperate areas.

environment. This was not the case in the northeastern United States. Since the 1930s, the *Zostera* beds have slowly returned to reoccupy their former habitat. More recently, however, Robblee et al. (1991) have reported the widespread demise of the eelgrass beds of *Thalassia testudinum* in Florida Bay. This catastrophic loss has the same characteristics as the wasting disease of *Zostera* and may have a similar cause. The extent and ecological results of this new outbreak are not known. During the last few decades, the *Zostera* beds have once again declined along the European and American Atlantic coasts, but this decline does not seem to be due to another episode of the wasting disease but to eutrophication of the shallow waters through human activity.

For some seagrass communities, den Hartog (1977) reports a successional pattern such as has been found for terrestrial communities (Figure 5.34). An example is the seagrass beds of the tropical Caribbean. Here, the climax community is dominated by *Thalassia testudinum*, but there is a series of stages leading up to *Thalassia* that are different, depending on the substrate. In the temperate Atlantic where the dominant seagrass is *Zostera marina*, no series of successional stages has been uncovered, and *Z. marina* colonizes directly; that is, it is both pioneer and terminal stage. In the temperate Pacific Ocean, the seagrass *Phyllospadix scouleri* occurs on hard substrates on waveswept shores (*Zostera* occurs on soft bottoms in protected areas). Turner (1983) has shown that there

FIGURE 5.35 A North Pacific sea pen (*Ptilosarcus*). (Photo courtesy of Randy Morse.)

is an obligate succession leading to the development of *P. scouleri* beds. Open areas in surf grass beds are settled first by *Ulva*, a green ephemeral alga. This is replaced by several red algae. Finally, *Phyllospadix* colonizes either by the rhizomes growing into the space or by seed recruitment. Rhizome invasion does not depend on any preexisting species and may occur directly, but for seeds to establish the surf grass, macroalgae must be present for the seeds to attach to. Furthermore, the seeds attach preferentially, or at least have a better survival, if the algae are turf forming with bushy branches and a narrow central axis; thus, a true facilitated succession occurs.

SOME SPECIAL COMMUNITIES

In addition to the widespread communities described, there are a few of more restricted occurrence that have received considerable study; hence, they deserve mention here as examples of how communities are organized, how they may vary temporally, and how they are maintained.

A Sea Pen Community

In Puget Sound, Washington, at depths of 10–50 m on soft bottoms, there exists a community dominated by dense stands of the long-lived cnidarian sea pen, *Ptilosarcus gurneyi* (Figures 5.35 and 5.36). In an extensive study, Birkeland (1974) found that *Ptilosarcus gurneyi* provides the major food source for seven predators: four starfish (*Hippasteria spinosa*, *Dermasterias imbricata*, *Crossaster papposus*, and *Mediaster aequalis*) and three opisthobranch mollusks (*Armina californica*, *Tritonia festiva*, and *Hermissenda crassicornis*). *Ptilosarcus* has a life span of about 15 years; takes several years to reach sexual maturity; has irregular, unpredictable recruitment, resulting in stands of differing age structure; and never reaches sizes at which it is immune to predation.

Adult *Ptilosarcus* are readily consumed by three starfish—*Hippasteria*, *Dermasterias*, and *Mediaster*—plus the opisthobranch *Armina*, which seems to prefer the largest sea pens (Figure 5.37). Of these four predators, *Hippasteria* has the most restricted diet, feeding only on *Ptilosarcus* (Figure 5.38). The others are capable of consuming other prey. In Puget Sound at least, *Armina* seems to feed exclusively on *Ptilosarcus*. Juvenile *Ptilosarcus* are preyed upon by *Hermissenda*, *Tritonia*, and *Crossaster*, but this is uncommon.

In this community, one might expect the starfish generalists (*Dermasterias*, *Mediaster*, and *Crossaster*), because they do not completely depend on *Ptilosarcus*, would maintain high population numbers and reduce the *Ptilosarcus* population, so the specialists (*Armina* and *Hippasteria*) would have nothing left. In other words, they could outcompete them. What is to prevent this? The answer lies with yet another starfish, the top predator *Solaster dawsoni*, which preys on *Mediaster*, *Dermasterias*, and *Crossaster*, reducing their numbers sufficiently

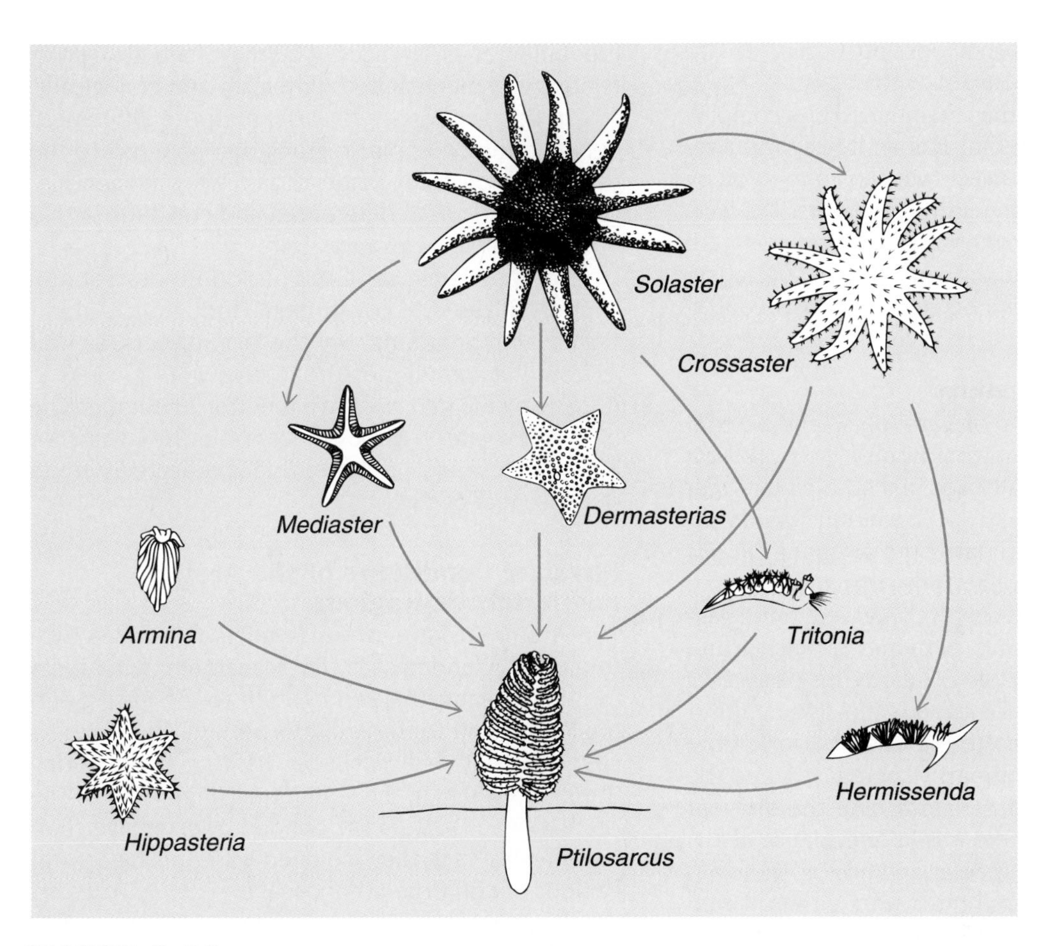

FIGURE 5.36 The food web of a North Pacific sea pen community.

FIGURE 5.37 The nudibranch *Armina californica* eating the sea pen *Ptilosarcus gurneyi*. (Photo courtesy of Dr. Ron Shimek.)

FIGURE 5.38 The sea star *Hippasteria* feeding on the sea pen *Ptilosarcus gurneyi*. (Photo courtesy of Dr. Ron Shimek.)

so that competitive exclusion does not occur (see Figure 5.36), thus allowing *Hippasteria* to persist.

This community is thus maintained by a complex food web relationship that has as its key industry species the sea pen *Ptilosarcus*, which is preyed on by several specialist and generalist predators. The generalists (*Mediaster*, *Dermasterias*, and *Crossaster*) are unable to dominate and outcompete the specialists (*Hippasteria* and *Armina*) because a top predator, *Solaster*, keeps their numbers in check.

A Sea Pansy Community

Off San Diego in 5–10 m of water on a sand bottom exists quite a different community. As noted on page 192, this community was first studied by Fager (1968) between 1957 and 1963, when it was dominated by cnidarians, primarily the sea pansy *Renilla kollikeri* and the anemones *Harenactis attenuata* and *Zaolutus actius*. Associated with these were three gastropods (*Nassarius fossatus*, *Nassarius perpinguis*, and *Polinices recluzianus*) and two echinoderms (*Amphiodia occidentalis* and *Astropecten armatus*). In this community, Fager found the pattern of distribution of the major organisms was primarily aggregated, and the populations remained constant over the six-year period. Furthermore, there was little evidence of any interaction among the species and no evidence that these large macroinvertebrates were either preyed on by mobile fishes or themselves preyed on the smaller infaunal organisms.

Eleven years later, in 1974, Davis and VanBlaricom (1978) repeated the study at the same site. Whereas Fager had observed essentially nonvariant densities of the major species over six years, Davis and VanBlaricom found both long- and short-term changes in the abundances and densities of seven of the dominant species. These studies suggest that shallow-water benthic communities are not completely stable over long periods, even though they may appear so for several years. It also appears that a much longer time interval is necessary to evaluate what is happening in such areas.

BENTHIC BIOLOGY OF POLAR SEAS

The Arctic and Antarctic regions of the earth appear outwardly similar in that both are characterized by the dominating presence of ice and snow, year-round cold temperatures, and drastic changes in photoperiod that prevent photosynthesis during a significant part of the year. Most of these outward similarities, however, mask fundamental differences that make the two polar regions very different both physically and biologically.

In this section we discuss, first, the differences between the two polar regions that give rise to the different biological associations. We will then discuss the biological differences and conclude with a brief look at some examples.

Perhaps because of the profound interest and support of the U.S. government through its Antarctic research programs over the last 30 years, as well as a similar commitment from other world powers to Antarctic research, we have a better understanding of shallow-water Antarctic communities than we have of those of the Arctic. This is reflected here in the coverage of each.

Physical Conditions of the Arctic and Antarctic Regions

The Arctic Ocean system is an isolated sea surrounded by landmasses that leave only two outlets to other oceans, the Bering Strait to the Pacific and the Fram Strait to the Atlantic. Both of these outlets have relatively shallow sill depths, 70 m in the Bering Strait and 400 m in the Fram Strait (Figure 5.39). The Arctic Ocean Basin is surrounded by extensive shallow shelves such that the deep water of the central basin is cut off from other oceans.

The circulation pattern consists of two major anticyclonic currents, the Beaufort gyre over the Canadian Basin and a transpolar current across the Eurasian part of the basin that exits through the Fram Strait (Figure 5.39).

The landmasses surrounding the Arctic Ocean have several large rivers that discharge significant amounts of sediment into the basin such that the basic substrate overlying the extensive continental shelves is particulate matter. The large freshwater discharges also lead to the development of a low-saline, stratified surface layer.

The Arctic Ocean pack ice is a persistent multi-year system that is harder and thicker than that of the Antarctic. Antarctic pack ice is usually seasonal and only a year old or younger. The whole central Arctic Ocean is permanently covered by ice (Figure 5.39).

By contrast, the Antarctic marine ecosystem is open to all oceans and is a circumpolar ring of water surrounding a central landmass. The Antarctic continental shelf is narrow and falls off rapidly to deep water. The northernmost marine boundary is defined by the circular Antarctic Convergence, which is situated over

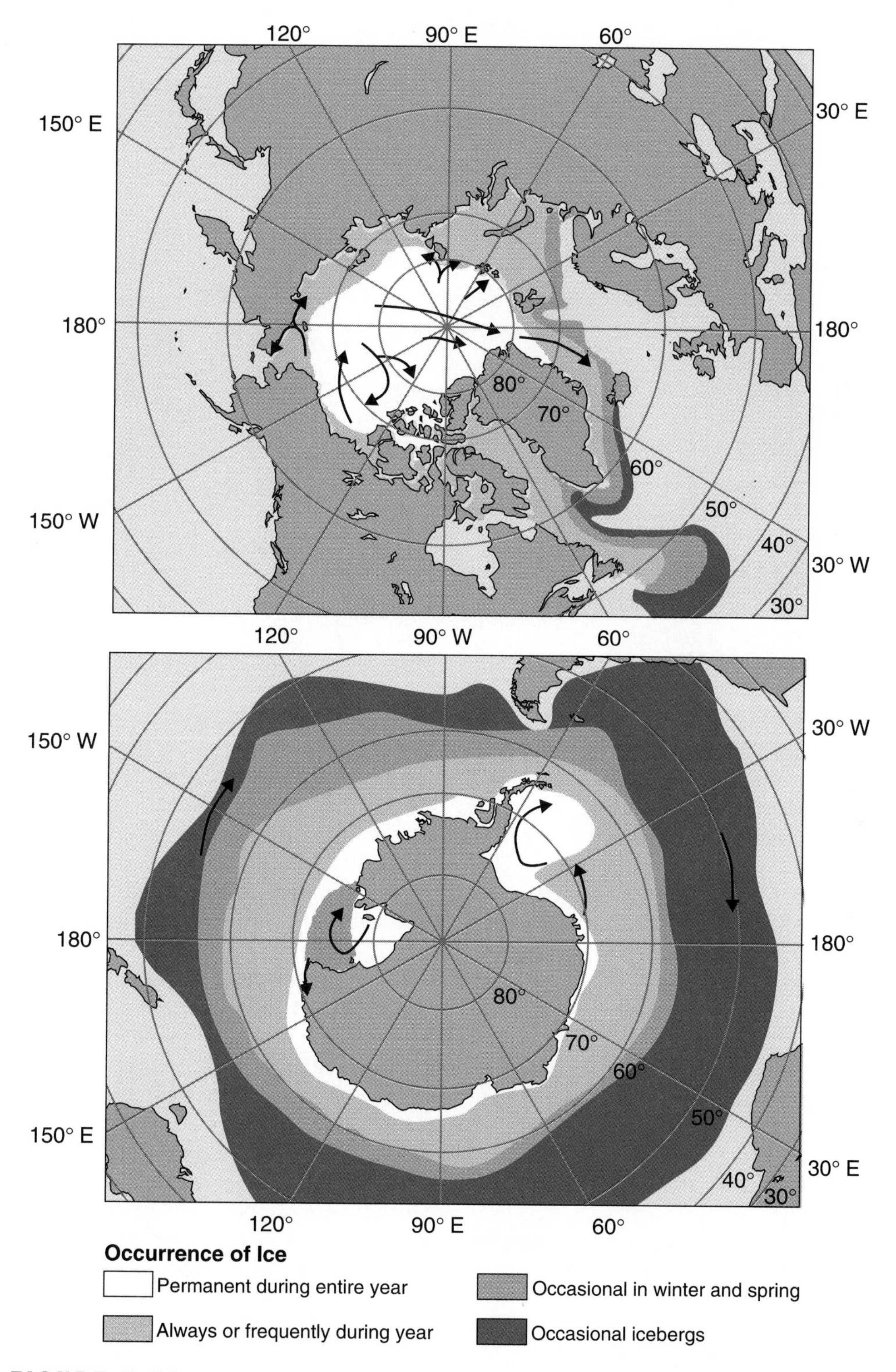

FIGURE 5.39 Geography of the Arctic and Antarctic regions, showing the seasonality of ice cover and major currents (black arrows). (From "On the Biology of Polar Seas" by G. Hempel in *Marine Biology of Polar Regions and Effects of Stress on Marine Organisms*, Gray & Christiansen, eds., © 1985 John Wiley & Sons Limited. Reproduced with permission.)

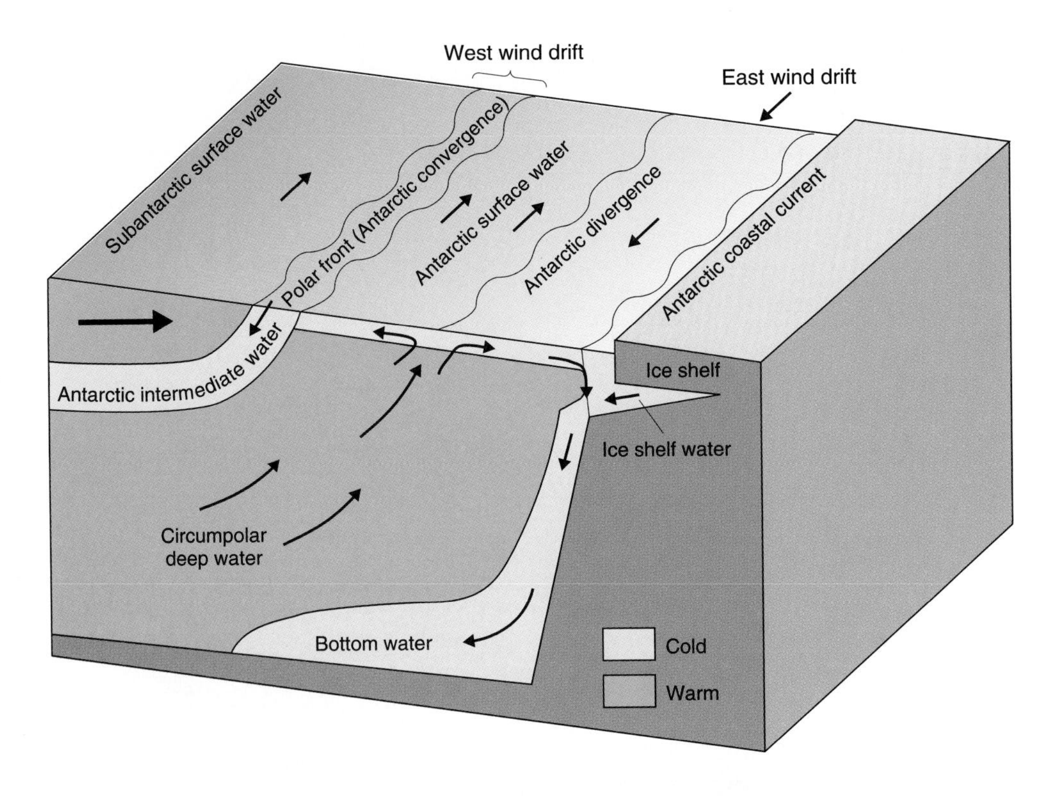

FIGURE 5.40 Diagrammatic representation of the currents off the Antarctic continent. (After "On the Biology of Polar Seas" by G. Hempel in *Marine Biology of Polar Regions and Effects of Stress on Marine Organisms*, Gray & Christiansen, eds., © 1985 John Wiley & Sons Limited. Reproduced with permission.)

deep water in a zone of persistent and strong westerly winds (Figure 5.40). Immediately south of the Antarctic Convergence, nutrient-rich circumpolar water rises toward the surface and fertilizes the Antarctic surface waters (Figure 5.40). Near the continent and beneath the ice shelves, seawater is cooled to freezing or below and flows downslope to form the bottom water of the earth's deep ocean basins. Since the Antarctic continent is permanently covered by ice and lacks any rivers, there is little or no sediment input and no formation of stratified surface layers.

Compared with the Arctic, seasonal variations in temperature are small in the Antarctic, both horizontally and vertically. In winter, pack ice covers more than half of the Antarctic marine area, but it is reduced to only 10% in late summer (see Figure 5.39). Antarctic pack ice is only 1–2 m thick, in contrast to the permanent 3–4 m thick Arctic pack ice. Both the Arctic and Antarctic have pronounced year to year variations in ice cover.

The Antarctic ocean occupies a ring around the earth between 50–60° and 65–70° south latitude, while the Arctic Ocean is an enclosed Mediterranean-type sea beyond 70° north latitude. Thus, the Antarctic ocean experiences strong, but less pronounced seasonal light conditions than the Arctic Ocean, and the plankton experience strong diurnal but weak seasonal vertical migrations. The Arctic plankton show weak diurnal vertical migrations but pronounced seasonal ones.

These physical differences between the two polar regions are summarized in Table 5.1.

Because the polar inshore waters are covered with ice most of the year, benthic organisms and communities are rarely disturbed by wave action. However, polar shallow-water areas are subject to a stress factor unknown in other areas—ice. During the winter months, ice forms a thick layer over the continental shelf areas, extending down several meters. Wherever this ice layer makes contact with the bottom, the fauna is destroyed, either by mechanical action of the ice grinding against the bottom or by freezing the animals. This means that in polar areas, the shallowest subtidal areas are devoid of permanent communities down to the level to which the ice extends. This area may, however, be inhabited during ice-free times of the year by mobile or transient organisms. Indeed, Conlan et al. (1998) studied ice scour disturbance in Barrow Strait in the high Arctic of Canada and found a gradient of high ice scour by shallow draft ice in inshore waters to less scour by less frequent icebergs and ice shelves at 30 m. They also found that ice scours at all depths were dominated by the same set of disturbance-associated fauna, which was quite different from the undisturbed benthic community outside the scours (Figure 5.41).

TABLE 5.1

Oceanographic Features of Arctic and Antarctic Oceans

	Antarctic	Arctic
Area (× 10^6 km)	35–38	14.6
Pack ice		
Mean maximum	22	12–13
Mean minimum	2.6	6–7
General features		
Form	Circumpolar ring between 50–60°S and 65–70°S	Mediterranean Sea around North Pole, almost enclosed by land at 70–80°N
Shelf	Narrow, open to all oceans, with large exchange in deep layers	Broad, two narrow openings (Bering Strait, Fram Strait)
Current systems	Circular currents (west wind drift, east wind drift), large eddies (Weddell Sea, Ross Sea)	Transpolar current, Beaufort gyre
Pack ice	High seasonality	Little seasonality
Cover in winter	c.50%	90%
Cover in summer	c.10%	80%
Age, thickness	Mainly one year, 1.5 m	Mainly multiyear, 3.5 m
Icebergs	Large, tabular icebergs abundant and carrying coarse material	Small, irregular icebergs, only in Greenland and Bering Sea, not in Arctic Basin
River input	None	Much
Stratification	Low vertical stability	Surface layer (20–50 m) stable all year
Nutrients in euphotic zone	High all year	Seasonal depletion

Source: Modified from "On the Biology of Polar Seas" by G. Hempel in *Marine Biology of Polar Regions and Effects of Stress on Marine Organisms*, Gray & Christiansen, eds., © 1985 John Wiley & Sons Limited. Reproduced with permission.

In Antarctic seas, in addition to the action of the thick layer of annual sea ice affecting the bottom, there is an additional area below that is subject to a different kind of ice action. In this area, which extends down to approximately 30 m, ice platelets begin to form on the bottom around any convenient nucleus. As these platelets increase in size, they may surround various sessile and sedentary invertebrates living on the bottom. Since ice is generally less dense than the surrounding water, as these platelets grow they lift off the bottom and rise toward the thick sea ice above, carrying with them the trapped invertebrates. Such organisms become trapped in the sea ice and are permanently lost from the community. Such ice formation is termed **anchor ice** (Figure 5.42). As a result of anchor ice formation and its tendency to remove organisms, a zonation of the benthic fauna with depth is created.

Biological Comparisons of Arctic and Antarctic Seas

As might be anticipated, the differences in physical characteristics also are reflected in differences in biology. Perhaps the most significant difference between the two polar regions is that the Antarctic region is far richer in species of benthic organisms (Table 5.2). It also has an extremely high degree of endemism (Table 5.3). The high diversity of Antarctic species is matched by a high biomass. By contrast, the Arctic fauna is impoverished and consists mainly of organisms derived from the Atlantic Ocean. In general, the biomass figures for any given depth in the Antarctic are one to two orders of magnitude higher than those of the Arctic. The high biomass of the Antarctic benthos is probably related to the nutrient-rich water and high offshore productivity, while the high species richness and endemism, according to Hempel (1985), are related to a long evolutionary history and isolation, as well as to the presence of both hard and soft substrates.

Arctic benthic communities are often dominated by one or only a few species, but Antarctic communities usually have several species that are common or dominant. Because of the extensive areas of sediments, the Arctic benthic fauna is mainly an infauna, whereas the Antarctic benthos is characterized by large numbers of epifaunal species. According to Hempel (1985), fish are less important in Antarctic benthic communities and mainly are restricted to the near bottom.

Most polar benthic invertebrates appear to be long-lived, slow-reproducing animals, but it is unclear

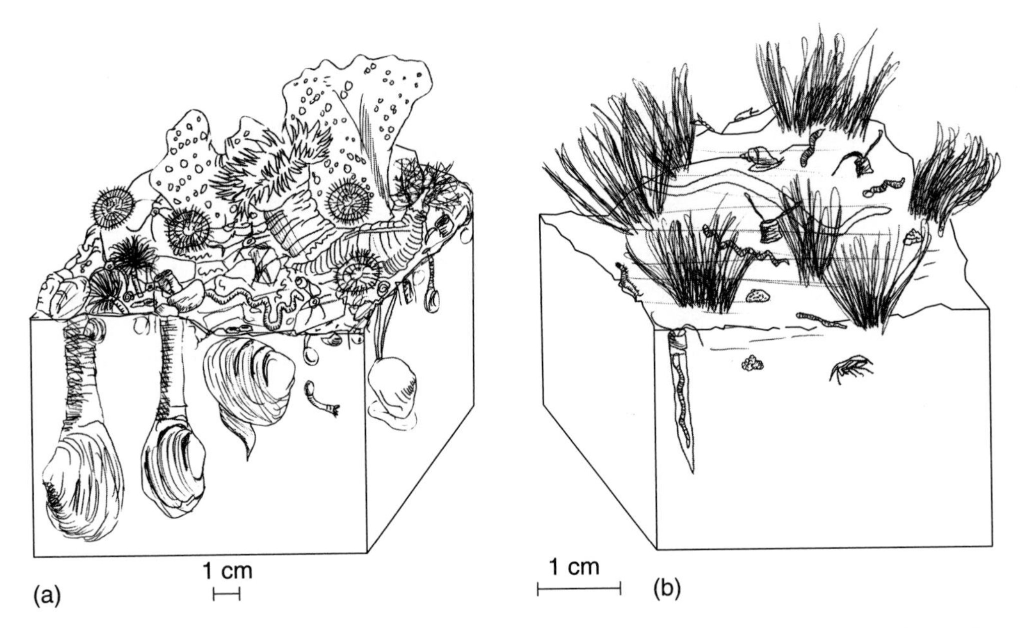

FIGURE 5.41 Dominant fauna (a) outside and (b) inside a one-year-old ice scour in Barrow Strait, Canadian Arctic, at 20 m depth. The undisturbed benthos (a) is dominated by the large bivalves *Mya truncata, Serripes groenlandicus*, and *Macoma calcarea*, the sea urchin *Strongylocentrotus pallidus*, the sea cucumber *Cucumaria frondosa*, the anemone *Tealia* sp., the polychaete *Phyllodoce groenlandicus*, and the kelp *Agarum cribosum*. Dominant occupants of a one-year-old scour (b) are tubicolous spionid polychaetes (vertical tubes) and *Capitella capitata* (horizontal tube), the mobile polychaetes *Ophryotrocha spatula* (spatulate posterior) and *Nereimyra punctata*, the amphipod *Monoculodes vibei*, and various cumaceans. Diatoms colonize in abundance (the tufts). (From "Ice Scour Disturbances to Benthic Communities in the Canadian Arctic" by K. E. Conlan et al, *Marine Ecology Program Series*, Vol. 166, pp. 1-16, 1998. Reprinted by permission of Inter-Research.)

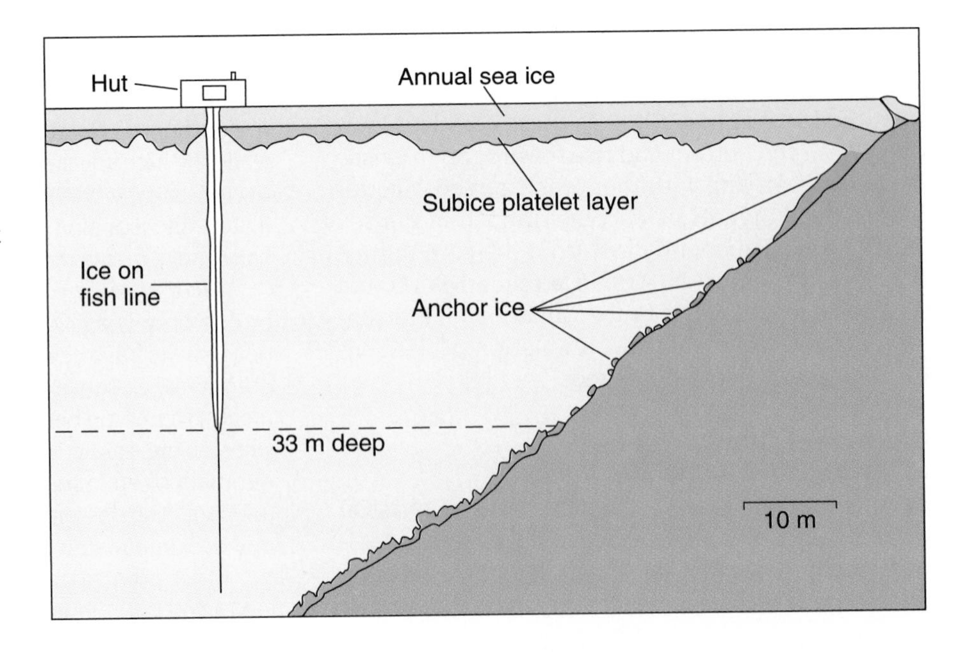

FIGURE 5.42 Anchor ice area in the Antarctic. (Reprinted with permission from "Anchor Ice Formation in McMurdo Sound, Antarctica, and its Biological Effect" P. K. Dayton et al., *Science*, Vol. 163, p. 273, 1969. Copyright © 1969 American Association for the Advancement of Science.)

TABLE 5.2

Comparison of the Number of Species of Macrobenthic Organisms from the Arctic and the Antarctic

	Arctic	*Antarctic*
Sponges	200	300
Polychaetes	c. 300	650
Amphipods	262 (Barents Sea)	470
Pycnogonids	29 (Barents Sea)	100
Ascidians	47	129
Bryozoans	200 (Barents Sea)	310
Mollusks	224 (Barents Sea)	875
Isopods	49 (Kara Sea)	299

Source: After "On the Biology of Polar Seas" by G. Hempel in *Marine Biology of Polar Regions and Effects of Stress on Marine Organisms*, Gray & Christiansen, eds., © 1985 John Wiley & Sons Limited. Reproduced with permission.

TABLE 5.3

Degree of Endemism in Selected Groups of Antarctic Marine Organisms

	Genera (%)	*Species* (%)
Isopoda and Tanaidacea	10	66
Fishes	70	95
Pycnogonida	14	90
Echinodermata	27	73
Echninoidea	25	77
Holothuroidea	5	58
Bryozoa	—	58
Polychaeta	5	57
Amphipoda	39	90

Source: "On the Biology of Polar Seas" by G. Hempel in *Marine Biology of Polar Regions and Effects of Stress on Marine Organisms*, Gray & Christiansen, eds., © 1985 John Wiley & Sons Limited. Reproduced with permission.

whether this is due to some inherent response to low temperatures or to some other feature of the polar environment, such as low food resources. Among the benthic invertebrates in the Antarctic, there is little evidence for any temperature adaptation with respect to reproduction, development, or growth. Pearse et al. (1991) also note that, contrary to "Thorson's rule," that cold-water benthic organisms should show more nonpelagic development, pelagic planktotrophic development is common. The only probable difference from temperate and tropical planktotrophic organisms is that the Antarctic pelagic larvae take a very long time to develop to metamorphosis. How these larvae survive the poor food conditions during the Antarctic winter is still not known.

While the shallow Antarctic seas have relatively few fish species—120 species according to Knox (1994)—most are endemic probably because they are isolated by deep water from all other continents. These fishes have generated considerable interest among physiologists because they have developed special mechanisms to keep their body fluids from freezing in the below-freezing water and because some of them (Channichthyidae) lack respiratory pigments in their blood. Such specialized endemic fish are not present in the Arctic.

There are also differences between the polar oceans with respect to the ecology of marine birds. In Antarctic seas, the distributions tend to be latitudinal, with the bird communities similar in all ocean basins at a given latitude. In the Arctic Ocean, distribution is longitudinal with strong differences between communities at similar latitudes in different parts of the ocean basin. Another difference is that in the Antarctic seas a whole community of birds is adapted to forage near the ice edge, and the open water near the ice margin is also a significant foraging area. In the Arctic, Hunt (1991) reports few birds forage in the marginal ice area or at the edge of the ice.

Ecological Communities of Polar Seas

Sea Ice Communities A marine environment unique to polar seas is that occupied by organisms inhabiting the sea ice. Sea ice provides an extensive habitat for a set of organisms ranging in size from bacteria to marine birds and mammals. We are here, however, mainly interested in the organisms that inhabit the ice on a long-term basis and that interact with each other.

In Antarctica the circumpolar ring of sea ice around the continent ranges in size seasonally from 4×10^6 to 20×10^6 km^2. The biological communities that develop in the sea ice are quite variable, reflecting the different physical, chemical, and biological features that characterize the physically distinct ice types and the seasonal differences that occur in the sea ice.

More than 200 species of marine organisms have been reported living in association with sea ice according to Garrison (1991). The major groups include bacteria, certain algae, heterotroph protistans, and small metazoans (Figure 5.43). There is still disagreement as to whether or not these ice associations are true communities, since many of the species are also found in adjacent open waters. It has been suggested that the ice may be a temporary habitat for planktonic species and may be important in providing a "seed" population for plankton blooms at the sea edge.

The algal species inhabiting the ice depend on light for photosynthesis and are, therefore, less

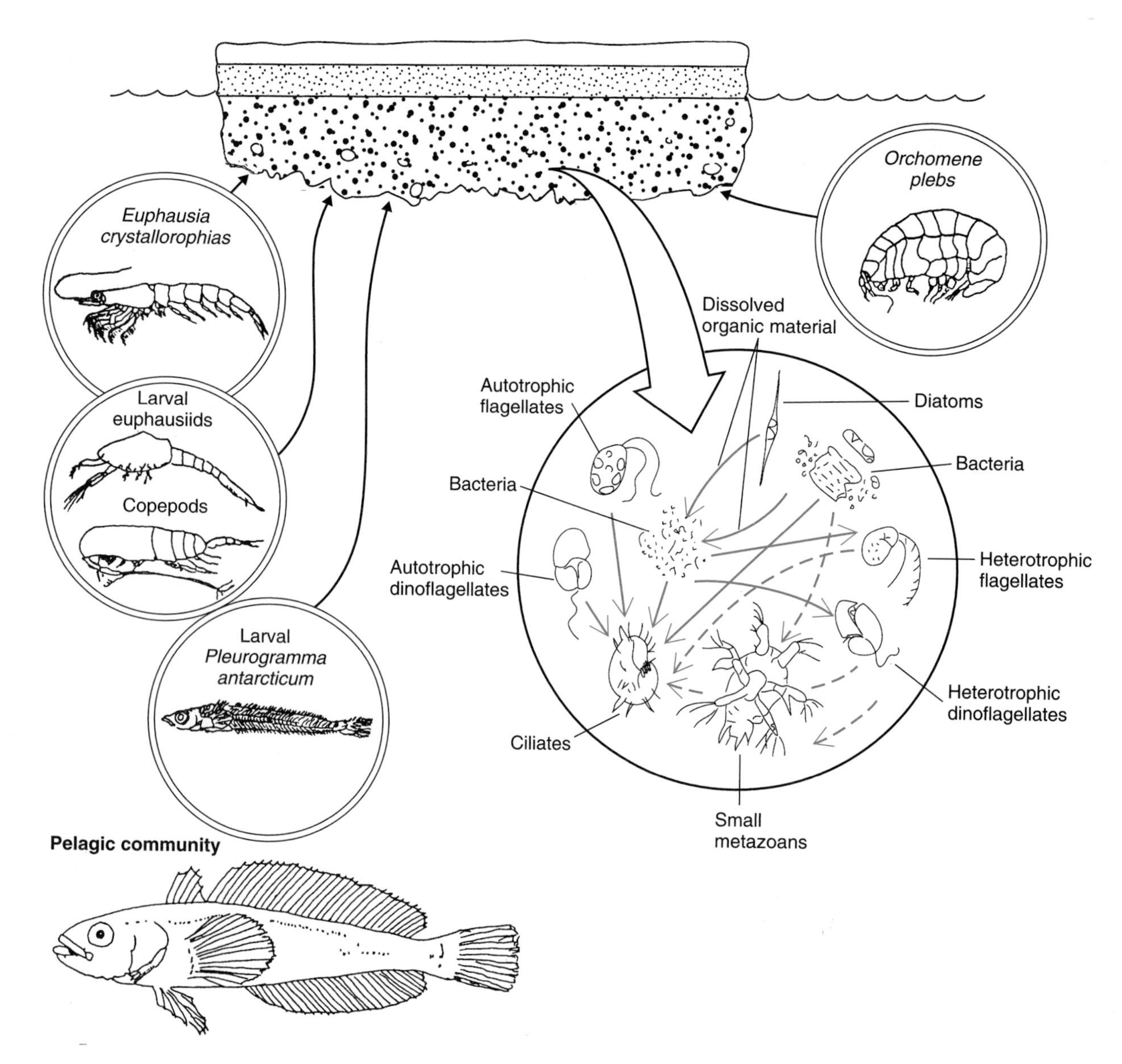

FIGURE 5.43 A conceptual food web for the pack ice microbial community. Solid arrows show feeding relationships documented. Dashed arrows show predator-prey relationships known in other systems but that have not been confirmed in sea ice. None of the feeding rates is known. (After "Antarctic Sea Ice Biota" D. L. Garrison, 1991, *American Zoologists*, Vol. 31, pp. 17-33, © 1991 American Society of Zoologists.)

abundant under the thick ice layers of the Arctic Ocean and are most abundant at the ice edges. In the Antarctic, where the ice is thinner and more seasonal, Stromberg (1991) reports these algae are more widespread. These algae are grazed by various zooplankton organisms of which the amphipods are the dominant forms in the Arctic. The large krill, *Euphausia superba*, is the dominant grazer in Antarctic seas.

There is a well-developed microbial food web in the sea ice. This microbial community or association may be significant not only to plankton, but to polar benthic communities as well because there is evidence that benthic organisms feed on the sea ice biota. A conceptual food web for the ice community is in Figure 5.43.

A final difference between the Arctic and Antarctic communities concerns the zooplankton. The zooplankton of the two oceans are similar with respect to the major taxa represented, but there is one major difference that has profound implications for the organization of perhaps all Antarctic marine com-

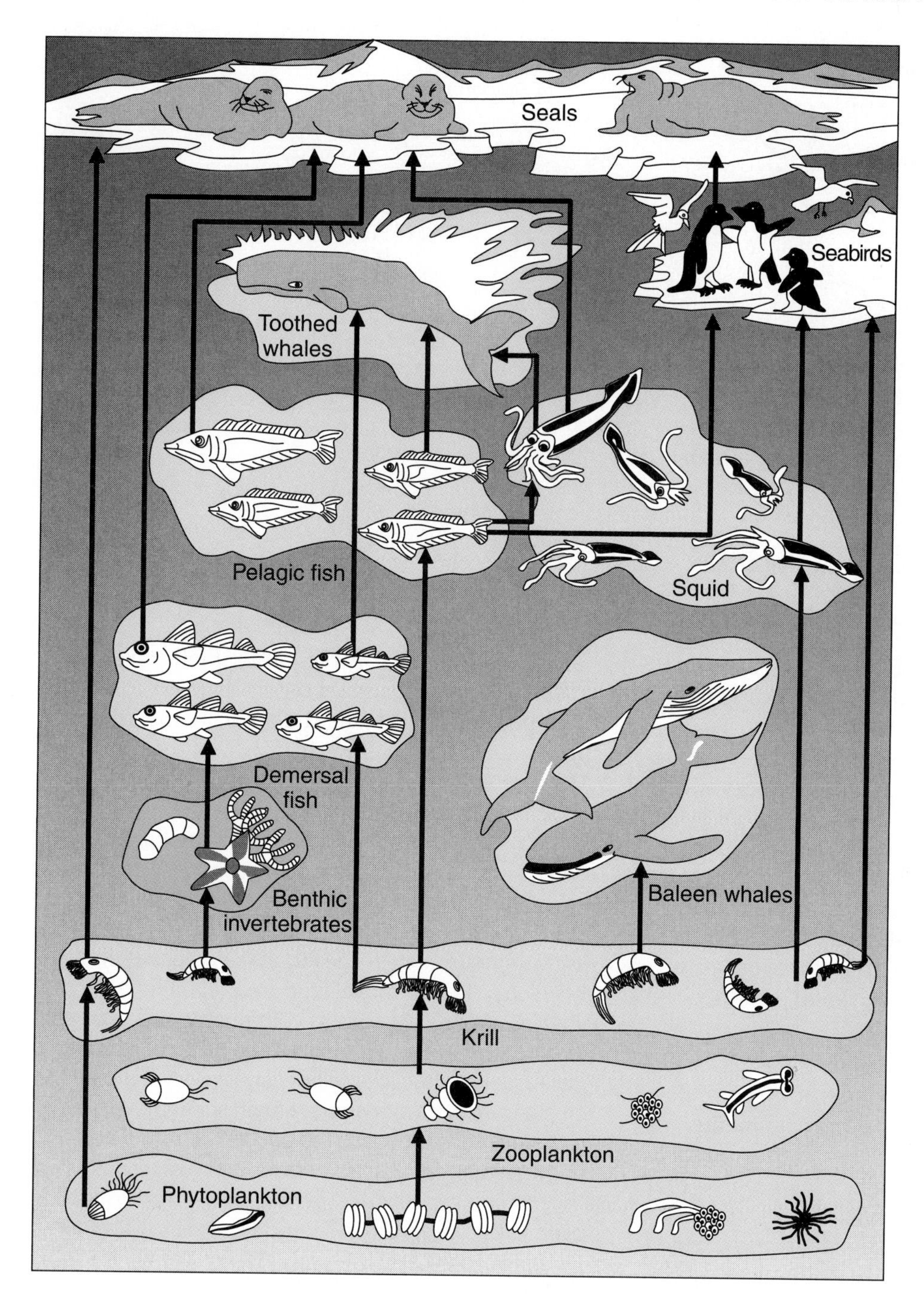

FIGURE 5.44
Diagrammatic representation of the central position of krill, *Euphausia superba*, in the Antarctic oceanic ecosystem. (After "Ecosystem Management and the Atlantic Krill." S. Nicol and W. de la Mare, *American Scientist*, Vol. 81, No. 5, pp. 36-47, 1993. Copyright © 1993 American Scientist/Elyse Carter. Reprinted by permission.)

munities. That is the presence south of the Antarctic Convergence of the euphausid *Euphausia superba* as the dominant herbivore in the pelagic system. No such dominating species occurs in the Arctic Ocean. The Antarctic krill, E. *superba*, is the largest of the euphausids, reaching 6 cm in size. It occurs in enormous numbers but is also subject to great interannual population fluctuations. Because E. *superba* is the staple food for a large number of Antarctic species, including such disparate forms as whales, seals, penguins, and fishes, any change in the populations of krill will affect the populations of all these vertebrate species as well (Figure 5.44). This is the reason for the central position of this species in the Antarctic seas. Recently, Loeb et al. (1997) have suggested that population fluctuations of krill are related

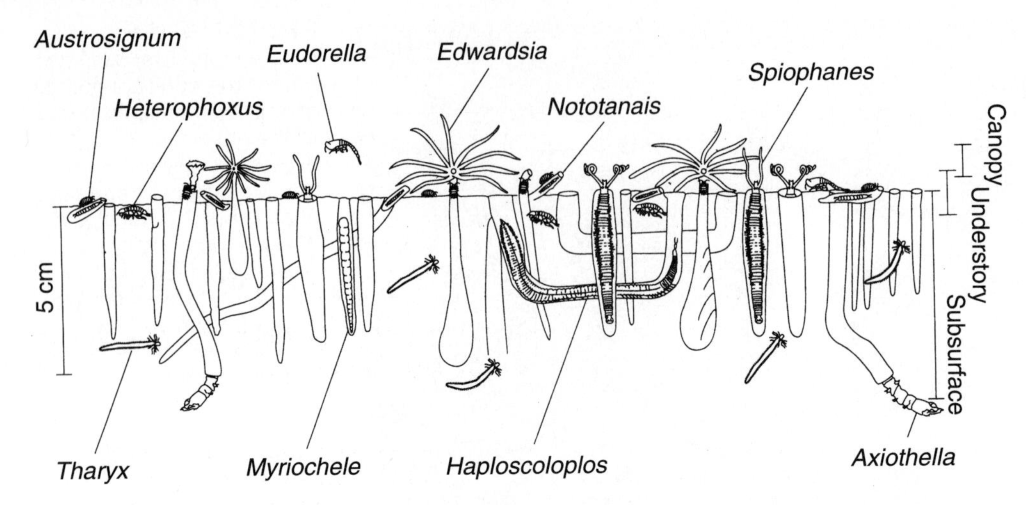

FIGURE 5.45 Schematic of a dense soft-bottom Antarctic assemblage, showing the three faunal groups: canopy, understory, and subsurface species. (From "Effects of Crustacean Predators on Species Composition and Population Structure of Soft-bodied Infauna from McMurdo Sound, Antarctica" J. S. Oliver & P. N. Slattery, *Ophelia,* Vol. 24, No. 3, pp. 155– 175, 1985. Copyright © 1985 Apollo Books. Reprinted by permission.)

to ice cover and abundance of the salp *Salpa thompsoni*. Salps are competitors of krill for phytoplankton but cannot feed on ice algae as can the krill. In years of extensive ice cover, the krill are favored, leading to good recruitment and high standing stocks, while salps decline. The reverse is true in years of low ice cover. Although we know a fair amount about *E. superba*, we are still unsure about many aspects of its biology. We do not, for example, know what the standing stocks are and what a safe level of harvest might be. These are important questions because in the last 25 years a fishery for the krill has developed (see also Chapter 11).

An Antarctic Soft-Sediment Community Oliver and Slattery (1985) studied a rare soft-bottom community in McMurdo Sound, Antarctica. This community was composed primarily of burrowing and tube-building polychaete worms and small crustaceans that occurred in densities of over 100,000 individuals per square meter. This dense assemblage was divided into three faunal groups of species that had similar characteristics. The three groups were canopy, understory, and subsurface (Figure 5.45). The canopy was dominated by the anemone *Edwardsia* and consisted primarily of suspension feeders. The understory included mainly mobile crustaceans that were deposit feeders and predators. The subsurface species group was composed primarily of tube-dwelling polychaetes that were mainly deposit feeders.

Experimental manipulations of this community suggested that the community structure was regulated by the understory predatory small crustaceans. These animals consumed the juveniles of the dominant polychaete species. The adults of the dominant polychaetes had a refuge in size from the predators. Therefore, the dense assemblage was dominated by large and long-lived species.

This community probably persists long periods because of several co-occurring factors. First, there are no large organisms or predators, such as walruses, gray whales, skates, or rays, that could introduce biological disturbances. These can, as we have seen earlier (see p. 123), remove large areas of infauna in other shallow-water areas. In this community, such a biological disturbance would destroy the assemblage, because the adults would be removed, while the motile crustacean predators would remain and remove the settling juveniles. Second, the adults are long-lived and immune to predation from the small crustaceans when they reach a certain size. This means there is a long time over which they reproduce, enough time to allow a few juveniles to escape the predation of the crustaceans and replace the few adults that may die.

In contrast to other soft-bottom infaunal communities that seem to be structured by competition (see pp. 190–194), there is no evidence of a competitive dominant.

An Antarctic Hard-Bottom Community Antarctic hard-bottom communities appear more common than soft-bottom communities. They have also had a longer period of investigation, beginning in the late 1960s. Indeed, the more than 30 years of marine ecological

FIGURE 5.46 Vertical zonation of invertebrates in Antarctic shelf areas. A few motile animals forage into Zone I, which is otherwise barren of sessile animals. Cnidarians dominate the sessile animals of Zone II, and sponges, Zone III. (From *Antarctic Biology*, Vol. 1, M. W. Holdgate 1970. Reprinted by permission of Academic Press Ltd.)

work in the McMurdo Sound area of Antarctica has given us a better understanding of community interaction in this remote area than we have for many temperate or tropical areas close to home.

In the shallow inshore waters of McMurdo Sound, Dayton, Robilliard, and Paine (1970) recognize three zones (Figure 5.46). Zone I extends 0–15 m and is essentially devoid of life, due to scouring by sea ice and the almost universal coverage by anchor ice. Only in ice-free periods can mobile organisms enter this zone.

Zone II extends from about 15 m down to the limit of anchor ice formation at approximately 30 m. This zone experiences less anchor ice and has a fauna of numerous sessile animals, mainly anemones and other cnidarians. The anemones here often feed upon rafted-in jellyfish or sea urchins, while the other cnidarians are suspension feeders. The anemones appear to be territorial.

Zone III begins abruptly where anchor ice ceases and continues down to an undetermined depth. The substrate changes from a cobbly-rocky bottom to one that is a thick mat of sponge spicules. Sponges dominate this area, but the area has a rich and diverse fauna of other invertebrates as well, of which starfish are the most conspicuous motile animals. This zone is noteworthy because of its extreme physical stability, which has resulted in a complex, highly diverse community that, according to Dayton et al. (1974), seems to be regulated by biological factors (predation) rather than by physical factors as in Zones I and II.

The dominant organisms in this area are very slow growing, so slow that in many cases the growth rates are too low to measure in one year. One exception is the sponge *Mycale acerata*, which has a very rapid growth rate and seems capable of outcompeting all other sessile invertebrates and taking over the substrate space. *Mycale* is prevented from dominating the substrate because it is preyed on by two sea stars, *Perknaster fuscus* and *Acodonaster conspicuus*, which keep it in check. *Acodonaster conspicuus* also preys on several other dominant sponges but is prevented from reaching population densities that would allow it to destroy the sponge community by yet another asteroid, *Odontaster validus*, which consumes the larvae, young, and even adults of *A. conspicuus* (Figures 5.47 and 5.48). Thus, we have a community that is regulated through finely tuned biological interactions, primarily predation.

FIGURE 5.47 Abundant starfish of the genus *Odontaster* on the bottom of McMurdo Sound, Antarctica. (Photo courtesy of Hunter Lenihan)

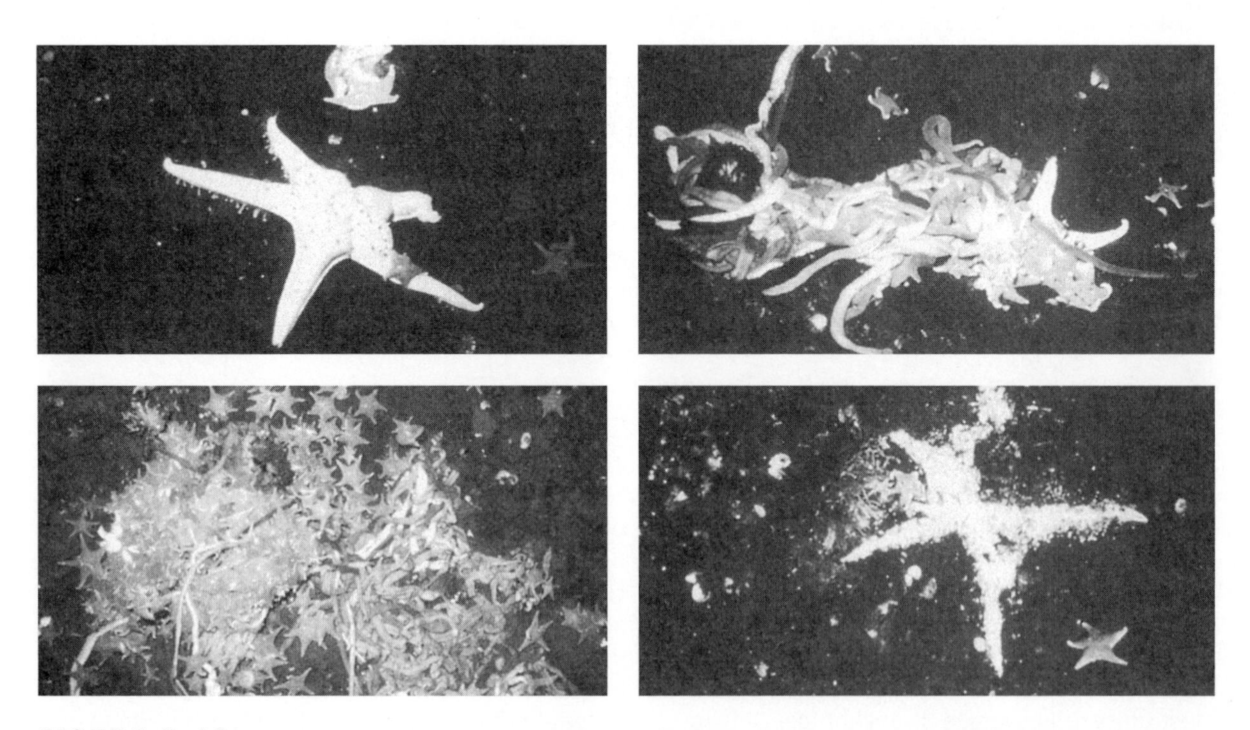

FIGURE 5.48 Sequence showing the consumption of the large Antarctic starfish *Acodonaster conspicuus* by the small starfish *Odontaster validus* and the nemertean *Lineus corrugatus*. (From P. K. Dayton, G. A. Robilliard, R. T. Paine, and L. B. Dayton, 1974, Biological accommodation in the benthic community at McMurdo Sound, Antarctica. *Ecol. Monogr.* 44(1):105–128.)

SUMMARY OF KEY CONCEPTS

- Continental shelf waters show more variability in environmental conditions than open sea or deep sea. Temperature and salinity are more variable, light penetration is reduced, and food supply is abundant.
- Unvegetated sedimentary environments were first sampled remotely by grabs, more recently using scuba and towed cameras and video systems.
- Unvegetated sedimentary systems are dominated by polychaetes, crustaceans, mollusks, and echinoderms.
- The first quantitative work in marine ecology was done by Petersen, who invented the quantitative grab. He was succeeded by Thorson, who established the concept of parallel bottom communities—that certain recurrent groups occupy the same substrate types over large areas of the continental shelves.
- Marine organisms have patchy distributions in time and space and may be categorized into functional groups such as sediment stabilizers, sediment destabilizers, detritivores, carnivores, omnivores, and herbivores.
- Changes in community structure occur through either physical factors or biological factors or some combination of the two. The simplest cases involve the dominance of physical factors. Biological factors are more complex and include predation, competition, and recruitment.
- Competition may be direct, involving either contact between organisms or interference in some aspect of the organisms life cycle, or it may be indirect, with one organism interrupting or preventing resource use by another.
- Burrowing deposit feeders are most abundant in mud, whereas suspension feeders are most abundant in sand.
- Trophic group amensalism is the exclusion of one trophic group via modification of the environment by another group.
- Competitive interference is the exclusion of one group of organisms via interference with their normal activities.
- Three categories of predators have been described: surface, burrowing, and digging.
- Caging out epifaunal predators in soft sediments usually results in increased numbers of infaunal organisms, suggesting that predation is an important process. However, this usually does not exclude infaunal predators, which may well be important.
- Recruitment may have the potential to determine community structure, but the difficulty of separating the effect of recruitment from various sources of postrecruitment mortality means that we can only speculate at present.
- Competition for space may occur in benthic infauna both vertically and horizontally and may affect the density or lead to more occurrences of commensalism.
- Community structure in open sedimentary environments changes over differing time periods and produces certain patterns or patches that recur on differing areal and time scales. These changes result from a complex interplay of physical and biological factors.
- Rocky subtidal communities are usually covered with low growing plants and sessile or sedentary invertebrates, leaving little open space and occupying a two-dimensional space.
- Most animals of subtidal rock are colonial encrusting species able to spread over surfaces by asexual means or single individuals that settle gregariously, forming large, space-occupying clumps. Most are suspension feeders.
- The organization and persistence of these areas and the high species diversity are explained by a combination of biological and physical factors, including storm disturbance, competition for space, grazing, recruitment, and mutualism.
- Disturbance generates patches of open space that are colonized by larvae or juveniles of nearby adults, or by spread of adjacent colonial organisms. Competition among these organisms for space coupled with seasonality of certain organisms and mutualistic interactions among certain species completes the cycle.
- In any area at any time, there is either a mosaic of patches of different ages with different sets or different abundances of species resulting in high species diversity, or a different zonation pattern.
- The existence in the same area of different groups of organisms or patches of organisms has been called alternate stable states.
- In cold temperate parts of the world, the subtidal hard substrates are occupied by large brown algae, giving rise to kelp forests and kelp beds.
- Kelps form extensive three-dimensional habitats composed of several vertical layers.
- The extent of kelp beds and forests is dependent upon a hard substrate for attachment, sufficient light, low temperatures, nutrients, and grazing.
- Kelp plants have a life cycle that alternates between an asexual macroscopic stage termed the sporophyte and a sexual microscopic stage called the gametophyte. Kelps have very high productivity.
- The causes of kelp mortality are wave action, nutrient depletion, warm temperatures, and grazing.
- Seagrasses have a worldwide distribution and form dense carpets of blades in shallow waters of temperate and tropical seas.
- All seagrasses have a similar appearance, with thin straplike blades arising from a creeping rhizome that stabilizes the sediment.
- Seagrass beds are most strongly influenced by water currents and wave action, which shape how the beds form.
- Seagrass beds are depositional environments, often with high organic content that can make the sediment under the bed anaerobic.
- Other physical factors affecting the beds include light, temperature, and desiccation.
- The productivity of seagrass beds is among the highest in the marine environment. Because they are flowering plants,

they extract nutrients from the sediment and can recycle otherwise trapped nutrients back into the ecosystem.

- Seagrass beds generally have only one or at most a few dominant species of grasses.
- Seagrass beds are complex systems with a narrow vertical three-dimensionality and complexity and contain large numbers of epiphytic and epizoic organisms, burrows, and motile animals.
- Seagrasses are grazed by relatively few animals, notably various birds, a few fish, sea urchins, and the green turtle, but these may consume a large fraction of the seagrass annual production.
- A wide variety of predators are associated with seagrass beds, but there is little evidence that they regulate the structure of the beds or abundance of the prey.
- Seagrass beds may offer some refuge from predators to some prey organisms.
- Much seagrass productivity may be exported in the form of detritus to other systems.
- Seagrass beds serve a number of functions in shallow inshore waters, including being a major source of primary productivity, stabilizing of the bottom, serving as a nursery grounds for many species, as a trap for sediment, as a protective canopy shielding inhabitants from strong sunlight, and as a substrate for various epizoic and epiphytic organisms.
- Seagrass beds are subject to catastrophic, episodic destruction by disease.
- Some seagrass beds show a successional pattern, whereas others do not.
- Examples of some special communities are the sea pen community in Puget Sound and the sea pansy community in southern California.
- Polar communities in shallow water differ from other communities due to the drastic seasonal changes in light, ice and snow, and the constant extremely low temperatures.
- The Arctic and Antarctic oceans have very different physical conditions, which lead to different communities. The Arctic Ocean is an isolated sea surrounded by landmasses with extensive continental shelf area and receiving the influx from many rivers, leading to mainly sedimentary benthic environments. In contrast, the Antarctic ocean is a circumpolar ring of water surrounding a central landmass with no rivers producing sediment and a narrow continental shelf that drops off quickly into deep water.
- A major disturbance factor in both the Arctic and Antarctic is ice, which can destroy benthos by removing the organisms by freezing and lifting them or by gouging them out.
- Biologically the Antarctic organisms are more diverse and show more endemism than those of the Arctic.
- Associated with the sea ice, particularly in Antarctica, is a whole set of organisms from bacteria to krill and a whole well-developed microbial food web that is significant to both the water column community and the benthos.
- Antarctic seas are characterized by the central position and importance of the krill *Euphausia superba*.
- The Antarctic soft-bottom community is composed primarily of burrowing and tube-building polychaete worms and small crustaceans dominated by large and long-lived species with no evidence of a competitive dominant.
- Antarctic hard-bottom communities are dominated by sponges with a rich and diverse associated fauna, of which various starfish are conspicuous and are regulated primarily by predation.

REVIEW QUESTIONS

ESSAY: Develop complete answers to these questions.

1. Why are there fewer niches available in the soft-bottom, sublittoral region than in rocky coasts? Discuss.
2. Some ecological wag has said, "Nichy nachy, the environment is patchy." Discuss three ways (one physical, one chemical, and one biological) in which the shallow-water subtidal benthos is (or becomes) patchy.
3. Compare a kelp forest with a terrestrial forest. How are they similar? How are they different? Consider, for example, light, nutrient uptake, vertical zonation, and general structure.
4. Sediments are profoundly three-dimensional, but they are not uniform. Present evidence that supports this statement and discuss its significance to the infauna.
5. Discuss examples whereby abiotic and biotic factors interact to structure the marine sublittoral into patches of organisms.
6. Explain the phenomenon called alternate stable states; how do these states develop and why do they change?
7. What important roles (functions) do seagrass beds serve in inshore regions? Discuss at least ________ of these. (Instructors should decide the number that their students are to discuss.)
8. Present an argument that would explain why the fauna of the Antarctic region has a high degree of diversity, high endemic rate, and high biomass as compared with that of the Arctic Ocean.

BIBLIOGRAPHY

Ackley, F. F., and C. W. Sullivan. 1994. Physical controls on the development and characteristics of Antarctic sea ice biological communities—a review and synthesis. *Deep-Sea Res.* I 41:1583–1604.

Ambrose, W. 1991. The importance of infaunal predators in structuring marine soft bottom communities. *Amer. Zoo.* 3:849–860.

Ayling, A. M. 1981. The role of biological disturbance in temperate subtidal encrusting communities. *Ecology* 62(3):830–847.

Barkai, A., and C. McQuaid. 1988. Predator-prey role reversal in a marine benthic ecosystem. *Science* 242:62–64.

Birkeland, C. 1974. Interactions between a sea pen and seven of its predators. *Ecol. Monogr.* 44:211–232.

Camp, D. K., S. P. Cobb, and J. F. Van Breedveld. 1973. Overgrazing of seagrasses by a regular urchin, *Lytechinus variegatus*. *Bioscience* 23:37–38.

Conlan, K. E., H. S. Lenihan, R. G. Kvitek, and J. S. Oliver. 1998. Ice scour disturbance to benthic communities in the Canadian High Arctic. *Mar. Ecol. Prog. Ser.* 166:1–16.

Cottam, C. 1934. Past periods of eelgrass scarcity. *Rhodora* 36:261–264.

Coull, B. C., ed. 1977. *Ecology of marine benthos*. Columbia: University of South Carolina Press.

Crisp, D. J., and P. S. Meadows. 1962. The chemical bases of gregariousness in cirripedes. *Proc. Roy. Soc. London* (*Biol.*) 150:500–520.

Davis, N., and G. VanBlaricom. 1978. Spatial and temporal heterogeneity in a sand bottom epifaunal community of invertebrates in shallow water. *Limn. Oceanog.* 23(3):417–427.

DeVlas, J. 1979. Annual food intake by plaice and flounder in a tidal flat area in the Dutch Wadden Sea with special reference to consumption of regenerating parts of macrobenthic prey. *Neth. J. Sea Res.* 13:117–153.

Dayton, P. K., G. A. Robilliard, and A. L. DeVries. 1969. Anchor ice formation in McMurdo Sound, Antarctica, and its biological effects. *Science* 163.

Dayton, P. K., G. A. Robilliard, and R. T. Paine. 1970. Benthic faunal zonation as a result of anchor ice at McMurdo Sound, Antarctica. In *Antarctic ecology*, Vol. 1. New York: Academic Press, 244–258.

Dayton, P. K., G. A. Robilliard, R. T. Paine, and L. B. Dayton. 1974. Biological accommodation in the benthic community at McMurdo Sound, Antarctica. *Ecol. Monogr.* 44(1):105–128.

den Hartog, C. 1977. Structure, function and classification in sea grass communities. In *Seagrass ecosystems*, edited by C. P McRoy and C. Helfferich. New York: Dekker, 89–121.

Deysher, L. E., and T. A. Dean. 1986. In situ recruitment of sporophytes of the giant kelp, *Macrocystis pyrifera* (L) C. A. Agardh: Effects of physical factors. *J. Exp. Mar. Biol. Ecol.* 103:41–63.

Ebeling, A. W., D. R. Laur, and R. J. Rowley. 1985. Severe storm disturbance and reversal of community structure in a southern California kelp forest. *Mar. Biol.* 84:287–294.

Elner, R. W., and R. L. Vadas. 1990. Inference in ecology: The sea urchin phenomenon in the Northwest Atlantic. *Amer. Nat.* 136:108–125.

Estes, J., and D. O. Duggins. 1995. Sea otters and kelp forests in Alaska: Generality and variation in a community ecological paradigm. *Ecol. Monogr.* 65(1):75–100.

Estes, J., and J. Palmisano. 1974. Sea otters, their role in structuring near-shore communities. *Science* 185:1058–1060.

Fager, E. W. 1968. A sand bottom epifaunal community of invertebrates in shallow water. *Limn. Oceanog.* 13(3):448–464.

Fenchel, T. 1977. Aspects of the decomposition of seagrasses. In *Seagrass ecosystems*, edited by C. P. McRoy and C. Helfferich. New York: Dekker, 123–146.

Fletcher, W. J. 1987. Interactions among subtidal Australian sea urchins, gastropods and algae: Effects of experimental removals. *Ecol. Monogr.* 57(1):89–109.

Foster, M. S., and D. R. Schiel. 1985. *The ecology of giant kelp forests in California: A community profile*. U.S. Fish and Wildlife Service Biological Report 85 (7.2).

Garrison, D. L. 1991. Antarctic sea ice biota. *Amer. Zoo.* 31:17–33.

Gray, J. S. 1981. *The ecology of marine sediments*. New York: Cambridge University Press.

Harrold, C., and D. Reed. 1985. Food availability, sea urchin grazing and kelp forest community structure. *Ecology* 66(4):1160–1169.

Hempel, G. 1985. On the biology of polar seas, particularly the southern ocean. In *Marine biology of polar regions and effects of stress on marine organisms*, edited by J. S. Gray and M. E. Christiansen. New York: Wiley, 3–33.

Holme, N. A., and A. D. McIntyre, eds. 1984. *Methods for the study of marine benthos*. 2d ed. Oxford: Blackwell Scientific Publications.

Hunt, G. L. 1991. Marine ecology of seabirds in polar oceans. *Amer. Zoo.* 31:131–142.

Jones, N. S. 1950. Marine bottom communities. *Biol. Rev.* 25:283–313.

Kastendiek, J. 1976. Behavior of the sea pansy, *Renilla kollikeri* Pfeffer (Coelenterata, Pennatulacea), and its influence on the distribution and biological interaction of the species. *Biol. Bull.* 151:518–537.

Keegan, B. F., P. O. Ceidigh, and P. J. S. Boaden, eds. 1977. *Biology of benthic organisms*. New York: Pergamon.

Kenworthy, W. J., J. C. Zieman, and G. W. Thayer. 1982. Evidence for the influence of seagrass on the benthic nitrogen cycle in a coastal plain estuary near Beaufort, North Carolina. *Oecologia* 54:152–158.

Knox, G. A. 1994. *The Biology of the southern ocean*. Cambridge, UK: Cambridge University Press.

Levin, L. A. 1981. Dispersion, feeding behavior and competition in two spionid polychaetes. *J. Mar. Res.* 39:99–117.

Levinton, J. S. 1977. Ecology of shallow water deposit feeding communities, Quisset Harbor, Massachusetts. In *Ecology of marine benthos*, edited by B. Coull. Columbia: University of South Carolina Press.

Loeb, V., V. Siegel, O. Holm-Hansen, R. Hewitt, W. Fraser, W. Trivelpiece, and S. Trivelpiece. 1997. Effects of sea-ice extent and krill or salp dominance on the Antarctic food web. *Nature* 387:897–900.

Mann, K. H. 1973. Seaweeds: Their productivity and strategy for growth. *Science* 182:975–983.

McRoy, C. P. 1966. The standing stock and ecology of eelgrass in Izembeck Lagoon, Alaska. Master's thesis, University of Washington.

McRoy, C. P., and C. Helfferich, eds. 1977. *Seagrass ecosystems*. New York: Dekker.

McRoy, C. P., and C. McMillan. 1977. Production ecology and physiology of sea grasses. In *Seagrass ecosystems*, edited by C. P. McRoy and C. Helfferich. New York: Dekker, 53–88.

Menge, B. A., and T. M. Farrell. 1989. Community structure and interaction webs in shallow marine hard bottom communities: Tests of an environmental model. *Adv. Ecol. Res.* 19:189–262.

Mills, E. L. 1969. The community concept in marine zoology, with comments on continua and instability in some marine communities: A review. *J. Fish. Res. Bd. Canada* 26:1415–1428.

Myers, A. C. 1977. Sediment processing in a marine subtidal sandy bottom community. II. Biological consequences. *J. Mar. Res.* 35:633–647.

Nicol, S., and W. de la Mare. 1993. Ecosystem management and the Antarctic krill. *Am. Sci.* 81(1):36–47.

North, W. 1971. Growth of individual fronds of the mature giant kelp, *Macrocystis*. In *The biology of giant kelp beds* (Macrocystis) *in California*, edited by W. North. Beihefte zur Nova Hedwigia Heft 32, Verlag von J. Cramer, 123–168.

Ockelmann, K. W., and O. Vahl. 1970. On the biology of the polychaete *Glycera alba*, especially its burrowing and feeding. *Ophelia* 8:275–294.

Ogden, J. C. 1980. Faunal relationships in Caribbean seagrass beds. In *Handbook of seagrass biology: An ecosystem perspective*, edited by R. C. Phillips and C. P. McRoy. New York: Garland STPM Press, 173–198.

Olafsson, E. B., C. H. Peterson, and W. G. Ambrose. 1994. Does recruitment limitation structure populations and communities of macro-invertebrates in marine soft sediments: The relative significance of pre- and post-settlement processes. *Oceanogr. Mar. Bio., Ann. Rev.* 32:65–109.

Oliver, J. S., and P. N. Slattery. 1985. Effects of crustacean predators on species composition and population structure of soft bodied infauna from McMurdo Sound, Antarctica. *Ophelia* 24(3):155–175.

Oliver, J. S., P. N. Slattery, L. W. Hulberg, and J. W. Nybakken. 1979. Relationships between wave disturbance and zonation of benthic invertebrate communities along a subtidal high energy beach in Monterey Bay, California. *Fish. Bull.* 78:437–454.

Orth, R. J. 1975. Destruction of eelgrass *Zostera marina* by the cownose ray *Rhinoptera bonasus* in the Chesapeake Bay. *Chesapeake Sci.* 16:205–208.

Pearse, J. S., J. B. McClintock, and I. Bosch. 1991. Reproduction of Antarctic benthic marine invertebrates: Tempos, modes and timing. *Amer. Zoo.* 31:65–80.

Petersen, C. G. Joh. 1918. The sea bottom and its production of fish food. A survey of the work done in connection with the valuation of the Danish waters from 1883–1917. *Rap. Danish Biol. Stat.* Vol. 25.

Petersen, C. G. Joh. 1924. A brief survey of the animal communities in Danish waters. *Amer. J. Sci., Ser.* 5 7(41):343–354.

Peterson, C. H. 1977. Competitive organization of the soft bottom macrobenthic communities of southern California lagoons. *Mar. Biol.* 43:343–359.

Peterson, C. H. 1982. Clam predation by whelks (*Busycon* spp.): Experimental tests of the importance of prey size, prey density and seagrass cover. *Mar. Biol.* 66:159–170.

Peterson, C. H., and M. L. Quammen. 1982. Siphon nipping: Its importance to small fishes and its impact on growth of the bivalve *Protothaca staminea* (Conrad). *J. Exp. Mar. Biol. Ecol.* 63:249–268.

Phillips, R. C., and C. P. McRoy, eds. 1980. *Handbook of seagrass biology*. New York: Garland STPM Press.

Pohle, D. G., V. M. Brieelj, and Z. Garcia-Esquiral. 1991. The eelgrass canopy: An above bottom refuge from benthic predators for juvenile bay scallops *Argopecten irradians*. *Mar. Ecol. Progr. Ser.* 74(1):47–59.

Rasmussen, E. 1973. Systematics and ecology of the Isefjord marine fauna (Denmark). *Ophelia* 11(12):1–495.

Reed, D. C., D. R. Laur, and A. W. Ebeling. 1988. Variation in algal dispersal and recruitment: The importance of episodic events. *Ecol. Monogr.* 58:321–335.

Reise, K. 1979. Spatial configurations generated by motile benthic polychaetes. *Helgol. Wiss. Meeresunters.* 32:55–72.

Rhoads, D. C., and D. K. Young. 1970. The influence of deposit feeding organisms on sediment stability and community structure. *J. Mar. Res.* 28(2):150–178.

Rhoads, D. C., and D. K. Young. 1971. Animal-sediment relations in Cape Cod Bay, Massachusetts. II. Reworking by *Molpadia oolitica*. *Mar. Biol.* 11(3):255–261.

Robblee, M. B., T. R. Barber, P. R. Carlson, Jr., M. J. Durako, J. W. Fourgurean, L. K. Muehlstein, D. Porter, L. A. Yarbro, R. T. Zieman, and J. C. Zieman. 1991. Mass mortality of the tropical seagrass *Thalassia testudinum* in Florida Bay (U.S.A.). *Mar. Ecol. Progr. Ser.* 71:297–299.

Robertson, A. I., and K. H. Mann. 1984. Disturbance by ice and life history adaptations of the seagrass *Zostera marina*. *Mar. Biol.* 80:131–142.

Rosenthal, R. J., W. D. Clarke, and P. K. Dayton. 1974. Ecology and natural history of a stand of giant kelp, *Macrocystis pyrifera*, off Del Mar, California. *Fish. Bull.* 72(3):670–684.

Scagel, R. F. 1947. *An investigation on marine plants near Hardy Bay*, B.C. Provincial Dept. Fisheries, Victoria, B.C., Canada, 1–70.

Sebens, K. 1985. The ecology of the rocky subtidal zone. *Amer. Sci.* 73:548–557.

Sebens, K. 1986. Spatial relationships among encrusting marine organisms in the New England subtidal zone. *Ecol. Monogr.* 56(1):73–96.

Setchell, W. A. 1929. Morphological and phenological notes on *Zostera marina*. *Univ. Calif. Pub. Botany* 14:389–452.

Short, F. T., L. K. Muehlstein, and D. Porter. 1987. Eelgrass wasting disease: Cause and recurrence of a marine epidemic. *Biol. Bull.* 173(3):557–562.

Stromberg, J. O. 1991. Marine ecology of Polar seas: A comparison Arctic/Antarctic. In *Marine biology its accomplishment and future prospect*, edited by J. Mauchline and T. Nemoto. New York: Elsevier, 247–261.

Sutherland, J. P. 1974. Multiple stable points in natural communities. *Amer. Nat.* 108:859–873.

Tegner, M. J., and P. K. Dayton. 1987. El Niño effects on southern California kelp forest communities. *Adv. Ecol. Res.* 17:243–279.

Tevesz, M. J. S., and P. L. McCall, eds. 1983. *Biotic interactions in recent and fossil benthic communities*. New York: Plenum Press.

Thayer, G. W., S. M. Adams, and M. W. LaCroix. 1975. Structural and functional aspects of a recently established *Zostera marina* community. In *Estuarine research*, Vol. 1., edited by L. E. Cronin. New York: Academic Press, 517–540.

Thayer, G. W., W. J. Kenworthy, and M. S. Fonseca. 1984. *The ecology of eelgrass meadows of the Atlantic coast: A community profile*. U.S. Fish and Wildlife Service. FWS/OBS-84/02.

Thorson, G. 1950. Reproductive and larval ecology of marine bottom invertebrates. *Biol. Rev.* 25:1–45.

Thorson, G. 1955. Modern aspects of marine level bottom animal communities. *J. Mar. Res.* 14:387–397.

Thorson, G. 1957. Bottom communities (sublittoral or shallow shelf). In *The treatise on marine ecology and paleoecology*, Vol. 1, *Ecology*. Geol. Soc. Amer. Memoir 67, 461–534.

Thorson, G. 1966. Some factors influencing the recruitment and establishment of marine benthic communities. *Neth. J. Sea. Res.* 3(2):267–293.

Turner, T. 1983. Facilitation as a successional mechanism in a rocky intertidal community. *Amer. Nat.* 121:729–738.

VanBlaricom, G. R. 1982. Experimental analyses of structural regulation in a marine sand community exposed to ocean swell. *Ecol. Monogr.* 52:283–305.

Van Veldhuizen, H., and D. Phillips. 1978. Prey capture by *Pisaster brevispinus* (Asteroidea, Echinodermata) on soft substrate. *Mar. Biol.* 48:89–97.

Virnstein, R. W. 1977. The importance of predation by crabs and fishes on benthic infauna in Chesapeake Bay. *Ecology* 58:1199–1217.

Warher, G. F. 1984. *Diving and marine biology, the ecology of the sublittoral*. New York: Cambridge University Press.

Wilkins, E. W. 1982. Waterfowl utilization of a submerged vegetation (*Zostera marina* and *Ruppia maritima*) bed in the lower Chesapeake Bay. Master's thesis, College of William and Mary.

Williams, S. 1987. Competition between seagrasses *Thalassia testudinum* and *Syringodinium filiforme* in a Caribbean lagoon. *Mar. Ecol. Progr. Ser.* 35:91–98.

Wilson, W. H., Jr. 1981. Sediment mediated interactions in a densely populated infaunal assemblage: The effects of the polychaete *Abarenicola pacifica*. *J. Mar. Res.* 39:735–748.

Wilson, W. H., Jr. 1991. Competition and predation in marine soft sediment communities. *Ann. Rev. Ecol. Syst.* 21:221–241.

Witman, J. D. 1985. Refuges, biological disturbance, and rocky subtidal community structure in New England. *Ecol. Monogr.* 55:421–445.

Witman, J. D. 1987. Subtidal coexistence: Storms, grazing, mutualism and the zonation of kelps and mussels. *Ecol. Monogr.* 57(2):167–187.

Woodin, S. A. 1976. Adult-larval interactions in dense infaunal assemblages: Patterns of abundance. *J. Mar. Res.* 34:25–41.

Woodin, S. A. 1983. Biotic interactions in recent marine sedimentary environments. In *Biotic interactions in recent and fossil benthic communities*, edited by M. J. S. Tevesz and P. L. McCall. New York: Plenum Press, 3–38.

Zieman, J. C. 1982. *The ecology of the seagrasses of South Florida: A community profile*. U.S. Fish and Wildlife Service Program FWS/OBS-82/25. 12.

INTERTIDAL ECOLOGY

Although it constitutes by far the smallest area of all in the world's oceans, the intertidal zone is perhaps the best known. It is an extremely narrow fringe area a few meters in extent between high and low water. Because it is the most accessible to humans, it has been studied more intensively than any other marine habitat. Only here can aquatic organisms be directly observed during low tide without special equipment. The intertidal zone has been of interest and use to humans since prehistoric time.

Despite being restricted, the intertidal has the greatest variations in environmental factors of any marine area, and these can occur within centimeters of each other. Coupled with this is a tremendous diversity of life, which may be as great as or greater than that found in the more extensive subtidal habitats. It should be emphasized that this area is an extension of the marine environment and is inhabited almost exclusively by marine organisms. Although it may be exposed to air as much as half the time, the terrestrial fauna and flora have not invaded it to any extent, except at the uppermost fringes.

The richness, the diversity of environmental factors, and the ease of access attract to this area a disproportionate amount of scientific attention. More knowledge is available for this small area and its organisms, and their interactions are better known, than in any other area. Thanks to such interest and knowledge, the intertidal has produced more unifying concepts with respect to the organization of marine communities than any other. As a result, a disproportionate amount of space in this book is devoted to a discussion of it.

ENVIRONMENTAL CONDITIONS

In part, the tremendous range of environmental factors found in the intertidal occurs because this zone is exposed to the air for a certain amount of time during a day and most physical factors show a wider range in air than in water (see Chapter 1, pp. 1–4).

Tides

The periodic, predictable rise and fall of the level of the sea over a given time interval is called a **tide.** This is the most important environmental factor influencing life in the intertidal zone. Without the presence of the tide or some other means of inducing a periodic rise and fall in the water level, this zone would not exist as it is, and many of the other factors would cease to be of influence.

With very few exceptions, most shore areas of the world experience tides. The only major seas that virtually lack tidal action are the Mediterranean, Black and the Baltic. In these areas, the water level on the shoreline fluctuates primarily due to wind action pushing water. This does not mean, however, that all seashores experience the same tidal range, or even the same type of tide. The reasons for the occurrence of different tides and different tidal ranges are complex and have to do with the interaction of the tide-generating forces—the sun and moon, the rotation of the earth, the geomorphology of the ocean basins, and the natural oscillations of the various ocean basins. Most of the explanation is beyond the scope of this text, but a simplified explanation will be given here.

Tides occur due to the interaction of the gravitational attraction of the sun and the moon on the earth and the centrifugal force generated by the rotating earth-moon system. As a result of these forces, the water in the ocean basins is pulled into bulges. Gravitational attraction of one body for another is a function of the mass of each body and their distance apart. In the case of the sun and the moon, the gravitational generating force of the moon on the earth is about twice that of the sun, even though the sun is many times more massive than the moon. This is due to the much greater distance of the sun from the earth.

The earth and moon form an orbiting system that revolves around their common center of mass. Because of the large size of the earth relative to the moon, this point is located inside the earth. The revolution of the earth-moon system creates a centrifugal force (acting outward) that is balanced by the gravitational force acting between the two bodies. Gravitational force, however, is much stronger than centrifugal force on the side of the earth facing the moon, and much weaker on the side opposite. As a result, the side facing the moon has the water pulled into a bulge (high tide). On the opposite side of the earth, the gravitational force of the moon will be the least, and the stronger centrifugal force will pull water into a bulge away from earth (another high tide; Figure 6.1); thus, we have the two high tides. These high tides circle the earth following the position of the moon as the earth revolves on its axis once every 24 hours. Low tides are about halfway between the high tides.

The system as described would give two high tides and two low tides of equal magnitude each day. As we know, however, tides are not equal, nor are there always four tides per day. How do these discrepancies arise? In part, this is because the earth is not vertical with respect to its plane of orbit around the sun. It actually is inclined 23½° from the vertical. As a result, during the earth's rotation on its axis, a given point on the earth's surface is subjected to different tidal heights, as Figure 6.1a shows. It should also be noted from Figure 6.1a that, at the highest latitudes (polar seas), only a single high tide would occur. Further changes in the height of the tides result from changes in the moon relative to the earth as the moon moves in its orbit around the earth. Since the orbit of the moon is not circular but elliptical, there are times when the moon is closer to the earth (the point where it is closest is called the **perigee**) and others when it is farther away (the point where it is most distant is called the **apogee**). Tides are greater at perigee and diminished at apogee.

The effect of the sun is seen in spring and neap tides (Figure 6.1b). Spring tides are tides that show the greatest range (both high and low) and result when the moon and sun are directly aligned, and these two forces are combined. Neap tides, on the other hand, are tides showing minimum range and result when sun and moon are at right angles to each other, and they counteract each other.

Differences both in number of high and low tides per day and in height in various parts of the world are due to peculiarities of the various ocean basins in which the tides occur. The extremely high tides of the Bay of Fundy in Nova Scotia and of Cook Inlet in Alaska result from the geometry of the basin in which the basic tide forces are acting.

Locations having a single low and high per day are said to experience **diurnal tides** (Figure 6.2). Those with two highs and lows per lunar day experience **semidiurnal tides**. Those having a mixture of diurnal and semidiurnal tides experience **mixed tides**. The heights of the highs and lows vary from day to day as the positions of the sun and moon change relative to each other, each being alternately fully aligned and 90° out of alignment every 14 days.

Changes in tides over time can have two direct effects on the presence and organization of intertidal communities. The first effect results from the time that a given area of the intertidal is exposed to the air relative to the time it is submerged in water. It is the *duration of exposure* to air that is most important, because that is when the marine organisms will be subjected to the greatest temperature ranges and the possibility of desiccation (water loss). The longer the duration of exposure, the greater the chance of encountering a lethal temperature or desiccating beyond tolerable limits. Since most of these animals also must wait until covered with water to feed, the longer the duration of emersion, the less opportunity to feed and

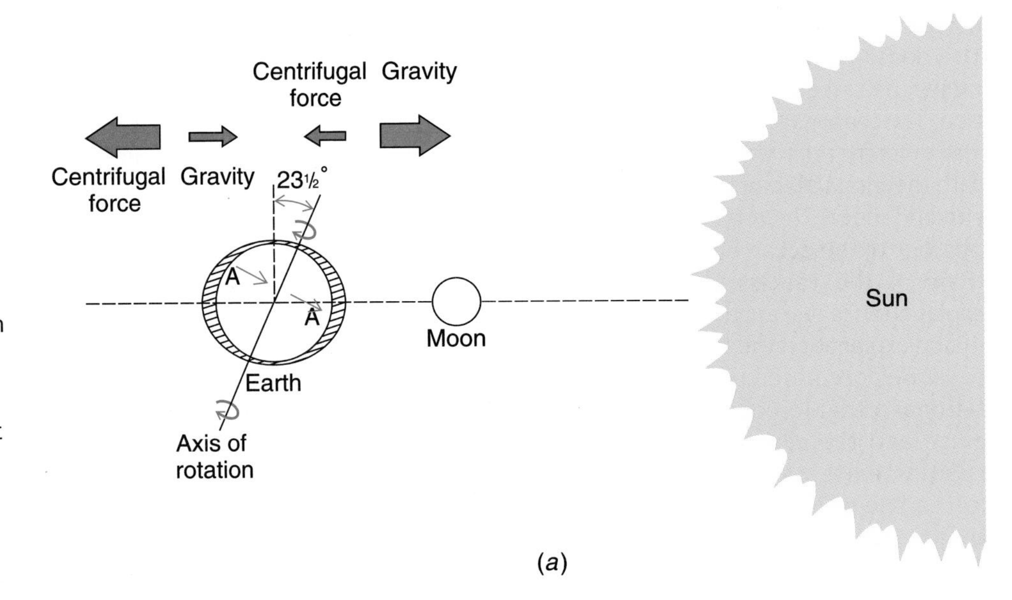

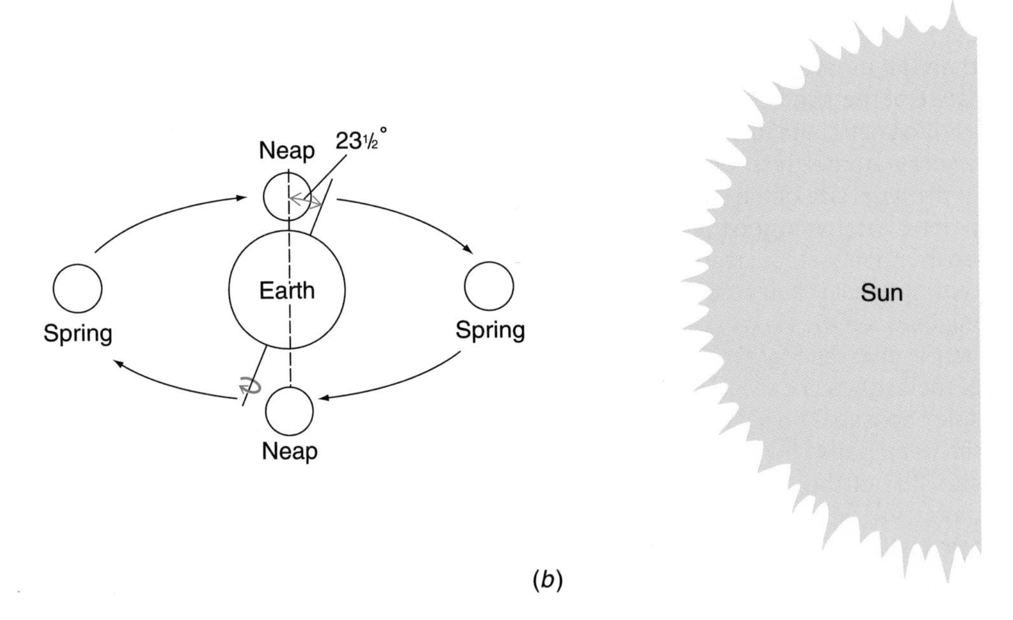

FIGURE 6.1 Origin of tides. (a) Moon raises a bulge on the side of earth nearest it due to greater gravitational force than counteracting centrifugal force. On the opposite side, the centrifugal force is stronger and throws another bulge outward. Because of the inclination of the earth on its rotational axis, point A, as it rotates, will experience two high tides of different height. (b) Position of the moon and sun at neap and spring tides.

obtain sufficient energy. Animals and plants of the tidal zone vary in their ability to tolerate immersion in air, and this differential is part of the reason for the different pattern of distribution of organisms observed on many rocky shores (see pp. 246–249).

In certain areas of the world there are two high and two low tides per day. Where the two highs or two lows are not equal (areas with mixed tides), the duration of immersion is not smooth going from low to high tide levels and there are break points where the duration changes greatly in a short distance. An organism living just a few centimeters above one of these points would be subjected to a greatly increased exposure to air relative to one just a few centimeters below. The existence of these **critical tide levels** was suggested by Doty (1946) as the reason for the sharp changes in fauna and flora distribution observed on rocky shores (see Figure 6.12b), since they were thought to delineate an important environmental break point.

The second effect of tide is the result of the time of day during which the intertidal is exposed to the air. As before, the potentially lethal ranges for physical parameters are encountered by marine organisms only when exposed to air. In any daily cycle, these vary greatly at different times of the day. If, for example, the low tides in the tropics occurred only during the hours of darkness, the intertidal fauna and flora would be exposed to much lower temperatures and less desiccation than if exposed at midday. As a result, we might expect a greater diversity of organisms in the intertidal of a tropical area where low tides regularly occur during the night or early morning or evening than in an area where they occur at midday. In the cold temperate zones, the reverse would be true, because in winter the lowest temperatures would occur in early morning or at night and might freeze the animals, whereas at midday the temperatures might be high enough to prevent freezing.

A final effect of the tide is that, because of its great predictability, it induces certain rhythms in the activities of shore organisms. These rhythms affect processes such as time of spawning as seen in the grunion, a fish of the Pacific coast of the United States that spawns on the beach only on certain nights of the highest spring tides (Figure 6.3), and the horseshoe crab (*Limulus polyphemus*) of the Atlantic and Gulf coasts. They may also affect feeding or other activities. Many intertidal organisms are quiescent when the tide is out and resume normal activity, such as feeding, only when the tide is in.

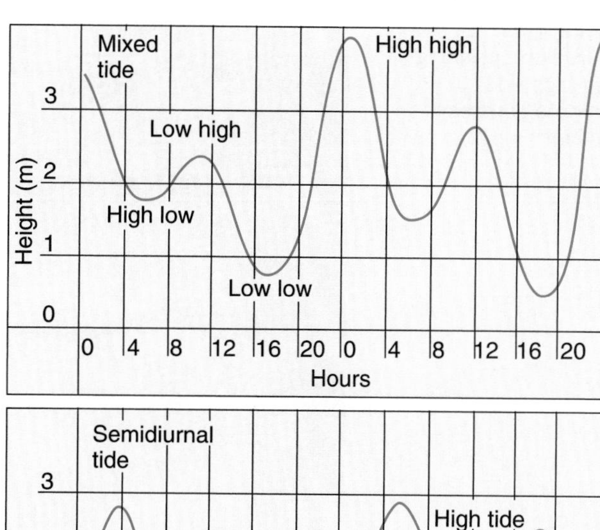

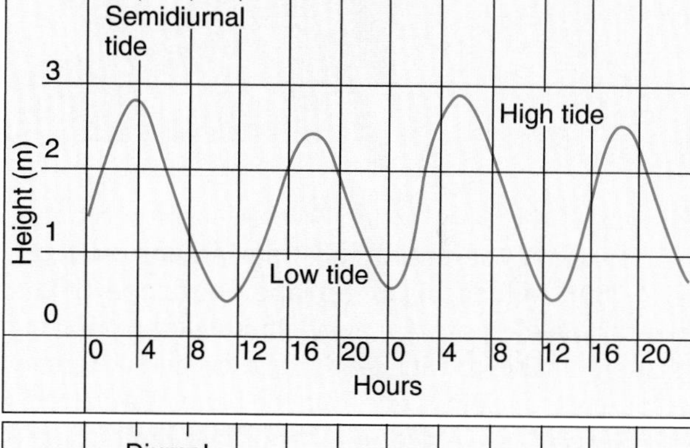

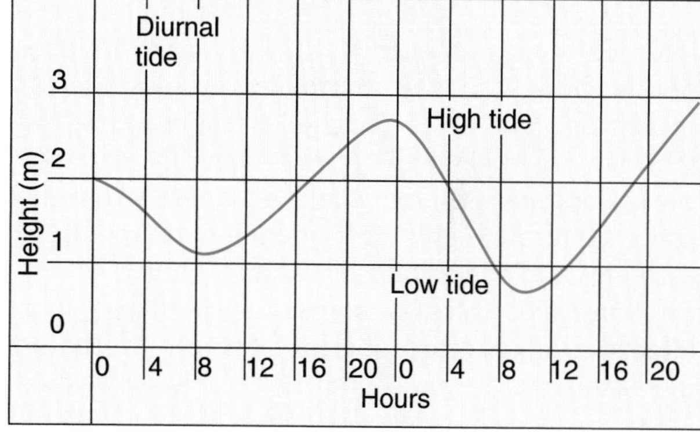

FIGURE 6.2 Mixed, semidiurnal, and diurnal tide curves. (Modified from *Introduction to Oceanography*, 4/e, D. A. Ross, 1988 p. 287, © 1988 Prentice Hall, Upper Saddle River, NJ.)

Temperature

Because of its inherent physical characteristics (see Chapter 1), water—particularly in a large body such as an ocean—shows a minimum range of temperature change. Furthermore, this range rarely exceeds the lethal limits of the organisms. Intertidal areas are, however, regularly subjected to aerial temperatures that may exceed lethal limits for varying periods. Even if death does not occur immediately, the organisms can be so weakened by the extreme temperatures that they cannot resume normal activities and will suffer mortality from secondary causes.

Temperature also has an indirect effect on marine organisms. They are subject to death by desiccation, which may be hastened by high temperatures.

Wave Action

In the intertidal zone, wave action exerts more influence on organisms and communities than in any other area of the sea. This influence is manifested directly and indirectly. Wave action affects shore life directly in two major ways. First, it has a mechanical effect that smashes and tears away objects. The many pictures and stories of the destruction of homes, harbors, and ships by storm waves are proof. The action is similar in the intertidal zone; hence, any creatures that inhabit this zone must be adapted to this persistent force in some way (see pp. 244–245). On shores of loose sand

FIGURE 6.3 Grunion spawning at high tide. (Photo courtesy of Jeff Foott.)

or gravel, the wave action is even more profound. Waves move the entire substrate around (see pp. 277–280), influencing the shape and slope of the beach. Wave exposure may be limiting for organisms that are unable to withstand the force, and necessary to others that cannot exist outside heavy wave areas.

Second, wave action extends the limits of the intertidal zone. It does this by throwing water higher on the shore than would normally occur as a result of the tide alone. This continual splashing allows the marine organisms to live higher in exposed waveswept areas than in sheltered areas within the same tidal range and often mediates temperature extremes and effects of desiccation (Figure 6.4).

Wave action also has some other, perhaps minor, effects. It mixes atmospheric gases into the water, increasing the oxygen content, so that wave-washed areas never lack oxygen. Because of the constant interaction with the atmosphere, formation of bubbles, and stirring of the substrate, waveswept areas may have decreased penetration of light. This is, however, of little ecological significance.

Salinity

Salinity changes may occur in the intertidal in two situations that may affect organisms. First, the intertidal may be exposed at low tide and subsequently flooded by heavy rains or runoff from heavy rains. This means organisms will be subjected to greatly reduced salinities. Since most intertidal organisms show a limited tolerance to decreased salinity, in extreme reductions of salinity the organisms may swell by osmosis and die. The second situation concerns tide pools, areas that retain seawater at low tide. These may either be flooded with freshwater runoff from heavy rains, thus reducing salinity, or they may show increased salinity due to evaporation during the day (see pp. 273–276).

Other Factors

Ice can have significant effects on intertidal organisms on far northern or southern shores. Wherever sea ice forms on the shore, or wherever large icebergs impinge on the intertidal zone, the result is catastrophic mortality by freezing and/or mechanical abrasion. Although the effect of sea ice on the intertidal is not well studied, Wethey (1985) has suggested that the presence of this disturbance might explain the absence from New England rocky shores of long-lived organisms otherwise common on intertidal shores unaffected by ice.

Different substrates—rock, sand, mud—have very different faunas and community structure in the intertidal, but these are discussed separately in this chapter.

ADAPTATIONS OF INTERTIDAL ORGANISMS

Since the intertidal organisms are primarily marine in origin, the adaptations we observe generally are for avoiding or minimizing the stresses of daily exposure to air. The major stress from the marine environment is wave action.

Resistance to Water Loss

As soon as marine organisms are moved from the water into air, they begin to lose water by evaporation.

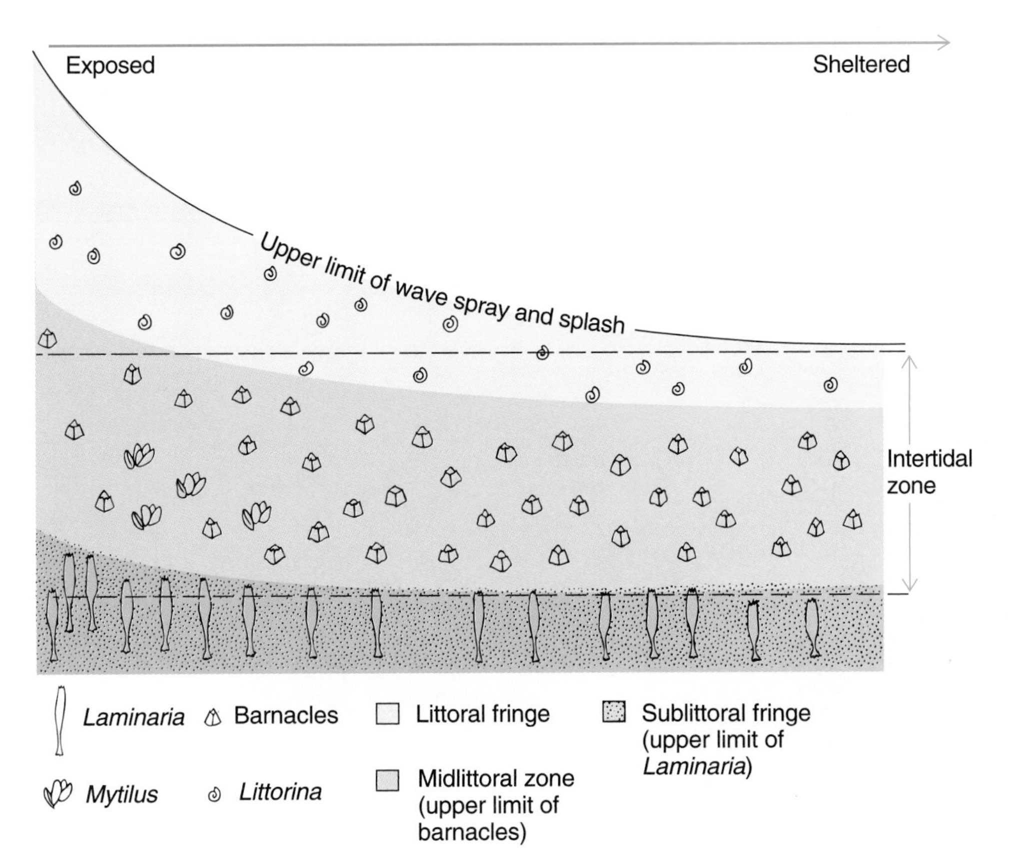

FIGURE 6.4 Changes in the extent of vertical zonation with change in exposure to wave action.

If they are to survive in the intertidal, this loss must be minimized, or the organisms must either have body systems that will tolerate considerable water loss during the hours of exposure to the air or have some mechanism to conserve body water loss until an external supply of water becomes available again.

The simplest mechanism for avoiding water loss is seen in mobile animals, such as crabs. These animals simply move from the exposed surface areas of the intertidal into very moist cracks, crevices, or burrows where water loss is reduced or absent. Alternatively, they may seek refuge under a covering of moist algae. These animals avoid the adverse environmental condition of the shore by actively selecting suitable microhabitats. A similar situation occurs in some species of anemones, such as *Anthopleura xanthogrammica* of the Pacific coast of North America (Figure 6.5). It is soft bodied with no way of preventing water loss. However, it normally is found only among barnacles or in crevices where water loss is reduced. Physiological adaptation is not needed. On the Atlantic coast, the anemone *Anthopleura varioarmatus* covers itself with shell fragments to shield itself from drying when exposed to air.

FIGURE 6.5 The large green Pacific coast anemone *Anthopleura xanthogrammica*. (Photo courtesy of Drs. Lovell and Libby Langstroth.)

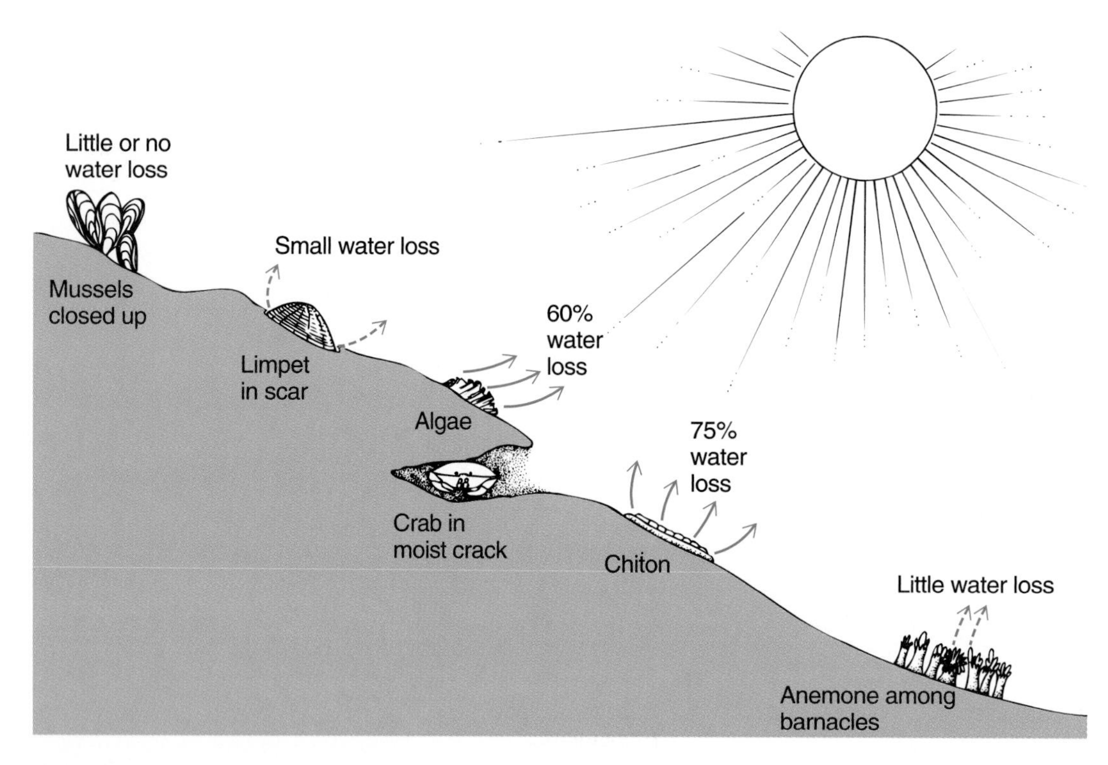

FIGURE 6.6 Diagrammatic representation of adaptations to water loss in intertidal organisms.

Another simple mechanism is that seen in several of the high intertidal algae genera, such as *Porphyra*, *Fucus*, and *Enteromorpha*. These plants cannot move and have no mechanism to avoid water loss. They simply are adapted to withstand a severe loss of water from their tissues. Specimens of these genera are dry and brittle after a long exposure at low tide, yet they will quickly take up water and resume normal body processes when the tide returns. Kanwisher (1957) found that they can tolerate as much as a 60–90% loss of water and still recover. A similar tolerance to great water loss is seen in some intertidal sedentary animals. Boyle (1969), for example, reported chitons can tolerate a 75% water loss (Figure 6.6). According to Davis (1969), limpets can lose 30–70% of their water, depending on the species.

Many species of intertidal animals, in contrast to those already mentioned, have mechanisms for preventing water loss. These mechanisms may be structural, behavioral, or both. Barnacles of many species are predominant members of the intertidal zone all over the world. These animals are sessile and avoid water loss by closing their shells at low tide. Here, the presence of an impermeable shell cuts down the evaporative water loss. Limpets of the genera *Patella*, *Acmaea*, *Lottia*, and *Collisella* are also dominant animals of the rocky intertidal. Certain species of these limpets have a "home scar" into which their shells fit exactly. At low tide, they return to these "homes." By fitting their shells into these grooves, they greatly reduce water loss (Figure 6.6). Other limpets that have no scar simply clamp down tightly against the rock so no tissue is exposed. Other gastropods, such as the periwinkles (*Littorina*), have opercula that completely seal off the aperture to their shells. At low tide, they may pull into the shell, closing the aperture with the operculum; thus, they reduce water loss. Certain bivalves, such as *Mytilus edulis*, and oysters, such as *Crassostrea virginica*, survive in the intertidal by closing their valves tightly to prevent water loss. Still other organisms, such as the anemone *Actinia* and the hydroid *Clava squamata*, produce mucus that reduces water loss. Many inhabitants of sand or mud simply burrow into the substrate to prevent desiccation.

Maintenance of Heat Balance

Intertidal organisms exposed to extremes of heat and cold show behavioral and structural adaptations to maintain their internal heat balance. Though deaths by freezing have been recorded for intertidal organisms, extreme low temperatures are less of a problem to shore organisms than high temperatures. This may be because intertidal organisms are often living much closer to their upper lethal temperature than to their lower lethal one. Hence, most mechanisms for heat balance concern the avoidance of too high temperatures. This may be accomplished by (1) reducing heat gain from the environment and (2) increasing heat loss from the body of the animal. Heat gain from the environment is reduced in several ways. One way is to have a relatively large body size as compared with similar species either lower in the intertidal or in the subtidal. Large body size means less surface area relative to volume and less area for gaining heat. At the same time, a larger body takes longer to heat up than a smaller one. It has been recorded for gastropod mollusks, such as *Littorina littorea* and *Olivella biplicata*, that the larger individuals occur higher in the tidal zone and the smaller ones lower. Yet another mechanism to reduce heat gain is to reduce the area of body tissue in contact with the substrate. This is, however, difficult to achieve for most intertidal animals, because they need the extra purchase provided by a large area of tissue in contact with the substrate to avoid being swept off by waves. There are some shore animals, however, that do have such a small area of attachment or none at all. These occur where wave action is not severe. For example, many high-shore tropical littorine snails attach to an overhang or ledge by means of a mucus thread, thereby eliminating all body contact with the substratum. In those animals that lose attachment, the ability to reattach as the tide comes in is well developed.

Heat loss is manifest in a number of adaptations. One mechanism, found in such hard-shelled organisms as mollusks, is greater elaboration of ridges and other sculpturing on the shell (Figure 6.7). These act as radiators and facilitate heat loss. Examples of such sculptured mollusks are commonly found in the tropics and include *Tectarius muricata* and *Nodolittorina tuberculata*. Also, dark-colored bodies gain and lose heat by radiation more rapidly than light-colored ones. Therefore, light-colored shelled gastropods lose heat more slowly than dark ones but they also heat up less as well, keeping cooler for a longer time when the sun shines on them. Many tropical and subtropical snails of the high intertidal, such as *Nerita peleronta* (Caribbean) and *Littorina unifasciata* (New Zealand), are much lighter in color than their lower-level relatives. Presumably, their coloring slows the heat gain.

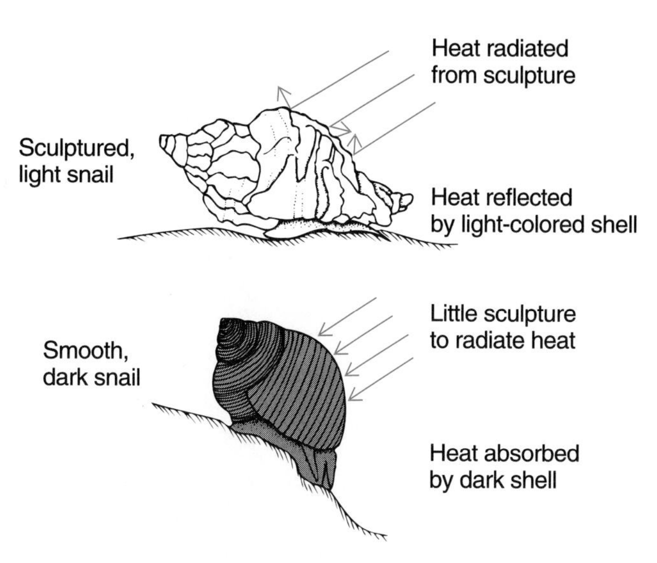

FIGURE 6.7 Differences in heat absorption between smooth, dark shells and sculptured, light shells.

Finally and most importantly, heat loss can occur through evaporation of water. Unfortunately, intertidal organisms must face the problem of desiccation and cannot afford to cool themselves through evaporation and thereby risk desiccation. Many intertidal organisms have adaptations that allow them to cool themselves by evaporating water while avoiding excessive desiccation (Segal and Dehnel, 1962). To facilitate this balance, many intertidal animals have an extra water supply for cooling. This extra water is held in the mantle cavity of barnacles and limpets, and it exceeds the amount the animal needs to survive desiccation. For example, the members of barnacle genus *Tetraclita* have a thick porous shell that traps water at high tide and slowly allows evaporation at low tide, providing evaporative cooling. Moreover, most high intertidal animals can withstand a remarkable amount of desiccation (see previous section). In addition, many animals can reduce the rate of water loss from their tissues when necessary. Those animals living in intertidal sedimentary environments escape most of the problems of water loss by living in the substrate that retains water even at low tide.

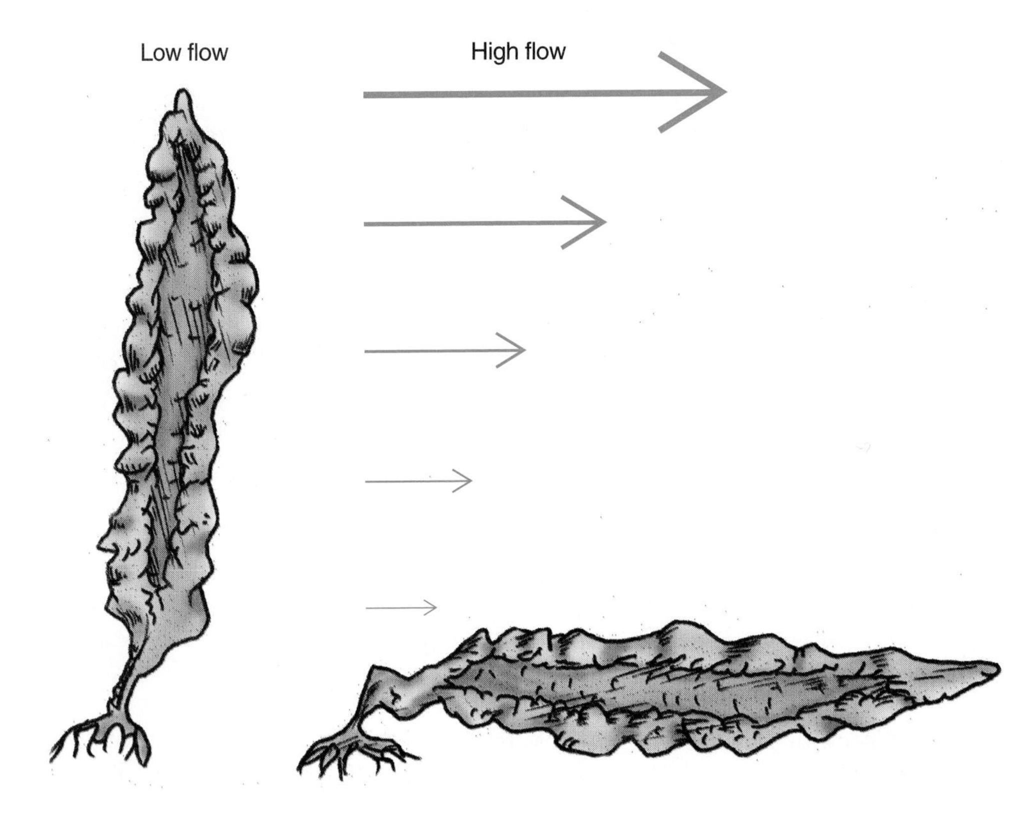

FIGURE 6.8 A large seaweed coping with high wave energy by bending into the flow, thus reducing area and drag to the wave. (Modified from *The Ecology of Atlantic Shorelines*, M. D. Bertness, p. 220. Copyright © 1999 Sinauer Associates, Inc. Reprinted by permission.)

Mechanical Stress

Wave action reaches a zenith in the intertidal zone. As a result, it is necessary for any organism that lives intertidally to adapt to resist the smashing and tearing effects of waves. Wave action has different effects on rocky shores than on sandy shores and requires different adaptations, discussed later. We discuss here some adaptations of dwellers on rocky substrates.

Perhaps the most common adaptation to wave stress among animals is the limitation of size and shape. Most common organisms that frequent the splash zone have small, squat bodies with streamlined shapes that minimize their exposure to the lift and drag of wave forces. This can be seen in animals such as limpets, barnacles, and chitons that are common in the rocky midintertidal. As animals increase in size, they become more vulnerable to wave destruction or removal unless they compensate with additional means of attachment. Unlike the animals, the intertidal algae are often large and cope with wave stress by being flexible and bending toward the substrate as the waves pass over, thereby minimizing the surface area in contact with the waves (Figure 6.8).

Another adaptation, observed in barnacles, oysters, and serpulid polychaete worms, is to become fixed to the substrate. Similarly, the majority of algae in the intertidal are attached via a holdfast to the bottom.

Other organisms develop strong but temporary attachments that give them limited movement. An example of such an attachment is the byssal threads of *Mytilus*, which anchor the animals securely but still can be broken and remade to allow limited, slow movement.

Most intertidal mollusks are adapted to wave shock by possessing thicker shells than their subtidal relatives. They also lack delicate sculpturing, which is easily broken in the waves. The dominant intertidal mollusks, such as the various limpets and chitons, resist wave action through a strong, enlarged foot that clamps them to the substrate and a shell that is low and flat, offering little resistance.

Motile organisms, such as crabs, have no structural mechanism to resist being swept away and survive only because they seek shelter from the waves in crevices or under rocks. Many, such as grapsids, are adapted for living in waveswept areas by being flattened, providing less "edge-on" resistance to waves.

Respiration

Since most animals inhabiting the intertidal are marine, they have respiratory surfaces, gills, adapted to extract O_2 from water. Usually, these are thin-walled extensions from the body surface. These respiratory organs are very susceptible to desiccation in air and usually do not function well unless immersed in water. Such organs would be at a disadvantage during aerial exposure in the intertidal. Among intertidal animals, there is a tendency to enclose the respiratory surfaces in a protective cavity to prevent them from drying. This is seen particularly in various mollusks where the gills are in a mantle cavity, which itself is protected by the shell. Another adaptation, especially of high intertidal gastropods, is the reduction of the gill and formation of a vascularized mantle cavity, which then serves as a lung for aerial respiration. A similar situation prevails in barnacles where the mantle tissue acts as a respiratory organ.

Because animals with protected respiratory organs must also conserve water at low tide, they often close up (operculum) or clamp down (chitons, limpets) such that gaseous exchange is reduced. To conserve O_2 and water, most animals are quiescent during low tide.

Intertidal fishes often are specialized for cutaneous respiration through the reduction of gills and the proliferation of blood vessels in the skin. Bridges (1988) suggests that many fishes may satisfy over half their oxygen needs by gas exchange through the skin.

Feeding

All intertidal animals must expose the fleshy parts of their bodies to feed. For those on hard substrates, this means exposing those parts most susceptible to desiccation. As a result, virtually all diurnal rocky intertidal animals are active only during the time the tide is in and they are covered with water. This is true whether the animals are grazers, filter feeders, detritus feeders, or predators. By contrast, organisms living in soft substrates are often protected by the water-filled substrate and may be active feeders at low tide. Nocturnal animals may also be active at low tides occurring at night.

Salinity Stress

As previously noted, the intertidal zone may be flooded by fresh water, which would create an osmotic stress on the intertidal organisms adapted only to seawater. This is particularly true for those intertidal organisms lacking adaptations to tolerate salinity changes, such as those seen in estuarine organisms (see Chapter 8). Intertidal organisms without mechanisms to control the salt content of their body fluids are **osmoconformers**. In such organisms adaptations to prevent salinity stress are the same as those for preventing desiccation, such as closing up valves or shells in barnacles and mollusks. Perhaps this is why there are records of catastrophic mortality of intertidal organisms following heavy rain or runoff. Apparently, such events are so rare or unpredictable that special osmoregulatory mechanisms have not evolved. On the other hand, tropical tide pools can reach salinities greater than 100 psu; this may be intolerable for some intertidal animals, but others, such as the gastropods *Batillaria* and *Nodolittorina*, are found abundantly in such extreme habitats.

Reproduction

Because so many of the intertidal organisms are sedentary or sessile, to disperse the species they must rely on fertilized eggs or larvae that are free-floating in the plankton for dispersal. As a result, we see a very high incidence of planktonic larvae among intertidal organisms. These larvae are of many types (see Figure 2.23 on meroplankton).

A second reproductive adaptation to the intertidal position is that most organisms' breeding cycles are synchronized with the occurrence of certain tides, such as spring tides, to ensure fertilization. An example is *Mytilus edulis*, in which gonads mature during periods of spring tides and spawning occurs on subsequent neap tides. In *Littorina neritoides*, the eggs are spawned during spring tides. In these cases, where the animals have external fertilization, synchronized spawning of all the individuals ensures the maximum number of gametes in the water at the same time to facilitate fertilization.

Semiterrestrial shore animals, such as the Caribbean crabs *Gecarcinus* and *Cenobita*, migrate from the shore to the sea to deposit their eggs.

ROCKY SHORES

Of all the intertidal shores, those composed of hard material, the rocky shores—particularly those of the temperate zones—are the most densely inhabited by macroorganisms and have the greatest diversity of animal and autotroph species. They contrast sharply with the almost barren appearance of the surface of sand and mud shores. It is these densely populated, topographically diverse, and species-rich rocky areas that have fascinated marine biologists and ecologists for many years. In the last 30 years, these areas have been the subject of many classic studies that have enhanced our understanding of how these associations of species interact to maintain or change the community. It should be noted at the outset that these intertidal rocky areas differ fundamentally from the subtidal rocky areas discussed in Chapter 5. Most of the dominant organisms of intertidal rocks are solitary or clonal animals, whereas subtidal rocks seem dominated by colonial encrusting animals.

Atlantic and Pacific Rocky Shores

There are significant differences in the geographical, physical, and biological factors shaping intertidal associations and communities among rocky intertidal areas in different parts of the world. These differences are perhaps best documented between the temperate Atlantic and Pacific rocky intertidal habitats of North America. These bear mentioning here because they contribute to differences in the organization and persistence of the intertidal communities in the two oceans as discussed later in this chapter. In the first place, the temperate Atlantic coast generally lacks rocky intertidal areas south of Cape Cod, where sedimentary intertidal areas dominate. In contrast, the Pacific coast intertidal is dominated by rock from California up to Alaska and sedimentary intertidal areas are uncommon. Thus, rocky shores occur over an extensive range of latitude on the Pacific coast but are confined to northern latitudes in the east. Second, the range of aerial temperatures over the Pacific coast is much narrower than over the Atlantic, both on a seasonal and daily scale. Thus, intertidal organisms in New England and eastern Canada can be exposed to winter temperatures as low as −20°C and summer temperatures of nearly 40°C, whereas on the Pacific coast, even in Canada and southern Alaska, the temperatures rarely go much below freezing in winter and persistent cloud and fog cover in summer reduce any potentially high temperatures. As a result, Atlantic organisms may be regularly killed by seasonal freezing or ice action in winter, which does not occur in the Pacific.

The species diversity in the intertidal of the Atlantic is lower than that of the Pacific. This is probably because during the ice ages the continental glaciers scoured the Atlantic shoreline, causing widespread extinction of the rocky intertidal organisms since the lack of suitable substrate prevented the populations from retreating southward. Since the glaciers retreated only 10,000 to 20,000 years ago, there has been insufficient time for much speciation. The Pacific intertidal, by contrast, was spared such extinctions and has a history of occupation by organisms that is much older geologically, allowing for more speciation.

In addition to differences in diversity, there are also differences in numbers of species. For example, neither chitons nor limpets are common herbivores in the temperate Atlantic, as they are in other areas. Whereas the Pacific coast has ten or more intertidal limpet species, the Atlantic has but two; in Puget Sound there are five common chitons of large size in the intertidal and these play an important ecological role, whereas in New England there are but three chitons, all small, uncommon, and primarily subtidal.

Finally, the dominant herbivore of the Atlantic intertidal, the littorine *Littorina littorea*, and one of the dominant predators, the green crab, *Carcinus maenus*, are both introduced species from Europe, whereas no introduced species dominates any Pacific coast intertidal. This suggests that it may no longer be possible to understand how the intertidal communities were organized before the advent of the Europeans on the eastern seaboard of North America.

Zonation

One of the most striking features of any rocky shore anywhere in the world at low tide is the prominent horizontal banding or **zonation** of the organisms. Each zone or band is set off from those adjacent by differences in color, morphology of the major organism, or some combination of color and morphology. These horizontal bands or zones succeed each other vertically as one progresses up from the level of the lowest low tides to true terrestrial conditions (Figures 6.9 and 6.10). This zonation on intertidal rocky shores is similar to the zonation pattern one observes with increasing elevation on a mountain, where the different horizontal zones of trees and shrubs succeed each other vertically until, if one progresses far enough, permanent snow cover is reached. The major difference between these two areas is the scale. Mountain zones are perhaps kilometers in extent as opposed to intertidal zones extending at most a few meters vertically.

Rocky intertidal zones vary in vertical extent, depending on the slope of the rocky surface, the tidal range, and the exposure to wave action. Where there is a gradual slope to the rock, individual zones may be broad. Under similar tidal and exposure conditions on a vertical face, the same zones would be narrow. In the same manner, exposed areas have broader zones than protected shores, and shores with greater tidal ranges have broader vertical zones (see Figures 6.4 and 6.10).

Of course, these striking bands may be interrupted or altered in various places wherever the rock substrate shows changes in slope, composition, or irregularities that alter its exposure or position relative to the prevailing water movement.

The fact that these prominent zones can be observed on nearly all rocky shores throughout the world under many different tidal regimes led Stephenson and Stephenson (1949) to propose, after some 30 years of study, a "universal" scheme of zonation for rocky shores (Figure 6.11). This universal scheme was really a framework using common terms that would allow comparison of diverse areas. It established zones based on the distributional limits of certain common groups of organisms and not on tides. It reflects the knowledge of the Stephensons, and other intertidal ecologists, that distribution patterns of the organisms and zones vary not only with tides, but with slopes and exposure. Therefore, under similar tidal conditions there could be different bandwidths due to different exposures or slopes of rocks. It was this universal scheme that established a standard format for describing shore zonation, replacing a bewildering host of schemes and names established by earlier biologists.

The Stephensons' scheme has three main divisions of the intertidal area. The uppermost is termed the **supralittoral fringe**. Its lower limit is the upper limit of barnacles, and it extends to the upper limit of snails of the genus *Littorina* (periwinkles). The dominant organisms are the littorine snails and black encrusting lichens (*Verrucaria* type). The extreme high water of spring tides reaches part

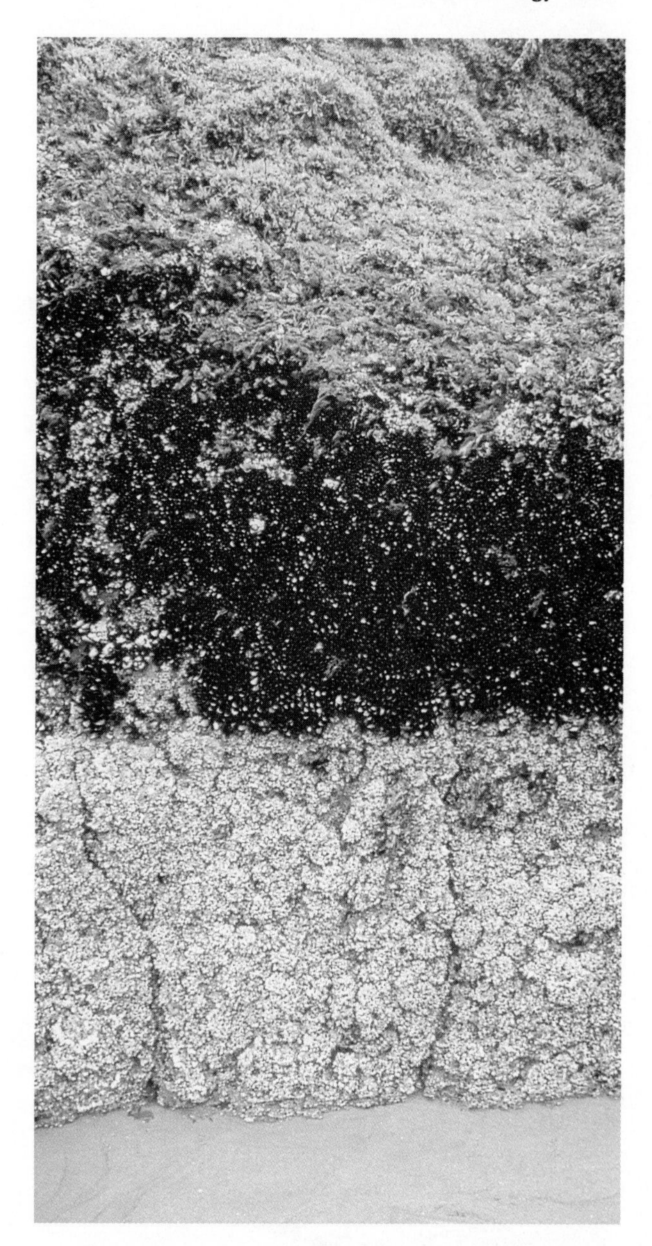

FIGURE 6.9 An intertidal area at low tide, illustrating the conspicuous banding or zonation of organisms. (Photo courtesy Norbert Wu.)

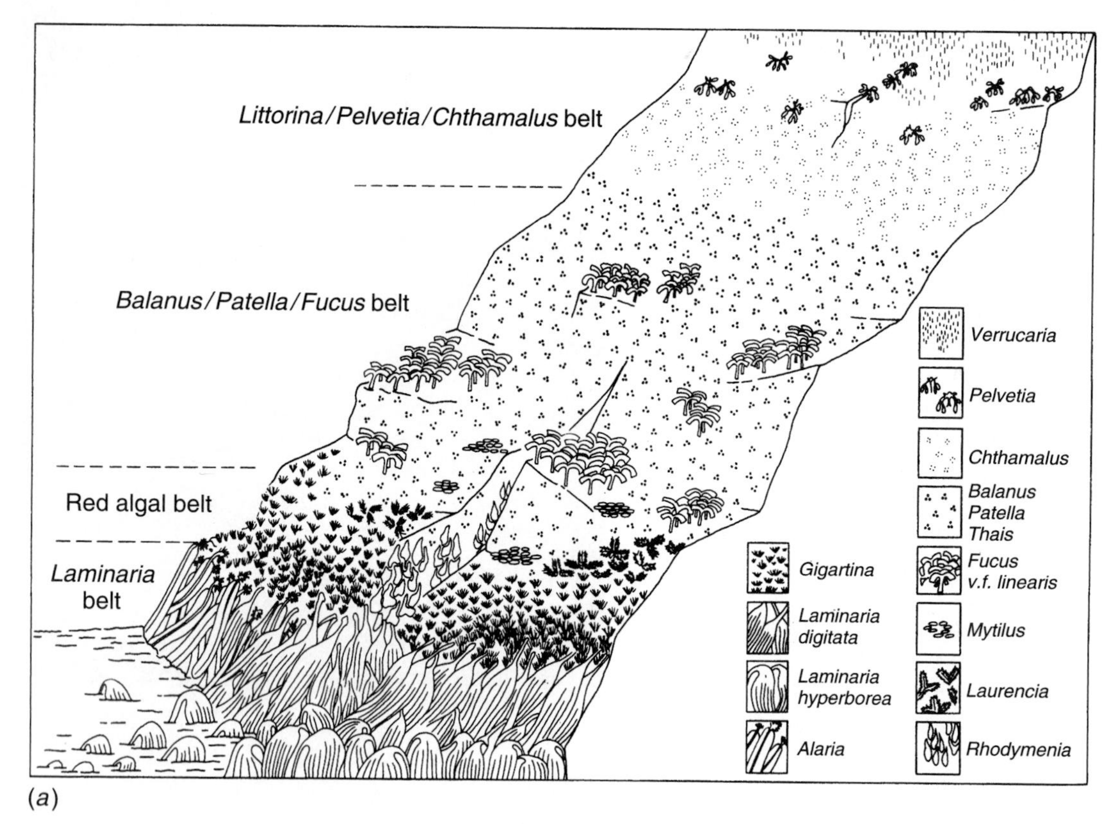

(*a*)

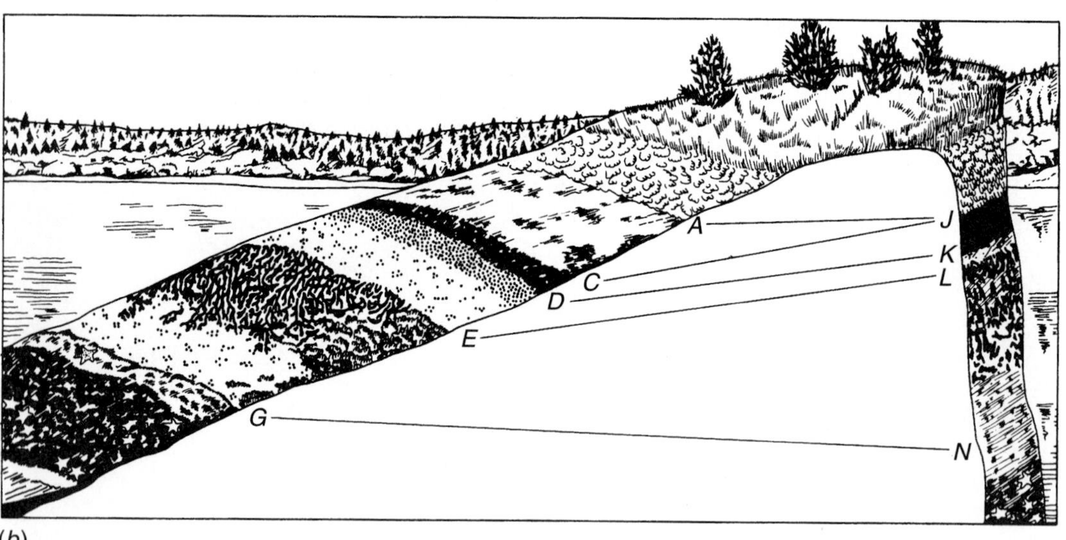

(*b*)

FIGURE 6.10 Zonation in the rocky intertidal. (a) An intertidal area in Scotland. (b) An intertidal area in British Columbia. Note in b the difference in the extent of zones. In b, line A–J indicates the lower limit of lichens; line C–J the upper limit of the "black" zone; line D–K, the upper limit of barnacles; line E–L, the upper limit of *Fucus*; line G–N, the lower limit of barnacles. (a, from J. R. Lewis, 1964, *The Ecology of Rocky Shores*, English Universities Press. b, from *Life Between the Tidemarks on Rocky Shores* T. A. Stephenson and A. Stephenson, © 1972 W. H. Freeman.)

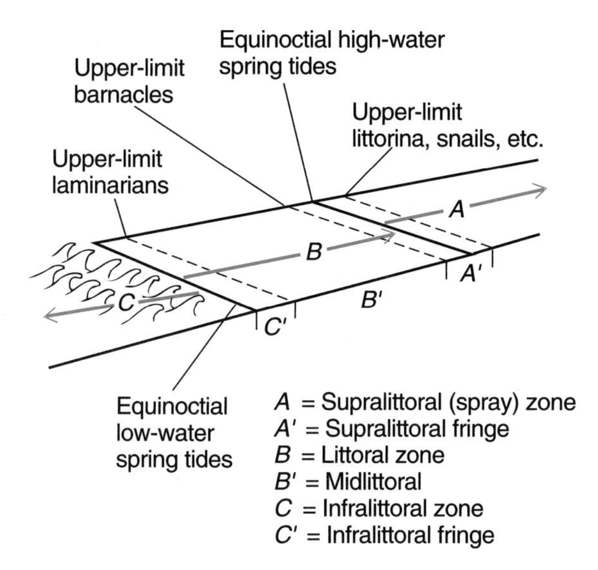

FIGURE 6.11 Stephensons' universal scheme of zonation for rocky shores. (From *Ecology and Field Biology*, 4/e, Robert L. Smith, p. 900. Copyright © 2000 Addison Wesley. Reprinted by permission of Addison Wesley Educational Publishers, Inc.)

of this zone, but most of its water comes from wave splash. Above this zone is the maritime terrestrial or supralittoral zone.

The middle part of the intertidal is termed the **midlittoral zone** and is the broadest in extent. Its upper limit coincides with the upper limit of barnacles, while its lower limit is the point where large kelps (*Laminaria*, etc.) reach their uppermost distribution. This zone is often subdivided and contains a host of different organisms. Perhaps the only universally present dominant group are the barnacles.

The lowest zone of this scheme is the **infralittoral fringe**, which extends from the lowest low tide up to the upper limit of the large kelps. This is an extremely rich zone composed of organisms that can tolerate only limited exposure to air. It is really an intertidal extension of the **infralittoral zone** (the Stephensons' term) or what we know as the **sublittoral** area.

Although the Stephensons' scheme did set forth a means for describing zonation on rocky shores, it does not offer an explanation of why the zonation occurs. It is this explanation of zonal patterns that intrigues many marine biologists.

Causes of Zonation

Whereas it is fairly easy to recognize and measure the extent of the zones on a rocky shore, it is more difficult to find suitable explanations for why organisms are distributed in these zones. Physical and biological factors can be considered to explain the phenomenon. We shall take up each in turn.

Physical Factors The most obvious explanation for the occurrence of the zones is that they are a result of the tidal action on the shore and reflect the different tolerances of the organisms to increasing exposure to the air and the resultant desiccation and temperature extremes. One difficulty with this explanation is that the rise and fall of a tide tend to follow a smooth curve with no obvious sharp breaks corresponding to the often sharp boundaries observed in the intertidal zones. If, however, one observes a whole series of such tidal curves, such a series of breaks does become apparent. For example, Figure 6.12a gives a typical tidal curve for a mixed tide area on the California coast. Figure 6.12b gives the maximum time of continuous submergence for various tide levels. The graph shows that there are certain points on this curve that reflect sharp increases in exposure to air. This can also be deduced from Figure 6.12a, which represents a single typical lunar tide cycle. Consider an organism at point X on the graph. At this point, the tide will cover the organism at least once every 6 hours. If, however, the organism moves up only a few inches, the tide will not cover it for 12 hours. There is a great change in exposure time with a very short vertical movement. Points exhibiting sharp increases in exposure time over short distances have been termed **critical tide levels** by Doty (1946) and were offered as one of the early explanations for the zonation patterns described previously. In this explanation, it is important to remember that it is not the tide alone that causes the limit but the fact that, at these critical points, the organisms are subjected to greatly increased time in the air; hence, they experienced greater temperature fluctuations and desiccation.

The critical tide hypothesis has been tested in various places by several scientists since its original promulgation by Doty in 1946. In general, it is difficult to find good correlations, particularly at the lower tide levels. This lack of good correlation can be partly attributed to the diverse topography of the different shores and to the variation in exposure. Thus,

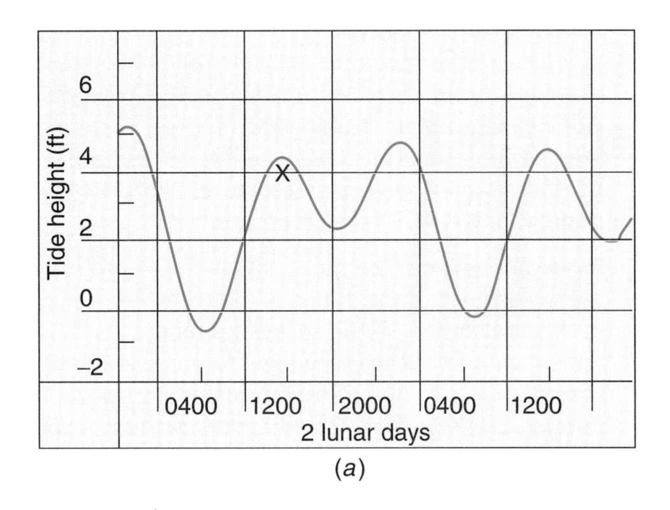

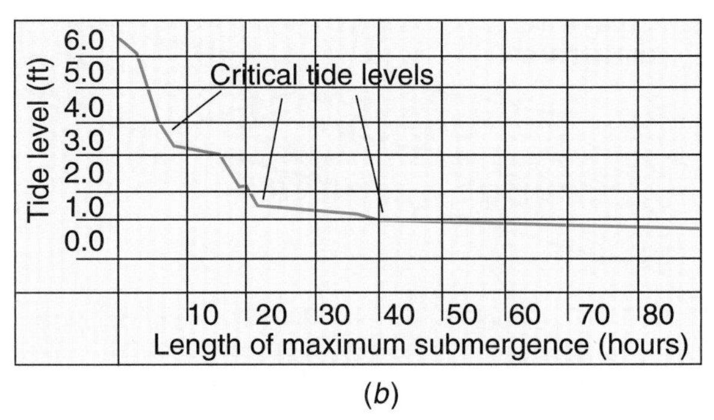

FIGURE 6.12 Typical tidal curve for a mixed tide area on the Pacific coast of North America. (b) Maximum time of continuous submergence for various tide levels on the Pacific coast of North America. (b from "Critical Tide Factors that are Correlated with the Vertical Distribution of Marine Algae and Other Organisms Along the Pacific Coast" Maxwell S. Doty, *Ecology*, Vol. 27, No. 4, October 1946, p.320, Figure 4. Reprinted by permission of the Ecological Society of America.)

a species may be able to exist above the critical tide level if the rocky shore is exposed to persistent violent wave action, which would throw water up higher and decrease desiccation. Similarly, caves, overhangs, and crevices remain moist when exposed areas are dried out; thus, organisms are able to persist above the critical tide level. The effects of differences in topography and exposure suggest, however, that the ultimate upper limits are set by certain physical factors: desiccation and temperature.

In the Gulf of Maine on the most wave-exposed headlands, the zonation is simple. The uppermost zone is dominated by the barnacle *Semibalanus balanoides* because it is the only primary space holder able to survive there (Figure 6.13a). *Semibalanus* is replaced at intermediate heights by *Mytilus* because the latter is competitively dominant. This zone grades into a kelp zone at the lowest level where the wave action precludes the presence of urchins (Figure 6.13a). This simple pattern is primarily due to the heavy wave action, which excludes most other organisms including the crabs, fishes, urchins, and snails that are the primary consumers of the dominant organisms. The situation is quite different on wave-protected shores (Figure 6.13b). On these shores below the barnacle zone, there is a profusion of algae, a number of mobile consumers, and few mussels. The lack of wave action allows the larger but unpalatable algae like *Ascophyllum* and *Fucus* to survive, forming a canopy that in turn protects a large assemblage of organisms from desiccation, thereby allowing them to persist. This assemblage includes herbivores such as *Littorina littorea* and carnivores such as the snail *Nucella*, the starfish *Asterias*, and crabs. These consumers limit the abundance of certain algae, as well as mussels and barnacles, so that there is often considerable open space (Figure 6.13b).

We have previously noted that desiccation is a serious problem. Several natural observations, as well as field experiments, have suggested that desiccation can set the upper limits to organisms and zones. In the North Temperate Zone, for example, rocks that have a north-facing slope often have the same organism occurring higher than do adjacent south-facing slopes. Since such rocks experience no difference in wave exposure or tides, the only reason for the height difference is that north faces dry out more slowly than south faces. Similarly, field experiments by Frank (1965) and others in which seawater was slowly dripped down slopes have resulted in the organisms extending higher up (Figure 6.14). In other experiments, organisms were transplanted above their normal positions. Notable in this respect are barnacles, which cannot move. When transplanted by Foster (1971) above their nor-

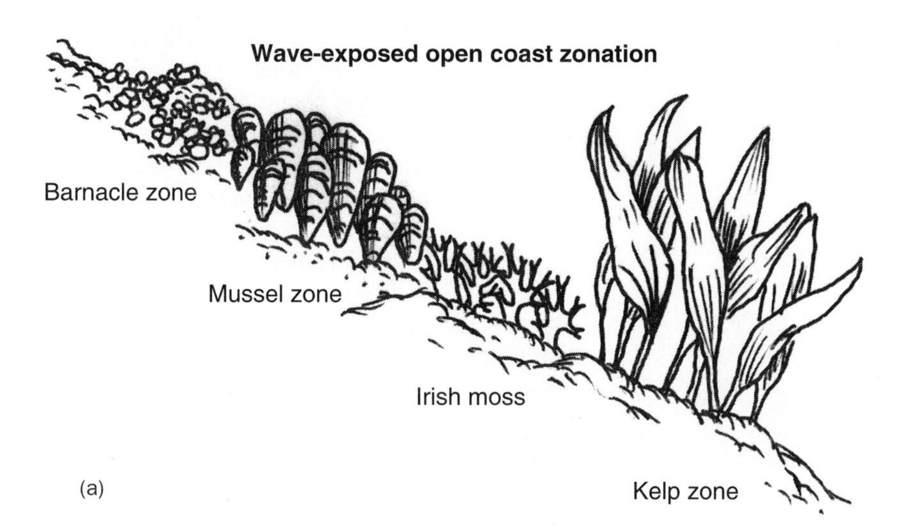

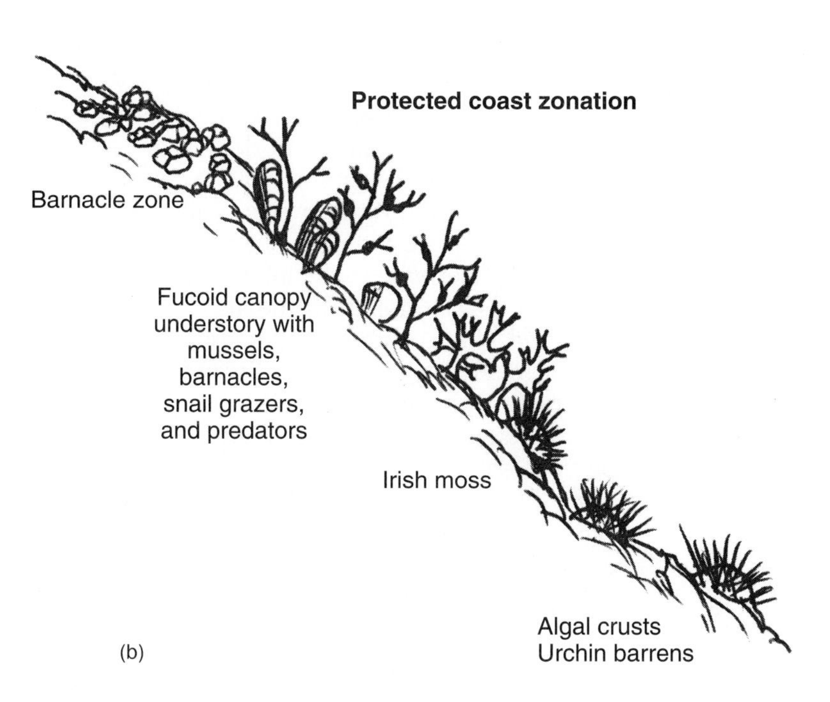

FIGURE 6.13
(a) Typical zonation on a wave-exposed open coast in New England. (b) Typical zonation on a wave-protected coast in New England. (Modified from *The Ecology of Atlantic Shorelines*, M. D. Bertness, p. 225. Copyright © 1999 Sinauer Associates, Inc. Reprinted by permission.)

mal tidal height, they died. Younger barnacles died more quickly than older ones. Again, desiccation can be suspected.

Often acting in concert with desiccation is temperature. Aerial temperatures have ranges that may exceed the lethal limits; hence, intertidal organisms may die from either freezing or "cooking." Upper limits of zones in part may be attributed to the tolerable temperature limits of the intertidal organisms. In addition, high temperature promotes desiccation, and the synergistic effect of these factors may be even more devastating than each acting alone. A good example of the importance of temperature comes from the New England intertidal. Here two barnacles, *Chthamalus fragilis* and *Semibalanus balanoides*, occur in the intertidal. *Chthamalus* occupies the high intertidal because it can tolerate higher temperatures than *Semibalanus*, but *Semibalanus* is competitively dominant over *Chthamalus* and outcompetes it throughout the rest of the intertidal.

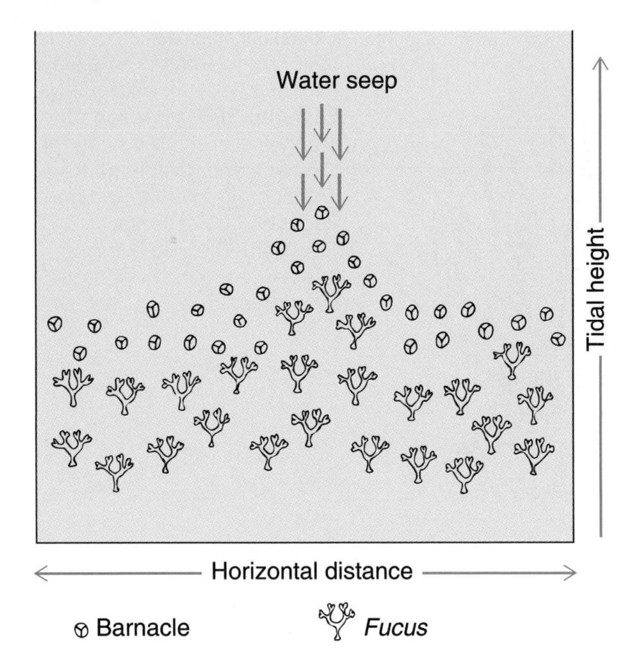

FIGURE 6.14 Zonation change in the presence of a water seep.

Wethey (1984) demonstrated that if *Semibalanus* was shaded to reduce the thermal stress in the uppermost intertidal, it would outcompete *Chthamalus* and take over (Figure 6.15). Indeed, north of Cape Cod the summer temperatures never become high enough to exclude *Semibalanus*, and it displaces *Chthamalus* throughout the high intertidal. At the other end of the temperature spectrum, differences in tolerance to freezing may also play a role in zonation. In New England rocky habitats the alga *Chondrus crispus* dominates many subtidal areas, but it is killed by freezing temperatures, which appear to limit the alga to subtidal areas. In contrast, the alga *Fucus* is common in the intertidal but suffers little mortality from freezing temperatures or encasement in ice. Of course, ice can also scour organisms from the rock, thus leaving many far northern Atlantic shores in a constant state of recovery.

Finally, sunlight may act adversely to limit organisms on the shore. Sunlight includes wavelengths in the ultraviolet (UV) region that can harm living tissue. Water absorbs these wavelengths and protects most marine animals. However, intertidal animals have direct exposure to such rays at low tide. The higher an organism is in the intertidal, the greater the exposure to these rays. At present, there is little information concerning whether UV radiation controls distribution of organisms, but the recent finding of UV-absorbing compounds in shallow-water organisms suggests its importance. Light also has been suggested as a regulator of the distribution of intertidal algae. This has to do primarily with the spectral quality of the light. As noted in Chapter 1, different wavelengths of light are absorbed differentially by water. In the case of intertidal algae, those needing light of longer wavelength (reds), which is absorbed most quickly by water, would tend to be found higher in the intertidal. When submerged, they would not be too deep for the penetration of red light (about 2 m). Since the main intertidal algae belong to three different groups—reds, browns, and greens—each of which has a slightly different absorption spectrum, it might be thought that they would be arranged along a depth gradient. In such a gradient, the green algae would be expected to be the highest, because they absorb mainly red light; the brown next; and the red algae deepest, because they absorb mainly the deep, penetrating green light. This, however, is not the case. The intertidal algae are a mixture of all types at most levels. This is likely due to the interaction of other factors and the physiology of the algae. It points out again the fallacy of attempting to explain patterns of species distribution with only single factors when a multitude are acting at all times.

Although there is a large body of evidence that physical factors are the strongest determinants of the upper limits of organisms, there may be alternative explanations, many of which have not been investigated. Underwood and Denley (1984) point out that other processes are at work in the high intertidal that may be confused with physical effects. For example, some animals in the high intertidal, particularly sessile filter feeders, may be killed by starvation during long periods of calm weather because they are not covered with water long enough to feed. Similarly, grazing animals may be absent, not because they cannot tolerate the physical factors, but because their algal food is absent, perhaps killed by severe physical factors. Finally, in the case of larvae settling out of the plankton, they may choose not to settle in the area, in which case the upper limit is set by processes affecting larval settlement. In one instance reported by Underwood (1980), the upper limit of several algal species in Australia was directly due to

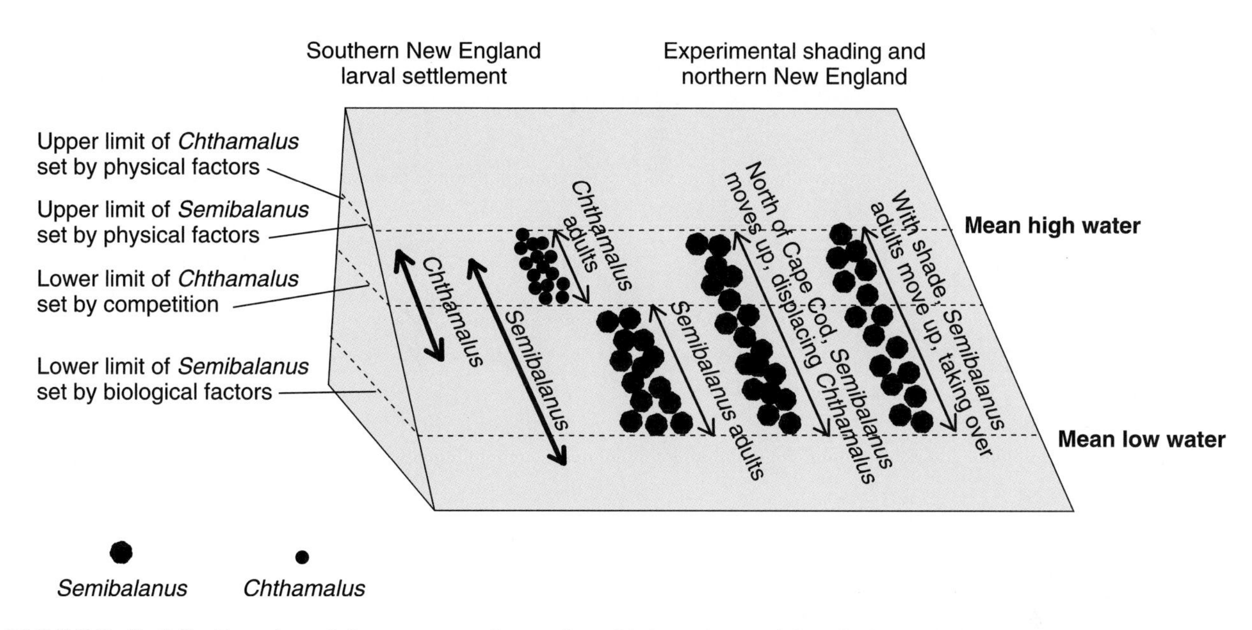

FIGURE 6.15 Zonation of the two acorn barnacles, *Chthamalus* and *Semibalanus*, in New England under different conditions. (Modified from *The Ecology of Atlantic Shorelines*, M. D. Bertness, p. 216. Copyright © 1999 Sinauer Associates, Inc. Reprinted by permission.)

grazing by gastropods. In each of these cases, the upper limits of the organisms are the result of biological and physical factors acting either independently or together.

■ We can summarize by stating that, although much current evidence suggests that physical factors are strong determinants of the upper limits of the distribution of intertidal organisms, other explanations are viable, and still others remain to be investigated. ■

Biological Factors Although the early work in the intertidal focused on the importance of physical factors in setting the zonal patterns, more recent work has begun to clearly establish the great importance of a number of biological factors in setting various observed distribution patterns. In general, these biological factors are more complex, often subtle, and closely linked to other factors. This is probably why we have only recently begun to understand how they act. The major biological factors are **competition, predation, grazing** (herbivory), and **larval settlement** (**recruitment**). We shall discuss them in order.

Competition for a certain resource does not occur if the resource is so plentiful that adequate supplies of it are available for all species or individuals. In the rocky intertidal zone, one resource commonly in limited supply is space. This is perhaps the most restricted area in the marine environment; at the same time, it is densely populated, at least in the temperate zone. As a result, there is an intense competition for space that has resulted in observed zonal patterns.

On the intertidal shores of Scotland, there is a distinct zonation of barnacles, with the small *Chthamalus stellatus* living in the highest zone and the larger *Semibalanus balanoides* occupying the major portion of the midintertidal. Studies done by Connell (1961) showed that *Chthamalus* larvae settled out throughout the zones occupied by both barnacles but survived to adulthood only in upper zones. The disappearance in the midlittoral region was due to competition from *Semibalanus balanoides*, which either overgrew, uplifted, or crushed the young *Chthamalus*. *Semibalanus* was prevented from completely eliminating *Chthamalus*, because unlike *Chthamalus*, *Semibalanus* did not have the tolerance to drying and high temperatures that prevailed at the higher tidal levels; hence, it could not survive there. Here, then, is a case where the

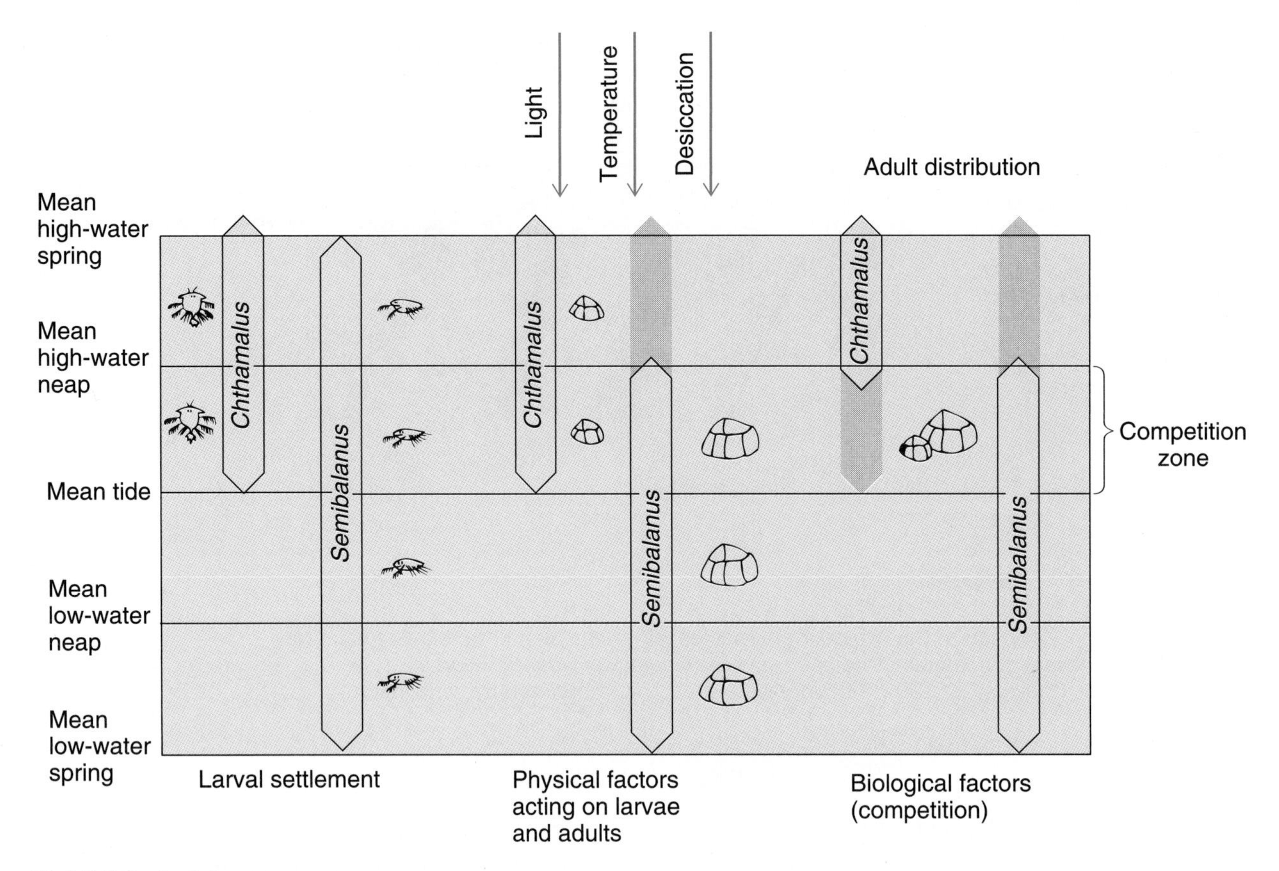

FIGURE 6.16 Intertidal zonation as a result of the interaction of physical and biological factors. The larvae of two barnacles, *Chthamalus stellatus* and *Semibalanus balanoides*, settle over a broad area. Physical factors, mainly desiccation, then limit survival of *S. balanoides* above mean high water of neap tides. Competition between *S. balanoides* and *C. stellatus* in the zone between mean tide and mean high water of neap tides eliminates *C. stellatus*.

zonation is at least partially a function of biological competition (Figure 6.16).

Intertidal algae, particularly those on temperate shores, also often show abrupt limits to their upper and lower distribution. Earlier, these limits were ascribed to critical tide levels, but it has also been suggested that this could be due to competition for space or access to light. In two experimental studies done on the Pacific coast of North America, evidence of competition was found. Dayton (1975) found that the dominant kelps—*Hedophyllum sessile*, *Laminaria setchelli*, and *Lessionopsis littoralis*—all outgrew and outcompeted certain smaller species in the lower intertidal. These smaller species were generally fast-growing species, which quickly colonized open areas. As noted in Chapter 1, they are **opportunistic** or **fugitive species**. Among the three dominants, *Hedophyllum* was outcompeted by the other two, and they dominated the areas. The second study was done in the subtidal, where Vadas (1968) found that the giant kelp *Nereocystis* outcompeted and overgrew the brown alga *Agarum*. On the New England coast, studies of tide pool algae by Lubchenco (1978) indicated that *Enteromorpha intestinalis* was a dominant space competitor as opposed to *Chondrus crispus*. In the absence of grazers, *E. intestinalis* would quickly outcompete *Chondrus* and take over the space. These few studies suggest that competition among algal species may be more widespread than originally assumed and may be a fertile ground for future ecological work.

Competition among the mussel *Mytilus californianus* and several species of barnacles on the Pacific coast of Washington is a more complex example. In this case, studies by Dayton (1971) and Paine (1966, 1974) have shown that *M. californianus* is the dominant space competitor on open coast shores. Given enough time and freedom from predators, the M. *californianus* eventually overgrew and outcompeted all other macroorganisms and took over substrates throughout most of the midintertidal. *Mytilus californianus* takeover is, however, slow. Wherever open space occurs, it may be rapidly colonized by other organisms, including three species of barnacles: *Balanus glandula*, B. *cariosus*, and *Pollicipes polymerus*. These, in turn, displace any rapidly growing algal species. The barnacles persist only until the mussels enter. The mussels outcompete and destroy the barnacles by settling on top of them and smothering them. Since nothing appears to be large enough to settle and smother the M. *californianus*, they remain in control of the intertidal space. Given this competitive edge, it would appear that eventually the rocky Pacific coast of Washington could be a monotonous band of M. *californianus*. It is also curious that M. *californianus* forms dense clumps or bands only in the intertidal, though it is perfectly capable of living subtidally. Since it is a premier space competitor, why this abrupt lower limit? The reasons for this have to do with other biological factors that prevent such resource monopolization, namely **predation.**

The role of predators in determining the distribution of organisms in the intertidal and the zonal patterns is best documented for the Pacific coast of Washington and is discussed here as an example of how complex biological interactions create the prevailing distributions.

The dominant abundant intertidal animals on the Pacific coast other than the space-dominating *Mytilus californianus* are the barnacles *Balanus cariosus* and B. *glandula*. These latter two species occur abundantly in the intertidal region even though they are competitively inferior to M. *californianus*, because a predatory starfish, *Pisaster ochraceus*, preferentially preys on M. *californianus*, preventing it from completely overgrowing the barnacles (Figure 6.17). *Pisaster ochraceus* is a voracious predator of mussels, consuming them at a rate that prevents them from occupying all the space.

At the same time, *Balanus glandula* is found primarily as a band of adults in the high intertidal, while B. *cariosus* occurs as scattered, large individuals or clumps in the midintertidal. This pattern, as Connell (1970) has shown, is also due to predation. *Balanus glandula*, like *Chthamalus* in Scotland, is capable of living throughout the intertidal zone and, indeed, settles throughout. The same is true for B. *cariosus*. That

FIGURE 6.17 A large mass of the mussel *Mytilus californianus* with large aggregation of its major predator, the starfish *Pisaster ochraceus* at Neptune Beach, Oregon. Notice the large areas of clean rock at the lower right where the starfish have already consumed the mussels. (Photo courtesy of Dr. Richard Mariscal.)

both show a restricted distribution is due to predation by three species of predatory gastropods of the genus *Nucella*: N. *lamellosa*, N. *emarginata*, and N. *canaliculata* (Figure 6.18). The abundance and motility of these predators is such that they are capable of completely consuming all the young B. *glandula* settling out in the midintertidal in 12–15 months. *Balanus glandula* survives only in a narrow band at the top of the intertidal, where the *Nucella* species are prevented from entering because of excessive desiccation. In the case of B. *cariosus*, however, the situation is somewhat different. There is no high-level refuge for this barnacle, and as with B. *glandula*, the young are consumed by the *Nucella*. Size is the defense for B. *cariosus*. Once it reaches two years of age, it is too large for *Nucella* to attack. The only mystery is how B. *cariosus* survives *Nucella* for two years. It may have to do with periodic events like big freezes at low tide or perhaps just plain random luck. As a result, the pattern of distribution for B. *cariosus* is random clumps of large, older barnacles.

It is not known what regulates *Nucella* populations, but they are preyed on by large *Pisaster* and are also vulnerable to periodic events, such as freezing during winter low tides. Such periodic events may reduce the population by such a significant amount that the *Balanus cariosus* could successfully establish themselves in an area. This may explain the differences among various shores with respect to the abundance of this barnacle.

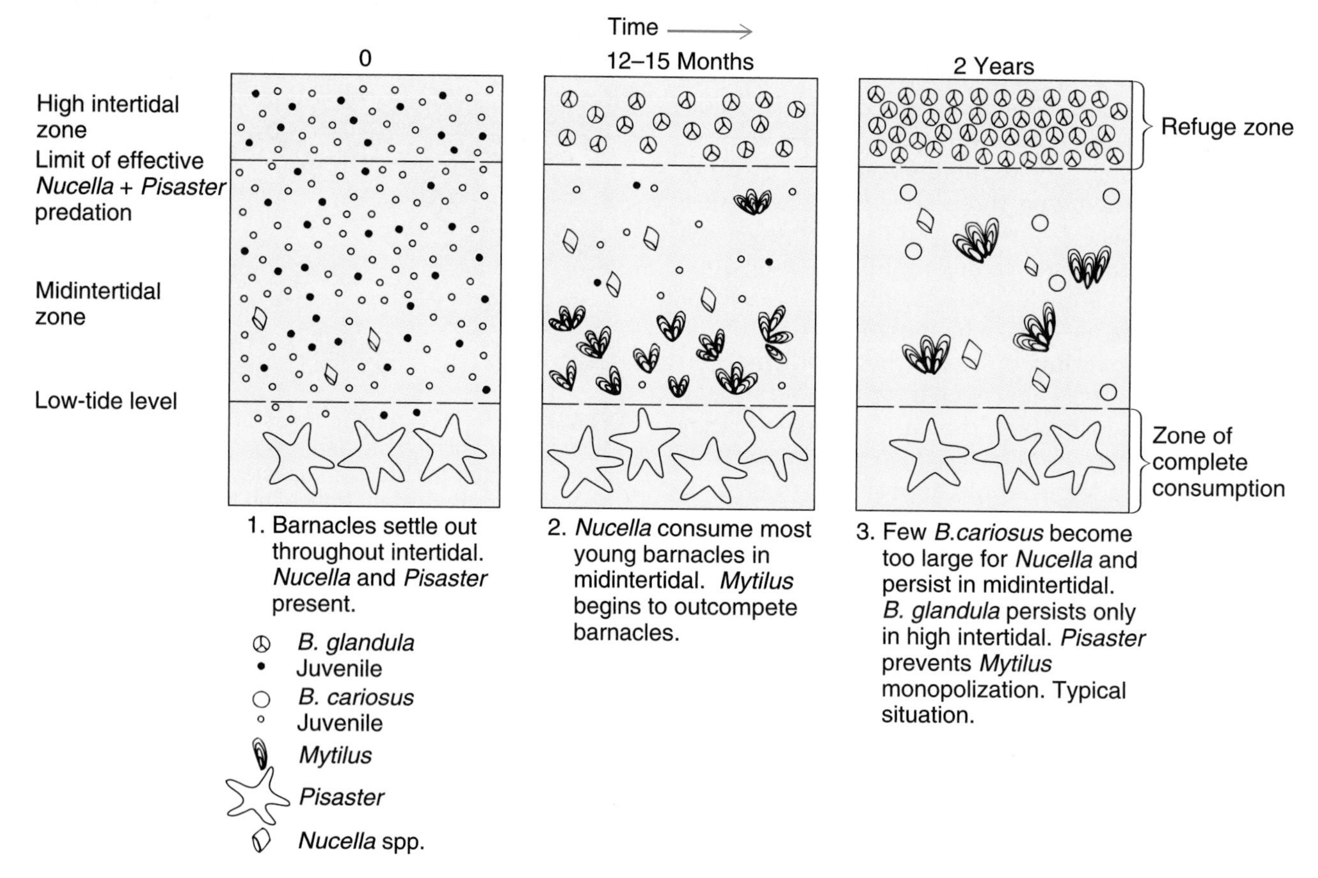

FIGURE 6.18 Interaction of predation and physical factors in establishing the zonation of the dominant intertidal organisms on the rocky shores of the northwest Pacific coast of North America.

Although it prefers mussels, *Pisaster ochraceus* can consume *Nucella* and barnacles of any size and is the primary predator of small- and medium-sized *Mytilus*. It is the top predator in the system. Because of its ability to influence the structure of the entire intertidal community by consuming *Mytilus* and preventing monopolization of the space, it has been called by Paine (1966) a **keystone species** as defined in Chapter 1.

The reason the upper intertidal is a refuge from predators is that *Pisaster* and *Nucella* can feed only when the tide is in, and they require a long period to attack their prey successfully. The short period of immersion of the upper intertidal does not allow sufficient time for them to make successful forays into that area. In subtidal areas, unlimited time is available; hence, the starfish have sufficient time to attack and consume their prey. It is probably for that reason that *Mytilus*, for example, do not extend into the subtidal areas, even though they can live there; hence, it is the reason for the sharp lower boundaries to the zonation of this species (Figure 6.18).

The situation regarding the importance of the interaction of mussels and starfish in structuring intertidal rocky communities of the Pacific coast of Washington has a confounding factor that suggests how difficult it is to understand how communities are structured and to extrapolate results from one area to another. Paine et al. (1985) have followed up on the earlier studies in Washington with similar studies of starfish removal from intertidal shores in Chile and New Zealand to assess the universality of the results obtained in Washington. In Chile, the dominant space occupier was the mussel *Perumytilus purpuratus*, and its predator was the sun star *Heliaster helianthus*. In New Zealand, the mussel was *Perna canaliculus*, and the predator was the sea star *Stichaster australis*. When predator removal studies were done in these areas and compared with Washington, a notable difference occurred. In Washington and New

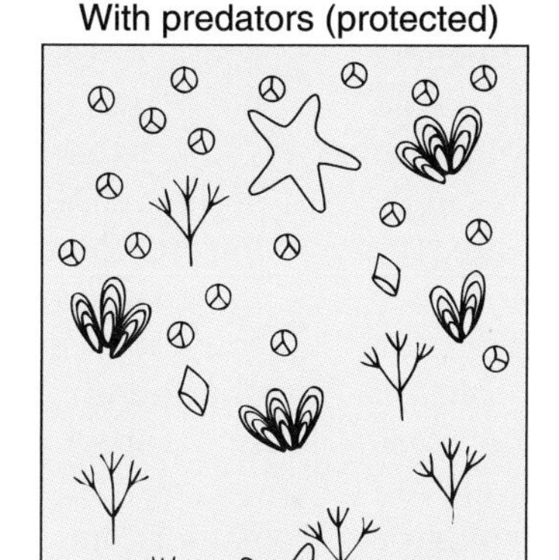

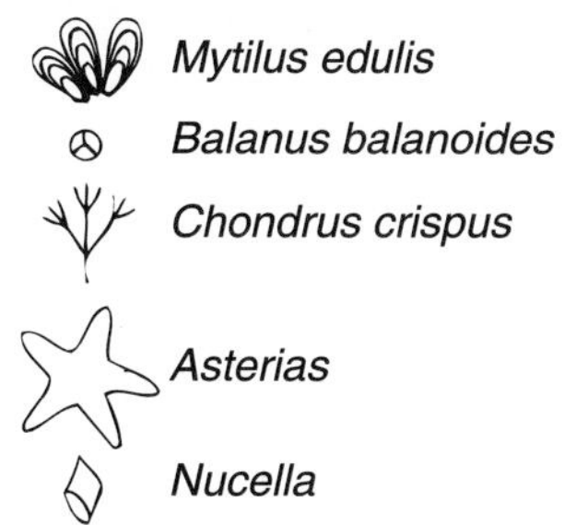

FIGURE 6.19 Composition of the low intertidal in the New England area with and without mussel predators.

Zealand, when the starfish were permitted to return to the areas from which they were formerly excluded, the mussel community persisted in monopolizing the space they had occupied when the starfish were excluded for as long as 14–17 years. In other words, there was no quick return to the conditions seen in control areas where predators were always present. By contrast, in Chile, when starfish were returned to the excluded areas, the community rapidly returned to the condition of the undisturbed control. What are the reasons for this difference? Paine et al. (1985) suggest that the main reason is that in Washington and New Zealand the mussels were able to attain a size during the absence of their main predator such that, when the predators were reintroduced, the mussels had grown to such a size that they were now immune to predation. In the case of Chile, *Perumytilus* never reaches a size at which it is immune to predation; hence, the quick return to a "normal" pattern. These results, in turn, suggest that in Washington and New Zealand the community will not return to the pre-exclusion state until the current large mussels either die or are removed by some other destructive event, such as wave action or crushing by wave-borne objects (see p. 268). This may take years. Whereas we may expect all three of these communities to eventually converge to their "normal" condition, the time involved may be markedly different. In the case of Washington and New Zealand, this may lead to an intertidal with a mosaic of patches—some completely dominated by mussels and others with scattered clumps, depending on the conditions and length of time the dominant predator was excluded.

Menge (1976) and Lubchenco and Menge (1978) have also demonstrated that predation is important in setting zonal patterns in the intertidal of the North Atlantic Ocean. In the low intertidal of New England, the competitive dominant species is *Mytilus edulis*. It is able to outcompete and eliminate the barnacle *Semibalanus balanoides* and the alga *Chondrus crispus*. It is prevented from doing so by the starfishes *Asterias forbesi* and *A. vulgaris* and the snail *Nucella lapillus*, all of which prey on *M. edulis*. Where these predators are absent—namely, in the most exposed, wave-beaten areas—*M. edulis* eliminates *Semibalanus* and *Chondrus* from the mid to low intertidal zone. The exposed, wave-beaten areas exhibit a zonal pattern in which the low intertidal is an *M. edulis* band, while in protected areas a diverse lower zone exists with *Mytilus*, *Semibalanus*, and *Chondrus* present (Figure 6.19).

An unexplained anomaly of the previous pattern is found in the very sheltered bays of New England. In these areas, Petraitis (1987) has demonstrated that the barnacle *Semibalanus balanoides* and the mussel *Mytilus edulis* are the most common organisms in the lower intertidal, where perennial algae are rare. This is a condition similar to the exposed outer shores, where predators are absent. However, here predators are abundant. The herbivorous gastropod *Littorina littorea* is thought to be responsible for the lack of algae, but the presence of the barnacles and mussels cannot be accounted for by a lack of asteroid predators nor by a refuge in size, since the starfish are capable of consuming all size classes.

The presence of the aforementioned anomalies in the keystone species concept has caused some researchers to question its usefulness, particularly how it should be defined and the generality of its occurrence. It is appropriate to briefly discuss these points.

At its inception by Paine (1969), the keystone species concept was defined as a single carnivore that preferentially preyed on and controlled the abundance of a prey species that, in turn, could competitively exclude other species and so dominate the community. It was subsequently used by other workers to refer to such noncarnivorous species as herbivores, mutualists, prey, and so forth; hence, confusion arose. Menge et al. (1994) have redefined the keystone species concept as originally stated and have clarified it by defining two other predator-prey relationships that are not keystone relationships. **Diffuse predation** is a condition where the total predation is strong and capable of controlling the abundance of a competitively dominant species but in which the predation is spread over several predators, not just one. If the total effect of predation on the competitive dominant prey is low, such that predation does not alone control the abundance, the condition is called **weak predation**.

The large geographical extent of the *Pisaster-Mytilus* community, extending from Alaska to Baja California, suggested to Paine (1969) that this keystone species concept, at least for this interaction, might hold true throughout the range. Subsequently, however, others such as Foster (1990) have questioned whether this generality could be maintained, particularly when *Pisaster* and *Mytilus* abundances are so variable in different geographic areas. Also, in southern California, it appeared that spiny lobsters, not *Pisaster*, were controlling the mussels.

In a more recent test of the generality of the keystone concept, Menge et al. (1994) investigated the high temporal and spatial constancy of the lower limits of *Mytilus* distribution at several sites in Washington and Oregon in which the abundance of *Pisaster* showed enormous variation. They compared sites that varied in size from meters up to kilometers, adding a question of scale that had not been posed earlier. They found that while *Pisaster* predation varied dramatically, they were able to conclude that *Pisaster* behaved as a keystone species, eliminating *Mytilus* at exposed sites. At more protected sites, however, *Pisaster* predation was much weaker and variable, and *Mytilus* was controlled by other factors, such as recruitment and burial by sand.

Recently Connolly and Roughgarden (1998) have provided evidence of a physical explanation for the differences in the *Pisaster-Mytilus* community structure observed along the extensive geographical range of the species. They point out that especially in the spring and summer along the Pacific coast, strong winds blow southward; this causes a surface layer of water to move offshore, resulting in a cold-water upwelling to replace the offshore moving surface water (Ekman transport). As a result, a nearshore circulation pattern is established with water moving shoreward at depth, surfacing as upwelling nearshore, and then moving offshore at the surface (Figure 6.20). Various larvae of intertidal organisms become entrained in this offshore moving "Ekman layer" and accumulate at offshore frontal zones where the offshore moving water meets the south-moving California Current. To replenish the intertidal zone, these larvae must end up on the shore to settle and metamorphose. However, this can only happen when the strong southward-blowing winds diminish and the front moves onshore. This happens periodically. However, the strength of the offshore currents is greater, the fronts are further offshore, the upwelling season is longer, and the relaxing of the winds less frequent in California as opposed to Oregon and Washington (Figure 6.20). As a result of these differences in circulation pattern, there is lower recruitment to the intertidal in California than in Washington. That in turn means lower adult abundances of the intertidal

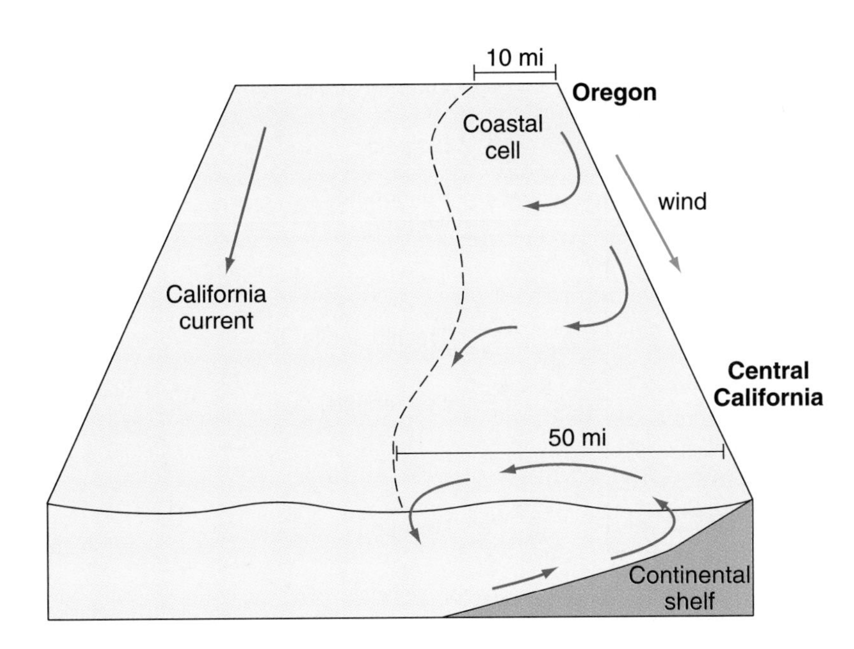

FIGURE 6.20 Diagrammatic representation of the California Current system. The dashed line represents the front between the upwelled water and the offshore California Current. (After "A Latitudinal Gradient in Northeast Pacific Intertidal Community Structure" S. R. Connolly & J. Roughgarden, *American Naturalist*, Vol. 151, p. 313, 1998. Copyright © 1998 University of Chicago Press. Reprinted by permission.)

invertebrates, more free space, and hence, weaker competitive and predatory interactions among the dominant intertidal invertebrates in California. The significant high frequent recruitment in the Pacific Northwest means competition, predation, and disturbance are the dominant factors structuring the *Pisaster-Mytilus* community in the Pacific Northwest, while the infrequent recruitment in California reduces such biological interaction and leads to community composition determined primarily by the infrequent recruitment events.

Finally, changes in temperature have now been demonstrated to affect keystone species predation and hence potentially change community structure. Sanford (1999) demonstrated in laboratory and field experiments that slight decreases in water temperature dramatically reduced the predation by *Pisaster ochraceus* on *Mytilus californianus* and *M. trossulus* as the stars consumed fewer mussels and decreased in density. Such results suggest that El Niño/La Niña events and global warming may have profound effects on intertidal community structure in the future.

It seems safe to conclude that the keystone species concept, at least for some temperate rocky shores, remains valid. At the same time, it is not a universal concept, and its occurrence is variable within the geographic range of the species or communities in question. It may be replaced under certain environmental conditions by other community regulatory mechanisms, such as diffuse or weak predation or physical factors.

A central theme of marine intertidal ecology, as argued by Connell (1975), is that wherever predation is reduced, competition will be increased. Indeed, the previous examples demonstrate this. It is well, however, to remember that this is not the inevitable result of predator removal. For example, Keough and Butler (1979) removed predatory starfish from pier pilings in Australia and found no changes in the invertebrate populations or any increase in competition. This indicates an important principle: the relative importance of biological factors and physical factors on rocky shores varies over relatively short distances. The variation is both

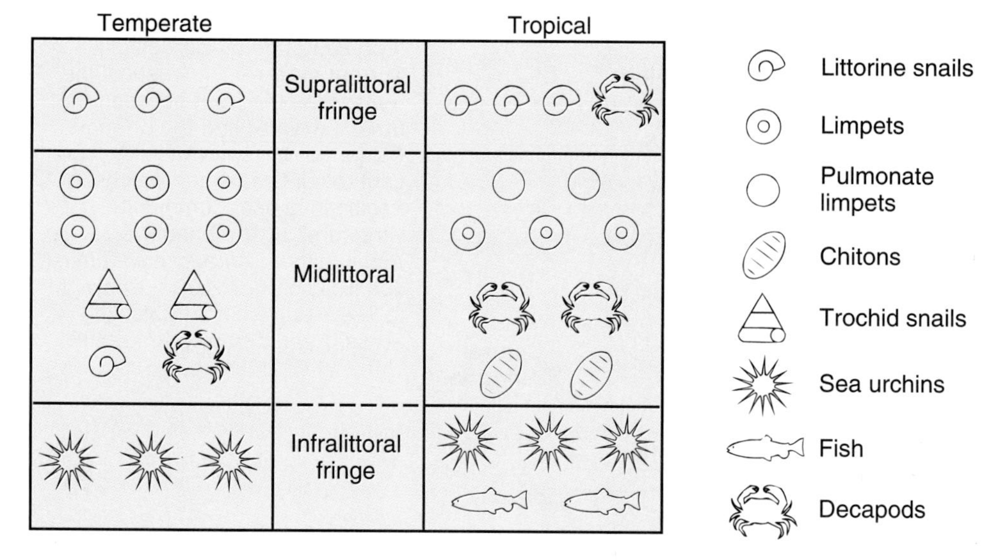

FIGURE 6.21 The main groups of algal grazers at different intertidal zones in temperate and tropical systems. Numbers of symbols indicate relative importance.

vertical and horizontal. Therefore, one cannot extrapolate from single experimental studies done on small areas to form models for the organization of large geographic areas of the intertidal.

The role of **grazers** or herbivores in regulating upper and lower limits of algal species is well studied and documented, and evidence suggests this process may also be important. Whereas a number of animals graze on intertidal algae, relatively few grazers seem abundant enough to significantly alter or determine community structure. The dominant grazers are various gastropod mollusks, certain crustaceans, sea urchins, and fishes. The relative importance of these groups varies latitudinally and vertically in the intertidal zone (Figure 6.21). Their relative abilities to keep algae in check may also vary.

Data on the importance of grazing were late in being reported (not until well after World War II). In the past 25 years, however, considerable evidence has been accumulated by various methods—experimental removals of grazers, caging experiments, natural disasters—so our understanding of the role of grazing is now fairly well advanced. Grazing affects a number of parameters, including algal zonation, species diversity, patchiness, and succession. Before the implementation of various experimental methods, it was assumed that in the intertidal the upper limits of the dominant algal species on rocky shores were a function of the species' ability to tolerate various physical factors. On the other hand, lower limits of algal species distribution seem not to be set by physical factors, because most algae grow faster when experimentally transplanted to lower levels or when they are grown constantly submerged.

Since most algae do grow better when set lower in the intertidal or submerged, why don't we find them there naturally? The answer coming from many researchers is that they are grazed out. For example, in various experiments that have removed grazing animals, the result has been the appearance lower on the shore of a number of algae otherwise restricted to higher levels, where grazing is less intense. This has been shown by Lubchenco (1978, 1980) to be true for *Chondrus* in New England, where the main grazer is the urchin *Strongylocentrotus droebachiensis*. Himmelman and Steele (1971) found the same thing for the alga *Alaria esculenta* in Great Britain, where *S. droebachiensis* is also responsible (Figure 6.22).

Grazing may also determine the community structure. For example, Dethier and Duggins (1988) removed the chiton *Katharina tunicata* from the low intertidal in Washington state, and the site developed from an area with few macroscopic plants into

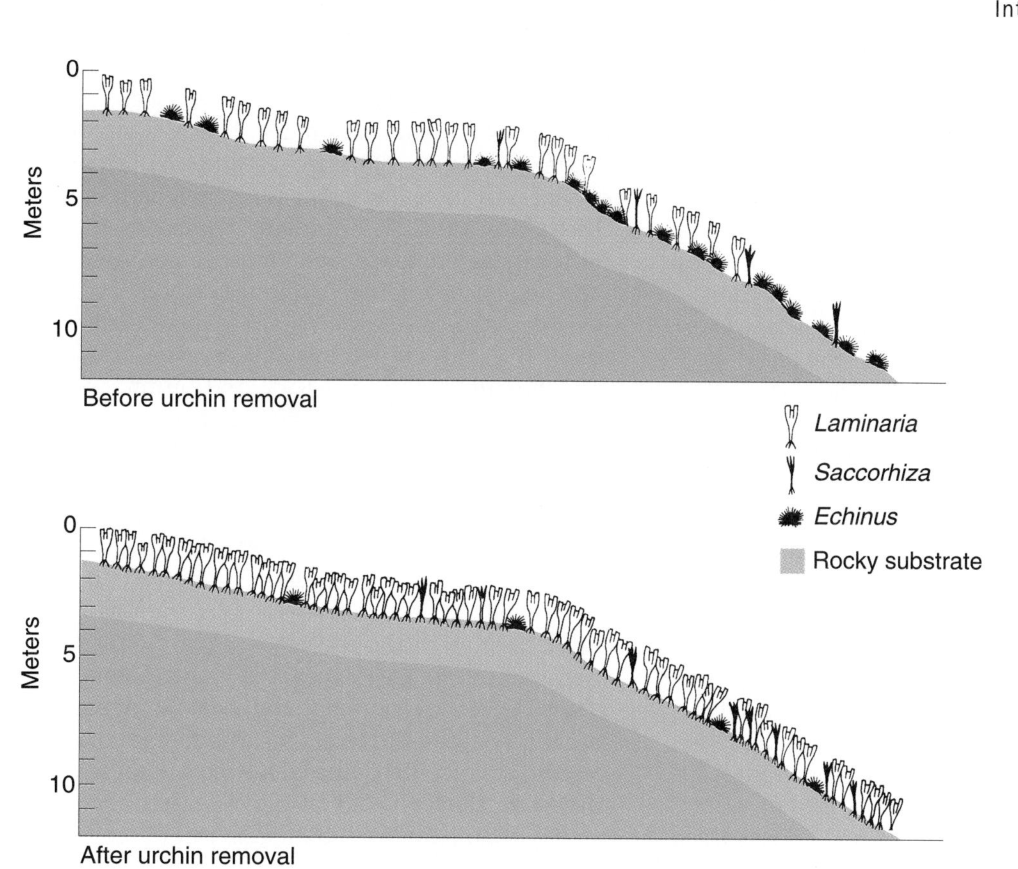

FIGURE 6.22 Effect of sea urchin removal on kelp growth on the Isle of Man, Great Britain. (Modified from *Pacific Seashores, A Guide to Intertidal Ecology*, T. Carefoot, 1977 Douglas & Mylntyre, Ltd.)

a kelp bed dominated by the kelps *Hedophyllum sessile*, *Alaria marginata*, and *Nereocystis luetkeana*. However, a similar removal of *Katharina* from a low intertidal area in Alaska did not result in development of a kelp bed, suggesting again that it is unwise to generalize from local experiments to broad geographical areas.

Grazing may also set the upper limits of algal distribution. One of the more interesting pieces of evidence supporting this comes from the Torrey Canyon oil spill disaster of 1968. This spill killed the main intertidal molluscan grazers on the shores of southern England. Southward and Southward (1978) documented a subsequent rise in the upper limits of a number of intertidal algae. Following the return of the limpets and other grazing gastropods, the algae were once again grazed down.

The mechanisms by which the removal of grazers facilitate the upward or downward extension of algae vary. In the simplest case, the algae are directly limited by the grazer. In other cases, the absence of the grazers may permit a rapid establishment of ephemeral algae, in turn, trapping moisture and permitting the newly settled sporelings of the dominant algae to survive. In still another scenario, the upward extension of algae may be seasonal due to changes in grazing pressure. For example, Castenholz (1961) showed that a summer decrease in high intertidal diatom cover was due to a summer increase in *Littorina* grazing activity in contrast to the less active winter condition. In a more complicated situation, grazing and physical factors interact. Hay (1979) found that in New Zealand the alga *Durvillea* extended its range upshore in the winter, when grazers were absent, but was cut back in the summer by a combination of increased grazing and desiccation at the higher levels.

Surprisingly enough, some of the limits of algae are set by the effect of the algae themselves on potential grazers. In New Zealand, for instance, Underwood and Jernakoff (1981) showed that the downshore extensions of the main gazing limpets,

Cellana and *Patelloida*, are limited by certain algae. In this case the algae settle and grow faster than the limpet grazers can remove them. As a result, the algae form a substrate that prevents the limpets from attaching, so they cannot enter the area and graze. Obviously, in this kind of situation there is a continuum. At some level, there is a balance point defining one area where grazers are more efficient at grazing than the algae are at growing, thus eliminating the algae. At the other extreme, the algae grow faster and exclude the limpets by creating an unsuitable surface for attachment.

Another consideration in the zonation of algae species is the ability of some algae to grow fast enough to escape grazing by becoming too large for the grazers to feed on. In these cases, the algae will mature only in areas where the grazers are low in density or restricted from grazing during some seasons; thus, the algae can reach a size at which they are immune to grazing.

So far we have considered grazing and its effects on vertical zonation patterns. Grazing may, however, also affect the horizontal patterns of zonation—patterns that have as the dominant physical factor the variation in wave action. The best studied examples come from the Atlantic coast of Europe. Here, the more exposed coastlines have the intertidal dominated by mussels, barnacles, and limpets, whereas the more sheltered shores are covered with dense stands of fucoid brown algae. From various removal experiments and natural experiments, such as the Torrey Canyon oil spill, a conceptual model to explain this has been developed by Southward and Southward (1978). Very simply, this model says that on sheltered shores limpets and other grazers are rare (but see preceding section on New England sheltered shores). This permits the algae to grow and flourish. Furthermore, the algal canopy reduces the ability of limpets to settle on the shore. On exposed shores, the limpets are abundant and graze out the algae, leaving the shore dominated by mussels and barnacles. On the opposite side of the Atlantic, in New England, limpets are rare and the dominant grazer, the snail *Littorina littorea*, decreases in abundance in waveswept areas. As a result, the community changes, going from exposed to sheltered conditions, are attributable not to grazing but to physical factors and predation, as noted previously.

We should not leave this issue of grazing without discussing a final subject, namely, the defenses of the algae to grazers. We cannot assume that algae are completely without protection from grazing. Indeed, they are not. The most obvious adaptations to grazing are morphological, and one of the most common is the laying down of calcium carbonate in the tissues. Calcium carbonate reduces the palatability of algae to grazers and makes the grazer exert more energy in feeding. Furthermore, Steneck (1982) has even shown for one coralline alga, *Clathromorphum*, that the meristem tissue and reproductive sites are all on the underside and protected from grazing. Growth form may also discourage grazing. Hay (1981) has shown, for example, that formation of short turfs in algae reduces grazing mortality. Still other algae, such as the Pacific coast *Egregia*, have considerable structural or "woody" tissue when mature; this may deter grazers.

Chemical defenses are also common. Many algae contain noxious or toxic compounds. For example, on the Pacific coast of North America the genus *Desmarestia* contains sulfuric acid concentrated enough to erode the calcium carbonate teeth of the grazing urchin *Strongylocentrotus franciscanus*. Other algae contain various alkaloids, phenolic compounds, and halogenated metabolites.

Finally, there are evolved defenses in the life histories of certain algae. Some algal species can exist in at least two growth forms, one a low "crust" and the other an upright frondose form. In areas with heavy grazing pressure, the slow-growing but grazer-resistant crusts predominate, but in areas of low grazing pressure or in areas from which grazers have been excluded, the frondose forms increase. Littler and Littler (1980) have suggested that having these two alternative growth forms in a single species is a strategy of **bet-hedging**. The frondose form has fast growth and high competitive ability but low grazing resistance. The crust form, on the other hand, has slow growth rates and competitive ability but high grazing resistance. Another life history feature that may have evolved in algae to cope with grazing is reproduction. Many algae, particularly those that are favored by grazers because of their high fraction of digestible material and low fraction of structural tissue, have evolved the opportunistic, or *r*-selected, habit of high reproductive output. This also makes them successful in colonizing newly available space. Quick maturity and high reproductive output mean that these species should be able to colonize an open area and reach maturity quickly before the grazers find and decimate them. Other species have evolved the opposite, or K-selected, features. They are relatively slower growing, perennial, and long-lived. To survive pressure from grazers, these algae either have evolved chemicals that make them unpalatable or grow large enough to be immune from grazers. Large size usually also means an enhanced competitive ability, at least for light.

We can conclude this section on grazing by looking briefly at the geographical variation in grazing and its effects on various kinds of communities. Beginning with the temperate zone, grazing is one of the main structuring agents of the rocky intertidal on the Atlantic shores of Europe and the Pacific shores of North America. On the Pacific coast of North America, a large number of species of limpets plus a few species of crabs and sea urchins are responsible for controlling community structure. Fewer grazers, primarily limpets and other snails, exist on the European shore, but they are efficient at structuring the communities. Grazing is less important on the Atlantic coast of North America, where limpets are rare and the major grazer, the snail *Littorina littorea*, seems less efficient than *Patella* in Europe. Grazing becomes less important as one moves poleward in the Atlantic. In the Southern Hemisphere, the intertidal zones of Australia and New Zealand are strongly structured by grazing and also have a high diversity of grazing species. Generally, as one moves toward the equator, the importance of grazing increases and so does the variety of grazing organisms. Crabs and fishes become more important as grazers in the tropics. The rocky intertidal in the tropics has been considered relatively barren due to the extremes of the physical factors acting there, but the large number of grazers suggests grazing may be more important (see following section on tropical shores). Additional work is needed to clarify this situation.

We can summarize the effects of grazing in the three intertidal zones defined by the Stephensons to create a general framework for the North Atlantic Ocean as a model or example of how grazing acts. In the supralittoral fringe, grazers are either uncommon or have little time to graze. The rigorous and stressful physical environment is thus the most important agency causing changes in algal composition. The midlittoral zone is dominated by barnacles and mussels but is strongly influenced by gastropod grazers, primarily limpets, and may vary in algal cover depending on exposure. This, in turn, may limit the numbers of gastropods. Barnacles, however, may also be present due to the ability of limpets and other grazers to keep the competing algae in check. In the sublittoral fringe, the algal growth exceeds the capabilities of the grazers to control it. Therefore, the competition for space and light among the various algae is the dominant interaction that structures communities at this level. These various factors and their relative importance are summarized in Figure 6.23.

The ability of larvae of various benthic invertebrates to select areas on which to settle has been known for more than half a century (see Chapter 1). However, the importance of this choice to the eventual structure of intertidal communities has remained obscure due to the difficulty of working with larvae under natural conditions. Several studies recently have, however, demonstrated the importance of this selection ability. In Australia, the larvae of the barnacle *Tesseropora rosea* settle only where adults occur; thus, the distribution of adults is not the result of competition with other barnacles outside their range. Moreover, the upper and lower limits on the shore are not set by physical factors, but by settlement of the larvae. Furthermore, the barnacle larvae were found by Denley and Underwood (1979) to avoid settling on substrates already occupied by other organisms. A more comprehensive example, also from Australia, includes a second barnacle, *Chamaesipho columna*. In New South Wales, Underwood and Denley (1984) report that shores protected from wave shock have grazing gastropods as dominants, but where there is moderate wave action, the midshore is dominated by *Chamaesipho*. On the most exposed areas, *Tesseropora* occupies midtide levels and *Chamaesipho* the upper reaches. Neither barnacle settles into the areas dominated by grazing gastropods, even if the grazers are removed. On exposed shores, neither settles into areas occupied by the adults of the other species. In other words, this whole intertidal zonation pattern is structured by the patterns of larval choice and not by predation, competition, or grazing.

Larval recruitment can be variable in time and in space. This variability has been documented, and its importance to the intertidal community structure can be significant. In some years, any given intertidal area may experience poor recruitment due to bad combinations of weather, reproduction by adults, larval

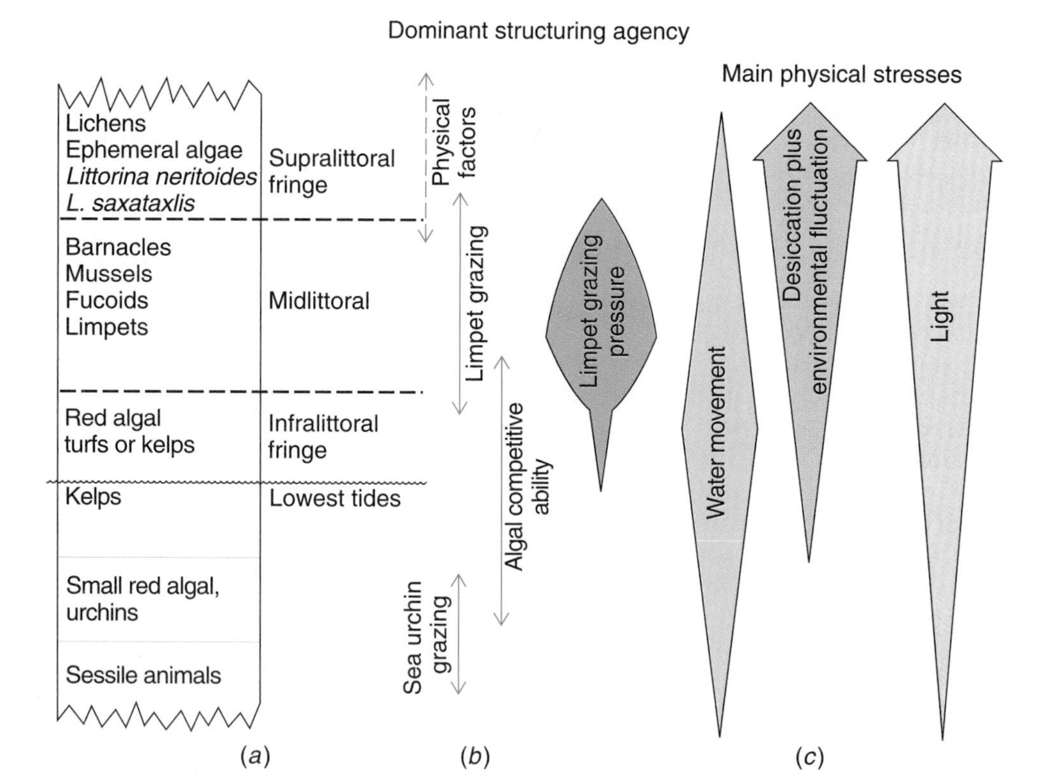

FIGURE 6.23 Structuring forces affecting the intertidal of the North Atlantic Ocean. (a) The main intertidal zones and their dominant organisms (after the Stephensons' universal scheme). (b) Dominant structuring agencies; overlap suggests a balance between factors. (c) Main physical stresses for marine algae and animals. (Modified from "Grazing of Intertidal Algae by Marine Invertebrates" S. J. Hawkins & K. G. Hartnoll, 1983, *OMBAR*, Vol. 21, pp. 195-282.)

mortality in the plankton, such water conditions as wave action, and so forth. The result is few survivors and a decrease in biological interactions, such as predation, competition, and grazing. In other years, favorable physical conditions may swamp an area with recruits, leading to increased competition or predation and probable changes in the adult population and the community. As a result, the same area may have a community that not only differs in composition over time, but may change in relative numbers of individuals. The importance of competition, predation, and grazing in determining the structure of the community will also, therefore, change.

Gaines and Roughgarden (1987) have demonstrated another significant factor that can lead to variation in larval recruitment. They were able to show that the juvenile rockfish that inhabit the kelp forest immediately offshore of the intertidal zone in central California are significant predators of the larvae of the intertidal barnacles. The fish are capable of reducing the number of recruits to 1/50 of the number that would settle in the absence of the fish. The number of barnacle larvae that actually make it across the kelp bed depends on the stock of rockfish, which, in turn, is a function of the extent of the kelp forest. Whenever the kelp forest is decreased, as in El Niño years, the rockfish also decrease, which leads to good barnacle recruitment and vice versa. In this scenario, the dynamics of the intertidal communities are directly coupled to another community, namely, the kelp forest (see Chapter 5 for factors influencing kelp forest dynamics).

This great spatial and temporal fluctuation in the settlement of larvae of the dominant intertidal organisms means that the importance of competition, predation, and grazing is likely to be variable. It is difficult to generalize about the effects of competition, predation, and grazing if the numbers of prey or potential competitors are also unpredictable. Although it is considered by some to be a "nuisance" in the setting up of experiments, this irregularity of appearance of various dominant intertidal invertebrates is a fact of nature and must be considered in any attempt to understand the structuring process in the intertidal zone. In other words, the irregular occurrence in a community of a numerically important, competitively dominant, or keystone species should lead to a variety of different outcomes from experimental studies done in the area because of variations in the intensity of biological interactions resulting from the larval recruitment patterns.

In an attempt to sort out the relative importance of competition, predation, and larval recruitment to the structure of rocky intertidal communities, Menge (1991) undertook a multiple regression analysis of these factors for New England and Panama shores. The results indicated that recruitment in New England explained 11% and predation and competition 50–78% of the structure, whereas in Panama recruitment explained 39–87% and predation and competition only 8–10% of the structure. If these figures hold true for other areas, this may indicate a fundamental difference in the structuring forces in temperate and tropical intertidal areas.

Tropical Intertidal Shores

Thus far, we have discussed the biological factors that determine community structure, employing examples exclusively from temperate regions of the world. What about the tropical intertidal? Are the communities and organisms similar? Unfortunately, there are few studies of the tropical intertidal from which to draw generalities. The most extensive are those of Menge et al. (1986) and Lubchenco et al. (1984) on the rocky intertidal of the Pacific coast of Panama, but less extensive studies exist for the Caribbean and Hong Kong that suggest there are marked differences among tropical areas. Considering first the Pacific coast of Panama, the general composition of the communities is similar to that of temperate rocky shores. The major differences are the lack of large algae and an abundance of herbivorous fishes and crabs, as well as predaceous fishes, in Panama. Other differences are evident. For example, studies indicate that the high intertidal of Panama has 90–98% bare substrate dominated by small barnacles. The mid to low intertidal is dominated by encrusting algae (26–90% cover), while sessile animals are scarce (1–10% cover). Only in the lowest parts of the intertidal is a diversity of organisms encountered. As might be expected from these figures, both species number and diversity increase with decreasing tide level, as does the maximum cover by organisms.

Working in the mid and low intertidal, Menge et al. (1986) found that space is dominated by crustose algae. Community structure is determined by complex interactions among a diverse assemblage of herbivores and predators that Menge et al. (1986) divided into four consumer groups: large fishes, small fishes and crabs, herbivorous mollusks, and predaceous gastropods (Figure 6.24). Experimental treatments involving deletions of various combinations of the four consumer groups and comparison with controls suggested that the effects of each consumer group on its prey species or group were often indirect and included both positive and negative effects. Both the small fishes and crabs group and the large fishes group reduced the abundance and cover of sessile colonial and solitary invertebrates, as well as fleshy algae, but were incapable of totally removing them. The molluscan herbivores, in turn, grazed down the fleshy algae until only the very resistant algal crusts survived, a group otherwise competitively inferior to the fleshy algae. The molluscan herbivores also inhibited recruitment of any sessile invertebrates. Finally, the predaceous gastropods also preyed on the sessile invertebrates and reduced their numbers. These group interactions and results were compounded by a combination of predator-prey interactions among the consumer groups and various competitive or symbiotic interactions among the prey.

The results of these studies indicate that there was no keystone species in this area that prevented overgrowth by a superior space competitor as we saw in certain cases of the mussel-dominated communities in the temperate intertidal. Instead, the continued dominance of the area by crustose algae is maintained by persistent intense predation and herbivory by a diverse assemblage of fishes and invertebrates on potentially dominant sessile invertebrates and fleshy algae. It is likely the dominance of this area by crustose algae and the failure of any organism or group of organisms to occupy a majority of the

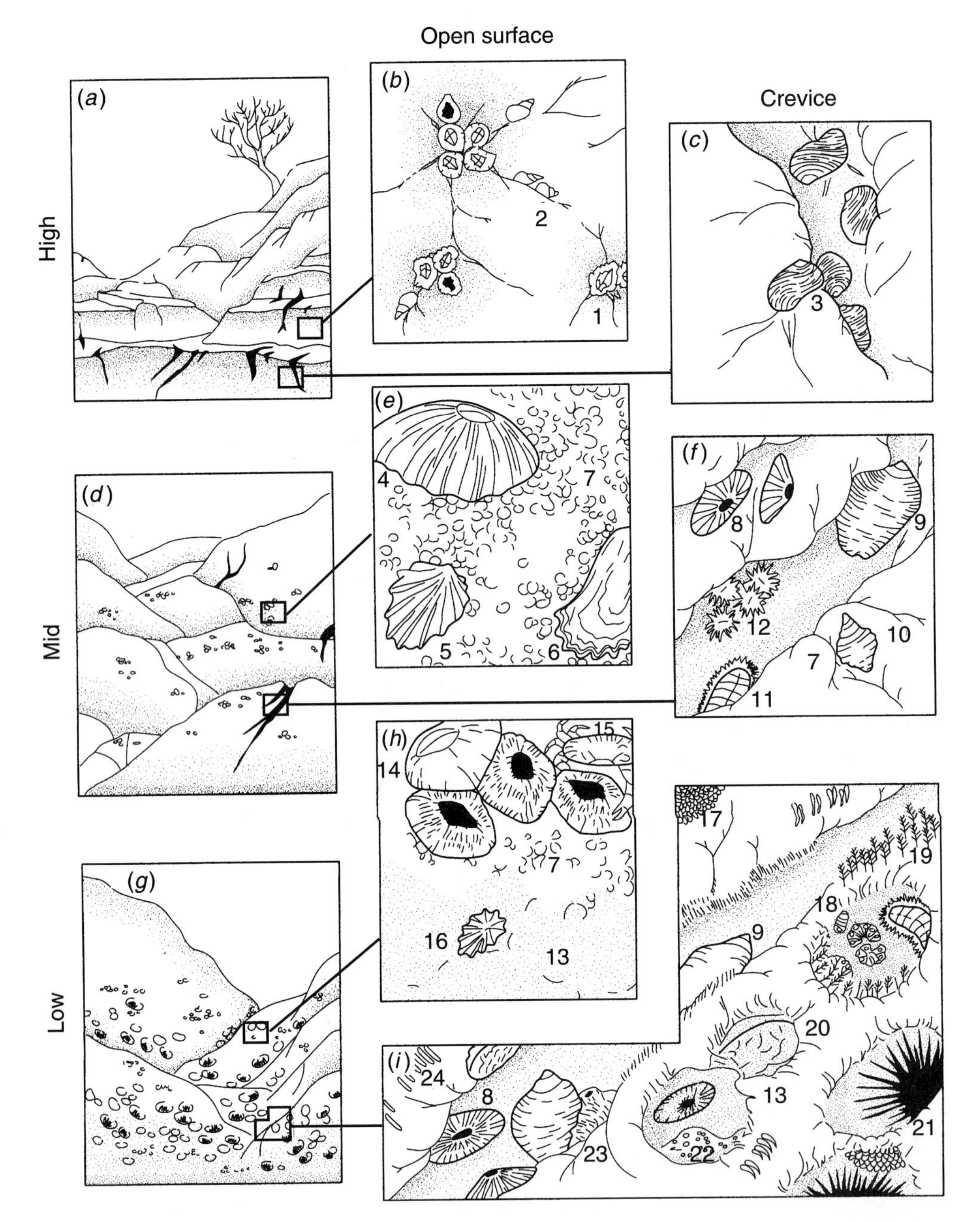

FIGURE 6.24 Diagrammatic illustrations of the organisms and physical heterogeneity at Taboguilla Island, Panama. The distant views are shown for the high (a), mid (d), and low (g) zones; each of these is <2 m across the foreground. Closeup views are illustrated for open surfaces (b) and crevices or holes (c) for the high, mid (e, f), and low (h, i), zones. Species identifications are (1) Chthamaloid barnacles (live and dead); (2) *Littorina* spp.; (3) *Nerita scabbricosta*; (4) *Tetraclita*; (5) *Siphonaria gigas*; (6) *Ostrea palmula*; (7) mixed algal crusts: *Schizothrix*, *Ralfsia*, and *Hildenbrandia*; (8) *Fissurella* spp.; (9) *Thais reelones*; (10) *Acanthina*; (11) *Ceratozona*; (12) anemones; (13) corallina curst; (14) *Balanus inexpectatus* (live and dead); (15) *Pachygrapsus*; (16) *Siphonaria maura*; (17) encrusting bryozoan; (18) colonial tunicate; (19) mixture of *Abietinaria* and filamentous algae; (20) *Chama*; (21) *Echinometra*; (22) sponges; (23) *Ascidia*; (24) rasp marks of grazing fishes. (After "Structure, Persistence, and Role of Consumers in a Tropical Intertidal Community" J. Lubchenco et al., 1984, *J. Exp. Marine Biology & Ecology*, vol. 78, pp. 23–73, 1984. Copyright © 1984 with permission from Elsevier Science.)

available space are maintained by a combination of extreme heat, desiccation, and ultraviolet ray stress that prevents occupation of much of the area by marine organisms.

In contrast to temperate rocky intertidal systems, in tropical Panama three-dimensional space in the form of holes and crevices appears to be the major refuge from predators and grazers. There does not appear to be a refuge in body size, time, or occupation of higher intertidal zones for any of the prey species.

In another study that encompassed Panama, Costa Rica, and the northern Gulf of California, Sutherland (1990) investigated the populations of the intertidal barnacle *Chthamalus fissus* and determined that in Costa Rica and Panama the populations of the species were regulated through unknown events in the water column that manifested themselves in the numbers of larvae available for recruitment. Predation was only important in localized areas where the predatory gastropod *Acanthina brevidentata* occurred, and competition was found sporadically with the mussel *Brachiodontes semilaevis*. In the northern Gulf of California, however, which is not a tropical area, recruitment was very high and competition and predation among the postsettlement barnacles were significant in structuring the community.

In the Caribbean area the intertidal rocky substrate is dominated by limestone, and littorinid snails are very abundant, but barnacles are conspicuous by their absence. Mussels are often absent or lower in the intertidal than mussels on temperate shores. Many of the high-zoned littorinids are grazers on endolithic algae and, as a result, become important bioerosive agents.

More studies are needed on tropical rocky shores before we can assess the importance of the various biological and physical factors to community structure and the amount of variation over time and space.

Patchiness

The major large-scale patterns of organism distribution in the rocky intertidal—the zonation—can be explained through one or more of the following factors acting separately or in various combinations: exposure, desiccation, temperature, competition, predation, larval settlement, or grazing. However, these vertical zonation bands are never perfect. Often, small-scale local variations alter the bands and affect the local distribution of certain organisms. Such small-scale **patchiness** or **local distribution** patterns can result from either biological or physical factors.

Patchiness is a basic feature of the structure of rocky intertidal areas just as it is of plankton (Chapter 2, pp. 85-87) and subtidal communities (Chapter 5, pp. 182-185). Patchiness is usually temporally and spatially variable. It is generated by any physical or biological activity that creates either an open space for recolonization, produces a different substrate, or alters the prevailing physical conditions of an area.

Perhaps the easiest of these local distribution patterns to explain are those caused by physical factors, primarily the variation in the nature, slope, and exposure of the rock surface. Existence of a different rock type in a local area as compared with the surrounding shore almost always means the presence of different organisms, or at least different abundances. An outcrop of sandstone among an otherwise granite rock shore usually means a different complement of organisms.

Similarly, slopes facing south and those facing north often have different organisms, or the densities and levels to which the organisms are found differ. This reflects the different amount of desiccation and harmful effects of direct exposure to the sun's rays (insolation). On the Pacific coast of North America, for example, two limpet species, *Collisella digitalis* and *C. scabra*, inhabit the high intertidal zone, but as Haven (1971) has shown, their distribution within the zone reflects their reaction to the small-scale changes in the substrate. *Collisella scabra* is found primarily on horizontal surfaces where it has a home scar or depression in which it remains during periods of low tide. *Collisella digitalis* has no scar and is found on the cooler, moist, shaded vertical slopes. *Collisella scabra* can survive on the drier horizontal rocks because of its scar, which is a tight enough fit to prevent desiccation at low tide. *Collisella digitalis*, on the other hand, without the increased desiccation protection of the scar, must live where desiccation is not as intense (i.e., north-facing slopes and crevices). Similarly, certain species do not prefer to settle on vertical faces or overhangs. These areas are then settled by other species unable to compete on the flat surfaces. The presence of cracks or crevices in the rock may help algal spores "escape" from grazers and create a patch of algae where others are absent due to grazing (Figure 6.25). Cracks and crevices also may shelter predators, which in turn affect the community. For example,

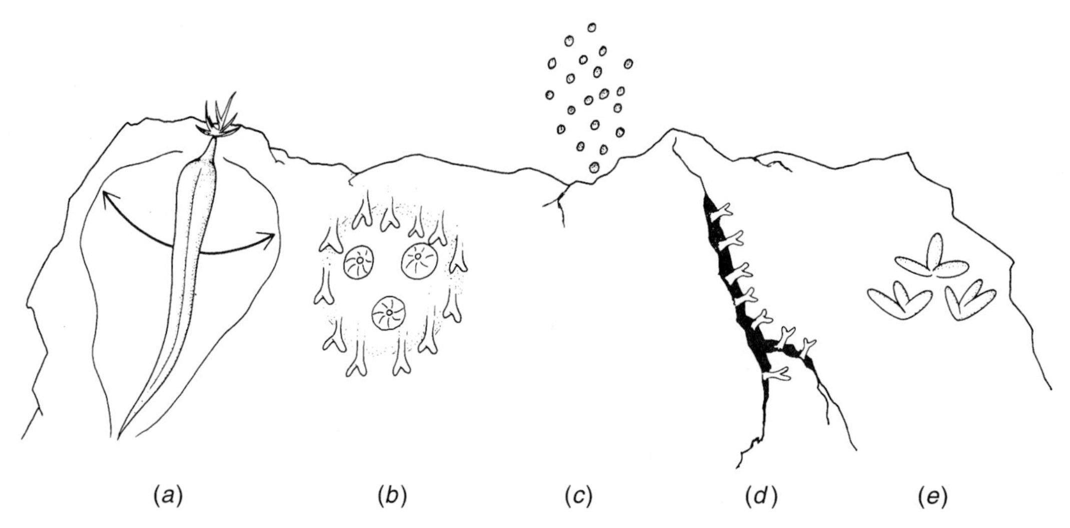

FIGURE 6.25 Causes of patchiness in algae on rocky shores. (a) Sweeping action of algal fronds. (b) Irregular spatial and temporal distribution of grazers. (c) Fluctuations in recruitment. (d) Refuge from grazing provided by pits and cracks in rock. (e) Escape of sporelings from grazers.

Moran (1980) filled the cracks in an intertidal area with concrete. This reduced the density of the predatory gastropod *Morula marginalba*, and led to increases in barnacle numbers and altered community structure. As previously noted, crevices and other three-dimensional openings are significant refuges for tropical intertidal organisms.

Much small-scale patchiness probably results from the destructive effects of wave action and such wave-borne objects as logs that may be hurled against the shore, smashing into small areas and clearing them of organisms. Such battering was studied by Dayton (1971) in the state of Washington, and he found that any point on an open shore had a 5–30% chance of being hit by a log within a three-year period. An additional effect of wave action is that it produces aggregations of mobile species, such as snails and crabs, in crevices or other protected areas. Heavy wave action from storms, even if unaccompanied by battering, also is responsible for removal of organisms from the intertidal. Menge (1976) found in the exposed intertidal in New England that winter storms could clear as much as 90% of the intertidal; whereas during summer and fall, less than 10% of the same space was free of organisms.

The premier space competitors, *Mytilus edulis* in New England and *M. californianus* on the Pacific coast of Washington, both form large clumps that grow out from the substrate. As the clumps get larger, fewer mussels are attached directly to the substrate through their byssal threads and more are attached to each other. This creates an unstable situation in which the battering waves are more likely to tear off a clump; hence, older and larger clumps become more vulnerable (Figure 6.26). Small-scale distribution anomalies may also result from biological interactions of various types. Some of these interactions are passive, such as many species of animals inhabiting moist areas to survive exposure to air; hence, they all shelter under the fronds of various intertidal algae. The patchiness of these forms follows the distribution of the algae, but it is not due to any direct action of the algae. Removal of the algal canopy also causes the disappearance of the sheltering organisms. Similarly, in the case of the larger algae with substantial fronds, the movement of the frond by the water causes a sweeping effect, which may remove larvae and spores from the rock near the alga, creating a patch (see Figure 6.25).

Other biological effects are more direct and lead to definite distribution patterns that can be observed. For example, the intertidal anemone *Anthopleura elegantissima* of the Pacific coast of North America tends to occur in large clumps, uniformly separated from each other. Each clump has resulted from the con-

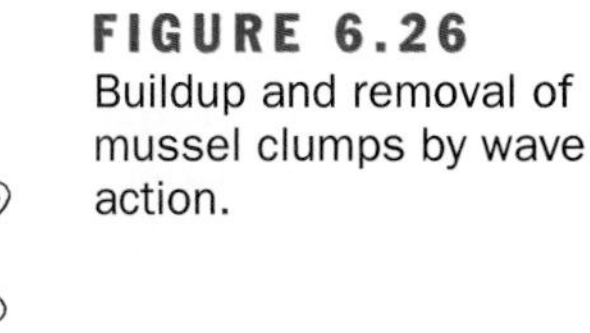

FIGURE 6.26 Buildup and removal of mussel clumps by wave action.

FIGURE 6.27 Clones of the anemone *Anthopleura elegantissima*, showing the separation caused by fighting between adjacent clones. (Photo by the author.)

tinued asexual division of the original anemone, and each clump consists of genetically identical individuals, or **clones**. Francis (1973) discovered that the sharp boundaries between adjacent clones are the result of active fighting between the outer individuals of each of the clones (Figure 6.27). In these fights, special tentacles called acrorhagi are used to inflict massive tissue damage to other anemones. The result of such clonal "border wars" is the separation of clones by anemone-free areas wide enough to prevent two anemones from reaching each other (Figure 6.28). As a result, clumps of anemones are always separated from each other.

Yet another example of biological interaction that produces spacing is **territoriality**. Territoriality is a pattern of behavior of an individual or group that actively keeps other members of the same or different

FIGURE 6.28 The aggregating anemone *Anthopleura elegantissima*, showing the anemone-free areas (bare brown area) between two genetically different clones. (Photo courtesy of Drs. Lovell and Libby Langstroth.)

FIGURE 6.29 The sea palm *Postelsia palmaeformis*, characteristic of areas of direct wave exposure and high wave shock on the Pacific coast of North America. (Photo courtesy of Dr. Richard Mariscal.)

species out of an area. This phenomenon is well known among various birds and mammals, but it also occurs among certain invertebrates. On the north Pacific coast of America, the largest limpet in the intertidal is *Lottia gigantea*. Stimson (1970) has shown that individuals of this species create a territory in the intertidal, which they actively defend against other *Lottia* and other limpets. Should other limpets invade this territory, the resident *Lottia* pushes the interloper out. In this way, each *Lottia* keeps a small patch of rock free of potential competition and ensures that all algae growing on the site will be consumed by itself.

Another more complicated combination of factors producing small patches is that observed by Dayton (1973) and Paine (1988) in the alga *Postelsia palmaeformis*, which occurs in isolated clumps in the most waveswept portions of the Pacific coast (Figure 6.29). *Postelsia* is an annual alga that inhabits areas dominated by the space-monopolizing mussel *Mytilus californianus*. The adult plants are immune to overgrowth by the mussels. The alga has a very limited dispersal of its propagules such that most settle within a few meters of the adult. These propagules may settle and begin to grow on almost any hard surface, but they are much more likely to survive to reproductive size if they settle on bare rock surfaces. *Postelsia*, although capable of settling, does not persist to adulthood in the presence of the common turf-like algae *Halosaccion* and *Corallina*. Therefore, the continued persistence and successful reproduction of *Postelsia* are dependent on a cyclical set of relationships. The heavy wave action is necessary to clear away patches of *Mytilus* back to the bare rock and to permit the settlement of the *Postelsia* propagules. The *Mytilus* are, in turn, necessary to eliminate the algal turf (*Halosaccion* and *Corallina*) too low to be removed by wave action and that outcompete *Postelsia*. The mussels do this by invading and overgrowing the lower-growing algae. If the mussels are not removed, the mussels will completely cover the intertidal. Removal of small patches of mussels by the violent wave action opens an area for *Postelsia* to occupy, provided its propagules colonize before the low-growing *Halosaccion* and *Corallina* take over the patches. The waveswept area, therefore, becomes a mosaic of patches of *Halosaccion* or *Corallina* and *Postelsia*. Patches of *Postelsia* result from its settling earlier than the two turflike algae, and *Halosaccion* or *Corallina* result from its first settlement of a patch. *Postelsia* may then persist in the patch it occupies only if the adult plants are ripped out after reproducing, leaving bare rock, or if another patch opens up within the dispersal range of the *Postelsia* propagules.

Predation and grazing can also create small patches or patchy distribution. Predation is the reason for the occurrence of *Balanus cariosus* (see p. 225) in small clumps in the intertidal. Grazing by sea urchins may result in tide pools free of algae, while others without urchins have an abundant growth. On the New England coast, as previously noted, tide pools without *Littorina littorea* are dominated by *Enteromorpha intestinalis*, whereas those with the snail have *Chondrus crispus*.

Another biological factor contributing to both large- and small-scale patchiness is the settlement preferences of various intertidal invertebrate larvae. A final biological factor to consider is succession, to which we have alluded and now take up in detail.

Succession

Creation of open spaces in any zone usually results in quick colonization by a series of **opportunistic** or **fugitive species**. These species quickly settle an open area, mature, and reproduce before the slower-growing dominant species retake the open space and force them out through competition. The most common of these opportunists are the smaller, filamentous green and red algae and such smaller animals as hydroids.

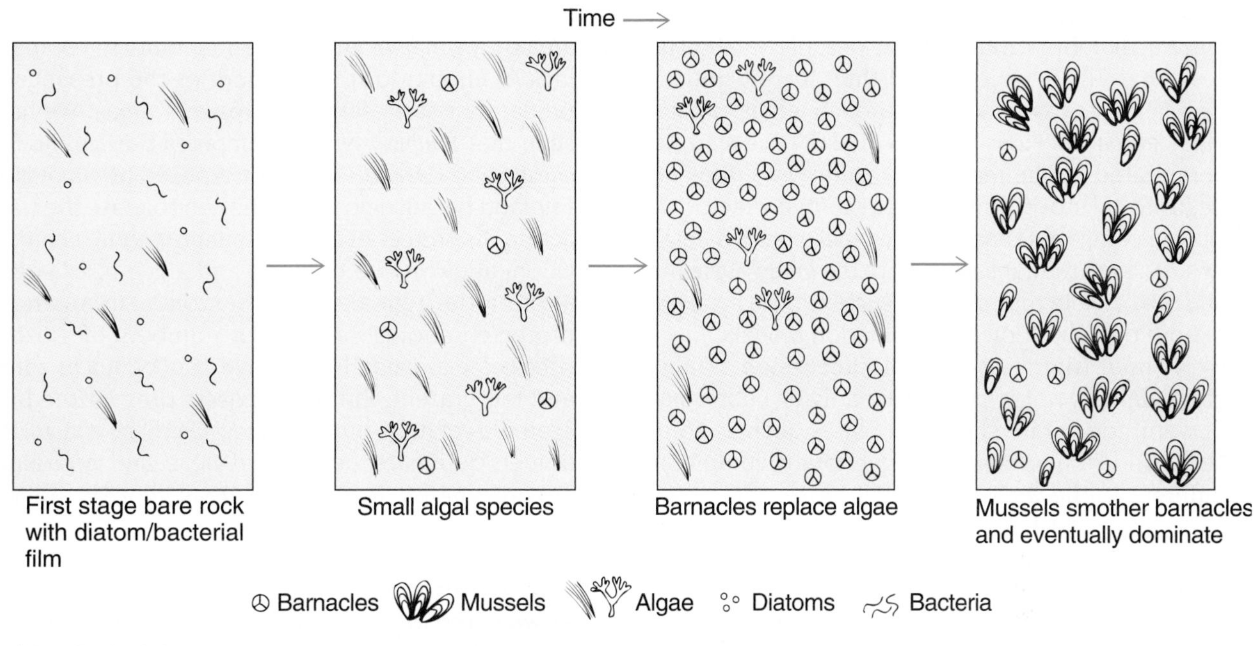

FIGURE 6.30 Succession in a northwest Pacific coast intertidal mussel bed in the absence of *Pisaster*.

These initial species are replaced successively by competitively better species until the competitive dominant in the area takes over (Figure 6.30). In any one zone, then, there may be many patches. Depending on the time since their clearance, they will be in differing stages of succession toward the dominant community characterizing the zone. This accounts for the different organisms in different patches in the same zone in the same area.

The ecological concept of succession, as we noted in Chapter 1, has several different models or hypotheses to explain the sequence of events following a disturbance in a community. In the rocky intertidal, succession seems to be a combination of the inhibition and facilitation models (see Chapter 1, pp. 21–22).

All successional sequences in the rocky intertidal depend on the time of the initial disturbance that frees the space. That is because the first colonizers of a space are usually those that have propagules ready to settle. Generally, the first organisms to settle on freed space are those with the life history characteristics of opportunistic species (see Table 1.2).

The sequence of events in succession is variable and dependent on various combinations of events in the rocky intertidal. On the Pacific coast of North America, Sousa (1980) found that succession in boulder fields followed the inhibition model. Once the early algal colonizer *Ulva* settled, it inhibited colonization by the perennial red algae, such as *Gigartina*. As long as *Ulva* remained healthy, it would prevent colonization by any other alga. However, this inhibition is broken by grazing of the crab *Pachygrapsus crassipes*, which permits the succession to proceed to a community of perennial red algae. These middle successional red algae, in turn, slow the invasion by the late successional dominant, *Gigartina canaliculata*, but eventually *Gigartina* colonizes because it is more resistant to desiccation and overgrowth.

In other instances, there does appear to be a successional sequence in which pioneer species are necessary before the later species can colonize (the facilitation model of succession). For example, Hawkins (1981), working on English rocky shores, showed that successional sequences were quite different on barnacle-covered rocks kept free of limpets than on areas scraped free of barnacles and also kept free of limpets. In the case of the barnacle-covered rocks, the colonization sequence was determined by the availability of propagules; but in the scraped areas, there was evidence that the larger green algae and *Fucus* could not settle until surfaces had been "prepared" by growth of diatoms and the green alga *Ulothrix*. The reasons for these differences in sequence are not clear, but it is suggested that the barnacle shells may have provided spatial refuges free from grazers, thereby facilitating "escapes" of the later successional algae, which could then recruit directly. Another suggestion is that although a mat of green algae enhances *Fucus* settlement, early *Fucus* growth is inhibited by the mat. As a result, *Fucus* does not begin to take over until either the mat dies or the slow-growing sporelings of *Fucus* finally reach a height or size where they are superior to the green algae mat and take over. In this example, succession is a mixture of both the inhibition and facilitation models.

A final topic is **seasonal succession** of algal communities, which occurs primarily on subtropical or warm temperate shores. In places such as south Texas and Hong Kong, the algal flora is completely different in the winter from that in the summer. In Hong Kong, which has a monsoonal climate, Kennish et al. (1996) found that foliose algae such as *Ulva*, *Porphyra*, and *Dermonema* dominated the shore during the winter but died off in the hot summer and were replaced by encrusting algae such as *Ralfsia*, coralline crusts, and encrusting cyanobacteria. In south Texas, Lowe and Cox (1978) described a winter algal community on Galveston Island dominated by *Enteromorpha*, *Bangia*, and *Gelidium* that shifted in the summer to a community dominated by *Cladophora*, *Bryocladia*, and *Ceramium*.

We can conclude this section by noting that continued physical disturbance of differing degrees of severity and physical extent coupled with seasonal variations in the types of propagules available, along with the fluctuations in the spatial distribution of dominant animals, ensure that succession is always occurring in the intertidal. This sequence—and on a larger scale—is one of the major sources of patchiness in the intertidal.

Horizontal Distribution Patterns

As we noted in the preceding sections, there are various changes in community structure as one moves from areas of high wave exposure to protected and quiet water. This is true for many geographical areas. Partly, these changes are caused by the action of water restricting certain species to or from waveswept areas. In other words, some species of algae or invertebrates cannot tolerate the wave action and are absent from waveswept areas. Usually, these are species with delicate structure or poor means of holding onto the substrate. In a similar fashion, certain species cannot tolerate quiet water areas. This may be due to their inability to compete, lack of enough or proper food, or the presence of predators not present in waveswept areas. We have seen that *Mytilus edulis* outcompetes *Semibalanus balanoides* and *Chondrus crispus* in exposed areas of New England because no predators can tolerate the wave action. In protected areas, predators reduce competition, allowing the barnacle and algae to exist and altering the appearance of the zones. In Australia, protected shores have larger numbers of grazers, whereas exposed shores have two barnacle zones and few grazers. European coasts present the best examples of horizontal changes. On these coasts, the limpets dominate the exposed areas, and large algae are uncommon. In sheltered areas, limpets are rare and large algae flourish.

Age Structure

As we noted in Chapter 1, terrestrial communities or associations consist of a matrix of long-lived plant species, which allows us to characterize terrestrial communities on the basis of the dominant plant species. In the marine environment, especially the intertidal, the long-lived species are various invertebrate animals, whereas the plants and the various algal species are more short-lived. We, in turn, name zones in the intertidal on the basis of animals: *Mytilus* zone, barnacle zone, and so forth. Even though the potential longevity is great for various marine invertebrates, it is well to consider under what conditions such longevity is achieved and what the age structure of the animals actually is.

As noted previously, on the Pacific coast of North America the high intertidal has a band of large *Balanus glandula*. Below that, any individuals of this species found are usually less than two years old. If the age structure of the large barnacles in the band is analyzed, it generally consists of a single age class. Why? Because the physical conditions are so harsh in this area that in most years the juvenile barnacles do not survive. As a consequence, the barnacles there result from a single good year in which conditions allowed survival through the vulnerable juvenile stage. As adults, the barnacles can survive succeeding harsh years. The reason all of the lower-level *B. glandula* are less than two years old is that, in two years, the carnivorous gastropods of the genus *Nucella* consume all midtide barnacles. Simi-

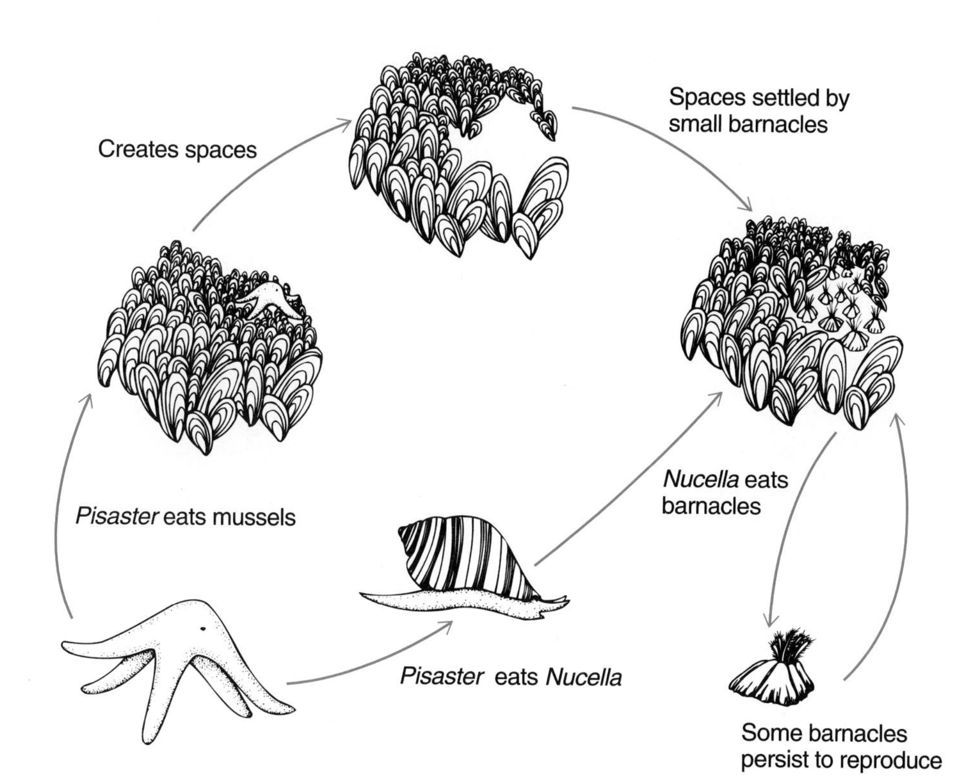

FIGURE 6.31
Interactions among mussels (*Mytilus*), barnacles, and their predators on the northwest Pacific coast of North America; these interactions allow barnacles to persist in the intertidal zone.

larly, *B. cariosus* populations in the midintertidal consist primarily of animals less than two years old. There are, however, a few that may be large and many years old. Again, the majority are less than two years old, because that is the time during which *Nucella* consume them. The few large, old individuals are those that somehow avoided predation until they were two years old and invulnerable to *Nucella*. Unless consumed by *Pisaster*, these individuals may live for many years. Thus, under very stringent environmental conditions or under optimal conditions low in the intertidal, the age structure of the dominant organisms is mainly of single-year classes.

Predators also seem to consist mainly of older individuals. Examples are *Pisaster ochraceus* and *Nucella lamellosa*. Herbivores, such as *Lottia gigantea* and the turban snail *Tegula funebralis*, also have populations dominated by older individuals. Why this should be the case is not known at present.

Interactions Among Factors: A Summary

We have seen how various physical and biological factors cause the large-scale vertical distribution patterns of algae and invertebrates that we call vertical zonation on rocky shores. We have also seen how some of these factors, plus others, introduce small-scale distribution anomalies into this pattern. In general, different factors operate in different areas and at different levels on the shore. The high intertidal tends to have the distribution pattern of its characteristic species set by physical factors, whereas biological factors become more important in setting limits in the low intertidal. It is important to realize that these factors do not act in isolation and the final distribution pattern observed, on either small or large scales, is often the result of the interaction of two or more factors (Figure 6.31).

Tide Pools

A characteristic feature of many rocky shores is the presence of tide pools of various sizes, depths, and locations. Certain conditions affecting life of tide pools differ markedly from the surrounding intertidal and necessitate a separate discussion here. Our remarks in this section will be restricted to those pools that undergo a complete interchange with the

FIGURE 6.32 An intertidal rock pool at low tide. (Photo by the author.)

ocean water during the tidal cycle (Figure 6.32). It should be pointed out at this juncture that the biotic communities of tide pools and the factors influencing their structure are less well understood and studied than those of the emergent rocky surfaces. It has even been suggested that tide pools do not represent an intertidal habitat because they are never exposed to the air during a tidal cycle. However, the fluctuations in physical and chemical factors in tide pools are a function of the tidal cycle; hence, it seems logical to consider them in this chapter. Furthermore, many of the organisms found in tidal pools are similar to those found on the adjacent exposed rock.

At first glance, tide pools appear more or less ideal places for aquatic organisms seeking to escape the harshness of the intertidal during its exposure to air. In reality, however, escape from such physical factors as desiccation may mean exposure to others that operate more severely in tide pools.

Tide pools vary a great deal in size and in the volume of water they contain. Since water is a great moderator of harsh physical conditions, the larger the pool and the greater the water volume, the less the fluctuations in physical factors. Other factors in addition to volume influence the physical and chemical conditions of the water held in a tide pool. These include the surface area, depth of the pool, height in the intertidal, exposure to wave action and subsequent splash, degree of shading, and drainage pattern. In addition, the physical and chemical environment of tide pools can vary vertically with depth and horizontally across the pool. Finally, all physical factors may fluctuate diurnally and seasonally. Given all these variables, Metaxas and Scheibling (1993) suggest that it is unlikely that any two tide pools will be similar in all characteristics; therefore, individual pools are unique physically. If this is true, it means ecologists cannot replicate experimental manipulations in different tide pools, and this may be a significant reason that there has been so little experimental ecological work done on pools. It may also be the reason that most studies have been primarily descriptive.

Three major physical factors are subject to variation in tide pools. The first is temperature. Whereas the ocean itself is a vast reservoir that heats and cools very slowly and usually within very narrow limits, the same is not true of tide pools. These relatively small bodies of water are subject to more rapid changes. Shallow tide pools exposed to the sun on warm days may quickly reach lethal or near lethal temperatures. Similarly, tide pools in cold temperate or subpolar regions may have temperatures in the freezing range in winter. An additional problem is temperature fluctuation. The pool may either heat up or cool down over a several-hour period while exposed to the air, but when the tide returns, it will be flooded with ocean water. This will suddenly change the temperature of the whole pool. Variations in daily temperature in a tide pool can be as much as

15°C, depending on its height in the intertidal, volume, degree of shading, and wave exposure. Finally, change in temperature during exposure at low tide may cause temporary thermal stratification of the pool. Organisms inhabiting such pools must still be adapted to considerable temperature fluctuation.

The second factor to vary in pools is salinity. During exposure at low tide, tide pools may heat up, and evaporation occurs, increasing salinity. Under hot tropical conditions, the salinity increase can be dramatic enough to reach the point of precipitating out salt. The opposite situation is the case when heavy rains occur at low tide and flood pools with fresh water, dramatically lowering the salinity. Fluctuations in salinity in tidal pools vary with the position of the pool on the shore and the other factors mentioned, but values that have been measured have ranged between 5 and 25 psu. Salinity stratification also may develop in the pool due to fresh water runoff during heavy rains or as a result of freezing in winter and evaporation in summer. Again, tide pool animals and plants may have to be adapted to wider ranges in salinity than typical marine or intertidal organisms. As before, when the tide returns, the pool will be flooded with seawater at some point, and there will be an abrupt return to normal conditions.

The final physical factor undergoing change in pools is oxygen concentration. Since the amount of oxygen that can be held in seawater is a function of temperature, it follows that tide pools that heat up during exposure to the air will lose oxygen. Under normal conditions, this may not be serious enough to produce oxygen stress, but if the pool is crowded with organisms, it may produce a stress situation. For example, a pool filled with algae that was exposed at night would produce a situation in which the lack of photosynthesis coupled with high respiration could reduce the oxygen level significantly. Oxygen level has been recorded falling to only 18% of saturation in tropical tide pools. It is also possible to develop an oxygen stratification in the water column of certain pools.

Tide pools are areas of refuge from desiccation for intertidal organisms, but in turn, these organisms suffer from rapid changes in temperature, salinity, and occasionally oxygen; thus, the fauna and flora is restricted to those organisms able to tolerate such ranges.

The organisms that inhabit tide pools are similar to those on the adjacent emergent substrates but often with differences in abundances between the pools and exposed surfaces. For example, several genera of algae, such as *Spongomorpha* and *Corallina* in the intertidal of Maine, *Prionitis* in Washington, and *Fucus* in Nova Scotia, are more abundant in pools. Other species are either absent or occur in lower numbers in pools than on exposed surfaces. An example is the alga *Ascophyllum nodosum* in New England. Finally, the occurrence of tide pools permits some organisms to extend their range upward in the intertidal beyond the levels to which they would be limited on the adjacent emergent surface. This is true for a great many algae and invertebrates and is particularly true for many fishes.

Zonation patterns have been described for tide pool organisms both within the intertidal zone and vertically within the tide pools. Along the intertidal gradient such green algae as the genera *Enteromorpha*, *Cladophora*, and *Chaetomorpha* dominate the upper tide pools, while such brown algae as *Fucus* and *Laminaria* and the red corallines, such as *Lithothamnion* and *Corallina*, are abundant in the lower intertidal pools. Benthic invertebrates also show zonation. High pools are inhabited by littorine snails, whereas low tide pools have a greater diversity of snails and all invertebrates. Fish zonation in tide pools has not been quantitatively documented according to Metaxas and Scheibling (1993), but qualitatively, there is a decrease in the number of species with increasing height of the pool. In general, both in number of species and biomass, high tide pools are depauperate compared with lower-level pools.

The role of biological interactions in determining the community structure of tide pools is poorly documented when compared to the studies for the exposed rocky intertidal. This is due not only to a relative lack of studies, but also to the variability among adjacent tide pools at the same intertidal level when subjected to experimental manipulation. Such variability has precluded obtaining statistically significant results in many cases, although correlative data often indicate a trend. We discuss some of these data here.

Herbivory has similar effects in tide pools as compared with those recorded for the emergent rock surfaces, namely, the altering of macroalgae abundance. For example, Paine and Vadas (1969) demonstrated that the removal of sea urchin grazers from tide pools in Washington resulted in increases in macroalgae. Similarly when Lubchenco (1978) added littorine snails to a pool, the density of the dominant

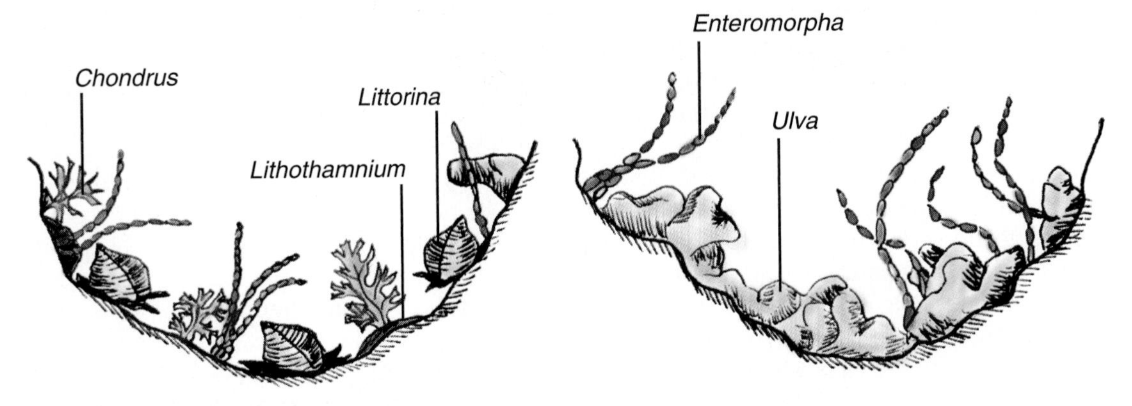

High-diversity pools with snails

Dominated by a mixture of palatable green seaweeds (*Ulva, Enteromorpha*) and slower-growing and well-defended species such as algal crusts (*Lithothamnium*) and Irish moss (*Chondrus*).

Low-diversity pools without snails

Dominated by fast-growing, competitively dominant green seaweeds such as *Ulva* and *Enteromorpha*. Green crabs colonize these pools and can limit the recruitment of herbivorous snails.

FIGURE 6.33 Diagrammatic representation of New England tide pools with and without grazers. (Modified from *The Ecology of Atlantic Shorelines*, M. D. Bertness, p. 234. Copyright © 1999, Sinauer Associates, Inc. Reprinted by permission.)

alga *Enteromorpha* was reduced, whereas when she removed littorines from a pool dominated by the alga *Chondrus crispus*, the density of *Chondrus* decreased (Figure 6.33). In Nova Scotia, Chapman and Johnson (1990) found that littorinid snails had a negative effect on the abundance of several species of the brown algal genus *Fucus*, a positive effect on the ephemeral algae, and no effect on the crustose alga *Hildenbrandia*.

Very few studies have investigated the role of predation in structuring tide pool communities. Indeed, there seem to be no well-documented manipulative studies. The few studies that have been done suggest that addition of predators to pools reduces the abundances of various prey organisms, but whether or not the predator exerts control over the community structure is uncertain.

The evidence for the importance of interspecific competition in regulating community structure in tide pools is also sparse. Lubchenco (1982) and Chapman (1990) have both demonstrated decreases in canopy cover in *Fucus* due to competition with ephemeral algae. Kooistra et al. (1989) have shown competitive dominance as indicated by overgrowth for the alga *Halichondria panicea* in tide pools in Brittany, France.

Although recruitment can be a significant factor in structuring tide pool communities, no studies have considered this factor directly. Similarly, there is little or no information about the role of disturbances, particularly large-scale episodic events, in the structuring of tide pool communities.

SANDY SHORES

Intertidal exposed sand beaches and protected sand flats are common throughout the world and are certainly better known than rocky shores to most humans, because they are used for various recreational activities. At the same time, they present a marked change from the previously described temperate rocky shores because, in contrast to the crowded life on the latter, exposed sand beaches appear devoid of macroscopic life. The environmental factors acting on these shores create conditions where virtually all organisms bury themselves in the substrate. Intertidal protected sand flats by contrast are more often populated, with large numbers of macroorganisms visible on the surface.

Exposed sand beaches are subject to more wave action and are usually facing the open sea, whereas sand flats commonly have less exposure to wave action and usually are facing a bay or lagoon protected from the open sea by a barrier island. Exposed sand beaches usually have a pronounced slope toward the sea, whereas sand flats have little or no gradient or slope.

In this section, we first explore the action of physical factors on sand beaches and intertidal protected

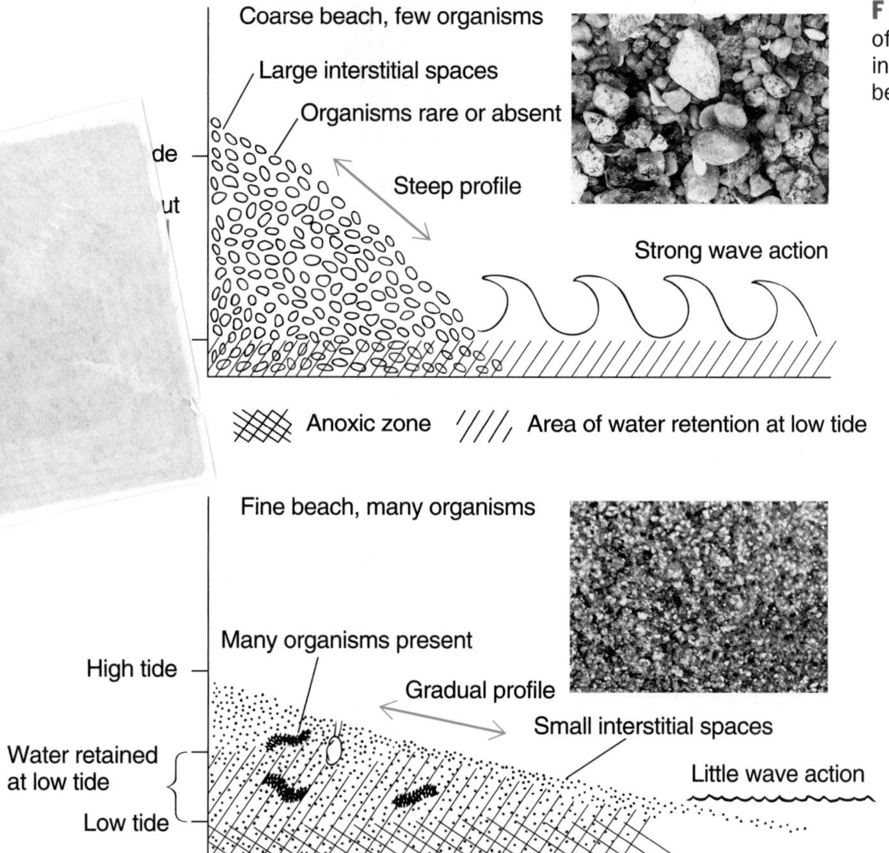

FIGURE 6.34 Comparison of the physical conditions found in fine-grained and coarse-grained beaches.

sand flats and see how these factors act to impose certain adaptations on the larger organisms found in this area. We then investigate the biological factors and how they act along with the physical factors to structure the communities. The specialized conditions and organisms existing in the tiny spaces between the sand grains are the subject of Chapter 7.

Environmental Conditions

In comparison with rocky intertidal shores, sand beaches and flats are subject to a similar array of physical factors, but the relative importance of these factors in structuring the community and their effect on the substrate differ.

Perhaps the most important physical factor governing life on exposed sand beaches is wave action and its attendant correlation with particle size and slope. Anyone who has visited such marine beaches is aware of the fact that the particle size and slope of the sand can differ among beaches and also among seasons on the same beach. Particularly in the temperate zone, the profile of the beach also changes between winter and summer. Beaches are defined by three factors: *particle size*, *wave action*, and *slope*. All three are interrelated such that defining two fixes the third. The importance of the particle size to organism distribution and abundance rests with its effect on *water retention* and its suitability for burrowing. Fine sand, through its capillary action, holds much water in its interstices after the tide has retreated. Coarse sand and gravel, on the other hand, allow water to drain away quickly as the tide retreats. Since the organisms inhabiting the intertidal are aquatic, they are well protected against desiccation in a fine sand beach but subject to desiccation in a coarse gravel beach. This makes the latter less hospitable. Fine sand is also more amenable to burrowing than coarse gravel (Figure 6.34).

The slope of a beach is the result of the interaction between the particle size, wave action, and the relative importance of swash and backwash water. **Swash** is water running up a beach after a wave breaks; this action carries particles with it, which may cause accretion of the beach if they remain there.

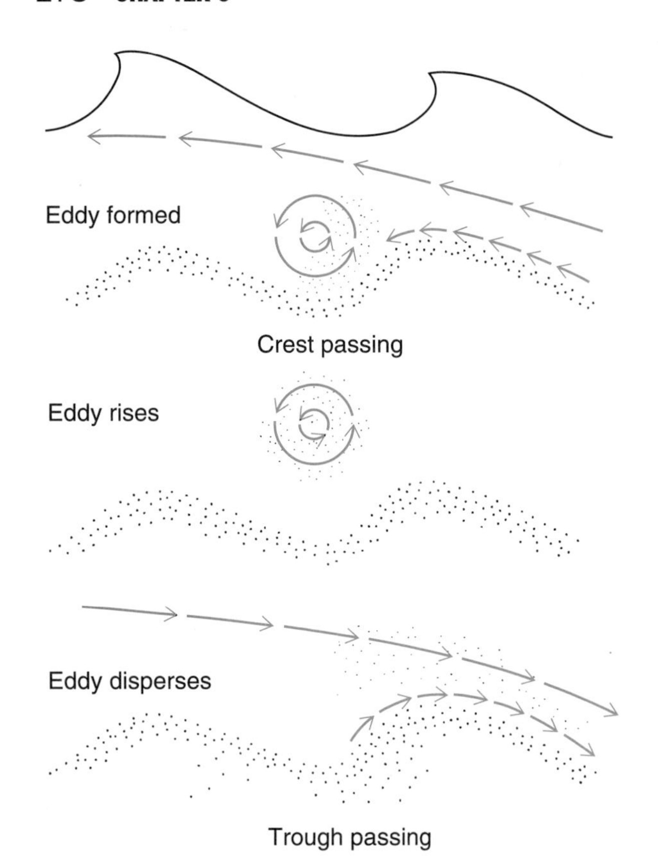

FIGURE 6.35 Transport of sand particles by wave action. (After "Treatise on Marine Ecology and Paleoecology" J. Hedgpeth, 1957, Vol. 1, *Ecology*, © 1957 Geological Society of America.)

Backwash is the water flowing back down the beach; this action removes particles from the beach, depending on the particle size.

Protected sand flats are much less seasonally variable, and they usually consist of finer-grained sand at all times. In protected sand flats, both waves and water currents affect grain size. The magnitude of the wind is largely a function of the size of the body of water over which the wind blows, and the large embayments on the East Coast of the United States, such as Chesapeake Bay and Pamlico Sound, are large enough to generate waves that can have an effect on grain size. In New England, because the embayments are all small, wind and wave action are less important.

A second important physical factor in sand beaches and tidal flats, which was not of concern in rocky shores, is also the product of wave action. That is wave-induced *substrate movement*. The particles on sand or gravel beaches are not large enough to be stable when waves strike the shore. As a result, with each passing wave, the substrate particles are picked up, churned in the water, and redeposited (Figure 6.35). Particles are being continually moved and sorted. The reason fine sand occurs only where wave action is light and coarse sand where it is heavy is that, in heavy wave action, the smaller particles remain in suspension so long that they are carried away from the beach, leaving only those heavy enough (coarse) to settle immediately. Light wave action also means that a smaller depth of particles is affected by the passage of a wave, whereas heavy wave action disturbs the substrate to a greater depth.

Within this continuum of grain size, slope, and wave action, a number of different beach types have been described. The extremes of these beach types are the dissipative and reflective (Figure 6.36). A **dissipative beach** occurs where wave action is strong but the wave energy is dissipated in a broad, flat surf zone located some distance from the beach face. This leads to a gentle swash and fine sediments. The beach face has a gentle slope and is maximally eroded, with much of the sediment occurring in the broad surf zone and in berms or bars parallel to the beach. A **reflective beach** is most fully developed where wave action impinges directly on the beach face, and the sediment is coarse. In these beaches, there is no offshore surf zone and waves produce large swashes up the beach face. The beach slope is steep, and the backwash and swash collide to deposit sediment. Wave energy is directed against the face of the shore and reflected off the face.

Studies by McLachlan (1983) and others have demonstrated that the abundance and diversity of macrofauna organisms are more strongly correlated with particle size and slope of open beaches than with wave action. Very exposed beaches often have more abundant faunas than less exposed beaches if the latter have coarse grains and steep slopes. Fine sand beaches usually have flatter slopes, and wave action is dissipated in the surf zone rather than on the beach itself. This means that a very exposed beach can support a rich macrofauna if it is dissipative and not reflective. Indeed, a high biomass of filter feeders is strongly correlated with dissipative conditions. Intertidal protected sand flats show much less local variability in grain size.

Since the substrate, at least the surface layers, is in constant motion and may be picked up and held in suspension for greater or lesser amounts of time, depending on the strength of the waves, it follows that any change in wave intensity on a given beach will

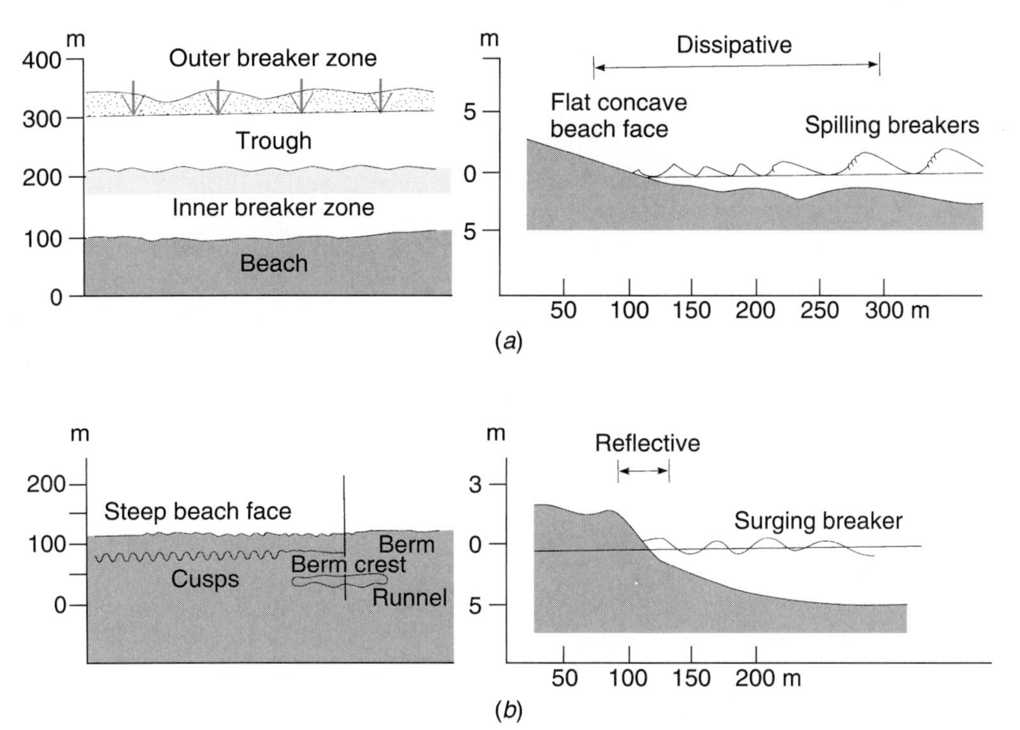

FIGURE 6.36 (a) Dissipative beach. The left side of the figure shows the beach from above; the right side is a cross-sectional view. (b) Reflective beach. The left side of the figure shows the beach from above; the right side is a cross-sectional view. (After *Ecology of the Sandy Shores*, A. C. Brown and A. McLachlan. Copyright © 1990 with permission from Elsevier Science)

mean a change not only in grain size but also in the profile or shape of the beach. This happens when gentle wave action is replaced by heavier waves, which dig deeper into the beach, pick up more and heavier grains, and hold them in suspension longer, so they are deposited some distance away, usually offshore. This changed profile is a common seasonal occurrence on many temperate zone beaches, where a gentle slope of fine sand occurs during the summer months and is replaced by a steep, coarse beach during winter storms (Figure 6.37). It is significant for organisms that the depth of substrate moved can be a meter or more!

Because of this ceaseless movement of surface layers of the sediment, few large organisms can permanently occupy the surface of open sand or gravel beaches. This is the reason for the barren appearance of such beaches.

In contrast to rocky shores, sand beaches and sand flats usually have a smooth, uniform profile; thus, they lack the great topographic diversity of rocky shores. There are no crevices, overhangs, permanent tide pools, or slopes facing different directions offering different moisture conditions. As a result, such environmental factors as temperature, desiccation, wave action, and insolation act uniformly at each tidal level on the beach or flat.

Because of its significant effects on particle size and slope of a beach, wave action is a dominant environmental factor acting on the exposed sand beaches, creating the special conditions that make it difficult, or impossible, for many organisms to inhabit this area. The much reduced wave action in protected sand flats, on the other hand, makes them suitable for a host of epifaunal and infaunal invertebrates. Both sand flats and sand beaches offer some advantages to marine organisms with respect to certain physical factors. Sand is an excellent buffer against large temperature and salinity changes. Measurements of temperature with depth on sand beaches at low tide have demonstrated that, below the first few centimeters, the temperature is very nearly that of the surrounding seawater. This amelioration of temperature is partly due to the insulating properties of sand and partly due to the water held in the interstices of the deeper layers. In a similar manner, salinity changes are also minimal below 10–15 cm on beaches, even if the upper layers have fresh water flowing over or falling on them. Again, this is because the water held in the interstices is salt water, which has a higher density than fresh water. As a result, the fresh water remains perched on the surface. Furthermore, sand is a barrier to any harmful effects of exposure to direct sunlight (insola-

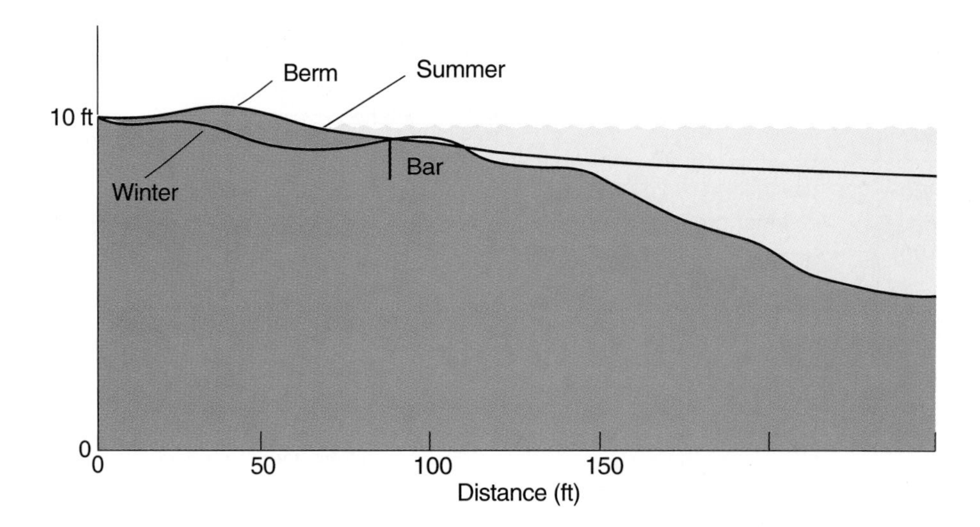

FIGURE 6.37 Beach profiles showing the difference between winter and summer.

tion) for any organism living in the sand, because the sand is opaque to light and reflects or absorbs it in the surface layers. Desiccation is not a problem as long as the beach sand is fine enough to hold water by capillary action during periods of low tide. Any organism burrowing into the sand will be constantly moist.

A final physical factor acting on beaches is oxygen content. Oxygen is never limiting in the water bathing the beach, because the turbulence of wave action ensures constant saturation. Where it might become limiting is in the substrate itself. The water held in the beach or sand flat ameliorates changes in temperature and salinity and also contains oxygen, which is available to the organisms. This supply, however, is used up by respiration of the organisms and must be replenished. This replenishment comes through interchange of the water with the sea above, and it depends on the fineness of the sediments. Fine sediments have a slow rate of exchange, and coarse sediments have a rapid exchange. In fine-grained beaches, particularly protected sand flats, water interchange is slow and may result in reduced oxygen or anoxic conditions below the uppermost layers. It is well to remember, however, that various organisms that live in the sediment, such as crabs and worms, may deepen the oxygenated layers through building and maintaining burrows.

Adaptations of Organisms

The dominant environmental factor acting on open sand beaches is wave action, which creates the unstable, constantly moving substrate. To inhabit this area, organisms must first be adapted to tolerate these features. Two routes may be taken by organisms in adapting. The first is for the organism to burrow deeply enough into the substrate such that the organism is deeper than the depth of sediment affected by the passing wave. This strategy is employed by many large clams, such as *Tivela stultorum*, the Pismo clam (Figure 6.38a). Such animals are usually also aided by developing a heavy shell, which helps to keep the animal situated in the substrate. They often have long siphons that enable them to burrow deeply. The only difficulty with this type of adaptation is that a severe storm may generate waves large enough to pull these animals out and throw them on the beach. That is the reason for the catastrophic destruction that occasionally occurs in these forms.

The second route of adaptation is for the organism to burrow very quickly as soon as the passing wave has removed the animal from the substrate. This is the more common mechanism and is employed by many annelid worms, small clams, and crustaceans. A good example of this type of adaptation can be found in the various sand crabs of the family Hippidae, which populate many open beaches around the world. These animals have a short body with limbs highly modified to dig quickly into wet sand. As soon as they are freed from the substrate by a passing wave or they leave by moving to the surface, they are able to burrow quickly before the water motion can carry them offshore (Figure 6.38c). Small clams of the genus *Donax* also do the same thing, and razor clams of the genera *Siliqua* and *Ensis* are extraordinarily fast at burrowing.

Other adaptations are correlated with the burrowing of these organisms. For example, some mollusks, such as *Olivella biplicata* (Figure 6.38b), have

(a)

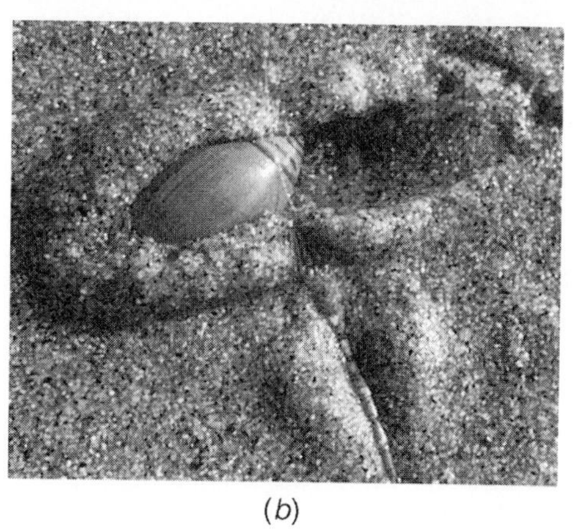

(b)

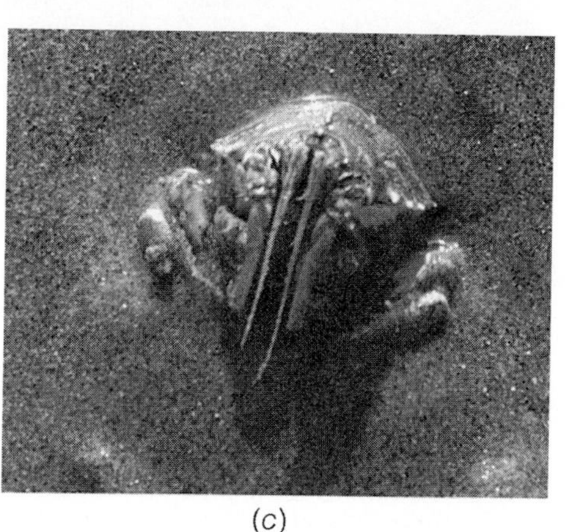

(c)

FIGURE 6.38 Some sandy beach animals of the Pacific coast of North America. (a) *Tivela stultorum* (Pismo clam). (b) *Olivella biplicata* (olive snail). (c) *Blepharipoda occidentalis* (sand crab). (Photos by the author.)

very smooth shells that reduce the resistance to burrowing into the sand, while others, as Stanley (1969) has noted, have developed special ridges on the shell to grip the sediment to aid in penetrating the substrate. Similarly, such shore echinoderms as sand dollars have much reduced spines to allow them to burrow into the sand.

A special adaptation observed by Chia (1973) in small sand dollars(*Dendraster excentricus*) is that they accumulate iron compounds in a special area of their digestive tracts. This iron serves as a "weight belt" to keep them down in the presence of wave action.

Reproductive problems facing sandy beach organisms include when to reproduce and whether to reproduce repeatedly (iteroparous) or only once (semelparous). A critical factor here is to coordinate gamete production or release of young with the tidal cycle so that the young are not stranded by the tide or consumed by predators. To this end, most sandy beach animals display lunar rhythms of reproductive activity. Many are also iteroparous, although at present we cannot demonstrate that the semelparous forms are at any particular disadvantage in this environment. The mode of development has a latitudinal gradient, with the tropical species having a majority of planktotrophic larvae and the temperate zone species having more lecithotrophic larvae.

A common aspect of many macrofaunal animals is their tendency toward gregariousness, about which we know very little. Perhaps the tendency is due to water movements. On the other hand, it may be biological, as when we see scavengers congregating on large food masses. At present, this is all speculation.

Sandy beach macrofaunal animals are vulnerable to predation, and many show adaptations to avoid predation. Chief among these is the ability to burrow deeply, as exhibited in the bivalves *Donax* and *Tellina* and the gastropod *Bullia*. Also, the tidal migrations of *Bullia* and the hippid crabs, which tend to keep them in the swash zone, may reduce predation by birds or fishes, which usually cannot feed in these areas. Crustaceans of the upper beach areas often avoid diurnal

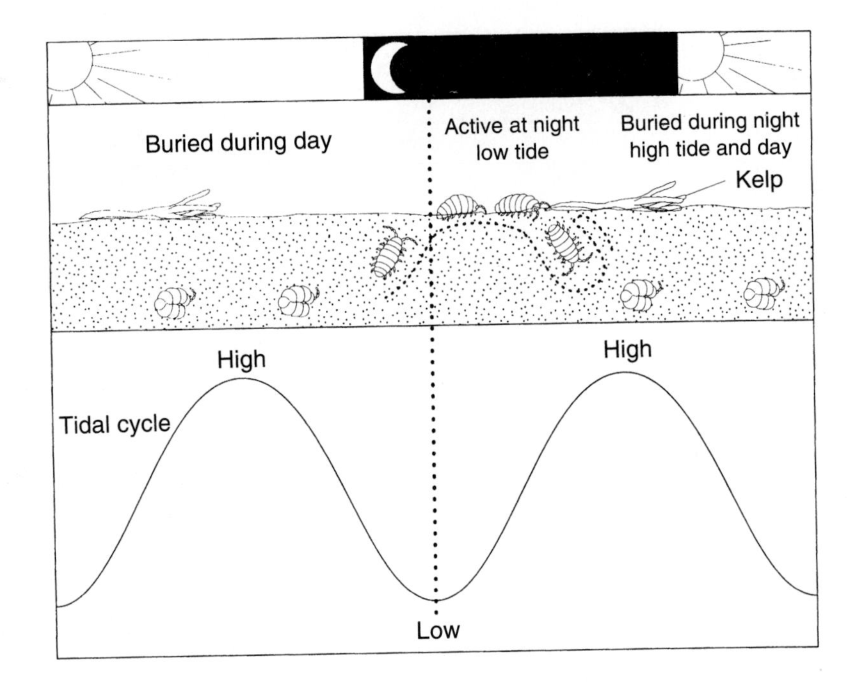

FIGURE 6.39 Endogenous activity rhythms of the semiterrestrial isopod *Tylos*. *Tylos* is active only during low tide at night, when it emerges from its burrow to roam the shore in search of the kelp on which it feeds. (After "Sandy Beach Ecology" A. McLachlan in *Sandy Beaches As Ecoystems*, eds. A. McLachlan & T. Erasmus, pp. 321-360. Copyright © 1983 Dr. W. Junk, with kind permission from Kluwer Academic Publishers.)

predators by residing in deep burrows during the day or by hiding under piles of wrack (Figure 6.39).

A final set of adaptations has to do with preventing clogging of respiratory surfaces by the resuspended sand. To prevent this from occurring, the intake siphons of sandy beach clams are often fitted with various screens that prevent entrance of sand but allow passage of water. Similarly, in sand crabs, the antennae, when held together, form a tube to the surface through which water enters the branchial chamber. These are densely clothed with closely spaced hairs that prevent entrance of sand.

Types of Organisms

The most conspicuous organisms absent from sand beaches are large plants. No macroscopic flowering plants or large algae occur on open sand beaches, presumably because there is no way for them to attach and maintain themselves in the wave action. Perennial macroscopic plants, primarily seagrasses, are common in the lowest intertidal areas of many protected sand flats of the North American Atlantic coast (see discussion of seagrass beds in Chapter 5) along with certain ephemeral algae, such as *Ulva* or *Enteromorpha*, which may be seasonally abundant. On open sand beaches the only primary producers are certain benthic diatoms and surf-living phytoplankton. These microflora organisms often exhibit vertical migration either within the sediments or between the sediment and the water column. Migrations may lead to concentrations of these autotrophs into highly productive patches. Protected sand flats, however, support a large and diverse microflora of benthic diatoms, dinoflagellates, and cyanobacteria, which typically appear as a brownish or greenish film on the sediment surface.

The second major group of organisms conspicuous by its absence is the sessile animals, such as barnacles and mussels, so dominant on rocky shores. Again, there is no place for them to attach. The sand beaches and intertidal flats are dominated by representatives from three invertebrate groups: polychaete worms, bivalve mollusks, and crustaceans. Various combinations of species from these three groups dominate sand beaches and protected sand flats throughout the world. There is a tendency for crustaceans to be more abundant on exposed and tropical beaches and bivalves to be the more abundant on protected and temperate shores. The total number of macrofaunal species of all groups increases with decreasing wave exposure, but the highest biomass comes from exposed beaches.

Feeding Biology

The absence of large multicellular plants and algal mats means that there is very little primary productivity on open sand beaches. Although diatoms are present, the opaqueness of the sand ensures that the

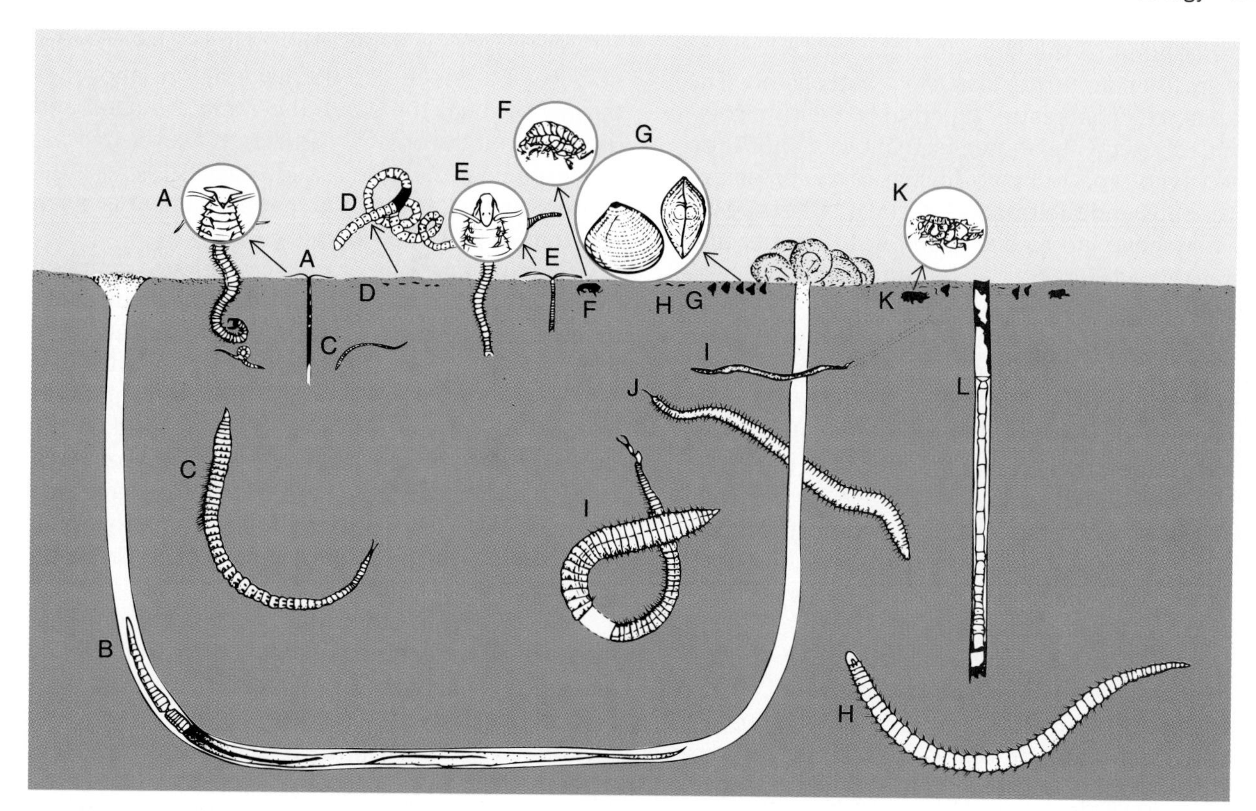

FIGURE 6.40 Some representative New England sand flat benthic invertebrates, indicating general life habits. (Not to scale.) Surface deposit feeders: (A) *Spiophanes bombyx* (spionid polychaete); (B) *Saccoglossus kowalewskyi* (protochordate); (E) *Pygospio elegans* (spionid polychaete). Burrowing deposit feeders: (C) *Aricidea* sp. (paraonid polychaete); (D) oligochaete; (H) *Exogone hebes* (syllid polychaete); (I) *Scoloplos* spp. (orbiniid polychaete); (J) *Nephtys* spp. (nephtyid polychaete). Suspension feeders: (G) *Gemma gemma* (venerid bivalve); (F) *Protohaustorius deichmannae* (haustorid amphipod); (K) *Acanthohaustorius millsi* (haustorid amphipod). Deep deposit feeder: (L) *Clymenella torquata* (malkanid polychaete). (From R. Whitlatch, 1982, *The ecology of New England tidal flats: A community profile*, U.S. Fish and Wildlife Service, Biological Services Program.)

diatom population is restricted to the surface layers. As a result, there are no macroscopic herbivores on a sand beach. Since there is virtually no primary productivity, the animals living on the beach must depend for food upon either the phytoplankton carried in seawater above the beach, the organic debris brought in by the waves, or consuming the other beach animals. These food sources dictate a benthic invertebrate association on open sand beaches dominated by filter feeders, detritus feeders, and scavengers. On protected sand flats, the productivity is confined to the microfloral films, seagrasses, and ephemeral macroalgae, but even the latter do not appear to be grazed extensively, leaving 90–95% of these macrophytes to be broken down into detritus. There are relatively few resident carnivores among open-beach animals, because to be such would require active movement across the substrate in search of prey. This is probably not feasible in the face of the heavy wave action. In addition, the erratic nature and unpredictability of large food items, either living or dead, mean that specialized feeders are rare, and most carnivores and scavengers are opportunists.

Suspension-feeding animals are those that filter particles out of the water column. These particles may be plankton organisms or, particularly on the sand beach, they may be various organic particles resuspended from the bottom by the passing waves. The dominant group of suspension feeders on sand beaches are the bivalve mollusks, such as razor clams (*Siliqua*, *Ensis*), surf clams (*Tivela*, *Spisula*), and coquinas (*Donax*). On protected sand flats, suspension feeders include small bivalves (*Gemma*) and amphipods (Haustoriidae; Figure 6.40). Filter-feeding bivalves in sand beaches and flats transfer a significant portion of the primary productivity in the water column to the benthic organisms wherever they are abundant. They also affect water clarity. The importance of this latter function to benthic communities was demonstrated by Ulanowicz and Tuttle (1992), who reported that the loss of the extensive oyster beds and their water filtering in Chesapeake Bay may have led to decreased water clarity and decreased productivity by benthic algae and seagrasses.

Open sand beaches tend to have less organic detritus than protected sand flats, but enough debris from various sources finds its way to the beach to be a reliable source of food for certain organisms. This detrital material is often carried up and down the beach, suspended in the wave wash, rather than being deposited on the bottom. Therefore, the mechanisms employed by detritus feeders on the open beach are often different from those used by detritus feeders on protected shores.

The sand crabs of the family Hippidae, such as *Emerita analoga* and *Blepharipoda occidentalis* of the Pacific coast of America, employ a unique mechanism to trap debris in the wave wash (see Figure 6.38). MacGinitie and MacGinitie (1949) observed that *Emerita* orient so that their heads point shoreward. They remain completely buried as the incoming wave passes over them shoreward. After passage, as the water begins to run down the beach, they stick out their very large second antennae. These are thickly clothed in hairs forming a net that, when spread into the backwash, intercepts all particles within a certain size range. The particles are then wiped off by the mouth and ingested. As the tide rises and falls, the position of the wave-wash front also moves. These crabs take advantage of this to maximize their feeding time by being moved up and down the beach with the tide. They move by surfacing as a wave rolls in, being carried shoreward some distance by the wave, and then quickly reburrowing. In a similar manner, as the tide recedes, they emerge and roll down the beach before reburrowing and commencing feeding.

A similar method of feeding is employed by the gastropod mollusk *Olivella columellaris* on the Pacific coast of Central America. In this case, a mucous net is produced by the animal and held across the wave backwash. This net filters and captures small particles. The animal feeds by consuming the entire net with the particles it has caught, and thus it must produce a new net at frequent intervals.

Most sand beaches also include one or more polychaetes. Much like earthworms on land, they burrow through the sand, ingesting the sand and digesting out the organic particles. However, they are rarely abundant on open sand beaches because the amount of organic material in the sediment is much less than in protected shores.

Sand dollars, one of the echinoderms common on sandy beaches, are generally detritus-feeding animals. These animals usually burrow through the sand. As they do so, the finer organic particles (but not the sand grains) fall down between the short spines, are trapped in mucus, and carried to the mouth. Some species, such as the Pacific coast *Dendraster excentricus*, according to Timko (1976), have even modified this basic feeding process to remove suspended particles by projecting part of the body up into the water column.

On protected sand flats, detritus feeders are more abundant in terms of numbers of both species and individuals. These deposit feeders play an important role by converting detrital organic material into macrofaunal biomass, which is subsequently available to higher trophic levels. Because of their abundance, size, and feeding mechanisms, deposit feeders are also significant bioturbators that disturb the upper layers of the substrate and thus influence the nature of the sediments. By far the dominant group is the polychaete worms. Two types of deposit feeders are recognized: surface deposit feeders and burrowing deposit feeders (see Figure 6.40). Most deposit feeders are generalized feeders whose diet is limited by the size of the particle they can ingest. Surface deposit feeders may live in vertical tubes (spionid and terebellid polychaetes) or burrow in the surface layer (amphipods). Still others, such as the mud snails *Ilyanassa* and *Hydrobia* and the fiddler crabs, collect particles from the surface and, according to Levinton and Lopez (1977), may affect the primary productivity by consuming the small surface-dwelling microalgae. Burrowing deposit feeders consist primarily of polychaete worms.

Resident carnivores are rare, particularly on open sand beaches, but the few found usually are polychaete worms, large crabs, or gastropod mollusks. The polychaetes, such as species of the genera *Nephtys* and *Glycera*, burrow through the sand and seize small prey items with their eversible proboscis. The carnivorous gastropods are represented most commonly by burrowing snails of the families Naticidae (moon

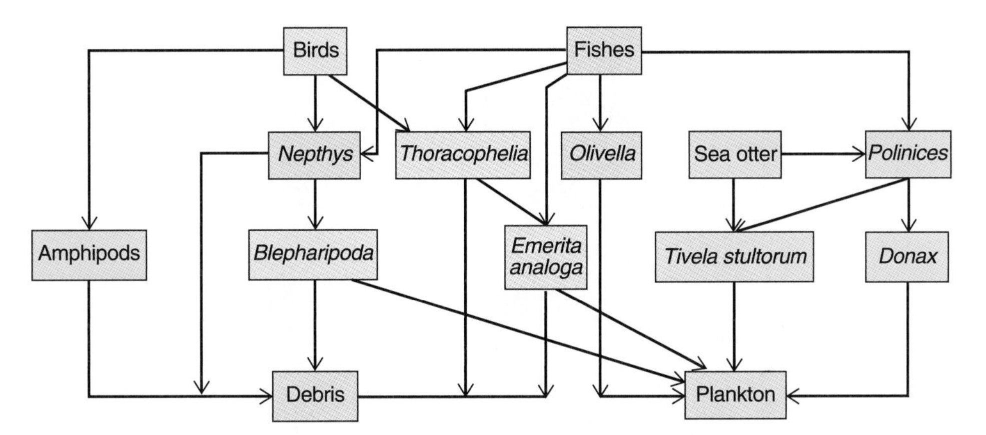

FIGURE 6.41
Generalized food web for a sand beach in California.

snails) and Nassariidae (dog whelks). Moon snails burrow slowly just below the surface layers of the sand in search of bivalve mollusks, which they usually consume by boring holes through the shell and eating out the contents. Nassariids of the genus *Bullia* are probably the best known of the larger open-sand gastropods. They are common on beaches all around Africa and as far east as India. They are extreme opportunists, consuming virtually all animal material available, live or dead. They detect food with their chemical sense organ, the osphradium, and move quickly to it by crawling or surfing. Once on the food, they attach via proboscis or foot, to avoid being separated from it by the continual movement of the food by the swash. On protected sand flats, epifaunal carnivorous crabs, such as *Callinectes* and *Carcinus*, various snails such as *Urosalpinx* and *Busycon*, and certain polychaetes are common predators of various infaunal invertebrates. On the Atlantic coast of North America, the horseshoe crab, *Limulus*, may be a seasonally abundant predator. A number of transient predators also occur. At low tide, various shorebirds actively feed on beach organisms, and at high tide, fishes invade the area. The rays, skates, and flatfishes are the most significant as predators of the benthic invertebrate communities (Figure 6.41).

Community Organization on Exposed Sand Beaches

Since there are few organisms visible on the surface of the intertidal sand beaches, these areas do not display the surface zonation patterns so obvious on rocky shores. Furthermore, since such physical factors as desiccation and temperature limit organisms on rocky shores but are far less important to organisms living in the sand, it might be suspected that zonation of organisms would be reduced or absent. There is, however, a zonation of sandy shore organisms, but it is neither as clearly defined nor as well understood as that of rocky shores. The lack of clear zonation patterns is due partly to the habit of some dominant organisms of moving up and down the beach for feeding and partly to the dearth of studies on what determines distribution patterns in this area. One of the major contributors to the lack of understanding is researchers' inability to set up field experiments on open sand beaches that exclude certain organisms. Wave action precludes such work.

From the few data available, it is possible to establish a three-zone division of most sandy beaches that closely follows the universal scheme proposed by the Stephensons for rocky shores. The highest parts of the sandy beach, corresponding to the supralittoral fringe, are usually inhabited by talitrid amphipod crustaceans (beach hoppers) in the temperate zone and by the fast-moving ghost crabs (*Ocypode*) in the warm temperate and tropics. Both groups excavate burrows and are scavengers. The broad midlittoral area, which corresponds to the area inhabited by barnacles and mussels on rocky shores, is much more variable. One group found here is isopods of the family Cirolanidae. Also coming into this zone when it is the area of wave wash are the various sand crabs and other animals that feed in a similar way (see previous section). The lowest zone, the infralittoral fringe or surf zone, is inhabited by the greatest number of species, including the large surf dams (*Tivela*, *Spisula*), sand dollars (*Dendraster*), various polychaete worms, a host of crustaceans, and the larger carnivorous snails (*Natica*, *Polinices*, *Bullia*). There is some tendency for this lowest zone to be occupied in the tropics by hippid crabs and in the

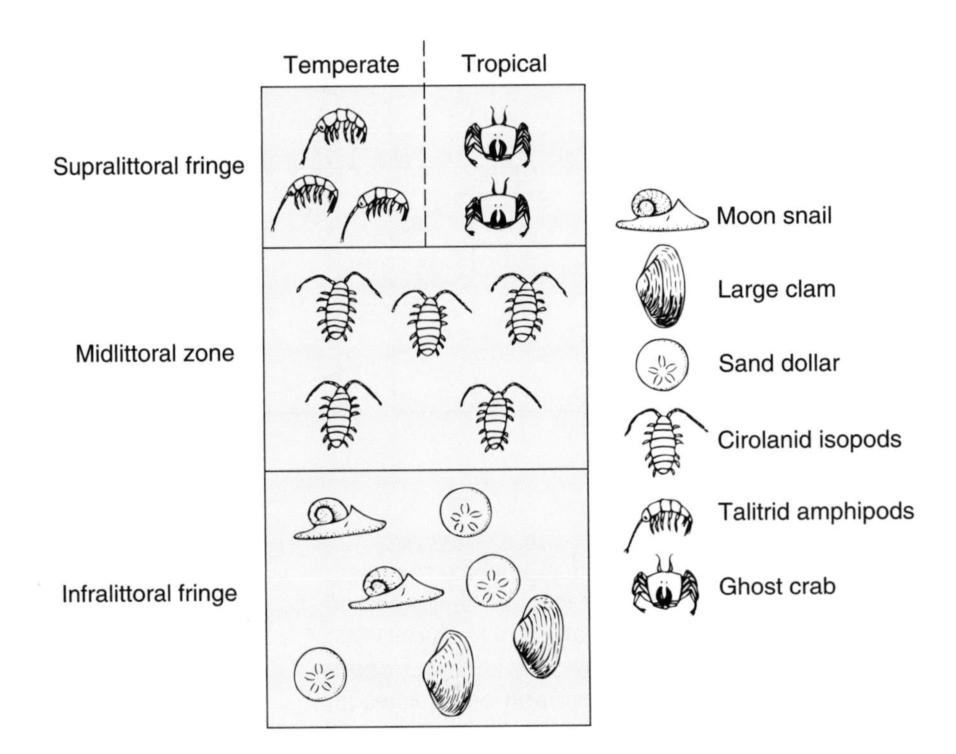

FIGURE 6.42 Generalized patterns of organism distributions for sandy beaches.

temperate zone by haustorid and other amphipods (Figures 6.42, 6.43, and 6.44).

In contrast to rocky shores where the fauna is sedentary or fixed, open sandy beaches are characterized by tidal migrations of many of the macrofauna. Any zonation pattern established by a researcher, therefore, reflects the faunal distribution at the time the samples were taken. Most macrofaunal migrations are simple movements of the animals that maintain them in the swash zone, where feeding conditions are good and predation pressure low. These tidal migrations compress the zonation pattern at high tide, when macrofaunal organisms are concentrated, and spread them at low tide.

In addition to tidal changes in faunal composition on open sand beaches, there are seasonal changes. The most dramatic of these changes are seen on the beaches of India where Ansell et al. (1972) and McLusky et al. (1975) have shown that most of the fauna disappears from the intertidal during the monsoon. Another seasonal change, occurring mainly on temperate zone beaches, is onshore and offshore movements in winter, which may be an adaptation to avoid winter storms. Long-term changes may also occur and may be dramatic in the case of periodic mass mortalities.

In direct contrast to rocky shores, where experimental work has demonstrated the importance of physical factors, competition, and predation in determining zonation patterns, the explanations for the establishment of the pattern in open sandy beaches have not been subjected to rigorous experimental analysis. As a result, we know little of the factors or their interactions that establish the sandy beach distribution patterns.

The extremely crowded conditions for organisms observed on rocky shores are not found on sand beaches. Indeed, the fauna is relatively sparse and does not occupy all available space. Also, the three-dimensional nature of the sedimentary substrate, coupled with the fact that many macrofaunal species burrow into it, minimizes interference competition. It is also difficult to conceive of organisms overgrowing others in this environment, where animals move freely vertically in the substrate, as well as horizontally on the surface. This immediately suggests that competition for space is not a major contributor to the observed distribution pattern. Competition for food also appears negligible, since the sparse populations and abundant plankton make competition unlikely. The fact that most sand-beach animals are opportunistic feeders also suggests competition is not important. Similarly, there are relatively few indigenous invertebrate predators, and it seems unlikely that they are responsible for any major

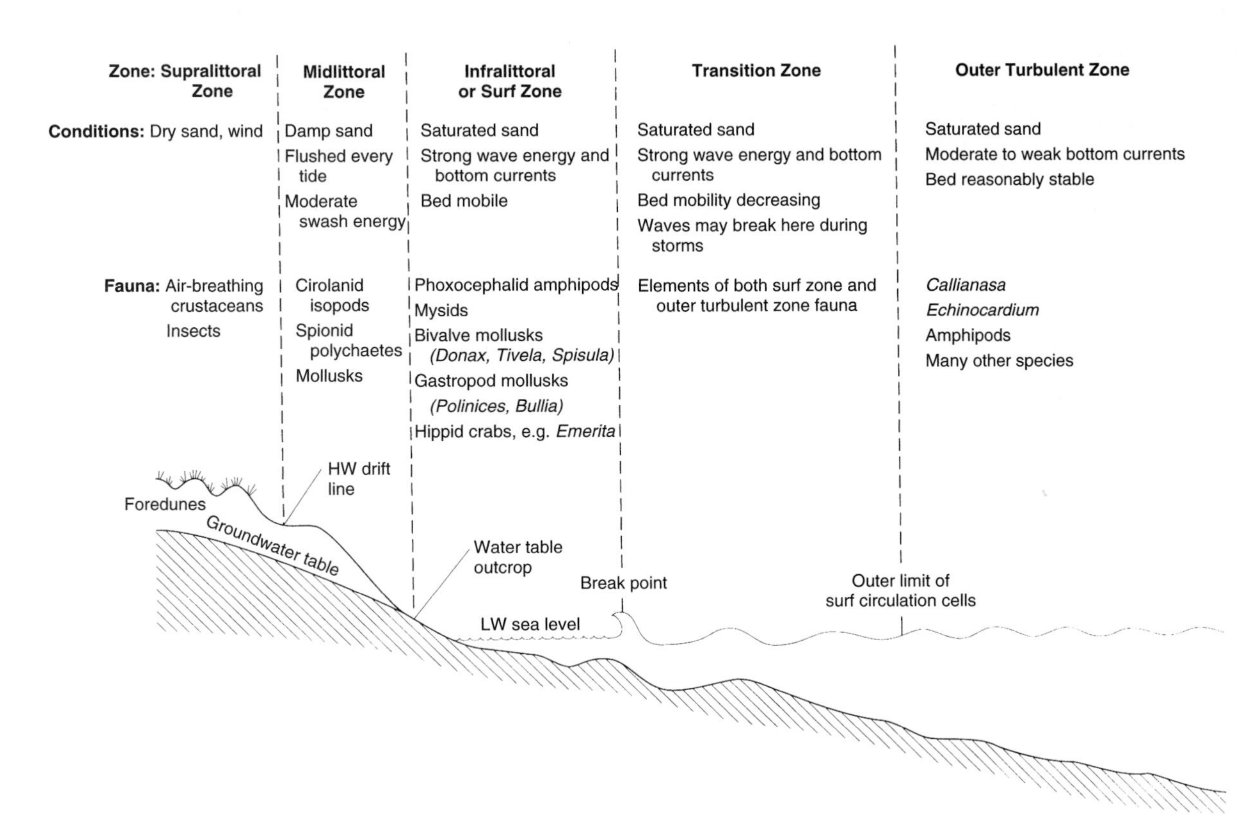

FIGURE 6.43 Generalized scheme of zonation on sandy shores. (After "Sandy Beach Ecology" A. McLachlan in *Sandy Beaches As Ecoystems*, eds. A. McLachlan & T. Erasmus, pp. 321-380. Copyright © 1983 Dr. W. Junk, with kind permission from Kluwer Academic Publishers.)

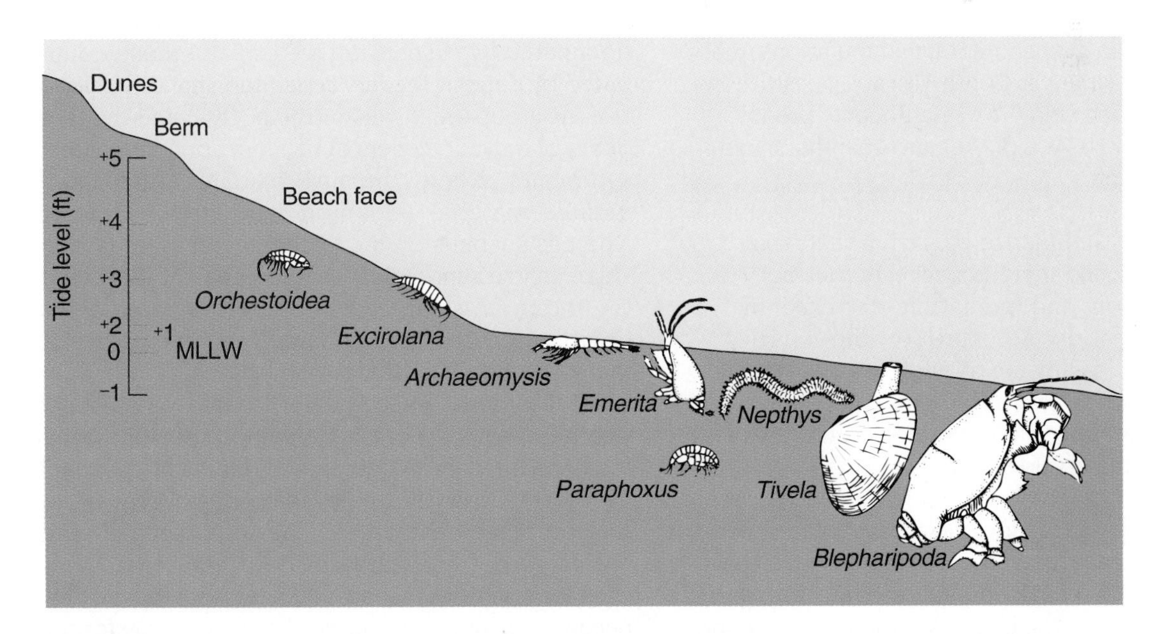

FIGURE 6.44 Zonation of the numerically dominant invertebrate genera on a central California intertidal sandy beach. *Orchestoidea*, *Paraphoxus* (Crustacea, Amphipoda); *Excirolana* (Crustacea, Isopoda); *Archaeomysis* (Crustacea, Mysidacea); *Emerita*, *Blepharipoda* (Crustacea, Decapoda); *Nepthys* (Annelida, Polychaeta); *Tivela* (Mollusca, Bivalvia). (Not to scale.)

distributional pattern. This is supported by the fact that, on the few occasions that predators have been experimentally excluded from sediments, the diversity has increased, not decreased, as in rocky shores. The effect, however, of large vertebrate predators, such as birds and fishes, remains to be investigated.

Little is known of the infralittoral or surf zone fish fauna and its potential effects on open-beach community structure. This is likely to remain true due to the danger and difficulty of sampling in the surf zone. What we do know is that these fish assemblages are dynamic, have few resident species, and are variable in time and space. Most species are opportunistic feeders. These generalities were confirmed in a four-year study by Gibson et al. (1993) on the fish fauna of a Scottish sandy beach. They found a clear seasonal cycle in abundance and species composition. Very few species were year-round residents, with the majority of species present only from spring to fall. There was also a marked annual cycle of species numbers and abundance. Recruitment of young of the year appeared to be the main reason for the increase of both number of species and numbers of individuals in the spring. Seasonal and annual changes in these parameters resulted from subsequent mortality of these recruits rather than either immigration or emigration by adults.

A similar situation prevails for birds. Here, again, our lack of knowledge stems primarily from a lack of direct studies and experimental manipulations. Indirect work, particularly in South Africa, suggests, however, that birds may be significant predators, removing from 10–49% of the macrobenthic invertebrate production.

■ We can summarize by saying that what little evidence there is suggests that both competition and predation decrease in importance as exposure of the beach increases, leaving the physical factors—primarily wave action, particle size, and beach slope—as probable major contributors to the patterns observed. ■

In the absence of indigenous large plants, open sand beaches have food sources mainly derived from the surrounding sea. The major determinants of the amount of food on the beach are the proximity and size of the food source and the characteristics of the surf zone delivering the food to the beach. The primary food types available on open sand beaches include particulate organic matter, carrion, dead large algae, dissolved organic material, plankton, and resident benthic microflora. Open-sand macrofaunal food chains generally include filter feeders, detritus feeders, scavengers, and carnivores. Filter feeders are more abundant on open sand beaches and detritus feeders on the more protected beaches. Scavengers and carnivores are rarely abundant. Fishes, birds, and certain large crustaceans dominate the top of open-sand food chains (Figure 6.45). In general, open sand beaches have a gradient in which terrestrial predators (mainly birds) dominate the upper shore zones and marine predators (fish) the lower shores. These distributions are summarized in Table 6.1.

Community Organization on Protected Sand Flats

In contrast to the situation on open sand beaches, protected sand flats are accessible for ecological investigations, including various experimental manipulations. As a result, we know more about the structuring of the communities of sand flats. Both physical and biological factors are important.

Physical factors play a significant role in the organization of sand flat communities in a number of ways. First, the grain size of the sand may set limits for the organisms that can inhabit the area.

There is a tendency to have those communities dominated by suspension feeders and those dominated by deposit feeders separated spatially due to trophic group amensalism (see p. 186), and the tendency of deposit feeders to be more abundant in fine sediments, where the fine particles have more organic material and more nutrients for them, whereas suspension feeders are most abundant in clean coarse sand. While this seems to hold true for the intertidal sand as well as for the subtidal, as described in Chapter 5, it is well to remember that many exceptions to this hypothesis exist.

Other direct effects are the result of seasonal weather changes. For example, in the New England intertidal sand flats, a number of dominant epifaunal invertebrates are affected by the seasonal changes in the water temperature. As the water temperature declines in the fall, Whitlatch (1982) reports that many crustacean species migrate to deeper water, where they become inactive over winter. The same is true for the dominant gastropod, the mud snail *Ilyanassa obsoleta*. Since *Ilyanassa* has an impact on algal mats, sediment

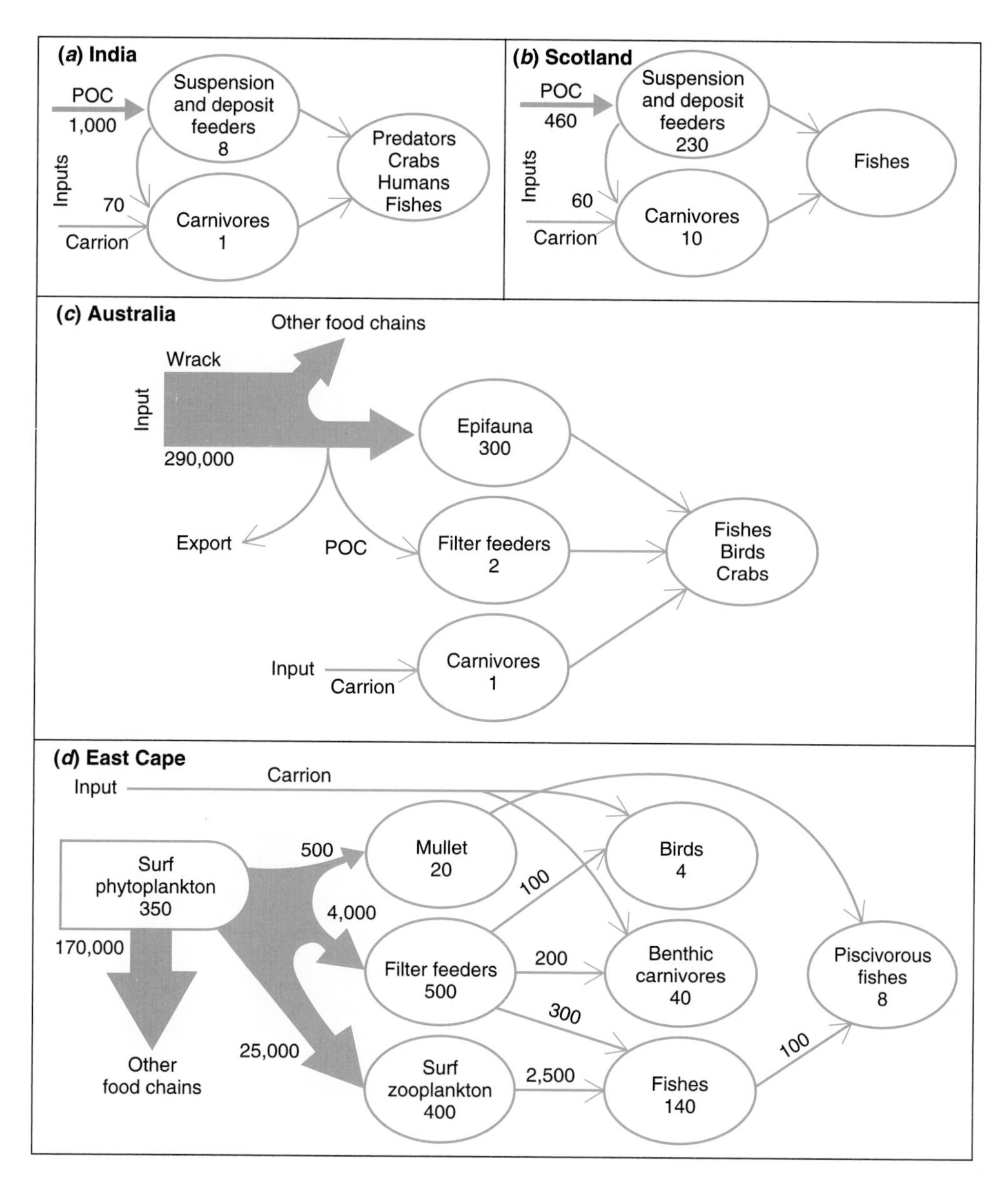

FIGURE 6.45 Macroscopic food chains and carbon flow in several different open sandy beach systems. (a) India. This is an example of a low-energy, mostly reflective beach with negligible surf and no drift line. Macrofauna are dominated by particulate feeders. Crabs, fish, and humans are the main predators. (b) Scotland. An example of a low-energy, dissipative beach. The macrofauna are dominated by particulate feeders, but the dominant predators are fishes, particularly juvenile flatfishes. (c) Western Australia. Example of a low-energy, reflective beach without a surf zone, but with large inputs of dead algae (wrack). The fauna here consists of large numbers of surface dwellers that consume wrack, small populations of hippid crabs and donacid bivalves in the midlittoral, and oxypodid crabs at the drift line. Predators are few. (d) South Africa, East Cape. Example of a high-energy beach tending toward the dissipative type and characterized by large numbers of surf diatoms supporting a rich fauna and more complex food chains. (All values are given in grams of carbon per meter squared per year.) (Modified from *Ecology of the Sandy Shores*, A. C. Brown and A. McLachlan. Copyright © 1990 with permission from Elsevier Science.)

TABLE 6.1

Major Beach Types by Climate Zone, Conditions, and Macrofaunal Composition

Type	*Latitude*	*Predominant Sources*	*Macrofaunal Biomass*	*Dominant Tropic Group*	*Predators*
High turnover, particulate-based food chain	Tropical	Particulates	Low to moderate	Filter feeders	Invertebrates, fishes
Low turnover, particulate-based food chain	Temperate	Particulates	Low to high	Filter feeders	Fishes, birds invertebrates
Macrodebris-based food chain	Temperate	Stranded macrophytes	Moderate	Scavengers, grazers	Birds
Sediment-based food chains in sheltered beaches	Temperate or tropical	Deposited detritus microorganisms and benthic microflora	Low to high	Deposit feeders	Birds, fishes, invertebrates
Carrion-based food chains in extremely exposed steep beaches	Temperate or tropical	Stranded carrion	Low	Scavengers, predators	Birds, crabs

Source: From "Sandy Beach Ecology" A. McLachlan in SANDY BEACHES AS ECOSYSTEMS, pp. 321-360, 1983.

stability, and subsequently on infaunal abundance, this seasonal migration can have significant effects on community structure (see Chapter 5, p. 185). In addition, particularly in high latitudes, the formation of sea ice and its scouring effect on the intertidal flats create changes in the community (see Chapter 5, pp. 220–224, on polar communities).

The prime biological factors that structure sand flat communities are predation, competition, and disturbance. Of the three, current information suggests that either predation or predation combined with disturbance is especially significant. There are a number of sand flat epifaunal predators that may have significant effects on the community. For example, on the East Coast, the moon snail *Polinices duplicatus* is an active predator of bivalves, particularly the soft shell clam *Mya arenaria*. Wiltse (1980) has demonstrated that when *Polinices* is excluded from areas, not only does the clam population rise, but so do the populations of other infauna. This snail affects community structure not only by feeding on clams, but also by disturbing the rest of the infauna as it plows through the substrate. Similar results have been reported by Peterson and Peterson (1979), who caged out predators in North Carolina tide flats. Another group of predators on the East Coast are various large crabs, such as *Carcinus maenas* in New England and *Callinectes sapidus* in the Middle Atlantic states. Virnstein (1977) has demonstrated that *Callinectes* is a voracious predator that also digs in the sediment; exclusion of the crab results in a great increase in infaunal density. A seasonal predator on eastern seaboard tidal flats is the horseshoe crab *Limulus polyphemus*. This animal digs distinctive pits when searching for infaunal prey, and Woodin (1978) has demonstrated that this also significantly reduces other infaunal organisms.

Other large East Coast invertebrate predators include several species of whelks (*Busycon* spp.) and tulip shells (*Fasciolaria*). The whelks, in particular, are devastating to the sand flat clam populations because they can consume all sizes. Unlike the barnacles and some mussels of the rocky intertidal, the clams have no refuge in size.

Despite the massive effects of the larger predators, there are areas that are refuges from predation and disturbance by these animals. Woodin (1978), for example, has demonstrated that the tube-forming polychaete worm *Diopatra cuprea* of the eastern United States intertidal sand flats, by forming upright tubes in the sand, effectively deters both *Limulus* and *Callinectes* to the extent that the infaunal abundances are greater in and around the tubes of *Diopatra*. All of the above strongly suggest that predation, and its accompanying disturbance, is the main biological structuring force in intertidal sand flats.

Another significant biological process determining community structure is, surprisingly, deposit feeding.

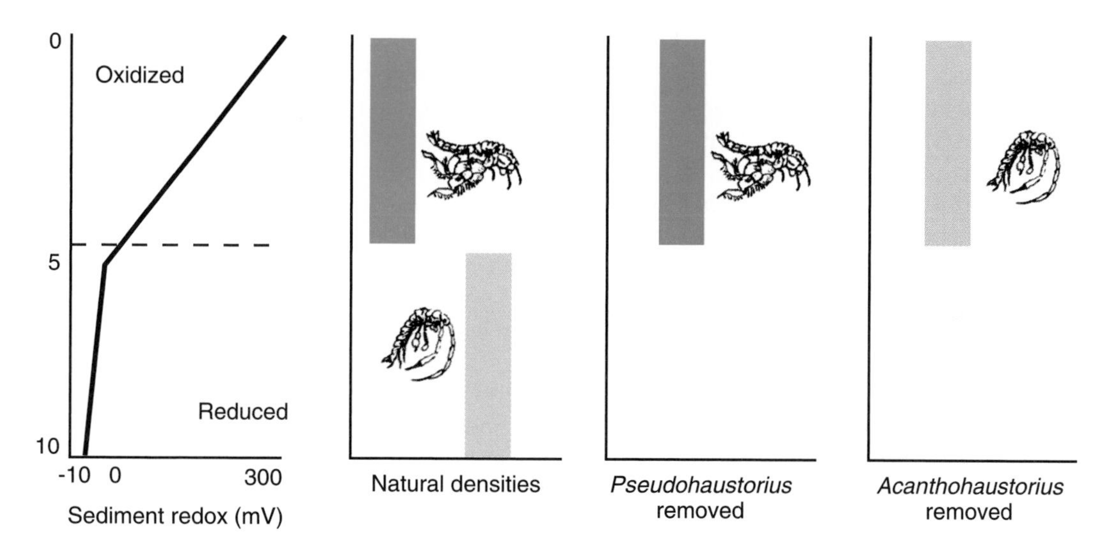

FIGURE 6.46 Competitive depth displacement of two species of amphipods under natural conditions and removal experiments. (Modified from *The Ecology of Atlantic Shorelines*, M. D. Bertness, p. 281. Copyright © 1999 Sinauer Associates, Inc. Reprinted by permission.)

In North Carolina, one of the more abundant deposit feeders is the hemichordate worm *Balanoglossus aurantiacus*. This animal, which Peterson and Peterson (1979) have called a "funnel feeder," digs a U-shaped burrow and then collapses the head end. Thus, sediment continues to fall down to form a funnel-shaped depression directed at the head of the worm. The worm then ingests the sediment, consuming the smaller infaunal organisms and causing the death of those that it does not digest. The densities of these large worms and their continual reworking of the sand mean that they keep numbers of infaunal organisms low, as well as exclude potential competitors. In North Carolina, the effect of *Balanoglossus* is heightened by the presence of yet another sediment processor, the sea cucumber *Leptosynapta tenuis*, which operates at the sediment surface.

Adult-larval interactions probably also have an effect on the composition of sand flat communities. These interactions should probably be categorized as predation because the interaction consists of deposit or suspension feeders ingesting the settling or newly settled larvae. Woodin (1976) suggests that when adults are in dense assemblages such interactions could be important, but we currently lack any experimental studies to document this.

The role of competition in structuring sand flat communities is different from that observed in rocky shores. The three-dimensionality of the protected sand shores, the dynamic relationship between the substrate and the animals living therein, and the interactions between juveniles and adults all tend to reduce the severity of direct competition. The best examples of direct competition in sand flats involve mobile animals that live on the sediment surface. In Long Island Sound, Levinton et al. (1985) have demonstrated that competition between two mud snails determines their intertidal distribution. *Hydrobia totteni*, which in Europe occupies the whole intertidal zone, is confined to the high intertidal in Long Island Sound because it is an inferior competitor to the native American mud snail *Ilyanassa obsoleta*. *Hydrobia* avoids *Ilyanassa* by living in the high intertidal, which *Ilyanassa* does not inhabit. A similar competitive interaction involves *Ilyanassa* and the introduced European snail *Littorina littorea*. According to Brenchley and Carlton (1983), *Ilyanassa* avoids contact with *Littorina* because *Littorina* destroys its egg capsules; this leads to spatial separation of the two.

Although less common, competitive interactions leading to different depth distributions do occur in protected sand beaches. One good example comes from South Carolina. Here, two amphipod species are found at different depths. *Acanthohaustorius* lives in the surface oxidized layer of sediment, while *Pseudohaustorius* lives in deeper, anoxic sediments. Through a series of experimental manipulations, Grant (1981) demonstrated that both species prefer to occupy the oxidized upper layer but that the presence of the superior competitor *Acanthohaustorius* displaces the *Pseudohaustorius* to the deeper sediments

(Figure 6.46). Although the above example of competition leads to physical depth displacement of one of the species resulting from competition for space, competition among the more sedentary infauna organisms such as bivalve mollusks is usually reflected in reduced growth rates and fecundity rather than displacement or death, suggesting that competition for space is not a significant factor in the distribution of these animals.

In direct contrast to the results of predator removal in the rocky intertidal, predator removal in soft substrates does not lead to increased competition and competitive exclusion. This is not to say that competition does not occur, for competition for space in sediments has been found by Levinton (1977) and Peterson (1977). The competition, however, does not seem to be a structuring force that determines a community makeup.

Another interesting contrast with rocky intertidal areas is provided by Peterson (1991), who notes that the major predators on rocky shores are the sedentary, slow-moving starfish and mollusks that are limited to the lower parts of the intertidal zone by their lack of mobility and need for coverage by water. In contrast, protected sand flats and mud flats are dominated by highly mobile predators, such as crabs, fishes, and birds that have feeding times less affected by the tides and that are capable of ranging throughout the intertidal.

■ We can summarize this section by stating that the macrofauna of intertidal sand flats displays high temporal and spatial variability. The total numbers of species and individuals may vary by several orders of magnitude within and between years. This high variability is due to the high predation pressure and biological disturbance that control the infaunal community structure to the extent that the majority of infaunal species are opportunistic. ■

MUDDY SHORES

Sharp boundaries exist between rocky shore and sandy shore, making them easily defined and recognizable. Boundaries between sand flats and muddy shores are not as easily defined. Indeed, it is not possible to draw sharp boundaries between the two, because as shores become more protected from wave action, they become finer grained and accumulate more organic matter; thus, they become more "muddy." Sand and mud shores are, therefore, the opposite ends of a continuum. The sand beaches have larger grain sizes, and the muddy areas have the finest grain sizes.

Since sharp boundaries do not exist and one blends into the other along a gradient of increasing protection from wave action, the fauna and flora of muddy shores also show a change from organisms typical of open sand beaches to those typical of muddy shores along the same gradient.

This section will discuss the features characteristic of typical muddy shores; that is, those at the opposite end of the continuum from the open sand beaches. It should be borne in mind, however, that various transitional communities exist that will show mixtures of organisms from both extremes. Since muddy shores are also composed of sediments, similar factors will apply here as they did with protected sandy shores and will not be repeated.

Muddy substrates are also characteristic of estuaries and salt marshes (Chapter 8). Most of the muddy shores of the world are associated with estuaries and similar embayments; hence, many of the types of organisms and adaptations are similar between the two areas. To avoid the problem of overlap in discussing these two areas, this section will treat the infaunal organisms of mud flats, their adaptations, feeding biology, and trophic structure. These considerations also will apply to the estuarine infauna of Chapter 8. The role of other organisms, productivity, and ecological relationships are discussed in Chapter 8.

Physical Factors

The major contrast with open sand beaches is that muddy shores cannot develop in the presence of significant wave action. Therefore, muddy shores are restricted to intertidal areas completely protected from open ocean wave activity. Muddy shores are best developed where there is a source of fine-grained sediment particles, and they are located in various partially enclosed bays, lagoons, harbors, and especially estuaries (see Chapter 8).

Since these areas develop where water movement is minimal, the slope of mud shores is much flatter than that of sand beaches. This is why these areas are often referred to as mud flats. Furthermore, these areas are more stable than sand substrates and more conducive to the establishment of permanent burrows.

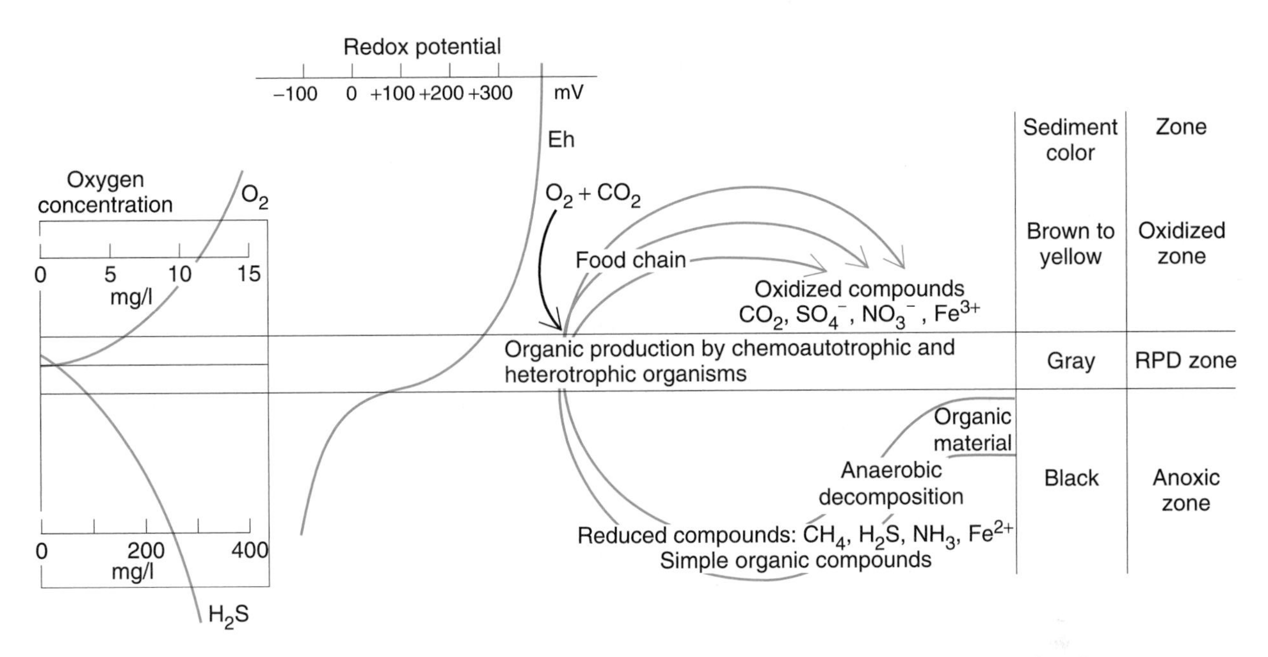

FIGURE 6.47 Diagrammatic representation of the physical and chemical characteristics of sediments across the redox discontinuity layer and the biological processes occurring in each. (From various sources.)

The very fine particle size, coupled with the very flat angle of repose of these sediments, means that water in the sediments does not drain away and is held within the substrate. This long retention time for water, combined with a poor interchange of the interstitial water with the seawater above, and a high internal bacterial population usually result in complete depletion of the oxygen in the sediments below the first few centimeters of the surface. **Anaerobic** conditions prevail within the sediment, and this is one of the most important characteristics of the muddy shore that separate it from sand flats.

Between the upper aerobic layer and the lower anaerobic layer is a transition zone called the **redox potential discontinuity (RPD) layer**. This zone is characterized by a rapid change from a positive redox potential (E_h), as measured by an electrode, to a negative redox potential. Below the RPD zone, sediments are anaerobic, and decomposition of organic compounds is by anaerobic bacteria. Above the RPD zone, decomposition is by aerobic means. The RPD layer is usually characterized by the gray color of the sediments; the oxidized layer above is usually brown or yellow; and the anoxic layer is black (Figure 6.47). The RPD zone is significant biologically for several reasons. First, reduced compounds diffuse upward and, as soon as oxygen is available, bacteria oxidize these compounds. The oxidized end products—including CO_2, NO_3, and SO_4—in turn are incorporated into bacterial biomass and form the basis of new food chains. Some compounds, however, diffuse downward below the RPD zone and are used by the anaerobic bacteria. These bacteria, in turn, produce more reduced compounds, which complete the cycle and release phosphate. This phosphate is important for new plant growth. Also present in the RPD zone are the chemoautotrophic bacteria. By oxidizing the reduced compounds and fixing CO_2, these bacteria produce more organic material. It is little wonder that high numbers of organisms inhabit the RPD zone.

Muddy shores tend to accumulate organic material. There is an abundant potential food supply for the resident organisms, but the abundant small organic particles "raining" down on the mud flat also have the potential to clog respiratory surfaces.

Adaptations of Organisms

As we noted for sandy shores, the surface of muddy shores is often barren, since few animals inhabit the surface of the mud flat. Most organisms inhabiting these areas show an adaptation to burrowing into and through the soft substrate or else inhabit permanent tubes in the substrate. In contrast to the open sand beaches, however, the presence of the organisms in

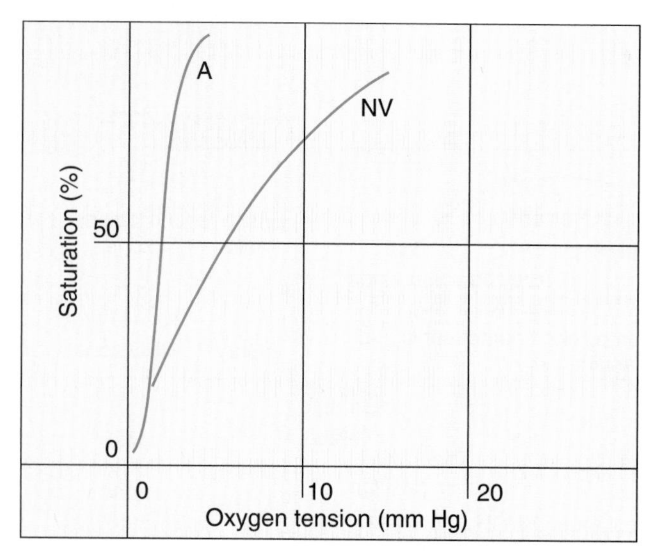

FIGURE 6.48 Oxygen dissociation curves for the blood of *Arenicola* (A) and *Nepthys* (NV). (Modified from "Observations on the Respiratory Physiology and on the Hemoglobin of the Polychaete Genus *Nephthys* with special reference to *N. hombergi*" J. D. Jones, *Journal of Experimental Biology,* Vol. 32, pp. 110-125, 1954. Reprinted by permission of the Company of Biologists, Ltd.)

the mud flat is advertised on the surface by the presence of various holes of differing sizes and shapes. One of the primary adaptations of organisms in mud flats is the ability to burrow into the substrate or to form permanent tubes. The latter is related to the increased stability of the fine sediments.

A second major adaptation concerns the anaerobic conditions in the substrate. If organisms are to survive while burrowed into the substrate, they must either be adapted to live under anaerobic conditions or they must have some way of bringing the overlying surface water with its oxygen supply down to them. Since most multicellular organisms cannot survive without oxygen, the latter is the more common adaptation. It is to obtain this oxygen-rich surface water and food that the various burrows, holes, and tubes appear on the surface of the mud flat.

Although most mud flat organisms are intolerant of completely anaerobic conditions, many have adaptations that permit them to exist at lower oxygen tensions than similar forms that live on open sand beaches. The most common adaptation to low oxygen supply is the development of carriers (e.g., hemoglobin) that will continue to pick up oxygen at concentrations well below that of similar pigments in other organisms (Figure 6.48). Others use glycogen stores for anaerobic metabolism during periods when there is little or no oxygen.

Since wave action is essentially absent on these mud flats, there is no need to develop either rapid burrowing or heavy bodies to maintain positions, as is the case on open sandy shores.

Types of Organisms

In contrast to sand beaches, muddy shores often develop a substantial growth of various plants. On the bare mud flats, the most abundant plants are diatoms, which live in the surface layers of mud and often give a brownish color to the surface at low tide. Other plants include large macroalgae, such as species of *Gracilaria* (red algae), *Ulva*, and *Enteromorpha* (green algae). These large algae often undergo seasonal cycles of abundance, becoming common in the warmer months and virtually disappearing in colder months. Other areas, particularly the lowest tidal levels, may be covered with a growth of various seagrasses, such as the genus *Zostera* (see Chapter 5). As a result of the occurrence of these primary producers, there is substantial primary productivity in the mud flats.

Mud flats contain large numbers of bacteria, which feed on the abundant organic matter. Bacteria are the only abundant organisms found in the anaerobic layers of the mud shore and constitute a significant biomass. Among the dominant bacteria inhabiting this area are a number capable of using the potential energy of the various reduced chemical compounds abundant here. These **chemolithoautotrophic** or **sulfur bacteria** obtain energy through the oxidation of a number of reduced sulfur compounds, such as various sulfides (for example, H_2S). These organisms are thus primary producers of organic matter analogous to green plants. They produce organic matter using energy obtained from the oxidation of the reduced sulfur compounds, whereas plants produce organic matter using energy obtained from sunlight (see also discussion of these bacteria at hydrothermal vents in Chapter 4).

Since these autotrophic bacteria are located in the RPD layer of the mud, mud flats are unique among marine environments in that they have two separate layers in which primary productivity occurs: the surface, where diatoms, algae, and marine grasses carry on photosynthesis; and a deeper layer, where bacteria conduct chemosynthesis.

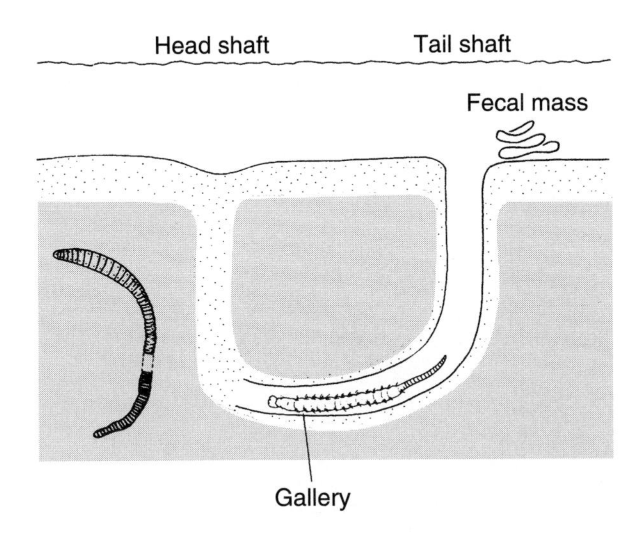

FIGURE 6.49 Two common polychaete worms of mud flats: *Arenicola* (right) in its U-shaped burrow and *Capitella* (left) burrowing through the substrate.

The dominant macrofaunal groups on muddy shores are the same as those encountered on sand beaches and sand flats—namely, various polychaete worms, bivalve mollusks, and small and large crustaceans, but of different sorts.

Feeding Biology and Trophic Structure

Because of the greater amount of organic matter present in and on muddy shores and because of increased productivity due to both bacteria and plants, there is vastly more food available on muddy shores than on sand beaches. This permits more large organisms to live on mud shores, and indeed, mud flats may be very densely populated.

The dominant feeding types on mud flats are deposit feeders and suspension feeders, a situation somewhat different from that on open sand beaches. Deposit feeders are particularly abundant due to the large amount of organic material and the large populations of bacteria in the sediments. Deposit-feeding polychaetes are represented by such genera as *Arenicola* and *Capitella*. Both feed by burrowing through the substrate, ingesting it, and digesting out the organic matter including bacteria, and passing out the undigested material through the anus. *Arenicola* spp., however, form a U-shaped burrow in which one arm of the U is a permanently open shaft to the surface, while the other arm is filled with sediment that the worm ingests (Figure 6.49). The worm consumes sediment from the filled shaft, passes it through its gut, and then moves up in the open burrow to defecate at the surface, leaving characteristic mounds. Capitellids, on the other hand, form no permanent tubes but move like aquatic earthworms through the surface layers of the substrate, ingesting it. Other deposit feeders include the surface-feeding terebellid polychaete worms and the burrowing hemichordates.

Deposit-feeding bivalves are also common in these areas. Temperate zone mud flats often have large numbers of small tellinid clams of the genus *Macoma* or *Scrobicularia*. These bivalves have siphons that are separated. Each clam lies buried in the substrate with its siphons extended to the surface. There, the intake siphon is moved over the surface like a vacuum cleaner, ingesting organic particles and bringing them down to the clam for digestion (Figure 6.50).

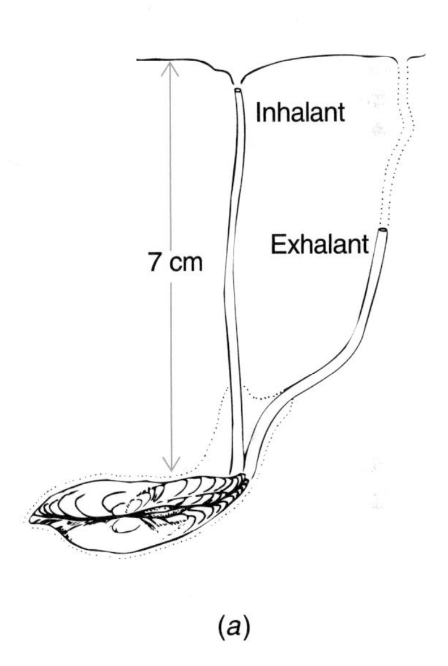

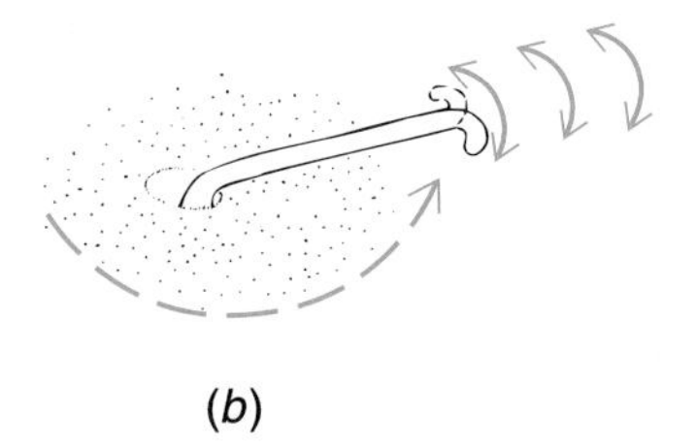

FIGURE 6.50 (a) *Macoma nasuta* in the substrate. (b) *Macoma nasuta* feeding with its in-current siphon. (Modified from "Life in sandy shores" A. E. Brafield, *Studies in Biology,* No. 89, 1978. Reprinted with the permision of Cambridge University Press.)

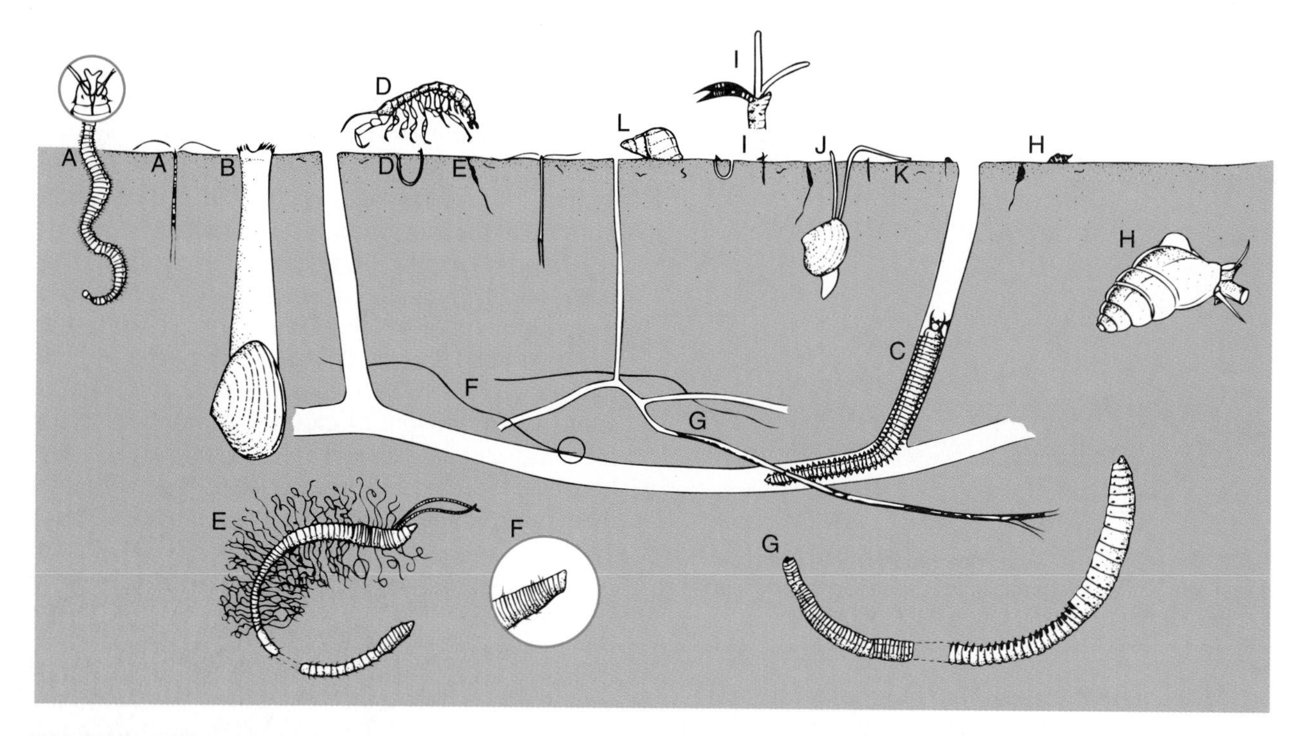

FIGURE 6.51 Some representative New England mud flat benthic invertebrates. Suspension feeder: (B) *Mya arenaria* (soft-shelled clam). Surface deposit feeders: (A) *Polydora ligni* (spionid polychaete); (D) *Corophium* spp. (gammaridean amphipod); (H) *Hydrobia totteni* (hydrobid gastropod); (I) *Streblospio benedicti* (spionid polychaete); (J) *Macoma balthica* (bivalve); (L) *Ilyanassa obsoleta* (mud snail). Burrowing omnivore: (C) *Nereis virens* (nereid polychaete). Burrowing deposit feeders: (E) *Tharyx* sp. (cirratulid polychaete); (F) *Lumbrineris tenuis* (lumbrinerid polychaete); (G) *Heteromastus filiformis* (capitellid polychaete); (K) oligochaete. (From R. Whitlatch, 1982, *The ecology of New England tidal flats: A community profile*, U.S. Fish and Wildlife Service, Biological Services Program.)

Suspension feeders include various other species of clams, some crustaceans, and numbers of polychaetes. Suspension-feeding mechanisms are not different from those observed in other areas. There is, however, some question as to how exclusively these forms are feeding on suspended plankton. The fine sediment particles are easily stirred up on mud flats and remain resuspended in the water. Suspension feeders may possibly take in amounts of these particles in addition to various plankton organisms. Most of the suspension feeders on mud flats also partially take in resuspended sediments; hence, they are actually feeding on both deposited and suspended material. In general, deposit feeders are more common in fine-grained shores, and suspension feeders become more abundant in coarser sediments, where there is little organic matter (Figure 6.51).

The major carnivores on mud shores are often the fishes, which feed when the tide is in, and the birds, which feed when the tide is out (Figure 6.52). Indigenous mud flat predators include a few polychaete worms (*Glycera* spp.), moon snails (*Polinices*, *Natica*), nemertean worms, and crabs.

Despite the relative abundance of plant material on muddy shores, there are very few herbivores. Most plant material finds its way into the food chains only after it has been broken down into small pieces to enter the deposit-feeding food web as detritus. Thus, the trophic structure of a mud flat is often built up from two bases: a detritus-bacteria base and an autotrophic base. The detritus base web is derived from plants and other organic sources and includes the bacteria that live on the detrital particles. Detrital particles can either be taken in directly by the large macrofaunal

FIGURE 6.52 Birds feeding on a mud flat. (Photo courtesy of David Weintraub.)

invertebrates or be consumed by bacteria. An additional bacterial component at the base of this trophic pyramid is the sulfur bacteria.

Bacteria are consumed commonly by various nematode worms, which occupy a position similar to that of herbivores in other areas. Bacteria are also consumed by deposit feeders as they ingest the organic particles on which the bacteria are found. Nematodes and deposit feeders are, in turn, consumed by various carnivores, including predatory invertebrates (moon snails, polychaetes such as *Glycera*), as well as birds and fishes (Figure 6.53).

The second food web is based on the microscopic diatoms as the autotrophic base. The diatoms are consumed by several different polychaetes, mollusks, and crustaceans (Figure 6.53). These, in turn, are consumed by the large predatory birds and fishes.

Zonation and Community Structure

Little information exists regarding the zonation of mud flats. The very gentle slope of these areas means that the intertidal is often very extensive, more so than rocky or sandy shores. The upper area, the supralittoral fringe, is often inhabited by various species of crabs, many of which burrow into the substrate. The very extensive midlittoral area is the home of most of the common species of clams and

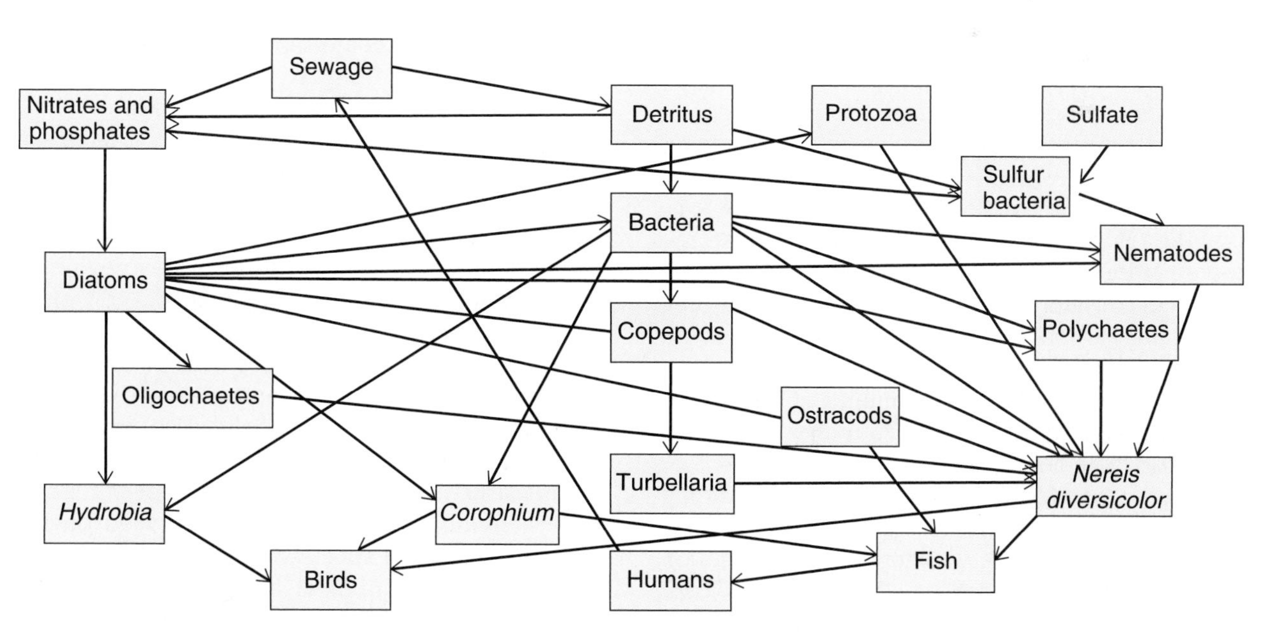

FIGURE 6.53 Generalized food web of a muddy shore. (From *Life in Mud and Sand*, S. K. Eltringham, 1971, Crane Russak.)

FIGURE 6.54 Examples of rocky intertidal fishes from the four dominant families: (a) Blenniidae (blennies). (b) Gobiidae (gobies). (c) Gobiesocidae (clingfishes). (d) Cottidae (cottids). (Modified from "The Biology of Behavior of Littoral Fish" R. N. Gibson, *OMBAR*, Vol. 7, pp. 367-410, 1969.)

polychaetes. There is no sharp boundary with the infralittoral fringe, and similar organisms are encountered.

The role of physical factors and biological factors in structuring these communities has been studied in a number of places. Structuring forces are similar to those acting in sand flats. Various biological factors seem significant, with predation being perhaps the most important. Since many of the predators are similar to those described for sand flats and since these areas grade into each other, it is not necessary to repeat that material here. Suffice it to say that both sand and mud flats seem to be similarly organized.

INTERTIDAL FISHES

Whereas a great deal of research has been concentrated on the ecology of invertebrates and plants of the intertidal zone on all three types of shores, there are relatively few accounts of the fishes of these areas or of the role they may play in community organization as grazers or predators. Since fishes are often present in considerable numbers in the intertidal and since it is known that they have a significant effect on other communities, such as coral reefs (see Chapter 9), and in various subtidal communities (Chapter 5), it is reasonable to assume that future research may clarify their role here as well.

Most rocky intertidal fishes, because of the turbulent environment, are small. The body shape is usually compressed and elongate (Blenniidae, Pholidae) or depressed (Cottidae, Gobiesocidae), which allows them to inhabit holes, tubes, crevices, or depressions for protection against both desiccation and wave action (Figure 6.54). Most also lack swim bladders and are closely associated with the substrate. Many of these fishes are adapted to withstand greater ranges of salinity and temperature than their subtidal relatives. A few are even adapted to spend some time out of the water (see Chapter 9 on *Periophthalmus*).

Most rocky intertidal fishes in the temperate zone are visual carnivores, again suggesting a potentially significant role in intertidal community organization.

The life history patterns of the few species that have been investigated are all similar. Eggs are demersal and laid on stones, rocks, or submerged vegetation. Often, the eggs are guarded by the male. The eggs hatch after a few weeks into planktonic larvae. The planktonic period varies, depending on the species. It may be as long as two months. During this period, the larvae gradually acquire the adult features and finally become demersal. Life spans of the adults are generally short, from two to ten years, and sexual maturity occurs in the first or second year.

On mud and sand flats, skates and rays as well as various species of flatfishes are significant predators of infaunal organisms. Rays are particularly common on intertidal flats. Their method of foraging, by excavating depressions or pits, causes widespread mortality among the infaunal organisms. Their effect extends beyond the species they consume. Digging uncovers many small infaunal species that are either killed by the disturbance or consumed by the smaller fishes that follow the rays. Flatfishes also may prey on various infaunal invertebrates, but their role in community structure seems more species-specific. Some have a minimal effect because they feed on few important infaunal species, whereas others reduce both abundance and diversity of species in a community. For example, Virnstein (1977) demonstrated in Chesapeake Bay that the hogchoker (*Trinectes maculatus*) had a minimal effect on the infauna, whereas the spot (*Leiostomus xanthurus*) significantly reduced the infaunal numbers. Some intertidal fishes are migratory, moving either tidally, diurnally, or seasonally.

FIGURE 6.55 New England tidal flat birds, showing general feeding positions of the various groups. (From R. Whitlatch, 1982, *The ecology of New England tidal flats: A community profile*. U.S. Fish and Wildlife Service, Biological Services Program.)

BIRDS

At low tide, a considerable variety of birds is often associated with the intertidal zone. They are particularly conspicuous on intertidal sand and mud flats, much less so in rocky intertidal areas. Sandy and muddy intertidal flats are exploited by many different shorebird species (Figure 6.55). These are mainly predators of infaunal invertebrates and, given their abundance and high food requirements, one might wonder how so many predatory species can coexist in a single flat without severely depleting the infauna or undergoing severe competition. There are several explanations. In the temperate zone, most of these birds are migratory and are only present during certain portions of the year. While they are present, they may locally deplete the infauna, but it has time to recover during the periods when they are absent. Another explanation is that tidal flats are not uniform in community structure, as we have seen. Different infaunal associations are found in different areas due to differences in grain size and various succession patterns that follow disturbances. The infauna is also segregated by depth in the sediment, as well as by tide level. A variety of different microhabitats exists. If the bird species differ in food preference or microhabitat preference, then the various species may be separated in space or time, thereby reducing competition and the chance that resources will become depleted. This seems to be

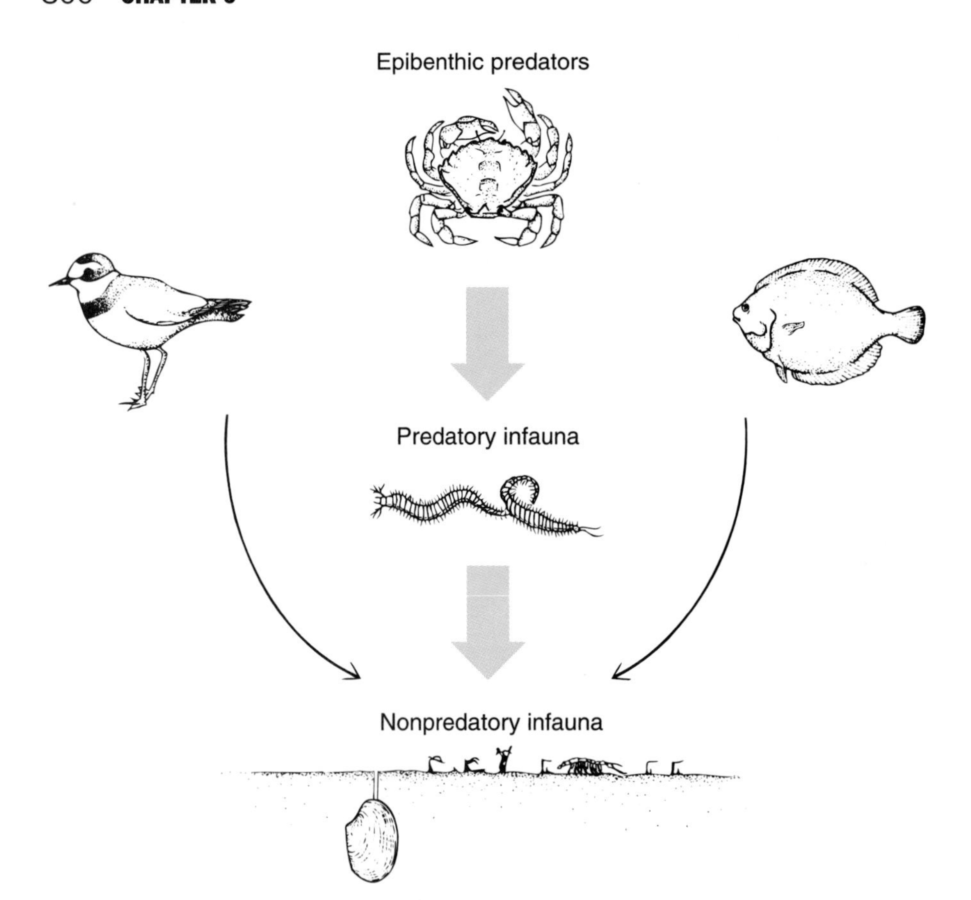

FIGURE 6.56
Diagrammatic representation of the interactions between various epibenthic predators and the infauna of sedimentary shores. Broad arrows indicate relations that may be particularly important. (From "Role of Predatory Infauna in Structuring Marine Soft-bottom Communities" G. Ambrose, *Marine Ecology Program Services,* Vol. 17, pp. 102–115, 1984. Reprinted by permission of Inter-Research.)

the case. For example, Harrington and Schneider (1978) have shown that sanderlings prefer sandy substrates, while dowitchers are more abundant on mud. Similarly, temporal segregation has been demonstrated among sanderlings and semipalmated plovers. Sanderlings follow the water's edge as the tide recedes, whereas semipalmated plovers forage in the middle intertidal zone (Harrington et al., 1974). Evans et al. (1979) have shown that competition between knots and dunlins is reduced because knots eat mollusks, whereas dunlins eat polychaetes. Morphological features of the birds, with respect to different leg and bill lengths, also influence where the birds can feed. Finally, behavior patterns dictate what food resources may be exploited.

Given that birds are present and abundant and that they do consume intertidal invertebrates, how important are they in structuring communities? It seems, on the basis of most studies, that they can be very important. Evans et al. (1979), for example, reported that wintering shorebirds were responsible for removing 90% of the *Hydrobia* snail population and 80% of the nereid polychaetes from an English mud flat. On the East Coast of the United States, Harrington and Schneider (1978) reported that the whole group of foraging shorebirds eliminated 50–70% of the invertebrate population. On the Pacific coast of North America, Quammen (1984) demonstrated that shorebirds had a significant effect on the densities of infaunal invertebrates on mud flat areas but little or no effect on sand flat invertebrate densities.

A special case may exist on "bird islands," those isolated small islands on which birds congregate in great numbers for breeding. In such situations, as Bosman and Hockey (1986) have noted, the greater nutrient input to the intertidal via bird guano leads to enhanced algal production and may, if there are predators of the herbivorous molluscan grazers, such as oystercatchers, lead to an intertidal dominated by foliose algae.

■ We can summarize by saying that bird predation may be intense and significant, particularly at certain times. It certainly plays an important role in structuring tidal flat communities (Figure 6.56). ■

SUMMARY OF KEY CONCEPTS

- The intertidal zone comprises the smallest area of the world's oceans but is the best known and most accessible to humans.
- The intertidal has the greatest variation in environmental factors of any marine area.
- The intertidal is inhabited primarily by marine organisms.
- The tide is a predictable rise and fall of sea level over a given time interval and is the most important factor influencing life in the intertidal.
- Tides occur due to the interaction of the gravitational attraction of the sun and the moon on the earth and the centrifugal force generated by the rotating earth-moon system.
- Spring tides occur when the sun and moon are directly aligned, whereas neap tides occur when sun and moon are at right angles to each other.
- Differences in the number of high and low tides per day and their heights are due to the peculiarities of the basins in which the tides occur.
- Tides expressed as a single high and low per day are diurnal tides; tides expressed as two highs and lows per day are called semidiurnal; and where the two kinds mix, we have mixed tides.
- Tides can have two direct effects on organisms. The first is the duration of exposure to the air. Since the organisms are primarily marine, the longer the exposure to air the greater the chance of desiccating beyond limits or encountering a lethal temperature. Intertidal organisms differ in their tolerance, and this contributes to their different patterns of distribution.
- The second effect is the result of the time of day that exposure to air occurs. Exposure to air during midday in the tropics could lead to lethal temperatures, whereas exposure at night in the cold temperate zones in winter could lead to freezing.
- The great predictability of tides induces certain rhythms, such as feeding and reproduction, in intertidal organisms.
- Air temperatures always have a greater range than water temperatures, and extremes may either kill organisms or weaken them, making them susceptible to death by other factors such as desiccation.
- Wave action directly affects organisms in two ways. First, it has a mechanical effect in smashing and tearing away organisms and on sedimentary shores by moving the entire substrate around. The second effect is waves throw water higher on the shore than would normally occur due to tides and therefore allows marine organisms to live higher than the tides would permit.
- Salinity changes in the intertidal that can affect organisms are primarily due to the flooding of the intertidal with fresh water from runoff or heavy rains.
- In the cold temperate zones in winter, catastrophic mortality can be caused by ice scour in the intertidal.
- Adaptations of intertidal organisms are primarily for avoiding stresses of daily exposure to air.
- Organisms show various methods of avoiding water loss, including moving into moist habitats at low tide and closing off soft tissues from the environment by closing shells, clamping down into home scars, or secreting a mucous covering. Others simply tolerate a great water loss during low tide and rehydrate during high tide.
- Intertidal organisms reduce heat gain from the environment in several ways, including large body size and reduction of area in contact with the substrate.
- Intertidal organisms lose heat from their bodies through radiation from ridges and sculpturing on shells, having light colors, and through evaporation of water from their bodies.
- Adaptations to reduce mechanical wave stress include streamlined shapes that minimize exposure to the lift and drag forces of waves, flexibility and bending in algae, and strong attachments to the substrate.
- Motile animals avoid wave stress by moving into sheltered areas.
- Most intertidal animals have respiratory surfaces that are adapted to extract oxygen from seawater and thus either protect these surfaces during aerial exposure or modify the organs to extract oxygen from air. Intertidal fishes may employ cutaneous respiration.
- Feeding in intertidal animals usually occurs only when submerged, which protects the fleshy parts of the animal from desiccation.
- Most intertidal animals have planktonic larvae, and breeding cycles are synchronized with the tides.
- There are significant differences between the rocky shores of the Atlantic and Pacific coasts of North America. In the Atlantic, rocky shores are generally absent south of New England and are replaced by sedimentary environments, whereas the Pacific coast is dominated by rocky shores and sedimentary environments are uncommon. Aerial temperature ranges are greater on the Atlantic shores than the Pacific, and Atlantic intertidal organisms in New England and Canada are regularly killed by freezing or ice action, which rarely affects the Pacific intertidal. The Pacific intertidal was also spared the widespread extinction that occurred in the Atlantic intertidal during the last ice age.
- The number of species and species diversity are greater in the Pacific intertidal than in the Atlantic, and the Atlantic intertidal is dominated by introduced species, which is not the case in the Pacific intertidal.
- A characteristic feature of the rocky intertidal throughout the world is that there is a distinct horizontal banding or zonation of the organisms, with these bands succeeding each other vertically on the shore.
- This zonation pattern led to the establishment of a "universal" scheme of zonation that established a standard

format for describing shore zonation. This scheme has three main divisions of the intertidal: the uppermost is the supralittoral fringe, the middle is the midlittoral zone, and the lowest is the infralittoral fringe. The boundaries of these divisions are defined by the limits of distribution of the dominant organisms.

- The most obvious explanation for the occurrence of zonal patterns on rocky shores is that they are the result of tidal action and reflect the different tolerances of the organisms to increasing exposure to the air and the resultant desiccation and temperature extremes. Many studies in different areas and under different conditions of tides, wave action, temperature, and season have produced evidence that suggests that physical factors are strong determinants of the upper limits of intertidal organisms, but that biological factors do act in some instances whereas other rocky shores remain to be investigated.
- Whereas early work on the causes of zonation focused on physical factors, more recent work has begun to clearly establish the great importance of the biological factors of competition, predation, grazing, and larval settlement in the establishment of zonal patterns.
- In the rocky intertidal there is intense competition among organisms for space, and several studies of competition among animals and plants have clearly demonstrated that zonation of intertidal organisms is at least partially a function of biological competition.
- Competition and the outcome of competitive interactions may be modified by another biological factor—predation.
- On both the Atlantic and Pacific rocky coasts of North America, there are competitive dominant animals that have the ability to outcompete all other animals and take over the substrate. These animals are *Mytilus californianus* on the Pacific coast and *Mytilus edulis* on the Atlantic coast. They are prevented from doing this by predation by starfishes, *Pisaster ochraceus* in the Pacific and *Asterias forbesi* and *A. vulgaris* in the Atlantic. Thus, predatory activity by top predators permits other less competitive animals to continue to exist in the intertidal. Because of their ability to influence the structure of the entire intertidal community, such predators have been called keystone species.
- Some prey animals escape predation in various ways that contribute to zonal patterns. The Pacific coast barnacle *Balanus cariosus* escapes predation by the snail *Nucella* by growing too large to be attacked, while *Balanus glandula* escapes by settling so high in the intertidal that *Nucella* cannot reach them due to excessive desiccation.
- Some anomalies in the keystone species concept have caused some researchers to question its usefulness. This has led to a redefinition of a keystone species as a single predator that controlled the abundance of a prey species that in turn could exclude all other species and dominate the community. If more than a single predator is involved in preventing competitive domination, then it is diffuse predation, and if predation alone cannot control the abundance of the competitive dominant, it is called weak predation.
- The keystone species concept remains valid for some temperate shores, but is not a universal concept and its occurrence is variable within the range of the species. It may be replaced under certain environmental conditions by other community regulatory mechanisms such as diffuse or weak predation, larval settlement, and physical factors.
- A central theme of intertidal ecology is that whenever predation is reduced competition will be increased, but this is not the inevitable result of predator removal.
- The relative importance of biological and physical factors on rocky shores varies over relatively short distances.
- The role of grazers in regulating the upper and lower limits of algal species is important, but relatively few grazers are abundant enough to determine community structure.
- Grazing affects a number of parameters, including algal zonation, species diversity, patchiness, and succession.
- Algae have defenses against grazers including laying down calcium carbonate in the tissues, clustering reproductive sites underneath away from grazers, production of toxic or noxious chemicals, and ability to exist in growth forms that are grazer resistant.
- Some algal species can exist in two growth forms that vary in their susceptibility to grazing; this exhibits the strategy of bet-hedging.
- The importance of larval settlement to the structure of the intertidal communities has remained obscure due to the difficulty of working with larvae under natural conditions.
- Larval recruitment is variable in time and space, and that variability can be significant to intertidal community structure.
- Recent studies have demonstrated that predation by offshore predators can significantly reduce the larval recruits that reach the shore.
- The great spatial and temporal fluctuations in the settlement of larvae means that the importance of competition, predation, and grazing are also likely to be variable.
- Studies of community structure on tropical intertidal shores are uncommon, and thus it is difficult to state generalities. On the Pacific coast of Panama, the upper shores are dominated by barnacles, the midintertidal is inhabited by encrusting algae, and the lowest intertidal has a high diversity of organisms. Community structure is determined by complex biological interactions among a diverse assemblage of herbivores and predators with no keystone species.
- Small-scale patchiness or local distribution patterns are basic features of the rocky intertidal, are temporally and spatially variable, and are caused by any physical or biological activity that creates either an open space for recolonization, produces a different substrate, or alters the prevailing physical conditions in the area.
- Successional sequences in the rocky intertidal seem to be governed by a combination of the inhibition and facil-

itation models and depend on the time of initial disturbance that frees the space.

- Continued physical disturbance of differing degrees of severity and physical extent coupled with seasonal variations in the types of propagules available, along with fluctuations in the spatial distribution of dominant animals, ensure that succession is always occurring in the intertidal and is a major source of patchiness in the intertidal.
- Community structure changes horizontally as one moves from areas of high wave exposure to protected quiet water.
- The age structure of the dominant organisms in the intertidal is mainly of single-year classes.
- In general the high intertidal tends to have the distribution pattern of its characteristic species set by physical factors, whereas biological factors become more important in setting limits in the low intertidal. However, these factors do not act in isolation, and the final distribution patterns observed on either a large or small scale are often the result of the interaction of two or more factors.
- Certain conditions affecting life in tide pools differ markedly from the surrounding intertidal.
- Tide pools are refuges from desiccation for intertidal organisms but can have rapid changes in temperature, salinity, and oxygen concentrations that may strongly affect the organisms.
- Zonation patterns have been described for tide pools both within single pools and among pools at different levels on the shore. In general, numbers of species or organisms decrease with increasing height of the tide pool.
- The role of biological interactions in determining community structure of tide pools is poorly documented due to lack of studies and also because of the variability among tide pools that precludes obtaining statistically significant results from manipulative studies.
- Perhaps the most important physical factor governing life on exposed sand beaches is wave action and its attendant correlation with particle size and slope.
- Beaches are defined by three factors: particle size, wave action, and slope. All are interrelated such that defining two fixes the third.
- Particle size is significant through its effect on water retention in the beach.
- The slope of a beach is the result of the interaction among particle size, wave action, and the relative importance of swash and backwash water.
- Protected sand flats are much less variable and usually consist of fine sand grains at all times.
- A second important physical factor on sand beaches and flats is the wave-induced substrate motion.
- Within the continuum of grain size, slope, and wave actions, a number of beach types has been described. The extremes of this continuum are the dissipative and reflective beaches.
- The abundance and diversity of macrofaunal organisms on open beaches are more strongly correlated with particle size and slope than with wave action.
- Because of the ceaseless movement of the surface layers of sediment on sand beaches, few large organisms occupy the surface.
- Sand beaches and flats have a smooth, uniform profile and lack the topographic diversity of rocky shores; hence, environmental factors act uniformly at each tidal level.
- Whereas the wave action on open sand beaches precludes most organisms from inhabiting the surface layers, the much reduced wave action on protected sand flats makes them suitable for habitation by epifaunal organisms.
- Sand is an excellent buffer against large temperature and salinity changes due to its insulating qualities and the fact that it holds large amounts of water in its interstices.
- Oxygen may become depleted in the sedimentary substrate, especially in fine-grained sediments where water exchange with the sea above is reduced.
- Two main adaptations to continued substrate movement by animals inhabiting open sand beaches are to either burrow deeper than the substrate is displaced or else to rapidly reburrow as soon as the wave has passed.
- Other adaptations include smooth shells and reduced spines to aid in burrowing, tidal migrations to keep the animals in the swash zone, avoidance of diurnal predators by becoming nocturnal, and preventing of clogging of respiratory surfaces through development of meshes that filter out sand.
- The types of organisms conspicuous by their absence on open sand beaches are large plants and sessile animals.
- The dominant animals on sand beaches and sand flats are polychaete worms, bivalve mollusks, and crustaceans.
- There is little primary productivity on open sand beaches, and the animals found there are mainly filter feeders, detritus feeders, and scavengers.
- On protected sand flats, there is substantial primary productivity and all feeding types of animals are present.
- The zonation pattern of organisms on open sand beaches is not clearly defined nor as well understood as that of rocky shores and this is largely due to the researcher's inability to set up field experiments on open sand beaches.
- Open sand beaches generally have three zones: an upper supralittoral fringe inhabited by amphipods or ghost crabs, a broad midlittoral with sand crabs and isopods, and a infralittoral fringe with large number of species.
- In contrast to rocky shores, sand beaches have tidal and seasonal migrations of their animals.
- The small amount of evidence suggests that both competition and predation decrease in importance as exposure of the beach increases. This leaves the physical factors, mainly wave action, particle size, and beach slope, as the probable major contributors to the patterns observed.
- On protected sand flats, there is a tendency to have a complementary distribution of communities dominated by suspension feeders and deposit feeders due to trophic group amensalism, with deposit feeders more abundant in fine sediments and filter feeders in coarser sediments.

- In higher latitude sand flats, there may be seasonal changes in the community due to migrations of the dominant animals or to the scouring effects of ice.
- The prime biological factors that structure sand flat communities are predation, competition, and disturbance. Of these, current information suggests that either predation or predation combined with disturbance is especially significant.
- The role of competition in structuring sand flat communities is different from that observed on rocky shores. The three-dimensionality of the substrate, the animals living therein, and the interactions between juveniles and adults all tend to reduce the severity of direct competition.
- Although less common, competitive interactions leading to different depth distributions do occur in protected sand beaches.
- In contrast to predator removal in rocky intertidal areas, predator removal in soft substrates does not lead to increased competition or competitive exclusion.
- In contrast to rocky shores, sand flats are dominated by highly mobile predators.
- The macrofauna of intertidal sand flats displays high temporal and spatial variability, and the numbers of species and individuals may vary by several orders of magnitude within and between years. This high variability is due to the high predation pressure and biological disturbance that control the infaunal community structure to the extent that the majority of infaunal species are opportunistic.
- There are no sharp boundaries between sand and mud shores, and they are the opposite ends of a continuum.
- Muddy shores are restricted to intertidal areas protected from ocean wave activity.
- Muddy shores have very fine particles that greatly restrict the water interchange and, with the high numbers of bacteria, create anaerobic conditions in the substrate below a shallow surface aerobic layer.
- Between the aerobic and anaerobic layers is a transition area, called the redox potential discontinuity layer, characterized by a rapid change from positive to negative redox potential.
- One of the primary adaptations of mud flat organisms is the ability to burrow into the substrate or to form permanent tubes that reach the surface.
- Another adaptation of mud flat organisms is being able to live under low oxygen conditions or having some way of bringing oxygenated water down to themselves.
- Mud flats often develop a substantial growth of various plants and contain large numbers of bacteria especially in the RPD layer. Among the bacteria in the RPD layer are the chemolithoautotrophic or sulfur bacteria that are primary producers of organic material using the energy of reduced sulfur compounds.
- The dominant large animals on mud flats are polychaete worms, bivalve mollusks, and crustaceans.
- The dominant feeding types on mud flats are deposit feeders and suspension feeders. The major predators are fishes and birds.
- Community structuring factors in mud flats are similar to those acting in sand flats.
- There are relatively few accounts of the role of fishes in the rocky intertidal but most appear to be visual carnivores, suggesting a potentially significant role in intertidal community organization.
- On sand and mud flats, skates and rays as well as flatfish are significant predators of infaunal animals. They also have wider community effects by excavating pits, causing widespread mortality among the other infaunal organisms.
- Bird predation on the intertidal may be intense and significant, particularly at certain times of the year, and it plays an important role in structuring tidal flat communities.

REVIEW QUESTIONS

ESSAY: Develop complete answers to these questions.

1. An important theme of intertidal ecology is that release from predation pressure permits an increase in competition. Discuss this concept, presenting evidence to support your essay.
2. Discuss how Menge's (1991) work indicates that there are fundamental differences in the forces that structure the rocky intertidal communities of temperate and tropical waters.
3. Why is the age structure of a group of barnacles often skewed toward a single age class? Discuss. OR Sessile intertidal rock shore organisms and similar sessile organisms that inhabit the surfaces of pilings often show marked variations in the age structure of the populations such that a single age class dominates. How can such a skewed age structure become established?
4. Why must a researcher who is studying zonation on a sandy beach take repeated samples at many different times over the course of the study? Discuss.
5. How does predation modify the structure of zonation in the rocky intertidal? Discuss.
6. How does small-scale patchiness develop? What forces are in action in developing patches in the rocky shore community?
7. What is meant by the term *succession*? Does succession occur in the rocky intertidal? Why or why not? Defend your answer.
8. What adaptations allow organisms to survive in the anaerobic mud flat regions? Discuss and give examples.

BIBLIOGRAPHY

Ambrose, W. G. 1984. Role of predatory infauna in structuring marine soft bottom communities. *Mar. Ecol. Progr. Ser.* 17:102–115.

Ansell, A. D., P. Sivades, B. Narayanan, and A. Trevallion. 1972. The ecology of two sand beaches in southwest India. II. Observations on the populations of *Donax incarnatus* and *Donax spiculurn*. *Mar. Biol.* 17:318–322.

Bertness, M. D. 1999. *The ecology of Atlantic shorelines*. Sunderland, MA: Sinauer Associates, Inc.

Bosman, A. L., and P. A. R. Hockey. 1986. Seabird guano as a determinant of rocky intertidal community structure. *Mar. Ecol. Progr. Ser.* 32 (2–3):247–257.

Boyle, P. R. 1969. The survival of osmotic stress by *Sypharochiton pelliserpentis*. *Biol. Bull.* 136:154–165.

Brafield, A. E. 1978. *Life in sandy shores*. Studies in biology no. 89. London: Edward Arnold.

Brenchley, G. A., and J. T. Carlton. 1983. Competitive displacement of native mud snails by introduced periwinkles in the New England intertidal zone. *Biol. Bull.* 165:543–558.

Bridges, C. A. 1988. Respiratory adaptations in intertidal fishes. *Amer. Zoo.* 28:79–96.

Brown, A. C., and A. McLachlan. 1990. *Ecology of sandy shores*. New York: Elsevier.

Carefoot, T. 1977. *Pacific seashores, a guide to intertidal ecology*. Seattle: University of Washington Press.

Castenholz, R. W. 1961. The effect of grazing on marine littoral diatom populations. *Ecology* 42:783–794.

Chapman, A. R. O. 1990. Effects of grazing, canopy cover and substratum type on the abundances of common species of seaweeds inhabiting littoral fringe tide pools. *Bot. Mar.* 33:319–326.

Chapman, A. R. O., and C. R. Johnson. 1990. Disturbance and organization of macroalgal assemblages in the Northwest Atlantic. *Hydrobiologia* 192:77–121.

Chia, F. S. 1973. Sand dollar: A weight belt for the juvenile. *Science* 181:73–74.

Connell, J. H. 1961. The influence of interspecific competition and other factors on the distribution of the barnacle *Chthamalus stellatus*. *Ecology* 42:710–723.

Connell, J. H. 1970. A predator–prey system in the marine intertidal region. I, *Balanus glandula* and several predatory species of *Thais*. *Ecol. Monogr.* 40:49–78.

Connell, J. H. 1975. Some mechanisms producing structure in natural communities: A model and evidence from field experiments. In *Ecology and evolution of communities*, edited by M. L. Cody and J. M. Diamond. Cambridge, MA: Harvard University Press, 460–490.

Connolly, S. R., and J. Roughgarden. 1998. A latitudinal gradient in Northeast Pacific intertidal community structure: Evidence for an oceanographically based synthesis of marine community theory. *Amer. Nat.* 151:311–326.

Davis, P. S. 1969. Physiological ecology of *Patella*. III. Desiccation effects. *Mar. Biol. Assoc.* U.K. 49:291–304.

Dayton, P. K. 1971. Competition, disturbance and community organization: The provision of and subsequent utilization of space in a rocky intertidal community. *Ecol. Monogr.* 41:351–389.

Dayton, P. K. 1973. Dispersion, dispersal and persistence of the annual intertidal alga *Postelsia palmaeformis* Ruprecht. *Ecology* 54(2):433–438.

Dayton, P. K. 1975. Experimental evaluation of ecological dominance in a rocky intertidal algal community. *Ecol. Monogr.* 45:137–159.

Denley, E. J., and A. J. Underwood. 1979. Experiments on factors influencing settlement, survival and growth of two species of barnacles in New South Wales. *J. Exp. Mar. Biol. Ecol.* 36:269–293.

Dethier, M. N., and D. O. Duggins. 1988. Variation in strong interactions in the intertidal zone along a geographical gradient: A Washington–Alaska comparison. *Mar. Ecol. Progr. Ser.* 50:97–105.

Doty, M. S. 1946. Critical tide factors that are correlated with the vertical distribution of marine algae and other organisms along the Pacific coast. *Ecology* 27:315–328.

Eltringham, S. K. 1971. *Life in mud and sand*. New York: Crane Russak.

Evans, P. R., D. M. Henderson, T. J. Knights, and M. W. Pienkowski. 1979. Short-term effects of reclamation of part of Seal Sands, Teesmouth, on wintering waders and shelduck. I, Shorebird diets, invertebrate densities and the impact of predation on the invertebrates. *Oecologia* 41:183–206.

Foster, B. A. 1971. On the determinants of the upper unit of intertidal distribution of barnacles (Crustacea, Cirripedia). *J. Anim. Ecol.* 40:33–48.

Foster, M. 1990. Organization of macroalgal assemblages in the Northeast Pacific: The assumption of homogeneity and the illusion of generality. *Hydrobiologia* 192:21–33.

Francis, L. 1973. Intraspecific aggression and its effect on the distribution of *Anthopleura elegantissima* and some related anemones. *Biol. Bull.* 144:73–92.

Frank, P. W. 1965. The biodemography of an intertidal snail population. *Ecology* 46:831–844.

Gaines, S. D., and J. Lubchenco. 1982. A unified approach to marine plant-herbivore interactions. II. Biogeography. *Ann. Rev. Ecol. Syst.* 13:111–138.

Gaines, S. D., and J. Roughgarden. 1987. Fish in offshore kelp forests affect recruitment to intertidal barnacle populations. *Science* 235:479–481.

Gibson, R. N. 1969. The biology and behavior of littoral fish. *Oceanog. Mar. Biol. Ann. Rev.* 7:367–410.

Gibson, R. N., A. D. Ansell, and L. Robb. 1993. Seasonal and annual variations in abundance and species composition of fish and macrocrustacean communities on a Scottish sandy beach. *Mar. Ecol. Progr. Ser.* 98:89–105.

Grant, J. 1981. Dynamics of competition among estuarine sand burrowing amphipods. *J. Exp. Mar. Biol. Ecol.* 49:255–265.

Harrington, B. A., and D. C. Schneider. 1978. *Studies of shorebirds at an autumn migration stopover area*. Final Report for U.S. Fish and Wildlife Service, Migratory Bird and Habitat Research Laboratory, Laurel, MD.

Harrington, B. A., S. K. Groves, and N. T. Houghton. 1974. *Season progress report. Massachusetts shorebird studies*. Contract 14-16-008-687. U.S. Fish and Wildlife Service, Manomet, MA.

Haven, S. B. 1971. Niche differences in the intertidal limpets *Acmaea scabra* and *Acmaea digitalis* (Gastropoda) in central California. *Veliger* 13:231–248.

Hawkins, S. J. 1981. The influence of season and barnacles on the algal colonization of *Patella vulgata* exclusion areas. *J. Mar. Biol. Assoc. U.K.* 61:1–15.

Hawkins, S. J., and R. G. Hartnoll. 1983. Grazing of intertidal algae by marine invertebrates. *Oceanog. Mar. Biol. Ann. Rev.* 21:195–282.

Hay, C. 1979. Some factors affecting the upper limit of the southern bell kelp *Durvillaea antarctica* (Chamisso) Hariot on the New Zealand shores. *J. Roy. Soc. N.Z.* 9:279–289.

Hay, M. E. 1981. The functional morphology of turf forming seaweeds: Persistence in stressful marine habitats. *Ecology* 62:739–750.

Hedgpeth, J., ed. 1957. *Treatise on marine ecology and paleoecology*. Vol. 1, *Ecology*. Washington, DC: Geol. Soc. Amer. Memoir 67.

Himmelman, J. H., and D. H. Steele. 1971. Food and predators of the green sea urchin *Strongylocentrotus droebachiensis* in Newfoundland waters. *Mar. Biol.* 9:315–322.

Jones, N. S. 1948. Observations and experiments on the biology of *Patella vulgata* at Port St. Mary, Isle of Man. *Liverpool Biol. Soc. Proc. Trans.* 56:60–77.

Jones, N. S., and J. M. Kain. 1967. Subtidal algal colonization following the removal of *Echinus*. *Helg. Wiss. Meeresunt.* 15:460–466.

Kanwisher, J. 1957. Freezing and drying in intertidal algae. *Biol. Bull.* 113:275–285.

Keough, M. J., and A. J. Butler. 1979. The role of asteroid predators in the organization of a sessile community on pier pilings. *Mar. Biol.* 51:167–177.

Kennish, R., G. A. Williams, and S. Y. Lee. 1996. Algal seasonality on an exposed rocky shore in Hong Kong and the dietary implications for the herbivorous crab *Grapsus albolineatus*. *Mar. Biol.* 125:55–64.

Kooistra, W. H. C. F., A. M. T. Joosten, and C. van den Hoek. 1989. Zonation patterns in intertidal pools and their possible causes: A multivariate approach. *Bot. Mar.* 32:9–26.

Kozloff, E. N. 1983. *Seashore life of the northern Pacific Coast*. Seattle: University of Washington Press.

Levinton, J. S. 1977. Ecology of shallow water deposit-feeding communities. In *Ecology of the marine benthos*, edited by B. C. Coull. Columbia: University of South Carolina Press, 191–227.

Levinton, J. S., and G. R. Lopez. 1977. A model of renewable resources and limitation of deposit feeding benthic populations. *Oecologia* 31:177–190.

Levinton, J. S., S. Stewart, and T. H. DeWitt. 1985. Field and laboratory experiments on interference between *Hydrobia totteni* and *Ilyanassa obsoleta* (Gastropoda) and its possible relation to seasonal shifts in vertical mudflat zonation. *Mar. Ecol. Progr. Ser.* 22:53–58.

Lewis, J. R. 1964. *The ecology of rocky shores*. London: The English Universities Press, Ltd.

Littler, M. M., and D. S. Littler. 1980. The evolution of thallus form and survival strategies in benthic marine macroalgae: Field and laboratory tests of a functional form model. *Amer. Nat.* 116:25–44.

Lowe, G. C., Jr., and E. R. Cox. 1978. Species composition and seasonal periodicity of the marine benthic algae of Galveston Island, Texas. *Contrib. Mar. Sci. Univ. Texas* 21:9–24.

Lubchenco, J. 1978. Plant species diversity in a marine intertidal community: Importance of herbivore food preference and algal competitive abilities. *Amer. Nat.* 112(983):23–39.

Lubchenco, J. 1980. Algal zonation in the New England rocky intertidal community: An experimental analysis. *Ecology* 61:333–344.

Lubchenco, J. 1982. Effects of grazers and algal competitors on fucoid colonization in tide pools. *J. Phycol.* 18:544–550.

Lubchenco, J., and S. D. Gaines. 1981. A unified approach to marine plant-herbivore interactions. I. Populations and communities. *Ann. Rev. Ecol. Syst.* 12:405–437.

Lubchenco, J., and B. Menge. 1978. Community development and persistence in a low rocky intertidal zone. *Ecol. Monogr.* 59:67–94.

Lubchenco, J., B. A. Menge, S. D. Garrity, P. J. Lubchenco, L. R. Ashkenas, S. D. Gaines, R. Eralet, J. Lucas, and S. Strauss. 1984. Structure, persistence, and role of consumers in a tropical intertidal community (Taboguilla Island, Bay of Panama). *J. Exp. Mar. Biol. Ecol.* 78:23–73.

MacGinitie, G. E., and N. MacGinitie. 1949. *Natural history of marine animals*. New York: McGraw-Hill.

Mann, K. H. 1972. Macrophyte production and detritus food chains in coastal waters. *Mem. Inst. Ital. Idrobiol.* 29(suppl.):353–383.

McLachlan, A. 1983. Sandy beach ecology. In *Sandy beaches as ecosystems*, edited by A. McLachlan and T. Erasmus. The Hague: Dr. W. Junk, 321–380.

McLusky, D. S., S. A. Nair, A. Stirling, and R. Bharoava. 1975. The ecology of a central west Indian beach with particular reference to *Donax incarnatus*. *Mar. Biol.* 30:267–270.

Menge, B. A. 1976. Organization of the New England rocky intertidal community: Role of predation, competition and environmental heterogeneity. *Ecol. Monogr.* 46(4):355–393.

Menge, B. A. 1991. Relative importance of recruitment and other causes of variation in rocky intertidal community structure. *J. Exp. Mar. Biol. Ecol.* 146:69–100.

Menge, B. A. 1995. Indirect effects in marine rocky intertidal interaction webs: Patterns and importance. *Ecol. Monogr.* 65:21–74.

Menge, B. A., and J. Lubchenco. 1981. Community organization in temperate and tropical rocky intertidal habitats: Prey refuges in relation to consumer pressure gradients. *Ecol. Monogr.* 51:429–450.

Menge, B. A., J. Lubchenco, L. R. Ashkenas, and F. Ramsey. 1986. Experimental separation of the effects of consumers on sessile prey in the low zone of a rocky shore in the Bay of Panama: Direct and indirect consequences of food web complexity. *J. Exp. Mar. Biol. Ecol.* 106:225–269.

Menge, B. A., E. L. Berlow, C. A. Blanchette, S. A. Navarrete, and S. B. Yamada. 1994. The keystone species concept: Variation in interaction strength in a rocky intertidal habitat. *Ecol. Monogr.* 64:249–286.

Metaxas, A., and R. E. Scheibling. 1993. Community structure and organization of tide pools. *Mar. Ecol. Progr. Ser.* 98:187–198.

Moran, M. J. 1980. The ecology and effects on prey of the predatory gastropod, *Morula marginalba*. Ph.D. dissertation, University of Sydney.

Newell, R. C. 1970. *Biology of intertidal animals*. New York: Elsevier.

Newell, R. C. 1979. *Biology of intertidal animals*. Faversham, Kent, U.K.: Marine Ecological Surveys Ltd.

Paine, R. T. 1966. Food web complexity and species diversity. *Amer. Nat.* 100:65–75.

Paine, R. T. 1969. A note on trophic complexity and community stability. *Amer. Nat.* 103:91–93.

Paine, R. T. 1974. Intertidal community structure: Experimental studies on the relationship between a dominant competitor and its principal predator. *Oecologia* 15:93–120.

Paine, R. T. 1979. Disaster, catastrophe and local persistence of the sea palm *Postelsia palmaeformis*. *Science* 205:685–687.

Paine, R. T. 1988. Habitat suitability and local population persistence of the sea palm *Postelsia palmaeformis*. *Ecology* 69(6):1784–1794.

Paine, R. T., and R. L. Vadas. 1969. The effects of grazing by sea urchins, *Strongylocentrotus* spp., on benthic algal populations. *Limn. Oceanog.* 14:710–719.

Paine, R. T., J. C. Castillo, and J. Cancimo. 1985. Perturbation and recovery patterns of starfish dominated intertidal assemblages in Chile, New Zealand and Washington State. *Amer. Nat.* 125(6):679–691.

Peterson, C. H. 1977. Competitive organization of the soft bottom macrobenthic communities of southern California lagoons. *Mar. Biol.* 43:343–359.

Peterson, C. H. 1991. Intertidal zonation of marine invertebrates in sand and mud. *Amer. Sci.* 79(3):236–249.

Peterson, C. H., and N. Peterson. 1979. *The ecology of intertidal flats of North Carolina: A community profile*. Slidell, LA: U.S. Fish and Wildlife Service, Office of Biological Services FWS/OBS-79/39.

Petraitis, P. S. 1987. Factors organizing rocky intertidal communities of New England: Herbivory and predation in sheltered bays. *J. Exp. Mar. Biol. Ecol.* 109:117–136.

Quammen, M. 1984. Predation by shorebirds, fish and crabs on invertebrates in intertidal mudflats: An experimental test. *Ecology* 65:529–537.

Ricketts, E., and J. Calvin. 1985. *Between Pacific tides*. 5th ed. Rev. by D. W. Phillips. Stanford, CA: Stanford University Press.

Sanford, E. 1999. Regulation of keystone predation by small changes in ocean temperature. *Science* 283:2095–2097.

Segal, E., and P. Dehnel. 1962. Osmotic behavior in an intertidal limpet, *Acmaea limatula*. *Biol. Bull.* 122:417–430.

Sousa, W. 1980. Experimental investigations of disturbance and ecological succession in a rocky intertidal algal community. *Ecol. Monogr.* 49:227–254.

Southward, A. J. 1965. *Life on the seashore*. Cambridge, MA: Harvard University Press.

Southward, A. J., and C. E. Southward. 1978. Recolonization of rocky shores in Cornwall after use of toxic dispersants to clean up the Torrey Canyon spill. *J. Fish. Res. Bd. Canada* 35:682–706.

Stanley, S. 1969. Bivalve mollusk burrowing aided by discordant shell ornamentation. *Science* 166:634–635.

Steneck, R. S. 1982. A limpet–coralline alga association: Adaptations and defenses between a selective herbivore and its prey. *Ecology* 63:507–522.

Stephenson, T. A., and A. Stephenson. 1949. The universal features of zonation between tidemarks on rocky coasts. *J. Ecol.* 37:289–305.

Stimson, J. 1970. Territorial behavior in the owl limpet, *Lottia gigantea*. *Ecology* 51:113–118.

Sutherland, J. P. 1990. Recruitment regulates demographic variation in a tropical intertidal barnacle. *Ecology* 71:955–972.

Timko, P. 1976. Sand dollars as suspension feeders: A new description of feeding in *Dendraster excentricus*. *Biol. Bull.* 151(1):247–259.

Ulanowicz, R. E., and J. H. Tuttle. 1992. The trophic consequences of oyster stock rehabilitation in Chesapeake Bay. *Estuaries* 15:298–306.

Underwood, A. J. 1980. The effects of grazing by gastropods and physical factors on the upper limits of distribution of intertidal macroalgae. *Oecologia* 46:201–213.

Underwood, A. J., and E. J. Denley. 1984. Paradigms, explanations and generalizations in models for the structure of intertidal communities on rocky shores. In *Ecological communities: Conceptual issues and the evidence*, edited by D. R. Strong, D. Simberloff, G. Abele, and A. B. Thistle. Princeton, NJ: Princeton University Press, 151–180.

Underwood, A. J., and P. Jernakoff. 1981. Effects of interactions between algae and grazing gastropods on the structure of a low shore intertidal algal community. *Oecologia* 48:221–233.

Vadas, R. L. 1968. The ecology of *Agarum* and the kelp community. Ph.D. dissertation. University of Washington, Seattle.

Virnstein, R. W. 1977. The importance of predation by crabs and fishes on benthic infauna in Chesapeake Bay. *Ecology* 58:1199–1217.

Voss, G. L. 1976. *Seashore life of Florida and the Caribbean*. Miami: E. A. Seeman.

Wethey, D. S. 1984. Thermal effects on the distribution of barnacle populations in New England. *Biol. Bull.* 59:160–169.

Wethey, D. S. 1985. Catastrophe, extinction and species diversity: A rocky intertidal example. *Ecology* 66:445–456.

Whitlatch, R. 1982. *The ecology of New England tidal flats: A community profile*. Washington, DC: U.S. Fish and Wildlife Service, Biological Services Program FWS/OBS-81/01.

Wiltse, W. I. 1980. Effects of *Polinices duplicatus* (Gastropoda: Naticidae) on infaunal community structure at Barnstable Harbor, Massachusetts. *Mar. Biol.* 56:301–310.

Woodin, S. A. 1976. Adult-larval interactions in dense infaunal assemblages: Patterns of abundance. *J. Mar. Res.* 34:25–41.

Woodin, S. A. 1978. Refuges, disturbance and community structure: A marine soft bottom example. *Ecology* 59:274–284.

MEIOFAUNA

In addition to the large macrofaunal organisms that inhabit sedimentary shores, there is another world that exists here made up of organisms that occupy the microspaces between particles or live on the individual particles. These are the ***meiofaunal*** *or* ***interstitial*** *organisms. The existence of this unique association of organisms was not realized by biologists until the twentieth century, when European scientists began investigations into these areas. Since then, many scientists throughout the world have made studies of this unique habitat. The first studies were primarily taxonomic, to determine what kinds of organisms were present and to name them. More recent ecological studies have given us a better understanding of the peculiar conditions under which these animals live.*

An interstitial faunal and floral assemblage has been found in intertidal and subtidal sedimentary substrates throughout the world, in both fresh water and salt water. Major interstitial groups have been found by Coull et al. (1977) at water depths as great as 5,000 m. Studies of these associations are, however, unequally distributed. The marine interstitial fauna is better known than the freshwater interstitial fauna, and the organisms and associations of the European seas are the best known of the various geographical areas. The Pacific coast of North America is practically unknown and unstudied with respect to interstitial fauna, whereas the Atlantic coast of North America has been the site of numerous studies.

In this chapter we will investigate the principles that govern meiofaunal life in marine sediments.

Definitions

As with any discipline or subdiscipline, there are a few technical terms to be learned. Interstitial is a general term that refers to the spaces between the sedimentary particles and is also used as a synonym to refer to the organisms living there. Meiofauna is the preferred term to refer to the organisms that live interstitially. It includes all animals that pass through a 0.5-mm screen but are retained on a 62-μm screen. A synonym is **meiobenthos**. Since the term meiofauna defines a size class, but does not specify a spatial location or how the organisms move within the habitat, and since the interstitial space is a function of the size of the particles, a few other terms have been used to refine the classification and indicate location. **Endobenthic** is a term used now to refer to meiofaunal-sized organisms that move within the sediment by displacing particles (i.e., they are bigger than the interstitial space). **Mesobenthic** refers to meiofaunal organisms living and moving within the interstitial space, and **epibenthic** refers to

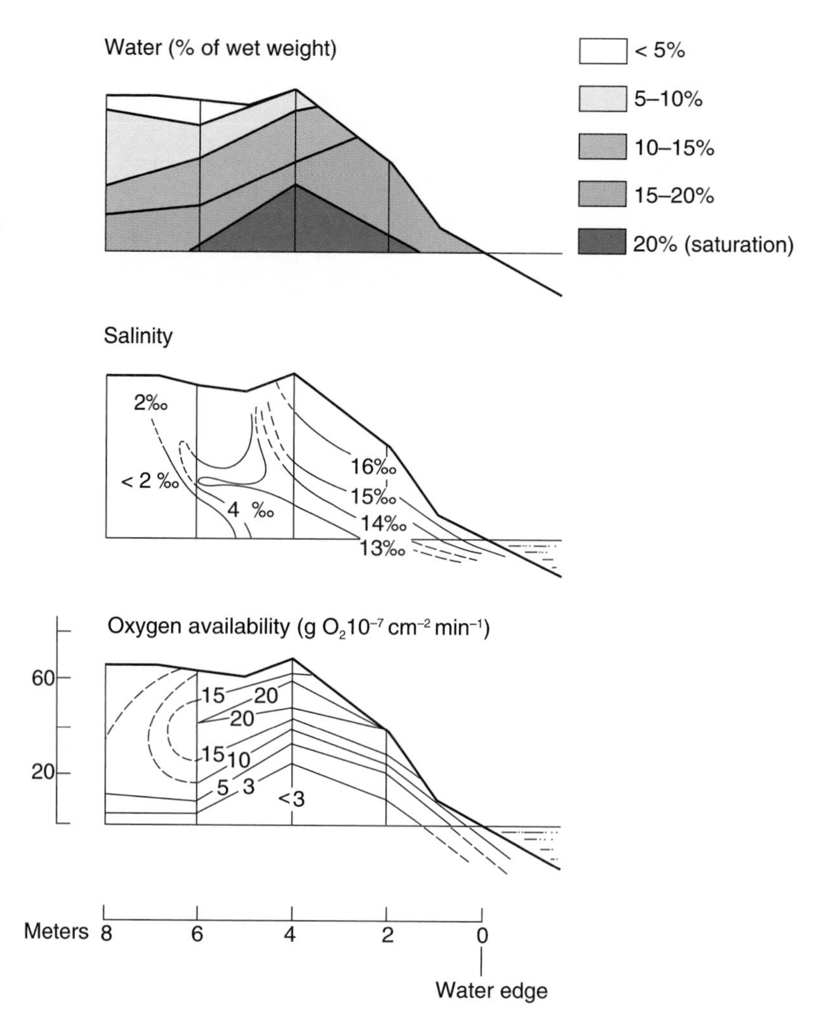

FIGURE 7.1 A transect of a Danish sand beach showing gradients of water, oxygen, and salinity at low tide. (From "Diurnal and Annual Variations of Temperature & Salinity of Interstitial Water in Sandy Beaches" B. O. Jansson, *Ophelia*, Vol. 4, pp. 173-201, 1967. Copyright © 1967 Apollo Books. Reprinted by permission.)

those meiofauna living at the sediment-water interface. In this scheme, mesobenthic is virtually equivalent to meiofauna. In this chapter, we deal primarily with meiobenthos operating as mesobenthic animals. In very fine sediments, however, meiobenthic organisms, such as nematodes, kinorhynchs, copepods, and others, will displace sediment grains as they move; thus, they operate as endobenthic animals.

ENVIRONMENTAL CHARACTERISTICS

The conditions that affect the meiofauna are somewhat different from those that affect the macrofauna in the same areas. Perhaps the most important factor influencing the presence, absence, and types of interstitial organisms is **grain size**. Grain size is of paramount importance in determining the amount of interstitial space available for habitation. The coarser the grain size, the greater the volume of interstitial space and the larger the interstitial organisms that can inhabit the area. Conversely, the finer the grain size, the lesser the space available and the smaller the organisms must be to inhabit the area. Grain size acts as a barrier to the movement of meiofaunal organisms. Certain organisms will be confined to sediments of certain grain size or areas of a given environment simply because they are too large to fit between the grains of adjacent areas and, therefore, cannot disperse through the whole area. Meiofaunal organisms show a definite zonation based on grain size. For example, sediments in which the median particle size is less than 125 μm are dominated by burrowing organisms, such as kinorhynchs, whereas sediments with particle sizes above 125 μm are dominated by forms that move in the interstices.

Grain size is also of great importance because it controls the ability of the sediment to retain and circulate water. Meiofaunal organisms are aquatic and require the presence of water in the interstices of the substrate to survive (Figure 7.1). If the grain size

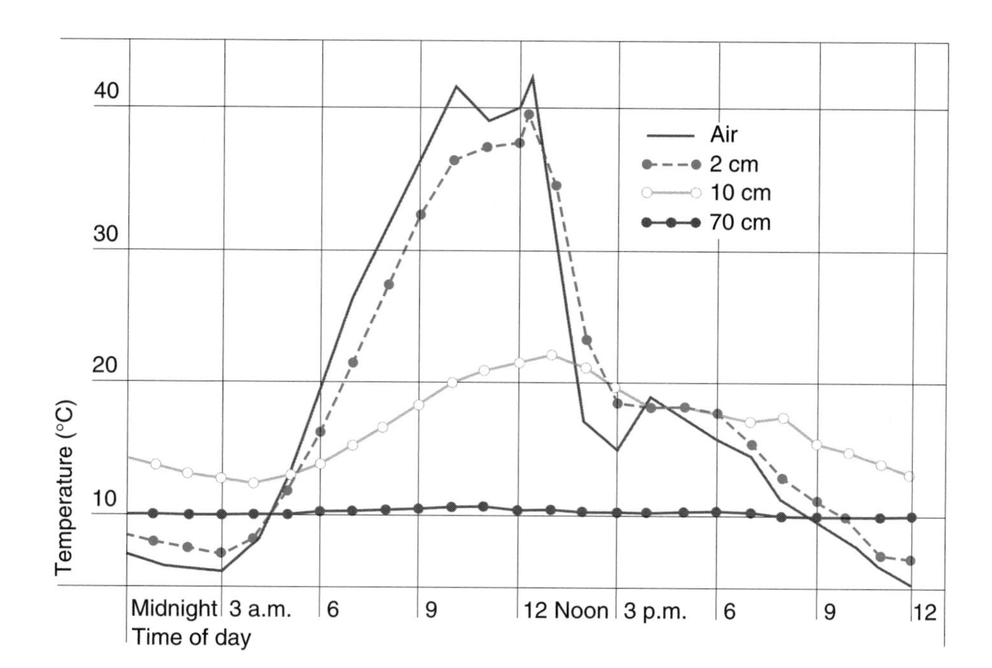

FIGURE 7.2 Graph showing the change in temperature at different depths in a Danish sand beach during a spring day. (From "Diurnal and Annual Variations of Temperature & Salinity of Interstitial Water in Sandy Beaches" B. O. Jansson, *Ophelia*, Vol. 4, pp. 173-201, 1967. Copyright © 1967 Apollo Books. Reprinted by permission.)

becomes too coarse, the water will not be held in the sand by capillary action when the tide ebbs and will drain away, leaving only a very thin layer coating the grains. On the other hand, fine-grained sediments can hold considerable water in the interstices through capillary action.

Circulation of water through the pore spaces in the sediment is important because this water movement is responsible for renewing the oxygen supply. Circulation is best in coarse-grained sediments and is reduced in fine-grained areas. If the grain size becomes really fine, as in mud shores, the circulation practically ceases. The result is the creation of anaerobic layers in the sediment (see discussion of RPD layer in Chapter 6, pp. 292–293). The maximum number of meiofauna have been found where sand grain diameters were between 0.175 and 0.275 mm. Fenchel (1978) reports that the interstitial fauna tends to disappear when the median grain size diameter drops below 0.1 mm.

In addition to grain size, the mineral nature of the grains is also of importance in determining the composition of the meiobenthos. Different associations of organisms may be found in siliceous sand as opposed to carbonate sand.

As suggested above, oxygen is an important factor in this environment. Practically all marine sediments have an oxidized layer at the surface, beneath which lies a completely anoxic layer of different chemical composition. Thus, meiofaunal organisms living below a certain depth will encounter oxygen-free conditions. The thickness of the oxygenated layer depends on a number of factors, such as grain size, amount of organic material, water turbulence, and bacterial metabolism.

Temperature is another environmental factor of importance in determining the presence or absence and distribution of meiofaunal organisms. Temperature ranges are most extreme in intertidal beaches and in the upper layers of the beach. Temperature changes are minimal in subtidal sediments and in intertidal sediments below 10–15 cm in depth (Figure 7.2). The surface layer of sand on a beach insulates the lower layers, dampening any significant temperature changes. Within the surface layers, however, the temperature may change markedly, depending on the air temperature, the effect of wind and rain, the amount of sunlight hitting the surface, and the temperature of the seawater. In cold temperate and polar regions, the temperature may be low enough to freeze the upper layers. Some meiofaunal organisms have adapted to survive being frozen (e.g., turbellarian *Coronhelmis lutheri*, gastrotrich *Turbanella hyalina*, oligochaete *Marionina southerni*, and harpacticoid copepod *Parastenocaris phyllura*). Because of the range of temperatures that may occur in the upper layers, the fauna inhabiting these areas may be different from those of lower layers. This includes those with a considerable tolerance for temperature change (eurythermal).

Another factor acting on meiobenthos, particularly of intertidal beaches, is **salinity**. Reduced salinities may occur in intertidal beaches as a result of freshwater runoff over the beach during low tide or

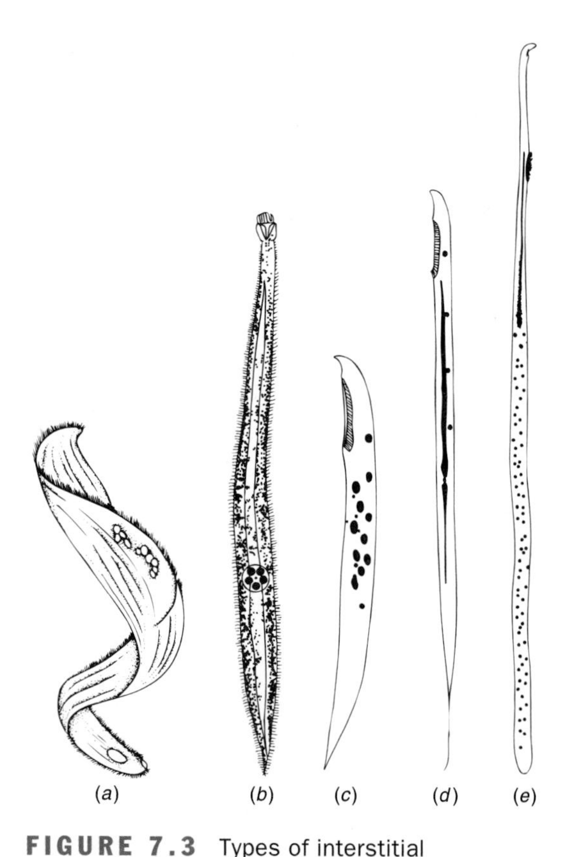

FIGURE 7.3 Types of interstitial ciliate protozoans, phylum Ciliophora.
(a) *Loxophyllum vermiforme* with flattened body.
(b) *Tracheloraphis remanei* with a cylindrical body.
(c) *Remanella faurei* with a flattened elongated body.
(d) *Remanella caudata* with an elongated body and tail.
(e) *Geleia gigas* with a threadlike body. (From "The Interstitial Fauna of Marine Sand" B. Swedmark, *iological Review*, vol. 32, pp. 1-42, 1964. Copyright © 1964. Reprinted by permission of Cambridge University Press.)

heavy rainfall. As noted in the last chapter, such salinity changes are usually confined to the upper layers of the beach because the lower layers retain, through capillary action, elevated levels of salt water. Since fresh water is less dense than salt water, fresh water cannot penetrate below the point where marine water is held by the capillary action. This means that only the uppermost layers are subject to salinity changes. Organisms inhabiting these layers are usually adapted to tolerate considerable salinity changes (e.g., down to 15 psu).

A final environmental factor is **wave action**. Wave action acts both intertidally and subtidally but is most dramatic intertidally. Breaking and moving waves resuspend the upper layers of sand and completely remove or deposit large amounts of sand seasonally. Whenever the beach is churned up, all internal space is rearranged, and the organisms themselves are moved about. This process has profound effects on organisms that depend on staying in the space between grains. For those organisms the interstitial space is being constantly rearranged, and they are in constant danger of being thrown out into the open water, where the risk of predation is much greater. Special adaptations for coping with this risk are known (see pp. 318–319).

Before leaving the realm of environmental factors, some mention should be made of light. Light rarely penetrates more than 5–15 mm into sediments. This means that microscopic plants should generally be restricted to the very surface layers. Diatoms, however, have been found at depths considerably below this, raising speculation about how they manage to survive. Most of the meioflora and microflora are attached to the surface of the sand grains and form an important food source for a variety of meiofaunal grazers.

COMPOSITION OF THE INTERSTITIAL ASSEMBLAGES

The types of organisms constituting the meiofauna are drawn from a broad range of invertebrate phyla. Some are represented by one or a few species, whereas others are abundant both in individuals and species. Their small size invites comparison to plankton organisms. In general, meiofaunal organisms have a size range similar to that of some of the smaller mesoplankton and the microplankton (see Chapter 2).

Those invertebrate phyla that normally have small bodies are, of course, preadapted to live in the small spaces between sand grains and are represented by many species and individuals. The protozoan phylum Ciliophora is represented abundantly by a large number of species. These are organisms diverse in form and, interestingly enough, often larger than the metazoan animals found with them (Figure 7.3). The class Turbellaria, of the phylum Platyhelminthes, is also abundantly represented by numerous small, flat, and elongate worms (Figure 7.4). Most abundant of all, perhaps, are the ubiquitous roundworms, the nematodes (Figure 7.5).

Several inconspicuous invertebrate phyla are well represented in the meiofauna, probably because they are normally small. These include the phyla Gastrotricha, Kinorhyncha, Tardigrada, and Rotifera (Figure 7.6). The Gastrotricha are particularly numerous in marine sediments (Figure 7.7) and Rotifera in fresh

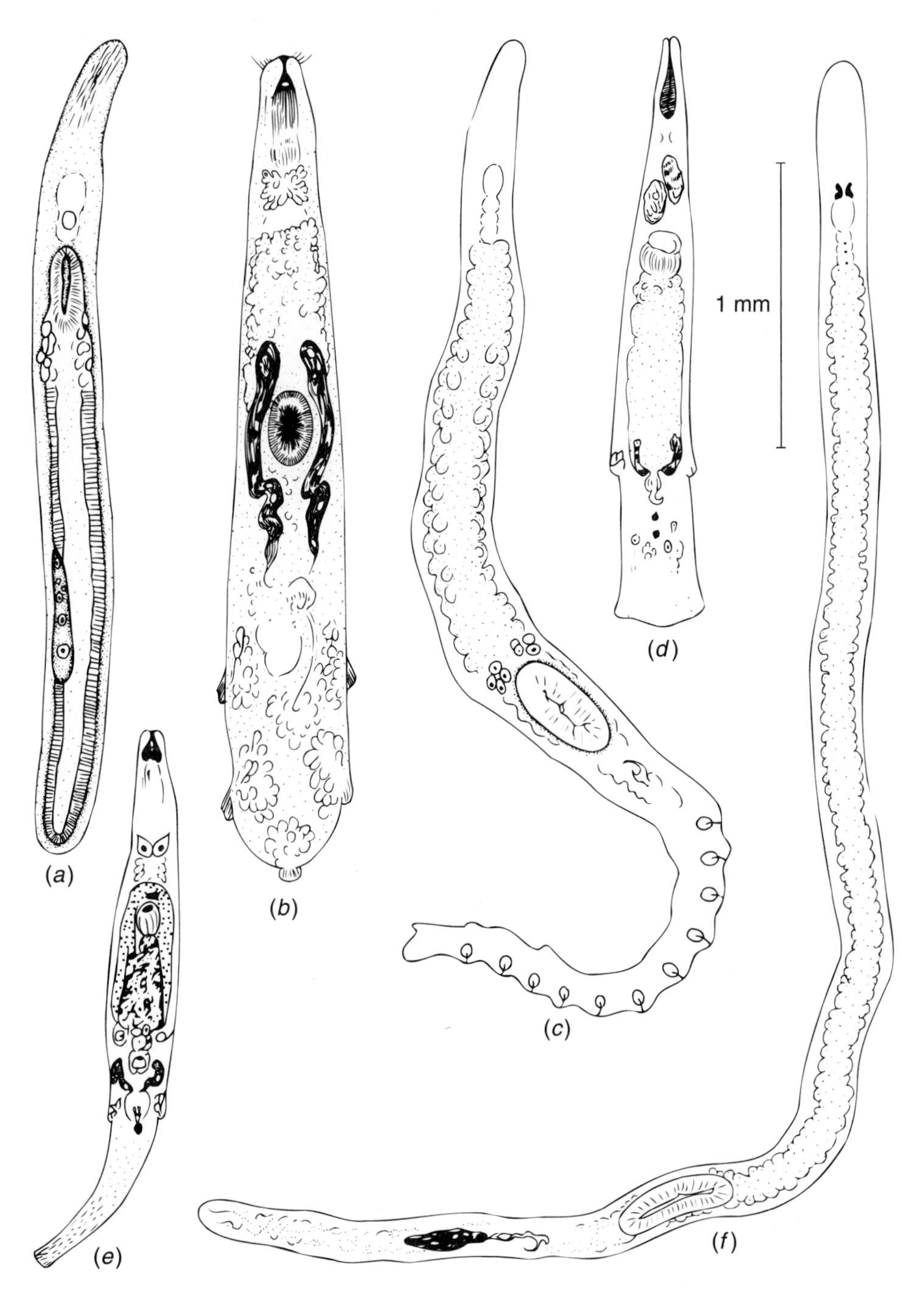

FIGURE 7.4
Examples of some turbellian flatworms from an interstitial marine sand beach in Florida.
(a) A member of the family Macrostomidae.
(b) A member of the family Kalyptorhynchidae.
(c) *Polystylophora* sp.
(d) *Proschizorhyncus* sp.
(e) *Cicerina* sp.
(f) *Nematoplana* sp.
(From Louise F. Bush, "Characteristics of Interstitial Sand Turbellaria" *Trans. Amer. Microscopical Society*, Vol. 87, No. 2, pp. 244-251, 1968. Reprinted by permission.)

water. Certain phyla with very few living representatives, such as Priapulida, Loricifera, and Entoprocta, are also found in the meiofauna (Figure 7.8).

The worms of the phylum Annelida with their elongated shape are well suited to the interstitial environment and are abundant. Representatives in the meiofauna are from both the classes Oligochaeta and Polychaeta (Figure 7.9).

Crustaceans are represented primarily by an abundance of harpacticoid copepods and ostracods and a few other small groups, such as Mystacocarida (Figure 7.10).

Phyla that normally consist of large organisms or of sedentary or sessile forms are poorly represented in the meiobenthos. In the case of large-bodied phyla, this is due to the body size, which does not allow

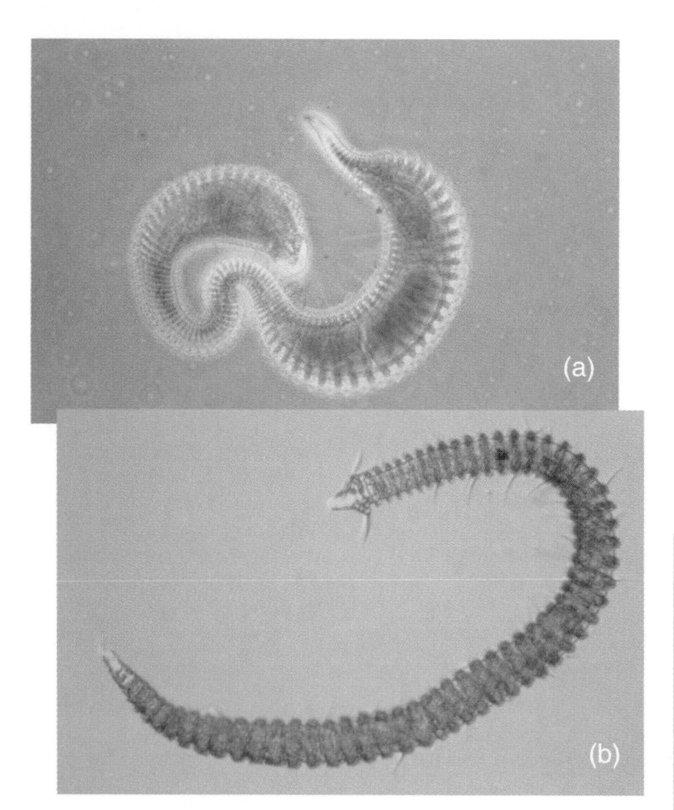

FIGURE 7.5 Two interstitial nematodes. (a) *Epsilonema* sp. (b) *Prototrichoma* sp. (Photos courtesy of Dr. Robert Higgins.)

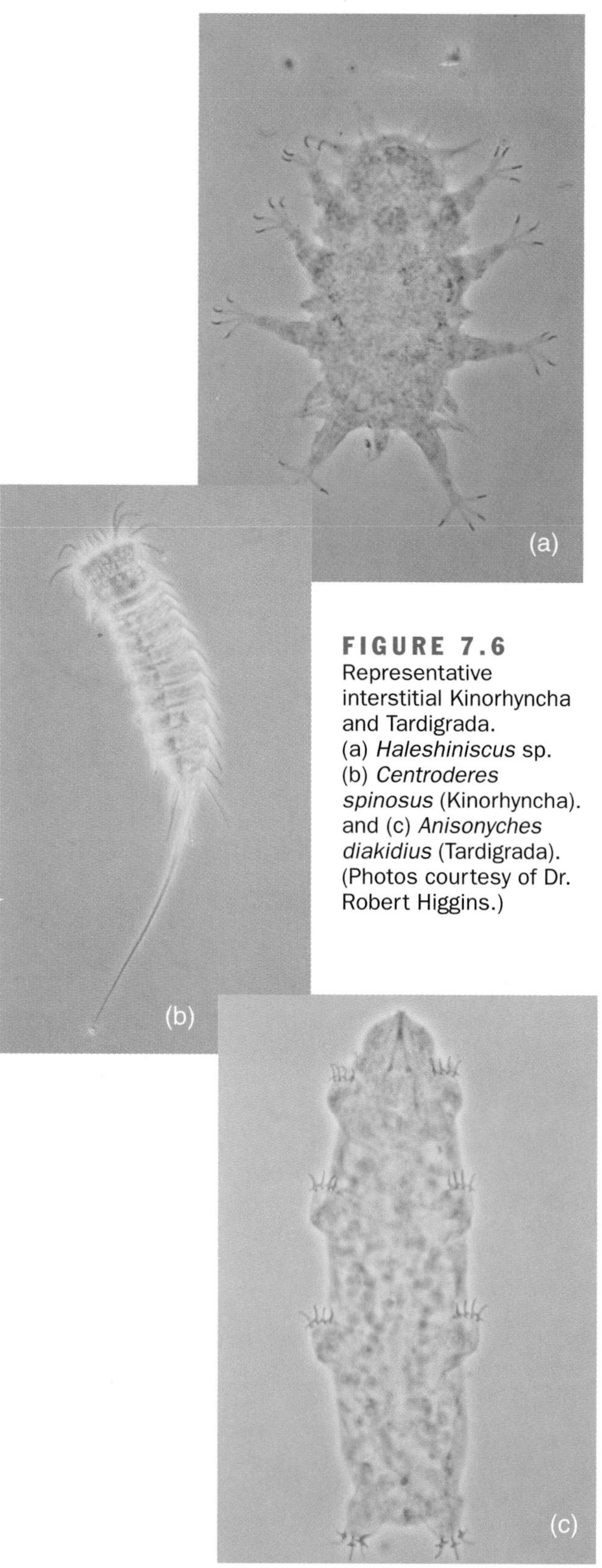

FIGURE 7.6 Representative interstitial Kinorhyncha and Tardigrada. (a) *Haleshiniscus* sp. (b) *Centroderes spinosus* (Kinorhyncha). and (c) *Anisonyches diakidius* (Tardigrada). (Photos courtesy of Dr. Robert Higgins.)

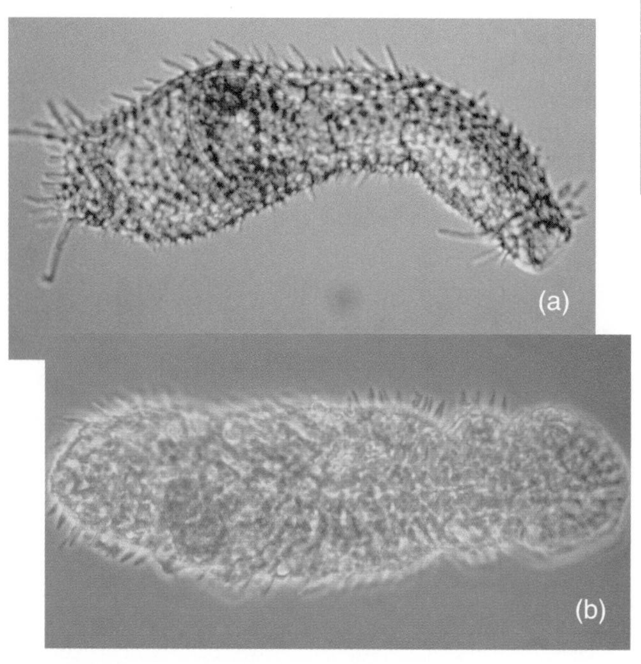

FIGURE 7.7 Two interstitial gastrotrichs. (a) *Pseudostomella* sp. (b) *Diplodasys* sp. (Photos courtesy of Dr. Robert Higgins.)

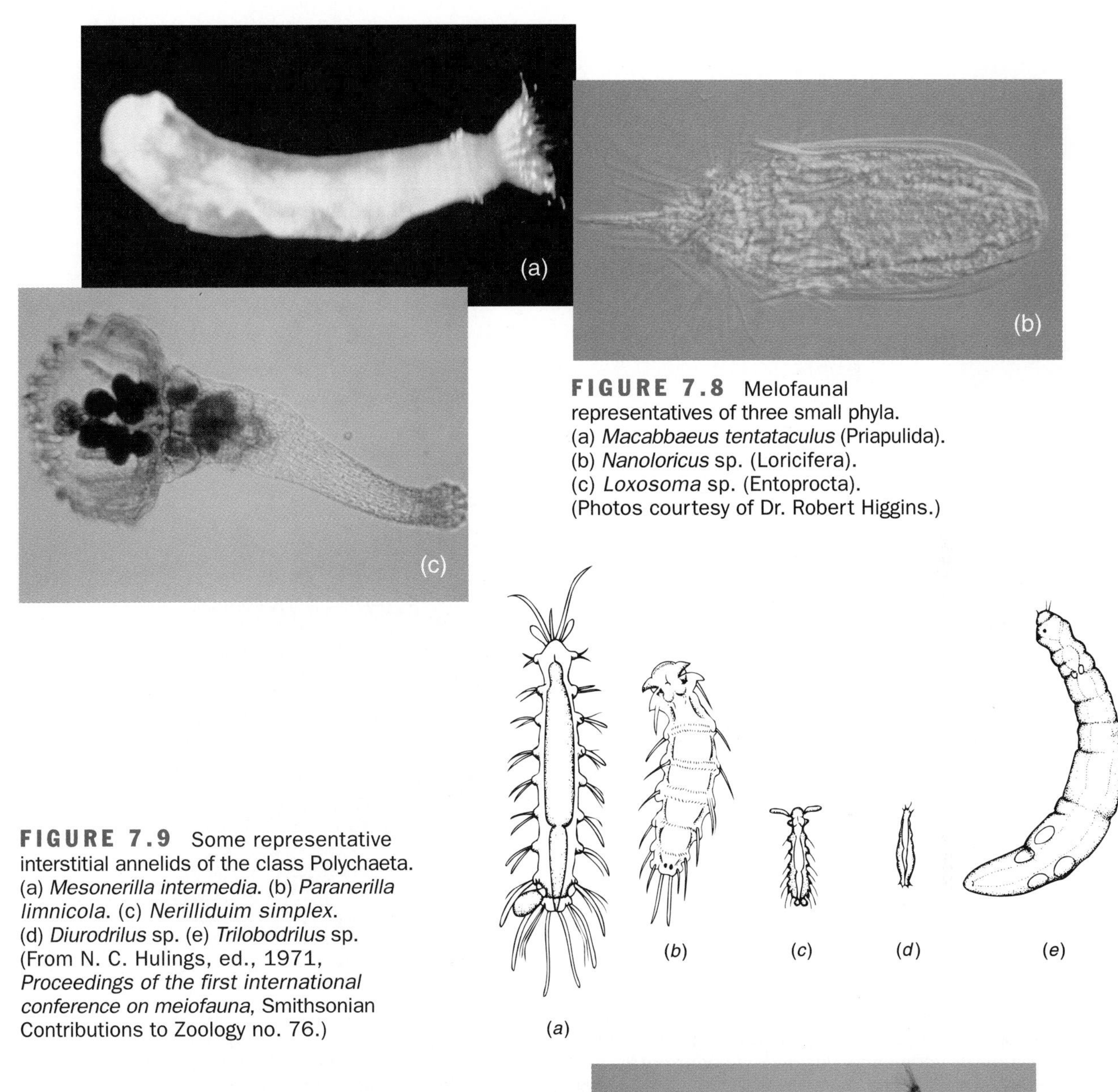

FIGURE 7.8 Melofaunal representatives of three small phyla. (a) *Macabbaeus tentataculus* (Priapulida). (b) *Nanoloricus* sp. (Loricifera). (c) *Loxosoma* sp. (Entoprocta). (Photos courtesy of Dr. Robert Higgins.)

FIGURE 7.9 Some representative interstitial annelids of the class Polychaeta. (a) *Mesonerilla intermedia*. (b) *Paranerilla limnicola*. (c) *Nerilliduim simplex*. (d) *Diurodrilus* sp. (e) *Trilobodrilus* sp. (From N. C. Hulings, ed., 1971, *Proceedings of the first international conference on meiofauna*, Smithsonian Contributions to Zoology no. 76.)

FIGURE 7.10 Some crustacea of the interstitial fauna. (a) *Derocheilocaris typica* (Mystacocarida). (b) Harpacticoid copepod. (Photos courtesy of Dr. Robert Higgins.)

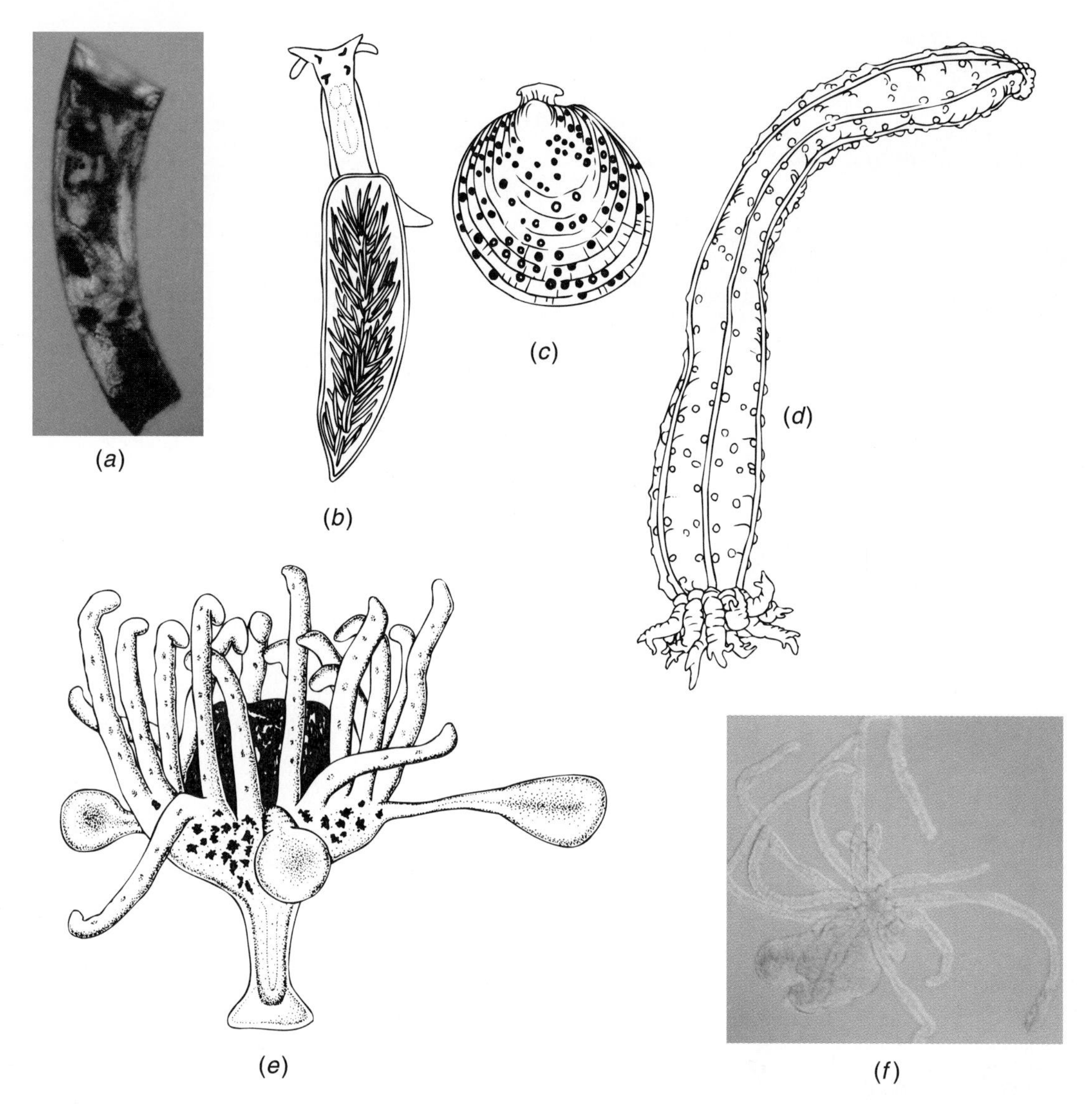

FIGURE 7.11 Representative interstitial Mollusca, Cnidaria, Echinodermata, and Brachiopoda. (a) *Caecum* sp. (Mollusca, Gastropoda). (b) *Hedylopsis brambeli* (Mollusca, Gastropoda). (c) *Gwynia capsula* (Brachiopoda). (d) *Labidoplax buskii* (Echinodermata). (e) *Stylocoronella riedli* (Cnidaria, Scyphozoa). (f) *Halammohydra* sp. (Cnidaria, Hydrozoa). (a, f, photos courtesy of Dr. Robert Higgins; b–e From Proceedings of the First International Conference on Meiofauna, *Smithsonians Contributions to Zoology*, No. 76, 1971.)

them to fit into the small spaces. These groups are, therefore, represented only by a few rather aberrant forms. Examples of such phyla are Echinodermata and Cnidaria (Figure 7.11). Phyla of sessile or sedentary habit have few representatives, because the dynamic nature of the substrate makes it difficult to remain permanently fixed in one place. Moreover, most of these phyla are suspension feeders, depending on filtering organisms out of the water. In the interstitial environment, there is relatively little water to filter and very few organisms in the water. Most organisms are attached to, or adhere to, the particles. The sessile groups represented by a very few species include the bryozoans and ascidians (Figure 7.12).

Absent from the interstitial fauna are the following phyla: Phoronida, Echiura, Porifera, Ctenophora, Hemichordata, and Chaetognatha.

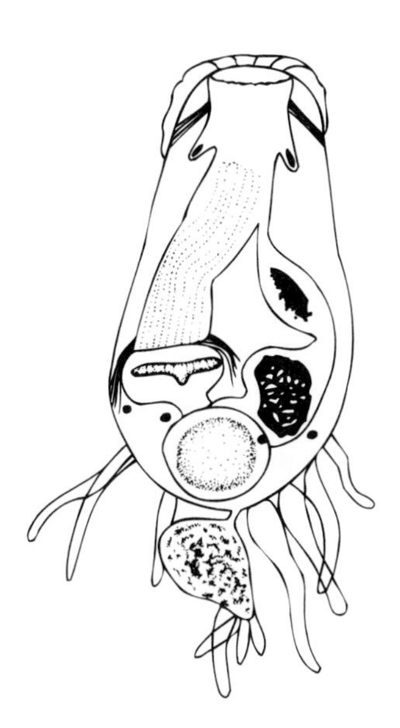

FIGURE 7.12 Sample interstitial Bryozoa. *Monobryozoon ambulans* (Bryozoa). (From Proceedings of the First International Conference on Meiofauna, *Smithsonians Contributions to Zoology*, No. 76, 1971.)

SAMPLING AND EXTRACTING MEIOFAUNA

Because interstitial organisms are very small and occupy a unique habitat, special techniques are needed to obtain samples. A great variety of sampling and extracting techniques have been employed by meiobenthologists over the years, and none is standard. Most techniques are, however, simple and may be grouped into general classes.

Because of the small size of the organisms, only a small sample of the substrate is required. This is most conveniently obtained by taking a core. Coring devices are usually hollow plastic or metal cylinders that may be pushed into the sand by hand, closed off at one or both ends, and removed from the beach (Figure 7.13). The enclosed sedimentary column may then be extruded and the organisms extracted. It is also possible to take a small volume or core from a larger grab or dredge sample that may be taken for macrofauna. Since the meiofauna exists throughout a number of vertical layers of the sand, it is best to take a core that extends some distance vertically into the substrate. The majority of the meiofaunal individuals will, however, be concentrated in the top 10–15 cm, so a depth of 25–30 cm is adequate to sample all but a few forms.

FIGURE 7.13 A simple coring device for obtaining samples of interstitial organisms. (Photos by the author.)

FIGURE 7.14 An elutriation apparatus for separating interstitial fauna from the sand. (Photo by the author.)

The diameter of the coring device is not critical, but those in common use range in size from 2–4 cm. Larger sizes produce such large samples that the extraction process is prolonged and difficult.

Once the sample is in hand, it is necessary to extract the organisms. Meiobenthic animals are reluctant to leave the sediment and must be removed by special methods in order to be studied. A number of different extraction techniques have been used over the years, and each gives somewhat different results. It should be noted that the fauna may be extracted alive or preserved. In most cases, it is preferable to extract the animals alive, because live animals are much more easily seen and because groups like flatworms and gastrotrichs are virtually unidentifiable when preserved.

The simplest extraction is to scoop up a bucket of sand and let it stagnate until organisms are driven to the surface by declining oxygen concentration. They may then be collected by pouring off the surface water through a 62-μm screen. Another simple technique is to treat a sediment sample with fresh water, to osmotically shock the animals to release from the particles, and then to pour off the water through a 62-μm screen. Perhaps the most used technique is to treat the sample with an isotonic solution of $MgCl_2$. This will anesthetize the animals, and they then can be collected by pouring the $MgCl_2$ solution through a fine net. Even more efficient is the method of **elutriation**. In this technique, a sediment sample is introduced into a container with openings at the top and bottom. The sediment is constantly stirred up by water entering from below. As the sediment stirs, the organisms are dislodged and carried over and out with the water through the upper opening. This water is filtered on a very fine mesh, which then catches the organisms (Figure 7.14). With this technique, one must remember to use only filtered water at the input. Otherwise, the sample will be contaminated by plankton organisms. This technique can be used with either living or preserved material.

ADAPTATIONS

Most of the morphological adaptations observed in the interstitial fauna correlate most closely with a few aspects of the environment, which can be reviewed here. The environment is dynamic, and the individual sediment grains of the upper layers are constantly being resuspended and deposited; thus, the location and amount of interstitial space is constantly changing. To survive here, meiofaunal organisms must remain in these spaces in the sediment and must avoid both being crushed by moving sand particles and being thrown out into the plankton, where their survival would be reduced.

Perhaps the most obvious adaptation to this environment is that of *size*. All meiofaunal organisms are very small. Furthermore, in this fauna, one finds the smallest representatives of most large phyla. For example, mollusks, echinoderms, and annelids are usually fairly large animals. All have representatives in the meiobenthos, but they are remarkable for their small size. The holothurian *Leptosynapta minuta* is 2 mm; the gastropod *Caecum glabrum* is 2 mm (see Figure 7.11); the polychaete *Diurodrilus minimus* is but 350 μm (see Figure 7.9); and the hydroid *Halamohydra* is 1 mm (see Figure 7.11). In the case of the ciliate protozoans, however, the size is generally larger than that of ciliates living in other environments. At the same time, however, these larger ciliates are considerably flattened and elongated, which allows them to fit more readily in the interstices (see Figure 7.3). In this case, larger really means longer.

Another adaptation concerns the *shape* of the body. Most of the meiofaunal organisms have elongated or vermiform bodies, even in those organisms that are normally not vermiform (see Figure 7.3). Another type of body shape observed is represented by organisms that are very flat. Flat organisms fit into the narrow spaces well, and they also give a greater surface area for hanging on to the grains. This latter adaptation has survival value in that it permits the organism to stay on the sediment grain during resuspension. Along with small body size, there is a reduction in the complexity and number of the body organ systems in metazoan animals. For example, in the polychaetes, the complex pharyngeal apparatus found in some macrofaunal polychaetes often disappears, as do the parapodial gills and even the kidneys (nephridia). In the meiofaunal hydroids, the number of tentacles is reduced from that observed in the forms not living interstitially.

One might expect that organisms living in an environment in which the substrate was often rearranged would be at risk of being crushed as sediment grains banged into each other. As a response to that, many meiofaunal animals have developed various types of reinforcement to their body walls. Skeletons of spicules are found in such ciliates as *Remanella*, such turbellarian flatworms as *Acanthomacrostomum spiculiferum*, and the acochlidiacean mollusks *Rhodope* and *Hedylopsis*. Gastrotrichs, on the other hand, have an armor of scales, while the nematodes depend on a thick, heavy cuticle. Those delicate animals lacking a protective armor have developed great abilities to extend and contract the body quickly, which also permits escape from crushing.

Perhaps because the environment is so dynamic, we find that most of the inhabitants are free-moving. This is true even for groups that are normally sessile, such as the Bryozoa and Ascidiacea. It is possible, however, that since all the common ways of extracting the meiofauna depend on the fauna leaving the substrate, there may be other sessile forms that we have not yet observed.

Two final types of adaptation may be noted. Both enable the meiofaunal organisms to remain in the sediment during the resuspension process. Many species of various phyla, such as flatworms and gastrotrichs, have adhesive organs to fix them to the sediment grains. Crustaceans and tardigrades use hooks or claws on their limbs. **Statocysts**, organs that detect gravity and help organisms differentiate up and down, are also common. Such sense organs enable any organism to determine quickly which way to move during resuspension; thus, they can return to the sediment rather than up into the open water.

Life History

As yet, we know relatively little of the life histories of many meiofaunal organisms. There are two groups of interstitial organisms. **Temporary meiofauna** consists of the newly settled juveniles of the macrofaunal organisms, and **permanent meiofauna** consists of organisms that spend their entire lives in the interstices of the sediment. The temporary meiofaunal organisms are not considered here.

Since the permanent meiofaunal organisms are so small, the number of gametes that they can produce at any time is also limited. In contrast to the macrofauna, which often produce thousands or hundreds of thousands of eggs, egg production by meiobenthic organisms is almost always below 100 per individual and usually ranges between one and ten!

When the number of eggs is that small, a species cannot afford to lose many or there will be no next generation. As a result, these animals have evolved a number of adaptations to ensure survival of the few propagules. First, there are adaptations that ensure that fertilization will occur. One way is to have copulation in which the sperm are directly transferred to the female. Such is the case with harpacticoid copepods. Another way is to package all the sperm in a single unit called a **spermatophore** and attach it to the female where it will be available to fertilize the eggs as they issue. Such a situation occurs in certain polychaete worms and the acochlidiacean mollusks. Finally, some species are hermaphroditic and have both male and female systems in the same individual, perhaps ensuring self-fertilization. This situation occurs in gastrotrichs, some hydroids (*Otohydra*), and some polychaetes (*Protodrilus*).

Once fertilization has occurred, it is important that the embryos be protected from loss. Usually, macrobenthic organisms produce larvae that enter the plankton in order to obtain maximum dispersal. However, the plankton community is notoriously dangerous, and most of the embryos are destroyed via predation. To compensate, the benthic animals using this route must produce large numbers of larvae. Since the meiofauna cannot produce large numbers, it would not be a good survival strategy to put larvae in the plankton. In fact, this is what is observed. Very few meiobenthic animals have planktonic larvae, and those that do are the ones that produce the largest numbers of eggs. Instead, most meiofaunal organisms have brood protection. The eggs are kept with the female until the young have reached a sufficient size to be self-sustaining. That selection pressure for brood protection is high in this environment can be deduced by observing that brood protection occurs here in groups of invertebrates in which it is otherwise virtually unknown, such as cnidarians (*Otohydra*) and gastrotrichs (*Urodasys viviparus*).

For those few that do not have brood protection, additional safeguards are found. Eggs that are spawned are either sticky so as to quickly attach to sand, or they are confined in cocoons that are sticky. In any case, the development is usually direct; that is, the egg hatches as a juvenile and does not have a free-swimming larva, or a larva is produced that is nonpelagic and remains in the interstices of the sediment.

Seasonal reproductive periods have been observed in many species. Studies by Gerlach (1971) suggest that generation times of interstitial organisms range from a few days to over a year. The average number of generations per year is three.

ECOLOGY

On a worldwide basis, Coull and Bell (1979) report that the average number of meiofaunal organisms per square meter is 10^6 and the biomass between 1 and 2 g/m^2. Both numbers of organisms and biomass are highest in intertidal areas and decrease with increasing depth in the oceans. The most numerically abundant taxa in all environments are nematodes and harpacticoid copepods.

The composition of the meiofauna varies both vertically and horizontally in the substrate. The factors creating this zonation are grain size differences and physiochemical factors, particularly temperature, oxygen, and salinity. Characteristic patterns of species assemblages are correlated with these physical factors. Most meiofaunal organisms are concentrated in the upper sediment layers, particularly in muddy sediments where Coull and Bell (1979) found that as much as 94% of all meiofauna were in the uppermost 1 cm. In sand beaches, meiofauna have been found as deep as 90 cm. Subtidally, perhaps 95% of the meiobenthos occur in the upper 7 cm. The variability in vertical extent seems related to the oxygen concentration; most taxa are present only in aerobic sediments. There is often a definite vertical zonation (Figure 7.15). Vertical zonation, however, may also be due to predation. Coull et al. (1989) reported that vertical cores taken where the fish *Leiostomus xanthurus* was feeding showed that the meiofauna was significantly reduced in the upper 2 cm.

There is a specialized faunal assemblage that is confined to the anoxic layers, the **thiobios**. Characteristic of this assemblage are ciliates of the orders Trichostomatida, Heterotrichida, and Odontostomatida, of which the latter contains only anaerobic species and is confined to this area. Zooflagellates, certain nematodes, turbellarians, gnathostomulids, rotifers, and gastrotrichs are also present, but it is not known whether these latter forms are permanent members of the community or transients from the upper oxidized zones.

Vertical migration of the organisms also occurs. This migration is usually triggered by temperature or salinity changes and is especially prominent in beaches in the temperate zone. The same organisms are often found deeper during the winter months and shallower during the summer. Strong tidal or wave action following storms may also trigger a migration, which may occur on a much shorter time scale, on the order of a few days or a tidal cycle.

There are also seasonal changes in the abundance of meiofaunal organisms. Certain species are more

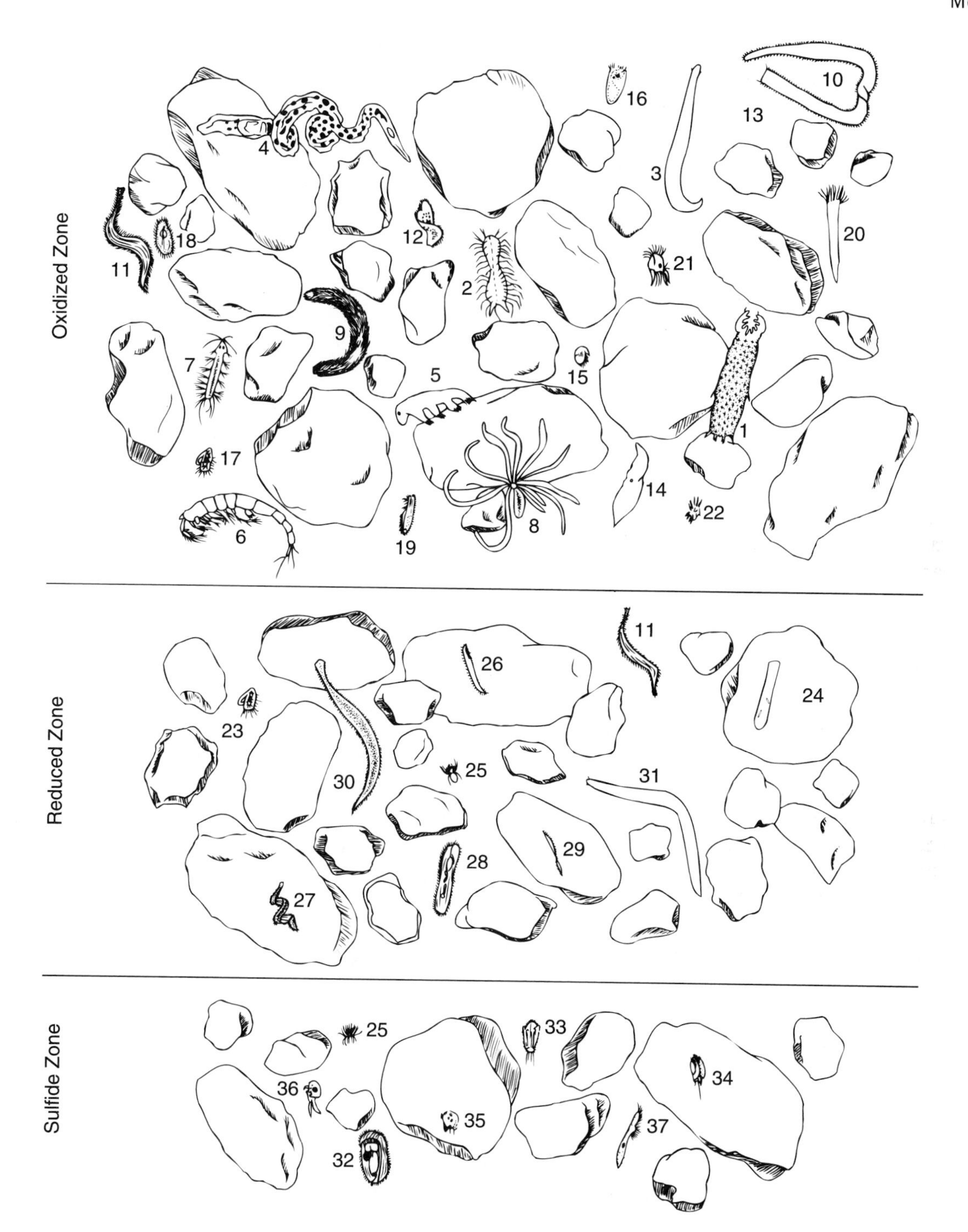

FIGURE 7.15 Vertical zonation of interstitial animals in a Scandinavian beach. (1) Microdasyoid, *Urodasys*; (2) gastrotrich, *Chaetonotus*; 3) nematode; (4) turbellarian, *Coelogynopora*; (5) tardigrades; (6) harpacticoid, *Cylindropsyllis*; (7) archiannelid, *Nemillidion*; (8) hydroid, *Halammohydra*. Cilates: (9) *Pseudoprorodon*; (10) *Helicoprorodon*; (11) *Tracheloraphis*; (12) *Loxophyllum*; (13) *Litonotus*; (14) *Dilepthus*; (15) *Lynchella*; (16) *Chlamydodon*; (17) *Prontonia*; (18) *Frontonia*; (19) *Blepharisma*; (20) *Condylostoma*; (21) *Diophrys*; (22) *Strombidium*; (23) *Aspidisca*; (24) *Paraspathidum*; (25) *Mesodinium*; (26) *Remanella*; (27) *Kentrophorus*; (28) *Cardiostomum*; (29) *Homalozoon*; (30) *Lacrymaria*; (31) *Geleia*; (32) *Sonderia*; (33) *Metopus*; (34) *Caenomorpha*; (35) *Saprodinium*; (36) *Myelostoma*; (37) *Parablepharism*.

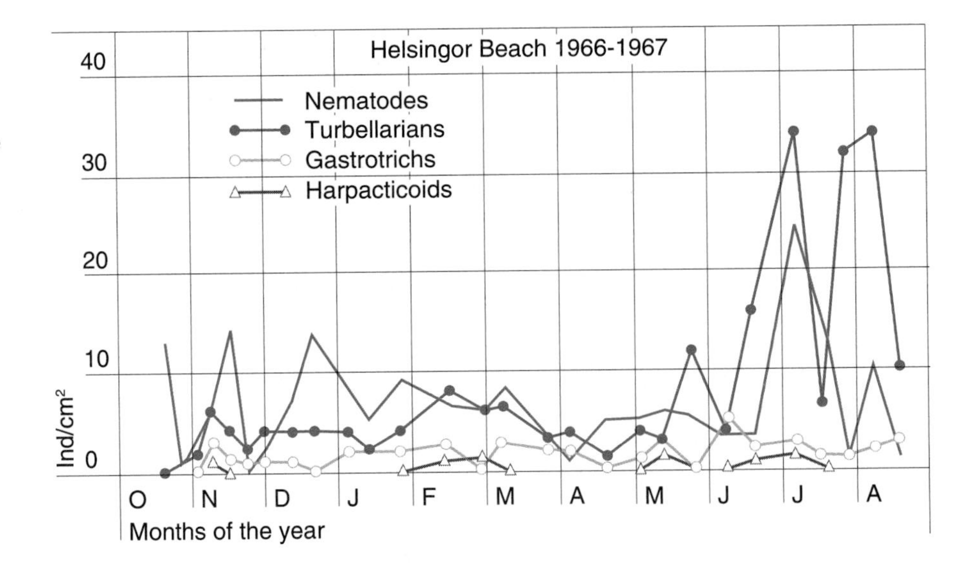

FIGURE 7.16 Changes in abundance of the major interstitial organisms with season in a Danish beach. (From "The Ecology of Marine Microbenthos" Fenchel et al., *Ophelia* Vol. 6, pp. 1–182, 1969. Copyright © 1969 Apollo Books. Reprinted by permission.)

abundant in one season than another (Figure 7.16). For example, on Danish beaches, Muus (1967) found that copepods were at a minimum in winter and peaked in numbers in the summer. Fenchel (1978) has noted that in temperate shallow-water areas, there is an annual change in the densities of the dominant groups. Copepods are succeeded by oligochaetes, nematodes, ostracods, and turbellarians. This temporal succession appears to be regulated by primary productivity and predation. Not only do abundances vary seasonally but so does the species composition, with different species becoming abundant at different times of the year. This is probably correlated with changes in the physiochemical factors mentioned previously, but it may also be due to migration.

Meiofauna are sensitive to human-induced or natural perturbations. Studies by Coull and Palmer (1984) recorded an immediate decrease in abundance and diversity of constituent meiofauna following a disturbance. In such perturbations, the crustaceans are most affected, and the nematodes are the least affected. Recovery rates from perturbations are variable, taking from only hours to months. Because of their rapid response to a disturbance, large numbers, stationary life habits, short generation time, and intimate association with sediments, meiofaunal organisms would seem to be ideal organisms to use as indicators of pollution. However, the great difficulty in identifying these tiny animals and the lack of experts coupled with the high expense of processing meiofaunal samples has precluded any widespread use of them for this purpose. In a review of the effects of pollution on meiofauna, Coull and Chandler (1992) noted that species diversity consistently decreased under polluted conditions, but the abundances of individuals of the major taxa increased in half the studies and decreased in the others.

Meiofaunal organisms are patchily distributed horizontally, even when the sediment grain distribution is homogeneous. We are ignorant of the causes of such patchiness, but various workers have suggested several reasons. Prominent among these suggestions is the idea that food sources are patchy, which leads to concentration of the meiofauna. Others have suggested that the occurrence of macrofaunal organisms or their constructions, such as burrows and tubes, increases the heterogeneity of the environment and the patchiness of distribution. Finally, since most meiofaunal organisms lack a larval dispersal mechanism, reproduction may result in aggregations of juveniles.

There is little information on the long-term variability of meiofauna. The few studies have produced conflicting results. The earliest, a nine-year study by McIntyre and Murison (1973), suggested that annual variations in meiofauna abundance were small, whereas Coull and Bell (1979) in a five-year investigation came to the opposite conclusion. More recently, Coull (1985a) in an 11-year study of two sites, one sand and one mud, demonstrated that the variability in the mud site was twice that of the sand, and the year-to-year variability was greater than the seasonality, but no long-term cycles were found.

Trophic relationships among the components of the interstitial fauna are not well known. It has been determined that certain flatfishes, such as *Pleuronectes platessa*, *Platichthys tiesus*, *Microchirus boscanion*, and *Limanda limanda* in European waters, actively prey upon meiofauna when they are young and recently metamorphosed. McIntyre (1969) reports that their stomachs contained numbers of harpacticoid copepods, ostracods, oligochaetes, and nematodes. More recently, Nelson and Coull (1989) demonstrated that the juveniles of the fish *Leiostomus xanthurus* preyed preferentially on the meiofaunal copepods in contrast to the more abundant nematodes. Other macrofaunal predators on meiofauna include shrimps, larger polychaetes, and hydroids. Macrofaunal deposit feeders, such as polychaetes and holothurians, must also destroy numbers of meiofauna existing on the particles they consume.

In addition to macrofaunal predators, there are interstitial predators, including certain nematodes, turbellarians, some tardigrades, and hydroids, such as *Protohydra*. The latter is a particularly voracious feeder on copepods and nematodes, according to Muus (1966). Predation by both meiofaunal and macrofaunal predators reduces the populations of other interstitial organisms, but it is not known how extensive this reduction is.

In addition to predators, the interstitial fauna contains three other feeding types: herbivores, detritus feeders, and suspension feeders. While some turbellarians are predators, others are scavengers, feeding on dead meiofaunal and macrofaunal organisms. Harpacticoid copepods are primarily herbivores, feeding on the few attached diatoms or detritus. Ostracods also ingest diatoms. Detritus feeders include gastrotrichs, some nematodes, and archiannelid polychaetes. Suspension feeders are quite rare, due to lack of an interstitial plankton.

A considerable degree of food specialization exists among the meiofauna. Fenchel (1968) studied the interstitial ciliates and demonstrated a surprising degree of food specificity for various bacterial and algal taxa. In addition, he showed that in certain habitats, whole groups of species subdivided their food resource according to size. A similar food niche separation has been reported by Tietjen and Lee (1977) for nematodes.

What controls meiofaunal communities? Again, we lack sufficient data to answer. Coull (1985b), in a study of two sites in South Carolina over an 11-year period, suggested that in fine-grained sediment (mud) organisms were controlled by predators, while in the nearby sand habitat the organism abundances were controlled by physical factors.

Meiofauna-Macrofauna Interactions

What is the significance of the meiofauna to the general ecology of benthic sedimentary systems and what, if any, is its relationship to the macrofauna? These questions have generated considerable interest and conflicting opinions over the years. The average meiofaunal biomass is from one-tenth to one-fifth of that of the resident macrofauna. In certain areas, such as the intertidal sand and mud flats and in the deep sea, the ratio is closer to one to one (Platt, 1981). This suggests that in these latter areas meiofauna plays an important role in energetics. Because of their small size, higher metabolic activity, and high turnover rate, productivity of meiofaunal organisms per unit of biomass is greater than that of the macrofauna. Therefore, even in systems dominated by macrofauna, total organic meiofaunal production approaches that of the macrofauna. Clearly, the meiofauna represents an important reservoir of organic material. Furthermore, meiofaunal animals stimulate bacterial production in the sediments by their metabolic secretions.

What is the ultimate fate of this production? Is it merely recycled within the meiofaunal system, or is there a linkage with the macrofauna and higher trophic levels? Earlier workers, such as McIntyre and Murison (1973), thought the meiobenthic system was separate, its production unavailable to higher trophic levels of the macrofauna. It now appears likely that these two systems are linked to some degree. For example, nonselective deposit feeders of the macrofauna cannot avoid consuming meiofauna, and if the digestion is equally nonselective, then meiofauna must contribute to their nutrition. Similarly, resuspension of bottom particles, which are captured by large suspension feeders, would lead to a similar macrofaunal ingestion. Macrofaunal predators that selectively feed on meiofaunal organisms seem rare, but Feller and Kaczynski (1975) have demonstrated that juvenile salmon are predators of meiobenthic copepods and Aarnio et al. (1998) have shown that the macrofaunal priapulid *Halicryptis spinulosa* significantly reduces meiofauna organisms in the Baltic Sea. Bell and Coull

FIGURE 7.17
Hypothetical benthic food web, illustrating possible connections between the meiofauna and macrofauna. (From "Meiofaunal Dynamics and the Origin of the Metazoa" H. M. Platt, 1981, pp. 207-216, Natural History Museum.)

(1978) have demonstrated that the grass shrimp *Palaemonetes pugio* not only feeds on meiofaunal organisms, but can control meiofaunal population numbers and community structure. Other fishes, particularly young stages of flatfishes, gobies, and mullet, have large quantities of meiofaunal organisms in their guts. Thus, the meiofaunal organisms probably fit into the macrofaunal food web, as illustrated in Figure 7.17.

The extent to which meiofauna enters the macrofaunal food web varies depending on the sediment type. It is likely that meiofauna is more significant as food to macrofauna in muddier sediments than in sands. This is partly because in muds the meiofaunal organisms are concentrated in the top layers, where they are more accessible, whereas in sand they may be much deeper and less accessible to many macrofaunal organisms. Since the macrofauna interact with the meiofauna, what effects do they have on meiofaunal community structure? Little is known at present, but the study by Bell and Coull (1978) has demonstrated that, in at least one case, meiofaunal numbers, particularly of the slower-moving wormlike meiobenthos, could be significantly reduced, suggesting a predator-controlled system.

Species Richness, Distribution, and Dispersal

One of the more remarkable features of the meiofauna is the species richness. Fenchel et al. (1967) found as many as 70 species of meiofaunal organisms in 50 cm^2 of sand, and they regularly obtained 30–50 species of ciliate protozoan alone. Even more striking is the discovery by Gerlach (1965) that the Arctic interstitial fauna is rich in contrast to the impoverished macrofauna at that latitude. Based on the limited studies available, it appears that there is no gradient in species richness in this fauna moving from polar regions to the tropics. This contrasts with the common distribution pattern of most macrofaunal organisms, where lower latitudes tend to have greater richness than higher latitudes.

Zoogeographically, the meiofauna have a pattern in which the common genera are often cosmopolitan, but the species are not. As a result, several different zoogeographical regions or provinces can be recognized.

Because most meiofaunal organisms lack a motile larval phase, they disperse mainly by the transport of adults by waterfowl, rafting on floating substrates (including icebergs), and turbulent water (Gerlach, 1977).

SUMMARY OF KEY CONCEPTS

- The meiofauna or interstitial organisms are those tiny animals that occupy the microspaces between particles or live on individual particles.
- Environmental conditions affecting the meiofauna are somewhat different from those that affect the macrofauna in the same area.
- Perhaps the most important factor influencing meiofauna is grain size.
- Coarse grain size means greater interstitial space; fine grain size means less interstitial space.
- Coarse-grained sediments hold less water by capillary action than fine-grained sediments.
- Circulation of water and oxygen is greater in coarse sediments and reduced in fine-grained sediments.
- In very fine grained sediments, such as mud, water circulation practically ceases, the interstitial habitat becomes anoxic, and the meiofauna disappears.
- Temperature is another factor of importance in determining presence, absence, and distribution of meiofaunal organisms. Temperatures are most variable in the surface layer of sediment, and the variability decreases with depth.
- Reductions in salinity due to flooding with fresh water usually affect only the uppermost layers of sediment, because the lower layers retain seawater by capillary action.
- Wave action churns up the sediment, constantly rearranging it and increasing the chances of throwing the meiofauna into the open water.
- The types of organisms constituting the meiofauna include a broad range of invertebrate phyla dominated by those phyla that normally have small enough sizes to fit between the sediment grains.
- Phyla that consist primarily of large or sedentary or sessile organisms are poorly represented in the meiofauna.
- Special techniques are required to extract the meiofauna from the sediment; these include osmotic shock, anesthetizing with magnesium chloride, seawater ice, and elutriation.
- The most obvious adaptation of meiofaunal organisms is that they are very small in size.
- Another adaptation is that these animals tend to be elongated or vermiform in shape and flattened.
- An adaptation to the dynamics of sediment movement is the presence of skeletons that offer some protection against crushing.
- Most animals in the interstitial are free-moving and have various adaptations to cling to the sediment grains.
- Temporary meiofauna consists of newly settled juveniles of macrofaunal organisms, whereas permanent meiofauna consists of organisms that spend their entire lives in the interstices of the sediment.
- Due to their small size, meiofauna produce very few eggs and ensure their fertilization and survival by various mechanisms that retain the embryos in the interstitial space.
- Worldwide, the average number of meiofaunal organisms per square meter is 10^6.
- The composition of the meiofauna varies both horizontally and vertically in the substrate, and most are concentrated in the upper layers.
- Vertical zonation of meiofauna in the sediment appears to be related to oxygen concentration and/or predation.
- Vertical migration triggered by temperature or salinity changes is especially prominent in temperate zone beaches.
- There are also seasonal changes in abundances of organisms and species composition, which may be regulated by primary productivity and predation.
- Meiofauna are sensitive to natural or human-induced perturbations.
- Meiofauna are distributed patchily, even where the sediment grain size is homogeneous; we are ignorant of the causes of this observation.
- There is little information on the long-term variability of meiofaunal communities, and the information that is available is conflicting.
- Trophic relations among the components of the interstitial fauna are not well known.
- Predators on the meiofauna include certain fishes and macrofaunal deposit feeders that kill meiofaunal when they ingest the sediment. There are also meiofaunal predators.
- The meiofauna also contains herbivores, detritus feeders, and suspension feeders.
- The significance of meiofauna to macrofaunal communities and food webs appears to depend on the sediment type. Meiofauna is more significant to macrofauna in finer sediments, probably partly due to the concentration of the meiofauna in the top layers, where they are more accessible.
- The meiofauna is species rich, and there is no gradient in richness with change in latitude.

REVIEW QUESTIONS

ESSAY: Develop complete answers to these questions.

1. How do temperature and salinity influence the distribution and types of interstitial organisms? Present evidence to support your answer.
2. What evidence supports the hypothesis that selection pressure for females to brood their offspring is high?
3. Why might one expect a diel rhythm of vertical migration of the meiofauna in the sand? How could you test to see if this rhythm is an endogenous behavior?
4. Discuss the evidence to support the hypothesis that meiofauna and macrofauna do interact.
5. What adaptations to the interstitial environment are found most commonly?
6. Why do meiofaunal organisms generally not produce planktonic larvae?

BIBLIOGRAPHY

Aarnio, K., E. Bonsdorff, and A. Norkko. 1998. Role of *Halicryptus spinulosus* (Priapulida) in structuring meiofauna and settling macrofauna. *Mar. Ecol. Progr. Ser.* 163:145–153.

Bell, S. S., and B. C. Coull. 1978. Field evidence that shrimp predation regulates meiofauna. *Oecologia* 35:141–148.

Bush, L. 1968. Characteristics of interstitial sand turbellaria: The significance of body elongation, muscular development and adhesive organs. *Trans. Amer. Microsc. Soc.* 87:244–251.

Coull, B. C. 1973. Estuarine meiofauna: A review: Trophic relationships and microbial interactions. In *Estuarine microbial ecology*, edited by L. H. Stevenson and R. R. Colwell. Columbia: University of South Carolina Press, 499–512.

Coull, B. C. 1985a. Long-term variability of estuarine meiobenthos: An 11-year study. *Mar. Ecol. Progr. Ser.* 24:205–218.

Coull, B. C. 1985b. The use of long-term biological data to generate testable hypotheses. *Estuaries* 8:84–92.

Coull, B. C., and S. S. Bell. 1979. Perspectives of marine meiofaunal ecology. In *Ecological processes in coastal and marine systems*, edited by R. J. Livingston. New York: Plenum Press, 189–216.

Coull, B. C., and G. T. Chandler. 1992. Pollution and meiofauna: Field, laboratory and mesocosm studies. *Oceanog. Mar. Biol. Ann. Rev.* 30:191–271.

Coull, B. C., and M. A. Palmer. 1984. Field experimentation in meiofaunal ecology. *Hydrobiologia* 19:1–19.

Coull, B. C., R. L. Ellison, J. W. Fleeger, R. P. Higgins, W. D. Hope, W. D. Hummon, R. M. Rieger, W. E. Sterrer, H. Thiel, and J. H. Tietjen. 1977. Quantitative estimates of the meiofauna from the deep sea off North Carolina, U.S.A. *Mar. Biol.* 39:233–240.

Coull, B. C., M. A. Palmer, and P. E. Meyers. 1989. Controls on vertical distribution of meiobenthos in the mud: Field and flume studies with juvenile fish. *Mar. Ecol. Progr. Ser.* 55:133–139.

Feller, R. J., and V. W. Kaczynski. 1975. Size selective predation by juvenile chum salmon (*Oncorhynchus keta*) on epibenthic prey in Puget Sound. *J. Fish. Res. Bd. Canada* 32:1419–1429.

Fenchel, T. 1968. The ecology of marine microbenthos. II. The food of marine benthic ciliates. *Ophelia* 5:73–121.

Fenchel, T. 1969. The ecology of marine meiobenthos. IV. Structure and function of the benthic ecosystem: Its chemical and physical factors and the microfaunal communities with special reference to the ciliated protozoa. *Ophelia* 6:1–182.

Fenchel, T. 1978. Ecology of micro- and meiobenthos. *Ann. Rev. Ecol. Syst.* 9:99–121.

Fenchel, T., B. O. Jansson, and W. von Thun. 1967. Vertical and horizontal distribution of the metazoan microfauna and of some physical factors on a sandy beach in the northern part of the Oresund. *Ophelia* 4:227–243.

Gerlach, S. A. 1965. Uber die Fauna in der Gezeitenzone von Spitzbergen. *Bot. Bothab. Acta Univ. Gothab.* 3:3–23.

Gerlach, S. A. 1971. On the importance of marine meiofauna for benthos communities. *Oecologia* 6:176–190.

Gerlach, S. A. 1977. Means of meiofauna dispersal. *Mikrofauna Meeresboden* 61:89–103.

Higgins, R. P., and H. Thiel, eds. 1988. *Introduction to the study of meiofauna*. Washington, DC: Smithsonian Institution Press.

Hulings, N. C., ed. 1971. Proceedings of the first international conference on meiofauna. Smithsonian Contributions to Zoology no. 76.

Hulings, N. C., and J. S. Gray, eds. 1971. A manual for the study of meiofauna. Smithsonian Contributions to Zoology no. 78.

McIntyre, A. D. 1969. Ecology of marine meiobenthos. *Biol. Rev.* 44:245–290.

McIntyre, A. D., and D. J. Murison. 1973. The meiofauna of a flatfish nursery ground. *J. Mar. Biol. Assoc. U.K.* 53:93–118.

Muus, B. J. 1967. The fauna of Danish estuaries and lagoons. *Meddr. Danm. Fisk.-og Havunders* 5:1–316.

Muus, K. 1966. Notes on the biology of *Protohydra leuckarti* Greef (Hydroidea, Protohydridae). *Ophelia* 3:141–150.

Nelson, A. L., and B. C. Coull. 1989. Selection of meiobenthic prey by juvenile spot (Pisces): An experimental study. *Mar. Ecol. Progr. Ser.* 53:51–57.

Platt, H. M. 1981. Meiofaunal dynamics and the origin of the metazoa. In *The evolving biosphere*, edited by P. L. Foray. British Museum (Natural History) and Cambridge University Press, U.K., 207–216.

Pollock, L. W. 1970. Distribution and dynamics of interstitial Tardigrada at Woods Hole, Massachusetts, U.S.A. *Ophelia* 7(2):145–165.

Sterrer, W., and P. Ax, eds. 1977. *The meiofauna species in time and space*. Mainz, West Germany: Akademie tier Wissenschaften und tier Literatur.

Swedmark, B. 1964. The interstitial fauna of marine sand. *Biol. Rev.* 39:1–42.

Tietjen, J. H., and J. J. Lee. 1977. Feeding behavior of marine nematodes. In *Ecology of marine benthos*, edited by B. C. Coull. Columbia: University of South Carolina Press, 21–35.

Vernberg, W., and B. C. Coull. 1981. Meiofauna. In *Functional adaptations of marine organisms*, edited by F. J. Vernberg and W. B. Vernberg. New York: Academic Press, 147–177.

ESTUARIES AND SALT MARSHES

Whereas the intertidal zone is the place where terrestrial and the marine habitats meet, an estuary is a place where fresh water and salt water meet. It is a transition zone between the two aquatic ecosystems. Estuaries have been and remain in close association with humans, because many of the major cities of the world are established on estuaries. We shall discuss the various special physical and chemical conditions in an estuary and how these create a rigorous environment in which relatively few species can persist.

TYPES OF ESTUARIES

There are many definitions of an estuary, because several geomorphological features of coastlines, such as lagoons, sloughs, fjords, and other shallow embayments, are often considered estuaries. A simple definition is that an **estuary** is a partially enclosed coastal embayment where fresh water and seawater meet and mix. This definition implies the free connection of the sea with the freshwater source, at least during a part of the year. It excludes permanently isolated coastal water impoundments, as well as such isolated brackish or saline bodies of water as the Caspian Sea, Aral Sea, and Great Salt Lake.

As a result of the geomorphology of an estuary, the geological history of the area, and the prevailing climatic conditions, there may be different estuarine types, each displaying somewhat different physical and chemical conditions. These may be grouped into a few basic types (Figure 8.1). Perhaps the most common type of estuary is the **coastal plain estuary**. Coastal plain estuaries were formed at the end of the last ice age when the rising sea level invaded low-lying coastal river valleys. Estuaries such as the Chesapeake Bay and the mouths of the Delaware and Hudson rivers in the United States and the Cornwall and Devon estuaries in Great Britain are examples of this type. Similar to this is the **tectonic estuary**. In this class of estuary, the sea reinvades the land due to subsidence of the land, not as a result of a rising sea level. A good example is San Francisco Bay. A third type of estuary is the **semienclosed bay** or **lagoon**. Here, sandbars build up parallel to the coastline and partially cut off the waters behind them from the sea. This creates a shallow lagoon behind the sandbars, which collects the freshwater discharge from the land. The water in such lagoons varies in salinity, depending on the climatic conditions, whether or not any major river flows into the lagoon, and the extent to which the bars restrict seawater access. Such estuaries are common in North Carolina, along the Texas and Florida Gulf coasts, in northwestern Europe (the Netherlands), and parts of Australia. A final category of estuary is the **fjord**. These are valleys that have been deepened by glacial action and are then invaded by the sea. They are characterized by a shallow sill at the mouth that greatly restricts water interchange between the deeper waters of the fjord and the sea. Often, these deeper waters are stagnant because of lack of circulation. Fjords are abundant on the coasts of Norway, Chile, Scotland, Alaska, New Zealand, and British Columbia.

Estuaries may be classified in yet another way, depending on the way the salinity gradients are formed. In most estuaries there is a gradient in salinity from full seawater (33–37 psu) at the mouth to fresh water at the upper reaches. Fresh water is less dense than seawater and, where the two meet, the fresh water will "float" on the seawater. Mixing occurs where the two come in contact, but the extent of the mixing varies with many other environmental factors, including basin shape, tide, river flow, and rainfall. In estuaries where there is substantial freshwater outflow and reduced evaporation (typical temperate zone estuaries), the fresh water moves out over the top of the salt water, mixing with it near the surface and reducing the salinity, leaving the deeper waters more saline. In such a situation, a cross section of the estuary shows isohalines (lines of equal salinity), which extend upstream at the bottom (Figure 8.2). At any given point on the estuary, a vertical column of water has highest salinity at or near the bottom and lowest at or near the surface. This is a **positive estuary** or **salt wedge estuary**. These estuaries are also called **river-dominated** or **stratified.** Such estuaries form a continuum—from those with little mixing and very prominent salt wedges, through those with partial mixing and lesser salt wedges, to **homogeneous**, **marine-dominated**, and **neutral estuaries** where either complete mixing or an evaporation rate equal to the freshwater inflow gives similar salinities from surface to bottom at any point. Where an estuary fits in this continuum depends not only on the amount of mixing of the water masses, but also on the tidal regime, the geometry of the estuarine basin, and the river flow. The tidal regime and river flow may be further altered seasonally. In general, positive estuaries are common, and neutral estuaries are rare. Galveston Bay, Texas, and Alligator Harbor, Florida, are examples of neutral estuaries.

In desert climates where the amount of freshwater input to the estuary is small and the rate of evaporation is high, a **negative** or **evaporite estuary** results. In a negative estuary, the incoming salt water enters at the surface and is somewhat diluted by mixing with the small amount of fresh water. The high evaporation rate, however, causes this surface water to become hypersaline. Hypersaline water is more dense than seawater, sinks to the bottom, and moves out of the estuary as a bottom current. A salinity profile of such an estuary is the reverse of the positive estuary, with highest values at the bottom and lowest at the top (Figure 8.2). A final category is the **seasonal** or **intermittent estuary**. These estuaries are formed in areas where there is a marked wet and dry season (Mediterranean climate). In the rainy season, they have freshwater input and are open to the ocean. In the dry season, they have no freshwater input, may become dry or stagnant, and are often cut off from the ocean by seasonal sandbars. Salinity in these estuaries varies not spatially but temporally.

PHYSICAL CHARACTERISTICS OF ESTUARIES

The physical-chemical regime of estuaries is one with large variations in many parameters, which often create a stressful environment for organisms. It is probably due to such stresses that the number of large species living in an estuarine area is small in comparison with other marine habitats.

Salinity

The dominant feature of the estuarine environment is the fluctuation in salinity. By definition, a salinity gradient exists at some time in an estuary, but the pattern of that gradient varies with the seasons, the topography of the estuary, the tides, and the amount of fresh water. We have already discussed the pattern of salinity distribution in salt wedge, partially mixed, and negative estuaries. There are, however, other factors that alter salinity patterns. The tide is one such force. Where the tidal range is significant, high tide

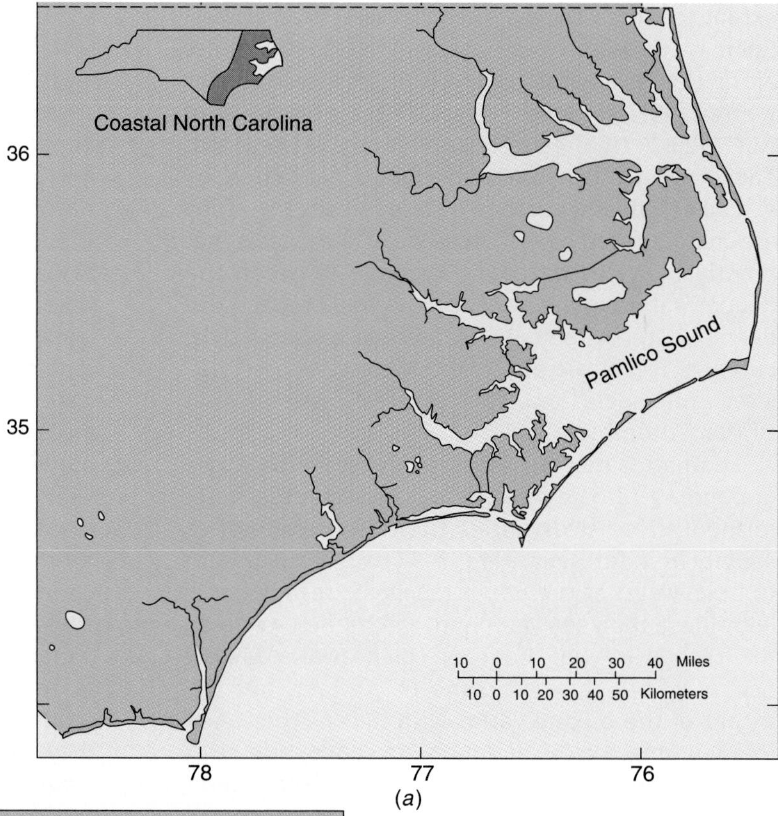

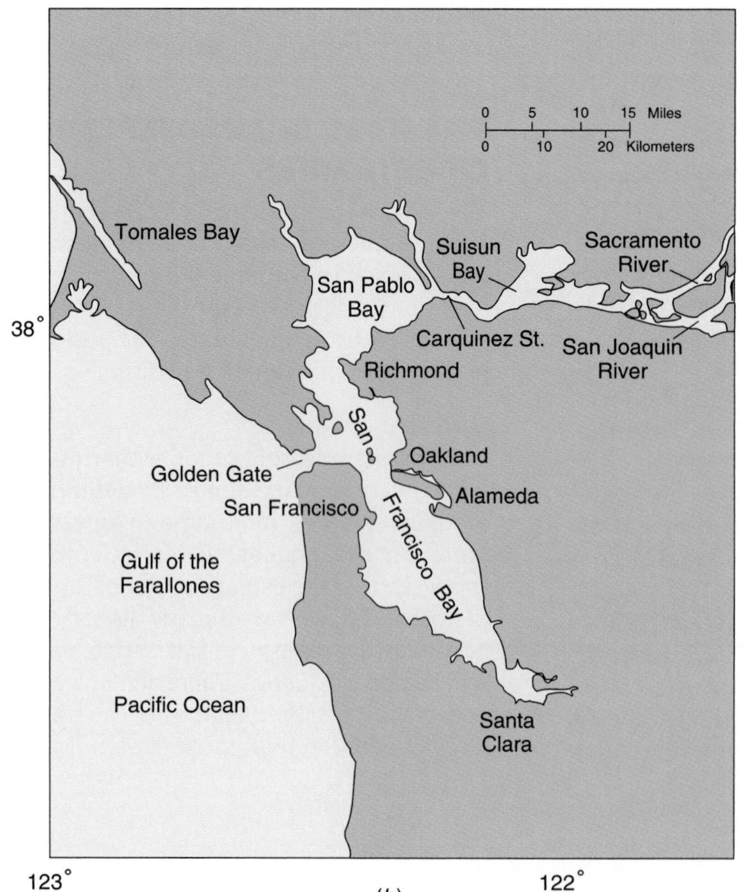

FIGURE 8.1 Types of estuaries. (a) Bay or lagoon lying behind a barrier beach; the major lagoons of the North Carolina coast. (b) Tectonic estuary; San Francisco Bay. (c) Fjord; Naeroy Fjord in Norway. (a, after C. Peterson and N. M. Peterson, 1979, *The ecology of intertidal flats of North Carolina: A community profile*, U.S. Fish and Wildlife Service, Biological Services Program FWS/OBS-79/39. b, after T. J. Conomos, ed., 1979, *San Francisco Bay, the urbanized estuary*, courtesy of Pacific Division—American Association for the Advancement of Science. c, photo by the author.)

(c)

FIGURE 8.1 (continued)

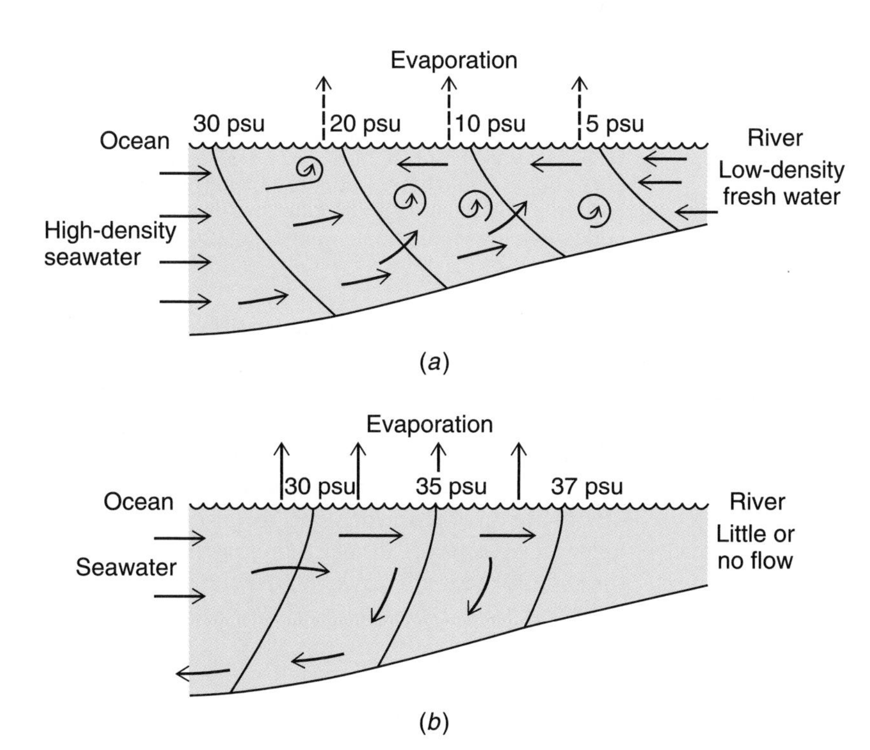

FIGURE 8.2 Positive and negative estuaries. (a) Positive (salt wedge) estuary. Freshwater runoff is significant throughout the year and greater than evaporation. Seawater enters along the bottom and gradually mixes with outward-flowing fresh water. (b) Negative estuary. Freshwater flow is diminished or absent during part of the year. Seawater enters along the surface. Evaporation is greater than runoff, so the salinity increases as one moves up the estuary. The hypersaline water sinks and flows out below the incoming seawater.

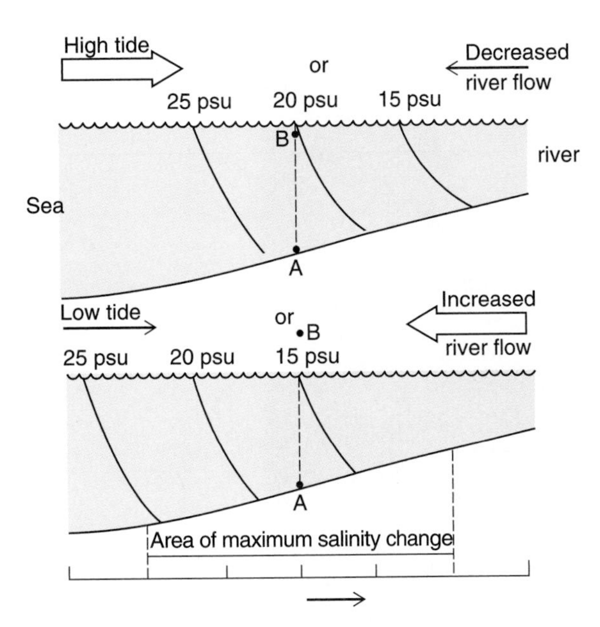

FIGURE 8.3 Change of salinity in an estuary with change in tide level or change in river discharge. Point A is on the bottom and B is intertidal. Because point B is covered only at high tide, it is inundated only with high-salinity water and is unaffected by low-salinity water at low tide. Point A, in contrast, is covered by water of different salinities at different tidal levels.

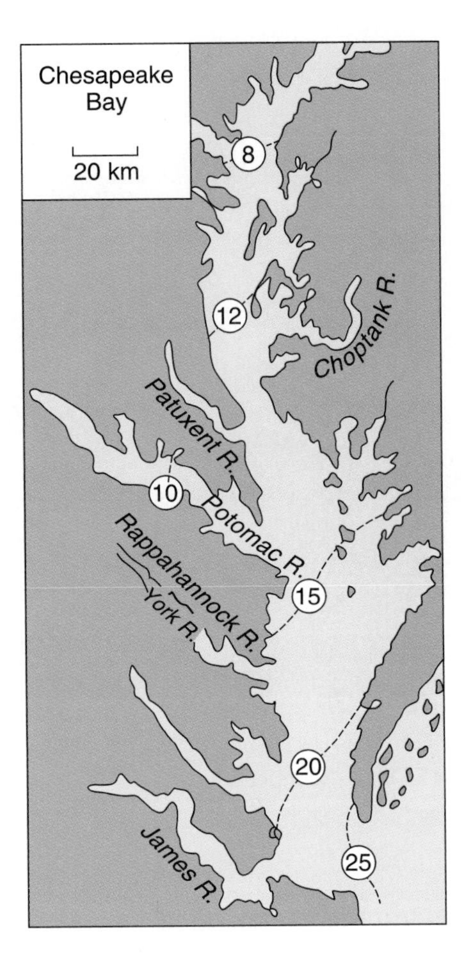

FIGURE 8.4 Result of Coriolis effect on salinity in the north-south oriented Chesapeake Bay estuary. Note the deflection of the isohalines to the right in this Northern Hemisphere estuary. Circled numbers refer to salinity values in parts per thousand. (After *Ecology of Estuaries*, D. S. McLuskey. Copyright © 1971 Heinemann Educational Publishers. Reprinted by permission of Heinemann Educational Publishers, a division of Reed Educational & Professional Publishing Ltd.)

drives the salt water further up the estuary, displacing the isohalines upstream. Low tides, by contrast, displace the isohalines downstream. As a result, a certain area of the estuary is subject to a salinity regime that changes with each tide (Figure 8.3). This area of an estuary has the maximum salinity fluctuation. The time scale of these salinity fluctuations is such that the salinity range over a 6- to 12-hour period equals or exceeds the entire annual range of salinities for some areas, even within an estuary.

A second force is the Coriolis effect. The rotation of the earth deflects flowing water. In the Northern Hemisphere, this effect deflects outflowing freshwater in north-south oriented estuaries to the right as one looks down the estuary toward the sea. Salt water flowing into the estuary from the ocean is also displaced to the right looking from the sea toward the estuary. The opposite is true in the Southern Hemisphere. As a result, two points, each on opposite sides of the estuary equidistant from the mouth, may have predictably different salinities (Figure 8.4).

Seasonal changes in salinity in the estuary are usually the result of seasonal changes in evaporation, freshwater flow, or both. In areas where freshwater discharge is reduced or absent for part of the year, higher salinities (salt wedge) may be found further upstream. With the onset of increased freshwater flow, the salinity gradients are moved downstream toward the mouth. At different seasons of the year, therefore, a given point in the estuary may experience different salinities.

Thus far, we have considered only the salinity changes in the water column itself. In the substrate, a different situation may prevail. Since estuarine substrates are sand and mud, water is held in the inter-

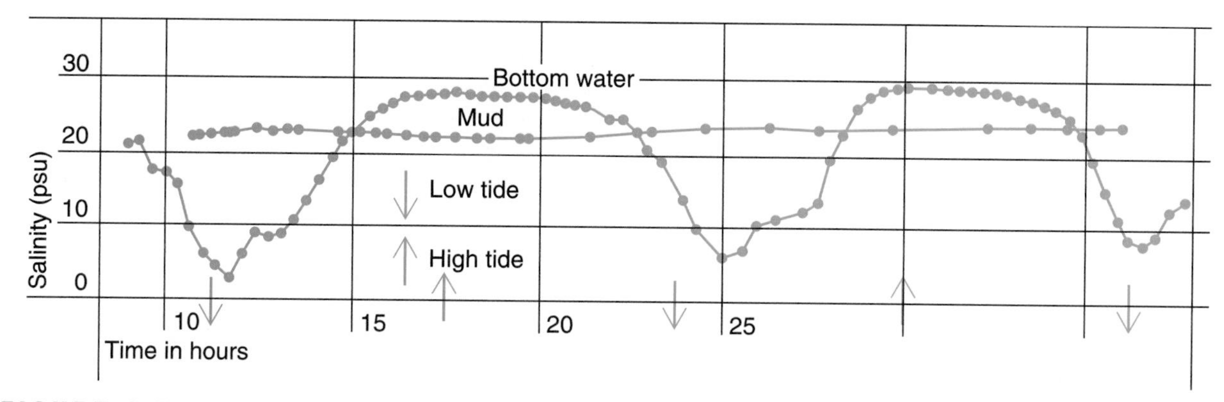

FIGURE 8.5 Comparison of salinity fluctuations in the water column with that interstitially in the bottom mud. Data from the Pocasset River estuary in Massachusetts. (Reprinted with permission from "Salinity Measurements in Estuaries" by P. C. Mangelsdorf, in *Estuaries*, Pub. No. 83, edited by G. H. Lauff, 1967, pp. 71-79. Copyright © 1967 American Association for the Advancement of Science.)

stices between the particles. This interstitial water originates from the overlying water. The interstitial water changes in salinity much more slowly than the overlying water because of the slow interchange between the two. The interstitial water and the surrounding mud and sand are, therefore, buffered with respect to the overlying water. Organisms dwelling within the substrate are subject to less drastic salinity changes than organisms in the water column at the same point. This buffering effect is probably most effective in the lower intertidal zone but not in the high intertidal. Variability in interstitial salinity increases with tidal height. This is because interstitial soil water at higher elevations has more opportunity for dilution during rains and is subject to more evaporation in dry weather (Figure 8.5).

Substrate

Most estuaries have soft, muddy substrates. These are derived from sediments carried into the estuary by both seawater and fresh water. Wind (aeolian) transport of larger sand particles into the estuary is often significant in certain areas, particularly coastal lagoons behind barrier beaches. In the case of fresh water, rivers and streams carry silt particles in suspension. When these suspended particles reach and mix with seawater in the estuary, the presence of the various ions in the seawater causes the silt particles to flocculate, creating larger, heavier particles. These particles settle out, forming the characteristic mud bottom. Seawater also carries much suspended material. When it enters an estuary, the sheltered conditions reduce the water motion that has kept the particles in suspension. As a result, the particles settle out and contribute to the forming of the mud or sand substrate. The relative importance of freshwater-borne or marine-borne particles to the development of the muddy substrate varies from estuary to estuary and also geographically.

The deposition of particles is also controlled by currents and the size of the particle. Large particles settle out faster than small particles, and strong currents keep particles in suspension longer than slow currents. Where strong currents prevail, the substrate will be coarse (sand or gravel), as only large particles can settle out; whereas where waters are calm and currents weak, fine silt will settle out. Thus, both seawater and fresh water drop their coarse sediments first, the former at the mouth of the estuary and the latter in the upper reaches of the river itself, so the area of mixing is dominated by fine silt (mud), resulting from decreased water movement and flocculation from the intermixing of the two water masses.

Catastrophic events may also play a significant role. Massive storms and their accompanying floods may produce large depositions or removal of sediment in estuaries, causing severe mortality of organisms. For example, Zedler (1982) reported 4 cm of silt deposited during a 1980 flood on the Tijuana estuary in southern California, and 40% of the low-tide volume of Magu Lagoon was lost during the same floods. Similarly, on the East Coast, Andrews (1973) reported that the massive flooding of fresh water into Chesapeake Bay following Hurricane Agnes in 1972 caused severe reduction in salinity and catastrophic losses of organisms.

The ecological conditions in the mud are similar to those already noted for muddy shores (see pp. 292–293). Indeed, many of the mud shores throughout the world are found in estuarine situations.

Among the particles that settle out in the estuary are terrestrial and marine organic matter. As a result, the substrate that accumulates is very rich in organic material. This material may serve as a large food reservoir for estuarine organisms. The large surface area relative to volume of the very small particles means that there is a very large area for bacteria to inhabit.

Temperature

Water temperatures in estuaries are more variable than in the nearby coastal waters. In part, this is because there is usually a smaller volume of water in an estuary and a larger surface area; therefore, it heats up and cools down more rapidly under prevailing atmospheric conditions (fjords, being deep and with a large volume, do not show this variation). Another reason for the variation is freshwater input. Fresh water in rivers and streams is more subject to seasonal temperature change than seawater. Rivers in the temperate zones are colder in winter and warmer in summer than adjacent seawater. When these fresh waters enter the estuary and mix with the seawater, they alter the temperature. As a result, estuarine waters are colder in winter and warmer in summer than surrounding coastal waters. The time scale is of interest in that with the changing tides, a given point in the estuary will show large temperature variation as a function of the difference between seawater and river water temperature.

Because the fresh waters show the greatest temperature ranges, it follows that as one progresses up the estuary, the temperature range on an annual basis becomes greater. Similarly, the temperature range is least at the entrance to the estuary, where mixing with fresh water is minimal. On a short-term basis, the central area of an estuary may show the greatest change with changing tide.

Temperature also varies vertically. The surface waters have the greatest temperature range and the deeper waters the smallest. In salt wedge estuaries, this vertical temperature difference also reflects the fact that surface waters are dominated by fresh water, whereas the deeper waters may be completely or predominantly marine.

Wave Action and Currents

Estuaries are surrounded by land on three sides. This means the water distance over which a wind can blow to create waves is minimal, at least in comparison with the oceans. Since the height of waves depends on the fetch or the open water distance over which the wind can blow, a small area of water can generate only small waves. The shallow depth of the water in most estuaries also precludes development of large waves. The narrowness of the mouth of the estuary, coupled with the shallow bottom, rapidly dissipate the effects of any waves entering the estuary from the sea. As a result of these processes, estuaries are generally places of calm water. Reduced wave action also promotes the deposition of fine sediments, allowing the development of rooted plants, which further stabilizes these sediments.

Currents in estuaries are caused primarily by tidal action and river flow. Currents are generally confined to channels, and within these channels, velocities up to several knots can occur. The highest velocities occur in the middle of channels, where the frictional resistance from the bottom and side banks is lowest. Although the estuary is an overall place of sediment deposition, the channels where currents are concentrated are often areas of erosion. Whenever currents change position, new channels are quickly eroded and old channels filled. This is particularly true in the intertidal areas, where a natural cycle of erosion and deposition occurs. However, in most estuaries deposition exceeds erosion, and there is a net accumulation of silt. In intermittent estuaries, where the mouth is often closed off from the sea during the dry part of the year, water movement is severely reduced at that time, leading to stagnation, reduced oxygen content, algal blooms, and fish kills.

For most estuaries there is a continual input of fresh water at the head. A given amount of this fresh water moves down the estuary, mixing to a greater or lesser degree with seawater. A volume of this water is eventually discharged from the estuary or evaporated to compensate for the next volume introduced at the head. The time required for a given mass of fresh water to be discharged from the estuary is the **flushing time**. This time interval can be a measure of the stability of the estuarine system. Long flushing times are important to the maintenance of estuarine plankton communities.

Turbidity

Because of the great number of particles in suspension in the water of estuaries, at least at certain times of the year, the turbidity of the water is high. Highest turbidities occur during times of maximum river flow. Turbidity is generally at a minimum near the mouth, where full seawater occurs, and increases with distance inland. The more an estuary approaches a lagoon type, the more the turbidity is a function of the plankton concentration or wind speed (producing resuspension).

The major ecological effect of turbidity is a marked decrease in the penetration of light. This, in turn, decreases photosynthesis by phytoplankton and benthic autotrophs, thereby reducing productivity. Under conditions of severe turbidity, phytoplankton production may be negligible, and the major production of organic matter is by emergent marsh plants (see pp. 358–360).

Oxygen

The regular influx of fresh water and salt water into the estuary, coupled with the shallowness, turbulence, and wind mixing, usually means there is an ample supply of oxygen in the water column. Since the solubility of oxygen in water decreases with increased temperature and salinity, the precise amount of oxygen in the water will vary as those parameters vary. In salt wedge estuaries, or any deep estuary during the summer when a thermocline can develop and where there is a vertical salinity stratification, there is often little interchange between the oxygen-rich surface waters and the deeper layers. This isolation of the deep waters from oxygen interchange, coupled with high biological activity and slow renewal or flushing rate, may deplete the oxygen of the bottom waters.

Oxygen is severely depleted in the substrate. The high organic content and high bacterial populations of the sediments exert a large oxygen demand on the interstitial water. Since the fine particle size of the sediments restricts the interchange of interstitial water with the water column above, oxygen is quickly depleted. Estuarine sediments are, therefore, anoxic below the first few centimeters—unless they have large particle size or large numbers of burrowing animals, such as the ghost shrimp *Callianassa* and the hemichordate worm *Balanoglossus*, which by their activities oxygenate lower sediment layers.

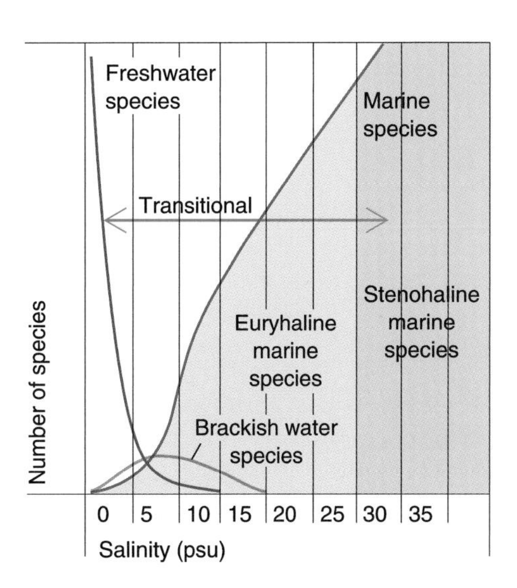

FIGURE 8.6 Numbers of species in each of the three major faunal components—freshwater, estuaries, and marine—of estuaries and their distribution with salinity. (After *Ecology of Estuaries*, D. S. McLuskey. Copyright © 1971 Heinemann Educational Publishers. Reprinted by permission of Heinemann Educational Publishers, a division of Reed Educational & Professional Publishing Ltd.)

THE BIOTA OF ESTUARIES

Faunal Composition

There are three types of fauna in estuaries: marine, freshwater, and brackish water or estuarine.

The marine fauna is the largest group in terms of numbers of species and includes two subgroups. The **stenohaline** marine animals are marine forms that either are unable, or barely able, to tolerate salinity changes. These organisms are usually restricted to the mouths of estuaries where salinity is generally 25 psu or above. These animals are often the same species found in the open sea. The second subgroup is the **euryhaline** marine animals. These marine animals are capable of tolerating varying amounts of salinity reduction below 30 psu. Such species can penetrate varying distances up the estuary. Most tolerate salinities down to 15–18 psu with a few hardy species tolerating levels down to 5 psu (Figure 8.6).

The brackish water or true estuarine species are found in the middle reaches of the estuary in salinities between 5 and 18 psu but are not found in fresh water or in full seawater. Examples of these animals include the polychaete *Nereis diversicolor*, oysters (*Crassostrea*, *Ostrea*), clams (*Scrobicularia plana*, *Macoma balthica*, *Rangia*

FIGURE 8.7 Some typical estuaries animals. (Not to scale.)

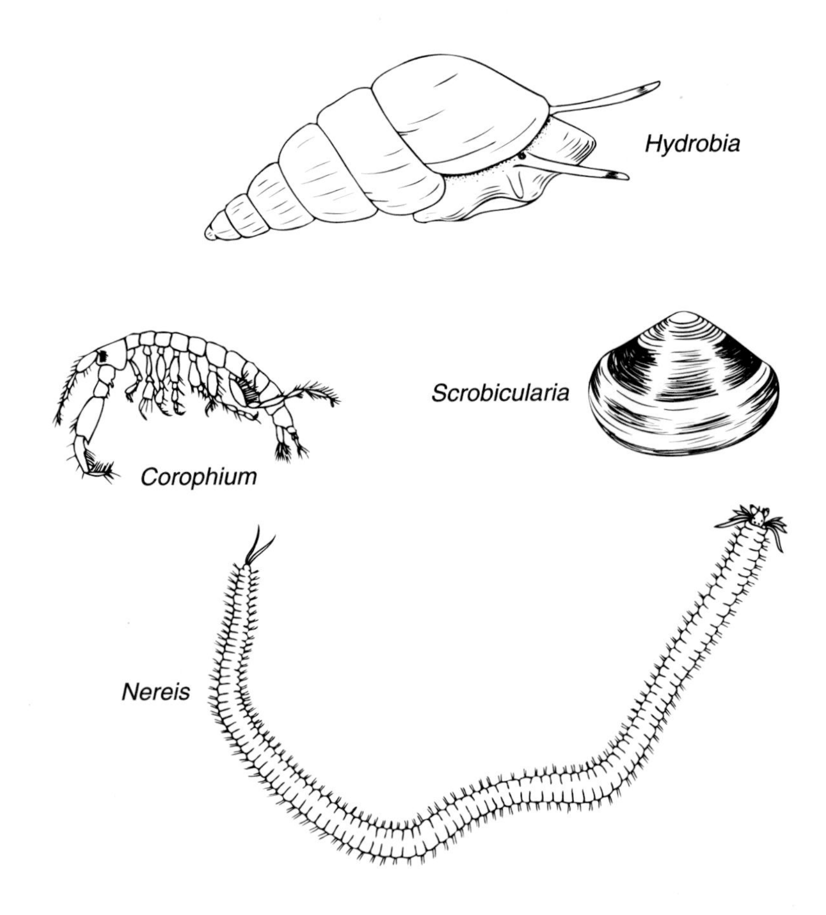

flexuosa), small gastropods (*Hydrobia*), crabs (*Callinectes*), and shrimps (*Palaemonetes*) (Figure 8.7). Some of these estuarine genera may be limited in a seaward direction not by physiological tolerances but by biological interactions, such as competition and predation; hence, this component may not be easily defined.

Another component is derived from fresh water. These animals usually cannot tolerate salinities much above 5 psu and are restricted to the upper reaches of the estuary (see Figure 8.6).

Finally, there is also a transitional component. This includes those organisms, such as migratory fishes, that pass through the estuary on their way to breeding grounds, either in fresh water or salt water. Common examples are salmon (*Salmo*, *Oncorhynchus*) and eels (*Anguilla*). Also included are forms that spend only part of their lives in the estuary—usually it is the juvenile stage, and the adults are found at sea. Good examples of this latter group are the various shrimps of the family Penaeidae (*Penaeus setiferus*, *P. aztecus*, *P. duorarum*), which form the basis of the Gulf of Mexico shrimp fishery. The young occur in estuaries. The transitional fauna also contains forms that enter the estuary only to feed and includes many birds and fishes.

The number of species of organisms inhabiting estuarine systems, as noted by Barnes (1974), is significantly lower than the number inhabiting nearby marine or freshwater habitats. This is probably because freshwater organisms cannot tolerate the increased salinities and marine organisms cannot tolerate the decreased salinities of the estuary.

The true estuarine organisms are derived primarily from marine stocks and not freshwater. This is similar to the situation in the other transitional zone, the intertidal, which is also populated mainly with marine organisms, not terrestrial forms. In contrast to the intertidal, however, the number of true estuarine species is very small, and the middle reaches of estuaries are depauperate (see Figure 8.6). Because the marine animals can tolerate a greater reduction in salinity than the freshwater species can endure salinity increases, and because the true estuarine organisms are primarily derived from marine stocks, most of the estuary is inhabited by marine animals.

Why are there so few estuarine species? The most common explanation is that the fluctuating environmental conditions, mainly salinity, are of such magnitude that only a few species have been able to evolve the necessary physiological specializations to exist there. Another explanation is that estuaries have not existed long enough in geological time to permit a complete estuarine fauna to develop. It should be noted that these two hypotheses are not mutually exclusive, so both could be operative. A final reason may be that estuarine areas have little topographic diversity, being mainly broad expanses of mud. There are fewer niches and, therefore, fewer species. It is not possible at this time to say if one, all, or none of the above are responsible for the fact that there are so few species, but apparently euryhalinity is a trait not easily acquired.

Estuarine Vegetation

The macroflora of estuaries is also limited. Most of the permanently submerged portions of estuaries consist of mud substrates unsuitable for macroalgal attachment. In addition, the highly turbid water restricts light penetration to a narrow upper layer. As a result, the deeper layers of the estuary are often barren of macrofloral life. The upper layers of water and the intertidal zone have a limited number of plants. In the lower reaches of the estuary at and below mean low water, there may be beds of seagrasses (*Zostera*, *Thalassia*, *Cymodocea*). These beds are considered as subtidal communities in Chapter 5.

The intertidal mud flats are inhabited by a limited number of green algal species. Common genera include *Ulva*, *Enteromorpha*, *Chaetomorpha*, and *Cladophora*. These are often seasonally abundant, disappearing during certain times of the year (Figure 8.8).

Estuarine mud flats often have an abundant diatom flora. As Lackey (1967) noted, the benthic diatoms are more abundant in estuaries than their planktonic relatives. Many are motile and undergo a rhythmic migratory pattern, moving up to the surface or down into the mud depending on the illumination. Estuarine muds also provide a suitable habitat for filamentous cyanobacteria of several types. These autotrophs may form mats up to 1 cm thick on the mud surface and on the bases of various emergent vascular plants.

The highly turbid water in estuaries means that the dominant vegetation in terms of biomass is emergent plants. These are generally long-lived flowering plants that root in the upper intertidal and form the characteristic salt marsh communities that fringe estuaries throughout the temperate zones of the world. The dominant genera include *Spartina* and *Salicornia* (see subsequent sections of this chapter for discussion). In the tropics, another community, the mangrove forest, is found (see Chapter 9).

A final component is the bacteria. Both the water and the mud of estuaries are rich in bacteria because of the abundance of organic matter to decompose. Estuarine waters have been shown by Wood (1965) to contain hundreds of times more bacteria than seawater, and the upper layers of mud contain more than a thousand times more bacteria than the overlying water. Densities of bacteria in estuarine muds of 100–400 million per gram have been reported.

Estuarine Plankton

The estuarine plankton is also reduced in number of species. It follows the same trend as the macrofauna and macrovegetation. Diatoms frequently dominate the phytoplankton, but dinoflagellates may achieve dominance during the warmer months and may remain dominant in some estuaries at all times. Dominant diatom genera include *Skeletonema*, *Asterionella*, *Chaetoceros*, *Nitzchia*, *Thalassionema*, and *Melosira*. Abundant dinoflagellate genera include *Gymnodinium*, *Gonyaulax*, *Peridinium*, and *Ceratium*. The phytoplankton may also be temporarily enriched by resuspension of characteristic bottom-dwelling diatoms. High turbidity and rapid flushing may restrict phytoplankton numbers and productivity in some estuaries. Where turbidity is low and flushing time long, however, diverse populations and relatively high productivity may result. Consequently, depending on prevailing conditions, estuaries may differ considerably in both phytoplankton numbers and productivity.

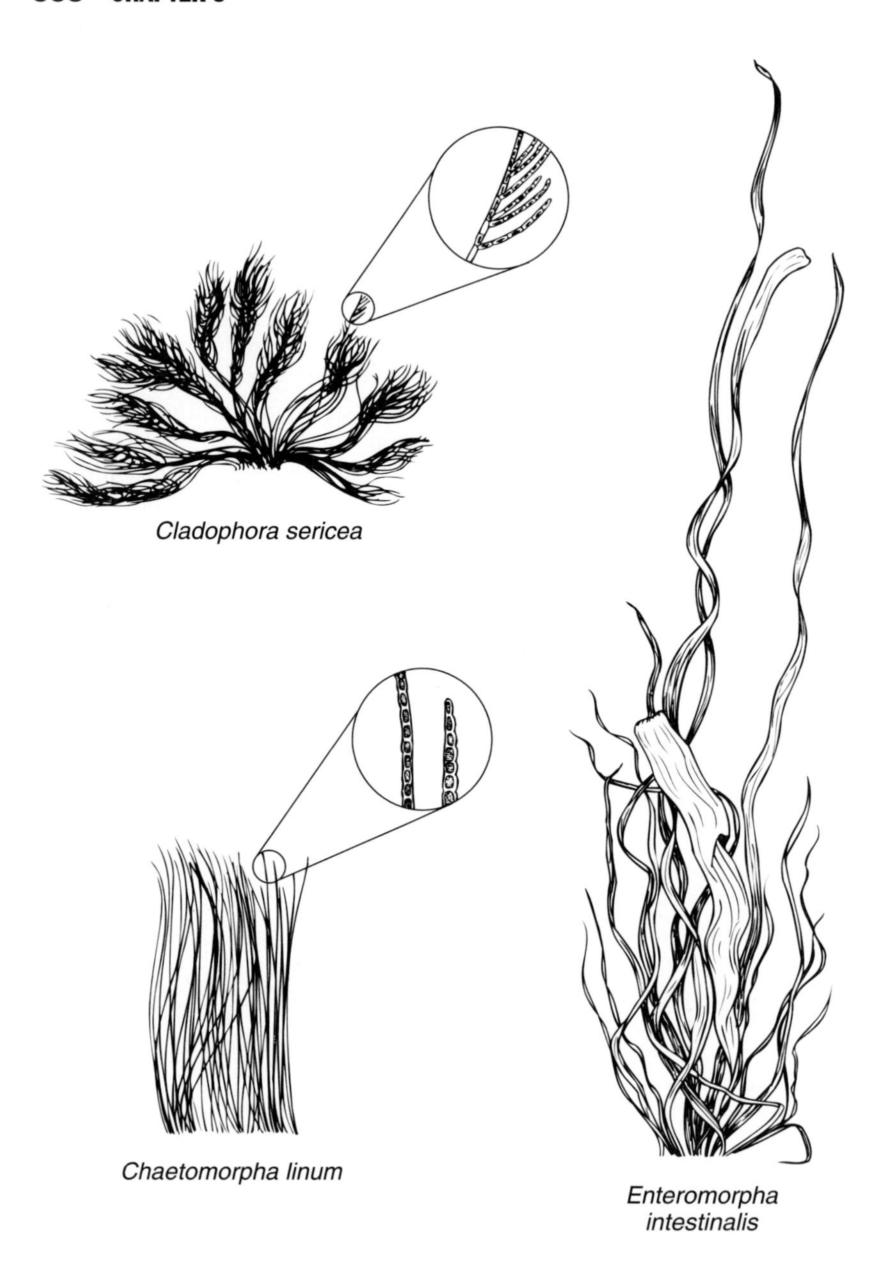

FIGURE 8.8 Typical estuarine algae. (Not to scale.)

Zooplankton in estuaries mirror the phytoplankton in being limited in species composition. The species composition also varies seasonally and with salinity gradients up the estuary. The few true estuarine zooplankters occur in larger, more stable estuaries where salinity gradients are less variable. Shallow, rapidly flushed estuaries are inhabited mainly by a typical marine zooplankton assemblage carried in and out with the tide. Characteristic estuarine zooplankters include species of the copepod genera *Eurytemora*, *Acartia*, *Pseudodiaptomus*, and *Centropages*; certain mysids, such as species of the genera *Neomysis*, *Praunus*, and *Mesopodopsis*; and certain amphipods, such as species of *Gammarus*. The estuarine zooplankton average about 1 ml/m^3 displacement volume, or somewhat greater than the concentration in adjacent coastal waters.

ADAPTATIONS OF ESTUARINE ORGANISMS

The variable nature of the estuarine habitat, especially defined by fluctuating salinities and temperatures, makes this a particularly stressful and rigorous habitat. For organisms to survive and successfully colonize this area, they must possess certain adaptations.

Morphological Adaptations

Few morphological adaptations among estuarine organisms can be attributed simply to living under conditions of fluctuating temperature and salinity. Most are adaptations to a given habitat, such as burrowing into mud. (See Chapter 6 for a discussion of adaptations of organisms living in mud.) Mud-dwelling organisms, whether estuarine or not, often have fine fringes of hair or setae, which guard the entrances to respiratory chambers to prevent clogging by silt particles. Such a situation prevails for estuarine crabs and many bivalve mollusks.

Other morphological changes in estuarine organisms, reported by Remane and Schlieper (1971), include a smaller body size than relatives living in full seawater and a reduced number of vertebrae among fishes. Reproduction is also affected. Marine species in an estuary often have a lowered reproductive rate and lowered fecundity. Freshwater species may have lower reproductive rates.

There are also certain morphological adaptations among vascular plants living in estuaries and salt marshes. All are relatively small, perhaps the result of long-term selection stemming from water stress. All have a special tissue, the **aerenchyma**, that supplies oxygen to the roots embedded in anoxic mud. This tissue may also bring oxygen indirectly to animals in the mud. Many plants also have mechanisms for dealing with salinity. Some, such as *Spartina alterniflora*, have **salt glands** for eliminating excess salt; thus, they maintain a stable water balance. Others have extensive stores of carbohydrates in the roots to provide a sugar source to cope with salinity fluctuations. Many also have high lignin content, which may be an adaptation to add strength to the plant to cope with high internal salinity. Abundant stomata and thin cuticles are suggestive of high transpiration rates. A final adaptation of vascular plants living in high-saline estuarine soils is to maintain high tissue water concentrations to buffer against water loss by osmosis, the **succulence strategy**, and to further reduce water loss by having reduced leaf area, few stomata, and photosynthetic stems.

Physiological Adaptations

The adaptations required for estuarine life are those that maintain the ionic balance of body fluids in fluctuating external salinities. **Osmosis** is the physical process in which water passes through a semipermeable membrane that separates two fluids of different salt concentration, moving from the area of lower to higher salt concentration. The ability to control the concentration of salts or water in internal fluids is called **osmoregulation**. Most marine organisms cannot control their internal salt content and are **osmoconformers**. Their ability to penetrate estuaries is limited by their tolerance for changes in their internal fluids. **Osmoregulators** are organisms that have physiological mechanisms to control the salt content of their internal fluids. Most estuarine animals either are osmoregulators or are osmoconformers able to function with fluctuating internal salt concentrations.

Penetrating into an estuary means encountering water with lowered salinity. Since the internal salt concentration of marine species is higher than that of the estuarine water, water tends to move across their membranes into their bodies to equalize the concentrations. Regulation means excreting the excess water without losing salts or excreting water and salts and replacing lost salts with active uptake of ions from the environment. For freshwater animals, which move from a more dilute to a less dilute medium when entering estuaries, the reverse is true.

Osmoregulatory ability is found in several taxa of estuarine animals, primarily polychaete worms, mollusks, and crustaceans. The field of osmoregulation has been the subject of intensive research for many years, resulting in a large literature. Detailed coverage is beyond the scope of this introductory book, but students interested in more in-depth studies may refer to the literature cited at the end of this chapter. Suffice it to say that osmoregulation can operate in three ways: (1) animals may move water; (2) they may move ions; or (3) they may adjust the internal water ion balance. In advanced invertebrates and vertebrates, the osmoregulatory organ is generally the kidney, where excess water is excreted and needed ions resorbed. Among many invertebrates, gills are the most common osmoregulatory structures. However, special

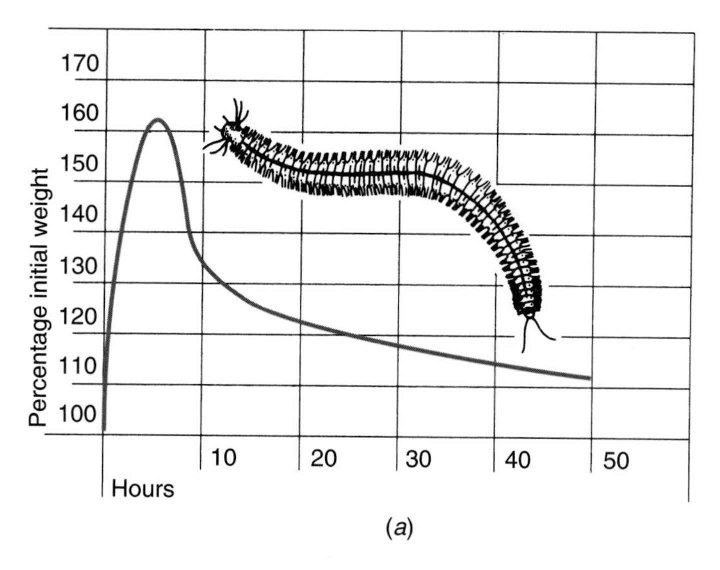

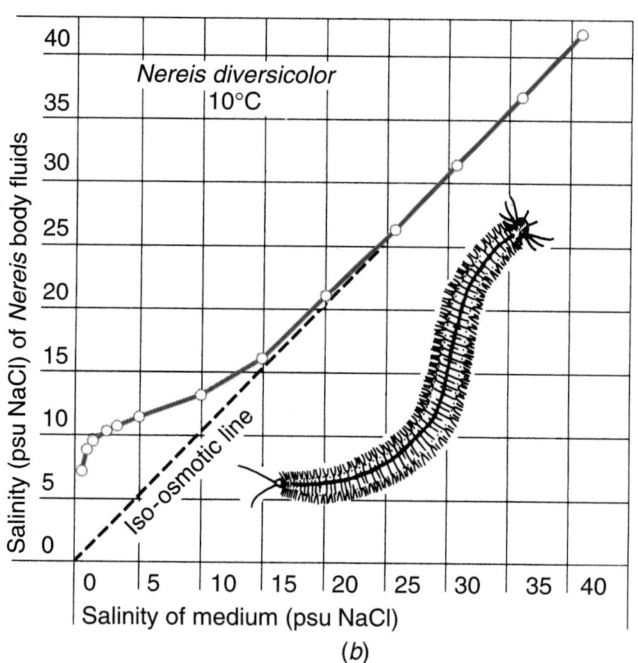

FIGURE 8.9 Changes in the body fluids of *Nereis diversicolor* with change in salinity. (a) Weight changes occurring after the animals are transferred to 20% seawater. Initial weight increase is due to the intake of water by osmosis. Weight decreases when the animal osmoregulates. (b) Osmotic concentration of the body fluids in relation to salinity change. The straight line on the graph indicates osmoconformity. (From *Ecology of Estuaries*, D. S. McLuskey. Copyright © 1971 Heinemann Educational Publishers. Reprinted by permission of Heinemann Educational Publishers, a division of Reed Educational & Professional Publishing Ltd.)

cells may also exist in other parts of the body, particularly for taking up or removing certain ions.

In certain soft-bodied forms, such as polychaete worms, an osmoregulatory mechanism is developed but is relatively slow to respond. These animals can tolerate wide ranges of internal ion concentrations, at least for certain periods of time. *Nereis diversicolor*, for example, when placed in dilute seawater takes up water due to osmosis. After some time, however, it begins to osmoregulate, and according to McLusky (1971), the internal water content falls to near that found when it was in full seawater (Figure 8.9). Not only is this osmoregulation delayed, it does not come into effect until the salinity of the external environment falls below a certain level. That is, N. *diversicolor* remains an osmoconformer as salinity decreases and regulates only when the external salinity falls below a certain level (Figure 8.9).

Bivalve mollusks are generally poor osmoregulators and respond to drastic salinity decreases by closing up in their shells to prevent excessive dilution of internal fluids. Certain gastropods, such as *Hydrobia*, have limited osmoregulatory ability, and the members of the genus are found in enormous densities in northern European estuaries where Muus (1963) reports they live in salinities of 5 psu or lower.

Among higher crustaceans, such as crabs, osmoregulation is well developed in some species. Greatly restricted body permeability due to the exoskeleton and the ability to regulate ion concentrations of internal fluids are probable reasons for their success in estuaries. Among vertebrates, osmoregulation is primarily through water removal by the excretory organs coupled with ion uptake from the environment to replace salts lost in water removal. In invertebrates, other mechanisms may be involved. For example, the cells of crustaceans engage in osmoregulation by changes in the concentration of amino acids. This "intracellular osmotic regulation" results from movements of free amino acids in and out of individual cells to equalize the osmotic concentration in the cells with the extracellular bathing fluid, preventing excessive water buildup and possible rupture of the cells (Figure 8.10).

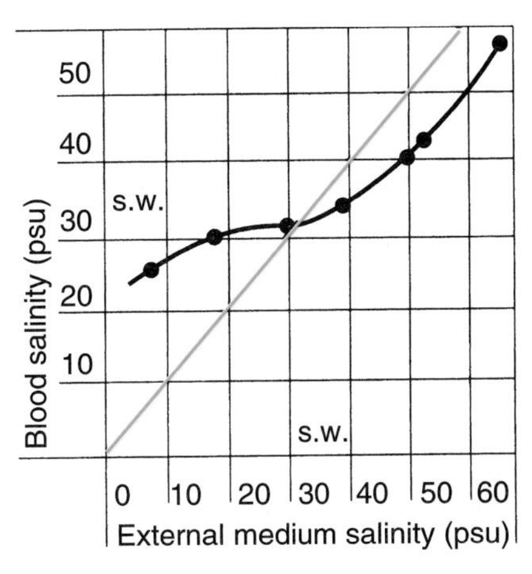

FIGURE 8.10 Graph showing the change in salinity of the blood of the crab *Australoplax tridentata* with change in the salinity of the external medium. (Modified from *Ecology of Estuaries*, D. S. McLuskey. Copyright © 1971 Heinemann Educational Publishers. Reprinted by permission of Heinemann Educational Publishers, a division of Reed Educational & Professional Publishing Ltd.)

Kinne (1963) noted that there is some correlation between temperature and the ability to osmoregulate. In the tropics where water temperatures are higher and the difference in temperature between fresh water and salt water is minimal, there are considerably more estuarine species, and stenohaline marine species penetrate further upstream than they do in the temperate zone estuaries. The reasons for this phenomenon are not yet clearly understood.

Vascular plants growing under estuarine conditions also display a set of physiological adaptations to osmotic stress. These rooted plants must maintain high concentrations of osmotically active substances to counter the water-retaining capacity of the surrounding medium. Generally, when the osmotic potential (increase in ion concentration) of the water in the soil around the roots increases, the plants respond by increasing the ion concentration of their tissues. Such an osmotic adjustment is the plant equivalent of osmoregulation in animals. In saline environments, salts are continuously transported into the plant. To prevent salts from reaching a lethal level, most plants have mechanisms to get rid of excess salts. Such mechanisms may include salt-secreting glands, leaching by rains of salts deposited on external surfaces by transpiration, and the shedding of leaves loaded with excess salts. Nonsucculent halophytes lose salts by some sort of secretion, whereas succulents release salts by shedding leaves.

Behavioral Adaptations

A certain number of behavioral adaptations are common to many estuarine organisms. Among the invertebrates, one such adaptation is burrowing into the mud. Although this adaptation is certainly not exclusive to estuarine animals, being common to many invertebrates in soft sediments throughout the world's oceans, it has two beneficial effects for an estuarine animal. First, in the case of those species with limited osmoregulatory abilities, existence in the mud means exposure to the interstitial water, which has much less salinity and temperature variation than the open water. It is a means to reduce salinity and temperature changes. Second, burying in the substrate means that these species are less likely to be consumed by surface- or water-dwelling predators, such as birds, fishes, or crabs.

Many adult estuarine crabs are able to live successfully in low salinities due to their developed osmoregulatory abilities. However, their eggs and young often lack these regulatory abilities. Therefore, many exhibit a specific migratory pattern, moving from the estuary to the adjacent sea for the

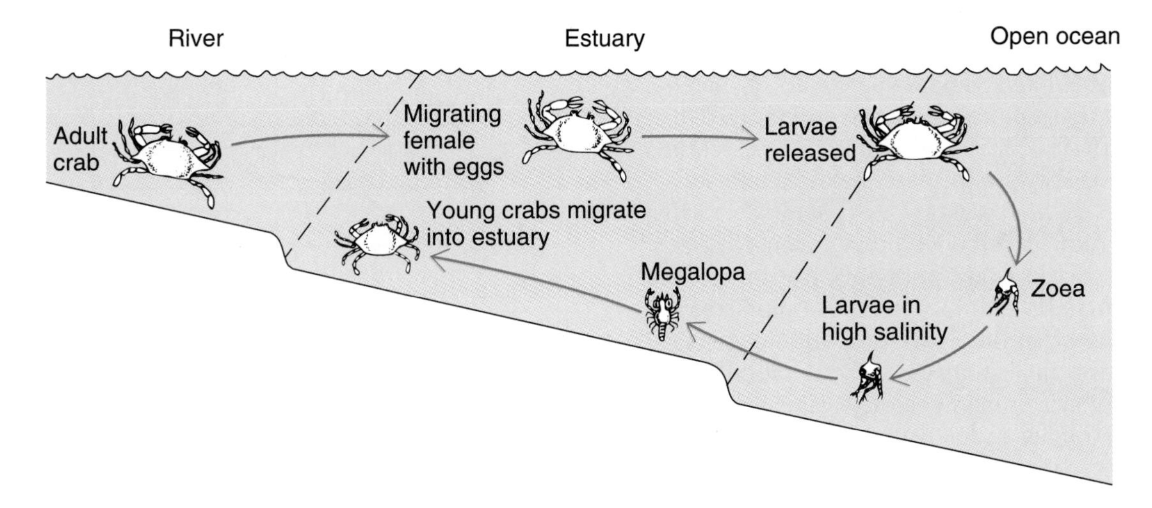

FIGURE 8.11 The life cycle of the blue crab *Callinectes sapidus* in estuaries of the Atlantic coast of the United States.

breeding season. One example of this type of behavior is the crab *Eriocheir sinensis*, which has been introduced into Europe from its native China. Although adults live hundreds of miles up rivers, *E. sinensis* migrates back to the sea to breed and to spend the early part of larval life. Another example is the blue crab *Callinectes sapidus*, which lives as an adult in estuaries of the American East and Gulf coasts, including Chesapeake Bay. The females migrate to waters of higher salinity to hatch the young. Once the larval stages are passed, the young crabs migrate back up the estuary (Figure 8.11).

On the other hand, certain other species take advantage of the great amount of food and the relative scarcity of predators to use the estuary as a nursery for their young. Most of these species are fishes. Examples include mullet (*Mugil* sp.), striped bass (*Roccus saxatilis*), and flounder (*Platichthys flesus*), which enter the estuary as juveniles and migrate back to the sea when mature. Various shrimp (*Penaeus*) also use the estuary as a nursery.

ECOLOGY OF ESTUARIES

Having considered the major physical-chemical features of estuaries, the common organisms, and the major adaptations of the organisms, it is now possible to consider certain aspects of the ecology of these transition areas.

Productivity, Organic Matter, and Food Sources

If we ignore for now the salt marsh, the primary productivity of estuaries resides in the phytoplankton, the benthic diatoms, seagrasses, and the various algal mats. Of these, the benthic diatoms and algal mats are significant. When all these sources are considered together, however, primary productivity by algae has traditionally been considered very low. Estuaries, however, have large amounts of organic material, large numbers of organisms, and high secondary productivity. If algal primary productivity is low, where does this organic material originate? Classically, it was believed that the major source of primary productivity in estuaries resided not in the estuaries, but in the emergent plants of the salt marsh surrounding the estuary (see pp. 358–360). Teal's (1962) oft-cited study of a Georgia salt marsh found there was a net productivity of 6,850 kcal/m^2/year, but the various algae contributed only 1,600 kcal/m^2/year. Algae may, however, be more important in some areas, such as the Pacific coast of North America, where Zedler et al. (1978) have shown that they contribute significantly to primary productivity. In addition, studies by Haines (1979) in the Georgia marshes suggested that the detritus from benthic and planktonic algae forms the bulk of the organic detritus. Haines found that most salt marsh plant detritus was accumulated and consumed in the marsh itself, leaving the algal detritus as the major material exported. In this scenario, the marsh is primarily a nursery for fishes, crabs, and shrimps, which are later exported to coastal waters.

Primary productivity in and around the estuary, however, is not the only source of organic material. As we noted previously, estuaries act as sinks for organic material brought down by rivers and in from the sea; thus, organic matter arrives from these sources as well. It is difficult to assess the role of primary productivity within the estuarine system in contributing to the total organic production for several reasons. First, there are few herbivores that feed directly on the plants. Most of the plant material, therefore, must be broken down to detritus before entering the various food webs. This process of breakdown involves bacterial action, which is an additional complicating factor. Also, once the plant material has become detritus, it is not easily distinguished from other organic detritus brought into the system by the river and the sea. Some fraction of the great amounts of organic material that move through the estuary is produced in the estuary (autochthonous), but some is transported in and is the result of primary production elsewhere (allochthonous). This includes organic matter transported from the surrounding salt marsh. Currently, the relative amount contributed by each source is under debate.

McLusky (1989) has considered the various sources of carbon "food" in the temperate estuaries of Europe and North America. From these studies he has separated estuaries into two major types that are at opposite ends of a spectrum and named for their primary geographical locations. The **European** (e.g., Britain, the Netherlands) **estuary** is dominated by large mud flats, is relatively barren of large plants, but has large populations of surface benthic diatoms that produce most of the primary productivity along with water column phytoplankton. These estuaries derive most of their energy from outside the estuary in the form of organic matter imported from the sea or river. They are net recipients of energy. Their ability to support large populations of consumers stems not from their productivity but from the fact that they are sinks for imported energy. At the opposite extreme is the **American estuary** (e.g., Louisiana, Florida, Georgia), dominated by extensive stands of *Spartina* grass and other salt marsh plants surrounding the less extensive mud flats. Phytoplankton and benthic diatoms are still significant primary producers but are overshadowed by the huge productivity of *Spartina* and the salt marsh plants. This type of estuary produces an excess of carbon, which is then exported to the surrounding sea, usually as detritus.

The detritus forms a substrate for a rich bacterial and algal growth, which is an important food source for various suspension- and detritus-feeding animals. When one refers to the detritus food source of estuaries, detritus is used in the widest sense to refer to organic particles, bacteria, algae, and even associated protozoans. This detritus forms an accumulation of food for estuarine organisms. As an example of the amount of food, Odum and de la Cruz (1967) noted that open seawater contains 1–3 mg of dry organic matter per liter, whereas drainage waters in estuaries may contain up to 110 mg dry weight organic matter per liter.

Food Webs

Low primary productivity in the water column, few herbivores, and the presence of large amounts of detritus have traditionally suggested that the basis of the food web in estuaries is detritus. This claim is currently under debate among scientists, but it is a convenient starting point.

To some extent to consider detritus as the basis of estuarine food webs is misleading, for it implies that detritivores actually digest the organic particles. Most likely they are digesting the bacteria and other microorganisms on the particles and excreting the particles without further degradation. This should be kept in mind in the following discussion.

Detritus may be consumed directly, either while in suspension in the water mass by various suspension-feeding benthic invertebrates, or more commonly by direct consumption of the material in or on the substrate. Suspension feeders, such as the clams *Cardium*, *Mya*, and *Mercenaria*, may obtain a small portion of their food from suspended detritus along with phytoplankton and zooplankton. Direct detritus consumers (deposit feeders) include certain clams, such as *Scrobicularia* and *Macoma*, which have separated siphons and feed by "vacuuming" up the detrital particles from the substrate via their inhalant siphons (see Figure 6.50).

Polychaete worms are very abundant in estuaries and are represented by many detritus feeders.

FIGURE 8.12 Mud snails, *Batillaria attramentaria*, on the surface of the mud in a California estuary. (Photo by the author.)

Members of the families Spionidae, Terebellidae, and Ampharetidae spread long tentacles over the surface of the mud and collect and transport the organic particles via ciliary-mucous tracts. Others, such as the families Capitellidae and Arenicolidae, are deposit feeders, ingesting the substrate directly, digesting out the organic material and bacteria during passage through their guts. Bacteria may be consumed directly by small protozoans and nematodes, both of which occur in large numbers in the upper layers of the mud.

Other animals, such as amphipods of the genus *Corophium*, sort particles using their appendages, while small gastropods, such as *Hydrobia* and *Batillaria*, rasp up the surface of the mud (Figure 8.12). Such sorting or digesting of the sediments reworks the substrate.

The large numbers of deposit feeders and suspension feeders are, in turn, consumed by a number of predators, both vertebrate and invertebrate. The main invertebrate predators are various species of crabs and shrimps; certain polychaetes, such as the genera *Glycera*, *Nephtys*, and perhaps some *Nereis*; and various predatory gastropod mollusks, such as *Polinices*, *Aglaja*, *Chelidonura*, and *Busycon*.

The dominant predators, however, are various species of fishes and birds. In European estuaries, Hartley (1940) found that the plaice (*Pleuronectes platessa*) feeds on polychaetes, while the flounder (*Platichthys flesus*) feeds on mysids, shrimps, and amphipods. In Louisiana estuaries, Darnell (1961) found the three most important larger invertebrate species eaten by fishes were the clam *Rangia cuneata* and two crabs, *Rithropanopeus harrisii* and *Callinectes sapidus*. Different fish species concentrate on different types of animals, and feeding habits change with age. A common developmental pattern in estuarine fishes is to progress from eating zooplankton, to detritus, to macroinvertebrates, to even other fishes. Much of the productivity of estuaries that ends up in fishes is ultimately lost to the estuarine system when the fishes move offshore as adults.

Estuarine birds include many ducks and geese, as well as a host of shorebirds, wading birds, gulls, and terns. The majority of the shorebirds feed on the infaunal invertebrates of the mud flats by probing at low tide (see also the discussion in Chapter 6, pp. 299–300). The variety of bill lengths among these birds enables the different species to feed at

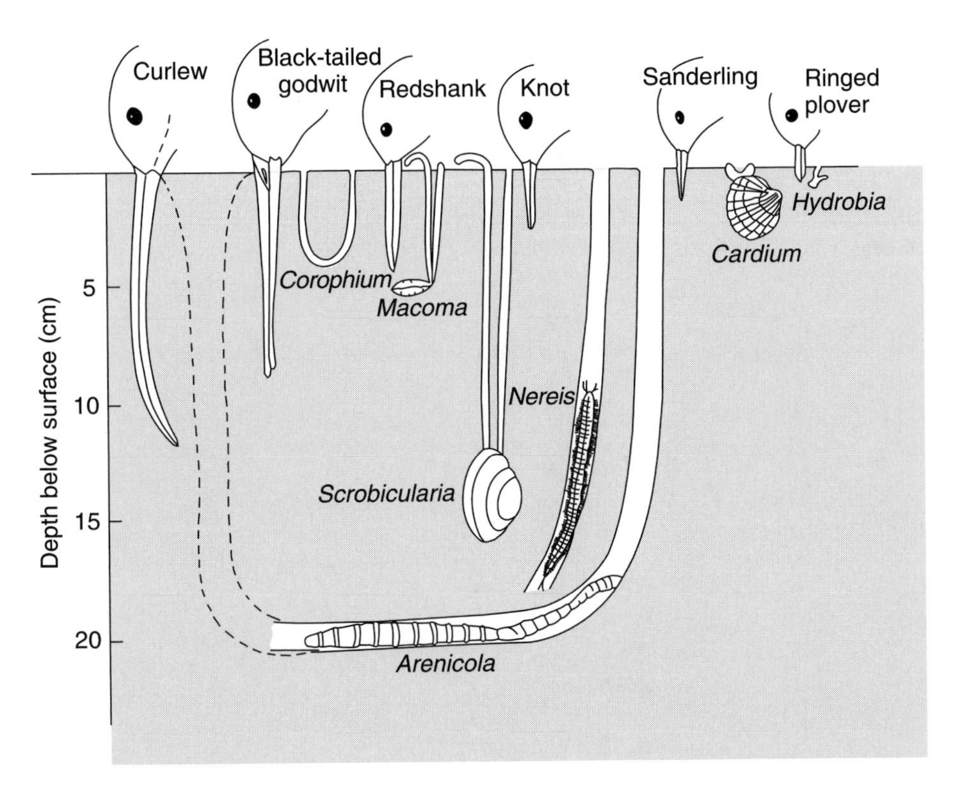

FIGURE 8.13 Diagram showing bill length of some common European shorebirds in relation to the depth of some common estuarine invertebrates. (After *The Biology of Estuarine Animals*, J. Green 1969.)

different levels in the mud (Figure 8.13). They follow the rise and fall of the tide during daylight hours to obtain their food. Observations and analyses of stomachs of these birds suggest that they are capable of consuming large numbers of invertebrates. Studies by Drinnan (1957), Goss-Custard (1969), and Prater (1972) have reported that populations of oystercatchers (*Haematopus ostralegus*) consume 315 *Cardium*, redshanks (*Tringa totanus*) consume 40,000 *Corophium*, and knots (*Calidris canutus*) consume 730 *Macoma* per day. The effect of this predation on the populations of infaunal invertebrates is not known, but various scientists have suggested that birds take 4–20% of the populations.

Over the past 20–25 years, there has been considerable interest in conducting field experiments that exclude large predators from the soft-bottom invertebrate communities of estuaries to determine the effect of predators on these communities. In summing up the results of these experiments, Peterson (1979) noted that when such systems are freed from predation the density of benthic organisms increases, but there is no corresponding tendency toward competitive exclusion by certain dominants. This contrasts with the results we have seen on rocky shores, where predator removal led to a competitive dominant occupying all space and forcing other species out. Why should such a different result occur here? Peterson (1979) has suggested several reasons. The initial one is simply that such experiments have not been conducted for a sufficient length of time. A second is that interference competition, which operates in the rocky intertidal, cannot operate in these soft muds because the organisms cannot obtain enough purchase to push away or crush another organism and because the organisms can exploit the third dimension (depth) of the soft substrate. Competitive exclusion

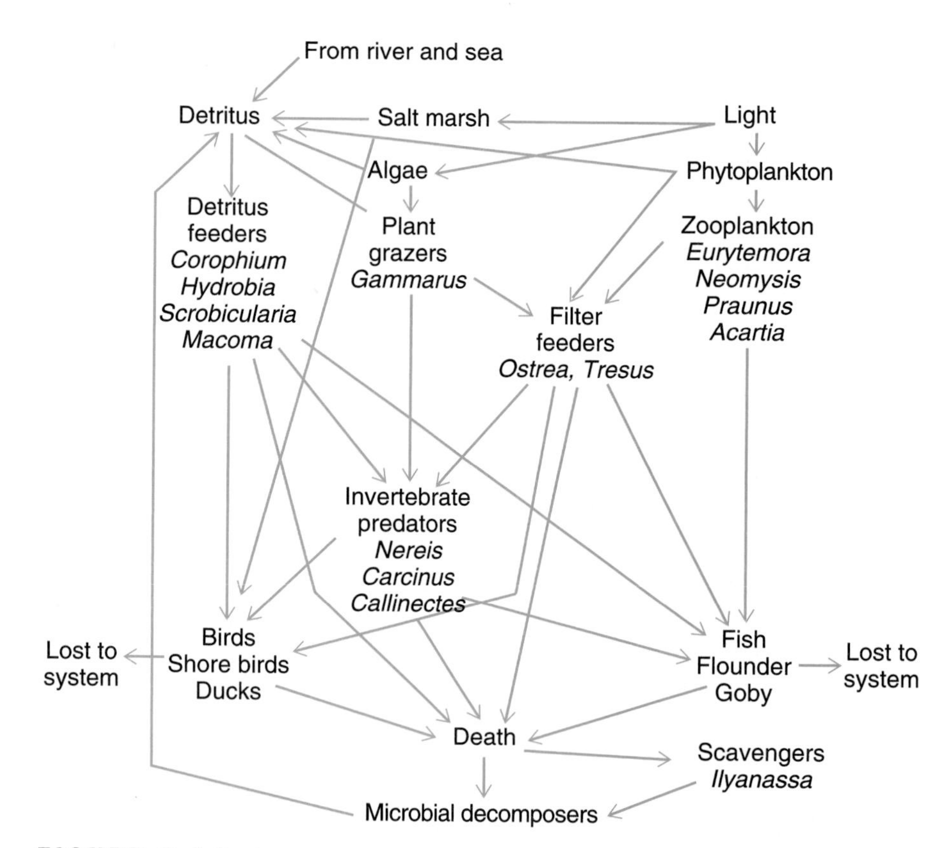

FIGURE 8.14 A generalized estuarine food web. (Modified from *Ecology of Estuaries*, D. S. McLuskey. Copyright © 1971 Heinemann Educational Publishers. Reprinted by permission of Heinemann Educational Publishers, a division of Reed Educational & Professional Publishing Ltd.)

by overgrowth, common on rocky shores, is also precluded because of high sediment mobility. This excludes species best able to use this mechanism, namely, colonial or clonal organisms. Adult-larval interactions also may be of sufficient intensity and development that they can hold the community at a density below its carrying capacity, preventing competitive exclusion without the necessity of predation (see also the discussion of intertidal mud flats in Chapter 6).

■ In summary, the estuarine food web has a number of energy pathways generated primarily from a detrital base and by a somewhat incomplete use of energy. This allows export to other areas primarily by fishes and birds. The incidence of omnivory is generally higher than in other systems. It is the abundance of food and the relative lack of predators that has permitted estuaries to be nurseries for the juveniles of many animals that spend their adult lives elsewhere. It is also the reason for their use as a feeding ground by migratory adults of many species of birds and fishes. The complex interactions of the estuarine food web are summarized in simplified form in Figures 8.14 and 8.15. ■

Plankton Cycles

Although the nutrient levels in estuaries are high, they are not well balanced; nitrogen concentration is often low and may even occasionally be limiting to estuarine phytoplankton. As we observed in the open sea (Chapter 2), there is also an annual cycle of phytoplankton in estuaries. In the temperate zones, low phytoplankton populations usually occur in late fall and winter due to reduced light and high

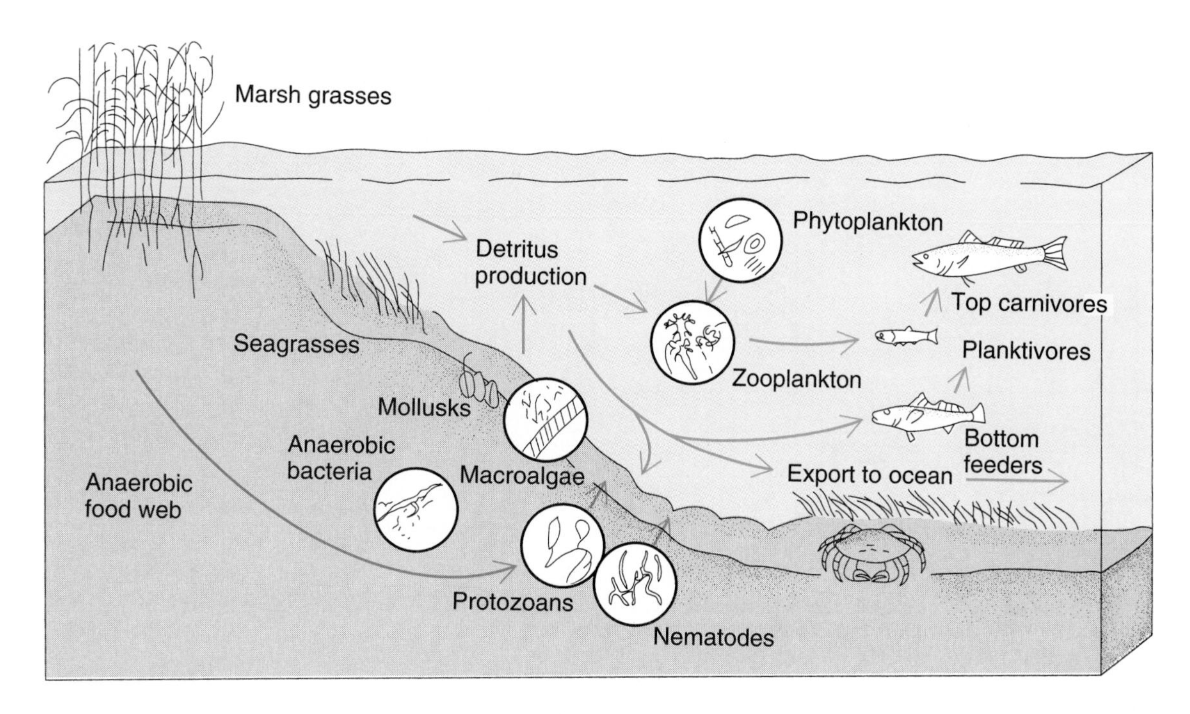

FIGURE 8.15 The food web of a typical estuary showing some of the major aquatic trophic groupings. (After *Estuarine Ecology*, Day et al., Copyright © 1989 John Wiley & Sons, Inc. Reprinted by permission.)

turbidity caused by high river runoff and turbulence. This is followed by a later winter diatom bloom. This bloom is ended in late spring, often not by zooplankton grazing, but probably by depletion of nitrogen sources, leading to the accumulation of diatoms on the mud surface. Populations remain low in the summer due to low nutrient levels and grazing, but there may be occasional blooms of dinoflagellates, leading to red tide situations. A large fall bloom occurs in some estuaries that rivals the spring bloom, whereas in others it is absent or reduced.

When one considers the relative abundance of nutrients, the phytoplankton in estuaries have a relatively low annual rate of production. McLusky (1989) suggests that this stems from three factors: (1) the shallow water, which is much less than the optimal depth for maximum photosynthesis; (2) the turbidity of the water, which limits light penetration; and (3) the continual removal of the phytoplankton caused by the rapid flushing rate of the estuary.

The zooplankton of estuaries may also be limited or reduced. This limitation is either caused by turbidity, which by limiting the phytoplankton limits the food available to the zooplankton, or currents. Currents carry the zooplankton out to sea. Elimination by currents can be reduced by the zooplankton remaining close to the bottom.

The zooplankton cycle in estuaries is variable and, as Deevey (1960) found, may change drastically from year to year. It does not closely follow the phytoplankton cycle. Populations of zooplankton peak later in the spring than the phytoplankton, and their numbers remain high throughout the summer, declining through the fall to winter lows. Zooplankton in estuaries consume only 50–60% of the net phytoplankton production. This leaves a significant portion available to the bottom suspension feeders. In shallow estuaries, the suspension-feeding benthos competes directly with the zooplankton for food, a situation not present in offshore waters.

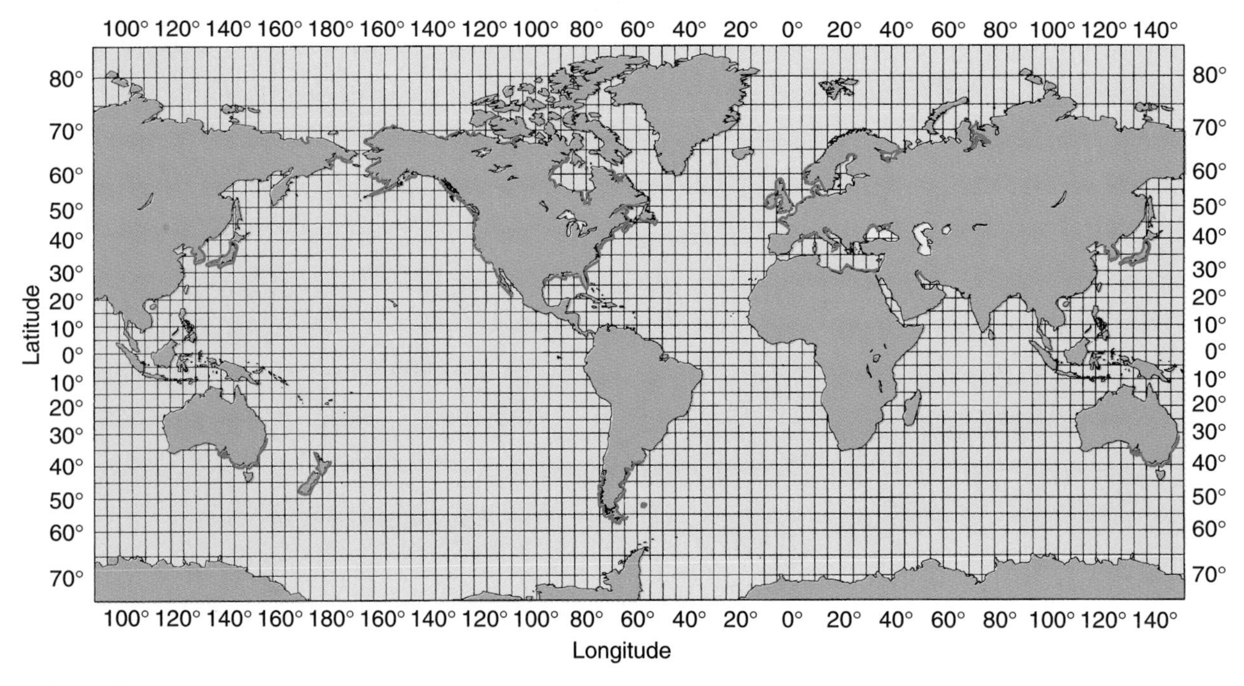

FIGURE 8.16 The distribution of salt marshes in the world.

SALT MARSHES

Bordering temperate and subpolar estuaries and protected marine shores and embayments throughout the world are special plant associations known as salt marshes (Figure 8.16). These marshes develop wherever sediment has accumulated and form a transition area between aquatic and terrestrial ecosystems. Such marshes dominate the intertidal Atlantic coast of North America south of New England as well as the northern Gulf of Mexico coast (Figure 8.17). They are a minor constituent of the intertidal coasts of the Pacific coast of North America.

Definition and Characterization

Salt marshes are communities of emergent herbs, grasses, or low shrubs rooted in soils alternately inundated and drained by tidal action. They occur mainly at the higher tidal levels in areas of protected water and most often in association with estuaries. Since the dominant plants are emergent flowering plants, they invade only the shallowest intertidal areas. They are **halophytes**, meaning they grow in soils with a high salt content. The dominant plants are herbaceous angiosperms. Because the upper portions of the plants are above water even during periods of high tide, this association has both terrestrial and aquatic components.

Environmental Characteristics

The salt marsh is a rigorous environment that shows wide variations in several environmental factors: salinity, temperature, and substrate. It shares with the estuary a fluctuating salinity that results from the interaction of river flow and marine water. Because it is intertidal, however, the salinity variation may be more sudden and more extreme than that experienced in the waters of the estuary. A sudden rainstorm at low tide may reduce surface salinities to near zero, whereas the return of the high tide may inundate the marsh with nearly full-strength seawater. It is not uncommon for marshes to experience salinities varying from 20 to 40 psu in a single tidal cycle.

Similarly, temperature undergoes wide fluctuations. Exposure at low tide opens the marsh and substratum to the extreme air temperatures of the terrestrial environment. Since these marshes occur in the temperate zone, this means air temperatures below freezing in winter and above 30°C in summer. Mud surface temperatures may vary 10°C during a single day.

The substrate is typically mud, similar to the estuarine sediments, with a high salt content caused by the perfusion of salt water coupled with a high evaporation rate. Salt marsh soils are often anaerobic. Within the marsh, the rate of sedimentation is higher in the lower part of the marsh than in the upper part.

The various environmental factors influencing the dominant vegetation are summarized in Figure 8.18.

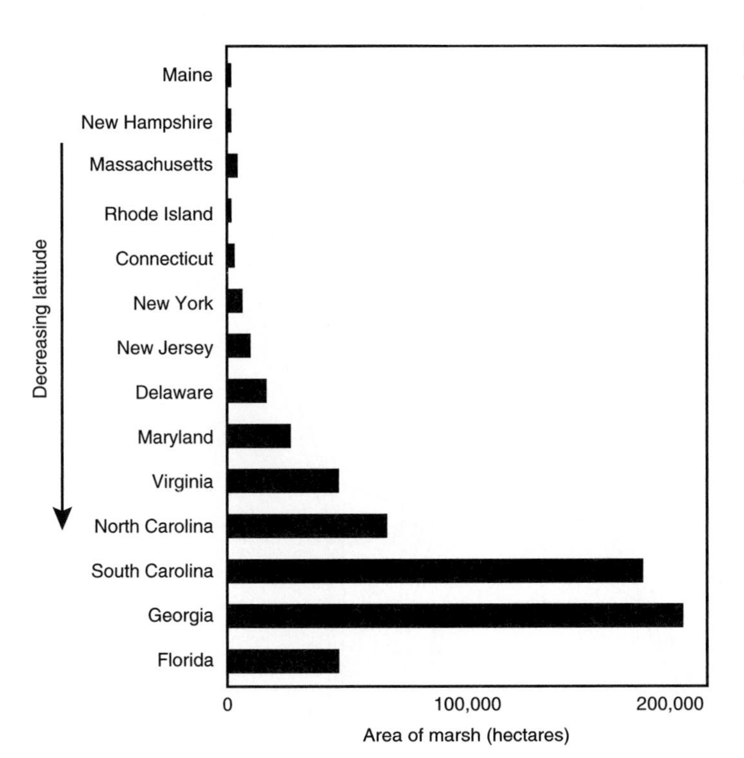

FIGURE 8.17 Areal extent of salt marshes on the Atlantic coast of the United States. (After "Mangals and Salt Marshes of the Eastern United States" R. J. Reimold, In *Wet Coastal Ecosystems, ed.*, V. J. Chapman, 1977, pp. 157–166. Copyright © 1977 with permission from Elsevier Science.)

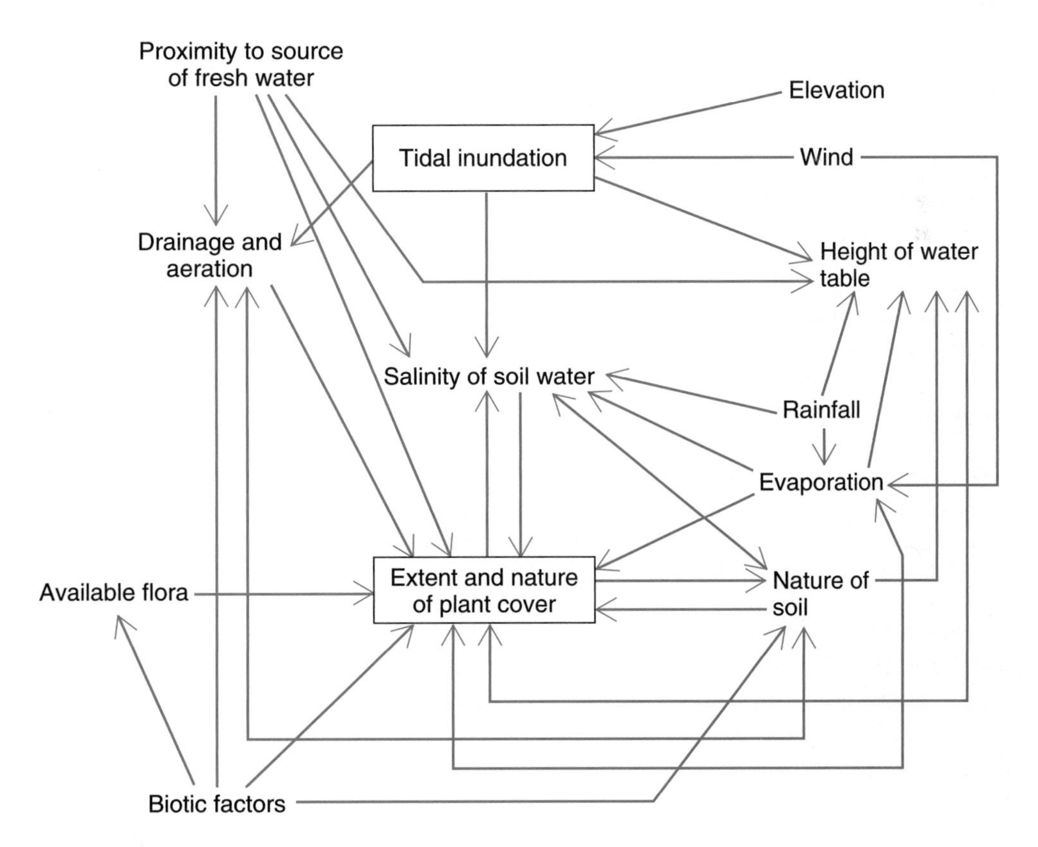

FIGURE 8.18 Environmental factors influencing slat marsh vegetation. (After *The Estuarine Ecosystem*, 2/e, D. S. McLuskey. Copyright © 1989 Chapman and Hall, with kind permission from Kluwer Academic Publishers.)

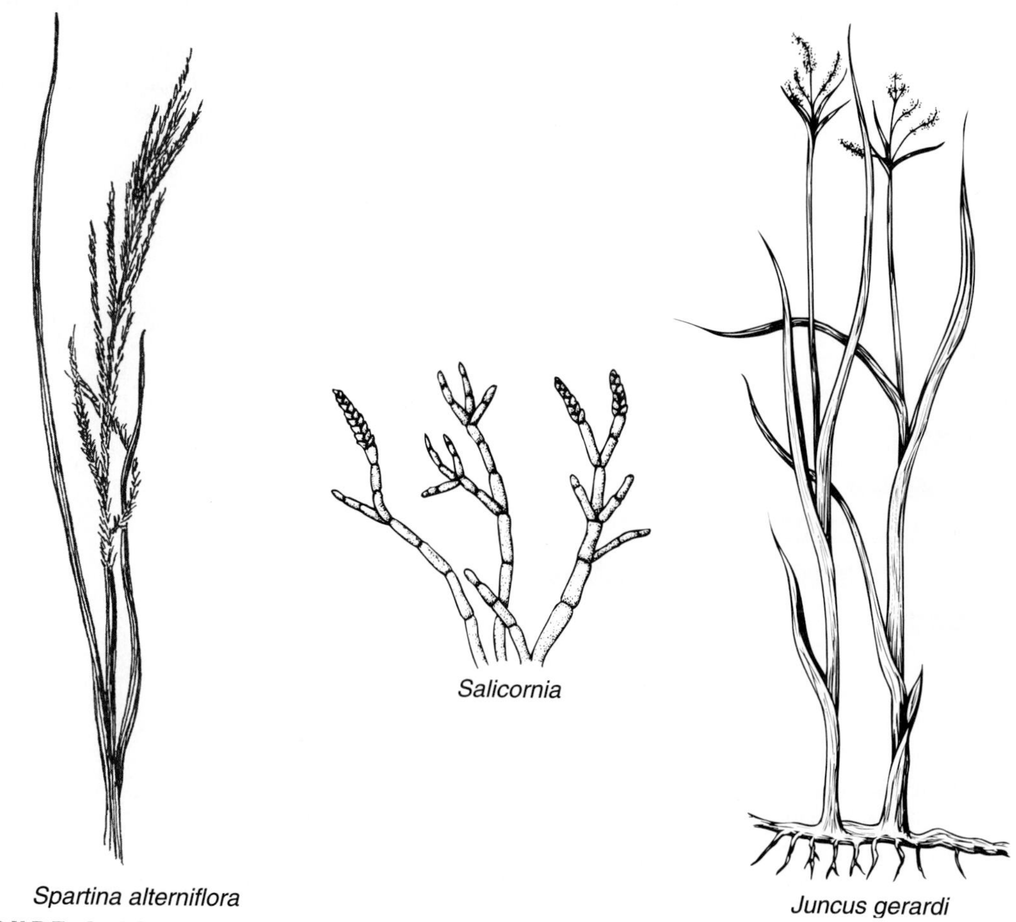

FIGURE 8.19 Some dominant salt marsh emergent plants.

Composition and Distribution

One of the most characteristic features of salt marshes is that they are species poor. Because of the rigorous environmental conditions and the high soil salt content, relatively few plants and animals are able to inhabit this area. Furthermore, these tolerant organisms show a high degree of taxonomic similarity over a wide geographic area. Dominant plants of salt marshes worldwide are grasses of the genus *Spartina* and species of the genera *Juncus* and *Salicornia* (Figure 8.19). Characteristic animals of marine origin are various species of crabs (*Uca*, *Hemigrapsus*, *Sesarma*), mussels (*Geukensia*), certain snails (*Littorina*, *Cerithidea*, *Melampus*), and smaller crustaceans such as amphipods, and shrimp.

Insects represent the major terrestrial animal component of the salt marsh and may live there permanently. There are also terrestrial animals, including raccoons, that enter the marsh only for feeding. At high tide, marine and estuarine animals enter the salt marsh, and at low tide, terrestrial species forage in the salt marsh (Figure 8.20). Despite the taxonomic similarity of the salt marsh species over broad geographic areas, certain combinations of topographic features and species distributions have produced several different marsh types. In the United States, there are two major groups of salt marshes, those of the Atlantic and Gulf coasts on the one hand and those of the Pacific coast on the other. On the East and Gulf coasts, salt marshes are far more extensive than West Coast marshes and occupy large areas of gently sloping

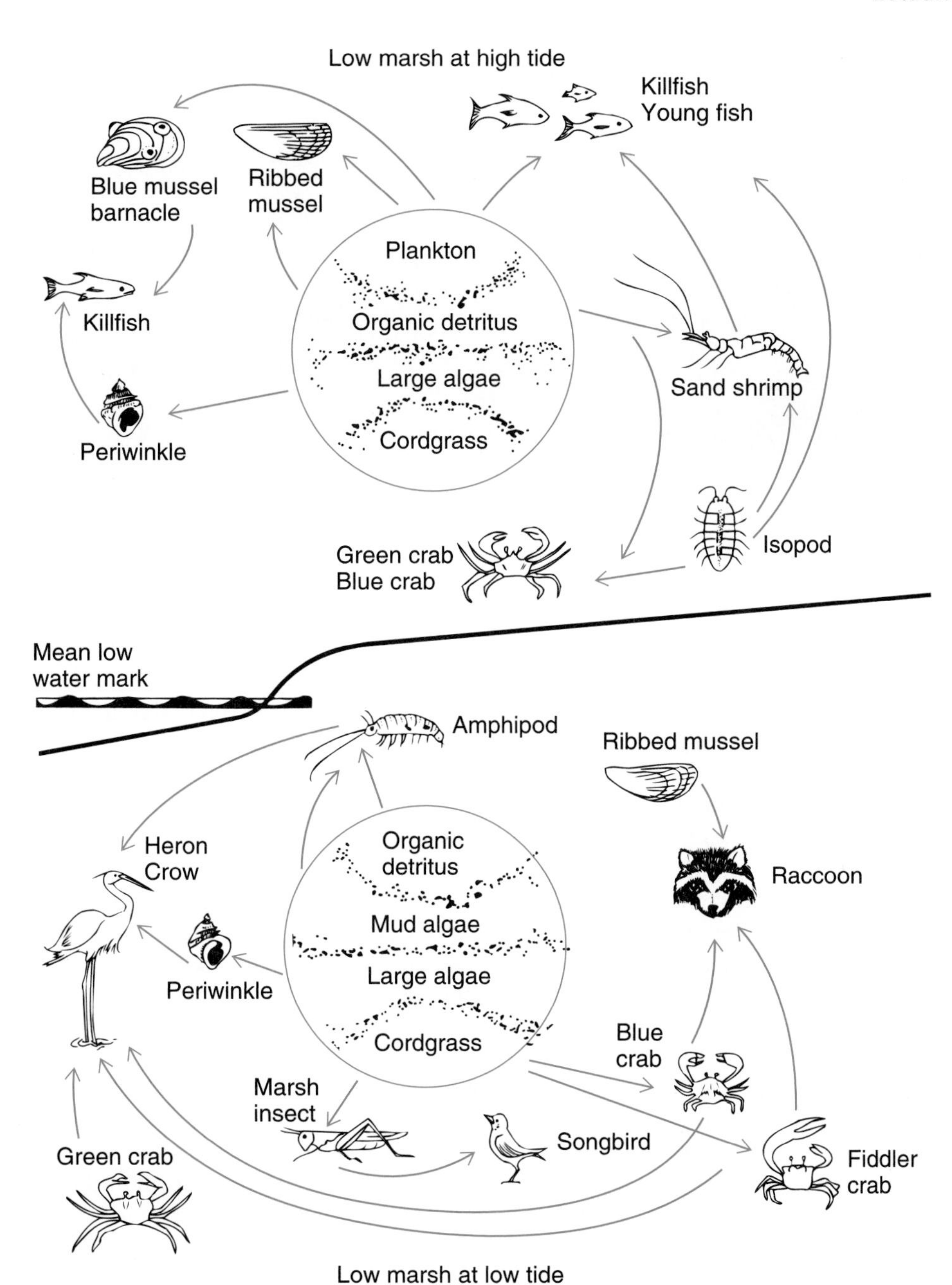

FIGURE 8.20 Characteristic animals present in a salt marsh at low and high tides on the Atlantic coast of North America.

coastline that surrounds the numerous broad estuaries and shallow bays associated with the extensive, shallow, offshore continental shelf. The Pacific coast, by contrast, has few rivers and bays, and steep coastal mountains lead directly to a narrow continental shelf. Consequently, river mouths are narrow and the extent of both estuaries and salt marshes is restricted. The East and Gulf coast marshes are dominated by various species of *Spartina*, which succeed each other in elevation and geographically. The more poorly developed West Coast marshes are characterized by broad bands of *Salicornia*. *Spartina* is of lesser importance or absent.

FIGURE 8.21 A tidal creek penetrating into a coastal salt marsh on the Pacific coast of North America. (Photo by the author.)

Physiognomy, Development, and Zonation

The general appearance of a salt marsh is similar throughout the world. It is a broad, flat expanse of low herbaceous, grassy, or shrubby plants of similar form, cut by a dendritic arrangement of channels leading to larger tidal creeks (Figure 8.21). These creeks lead to mud flats devoid of macrophytes or to the open waters of the estuary. Within the marsh there may be scattered open areas (pans) and shallow pools. At the upland side of the marsh, a variety of terrestrial shrubs and trees replace the halophytes. In contrast to the rocky intertidal where the majority of organisms are marine, salt marsh plants are mainly terrestrial in origin.

On the Atlantic coast of North America, marsh development begins with colonization of the intertidal sediment by the smooth cordgrass *Spartina alterniflora*. Cordgrass spreads vegetatively by rhizome and dampens waves, slows water movement, enhances sedimentation, and binds sediment particles into a dense mat with its roots. Young marshes therefore usually consist of a monoculture of *S. alterniflora* (Figure 8.22). With increasing time and sedimentation, the marsh grows vertically such that mature marshes consist of a high marsh and a low marsh cut by tidal creeks or drainage channels. The high marsh area is flooded only by the highest tides, whereas the low marsh and drainage channels are flooded daily by the tides. Other salt marsh plants become established, and a vertical zonation pattern emerges due to the changing physical conditions across the intertidal zone. In the northern marshes of the Atlantic coast, maturation of the marsh leads to the accumulation of marsh plant debris or **peat**, which adds to the vertical accretion of the marsh along with sedimentation. Peat does not occur in the more southern marshes of the Atlantic coast, and hence they accrete only by sedimentation.

A cross section of a typical mature marsh from the tidal creek to the true terrestrial vegetation reveals a zonation pattern. This pattern varies in detail and species composition from one geographical area to another because of tidal regime, drainage, slope, climate, and other factors, but a basic common zonation pattern for the Atlantic coast of North America exists and is described here.

The lowest zone is the creek bank and bottom, usually barren of macrophytic vegetation and containing mainly infaunal estuarine and marine animals and many mud snails. In the New England area the low marsh, the area flooded daily by the tides, is dominated by the cordgrass *Spartina alterniflora* (Figure 8.23a). This cordgrass zone is usually characterized by lush, tall cordgrass on the seaward margin and a shorter growth form of the same species immediately landward, where the soil is anoxic and accumulating peat. The next zone, the seaward border of the high marsh, is dominated by salt marsh hay, *Spartina patens*. Progressing landward, the next high marsh zone is dominated by the black rush *Juncus ger-*

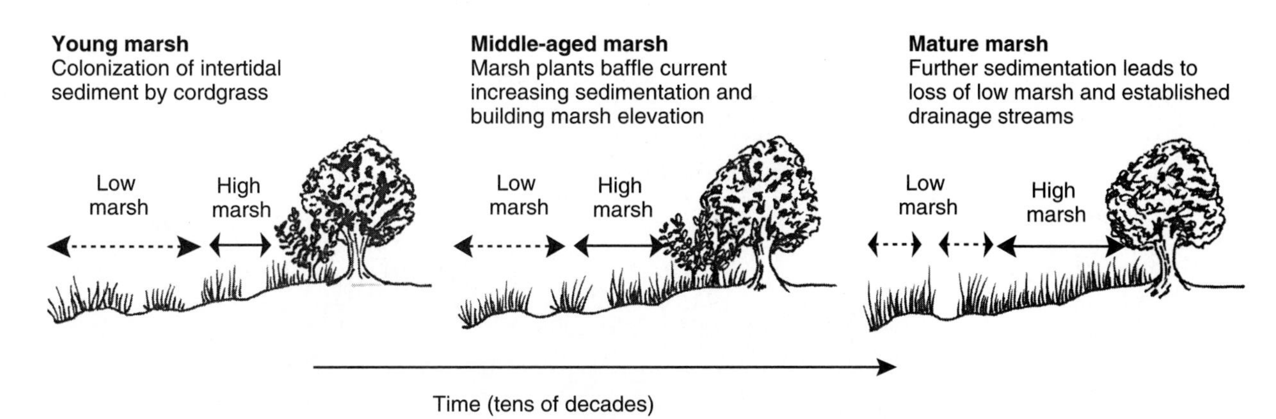

FIGURE 8.22 Marsh development. (After "Coastal Salt Marshes" R. W. Frey & P. B. Basan in *Coastal Sedimentary Environments*, ed., R. A. Davis, pp. 225–301, 1985. Copyright © 1985 Springer-Verlag. Reprinted by permission.)

ardi. Black rush is a perennial with a dense root mass, and it is more intolerant of salty soils and flooding by marine water than S. *patens*. Bordering the terrestrial vegetation at the upper limit of the marsh in New England is the marsh elder *Iva frutescens*. This species is very intolerant of tidal flooding and only grows where the soil is well drained. In addition to these dominant species, there are some less abundant species that are found in small patches, usually inhabiting disturbed areas within the above described zones. These species include the spikegrass *Distichlis spicata* and the glassworts *Salicornia europaea* and *Salicornia biglovii*. All these species are highly tolerant of high salinity and anoxic soils.

Southern Atlantic marshes have a somewhat different set of marsh plants and zonation pattern (Figure 8.23b). These marshes are subject to higher rates of evaporation, and hence higher salinities, as a result of their exposure to more intense solar radiation. Hence, Pennings and Bertness (1998) report that these marshes are characterized by a plant assemblage that is more salt tolerant. The high salinities also foster the development of vegetation-free areas in the marsh called **salt pans**. In the New England marshes vegetation growth is highly seasonal, from May to September, and during the fall and winter the emergent portions of the vegetation are composed of the dead remains from the growing season. This seasonally produced debris can form into mats over winter and either be rafted onto the upper marsh by the tides, where it can smother other marsh plants, or be exported from the marsh in a seasonal pulse. In the southern marshes vegetation growth occurs throughout the year and the plant debris has no seasonal pulse.

The seaward border of the southern Atlantic low marsh that is flooded daily is dominated by tall cordgrass, *Spartina alterniflora*, similar to the pattern in New England marshes (Figure 8.23b). Landward from this is an extensive stand of the short form of the cordgrass, which displaces the marsh hay *Spartina patens*. Although marsh hay is no longer the dominant that it was in New England, it is still found in patches. The highest parts of the southern salt marshes are dominated by the black needle rush *Juncus romerianus*, which displaces both *Spartina patens* and *Juncus gerardi* from these high marsh habitats, which they occupied in New England. Between the short cordgrass zone and the J. *romerianus* zone is a distinct intermediate zone, which typically contains permanent salt pans that have extremely high soil salinities (up to 100 psu), precluding plant colonization except on the fringes. Whereas tidal flushing at low elevations and rain and runoff at high elevations in the marsh prevent accumulations of soil salts in those areas, the lack of such flushing combined with high evaporation lead to the high soil accumulations and salt pans. Around the borders of the salt pans is a characteristic suite of usually very salt tolerant plants forming the salt meadow zone (Figure 8.23b). These include

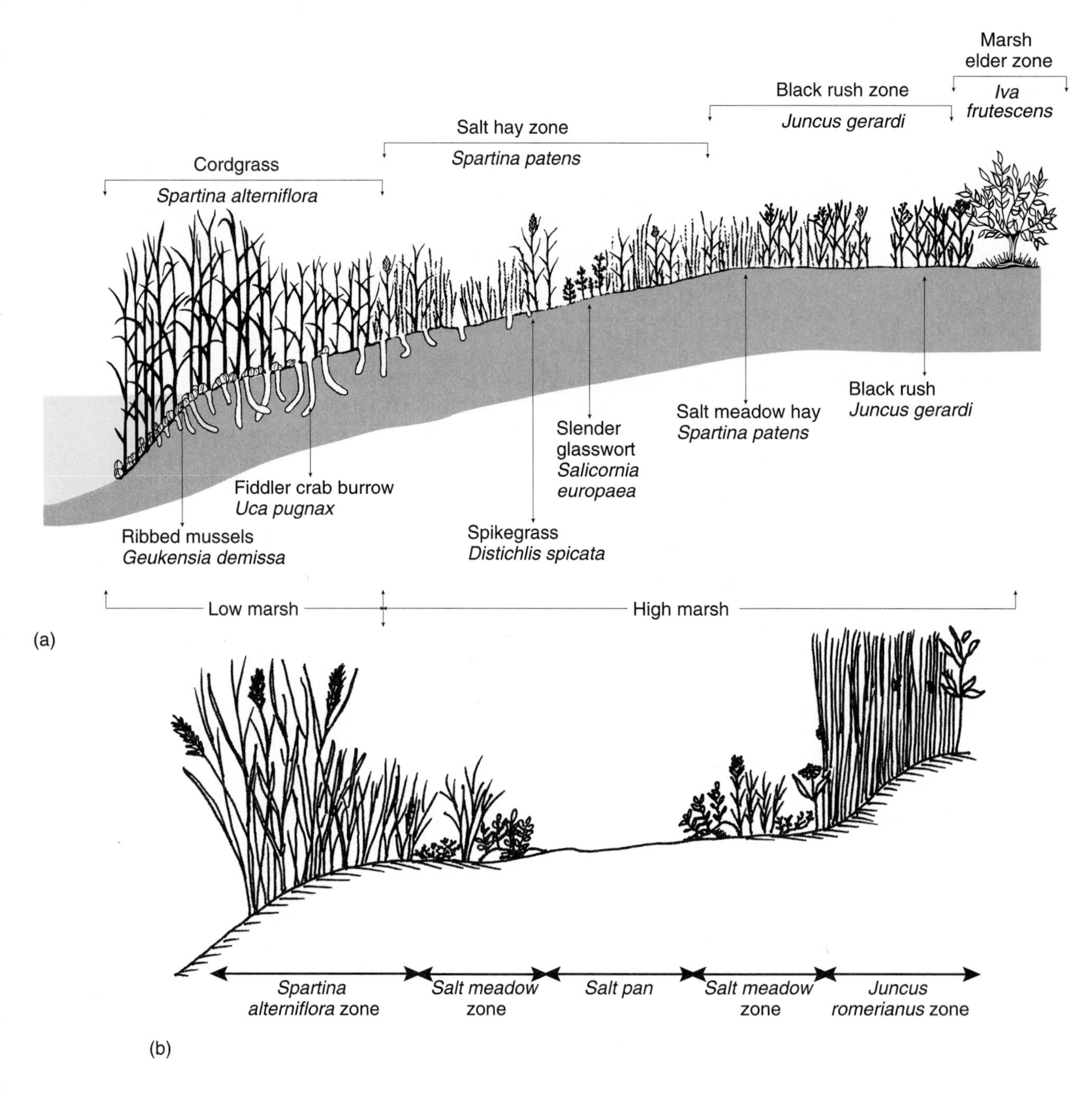

FIGURE 8.23 Diagrammatic representation of the zonation patterns of salt marshes of the Atlantic coast of the United States. (a) Zonation in a New England marsh. (b) Zonation in a southern marsh. (a, Copyright © Vergi Kask. Reprinted by permission. b from *The Ecology of Atlantic Shorelines,* M.D. Bertness, p.321. Copyright © 1999 Sinauer Associates, Inc. Reprinted by permission)

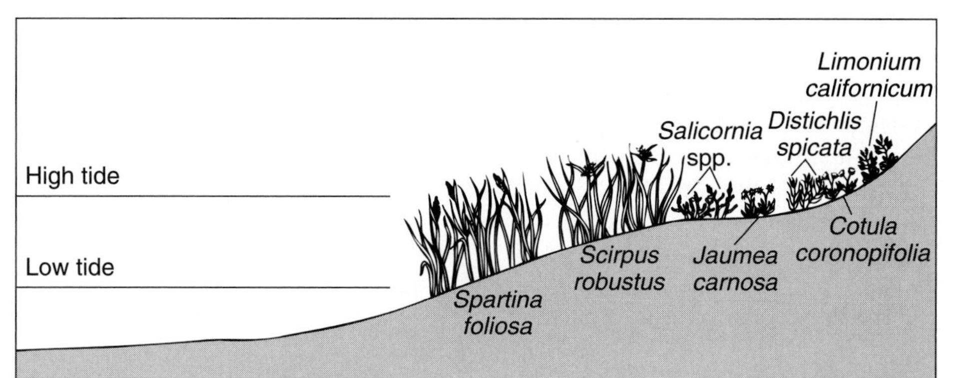

FIGURE 8.24 Generalized zonation of a Pacific coast salt marsh based on the marshes of San Francisco Bay.

stunted forms of the black needle rush, spikegrass (*Distichlis spicata*), and the glassworts *Salicornica europaea*, *S. virginica*, and *S. biglovii*, the latter of which seems to be the most tolerant and can be found growing in the middle of salt pans.

On the West Coast, marshes tend to have a narrow fringe of *Spartina* in the lower zone, followed by a broad expanse of *Salicornia* occupying most of the midintertidal zone. The highest intertidal area is occupied by a more diverse assemblage of which *Jaumea*, *Distichlis*, and *Limonium* are common (Figure 8.24). In contrast to intertidal rocky shores, where the reverse is true, Adam (1990) notes that competition among the various plant species is a significant factor governing the uppermost zonal boundaries of plants in salt marshes, while the lower limits are set by physiological tolerances for submersion in salt water (for comparison see pp. 249–253 in Chapter 6).

Causes of Zonation

The various physical and biological interactions that establish the zonation patterns observed in salt marshes have not been investigated as systematically throughout the world as they have been in the rocky intertidal, and so our understanding is imperfect. The greatest amount of research on the ecology of salt marshes has been done on the extensive marshes of the Atlantic and Gulf coasts of North America. They are discussed here as an example of how biological and physical factors interact to cause zonal patterns. The great similarity of salt marshes throughout the world suggests that it is reasonable to assume the generality of these examples for the organization and community structure of salt marshes in other parts of the world.

As noted above, in New England the tidal salt marshes are characterized by a distinct zonation (see Figure 8.23a). The highest zone is dominated by the marsh elder *Iva frutescens*. The next highest zone is occupied by the black rush *Juncus gerardi* and is flooded only at the highest tides. Beginning at the mean highest tide line the black rush gives way to a broad zone of *Spartina patens*, salt meadow hay, that extends down to the mean high water line. Also found in this zone are patches of the spikegrass *Distichlis spicata* and the glasswort *Salicornia europaea*. Below the mean high water line the *Spartina patens* is replaced by *Spartina alterniflora*, the cordgrass. Surrounding the roots of the cordgrass are large numbers of burrows of the fiddler crab *Uca pugnax*. The cordgrass finally gives way to open mud flats marking the end of the marsh. At this edge, large numbers of ribbed mussels, *Geukensia demissa*, live at the bases of the cordgrass.

In Chapters 5 and 6, we saw that much of the zonation pattern of intertidal and subtidal organisms on rocky substrates could be explained by biological interactions, mainly competition, grazing, and predation. Salt marshes differ from the rocky subtidal and intertidal in that grazers have a limited effect on the dominant plants. Since the dominant organisms are all large plants, predation also is rendered unimportant. This, in turn, means the keystone species concept does not apply in these communities. This would leave us with only competition as a possible biological determinant of salt marsh zonation. However, salt marshes have large variations in physical factors, primarily temperature and salinity, and these factors should enter the equation. Let us see how all of these factors interact to give the described zonal patterns and how they may change with geographical location.

In New England marshes at the uppermost part of the high marsh, the marsh elder *Iva frutescens* outcompetes *Juncus gerardi*, displacing it to a lower position in the high marsh. However, *J. gerardi* in turn outcompetes all other marsh plants. It does this in two ways, as Bertness (1991a) determined with removal and transplant experiments. First, *J. gerardi* has a very dense root mat, whereas *Distichlis* and *Spartina alterniflora* both have belowground runners. Various studies have demonstrated that dense root mats win out in competition with runners. Second, *J. gerardi* emerges early in the spring and begins growing before any other marsh plant, thereby defeating the others including *S. patens*, which also possesses a dense root mat. *Spartina patens* is restricted to the lower parts of the high marsh by competition with *J. gerardi*, but *S. patens*, in turn, is competitively dominant over *S. alterniflora*.

Since *S. alterniflora* is competitively inferior to both *J. gerardi* and *S. patens*, how does it manage to survive? The answer has to do with the sediment conditions and the morphology of *S. alterniflora*. The low marsh where *S. alterniflora* is dominant has a sediment that is nearly anoxic, and *S. patens* cannot grow there according to Bertness (1991b). *Spartina alterniflora*, however, is able to thrive on this anoxic sediment because it possesses aerenchyma tissue that provides oxygen to its roots. Cordgrass is usually taller in a band nearest the water and open mud flat, and stunted at the upper reaches where it meets the *S. patens* band. This is the result of the accumulation of peat in the soil. Near the water there is little peat, and the cordgrass grows well. Further inshore, peat has accumulated in the soil from the root masses of the cordgrass and stunts growth. *Spartina alterniflora*, therefore, destroys its own habitat behind it as it moves outward to colonize the open mud flat.

The fiddler crabs, *Uca pugnax*, that burrow in the mud around the roots of the cordgrass enhance the productivity of the cordgrass because their burrows aerate the soil and slow down peat accumulation. At the same time, the presence of the root masses of the cordgrass in the soft mud stabilizes it enough to permit the crabs to burrow. The relationship between the cordgrass and the crabs is best described as a facultative mutualism.

Another mutualistic interaction is that between the cordgrass and the mussel *Geukensia demissa*. The mussels filter food from the water column and deposit high-nitrogen fecal matter in the sediment that can increase the production of the cordgrass by 50% in a single season according to Bertness (1992). In addition, the mussels buffer the seaward edge of the cordgrass against physical disturbance by binding sediment and cordgrass roots with their byssal threads; thus, the seaward edge of the marsh is protected from wave- and current-induced erosion. Indeed, the mussels could not exist here if the cordgrass did not provide the anchoring material of roots for the byssal threads of the mussels.

The previous discussion accounts for the zonation of the major dominants. However, we have not accounted for the patchy occurrence of two other plants, *Distichlis spicata* and *Salicornia europaea*, that are competitively inferior to the dominants. How can they exist when outcompeted by the two species of *Spartina* and the *Juncus*? The answer lies with disturbance of the dominants and the salinity tolerances of the two fugitive marsh plant species.

During the winter, ice forms on the cordgrass in the lowest part of the marsh. These ice sheets are often moved by the tides, and their motion clips off the cordgrass, creating rafts of matted, broken cordgrass pieces. Alternatively, the ice sheet may be lifted by high tides and uproot the large masses of cordgrass to which they are attached. In either case, the very high tides of spring then carry these mats into the high marsh and deposit them on the existing dominant plants, killing them. After decaying completely or washing away, they leave a bare patch of sediment. With the increasing temperatures of late spring and summer, the sediment in these bare patches heats, causing the water in the surface layers to evaporate and the salinity to increase as much as 30 times that of the surrounding sediment that is densely covered with marsh plants. This increase is restricted to patches in the *S. patens* zone, because in the black rush zone rainwater dilutes the salinity, and in the *S. alterniflora* zone the frequent inundation by seawater prevents the salt from increasing significantly. In both of these zones, the competitive dominants quickly reoccupy the patches.

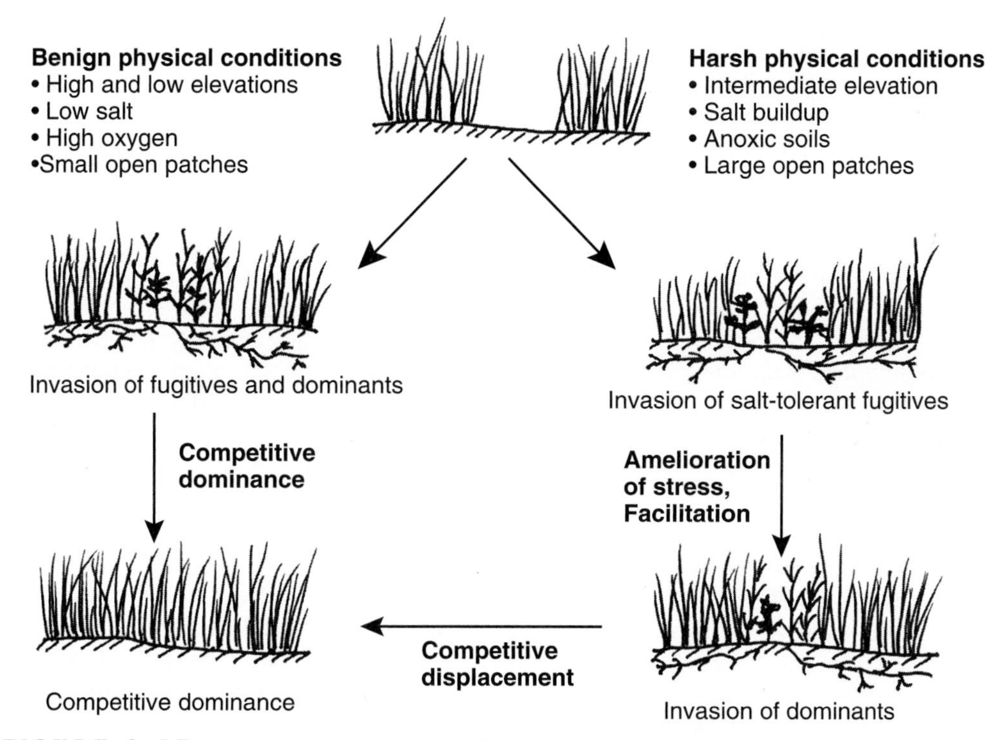

FIGURE 8.25 Alternative succession patterns in bare patches of the high marsh as a result of differing physical conditions. (After *The Ecology of Atlantic Shorelines*, M. D. Bertness, p. 348. Copyright © 1999 Sinauer Associates, Inc. Reprinted by permission.)

The hypersaline conditions of the sediment in the bare patches of the S. *patens* zone prevent the competitive dominants, S. *patens* and *Juncus*, from invading the patch. *Salicornia europaea*, however, thrives in such hypersaline conditions and quickly germinates and takes over such patches. Its dominance of the patch is short-lived because *Distichlis* begins to invade the patch by sending out runners from plants in nearby less saline environments. *Distichlis* can invade not only because it is salt tolerant, but because it receives its water through the runners from plants outside the hypersaline patch.

As the spikegrass takes over the patch, it shades the sediment, reducing the evaporation rate and decreasing the salinity. After a few years when the salinity has returned to normal, the competitive dominants will replace the *Salicornia* and/or *Distichlis* (Figure 8.25).

The salt marshes of the southern Atlantic and Gulf coasts are not subject to destruction by winter ice nor do they produce rafts of wrack that may settle and destroy patches of high marsh plants. They are also subject to higher solar radiation, leading to the presence of more or less permanent salt pans. Thus, the differences in zonal patterns observed between the southern marshes and northern marshes seem to be tightly linked to the differing climatic conditions.

Whereas competition and physical stresses certainly have the major roles in generating zonal patterns in the salt marshes, there are also positive interactions that ameliorate the physical environment and contribute to the zonation observed. For example, in the intermediate elevations of the salt marsh where tidal flushing and rainwater runoff are reduced, high soil salinities can occur in the absence of plant cover. Plant cover reduces the evaporation of water from the soil surface, and thus there is a positive feedback among plant neighbors that buffers them from salinity stress. Also, positive interactions between the dominant *Iva frutescens* and its neighbors at the seaward edge of the zone ensure its survival. In this scenario, adult *Iva* depend upon the presence of neighboring turf plants to prevent the decrease of soil oxygen and the increase of soil salinity that together will kill the adult *Iva*.

■ In summary, the New England salt marshes have their primary space dominated by dense stands of perennial, turf-forming plant species that through competitive interactions form monospecific bands that succeed each other across the intertidal. Competitively inferior, fugitive, plants species are rare. They only occur in patches created in the turf species when mats of rafted vegetation have been deposited on the dominants, killing them and creating hypersaline conditions through increased exposure of sediment to evaporation of water. The very harsh physical environment of the hypersaline bare patches cannot be tolerated by the competitively superior plants until that environment has been altered by the fugitive species. Spartina alterniflora forms a mutualistic association with both mussels and fiddler crabs. ■

Productivity

The primary producers of the marsh are the emergent marsh plants and the various microalgae that live on the surface of the marsh plants and the mud. In contrast to the water and mud flats of estuaries, productivity is very high in marshes, especially for the rooted emergent plants.

Although salt marshes have been extolled as one of the most productive of all plant communities, it is now clear that the methods and assumptions used to obtain the primary production figures were often flawed or had methodological limitations, making them suspect or causing widely varying estimates. The figures presented here may be revised in the future and are only presented to approximate the relative production among different areas.

It is most convenient to consider primary productivity in two groups: vascular emergent plants and various algae. The most complete data on productivity exist for the vascular plants and especially for the East Coast marshes. The major primary producers in these marshes are the various species of *Spartina*. Teal (1962) has shown that Georgia marshes have an average net productivity of 1,600 g C/m^2/year. This decreases to the north, such that in New Jersey, Good (1965) reported only 325 g C/m^2/year. On the Gulf coast marshes, the primary productivity of vascular plants ranges from 300–400 to more than 3,000 g C/m^2/year, according to Turner (1976). On the Pacific coast, Atwater et al. (1979) reported San Francisco marshes at 50–1,500 g C/m^2/year of net production (compare with Table 2.2, p.68), and Seliskar and Gallagher (1983) estimated 100–1,000 g C/m^2/year for the marshes of the Pacific Northwest.

Algal productivity estimates are fewer, more variable, and unavailable for many areas. On the Atlantic coast, studies by Teal (1962), Gallagher and Daiber (1974), and Van Raalte et al. (1976) have provided productivity estimates from 20 to 25% of that of the emergent vegetation. On the Pacific coast, Zedler (1982) has shown, however, that algal productivity in southern California marshes is 40–60% of that of the emergent vegetation. In the Pacific Northwest, Seliskar and Gallagher (1983) estimate algal production to be considerably less, 100 g C/m^2/year.

According to Bertness (1999) salt marsh primary productivity seems to be nitrogen limited, because nitrogen fertilization can entirely change the competitive interactions described above. Under experimental conditions of high nitrogen, *Distichlis spicata* outcompetes *Juncus gerardi* and *Spartina patens* invades the high marsh as does *Spartina alterniflora*. These results suggest that the typical pattern of zonation described here is due to persistent low nitrogen levels in the marshes (Figure 8.26). It also suggests that these marshes would undergo considerable change under conditions of human-induced eutrophication.

Zedler (1982) has suggested that the reduced productivity of vascular plants and enhanced productivity of algae observed in southern California marshes, but not in the East Coast marshes, may be because these California marshes surround intermittent estuaries where the soils are hypersaline during the summer growth period. The hypersalinity reduces the growth of the vascular plants, leading to a relatively open canopy; hence, more light penetrates to the substrate, enhancing algal production. Zedler also points out that this system may be more energy efficient, since algae can be consumed directly, whereas the vascular plants must be attacked by decomposers first, thereby adding another step to the food chain and reducing the total energy available to higher trophic levels.

Because there are few herbivores, most of the productivity of the marsh is not consumed directly. In the North Temperate Zone, the major natural grazers on the emergent salt marsh vegetation are the

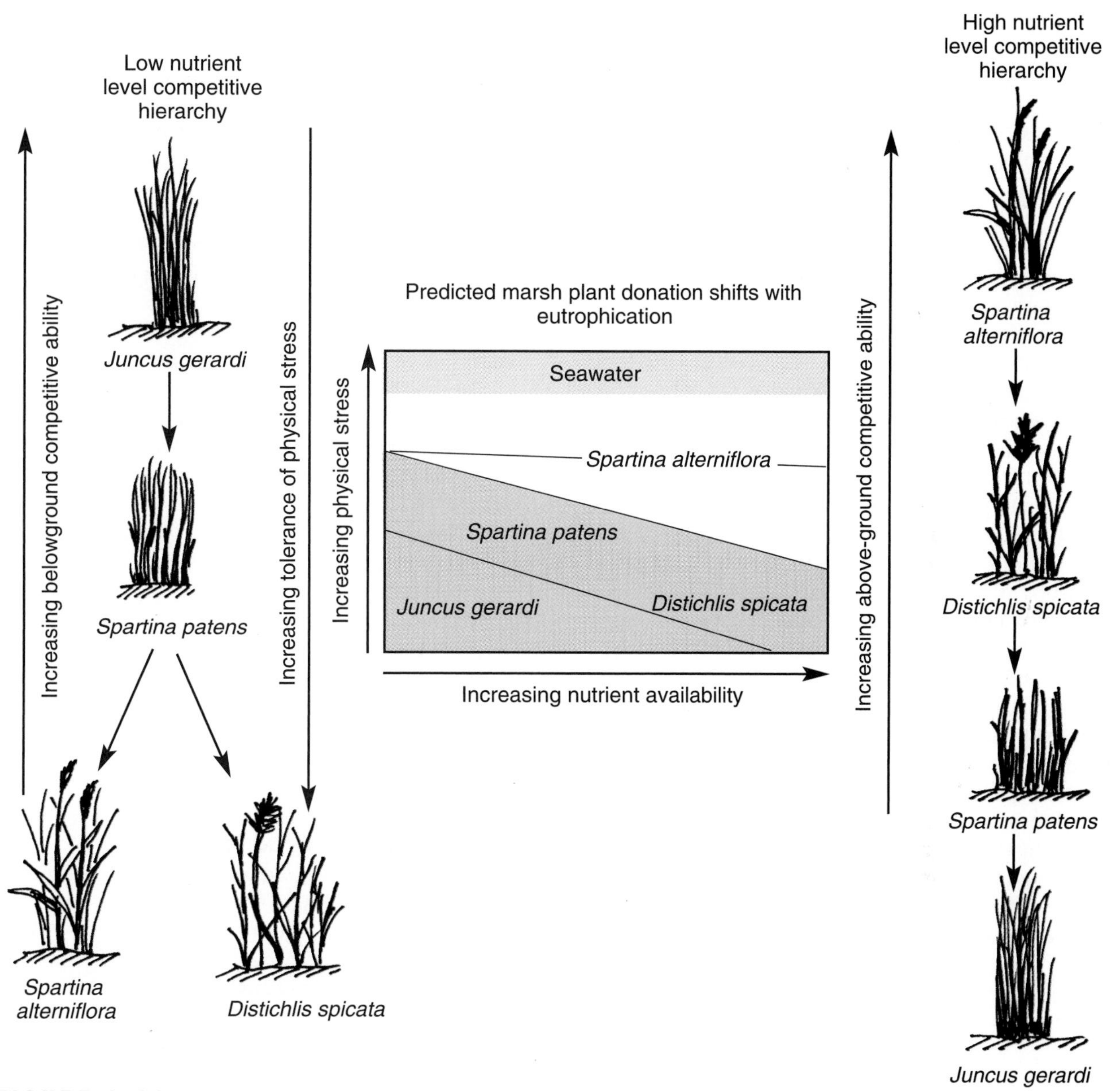

FIGURE 8.26 Change in competitive relations among marsh plants under high and low nitrogen levels. Competitive dominance at low nitrogen levels falls to plants that exploit belowground resources, whereas competitive dominance at high nitrogen levels falls to plants that are the better aboveground competitions. (After "Nutrient Availability and the Zonation of Marsh Plant Communities" J. S. Brewer & M. D. Bertnes, *Journal of Ecology*, Vol. 86, pp. 285–292, 1998. Copyright © 1998 Blackwell Science, Ltd. Reprinted by permission.)

various ducks and geese. These waterfowl may have a significant effect where they occur in large numbers (see pp. 213–214). Omnivorous crabs such as the warf crab *Armases cinereum* and the marsh crab *Sesarma reticulatum* are common in Atlantic coast marshes and are known to consume marsh plants. However, little is known about their natural diet and if they might have any effect on the vegetation. Why there are few herbivores is not known. Since marsh vascular plants have high salt content and low nutritional value, it may be that they are unpalatable, difficult to break down, or both or else cause water balance problems for the potential herbivores. In addition, most salt marsh plants are structurally defended by large amounts of cellulose and silica. Still others have chemical defenses, especially marshes at lower latitudes. In New England, salt hay has high phenolic concentrations that deter herbivory. The majority of the vascular plant production, perhaps 90% of it, is channeled into detritus and is either decomposed in place or washed out of the marsh. The dead plants are broken down by bacteria on the surface of the mud or in the water. During this process, the food value of the plant parts is increased as the indigestible cellulose is broken down into digestible carbohydrates. The protein content rises as the microorganism biomass increases on the pieces of decomposing plant. On the other hand, the algal mats are often subject to heavier grazing pressure, probably in part because of their higher nutritional value and easier digestibility. For example, Pace et al. (1979) excluded the snail *Ilyanassa obsoletus* and observed a rapid increase of the algal microflora.

Over the years a prevailing view has developed among biologists that, because of the high productivity of marshes and because the productivity is little used or even underused by resident animals, salt marshes are an outwelling nutrition source that sustains the productivity of estuarine and nearshore waters, particularly fisheries. This concept was widely supported by the statement of Teal (1962) that 45% of marsh production was removed, permitting that fraction to be used to support organisms outside the marsh. Since that time, additional studies and the use of new methodologies have led to a modification of this view. According to Nixon (1980), there now seems to be agreement that salt marshes do export a certain amount of dissolved and particulate organic matter, but the magnitude of this exported material and its importance to productivity in adjacent waters are variable and depend on a number of biological and geographical features. For example, Nixon (1980) showed that, where marsh area is large relative to the area of nearshore water, the organic supplement from the marsh may amount to about 50% of the open-water phytoplankton production. Such areas would be like those found off Georgia. In other cases, such as Chesapeake Bay and most New England marshes, the organic supplement is 10% or less. Furthermore, there is no evidence that the exported carbon provides significantly greater production of fish or shellfish than is found in coastal areas without marshes. We do not know why the export of organic matter does not lead to increased production at the higher trophic levels, but Nixon suggests it may be due to such factors as rapid sedimentation, the low nutrition of outwelled particles, or the rapid respiration of carbon by bacteria.

Marshes also export dissolved nitrogen and phosphorus. According to Nixon (1980), however, these nutrients do not have significant effects on primary productivity in nearshore waters, since primary productivity is no higher in areas where marshes occur than in areas where they do not.

On the Pacific coast, the marshes are small and adjacent to a nutrient-rich, upwelling ocean. Any outwelling of carbon or nutrients, therefore, would be insignificant to ocean productivity in these areas.

■ We can summarize this section by saying that our understanding of the productivity within salt marshes and the fate of that production including export to estuaries and nearshore coastal waters has advanced in the past 25 years. There is agreement that marshes export carbon, but this export fraction is variable depending on various geographical, biological, and physical features. ■

Interactions and Food Webs

The salt marsh is an ecotone between estuarine and terrestrial systems. There are numerous interactions between the animals of both areas and the marsh plants and their environment. In general, salt marsh animals live in a stressful environment of widely varying physical conditions. Few animals possess tolerance limits that include the range of temperature and salinity experienced in a marsh; therefore, marsh animals have evolved strategies that minimize the environmental variations to which they are exposed. As a result, there are more aquatic species found in low areas seldom exposed to the air and more terrestrial forms in areas above the effects of the highest tides. The fewest species occur in the intermediate area.

Marsh animals minimize the environmental changes in a variety of ways. Crabs commonly burrow into the mud to escape desiccation and surround themselves with soil water of higher salinity. Coffee bean snails (*Melampus*), marsh periwinkles (*Littoraria irrorata*), and certain insects escape inundation at high tide by climbing up the marsh plants. This may also

FIGURE 8.27 The migration of the marsh periwinkle *Littoraria irrorata* up the cordgrass blades to avoid predation by blue crabs (*Callinectes sapidus*). (After *The Ecology of Atlantic Shorelines*, M. D. Bertness, p. 355. Copyright © 1999 Sinauer Associates, Inc. Reprinted by permission.)

be a strategy to avoid predators. In the case of *Littoraria* it has been demonstrated by Hamilton (1976) that the migration limits their contact with blue crabs (*Callinectes*), their most significant predator (Figure 8.27). A great many animals simply migrate in and out of the marsh with the tide—aquatic animals in at high tide and terrestrial animals in at low tide.

The importance of the marsh periwinkle *Littoraria irrorata* to marsh productivity has been demonstrated by Silliman (1998), who experimentally removed the snails from penned areas of cordgrass and found that the production increased up to 25%; other penned areas, where densities of *L. irrorata* were artificially enhanced, showed production decreases up to 40%. It would appear that the periwinkles are capable of exerting significant top-down control of cordgrass, but are prevented from doing so, probably by the crab predators.

The marsh plants themselves are beneficial to the animals in a number of ways. They provide cover that reduces predation. The various small invertebrates, such as crabs, snails, and insects, are less subject to predation from birds and fish if they are beneath the canopy of marsh plants than if they are exposed on the open mud flat. For example, Willason (1980) found that shelter of the marsh was important to the survival of the crab *Pachygrapsus crassipes*, and Boland (1981) showed that most predatory birds used open mud flats rather than marsh areas. In southern California, there is evidence that the native hornsnail (*Cerithidea californica*) is more heavily preyed on in open mud flats than in areas within the marsh. The plants also stabilize the mud and provide a substrate to which certain animals may attach. Without the presence of *Spartina*, for example, the mussel *Geukensia* would have nothing to which to attach its byssal threads; barnacles would be unable to exist due to lack of a surface to cement to; and the terrestrial insects and pulmonate snails would have no vertical substrate to climb to escape inundation at high tide. The marsh plants also act indirectly by reducing wave and current action, which might otherwise remove or destroy some animals. They also reduce light penetration to the mud surface and ameliorate the temperatures and salinities on the mud surface, decreasing the range of variation to which the animals are exposed.

In turn, the animals have effects on the plants. The direct effects are minimal because only a few grasshoppers and leafhoppers consume the living plants; and these do not appear to have any significant effect on the plants. The burrowing of various crabs aerates otherwise stagnant mud and benefit the roots of the marsh plants. On the other hand, burrowing also may lead to harmful cropping of the roots. Burrowing by crabs and isopods, such as *Sphaeroma*,

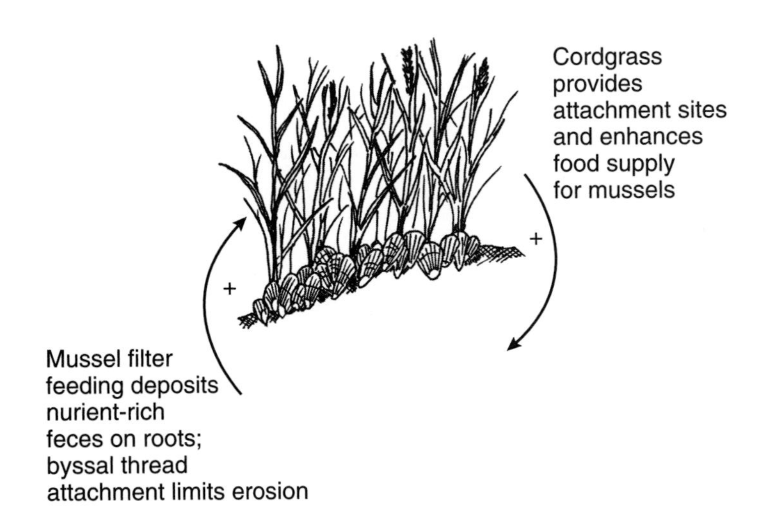

FIGURE 8.28 The mutual interaction between cordgrass and mussels in a salt marsh. (After "Ribbed Mussels and the Productivity of *Spartina alterniflora* in a New England Salt Marsh" M. D. Bertness, *Ecology*, Vol. 65, pp. 1794–180, 1984. Copyright © 1984 Ecological Society of America. Reprinted by permission.)

may also lead to bank erosion and the loss of parts of the marsh as they collapse into the tidal creeks. Such destruction of marshland has been reported by Josselyn (1983) for San Francisco Bay. Invertebrate fecal matter provides nutrients to the marsh plants, as we already noted. Certain snails, such as *Melampus* and *Littoraria*, graze the algae that settle on the stems of the marsh halophytes and enhance their growth by eliminating or reducing this algal cover.

Filter-feeding bivalve mollusks are common inhabitants of the seaward edges of marshes and help to couple water column and marsh food webs. Both oysters, *Crassostrea virginica*, and ribbed mussels, *Geukensia demissa*, historically formed dense aggregations or beds in many Atlantic coast marshes, where their large numbers were capable of filtering all of the incoming water at each tidal cycle. As a result nutrients extracted in the form of particulate plankton were transferred via fecal material to the marsh sediment, providing additional nutrients to the nutrient-limited marsh plants (Figure 8.28). Mussels and oysters also stabilized the sediments, reducing erosion and stabilizing the plants. In this century the overharvesting of oysters, diseases, and the increasing eutrophication of salt marshes have led to a steep decline in the oysters and a resultant deterioration of the salt marsh habitats.

The most common large deposit feeders in salt marshes are several species of fiddler crabs of the genus *Uca*. There is a positive association between the fiddler crabs and the marsh plants. The fiddler crab burrows allow oxygenation of the otherwise anoxic sediments, increase drainage of the waterlogged soil, and add nutrients with their waste products. All of these help to increase marsh plant production. In turn, the marsh plants' roots structurally support the crab burrows, the plant debris is a source of nutrition to the crabs, and the presence of the plants provides cover to the crabs from their predators.

The role of large predators in marsh habitats is not well studied. The two best-studied predators are the blue crab *Callinectes sapidus* and the killifish *Fundulus heteroclitis*. Blue crab predation restricts *Geukensia* to high marsh habitats in southern marshes and excludes marsh periwinkles from places where they cannot migrate out of the water at high tide. Similarly, killifish predation limits the snail *Melampus*, juvenile fishes, and various amphipods to the high marsh (Figure 8.29).

Competitive interactions among animals in the salt marsh have been little studied and remain a

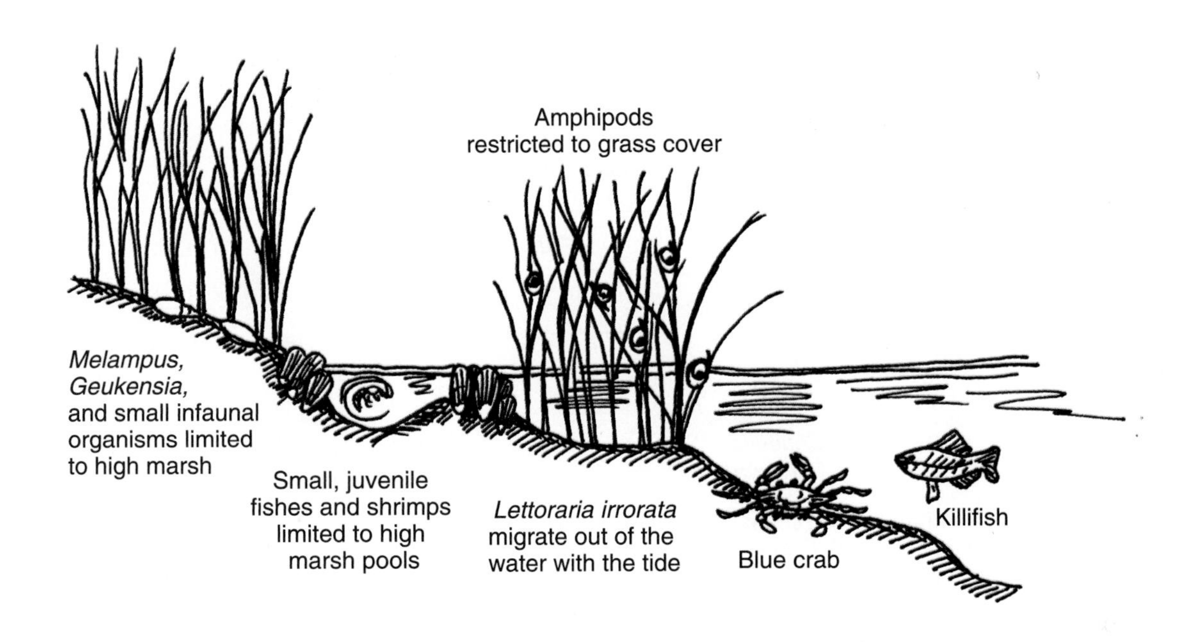

FIGURE 8.29 Representation of the effects of predation on some of the salt marsh inhabitants. (After *The Ecology of Atlantic Shorelines*, M. D. Bertness, p. 370. Copyright © 1999 Sinauer Associates, Inc. Reprinted by permission.)

fertile area for future work. The few studies that exist suggest that this process may be important. For example, Race (1982) has demonstrated that in San Francisco Bay the introduced snail *Ilyanassa obsoletus* has eliminated the native snail *Cerithidea californica* from mud flats. The latter is now found only in the upper portions of the marsh, because it is more tolerant of desiccation than *Ilyanassa*. Similarly, Sliger (1982) found that the two common marsh crabs *Pachygrapsus crassipes* and *Hemigrapsus oregonensis* were segregated vertically. *Pachygrapsus* appeared to be the dominant crab and had eliminated *Hemigrapsus* from higher tidal areas. *Hemigrapsus*, however, had a refuge in the lower intertidal areas because it could tolerate silty water. Its gills were not so easily clogged by silt as those of *Pachygrapsus*.

Marshes also experience a number of disturbances, both human induced and natural. Natural disturbances, such as those caused by wave or current action or by biological agents, including predatory activity, burrowing or excavating, and accumulations of drift plant and animal debris, among others, will lead to patchy distributions of organisms in the marsh. Human-induced disturbances, such as diking, dredging, channeling water, pollution, introduction of exotic species, particularly such large grazers as cattle, and construction of structures, often have a more profound effect on the marshes. They not only can lead to large-scale changes in the composition of a marsh but also can threaten its very existence. For example, San Francisco Bay, the largest estuarine salt marsh system on the Pacific coast, has, according to Nichols et al. (1986), been reduced from an areal extent of 2,200 km^2 in 1850 to less than 125 km^2 today. All of this is due to the effects of humans. In addition, deliberate and inadvertent introductions of exotic species have led to the current condition in which nearly all common macroinvertebrate species in the bay are alien (Nichols et al., 1986). Of the major estuaries of the United States, San Francisco Bay is now considered the one most changed by human activity.

In some cases, however, human-induced disturbances have given us better insights into how marshes operate. For example, Valiela et al. (1975) investigated the results of placing sewage on marshes

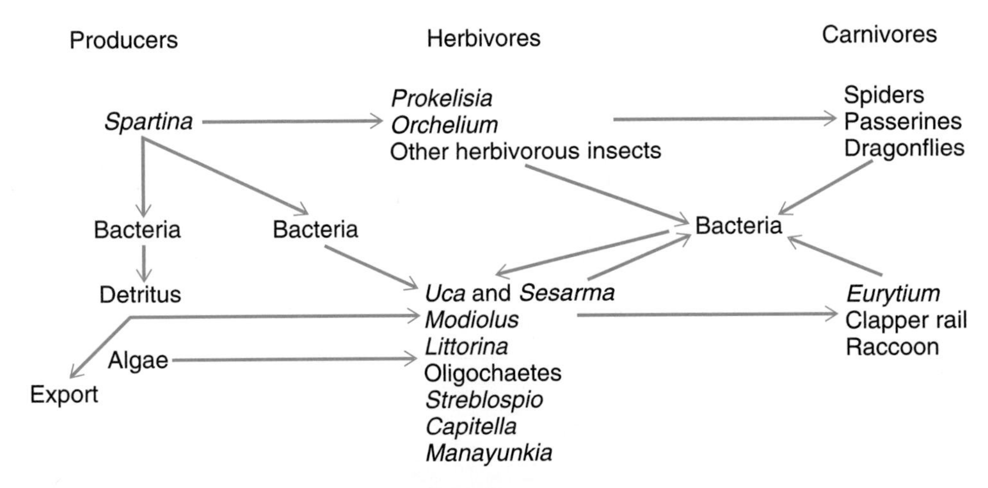

FIGURE 8.30 The food web of an Atlantic coast salt marsh. (After "Energy Flow in the Salt Marsh Ecosystem of Georgia" J. Teal, *Ecology*, Vol. 43, pp. 614–624, 1962. Copyright © 1962 Ecological Society of America. Reprinted by permission.)

on the East Coast. They discovered that the growth of *Spartina* was stimulated to such an extent that other plants were eliminated from the area. They also discovered that the *Spartina* plants accumulated high levels of toxic metals, such as cadmium and lead.

The food webs of salt marshes are based primarily on the consumption of detritus (and associated bacteria) and perhaps various algae. Since most of the detrital consumers distinguish only by particle size, a certain amount of live benthic algae is also consumed. These detrital consumers plus the few insects feeding directly on the marsh plants are, in turn, consumed by a variety of both aquatic and terrestrial carnivores (Figure 8.30).

SUMMARY OF KEY CONCEPTS

- An estuary is defined as a partially enclosed coastal embayment where fresh water and seawater meet and mix.
- Based on geology and geomorphology, there are four basic estuarine types: coastal plain, tectonic, semienclosed bay or lagoon, and fjord.
- Based on salinity gradients, there are three groups of estuaries: positive or salt wedge, negative or evaporite, and seasonal or intermittent.
- The dominant feature of the estuarine environment is the fluctuation in salinity.
- The pattern of water column salinity distribution in an estuary may be altered by the tidal regime, Coriolis effect, and seasonal changes in evaporation, freshwater flow, or both.
- Salinity changes in the substrate are less than in the water column, and so organisms in the substrate are partially buffered.
- Most estuaries have soft, muddy substrates derived from sediments carried into the estuary by both seawater and fresh water.
- Reduced water motion and the presence of various ions in the water causes silt particles to flocculate and settle out.
- Catastrophic events such as massive storms and accompanying floods may produce large depositions or removal of sediments in an estuary.
- Large amounts of particulate organic matter settle out in an estuary, which can serve as a food reservoir for estuarine organisms.
- Water temperatures in an estuary are more variable than in nearby coastal waters due to a smaller volume of water and a larger surface area for evaporation.
- Temperature range increases as one progresses up the estuary.
- Temperature also varies vertically, with the temperature at the surface showing a greater range than deeper waters.
- Wave action is minimal in estuaries due to the shallow depth and short fetch.
- Currents in estuaries are caused primarily by tidal action and river flow, with the highest velocities occurring in the

middle of channels where the frictional resistance from bottom and side banks is lowest.
- The time required for a given mass of fresh water to be discharged from the estuary is the flushing time and provides a measure of the stability of the system.
- Because of the great number of particles in suspension, turbidity is high in estuaries; this, in turn, decreases light penetration, reducing primary productivity.
- Oxygen is usually in ample supply in the water column but is severely depleted in the substrate.
- There are four types of fauna in estuaries: marine, freshwater, estuarine, and transitional.
- The largest group of species are marine and includes both stenohaline and euryhaline species.
- True estuarine organisms are found in the middle reaches of the estuary where the salinity is between 5 and 18 psu.
- The freshwater species are intolerant of salinities above 5 psu and are confined to the upper reaches of the estuary.
- The transitional fauna includes those animals that pass through the estuary to breeding grounds either in fresh or salt water and those that spend only part of their life cycle in the estuary.
- The number of species inhabiting estuaries is significantly lower than the number inhabiting nearby freshwater or marine habitats.
- The deeper substrates of estuaries are often barren of plant life, but the shallow and intertidal substrates may be occupied by seagrass beds, a number of macroalgae, and an abundance of diatoms.
- The dominant vegetation in the intertidal regions of estuaries are emergent flowering plants that form the characteristic salt marshes.
- Bacteria are extremely abundant both in the water column and the substrate of estuaries.
- Estuarine plankton is also reduced in number of species, with diatoms dominating the phytoplankton and copepods the zooplankton.
- There are few morphological adaptations among estuarine animals that can be attributed simply to living under conditions of fluctuating temperature and salinity.
- Among plants, morphological adaptations include aerenchyma tissue, salt glands, high lignin and carbohydrate content, abundant stomata, thin cuticles, and reduced leaf area.
- Osmosis is the physical process in which water passes through a semipermeable membrane that separates two fluids of different salt concentration, moving from the one of lower to higher salt concentration and thereby tending to equalize the salinities.
- Organisms that can control their internal salt concentrations are called osmoregulators; those that cannot are called osmoconformers.
- Osmoregulation can operate in three ways: animals may move water, they may move ions, or they may adjust their internal water-ion balance.
- Rooted plants maintain high concentrations of osmotically active substances to counter the water-retaining capacity of the surrounding medium.
- Most plants have mechanisms to get rid of excess salts, such as salt-secreting glands, deposition of salt on external surfaces, and shedding of leaves loaded with excess salt.
- A common behavioral adaptation in estuarine organisms is burrowing into the substrate, which exposes them to much less variation in salinity and temperature and decreases their exposure to surface predators.
- Many adult estuarine crabs can live successfully in estuaries but the eggs and young cannot, necessitating a migration back to the sea for breeding purposes.
- Other species take advantage of the estuary's rich food supply and relative scarcity of predators to use the estuary as a nursery for their young.
- The primary productivity of estuaries resides in the phytoplankton, benthic diatoms, seagrasses, and various algal mats; of these, the diatoms and algal mats are most significant.
- Classically it was believed that the major source of primary productivity in estuaries resided not in estuaries themselves but in the emergent plants of the surrounding salt marsh.
- Algae may be more important as primary producers in certain areas, particularly in southern California.
- Estuaries serve as a sink for organic material brought within by the rivers and the sea. Thus, some of the great amount of organic material that moves through the estuary is produced in the estuary and some is transported in from elsewhere.
- A European estuary is dominated by relatively barren mud flats, with the primary productivity centered in the benthic diatoms and phytoplankton. An American estuary is dominated by extensive stands of emergent salt marsh plants around the mud flats, which are the major primary producers.
- The estuarine food web has a number of energy pathways. These are generated primarily from a detrital base and have a somewhat incomplete use of energy that allows export to other areas, primarily by fishes and birds.
- The incidence of omnivory is generally higher than in other systems, and it is the abundance of food and relative lack of predators that has permitted estuaries to be nurseries for the juveniles of many animals that spend their lives elsewhere.
- There is an annual cycle of phytoplankton in estuaries, with a late winter diatom bloom ending in late spring by depletion of nitrogen. Populations remain low through the summer due to low nutrient levels and grazing, but the fall may bring another bloom.
- The zooplankton cycle in estuaries is variable and does not closely follow the phytoplankton cycle, as their

populations peak later than the phytoplankton and remain high during the summer.

- Salt marshes are communities of herbs, grasses, and shrubs rooted in soils alternately inundated and drained by tidal action. They are halophytes growing on soils with high salt content.
- The salt marsh is a rigorous environment with wide variations in salinity and temperature.
- Salt marshes are species poor, and the dominant plants show a high degree of similarity over a wide geographic area.
- The major terrestrial animal component of marshes are insects, and the marshes are invaded at high tide by marine and estuarine animals.
- Two major groups of marshes are found in North America. The Atlantic and Gulf coast marshes are extensive, occupying gently sloping substrates, and are dominated by various species of *Spartina*; the Pacific coast marshes occupy much smaller areas, characterized by broad bands of *Salicornia* with *Spartina* of lesser importance.
- The general appearance of a salt marsh is similar throughout the world. It is a broad, flat expanse of low herbaceous, grassy, or shrubby plants of similar form cut by dendritic channels leading to tidal creeks that in turn lead to bare mud flats or open waters of an estuary.
- On the Atlantic coast of North America, marsh development begins with the colonization of the intertidal sediment by the smooth cordgrass *Spartina alterniflora*; with time, other plants become established and a vertical zonation pattern emerges due to the changing physical conditions across the intertidal zone.
- The zonation pattern in the New England marshes has the lowest zone dominated by *Spartina alterniflora*, followed by a zone of *Spartina patens*, and next a zone of *Juncus gerardi* followed at the uppermost edge bordering the terrestrial vegetation by a band of *Iva frutescens*.
- Southern Atlantic marshes have the seaward edge of the marsh dominated by *Spartina alterniflora* extending landward displacing most of the *Spartina patens*, which is still found in patches. The highest parts of the southern marshes are dominated by *Juncus romerianus*, and between this zone and the *Spartina* zone is a distinct intermediate zone with permanent bare salt pans of exceedingly high salt content bordered by a suite of very salt tolerant plants.
- Pacific coast marshes tend to have a narrow seaward fringe of *Spartina* followed by a broad expanse of *Salicornia* occupying most of the midintertidal zone. The highest zone is occupied by a more diverse assemblage.
- The various physical and biological interactions that establish zonation patterns have not been investigated as thoroughly in salt marshes as in the rocky intertidal, but are perhaps best understood for the extensive marshes of the Atlantic and Gulf coasts of North America, which can serve as a model.
- The New England salt marshes have their primary space dominated by dense stands of perennial, turf-forming plant species that through competitive interactions form monospecific bands that succeed each other across the intertidal.
- Competitively inferior, fugitive plant species are rare in New England salt marshes. They only occur in patches created in the turf species when mats of rafted vegetation have been deposited on the dominants, killing them and creating hypersaline conditions through increased exposure of sediment to evaporation of water. The very harsh physical environment of the hypersaline bare patches cannot be tolerated by the competitively superior plants until that environment has been altered by the fugitive species.
- *Spartina alterniflora* forms a mutualistic association with both mussels and fiddler crabs.
- The primary producers of the marsh are the emergent marsh plants and the various microalgae that live on the surfaces of the plants and substrate.
- Vascular emergent plant productivity in marshes is high but varies with geographic locality, while algal productivity is less in all regions.
- Marsh productivity may be nitrogen limited, since under experimental conditions of enhanced nitrogen the whole pattern of plant competition changes.
- The reduced productivity of vascular plants and enhanced algal productivity seen in southern California marshes may be due to the open canopy that permits more light to penetrate to the algae on the substrate.
- Few herbivores consume marsh vegetation, possibly because marsh plants have high salt content, large amounts of cellulose and silica, and chemical substances that make them unpalatable or difficult to break down, or cause water balance problems for potential herbivores.
- Salt marshes do export a certain amount of dissolved and particulate organic matter, but the magnitude of this exported material and its importance to the production in adjacent waters are variable and depend on a number of geographical, biological, and physical features.
- Salt marsh animals live in a stressful environment and have evolved strategies that minimize the environmental variations to which they are exposed.
- Marsh plants are beneficial to the animals in a number of ways including providing cover that reduces predation, stabilizing the substrate and providing a surface for attachment, and reducing wave and current action.
- Animals, in turn, have effects on the plants including aerating the substrate by burrowing, grazing off of epiphytic algae, providing nutrients through fecal deposition, and in the case of mussels and oysters, stabilizing the substrate.
- The role of large predators in marsh habitats is not well studied.
- Competitive interactions among animals in the salt marsh have been little studied and remain a fertile area for future work.
- Although marshes experience a number of natural and human disturbances, those by humans often have the most profound effects and can threaten the existence of these communities.

REVIEW QUESTIONS

ESSAY: Develop complete answers to these questions.

1. Why are there relatively few truly estuarine species? Discuss.
2. Compare and contrast the European estuary with the American estuary in terms of the plants that dominate the mud flats and relative inputs and outputs of carbon.
3. What generally occurs when predators are excluded from soft-bottom invertebrate communities? Discuss.
4. Why is the keystone species concept nonfunctional in a salt marsh?
5. How do physical and biological factors interact to create zonal patterns in the New England salt marshes?
6. Why is it that stenohaline animals are located deeper in the water column the farther up an estuary they are found?
7. How can the disruption of habitat in coastal wetlands and estuaries affect organisms that live in the subtidal or intertidal regions? In coastal waters? In continental shelf waters? In the open ocean?

BIBLIOGRAPHY

Adam, P. 1990. *Saltmarsh ecology*. New York: Cambridge University Press.

Andrews, J. D. 1973. Effects of tropical storm Agnes on epifaunal invertebrates in Virginia estuaries. *Chesapeake Sci.* 14:223–234.

Atwater, B. F., S. G. Conard, J. N. Dowden, C. W. Hedel, R. L. McDonald, and W. Savage. 1979. History, land forms and vegetation of the estuary's tidal marshes. In *San Francisco Bay, the urbanized estuary*, edited by J. Conomos. San Francisco: Pacific Division AAAS, 347–386.

Barnes, R. S. K. 1967. The osmotic behaviour of a number of grapsoid crabs with respect to their differential penetration of an estuarine system. *J. Exp. Biol.* 47:535–551.

Barnes, R. S. K. 1974. *Estuarine biology*. Studies in Biology, no. 49. London: Edward Arnold. (Distributed in U.S. by Baltimore: University Park Press.)

Bertness, M. D. 1984. Ribbed mussels and the productivity of *Spartina alterniflora* in a New England salt marsh. *Ecology* 65:1794–1807.

Bertness, M. D. 1991a. Interspecific interactions among high marsh perennials in a New England salt marsh. *Ecology* 72(1):125–137.

Bertness, M. D. 1991b. Zonation of *Spartina patens* and *Spartina alterniflora* in a New England salt marsh. *Ecology* 72(1):138–148.

Bertness, M. D. 1992. The ecology of a New England salt marsh. *Amer. Sci.* 80:260–268.

Bertness, M. D. 1999. *The ecology of Atlantic shorelines*. Sunderland, MA: Sinauer Associates, Inc.

Bertness, M. D., and S. W. Shumway. 1993. Competition and facilitation in marsh plants. *Amer. Nat.* 142(4):718–724.

Boland, J. M. 1981. Seasonal abundance, habitat utilization, feeding strategies and interspecific competition within a wintering shorebird community and their possible relationships with the latitudinal distribution of shorebird species. Master's thesis, South Dakota State University.

Chapman, V. J., ed. 1977. *Ecosystems of the world. 1. Wet coastal ecosystems*. New York: Elsevier.

Conomos, J., ed. 1979. *San Francisco Bay, the urbanized estuary*. San Francisco: Pacific Division AAAS.

Darnell, R. 1961. Trophic spectrum of an estuarine community based on studies of Lake Ponchartrain, Louisiana. *Ecology* 45(3):553–568.

Day, J. W., C. A. S. Hall, W. M. Kemp, and A. Yañez-Arancibia. 1989. *Estuarine ecology*. New York: Wiley.

Deevey, G. B. 1960. The zooplankton of the surface waters of the Delaware Bay region. *Bull. Bing. Ocean. Coll.* 17(2):54–86.

Drinnan, R. E. 1957. The winter feeding of the oyster catcher (*Haematopus ostralegus*) on the edible cockle (*Cardium edule*). *J. Anim. Ecol.* 26:441–469.

Frey, R. W., and P. B. Basan. 1985. Coastal salt marshes. In *Coastal sedimentary environments*, edited by R. A. Davis. New York: Springer-Verlag, 225–301.

Gallagher, J. L., and F. C. Daiber. 1974. Primary production of edaphic algal communities in a Delaware salt marsh. *Limn. Oceanog.* 19:390–395.

Good, R. E. 1965. Salt marsh vegetation, Cape May, New Jersey. *N.J. Acad. Sci. Bull.* 10(1):1–11.

Goss-Custard, J. D. 1969. The winter feeding ecology of the redshank *Tringa totanus*. *Ibis* 111:338–356.

Green, J. 1968. *The biology of estuarine animals*. Seattle: University of Washington Press.

Haines, E. B. 1979. Interactions between Georgia salt marshes and coastal waters: A changing paradigm. In *Ecological processes in coastal and marine systems*, edited by R. J. Livingston. New York: Plenum Press, 35–46.

Hamilton, P. V. 1976. Predation on *Littorina irrorata* by *Callinectes sapidus*. *Bull. Mar. Sci.* 26:401–409.

Hartley, P. H. T. 1940. The saltash tuck-net fishery and the ecology of some estuarine fishes. *J. Mar. Biol. Assoc.* 24:1–68.

Josselyn, M. 1983. *The ecology of San Francisco Bay tidal marshes: A community profile*. U.S. Fish and Wildlife Service Division of Biological Services FWS/OBS-83/23.

Kinne, O. 1963. The effects of temperature and salinity on marine and brackish water animals. *Oceanog. Mar. Biol. Ann. Rev.* 1:301–340.

Lackey, J. 1967. The microbiota of estuaries and their roles. In *Estuaries*, edited by G. H. Lauff. Publ. no. 83. Washington, DC: AAAS, 291–302.

Lauff, G. H., ed. 1967. *Estuaries*. Publ. no. 83. Washington, DC: AAAS.

Levine, J., S. Brewer, and M. D. Bertness. 1998. Nutrient availability and the zonation of marsh plant communities. *J. Ecol.* 86:285–292.

Long, S. P., and C. F. Mason. 1983. *Saltmarsh ecology*. London: Blackie and Son, Ltd.

McLusky, D. S. 1971. *Ecology of estuaries*. London: Heinemann.

McLusky, D. S. 1981. *The estuarine ecosystem*. New York: Wiley.

McLusky, D. S. 1989. *The estuarine ecosystem*. 2d ed. New York: Chapman and Hall.

Muus, B. J. 1963. Some Danish Hydrobiidae with a description of a new species, *Hydrobia neglecta*. *Proc. Malac. Soc. London* 35:131–138.

Nichols, F. H., J. E. Cloern, S. N. Luoma, and D. H. Peterson. 1986. The modification of an estuary. *Science* 231:567–573.

Nixon, S. W. 1980. Between coastal marshes and coastal waters—a review of twenty years of speculation and research on the role of salt marshes in estuarine productivity and water chemistry. In *Estuarine and wetland processes*, edited by P. Hamilton and K. B. MacDonald. New York: Plenum Press, 437–525.

Nixon, S. W. 1982. *The ecology of New England high salt marshes: A community profile*. Biological Report 81(55). U. S. Fish and Wildlife Service, Washington, DC.

Odum, E. P., and A. A. de la Cruz. 1967. Particulate detritus in a Georgia salt marsh-estuarine ecosystem. In *Estuaries*, edited by G. H. Lauff. Publ. no. 83. Washington, DC: AAAS, 383–388.

Odum, H. T., B. J. Copeland, and E. A. McMahan. 1974. *Coastal ecosystems of the United States*. 4 vols. Washington, DC: The Conservation Foundation.

Pace, M. L., S. Shimmel, and W. M. Darley. 1979. The effect of grazing by a gastropod, *Nassarius obsoletus*, on the benthic microbial community of a salt marsh mudflat. *Est. Coast. Mar. Sci.* 9:121–134.

Pennings, S. C., and M. D. Bertness. 1998. Using latitudinal variation to examine effects of climate on coastal salt marsh pattern and process. Proceedings of the OECD Workshop on Global Change and Wetlands. To appear in *J. Wetl. Geochem.*

Peterson, C. H. 1979. Predation, competitive exclusion and diversity in the soft bottom benthic communities of estuaries and lagoons. In *Ecological processes in coastal and marine systems*, edited by R. J. Livingston. New York: Plenum Press, 233–264.

Peterson, C., and N. M. Peterson, 1979. *The ecology of intertidal flats of North Carolina: A community profile*. U.S. Fish and Wildlife Service, Biological Services Program. FWS/OBS-79/39.

Pomeroy, L. R., and R. G. Wiegert, eds. 1981. *The ecology of a salt marsh*. New York: Springer-Verlag.

Prater, A. J. 1972. Ecology of Morecambe Bay. III. The food and feeding habits of knot (*Calidris canutus*) in Morecambe Bay. *J. Appl. Ecol.* 9:179–194.

Race, M. 1982. Competitive displacement and predation between introduced and native mud snails. *Oecologia* 54:337–347.

Reid, G. K., and R. D. Wood. 1976. *Ecology of inland waters and estuaries*. 2d ed. New York: Van Nostrand.

Reimold, R. J. 1977. Mangals and salt marshes of the eastern United States. In *Wet coastal ecosystems*, edited by V. J. Chapman. Amsterdam: Elsevier, 157–166.

Reimold, R. J., and W. H. Queen, eds. 1974. *Ecology of halophytes*. New York: Academic Press.

Remane, A., and C. Schlieper. 1971. *Biology of brackish water*. 2d ed. Stuttgart: Schweizerbartsche.

Seliskar, D. M., and J. L. Gallagher. 1983. *The ecology of tidal marshes of the Pacific Northwest Coast: A community profile*. U.S. Fish and Wildlife Service Division of Biological Services, Washington, DC FWS/085-82/32.65.

Silliman, B. R. 1998. Top down control of *Spartina alterniflora* growth: Salt marsh periwinkles eat more than just detritus. Masters thesis, University of Virginia, Charlottesville.

Sliger, M. 1982. Distribution and microhabitat selection of *Hemigrapsus oregonensis* (Dana) and *Pachygrapsus crassipes* (Randall) in Elkhorn Slough, Monterey County, California. Master's thesis, California State University, Hayward.

Stickney, Robert R. 1984. Estuarine ecology of the southeastern United States and the Gulf of Mexico. College Station: Texas A and M University Press.

Stout, J. P. 1984. *The ecology of irregularly flooded salt marshes of the northeastern Gulf of Mexico: A community profile*. U.S. Fish and Wildlife Service Biological Report 85 (7.1).

Teal, J. 1962. Energy flow in the salt marsh ecosystem of Georgia. *Ecology* 43:614–624.

Teal, J. 1985. *The ecology of New England low marsh habitats: A community profile*. Washington, DC: U.S. Fish and Wildlife Service.

Turner, R. E. 1976. Geographic variations in salt marsh macrophyte production: A review. *Contrib. Mar. Sci.* 20:47–68.

Valiela, I., J. M. Teal, and W. Sass. 1975. Production and dynamics of salt marsh vegetation and the effects of experimental treatment with sewage sludge: Biomass, production and species composition. *J. Appl. Ecol.* 12:973–982.

Van Engel, W. A. 1958. The blue crab and its fishery in Chesapeake Bay. *Commercial Fisheries Rev.* 20(16):6–17.

Van Raalte, C. D., I. Valiela, and J. M. Teal. 1976. Production of epibenthic salt marsh algae: Light and nutrient limitation. *Limn. Oceanog.* 21:862–872.

Waisel, Y. 1972. *Biology of halophytes*. New York: Academic Press.

Wiegert, R. G., and B. J. Freeman. 1990. *Tidal salt marshes of the southeast Atlantic coast: A community profile.* Biological Report 85(7.29). Washington, DC: U.S. Fish and Wildlife Service.

Willason, S. W. 1980. Factors influencing the biology and coexistence of grapsoid crabs, *Pachygrapsus crassipes* (Randall) and *Hemigrapsus oregonensis* (Dana) in a California salt marsh. Master's thesis, University of California, Santa Barbara.

Wood, E. J. F. 1965. *Marine microbial ecology.* London: Chapman and Hall.

Zedler, J. B. 1982. *The ecology of southern California coastal salt marshes: A community profile.* U.S. Fish and Wildlife Service Biological Services Program FWS/OBS-81/54.

Zedler, J., T. Winfield, and D. Mauriello. 1978. Primary productivity in a southern California estuary. *Coastal Zone* 3:649–662.

Zobell, C. E., and C. B. Feltham. 1942. The bacterial flora of a marine mud flat as an ecological factor. *Ecology* 23:69–78.

TROPICAL COMMUNITIES

For color, sheer beauty of form and design, and tremendous variety of life, perhaps no natural areas in the world can equal coral reefs. Their beauty has fascinated generations of people, both scientific and lay, down through the years. By the same token, few areas of the marine environment have less initial aesthetic appeal than the dark, mud-wreathed areas known as mangrove forests, where passage is barred by a maze of tangled roots standing above a soft mud surface into which one quickly sinks, yet these two very different associations are characteristic of vast tropical regions of the world. They are unique, and must be considered if one is to understand the functioning of shallow-water areas in the tropics. In this chapter, we discuss first the distribution, zonation, and structure of these assemblages and then attempt to account for the maintenance and continuation of these communities by considering the component organisms and their interactions.

CORAL REEFS

Over a vast region (millions of square miles) of the tropics, the shallow inshore waters are dominated by the formation of coral reefs, and such reefs are often used to define the limits of the tropical marine environment (Figure 9.1). Coral reefs have been estimated to occupy about 600,000 square miles of the surface of the earth, which represents about 0.17% of the total area of the planet.

Coral reefs have the greatest diversity per unit area of any marine ecosystem with respect to higher taxa, and it has been estimated by Karlson (1999) that perhaps 4–5% of all species, or about 91,000, are found on coral reefs. Coral reefs are globally significant in that about half the calcium that enters the world's oceans each year is taken up and bound into coral reefs as calcium carbonate. Since each bound calcium atom requires the incorporation of a molecule of carbon dioxide, reefs also remove about 700 billion kilograms of carbon per year. These facts mean that coral reefs are very significant to the continuing health of the marine environment and the planet.

Coral reefs are also among the most productive systems in the marine environment and have existed on the planet for hundreds of millions of years. Despite their long history and ability to create the most massive structures built by living organisms, the thin layer of living coral tissue seems particularly sensitive to a number of natural and human-made disturbances that today seem to be converging and challenging their continued survival.

Coral reefs are unique among marine associations or communities in that they are built up entirely by biological activity. The reefs are essentially mas-

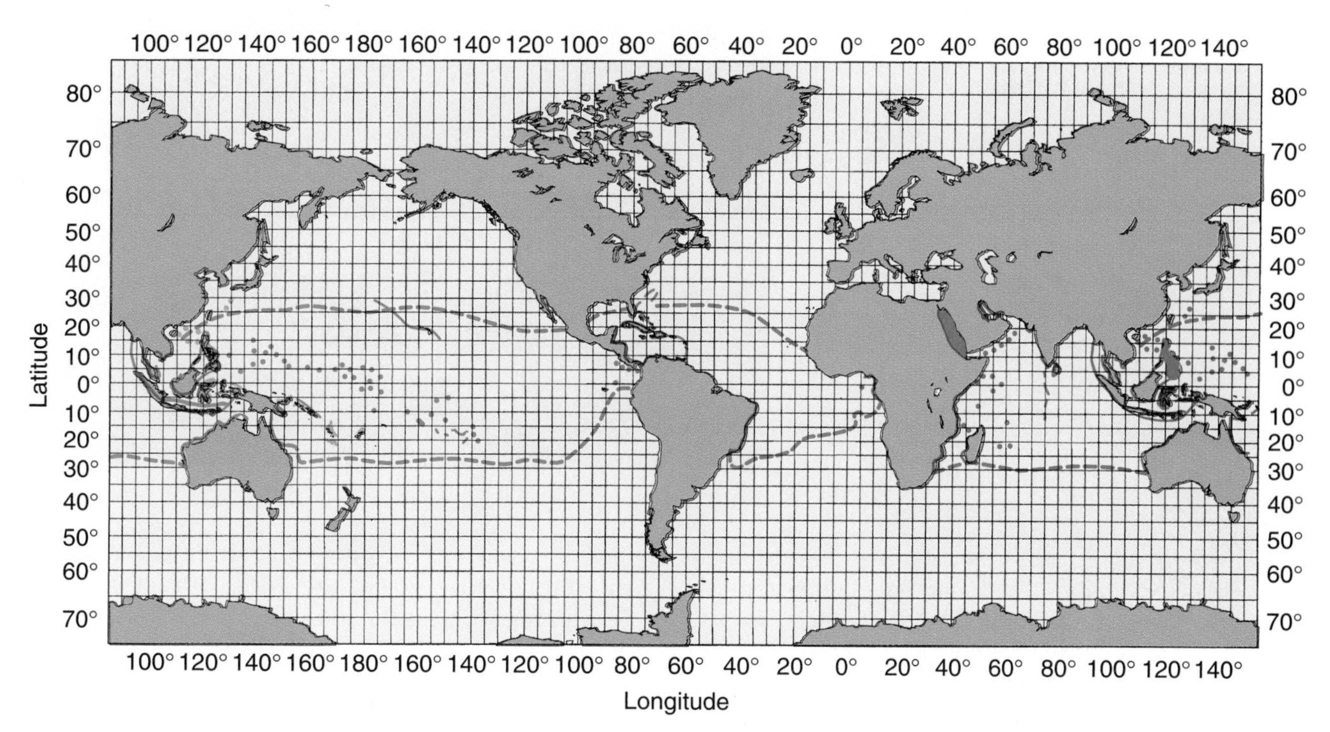

FIGURE 9.1 Distribution of coral reefs in the world; 20°C isotherm is indicated by the dashed lines.

FIGURE 9.2 Underwater view of a Pacific Ocean coral reef. (Photo courtesy of Dr. Gary Williams.)

sive deposits of calcium carbonate produced primarily by corals (phylum Cnidaria, class Anthozoa, order Scleractinia) with lesser additions from calcareous algae and other organisms that secrete calcium carbonate (Figures 9.2 and 9.3). Although corals are found throughout the oceans of the world in polar and temperate waters as well as in the tropics, it is only in the tropics that reefs achieve their greatest development and diversity. This is because there are two different groups of corals, one called **hermatypic** and the other **ahermatypic**. Hermatypic corals are those that produce reefs; ahermatypic do not form

FIGURE 9.3 Photo showing the crowding of corals on a coral reef in Palau. (Photo courtesy of Dr. Gary Williams.)

reefs. Ahermatypic corals are distributed worldwide, but hermatypic corals are found only in the tropical regions. The distinguishing feature between the two is that most hermatypic corals have in their tissues small symbiotic (living together) plant cells called **zooxanthellae**, whereas most ahermatypic corals do not have these cells. The role of these plants will be considered later (see p. 387).

Reef Distribution and Limiting Factors

Six major physical factors limit coral reef development: temperature, depth, light, salinity, sedimentation, and emergence into air. A glance at a map of coral reef distribution in the world immediately shows that nearly all coral reefs are found only within the 20°C surface isotherm (see Figure 9.1). Hermatypic corals can maintain themselves for periods at temperatures somewhat below this; however, as Wells (1957) notes, no reefs develop where the annual mean minimum temperature is below 18°C. Optimal reef development occurs in waters where the mean annual temperatures are about 23–25°C. Some coral reefs can tolerate temperatures up to about 36–40°C. It may be noted that reefs are reduced or absent from large areas on the west coast of South and Central America and also from the west coast of Africa; both areas lie well within the tropical zone. The reason for the great reduction or absence of reefs in these areas is that the west coasts of both of these continents are areas of strong upwelling of cold water, which reduces the temperature of the shallow inshore waters below that required for reef development in many places. Furthermore, both of these coasts have strong, north-flowing cold currents that keep the temperature down, the Peru Current on the South American coast and the Benguela Current off West Africa. The situation in West Africa is complicated also by the influx of large amounts of fresh water from the Niger and Congo rivers.

Coral reefs are also limited by depth. Coral reefs do not develop in water that is deeper than about 50–70 m. Most reefs grow in depths of 25 m or less. This explains why these structures are restricted to the margin of continents or islands. The peculiar feature of the development of atolls, which rise out of water many kilometers deep, will be discussed later (see pp. 375–377). The depth restriction is due to the hermatypic corals' requirement for light. Light is one of the most important factors limiting coral reefs. Why? Sufficient light must be available to allow photosynthesis by the symbiotic zooxanthellae in the coral tissue. Without sufficient light, the photosynthetic rate is reduced and, with it, the ability of the corals to secrete calcium carbonate and produce reefs. The compensation point for most reef corals seems to be the depth where light intensity has been reduced to 1–2% of surface intensity, according to Yentsch (1966).

Another factor that restricts coral reef development is salinity. Hermatypic corals are true marine organisms and are intolerant of salinities deviating significantly from that of normal seawater (32–35 psu). Wherever inshore waters are subject to continuing influxes of fresh water from river discharge so that the salinity is lowered, reefs will be absent. Such is the case along large portions of the Atlantic coast of South America, where the Amazon and Orinoco

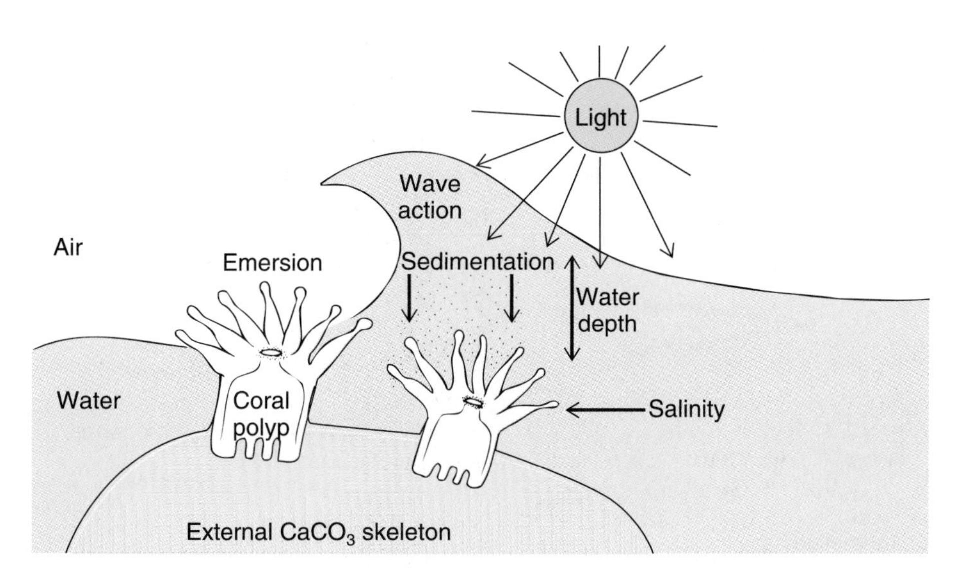

FIGURE 9.4 Summary of physical factors acting on coral polyps and coral reefs that may limit their distribution.

rivers discharge a huge volume of fresh water, and reefs are absent (see Figure 9.1). On a smaller scale, this occurs in many areas of the tropics where rivers and streams running out to sea cause breaks in reef development. At the other extreme, coral reefs do occur in regions of elevated salinity, such as the Persian Gulf, where reefs flourish at 42 psu.

Often correlated with freshwater runoff is the factor of sedimentation. Sediment, both in the water and settling out on the coral reefs, has an adverse effect on the corals. Many corals can remove limited amounts of sediment by trapping it in mucus and carrying it off by ciliary action. Most hermatypic corals, however, cannot withstand heavy sedimentation, which overpowers their ciliary-mucus cleansing mechanism, clogs their feeding structures, and smothers them. Sediment in the water (turbidity) also reduces the light necessary for photosynthesis by the zooxanthellae in the coral tissue. As a result, coral reef development is reduced or eliminated in areas of high turbidity. When this sediment is carried by rivers or streams, the combination of reduced salinity and excess sediment is responsible for the absence of reefs. While hermatypic coral species may vary in their tolerance for sedimentation, few can tolerate high sedimentation rates, and these are found as isolated colonies in sedimentary areas.

In general, coral reef development is greater in areas subject to moderate wave action. Coral colonies with their dense, massive skeletons of calcium carbonate are very resistant to damage by wave action. At the same time, the wave action provides a constant source of fresh, oxygenated seawater and prevents sediment from settling on the colony. Wave action is also responsible for renewing the plankton, which is food for the coral colony.

Finally, coral reefs are limited in an upward direction by emergence into air. Whereas abundant secretion of mucus may prevent dehydration for a short time, an hour or two, most corals are killed by long exposure to air; thus, their upward growth is limited to the level of the lowest tides. During an unusual and extremely low tide series over a five-day period in the Gulf of Aquaba, Loya (1976) found that coral mortality was between 80 and 90%, which suggests this factor can be of considerable importance. On very low spring tides, coral reef areas may, however, be exposed for short periods up to a few hours, which does not seem to harm them, at least in the Indo-Pacific region.

All these factors are summarized diagrammatically in Figure 9.4.

Structure of Corals

Since the prominent members of coral reefs are the corals, it is necessary to understand their anatomy. Corals are members of the phylum Cnidaria, which includes such diverse forms as jellyfish, hydroids, the freshwater *Hydra*, and sea anemones. Corals and sea anemones are members of the same taxonomic class, Anthozoa. They differ primarily in that corals secrete an external calcium carbonate skeleton and anemones do not. Corals may be colonial or solitary,

FIGURE 9.5 Anatomy of a coral polyp.

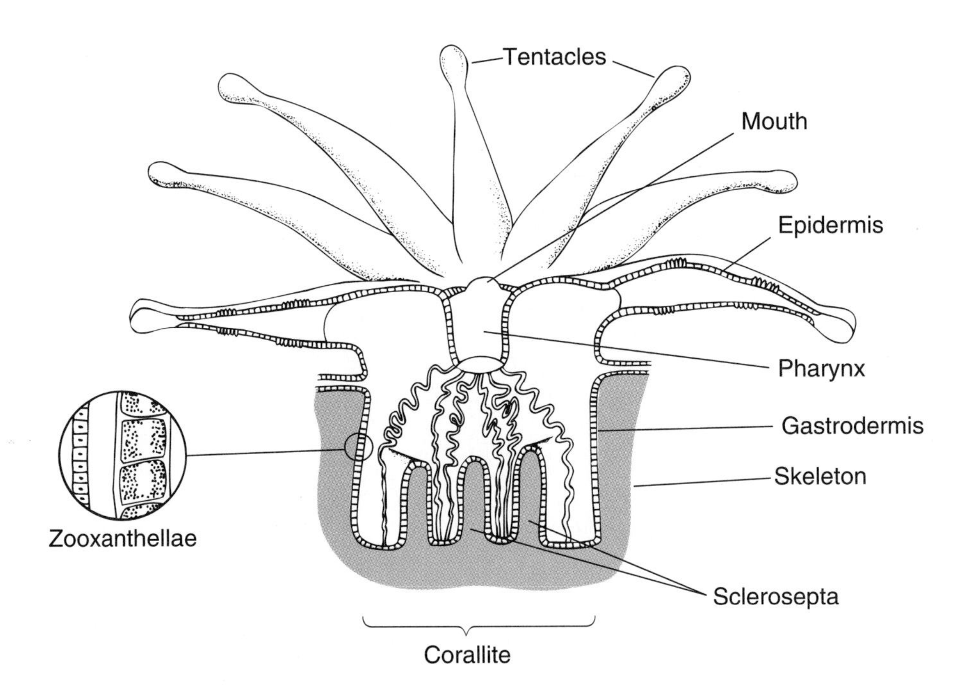

but almost all hermatypic corals are colonial, with the various individual coral animals or **polyps** occupying little cups or **corallites** in the massive skeleton (Figures 9.5, 9.6, and 9.7). Each cup or corallite has a series of sharp, bladelike **sclerosepta** rising from the base. The pattern of these septa differs from species to species and is often used as a taxonomic tool to distinguish among coral species. The sclerosepta of the skeletal cup alternate with the internal septa of the gastrovascular cavity of the coral polyp. Each polyp is a three-layered animal with an outer epidermis separated by a fibrous mesoglea from an internal gastrodermis. Around the mouth is a series of tentacles that have batteries of stinging capsules or nematocysts that the animals use to capture their zooplankton food (Figure 9.8). The symbiotic zooxanthellae, actually dinoflagellates (see Chapter 10), are found intracellularly in the gastrodermal layer.

FIGURE 9.6 A single coral colony, showing the characteristic appearance when the polyps are expanded. (Photo courtesy of Drs. Lovell and Libby Langstroth.)

FIGURE 9.7 Expanded coral polyps. (Photo courtesy of Drs. Lovell and Libby Langstroth.)

FIGURE 9.8 Orange tube coral, *Tubastrea coccinea*, showing the clusters of nematocysts in the form of dots on the tentacles. (Photo courtesy of Dr. Diane Nelson.)

Coral colonies grow by having the polyps bud off new polyps asexually. New colonies are established by the fragmentation of skeletal pieces or through the settlement of a planktonic **planula** larva, which is the result of sexual reproduction.

Types of Reefs

Coral reefs occur in many different sizes and shapes, resulting from the particular hydrological and geological situations that recur in different areas of the tropics. This has led scientists to classify reefs into various schemes, often generating a bewildering array of names and terms. In an attempt to keep things simple and prevent excessive terminology, we choose to limit the number of types to those that are most common. In general, coral reefs are grouped into one of three categories: **atolls, barrier reefs,** and **fringing reefs** (Figure 9.9). Atolls are usually easily distinguished because they are modified horseshoe-shaped reefs that rise out of very deep water far from land and enclose a lagoon. This lagoon may contain **lagoon reefs** or **patch reefs**. With few exceptions, atolls are found only in the Indo-Pacific area. Barrier reefs and fringing reefs, on the other hand, tend to grade into each other and are not readily separable. Some scientists would prefer to group them into a single category. Both types occur adjacent to a landmass, with a barrier reef being separated from the landmass by a greater distance and deeper water channel than the fringing reef (Figure 9.9b). Fringing reefs and barrier reefs are common throughout the coral reef zones in all oceans. The largest barrier reef is the Great Barrier Reef of Australia, which stretches for almost 2,000 km along the eastern coast of Australia, from near New Guinea to just north of Brisbane. The second largest barrier reef is that off the Yucatan Peninsula in Belize.

Origin of Reefs

Different types of reefs and reefs in different oceans may have diverse origins and histories. The greatest interest in the origin of reefs has centered on atolls.

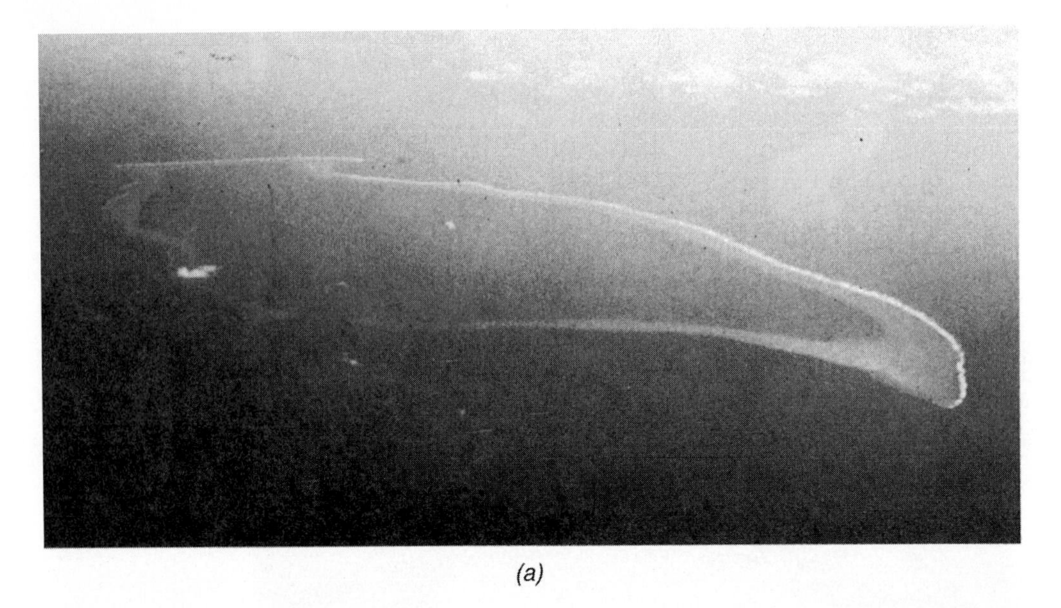

(a)

(b)

FIGURE 9.9 Aerial photo of an atoll (a) and a barrier reef (b). (Photos by Dr. Richard Mariscal.)

For many years, humans speculated as to how such reefs could develop out of such deep water, miles from the nearest emergent land. This interest was heightened when it was discovered that reef corals could not live deeper than 50–70 m. This led to the development of several theories concerning the origin of atolls. Only one need be discussed here—the **subsidence** or **compensation theory**. This explanation for the origin of atolls was first promulgated by Charles Darwin following his five-year voyage on the *Beagle*, during which time he had the opportunity to study reefs in several areas. According to Darwin's subsidence theory, atolls are created when fringing reefs begin to grow on the shores of newly formed

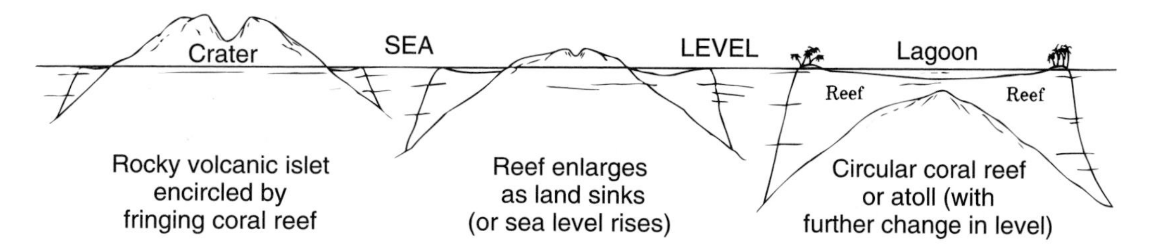

FIGURE 9.10 Geological evolution of a coral atoll according to the subsidence hypothesis of Darwin. (After *General Zoology*, 6/e, Stoyer et al., 1979.)

volcanic islands that have pushed to the surface from deep water. These islands often begin to subside, and if the subsidence is not too fast, reef growth will keep up with the subsidence, forming next a barrier reef and finally an atoll as the island disappears beneath the sea (Figure 9.10). When the island has disappeared, corals continue to grow on the outside and keep the reef at the surface. On the inside, where the island used to be, quiet water conditions and high sedimentation prevail. These conditions prevent continued vigorous coral growth; hence, a lagoon develops. This theory links all three reef types in an evolutionary sequence but is not an explanation for all fringing and barrier reef types.

Since the current surface features of atolls give no evidence of a volcanic base, in the years after the development of Darwin's theory other explanations were offered, and the whole concept of the origin of atolls became embroiled in the "coral reef problem." If Darwin's theory was correct, it must be assumed that drilling down through the current atoll reefs would yield layer after layer of reef limestone until, finally, volcanic rock would be encountered. The ability to drill to the base of atoll reefs and resolve the problem had to wait until the mid-twentieth century. In 1953, Ladd et al. reported borings at Eniwetok Atoll in the Marshall Islands that penetrated 1,283 m of reef limestone and then hit volcanic rock. This was the evidence that Darwin's theory was substantially correct. Correctness of this theory has been strengthened by the discovery of flat-topped submerged mountains or **guyots** that, at present, have their tops many hundreds or thousands of meters below the ocean surface but have on their surface the remains of shallow-water corals. Evidently, these mountains sank too fast for reef growth to keep within the lighted zone.

Although the subsidence theory links all three reef types in a successional sequence, not all barrier reefs and fringing reefs can be explained by this mechanism. Indeed, the reasons barrier and fringing reef types occur around continental margins and high, nonvolcanic islands are simply that these areas offer suitable environmental conditions for the growth of reefs and a suitable substrate on which to begin growth. The extensive reefs around the Indonesian Islands, the Philippines, New Guinea, Fiji, and most of the Caribbean islands and Florida are there because a suitable substrate in shallow water existed on which they could initiate growth. In none of these areas are large land areas subsiding, nor will these reefs ultimately become atolls.

Whereas atolls are often very old structures (for example, the 1,283 m of reef limestone at Eniwetok represent reef growth for about 60 million years), other reefs may be much younger. This is particularly true for the reefs in the Atlantic Ocean. They are all geologically very recent, most dating only from the last glacial age, or 10,000–15,000 years before the present. Perhaps even more remarkable is the more recent determination, as reported by Sale (1991) and Anderson (1990), that the world's largest reef system, the Great Barrier Reef of Australia, is at least 40 times younger than previously thought with new estimates of its age ranging from 500,000 years to as young as 9,000 years.

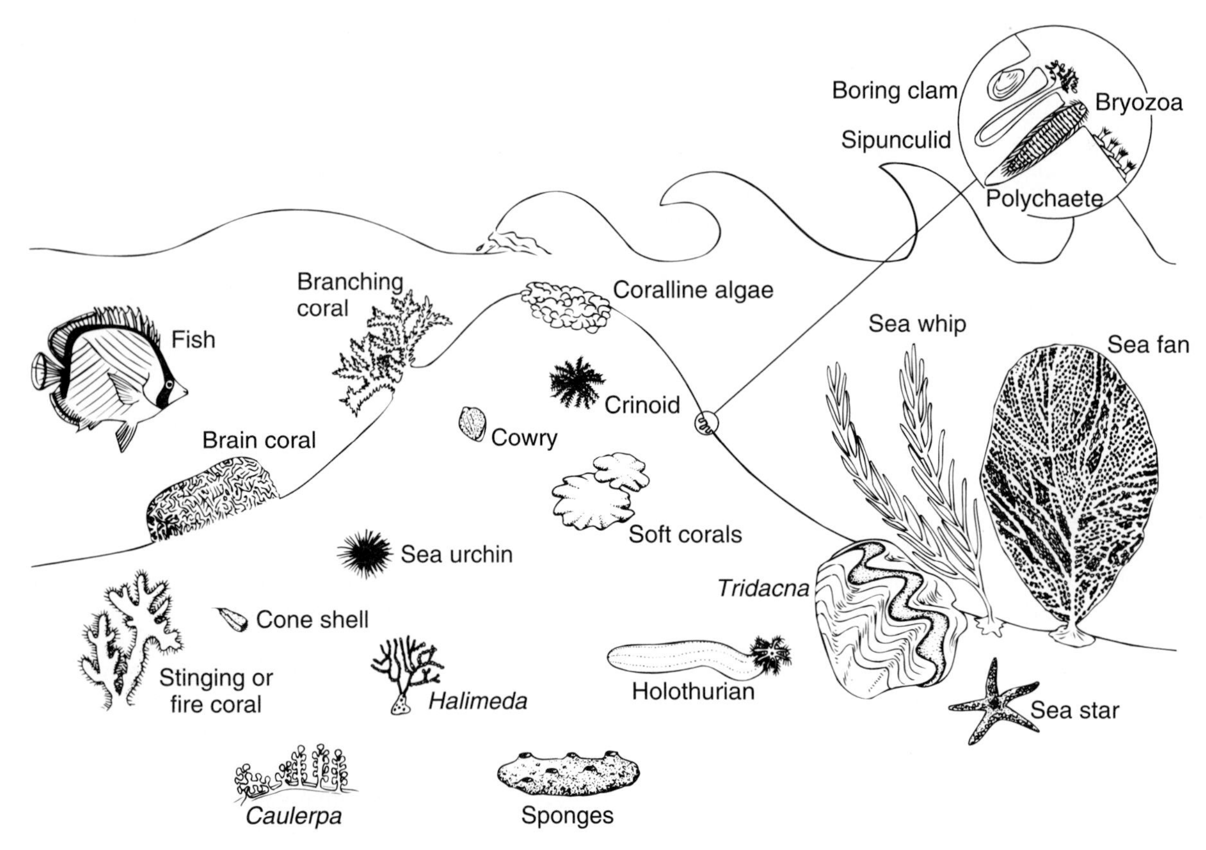

FIGURE 9.11 Some of the dominant and conspicuous components of a coral reef.

Composition of Reefs

Although the corals are the major organisms that form the basic reef structure, there is a bewildering array of other organisms associated with reefs, such that these areas are perhaps the most diverse and species-rich areas that exist in the marine environment today. Members of practically all phyla and classes may be found on coral reefs. It is not practical to mention here all of these forms, nor do we even know the roles played by many of them. Before considering the ecological structure of reefs, however, it is worth pointing out some of the more important and abundant groups.

In addition to the stony corals (order Scleractinia), certain other cnidarians contribute strongly to the reef. These include those anthozoan relatives of the corals called gorgonians (order Gorgonacea), commonly called sea fans and sea whips (Figure 9.11; see also Figure 9.3). These organisms have internal skeletons consisting of an axial, flexible protein rod and spicules. They are particularly conspicuous on the Atlantic reefs. The soft corals (order Alcyonacea) are very common on Indo-Pacific reefs and may be more abundant than the stony corals in some areas of the reef. They are often rare on Atlantic reefs. The hydrocoral *Millepora* (order Hydrocorallina), often called "fire coral" for its powerful nematocysts, is a conspicuous member of Atlantic reefs.

The **coralline algae** are an extremely important group in constructing and maintaining reefs. These red algae precipitate $CaCO_3$ as do the corals, but the algae are encrusting, spreading out in thin layers over the reef, literally cementing the various pieces together. By this action, the various individual calcium carbonate pieces are welded together to form a strong bulwark that resists wave destruction. These plants form the **algal ridge** on reefs, which is the most rapidly calcifying zone on the reef. Still other green algae are nonencrusting, grow erect, and secrete calcium carbonate but are not coralline. Together with the corals, the coralline and calcareous green algae make up the three dominant calcium carbonate-secreting organisms of the reef. Much of the sand in coral reef systems is derived from breakup of these erect algae, usually of the green alga genus *Hal-*

imeda. Other free-living, noncoralline algae are inconspicuous on reefs, but a significant algal flora also exists just below the surface layers of calcium carbonate in the coral colonies themselves.

Significant contributions to the calcium carbonate deposits on reefs are also made by mollusks of various types. The most conspicuous and important of these are the various giant clams (*Tridacna*, *Hippopus*), which Salvat and Richard (1985) have recorded occurring in numbers as great as 60 per square meter in some atoll lagoons in the Tuamotu Archipelago. Generally, McMichael (1974) records their density at 1 per square meter or less (Figure 9.12). Giant clams are absent from the Atlantic reefs. Various small gastropods are also abundant on reefs, but they are usually inconspicuous due to their size or because they are hidden.

Echinoderms, primarily sea urchins, sea cucumbers, starfish, and feather stars (crinoids), are another abundant and conspicuous group on reefs, but their role in reef ecosystems (except for the sea star *Acanthaster* and certain urchins) is incompletely understood (Figures 9.13 and 9.14).

Although various crustaceans and polychaete worms are very abundant on reefs, they are often inconspicuous, and little is known about their functions in the reef ecosystem. Various sponges are common on reefs, but they seem to play a minor role in modern reef construction. The only sponges contributing to the construction are the sclerosponges, and they are restricted to deeper slopes and in caves or overhangs, where coral cover and calcification are low, according to Wilkinson (1983). Siliceous sponges (Demospongiae) may, however, be important in holding coral and rubble together, preventing loss from the reef until it can be fused together by coralline algae. Wilkinson (1983) has also demonstrated that certain of the most common barrier reef sponges have cyanobacteria symbionts responsible for considerable primary productivity.

The final large group, which is very conspicuous on reefs, is the various fishes (Figures 9.15–9.18). The number of fish species present on coral reefs can number in the hundreds or occasionally thousands and

FIGURE 9.12 A giant clam, *Tridacna maxima*, on a Palau coral reef. The patterned mantle tissue is expanded. The mantle tissue contains the symbiotic zooxanthellae that furnish the clam with a portion of its nutrition. (Photo courtesy of Drs. Lovell and Libby Langstroth.)

FIGURE 9.13 The starfish *Fromia monilis* on a soft coral on the Great Barrier Reef. (Photo courtesy of Dr. Diane Nelson.)

FIGURE 9.14 A crinoid of the genus *Comanthus* on a Great Barrier coral reef. (Photo courtesy of Dr. Diane Nelson.)

FIGURE 9.15 Moorish idols, *Zanclus canescens*, on a coral reef in Palau. (Photo courtesy of Drs. Lovell and Libby Langstroth.)

FIGURE 9.16 A hawkfish swimming among the branches of a coral reef in Palau. (Photo courtesy of Drs. Lovell and Libby Langstroth.)

FIGURE 9.17 Queen angelfish, *Holocanthus ciliaris*, on a Caribbean coral reef. (Photo courtesy of Dr. Diane Nelson.)

FIGURE 9.18 A lionfish, *Pterios volitans*, on a Pacific coral reef. (Photo courtesy of Drs. Lovell and Libby Langstroth.)

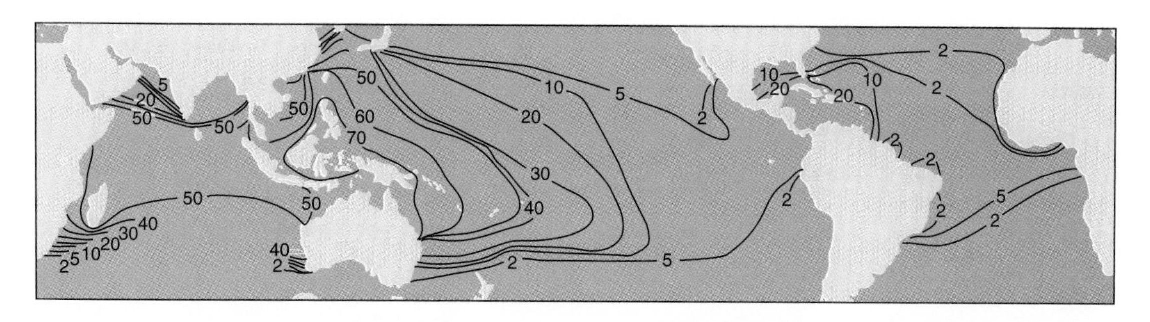

FIGURE 9.19 The number of genera of reef corals occurring in various tropical regions (After "Distribution of Reef-Building Corals" J. E. N. Veron, *Oceanus*, Vol. 29, No. 2, p. 27, 1986. Copyright © 1986 Woods Hole Oceanographic Institution.)

makes the coral reef one of the richest environments for fish on earth. The fish contribute considerably to the biological structure and integrity of the reefs.

Bacteria are also very abundant on reefs and are responsible for part of the high productivity of the reef, as well as the decomposition and quick cycling of organic matter.

Coral Distribution and Reef Zonation

The largest number of reef coral species and genera occur in an area of the Indo-Pacific that includes the Philippine Islands, the Indonesian Archipelago, New Guinea, and northern Australia (Figure 9.19). Within this area, Veron (1986) lists 70 genera and several hundred species. Considering the whole Indo-Pacific region, Veron (1986) lists 88 genera and 500 species. The number of genera and species declines in all directions away from this central diversity area, but most of the central Indo-Pacific has reefs with 20–40 genera of corals. This contrasts with the Atlantic Ocean, where Goreau and Wells (1967) report only 36 genera and 62 species. A dominant reef-building genus in both oceans, *Acropora*, has as many as 200 species in the Indo-Pacific but only three in the Atlantic. Some important reef-building genera of the Pacific, such as *Pocillopora*, *Pavona*, and *Goniopora*, are absent from the Atlantic. Thus, the species richness of Atlantic reefs is much less than that of Pacific reefs. Whether in the Atlantic or Pacific, as one moves away from the centers of the distribution, the number of coral species declines and is dramatically reduced at the edges of the coral reef zone. For example, the northernmost Pacific reefs, at Midway Island in the Hawaiian chain, have but seven genera of corals according to Grigg (1983). Similarly, in the Atlantic the most northern reefs in Bermuda have but 13 coral genera while those in Jamaica have 24 genera.

Coral reefs are large and complex associations of organisms that have a number of different habitat types all present in the same system. For example, there are areas of hard substrate on which many sessile organisms attach and that are analogous to rocky shores; at the same time, there are areas of sand that require a different set of adaptations, much as we saw for soft bottoms (see Chapter 5). Similarly, there are areas of heavy wave action and strong currents and areas of virtual calm, where water movement is minimal. Still other areas may have a lush growth of calcareous green algae (*Halimeda*) mimicking seagrass beds. In other areas, we may have mangroves or seagrass beds. All of these habitat types are found in a relatively small area. Therefore, one of the reasons for the great diversity of life in coral reefs is the great diversity of habitats. As might be expected, these different habitat types, as well as others not mentioned here, vary in their extent or presence on reefs in different geographic areas; hence, it is difficult to describe a typical zonation pattern for reefs that will be generally applicable. On all reefs, gradients in water motion (waves, currents) and light penetration (depth, turbidity) are the two most significant physical factors determining zonation. When these are coupled with the various biological interactions (competition, predation, grazing), the result is the various zonal patterns. Broad, large-scale zonation patterns are generally the result of physical factors. Small-scale "patchiness" is usually the result of biological interaction. We discuss first the broad-scale zonation patterns imposed by physical factors. Biological factors are taken up subsequently. The pattern of zonation, though complex, is most consistent for atolls, and it can be used as a general guide.

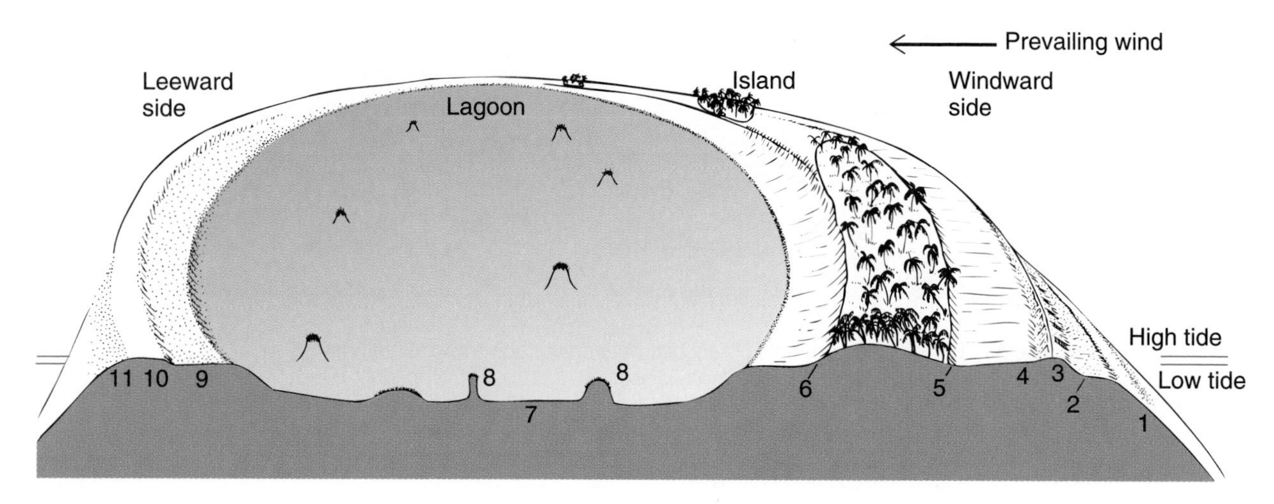

FIGURE 9.20 Diagrammatic cross section of a typical atoll: (1) outer seaward slope; (2) windward reef margin with spur and buttress zone; (3) algal ridge; (4) the reef flat; (5) seaward beach of the island; (6) lagoon beach of the island; (7) lagoon floor; (8) lagoon reefs; (9) leeward reef flat; (10) leeward reef margin; (11) leeward reef slope. (Modified from "The Biology of Coral Reefs" C. M. Yonge, 1963, *Adv. Mar. Biol*, Vol. 1, pp. 209–260. Copyright © 1963 Academic Press, Ltd. Reprinted by permission.)

Reef development and zonation on atolls are strongly influenced by wave action, a product of the prevailing wind. Therefore, even though these reefs are essentially horseshoe shaped, a cross section shows different zonation, depending on whether the reefs face the prevailing wind and waves or lie in the lee of them. Figure 9.20 is a diagrammatic cross section of a typical atoll and indicates the major features of zonation.

Approaching an atoll from the sea on the windward side, the first zone on a reef is the **outer seaward slope**, where living coral becomes abundant at a depth of about 50 m. Corals in this area are few and often delicate. At about 15 m, there is often a terrace or step in the otherwise steep outer slope; from here to the surface, coral growth is lush. This is also the area where wave action is the most severe; hence, we know little of this area. At the surface of the water is the **windward reef margin,** which is characterized by the development of coral-algal spurs or buttresses extending into the waves and down to the terrace. These spurs or buttresses alternate irregularly with deep channels or grooves into which water surges onto the reef and through which water and associated debris leave the reef (Figure 9.21). This is the so-called **spur and**

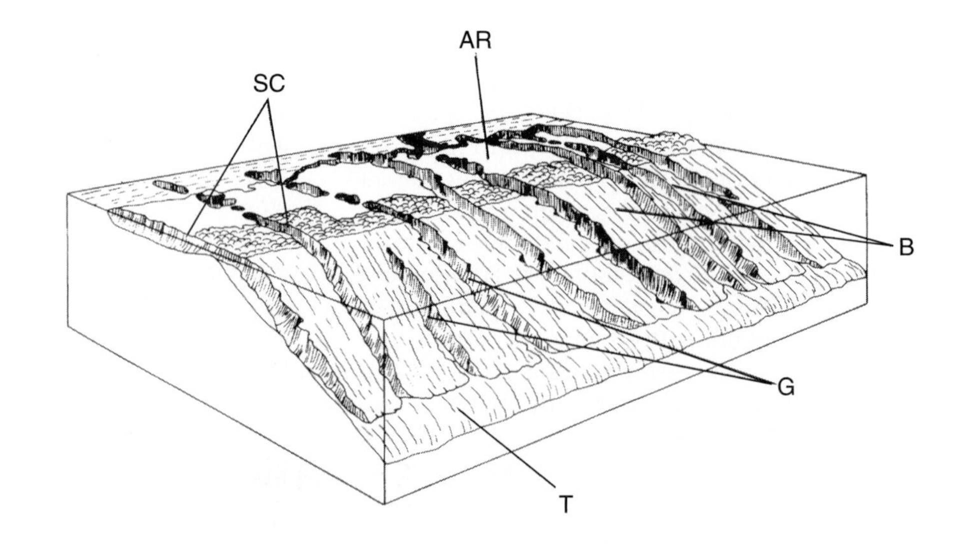

FIGURE 9.21 Generalized sketch of the spur and buttress zone of a windward reef. (AR) Algal ridge; (B) buttresses or spurs; (G) grooves; (SC) surge channels; (T) 18-m terrace. (After "The Biology of Coral Reefs" C. M. Yonge, 1963, *Adv. Mar. Biol*, Vol. 1, pp. 209–260. Copyright © 1963 Academic Press, Ltd. Reprinted by permission.)

FIGURE 9.22 A micro-atoll on the Great Barrier Reef. (Photo by the author.)

groove or **buttress zone.** The sides of the grooves or channels and the spurs themselves support a lush growth of the dominant reef-forming corals, such as *Acropora*. Because environmental conditions are optimal, it is here that coral reef accretion occurs most rapidly. Immediately behind the buttress zone is a smooth, coral-free ridge of encrusting coralline algae called the **algal ridge**. This ridge receives the full impact of the wave energy and, therefore, is virtually free of any organisms except the encrusting coralline algae. The coralline algae present are usually of several genera, including *Porolithon*, *Hydrolithon*, and *Lithothamnion* or related genera, hence, the origin of the alternative name for this zone, the **lithothamnion ridge**. Algal ridges are responsible for the asymmetry characteristic of many atolls. Typically the windward side exhibits the widest islands and has the algal ridge. There is a correlation between the development of an algal ridge and the degree of fetch and wind duration, so that many Pacific reefs have them while many Caribbean reefs do not. The grooves of the buttress zone, often called **surge channels**, penetrate behind the algal ridge as deep channels or often as blow holes where the channel has been roofed over by algal or coral growth, leaving only one or a few openings on the inside of the ridge through which water may enter or leave. These blow holes may offer a spectacular display under certain wave conditions, as the water is forced into the tubes at the outer slope of the reef and explodes as a geyser on the inside. This whole area, algal ridge and buttress zone, is extremely dangerous. Wave action is violent, and a step into a blow hole or surge channel could mean disaster.

Behind the algal ridge is a very shallow **reef flat** zone. This is a complex area with strong gradients in such environmental factors as temperature, turbidity, and exposure moving from the algal ridge to the shore of the atoll island. These gradients, coupled with differences in depth and substrate type (coral rock, sand), provide many habitats that subdivide this zone. Near the algal ridge, conditions for coral growth are good, and a variety of corals is found. This is another area where active accumulation of reef material occurs. Conditions for coral growth depart from optimal as one approaches the island, and corals become less and less abundant. Micro-atolls often develop in deeper areas of the middle reef flat. They are larger colonies of a massive coral, such as some species of *Porites*, in which the top of the colony has been killed by exposure to air or excess sedimentation. This is followed by erosion of the skeleton, so it becomes a small basin filled with water in which other organisms may grow (Figure 9.22). These micro-atolls support a variety of organisms. Also present in this reef flat area are the giant clams (*Tridacna*, *Hippopus*).

The reef flat zone ends at the **seaward beach** of the atoll island. On the lagoon side of the island is a narrow beach, followed by the quiet waters of the lagoon. There are two major zones in the lagoon, the **lagoon reefs** and the **lagoon floor**. Lagoon reefs are found around the perimeter of the lagoon and also exist as patch reefs or pinnacles rising from the lagoon floor to the surface. The depth of the lagoon is usually less than 50 m and, therefore, within the depth range of coral growth. Conditions within the lagoon are not as good for coral growth as on the windward side, because wave action and circulation are not as great and sedimentation is greater. As a result, coral growth is often restricted to areas where conditions are better, leading to the existence of patch reefs and pinnacles and leaving large areas of the lagoon floor free of coral growth. Sedimentation leads to large areas of sand, and the lagoon floor may have extensive beds of seagrasses (*Thalassia*, *Cymodocea*) or of green algae, such as *Caulerpa* and *Halimeda*.

The leeward side of the lagoon is bordered by the **leeward reef flat**. It is generally narrower than the seaward reef flat and separated from the lagoon reefs by a barren boulder zone or by an area of poor coral growth, dominated by isolated colonies of *Porites*. The seaward margin of this flat is similar to the windward flat. At the leeward reef margin, the algal ridge is usually poorly developed or absent, and there are no surge channels or buttresses. At the seaward edge, extending down to 15–20 m, coral growth is lush and diverse, dominated by the branching arms of *Acropora*. In addition to the massive corals seen on the seaward reef margin, numerous branching species are found here, due to the absence of violent wave action. Below this, the leeward reef slope is similar to the outer seaward slope.

Fringing reefs and barrier reefs can be considered truncated portions of the atoll zonation scheme. Fringing reefs and barrier reefs on the windward sides of islands have a zonation like that of the windward side of atolls, and those on the leeward side are similar to the leeward reef of atolls. The major difference between the two is that barrier reefs, especially large ones such as the Great Barrier Reef, may have a very extensive lagoon between the reef flat and the shore of the continent or island, and that lagoon may develop additional coral reefs.

Atlantic and Indo-Pacific Reefs

As we have already indicated, there are differences between the Atlantic and Indo-Pacific reefs. We will now examine them in detail. They include not only structure and zonation features, but presence, absence, and abundance of various fauna and flora.

The physical conditions under which reefs grow in both the Atlantic and Pacific are quite similar with respect to temperature, salinity, and turbidity, but the number of coral genera and species is very different between the two areas as described in the preceding section. This reduction in number of species between Pacific and Atlantic reefs is not limited to corals. It also occurs among most of the reef components, including mollusks, fishes, and crustaceans. For example, there are perhaps 5,000 species of mollusks on Indo-Pacific reefs, but there are only 1,200 on the Atlantic reefs. Similarly, there are 2,000 species of Indo-Pacific reef fish compared with 600 Atlantic reef fish. Atlantic reefs commonly have large numbers of sea fans and whips (gorgonians), whereas these are much less in evidence in Indo-Pacific reefs. Indo-Pacific reefs have large numbers of soft or alcyonacean coral, such as *Sarcophyton* and *Lobophyton*, whereas these are inconspicuous in the Atlantic. Algal ridges and coralline algae were long thought to be of minor significance on Atlantic reefs, but work by Adey et al. (1977) and Steneck and Adey (1976) now suggests that they are important. Atlantic reefs also lack the giant clams (*Tridacna*, *Hippopus*) and the octocorals *Heliopora* and *Tubipora*. There are no anemone fish (*Amphiprion*) in the Atlantic, though the Atlantic does have fishes that associate with anemones. Similarly, crabs and shrimp that guard certain corals against predators are absent from the Atlantic, as is the coral-eating starfish *Acanthaster planci*.

Dominance is also different in the two regions. Atlantic reefs are dominated by *Acropora palmata* (elkhorn coral) and *Millepora complanata*, a hydrocoral. *Millepora* does not dominate any Indo-Pacific reef. The amount of substrate covered by corals differs in the two oceans. Rhythms differ in that most Atlantic corals are nocturnal, but in the Pacific many, such as *Porites*, *Goniopora*, and *Favia*, are diurnal.

There are also differences in relative abundances of some of the noncoral species. Wilkinson (1987b) has noted that sponge biomass is 2–10 times as great on Caribbean reefs as on the Great Barrier Reef. Few, if any, Caribbean sponges have the phototrophic symbionts present in many Pacific sponges. Similarly, Sammarco (1987) has noted that echinoids are the major grazers in the Atlantic, whereas fishes fill that role in the Pacific.

Not only are there differences in species composition and relative abundances between these two areas, there are also differences in age and reef morphology (construction). Atlantic reefs usually rest on shallow banks or platforms, probably the result of erosion by wave action during the Pleistocene period when sea level was much lower. Borings through Atlantic reefs, therefore, do not show the great depth of reef limestone that indicates the great age of the Pacific atoll reefs. Why this great discrepancy in age? This is not known for certain, but two explanations may be offered. First, the Atlantic is a geologically more recent ocean; hence, there has been less time for reefs to develop. Second, in the ice age, the Atlantic may have become too cold for reefs to survive; hence, they died and were replaced only when the seas warmed following the retreat of the glaciers. Glynn (1973) states that Atlantic reefs, though of more recent origin, do develop massive structural frameworks similar to Pacific reefs and show similar vigorous growth.

Zonation is somewhat different. Atlantic reefs have the buttress zone deeper than the Pacific reefs, and active coral growth extends to 100 m in the Atlantic but only to 60 m in the Pacific. Atoll islands are often absent in the Atlantic, but reef flats and shallow lagoons are common. Also, the areal coverage by corals differs. In the Atlantic, corals cover less than 60% of the substrate as opposed to 80–90% in the Pacific.

Finally, there are some subtle differences in the life histories of organisms and ecological processes. For example, Sammarco (1987) reports that coral recruitment in the Caribbean, at least for the dominant genus *Acropora*, is primarily asexual via branch breaking. In the Pacific, recruitment is by sexually reproduced planulae. Similarly, Vermeij (1987) reports that Pacific invertebrates are more specialized, particularly in dealing with predators and symbiotic relationships, than their Atlantic relatives. Birkeland (1989) also notes that no asteroid species are influential predators of corals in the Atlantic.

Productivity

Coral reefs are truly oases in a watery desert. As we noted in Chapter 2, tropical marine waters are extremely poor in nutrients; hence, they have very low productivity (see p. 68). Coral reefs, however, abound with life, and all studies done on them have indicated a very high productivity. Most studies—such as those by Kohn and Helfrich (1957), Odum and Odum (1955), Johannes et al. (1972), Smith and Marsh (1973), and Atkinson (1992)—have given gross primary productivity estimates of 1,500–5,000 g C/m^2/year. This is in contrast to a gross primary productivity in the open tropical oceans of 18–50 g C/m^2/year. How can coral reefs maintain this production in such nutrient-poor areas? This is less hard to explain for reefs bordering high islands or continents, where much nutrient material might become available through runoff from the land, than it is for atolls isolated far from land. In the latter case, we do not yet know all the reasons for the high productivity. While there is little or no disagreement among scientists that coral reefs are extremely productive, there is still no consensus about the underlying factors that sustain this high productivity. The problem revolves mainly around the necessary source or sources of the limiting nutrients, nitrogen and phosphorous. Some researchers have suggested that the reefs need a sustained external source of these nutrients and have invoked such processes as upwelling in the water column, and groundwater leaching. More recently, Rougerie et al. (1992) suggested that geothermal endo-upwelling of nutrients from the interstitial water in the reef limestone maintained sufficient fluxes of these nutrients to support productivity. Other researchers dispute the need for external sources of nutrients and suggest that tight recycling by biological processes can maintain high enough fluxes of the nutrients to sustain the productivity. Still others, such as Atkinson (1992), have suggested that a high flow rate of the tropical oligotrophic water across the reef is sufficient to account for the high productivity without invoking outside sources or even recycling.

Sorokin (1990) has invoked a combination of factors that would explain the maintenance of high productivity. In this scenario, nitrogen and phosphorous are assumed not to be limiting for several reasons. First, nitrogen is assumed not to be as limiting as phosphorous, since it can be fixed from atmospheric nitrogen by the nitrogen-fixing cyanobacteria. Second, both nitrogen and phosphorous exist in the surrounding waters as dissolved substances and particulate organic matter (POM). The amount of nitrogen and phosphorous in the POM is much greater than that in the dissolved inorganic matter. This organic matter is removed from the water passing over the reef by the large numbers of bacteria on

TABLE 9.1

Primary Productivities of Component Autotroph Types on Coral Reefs and the Distribution of These Communities on Reefs

Autotroph Type	*Range of Productivity* ($g\ C/m^2/day$)	*Approximate Area/ Cover on Reefs* %
Benthic algae	0.1–4	0.1–5
Turf algae	1–6	10–50
Zooxanthellae	0.6	10–50
Sand algae	0.1–0.5	10–50
Phytoplankton	0.1–0.5	10–50
Seagrasses	1–7	0–40

Source: From Larkham, 1983.

the reef, sequestering significant amounts of the limiting nutrients. The removal of nutrients is further aided by the biofiltering activity of such reef fauna as sponges, bivalve mollusks, polychaete worms, the corals themselves, and possibly the water column zooplankton. These organisms remove not only the POM, but the bacterioplankton as well. In Sorokin's view, this large biomass of reef organisms is an extremely effective biofilter that scavenges dissolved and particulate organic material in the oceanic offshore waters that pass onto the reef. In this way, the reef system derives a positive input of limiting nutrients from the surrounding oligotrophic ocean. A final source of nutrients is in the form of the accumulation of wind-driven cyanobacteria from the open ocean onto the reef. Tropical oceans often contain cyanobacteria, such as the genus *Trichodesmium*, in suspension, and when these masses are thrown on reefs or accumulate in the lagoon, their decomposition adds further supplies of nutrients. All these processes provide a positive balance of the potentially limiting nutrients, even when the surrounding waters are oligotrophic. The reef, thus, avoids nutrient limitation, and the result is high productivity.

In the absence of any consensus about nutrient fluxes, it is still possible to suggest other reasons that reefs are so productive. First, the amount of plant tissue present on a reef and capable of photosynthesis is very large, much larger than in a similar area of the open ocean. Photosynthetic organisms present on a reef include zooxanthellae, prokaryotic symbionts in sponges, crustose coralline algae, turf algae, large macroalgae, sand algae, seagrasses, and phytoplankton. Each of these components varies in productivity and in the amount of areal cover on reefs (Table 9.1), but together they form a significant biomass. According to Odum and Odum (1955), the biomass of these organisms is greater than the animal biomass, even if less conspicuous. Of these autotrophic components, it appears that the ubiquitous small turf algae are the most significant primary producers (Table 9.1). Reefs also act as a sink for any nutrients brought in from the outside. This ensures that nutrients remain within the reef system and are not lost to deeper offshore waters. It also means that any plankton from the open sea that impinges on the reef remains there, as do the nutrients brought with them.

Another factor that aids in productivity in coral reefs is the close coupling of photosynthetic organisms with either the animals or the matrix of the reef. The best example is the location of the zooxanthellae in the coral polyps. Placing the plants in the coral tissue means that the algae will not be washed from the reef. It also means that any nutrients produced by the coral (NO_3, PO_4) as a result of its metabolism will be available directly to the algae without cycling it outside into the water where it might be lost. The same is true of the cyanobacteria located in sponges. Similarly, most of the free-living algae on the reef contain calcium carbonate. They are too heavy to be moved easily off the reef, or they are fused into the reef structure by the coralline algae; hence, their nutrients remain on the reef to nourish additional generations. Rapid cycling is also aided by the large bacteria populations in coral reefs. These bacteria quickly break down dead material, making enclosed nutrients available. For all these reasons and probably for others yet to be discovered, coral atolls have very high productivity, as do fringing and barrier reefs. Given the controversy regarding the necessity and source of nutrients, however, it is not yet understood how such systems are maintained. Much of this explanation will come from a more complete understanding of how the components of the reef ecosystem interact. Of these components, the first and most significant are the corals themselves.

Biology of Hermatypic Corals

Hermatypic corals are the dominant group involved in reef formation and maintenance; therefore, a knowledge of certain biological aspects of the group is crucial to our understanding of the ecology of reefs.

Nutrition Reef corals have long been known to be carnivorous animals, similar in that respect to most other members of their phylum. They have tentacles studded with stinging capsules, nematocysts, used to sting and capture small plankton organisms. Since they live in colonies of hundreds or thousands of individuals and the number of colonies is also huge, the total area for feeding is enormous. In addition to the tentacles and their nematocysts, the outer epidermis of corals is ciliated and produces mucus. This ciliary mucous mechanism is generally used by corals to rid themselves of sediment settling on the surface. In some corals, it has been modified for feeding as well. Those species using mucous mechanisms to capture plankton tend to have smaller, shorter tentacles than those that feed directly on plankton by snaring them with their tentacles. Food organisms are detected by chemoreception.

Since corals exist in such numbers on reefs and the surrounding ocean is notoriously poor in plankton, how is it possible for these animals to obtain enough food? This question has puzzled generations of coral reef biologists, and the answer is now becoming clear. Rigorous plankton sampling just offshore on reefs by Johannes et al. (1970), Porter (1974), and others, both during the day and at night, has revealed an abundant plankton population. A large fraction of this plankton population, however, is not from the open ocean but is a population indigenous to the reef itself. This indigenous plankton population is composed primarily of demersal meroplankton, which spends the daylight hours on the bottom and rises into the waters over the reef at dusk. This may also explain why many corals are closed during the daylight hours, only expanding to feed during the hours of darkness. In spite of this discovery of enhanced abundances of plankton over the reefs, Johannes et al. (1970) have determined that the amount of plankton available to the corals is sufficient to satisfy only 5–10% of their total food requirements. Where, then, does the remainder of their food come from? Only one source seems possible: the zooxanthellae found in the coral tissues.

The role of zooxanthellae in coral biology has long puzzled biologists. Long ago, it was learned that corals were carnivores and could not digest plant tissue. Furthermore, early experimental studies showed that, if large tentacled corals were starved or kept in the dark, they would expel their zooxanthellae but could continue to live for a few weeks without them. This was possible because these corals could capture zooplankton and use them. These observations suggested that the zooxanthellae were really unnecessary to the coral. In 1969, however, Muscatine and Cernichiari used radioactive tracers to prove that organic compounds fixed by the zooxanthellae in photosynthesis are transferred to the coral, where they serve as food for the coral. Franzisket (1969) and Johannes (1974) also demonstrated that, if corals are not fed but kept in the light, they gain weight. This was further established in field studies by Wellington (1982b). This could happen only if the zooxanthellae furnished food to them. Whereas the role of zooxanthellae in nutrition seems proven, we do not know how important this role is for all coral species. Studies by Porter (1976) and Wellington (1982b) suggest that corals are primarily **phototrophic** (meaning dependent on zooxanthellae), but corals with large polyps and tentacles are less dependent on zooxanthellae for nutrition than species with smaller polyps.

Zooxanthellae are acquired by the coral either directly from the parent or indirectly from the environment. In the case of asexually reproducing corals, the zooxanthellae are directly transmitted during the division or fragmentation process. Sexually reproducing corals, however, may transmit the parent zooxanthellae directly to the eggs or brooded larvae; alternatively, the larvae or young coral may have to acquire the zooxanthellae from the environment after release from the parent. Which mechanism is employed is generally species specific.

Originally, it was thought that there was only a single species of zooxanthellae, but recent molecular taxonomic work has revealed 16 species. Zooxanthellae in corals grow rapidly and fix carbon at a rapid rate. This rapid growth continues until they reach a certain density. At this point, they stop reproducing and growth slows as they release carbon compounds to the coral. If at this point higher concentrations of N and P are added, the zooxanthellae once again begin to reproduce and reach even higher densities. This suggests that corals are nitrogen limited. Corals seem able to limit the growth and densities of zooxanthellae through some unknown "host factor," causing the zooxanthellae to release the organic molecules they use for food.

Growth and Calcification Although there has been much interest in the growth rate of both corals and coral reefs for many years, it is only in recent years that we have come to understand the growth of individual corals and reefs. The first requirement for active coral growth is light. As Connell (1973) has shown, if corals are shaded or denied access to light, they stop growing. If light is withheld long enough, they will die. This requirement for light is based on the needs of the zooxanthellae, and it suggests the other important role of zooxanthellae in the life of corals. Goreau (1961) found that zooxanthellae significantly increase the calcification rate of corals and the growth rate of the coral colonies. Exactly how the zooxanthellae increase this rate of skeletal accretion is not fully understood, but it seems that light enhances oxygen production, which, in turn, stimulates coral metabolism and leads to increased calcium carbonate deposition. This again indicates the important role of the zooxanthellae in the coral reef ecosystem, where rapid calcification is needed to maintain the reef against the various destructive forces acting against it.

The most direct way to measure coral growth is to record a particular dimension of the coral and then repeat the measurement after an appropriate time interval. Dimensions measured according to Buddemeir and Kinzie (1976) have included length of branch, diameter of the colony, volume, area of the colony, and dry weight. Other means of measuring coral growth have included monitoring the rate of calcification by radioisotope uptake. A rapid and simple way of recording coral growth has been employed by Connell (1973). It is to take photographs of an area of the reef and then measure the colonies later from the developed print (Figure 9.23).

Coral growth rates vary with species, colonial age, and location on the reef. Young, small colonies tend to grow more rapidly than older, larger colonies, and branched or foliaceous corals tend to grow more rapidly than massive (brain) corals. In growth, corals increase their diameters rapidly. *Acropora*, for example, is a branched genus; measurements of the Atlantic species reported in Davies (1983) suggest that the species could grow 2.5–26.6 cm/year in length. *Montastrea annularis*, a massive type, had a linear growth of 0.81–2.5 cm/year. Despite much recent work that has given us good estimates of growth rates in some species, the fact that there is growth throughout the life of the colony and that growth rates in the larger, older colonies are variable and difficult to estimate means we still do not understand the age structure of the colonies seen on a reef, nor do we know how long corals can live. What little information there is, primarily that of Connell (1973), suggests that most coral colonies on reefs are ten years of age or younger. However, some of the large massive corals, such as those forming micro-atolls, may be 100 years old or even older, suggesting that corals can live much longer than the ages generally attained by most colonies on the reef.

It is also difficult to translate the growth rates of individual species into growth of a whole reef system. As a result, most estimates of reef growth have been made by extrapolating from changes in reef topography over a period of years or from age of the reef as determined from geological data, such as thickness of reef limestone deposits. These estimates vary widely and also vary for reefs in different areas of the world. The range reported by Stoddard (1969) is 0.2–8 mm of upward growth per year.

The growth form of a given coral species may also vary, depending on its location on a reef. In contrast to the same species in shallow water, a coral species at greater depth is often thinner and more fragile, probably as a result of less calcification. Wave action forces branched species to have shorter and stubbier branches, and currents cause branching forms to have a definite alignment of the branches. This ability of coral species to have different skeletal shapes has resulted in much confusion among coral reef taxonomists, such that different growth forms of the same species have often been named as different species.

Sexual Maturity, Reproduction, and Recruitment Corals perform both sexual and asexual reproduction. Asexual reproduction is generally accomplished by a new individual budding off the parent, and continued budding is the mechanism of increasing the size of the colony but generally not for producing new colonies. Highsmith (1982), however, has reported fragmentation as a means of starting new colonies in some species. Jackson and Hughes (1985) have shown in their study site in the Caribbean Sea that over a three-year period 38 new colonies of the coral *Leptoseris cucullata* were started by fragmentation but only 16 by larval recruitment. Sexual reproduction results in a free-swimming planula larva, which initiates development of a new

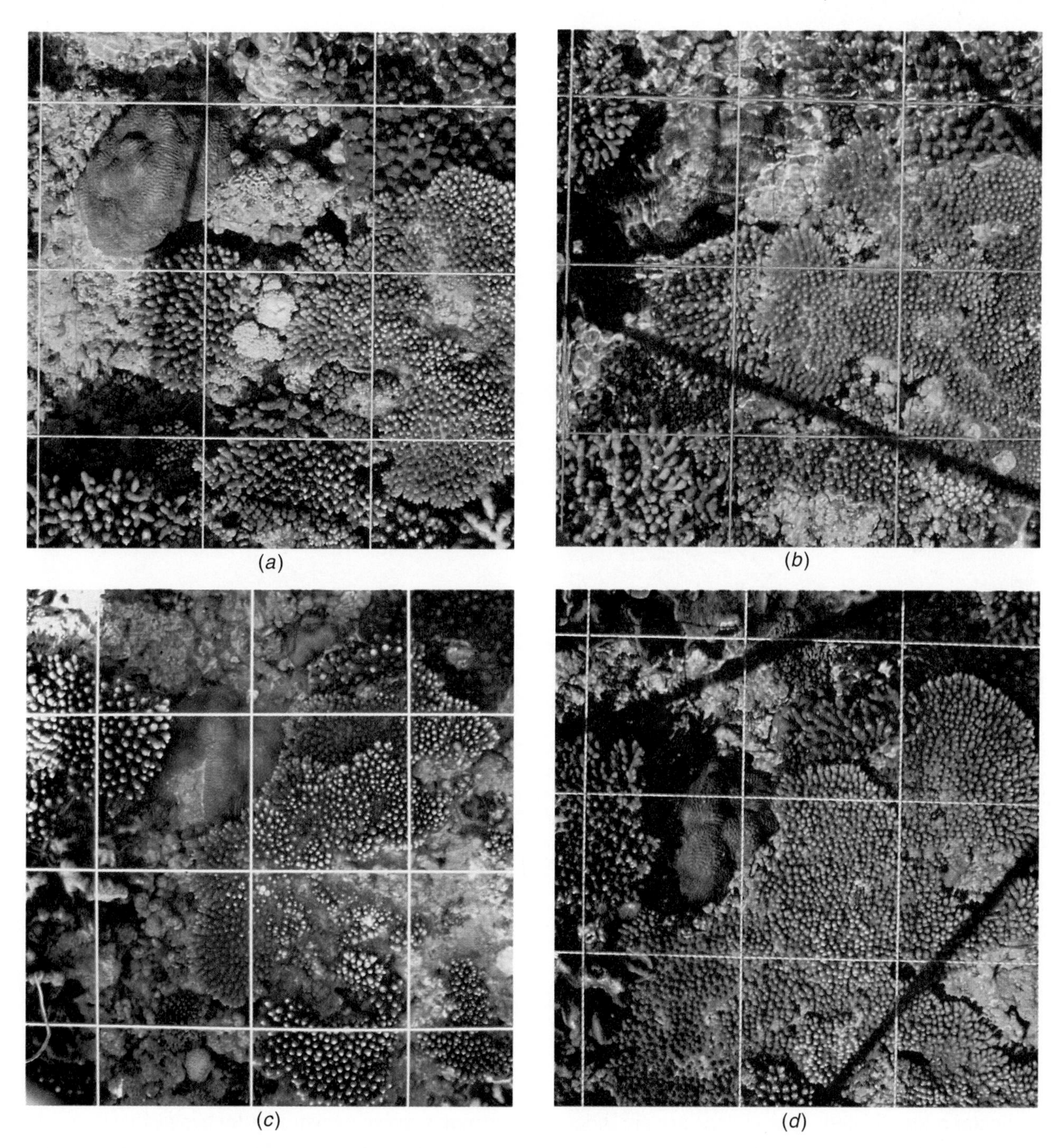

FIGURE 9.23 Photographs of the same areas of a 1-m quadrat showing the changes in growth and occurrence of coral colonies over an eight-year period on the Great Barrier Reef at Heron Island. For each photograph the square meter frame, divided into 20 × 20 cm squares, was positioned on permanent stakes (only a portion of the 20 × 20 squares are shown). Note the great changes that have occurred in growth and in presence of corals over short time intervals. (a) 1963. (b) 1965. (c) 1969. (d) 1971. (Photos courtesy of Dr. Joseph Connell, Dept. of Biological Sciences, University of California, Santa Barbara.)

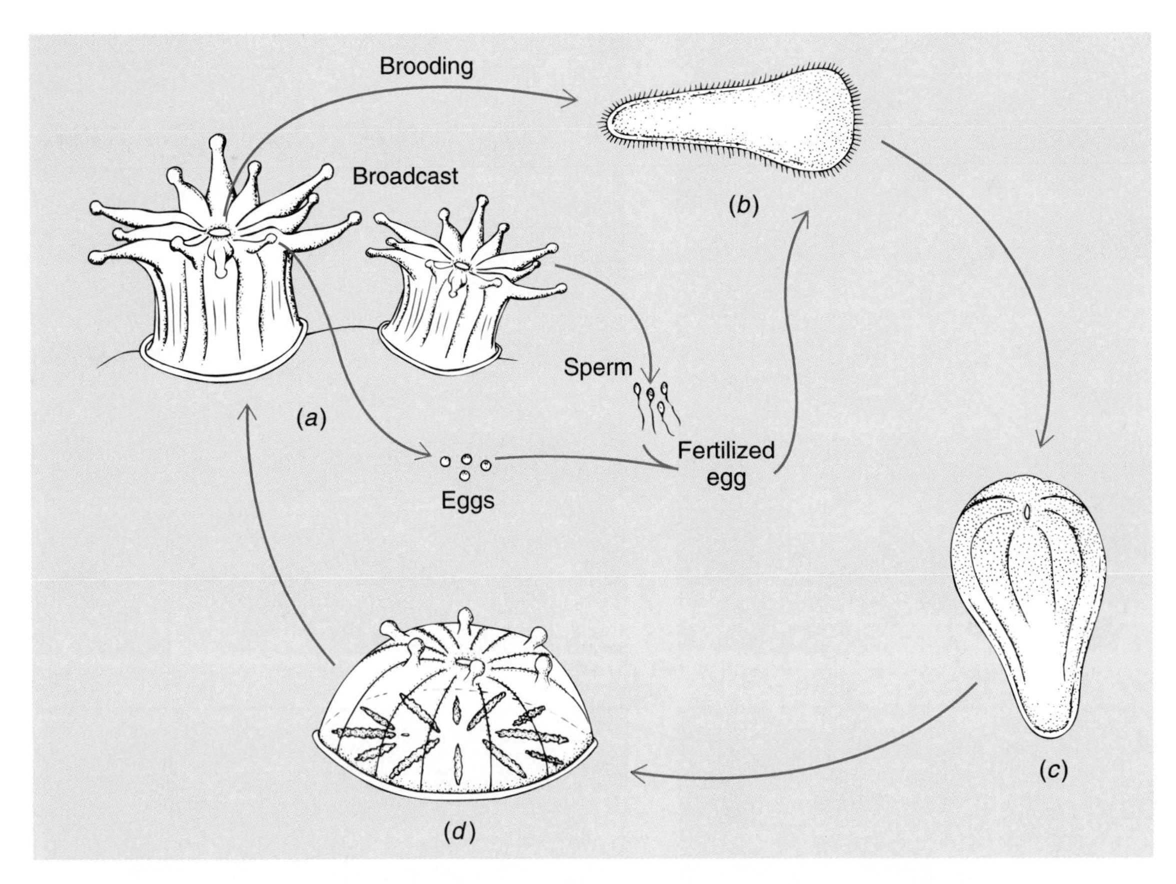

FIGURE 9.24 Reproduction of corals. (a) Adult polyp. (b) Planula larva. (c) Later planula with developing septa. (d) Young polyp after attachment. (Not to scale; planula and young polyp are much smaller than the adult.) (Modified from L. Hyman, 1940, *The Invertebrates, Vol. 1, Protozoa Through Ctenophora*, McGraw-Hill. Copyright © 1940; used with the permission of McGraw-Hill Book Company.)

colony when it settles down (Figure 9.24). It has now been demonstrated, however, that some planulae are produced asexually, so that asexual reproduction may be more important than first believed in the establishment of new colonies. This may be especially true if fragmentation is widespread.

From limited data, it appears that most corals reach sexual maturity at seven to ten years of age. Corals are primarily hermaphroditic; a few are dioecious (gonochoric).

The time at which corals form gametes depends on the size and age of the parent colony. Fragmentation or partial destruction of a colony can affect its fecundity. Most corals are broadcast spawners, releasing eggs and sperm into seawater, but a smaller number are brooders, retaining fertilized eggs in the gastrovascular cavity until development has proceeded to the planula stage (Figure 9.24). Spawning is usually seasonal, with a single brief annual event. Broadcast spawners are regulated by the lunar cycle, and there are some interesting regional differences. According to Szmant-Froelich (1985), many species on the Great Barrier Reef spawn in late spring to early summer, after the October and November lunar cycles, and are often synchronous. In the Caribbean, the major spawning period is longer, lasting from July through September, and is nonsynchronous. Among the brooders, release of planulae is also coordinated by the lunar cycle. The cue, according to Szmant-Froelich (1985), is nocturnal illumination.

The planulae are released and swim in the open water for an undetermined time, generally for a few days to a week or more, before settling down and starting a new colony. There is still disagreement among

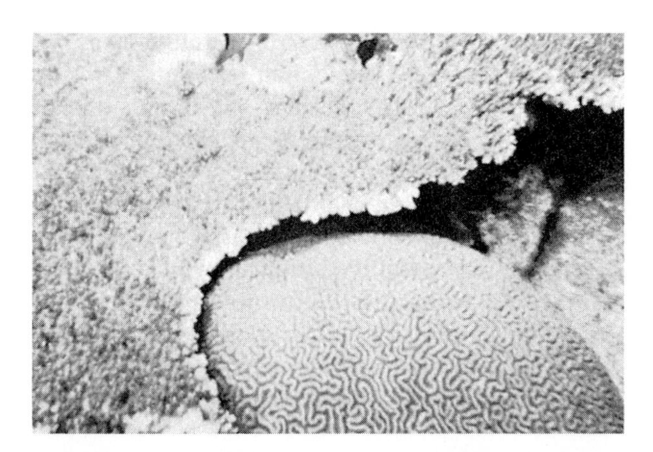

FIGURE 9.25 Photo showing the killing of portions of a tabular *Acropora* colony (left) by the brain coral *Ctenella chagius* (right), thereby preventing overgrowth by the faster-growing *Acropora*. (Photo from C. R. C. Sheppard, 1979, Interspecific aggression between reef corals with reference to their distribution. *Mar. Ecol. Progr. Ser.* 1:237–247.)

scientists about the amount of time planula larvae can remain in the plankton and, therefore, about the dispersal range of a given coral species. Since adult corals are fixed in place, the planula larvae are the only means of dispersal of the various coral species.

The rate of recruitment of new coral colonies to the reef system is variable in time and in space. Sammarco (1985) found that in both the Great Barrier Reef and the Caribbean, where *Acropora* was the dominant coral genus, recruits of this genus were rare in the Caribbean but among the most abundant in the Great Barrier Reef. There appears, therefore, to be a major difference between the two geographic areas in that there is marked diminishing of recruitment success by broadcast species in the Caribbean. This could, as Sammarco (1985) notes, have implications for reef recovery from catastrophic events. In the Great Barrier Reef system, the higher number of recruits and seemingly higher survival rate would lead to faster reef recovery, whereas in the Caribbean, recovery would be much slower. A study by Connell (1973) has indicated that the rate of recruitment on the Great Barrier Reef was 0–13 new colonies/m^2/year. The more common coral species were the most abundant of the new colonizers, but there were great differences in species that settled among different areas. It also appeared that mortality was high among the newly settled colonies. According to Babcock (1985), growth rates of newly settled broadcast corals were slower than those of juveniles resulting from brooding corals.

Species Interactions and Ecology of Reefs

At this point, we have covered the basic biology of the dominant organisms of the reefs—namely, the scleractinian corals—and have outlined the general zonation pattern of reefs and the physical conditions acting on them. Reefs, however, are much more than simple static systems. They are living systems that increase or decrease in size as a result of the complex interactions among various biological and physical forces. We can now consider how these interacting factors produce the reef and ensure its persistence. Since coral reefs are among the most complex systems in the marine environment and since they have not received as much detailed ecological study as rocky shores, our knowledge of this aspect of the coral reef system is not as complete as we would like. However, it is still possible to suggest how some factors interact to maintain the system. The general picture emerging is that the interactions that account for the reef are similar to those responsible for the patterns we observed on rocky shores.

Competition One of the most obvious features of a coral reef in an area of active coral growth is that there is virtually no open space. The primary space is almost completely covered by the dominant corals or an algal turf (all the small, short macroalgae growing a few centimeters from the substrate). Since all space is occupied and since corals must have access to light to survive, we might expect competition among corals for space and light.

Such competition does occur among coral colonies. Upright, branching coral colonies grow more rapidly than encrusting or massive corals, and they often extend themselves up and over the encrusting forms, shutting them off from the light. Where this occurs, the part of the encrusting colony in the shade dies. This is **exploitative competition**. Since these branching corals can grow faster and shade out the slower-growing massive and encrusting corals, how do these slow-growing forms manage to persist? There are a number of reasons, but the most intriguing was the discovery by Lang (1973) that the slow-growing species can extend digestive filaments from their gastrovascular cavities, which kill the tissues of adjacent, competing coral species (Figure 9.25). This kind of interaction, called **interference**

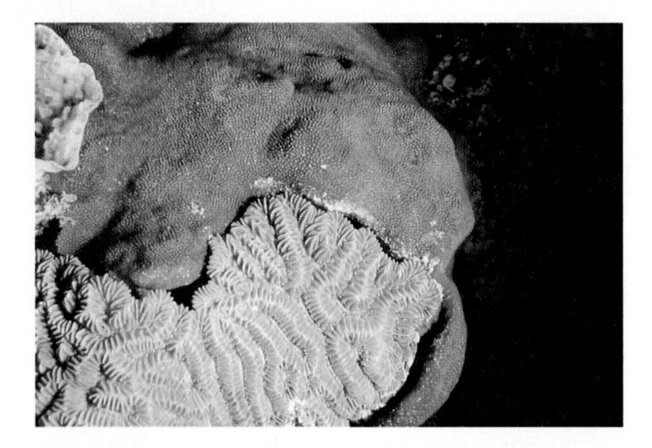

FIGURE 9.26 Competitive interaction between two coral species of the genus *Diploria*. (Photo courtesy of Dr. Ron Shimek.)

competition, is a means by which slower-growing but more aggressive coral species are capable of maintaining space while competing with more rapidly growing neighbors (Figure 9.26).

Lang found that corals capable of utilizing their digestive filaments this way could be arranged in specific order so that when different species were paired, the winner could be predicted. She suggested that this aggressive pecking order was a means by which diversity may be maintained locally on coral reefs. Subsequent testing has, however, revealed that this may be a far more complicated interaction than first believed. For example, in the eastern Pacific, Glynn (1974), in laboratory experiments, found that the slow-growing, massive coral *Pavona* dominated the faster-growing, foliaceous coral *Pocillopora*. This was anticipated from Lang's studies. In the field, however, *Pocillopora* seemed to damage *Pavona* more often. Wellington (1980) showed that during the initial contact between the two genera, *Pavona* did extend its mesenterial filaments and partially kill the *Pocillopora*. After 7–60 days, however, the undamaged tentacles of *Pocillopora* polyps at the edge of the area killed by *Pavona* began to elongate to 30 times their normal length and changed their nematocyst batteries to be more powerful. These "sweeper tentacles" were passively carried by water currents into the *Pavona* colony, where they inflicted damage. This retaliation permitted the *Pocillopora* to repair the original damage and continue growth toward the *Pavona*. The outcome is that *Pocillopora* wins the direct conflict. A similar system of sweeper tentacles has been reported by Richardson et al. (1979) for two corals in the Caribbean, suggesting that this phenomenon is more widespread. The competitive superiority of *Pocillopora* helps explain why it is the dominant coral on shallow reefs of the eastern Pacific.

At the shallowest depths on coral reefs, cover by coral or other organisms may approach 100%, according to Huston (1985). Under such crowded conditions, the rapidly growing, branching corals reduce or eliminate the slower-growing, massive corals by overtopping them, thereby reducing their access to light and water movement. Such exploitative competition may explain the dominance of *Pocillopora* at the shallowest depths in the eastern Pacific and *Acropora* in the Caribbean. If the slow-growing corals cannot gain access to full sunlight, how do they survive? One mechanism, according to Huston (1985), is that they are more shade tolerant and grow at greater depths, where reduced competition under the low-light conditions permits their survival. This, in turn, leads to increased species diversity with depth. We see this not only in corals but in other members of the reef fauna as well.

As more studies were done on competitive interactions among coral species, it became apparent that other mechanisms were employed by corals in competitive interactions among the species along with the mesenterial filaments and sweeper tentacles. Lang and Chornesky (1990) reviewed these and expanded the number of categories of competitive mechanisms used by corals in interactions to eight. These were mesenterial filaments, sweeper tentacles, mucous secretions, overgrowth, overtopping, sweeper polyps, water-borne chemicals, and histoincompatibility. Furthermore, the interspecific competition is made even more complex because it can also vary geographically and temporally and can be influenced by various other factors such as predation, disturbance, and changing environmental conditions. For example, a hierarchical ranking of coral species based on aggressive use of mesenterial filaments is different in Jamaica and Bermuda; and in the above *Pavona*-*Pocillopora* interaction, the hierarchical ranking by mesenterial filaments puts *Pavona* on the top, but if the ranking is by sweeper tentacles then *Pocillopora* is first ranked. These categories of competition may also describe the competition between corals and other space-occupying invertebrates on the reef. For example, Sammarco and Coll (1992) describe the inhibition of growth and tissue necrosis among scleractinian corals as a result of competitive encounters with soft corals, which seems to be mediated by water-borne chemicals. Similarly, sponges have been known to overgrow corals, and chemicals produced by the sponges have been implicated as the competitive mechanism permitting the sponges to displace the corals.

Competition between corals and other reef taxa, particularly various fast-growing algal species, is reduced by diverse grazing fishes and invertebrates, which is advantageous to the corals. Also, whereas algae are competitively superior and overgrow juvenile coral colonies in shallow water, Birkeland (1977) has shown that the corals can maintain their space in deeper waters. The importance of algal grazers to corals is shown by grazer exclusion experiments that demonstrate, according to Brock (1979), Hatcher (1983), and Sammarco (1980), that the algae are competitively dominant over corals.

Competitive interactions are not limited to those among corals or between corals and other reef organisms; there are also competitive interactions among other invertebrate taxa. For example, Wahle (1980) described a situation in which two hydrocorals of the genus *Millepora* directed growth toward adjacent gorgonians *Plexaura homomalia* and *Briareum asbestinum* and over a period of months completely overgrew the gorgonians. Similarly, Karlson (1980) has described the overgrowth of the zoanthid *Zoanthus solanderi* by another zoanthid, *Palythoa caribaeorum*, and an aggressive gorgonian, *Erythropodium caribaeorum*.

It seems clear now, as Karlson (1999) has pointed out, that various competitive interactions among corals and with other clonal and colonial sessile invertebrate taxa are very complex and involve multiple mechanisms. Furthermore, these interactions can be variable in time and space and may be modified by disturbances, geography, and predation such that the resultant community structure is difficult to predict. Finally, many of these competitive interactions can also result in standoffs, reversals, and even competitive coexistence as well as competitive exclusion. Thus, the degree to which competition between coral species, or between corals and other space-occupying invertebrates, controls community structure on reef is presently unknown. Furthermore, competition is only one mechanism contributing to coral reef diversity, structure, and zonation patterns.

Finally, there is one other set of organisms among which competition for space seems common. That group is the **cryptofauna**, the smaller colonial encrusting invertebrates, such as sponges, bryozoans, polychaetes, brachiopods, bivalves, ahermatypic corals, and tunicates, that occupy the undersides of coral heads, overhangs, cliffs, and caves. As Jackson (1977, 1979) has demonstrated, individual clones of these small, fast-growing invertebrates often undergo intense competition among themselves for space but also can overgrow various solitary invertebrate taxa. The overgrowth relationships among these cryptofaunal organisms form competitive networks rather than hierarchical sequences according to Jackson (1979), and they are influenced by other factors that insert a certain variability and unpredictability to the outcome of the competitive interactions. Due to indeterminate and rapid asexual growth, the cryptofauna have an advantage over the slower-growing ahermatypic corals. Their competitive abilities are offset, as Jackson (1977) has noted, by their increased susceptibility to predation and disturbance. One effect of the intense space competition on reefs is the tendency of various organisms to use the same space in some close association. This is commensalism. Commensal relationships are more common among reef organisms than elsewhere (see Chapter 10 for complete discussion).

Predation Although large numbers of species of various other invertebrates exist on coral reefs, one sees virtually no invertebrates on the reef except for a few large echinoderms (sea cucumbers, urchins, feather stars) and large mollusks (*Tridacna* and others). The reef seems dominated by the corals and by the abundant fishes. This is because the other invertebrates are hidden. It has been suggested by Bakus (1964) that the reason for this cryptic pattern is the intense predation pressure on the reef. Any soft-bodied invertebrate in the open would be quickly consumed. This can be verified by turning over coral heads and exposing the fauna, which immediately attracts fishes to the area to consume the exposed organisms. Presumably, the larger echinoderms and mollusks are more immune to predation.

The role of predation in determining the structure and composition of coral reefs is not as well studied or understood as it is for rocky intertidal areas, but we do have an increasing body of evidence of its importance. A surprising number of animals feed on live coral and can be classified as predators. Most of these predators, however, are small in relation to the coral colony. The process of predation in

FIGURE 9.27
Photo showing the large numbers of *Acanthaster planci* on a heavily infested reef in Palau in 1979. (Photo courtesy of Dr. Diane Nelson.)

these forms resembles the process of grazing by herbivores, whereby patches of coral polyps are removed but the entire colony is not destroyed. If not too much tissue is removed, the coral may regrow polyps to cover the area, very much as grass regrows after grazing. These small predators include a number of gastropod mollusks (families Architectonicidae, Epitoniidae, Ovulidae, Muricidae, and Coralliophilidae), at least one nudibranch mollusk (*Phestilla*), amphinomid polychaete worms (*Hermodice*), certain barnacles (*Pyrgoma*), and several crabs (*Mithraculus, Tetralia*). Where these predators have been studied in detail, it appears that most are dietary specialists with respect to the types of corals consumed. In general, the mollusk and echinoderm corallivores seem to prey on the fast-growing and dominant corals. Taken together, these predators do not appear to have a significant effect on coral colonies, nor do they seem able to affect the community structure. However, according to Glynn (1990), if the living corals on a reef are greatly reduced as a result of massive destruction by a physical event, such as a hurricane or an El Niño episode, the remaining living colonies may concentrate the now-starving corallivores, resulting in massive secondary mortality, which could have an even greater effect than the initial destruction. Such a delayed mortality was reported by Knowlton et al. (1981) for the dominant Caribbean coral *Acropora cervicornis* following the destructive passage of a hurricane on the island of Jamaica.

Two other taxa of predators are, however, capable of destroying coral colonies and modifying the structure of reefs. These are the crown-of-thorns starfish *Acanthaster planci* and various fishes. *Acanthaster planci* is a very large, multiarmed starfish that feeds only on living coral (Figure 9.27). Because of its size, it is capable of destroying an entire colony during feeding. Normally, *Acanthaster* is rare on coral reefs in the Indo-Pacific (not found in the Atlantic), but when present in moderate to large numbers, it can exert an influence on the composition of reefs. For example, on eastern Pacific reefs, Glynn (1976) found that *Acanthaster* preferred to feed on the abundant, fast-growing *Pocillopora* colonies, but in the field, it fed most of the time on the slow-growing, nonbranching colonies in the deeper parts of the reef. This selective predation on nonpocilloporid corals reduced coral cover and diversity. The reason the preferred *Pocillopora* was not eaten was that the colonies harbored symbiotic shrimp (*Trapezia, Alpheus*) that repulsed the starfish and protected the colonies. *Acanthaster* may also cause severe changes in reef structure when a preceding event gives it access to otherwise inaccessible corals. On coral reefs of the eastern Pacific, the presence of the pocilloporid corals prevented access to the other corals by *Acanthaster* because the shrimp symbionts living in *Pocillopora* plus the nematocysts of the *Pocillopora* prevented the star from crossing the coral barrier. However, as Glynn (1985) noted, the 1982–83 El Niño event killed most of the pocilloporids and allowed the *Acanthaster* access to the non-pocilloporid corals, which *Acanthaster* then destroyed.

The second influence on the reef structure by *Acanthaster* comes when conditions bring about high starfish densities. Starting in the 1960s, *Acanthaster planci* underwent a population explosion on certain reefs in the western Pacific such that the normal population density reported by Endean (1973) of one to three starfish observed in four to five hours of search increased to more than 100 starfish seen in 20 minutes of search. The latter figure represents tens of thousands of starfish on each reef (Figure 9.27). In these cases, the entire reef is destroyed by the voracious starfish (see pp. 411–413), and nearly all corals are consumed. Even under circumstances when the *Acanthaster* populations are normal (20–40 individuals per hectare), the fact that *Acanthaster* feeds preferentially on fast-growing corals means that the relative abundance of corals is changed as is the community structure.

We should point out that *Acanthaster* is not without predators. For example, in the Red Sea, puffer fish and trigger fish kill as many as 4,000 of the stars per hectare per year and may be responsible for the low densities of *Acanthaster* there. Other small predators include the shrimp *Hymenocera* and the polychaete worm *Pherecardia*. Such predators may be able to prevent the population outbreaks seen in the western Pacific, but there is currently no evidence for this.

Two groups of fishes actively graze over coral colonies: those species that consume the coral polyps themselves (corallivores), such as puffers (Tetraodontidae), file fish (Monacanthidae), triggerfish (Balistidae), and butterfly fish (Chaetodontidae)(Figure 9.28a); and a group of multivores (omnivores), such as surgeonfish (Acanthuridae) and parrotfish (Scaridae), that remove the coral polyps to obtain either the algae in the coral skeleton or various invertebrates that have bored into the skeleton (Figure 9.28b).

Corallivores that feed by biting off and ingesting portions of the coral skeleton tend to feed on the faster-growing, branching species, such as *Porites* and *Acropora* in the Atlantic and on *Pocillopora*, *Acropora*, and *Montipora* in the Pacific. There is evidence that these corallivorous fish can influence community structure. For example, Neudecker (1977) found that transplanted *Pocillopora damicornis* colonies were severely damaged if transplanted to depths of 15–30 m but were unaffected by corallivorous fish at depths of 2–3 m. Similarly, Wellington (1982a) found that *P. damicornis* grew better in damselfish territories where it was protected from corallivorous fish by the aggressive territorial defense of the damselfish. These two cases—plus the observations that *Pocillopora* is rarely found deep on reefs, and that their colonies when caged away from predators can grow faster at 15 m than at 2 m—are strong evidence that corallivorous fish determine community structure. The extent to which this activity restricts corals to certain regions or is responsible for coral distribution patterns is, however, unknown at present. What is certain is that the two families of fishes, Acanthuridae (surgeonfishes) and Scaridae (parrotfishes), that do consume living and dead coral are among the more abundant fishes on reefs. By passing coral skeletons through their guts, they are significant producers of sand and sediment on reefs.

Other carnivores may affect reef structure by preying on groups of organisms that may compete with the corals. If carnivores preyed upon such groups as sponges, alcyonarians, gorgonians, and tunicates more than on the reef corals and coralline algae, this would give the latter an advantage in sequestering space. This currently remains speculative because of a lack of evidence.

Grazing Aside from the abundant and conspicuous coralline algae on reefs, there is little evidence of other large algae or other plants. If you look closely, however, you will find in many parts of the reef a short algal turf or small clumps of algae. Algae can be major space competitors with corals on reefs, invading and establishing themselves more quickly than corals. Since the algae are not dominant on the reef and obviously do not outcompete the corals, what restrains them? The answer appears to be intense grazing pressure by fishes and certain invertebrates, primarily sea urchins. Many families of fishes contain herbivorous grazers, including Siganidae and Pomacentridae, as well as Acanthuridae and Scaridae. Continued grazing by these fishes, plus the grazing by abundant sea urchins, such as the ubiquitous genus *Diadema*, keep the algae reduced to a minimum and enhance the survival of coral recruits (Figure 9.28b). As Huston (1985) points out, if the herbivores are either removed from reefs or prevented from grazing by

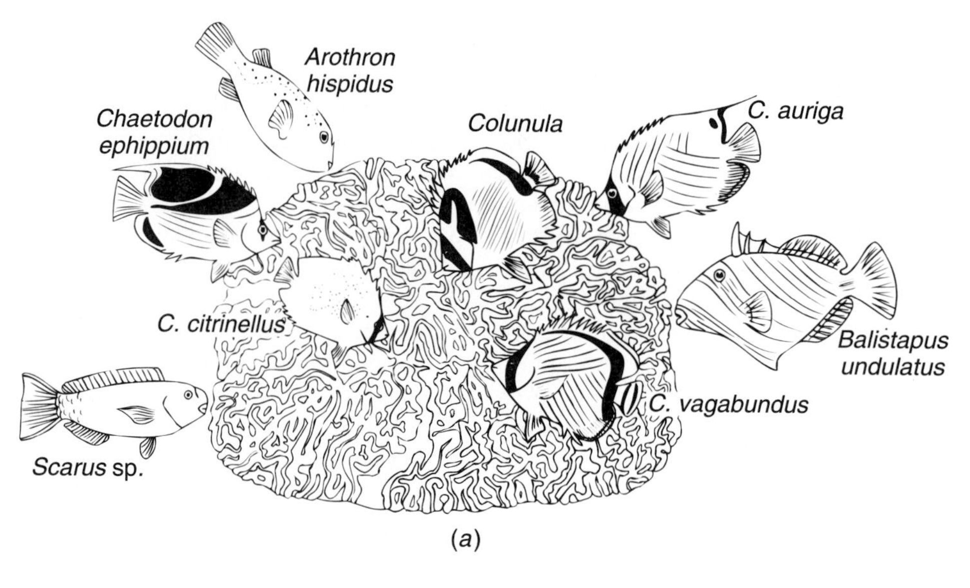

(*a*)

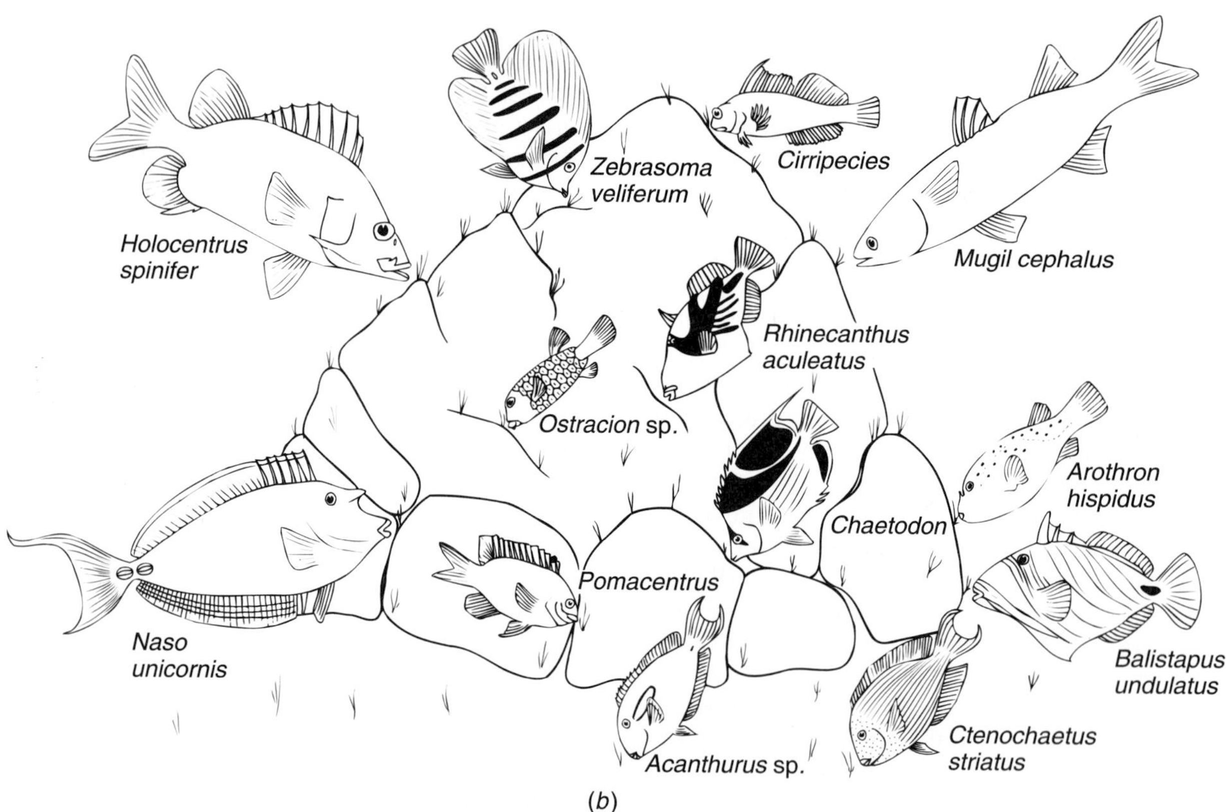

(*b*)

FIGURE 9.28 (a) Fishes that graze coral colonies; butterfly fish (Chaetodontidae), *Chaetodon*, *Colunula*; parrotfish (Scaridae), *Scarus*; triggerfish (Balistidae), *Balistapus*; puffers (Tetraodontidae), *Arothron*. (b) Herbivorous fishes of the reef; triggerfish (Balistidae), *Balistapus*, *Rhinecanthus*; squirrel fishes (Holocentridae), *Holocentrus*; surgeonfish (Acanthuridae), *Acanthurus*, *Ctenochaetus*, *Zenbrasoma*, *Naso*; damselfishes (Pomacentridae), *Pomacentrus*; butterfly fishes (Chaetodontidae), *Chaetodon*; puffers (Tetraodontidae), *Arothron*; mullet (Mugilidae), *Mugil*; boxfishes (Ostraciontidae), *Ostracion*; blennies (Blenniidae), *Cirripecies*. (Not to scale.) (Modified from "Ecological Relationships of the Fish Fauna on Coral Reefs of the Marshall Islands" R. W. Hiatt & D. W. Strasburg, *Ecological Monographs*, Vol. 30, pp. 65-127, 1960. Copyright © 1960 Ecological Society of America. Reprinted by permission.)

caging, coral is rapidly overgrown and killed by algae. Indeed, Hay (1991) has noted that not only are there more herbivorous fishes on coral reefs than in any other marine habitat, but their grazing rates on certain portions of the coral reef exceed the rates of grazing for any other habitat, terrestrial or marine. Herbivorous fish appear to have the ability to remove anywhere from 50 to 100% of total algal production. It has been suggested that the reason for the removal of such vast quantities of algae is that the algae are actually low in nutritive value for the fish needs; hence, the fish must consume large amounts. It is also true that herbivorous invertebrates, primarily sea urchins and gastropod mollusks, are a greater proportion of the invertebrate fauna in tropical seas than in temperate seas.

Grazing by herbivorous fish and urchins directly affects the algae, but it only affects the corals indirectly. The effect of the grazers on the algae is the reduction of the algal biomass. This reduction can be enormous. In the Virgin Islands, when Sammarco et al. (1974) removed all the *Diadema antillarum* from an isolated reef, they had a tenfold increase in algal biomass within four months. There may also be more complex and subtle effects. Grazing, for example, may prevent a single competitively dominant alga from taking over all the space and reducing the diversity. For instance, Sammarco et al. (1974) reported higher algal diversity in grazed patch reefs than in ungrazed reefs in the Virgin Islands; the latter were dominated by a single algal species. The universality of this may be questioned, however, because the same investigator did not find that situation in a similar study in Jamaica.

Grazing can also result in shifts in abundance of groups of algal species the grazers prefer. This is best exemplified by the coralline algae and the fleshy filamentous species. The latter group grows more rapidly and may eliminate the slower-growing corallines if ungrazed. However, the fleshy filamentous algae are much more palatable and so are preferred by the grazers. In areas of heavy grazing, therefore, the coralline algae increase at the expense of the fleshy forms.

Given the intensity of the grazing pressure, how do so many algae survive in these reef systems? There are a number of ways that algae deter various herbivores. One way is to become less palatable. This can be done, as in the case of the coralline algae, by laying down calcium carbonate in the tissues; thus, the algal species are difficult to graze. Another is to produce chemicals that are unpalatable or noxious to the grazers. Some algal species reduce or avoid grazing by settling into habitats on the reef that provide spatial or temporal refuges from the grazers. In this respect, it is known that herbivory is highest in the shallow forereef areas and decreases markedly in the deeper forereef area, as well as on the sandy lagoon and deep sand plains. Residing in the deeper areas means that the water temperatures are lower as well, and temperature affects the rate of grazing; thus they survive. The existence of such spatial refuges mean that the reef is usually a mosaic of such refuges interspersed with heavily grazed areas, giving rise to different microhabitats that support different algal communities and lead to greater diversity of organisms.

The indirect effects of grazing on corals and coral growth produce a more complex situation. Although it is undoubtedly true that herbivores, in general, permit the survival of coral that would otherwise be eliminated by competition with the more rapidly growing fleshy algae, it is possible that they may have certain deleterious effects. For example, sea urchins seem to be rather unselective grazers in contrast to fishes and may destroy young coral colonies as they graze. The studies done by Sammarco et al. (1974) with *Diadema* suggest that at very high densities, this urchin will graze on all organisms, not just algae; thus, coral growth is prevented as well. At lower densities, the urchin removes algae and permits coral colonies to become established. Grazing by moderate densities of *Diadema* clears areas for coral planulae to settle into and contributes to coral reef maintenance.

Within the tropics, the relative importance of grazing fishes and urchins to the maintenance and structure of reefs varies with human fishing pressure. Fishing removes fishes but, according to Hay (1984), it removes both herbivorous fishes and those that prey on urchins. This means that urchins become relatively more important on reefs that are subject to heavy fishing pressure.

The relative importance of fishes and urchins on reefs has been further underlined as a result of the catastrophic die-off of virtually all of the *Diadema antillarum* in the Caribbean. Beginning at the Atlantic entrance to the Panama Canal and at Barbados, *D. antillarum* populations began to mysteriously disappear starting in 1983. Over the next 13 months, the populations died off throughout the Caribbean. According to Lessios (1988), no tropical western

FIGURE 9.29 Threespot Damselfish in its territory. (Photo courtesy of Charles V. Angelo/ Photo Researchers Inc.)

Atlantic area was spared, and mortality was 93–99%. Mortality was followed by nearly complete failure of urchin larval recruitment in 1984.

The immediate response to this loss of a major grazer was the increase in the algal biomass, and this was most dramatic in areas where herbivorous fish had been reduced by fishing pressure. In areas that experienced the largest algal blooms, the fleshy algae increased at the expense of the encrusting coralline forms. In the years since the die-off, studies strongly suggest that the relaxing of grazing pressure by *Diadema antillarum* has had a major effect on algal cover, biomass, and composition, and this may be much modified if herbivorous fish are abundant enough to compensate for the urchins.

In general, moderate grazing prevents monopolization of substrates by dominant macroalgae and increases diversity. On the other hand, heavy grazing leads to the destruction of not only the algae, but the corals as well and decreases the diversity of the reef system.

A special case of the complex effect of grazers on coral is that of the damselfish (Pomacentridae). Damselfish are territorial fish that graze either selectively or nonselectively on the algae within their territory (Figure 9.29). They defend their "algal gardens" against other grazing fishes, and they may even control the algal species composition in their territory by selective "weeding," according to Ogden and Lobel (1978). Since the algae in the territory are protected from heavy grazing by other grazers, Potts (1977) found that they grow rapidly enough to kill the small settling corals by overgrowth and shading. Corals are, therefore, reduced or absent. Damselfish "gardens" may occupy significant portions of reef surfaces. According to Wellington (1982a), up to 60% of the area of reef flats may be occupied by such damselfish territories; hence, they are important contributors to community structure, as well as providing refuges for juvenile invertebrates and demersal plankton. All is not, however, as simple as just indicated. Since damselfish protect their territory from all grazing fishes including grazing corallivorous fishes and given that massive corals recover from fish grazing and grow more slowly than branching foliose corals, the damselfish contribute to the selective dominance of the reef areas by branching corals in the areas they occupy. According to Kaufman (1977), damselfish in the Caribbean eliminate the massive coral *Montastrea annularis* from the upper parts of the reef where damselfish are most common. This leads to the dominance by the fast-growing coral *Acropora cervicornis*. In the eastern Pacific, there is an even more dramatic example of the effects of damselfish on reef structures. Here, Glynn and Wellington (1983) have demonstrated that the damselfish also remove the dominant urchin, *Eucidaris thouar-*

sii, from their territories; thus, they contribute to the survival of corals that might otherwise be destroyed by the grazing urchin.

Finally, though we have emphasized that heavy algal growth will kill corals, Huston (1985) thinks that a certain amount of algal growth may promote coral diversity by reducing the interspecific contact and subsequent competition between coral colonies.

Regional Differences There are some interesting regional differences in species interactions that can be noted at this time. It is not known yet, however, how they may relate to community structure and coral reef development. First, it is likely that herbivory is more intense on the Pacific reefs than on those of the Atlantic. Fishes remove more algae from settling plates in the Pacific, and there are more numerous bite marks per unit area of substrate in the Pacific. A second observation is that herbivorous sea urchins are more common in the Atlantic than in the Pacific. Comparison of the frequency of shell injuries to gastropod mollusks between the Pacific and the Atlantic shows that these injuries are more common in the Pacific and suggests this is the result of a higher rate of predation on those species in the Pacific.

Bacteria in Reef Systems

The role of bacteria in coral reef systems is poorly understood even today. There have been only a few studies undertaken, primarily in the Great Barrier Reef system, and none on atolls.

On the Great Barrier Reef, bacteria have been investigated in both the water column and in the sediment. Bacterial production is dependent on the dissolved organic carbon (DOC) from the water flowing over the reef and the particulate organic carbon (POC) from the breakdown of reef organisms. The fluxes of DOC and POC are great enough to permit large populations of bacteria to exist in both the water columns of the lagoons and the sediments. Ducklow (1990) reports that bacterial biomass and production are orders of magnitude larger in the sediments than in the water column, and the ratio of total reef bacterial production to total reef primary production ranges from 0.6 to 1.0. The bacteria represent a huge energy resource for the reef community and suggest that a microbial loop system could be present, similar to that existing in the plankton (see Chapter 2), which could conserve energy in the reef system. Sorokin (1990) has proposed that such a loop system exists and has claimed it is one of the main sinks for nutrients. This would explain the maintenance of high productivity of coral reefs, but Ducklow (1990) indicates that such a loop system has not been discovered, and that this large bacterial energy source is either not used by reef filter feeders or is rarely used by filtering organisms.

We are left with a controversy. If one believes Sorokin, coral reefs are closely coupled systems adapted to tightly hold all inorganic nutrients to prevent loss to a "barren" surrounding ocean and making them essentially self-sustaining. On the other hand, if one believes Ducklow, the large carbon and nutrient source represented by the POC and bacterial populations are ignored. It is clear that we are in need of more studies to clarify the role of bacteria in reef systems.

Bacteria may also be important in the bioerosion of coral reefs. Glynn (1997) reports that several species of cyanobacteria are capable of eroding reef rock and have been found on reefs. Cyanobacteria are also important in coral reef diseases such as black-band disease.

Role of Algae in Reef Systems

As we have noted, algae tend to be small and inconspicuous on coral reefs, in marked contrast to the great kelp forests of temperate shores. However, through various biological interactions, algae are ecologically important to the community structure of the reef system. We discuss now some additional, primarily physical, roles of algae.

First, the encrusting red coralline algae, such as *Lithothamnion*, maintain the integrity of the reef by constantly cementing together various pieces of calcium carbonate; thus, they reinforce the reef against destruction by wave action and prevent individual pieces from being carried off the reef. The algal ridge is further responsible for breaking the velocity of the waves and producing calmer conditions that allow growth of other organisms in the reef flat behind it. The second group of calcareous algae are the greens, dominated by the genus *Halimeda*. These algae are significant contributors to the sand found in reefs, particularly in the lagoon area. They, therefore, create a special habitat.

Certain algae also bore into coral skeletons, but they do not appear to be significant in bioerosion. Algae also create habitats in and around themselves and may furnish necessary shade from the hot tropical sun for certain organisms.

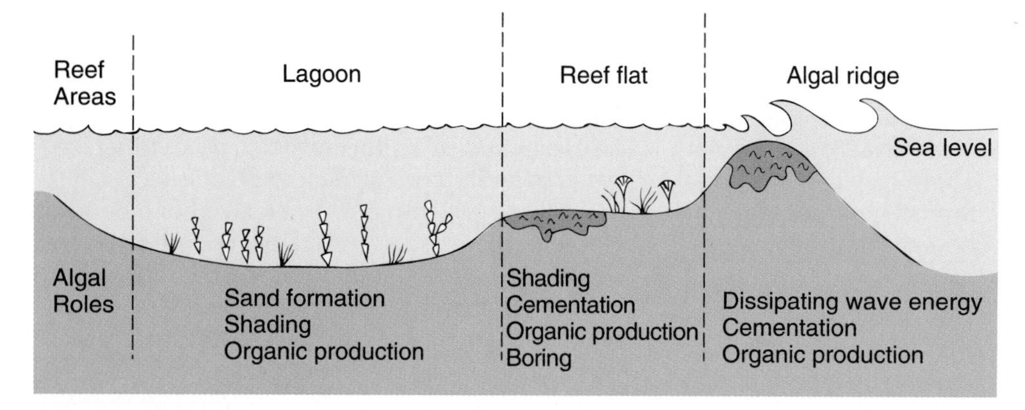

FIGURE 9.30 The various roles of free-living algae in different zones of a reef. (Modified from *Proceedings of the Second International Symposium on Coral Reefs*, Vol. 1, M. Doty, 1974, Great Barrier Reef Committee, Brisbane. Used by permission of the Great Barrier Reef Committee.)

Algae are important as primary producers in the reef system and as a food for various herbivores. These roles are summarized in Figures 9.30 and 9.31.

Ecology of Reef Fishes

Coral reefs harbor more species and diverse fish communities than any environment on earth. It is impossible to conceive of a coral reef without the myriad brilliantly colored fishes gracefully moving among the various corals and associated organisms. Indeed, fishes are the most abundant and conspicuous large organisms encountered on a reef. Because of their abundance and all-pervading presence, it seems obvious that they must contribute in significant ways to the reef ecosystem. We have noted some of these important structuring relationships in previous sections, and in this section we will review other aspects of their ecology.

The species richness of coral reef fishes is similar to that observed in corals. The central Indo-Pacific area of the Philippine and Indonesian archipelagos has the greatest number of species, and the number decreases in all directions away from this center (Table 9.2). The Atlantic reefs are also relatively species poor in this comparison. The number of species found on a single reef is remarkable—500 for one reef in the Great Barrier Reef system. How does this great diversity of species come about, and how is it maintained? This is the question that has intrigued coral reef biologists for many years, and it is still generating debate. We can only suggest some of the reasons here; more remain to be discovered.

One of the reasons for the high diversity of species on reefs is the great variety of habitats that exists on

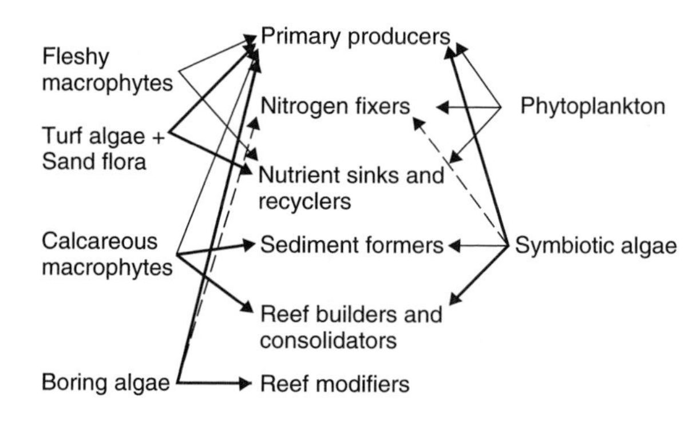

FIGURE 9.31 The main algal groups and their roles in the coral reef ecosystem. The thickness of the lines indicates the relative importance of that algal group to the processes listed. Dashed lines indicate a possible role for that group. (From "Reef algae" M. A. Borowitzka and A. W. D. Larkum, *Oceanus* Vol. 29, No. 2, p. 52, 1986. Reprinted by permission.)

TABLE 9.2

Number of Fish Species in Several Coral Reef Areas

Geographical Area	*Number of Fish Species*
Philippine Islands	2,177
New Guinea	1,700
Great Barrier Reef	1,500
Seychelles Islands	880
Marshall and Mariana Islands	669
Bahama Islands	507
Hawaiian Islands	448

Source: After Goldman and Talbot, 1976.

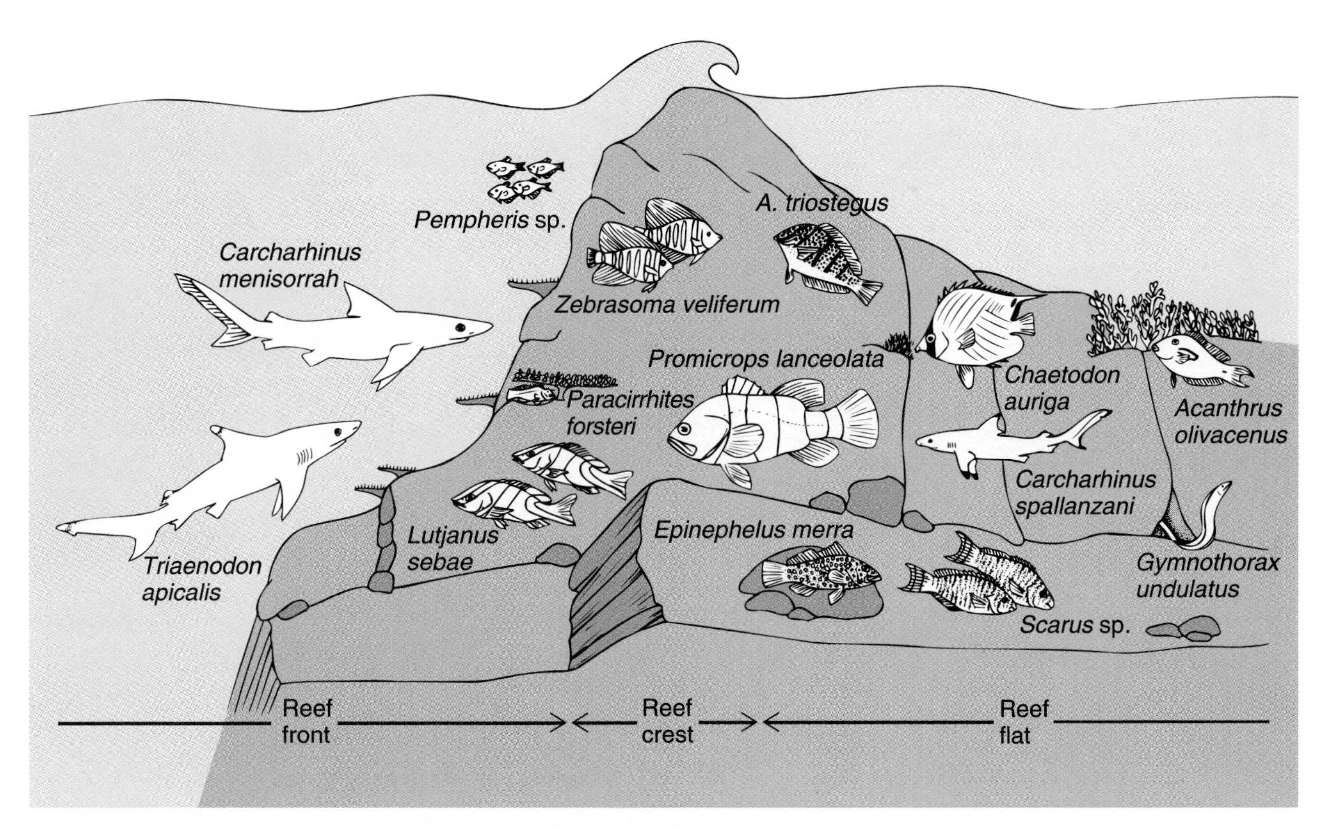

FIGURE 9.32 Generalized picture of the characteristic fish species and their habitats on a seaward coral reef in the Marshall Islands. Requiem sharks (Carcharinidae), *Triaenodon*, *Carcharhinus*; moray eels (Muraenidae), *Gymnothorax*; hawk fishes (Cirrhitidae), *Paracirrhites*; groupers (Serranidae), *Epinephelus*, *Promicrops*; snappers (Lutjanidae), *Lutjanus*; surgeonfishes (Acanthuridae), *Acanthurus*, *Zebrasoma*; parrotfishes (Scaridae), *Scarus*; butterfly fishes (Chaetodontidae), *Chaetodon*; sweepers (Pempheridae), *Pempheris*. (Not to scale.) (Modified from "Ecological Relationships of the Fish Fauna on Coral Reefs of the Marshall Islands' R. W. Hiatt & D. W. Strasburg, *Ecological Monographs*, Vol. 30, pp. 65-127, 1960. Copyright © 1960 Ecological Society of America. Reprinted by permission.)

reefs. Coral reefs encompass not only coral, but also areas of sand, various caves and crevices, areas of algae, shallow and deep water, and different zones progressing across the reef. This diversity of habitats alone can go far to explain the increased number of fishes. These habitats and their associated fish inhabitants are summarized in Figures 9.32 and 9.33.

The numerous habitats, however, are not enough to explain the high diversity of coral reef fishes, particularly in local areas. Numerous experimental and theoretical studies have attempted to explain how so many species of reef fish can exist within local communities. Four opposing theories based on different aspects of recruitment or competition have arisen over the last 25 years (Figure 9.34). The more classical view is that high diversity is the result of strong competitive interactions following recruitment that lead to a high degree of specialization. Each species has a specific set of adaptations that give it the competitive edge in at least one situation on a reef. That is to say, these fishes have narrower ecological niches, so more species can be accommodated in a given area. This model of Smith and Tyler (1972) is called the **competition model**. It is partially supported by correlations between fish species diversity and habitat complexity or "rugosity" by Risk (1972) and Luckhurst and Luckhurst (1978). In this model, recruitment is not significant in structuring the adult population because it is assumed there is always an excess of potential recruits. A second model, promulgated by Sale (1977, 1980), is the **lottery hypothesis**. This hypothesis rests on the fact that most coral reef fishes produce large numbers of larvae that are dispersed into the plankton. It states further that fishes are not specialized (many similar species having

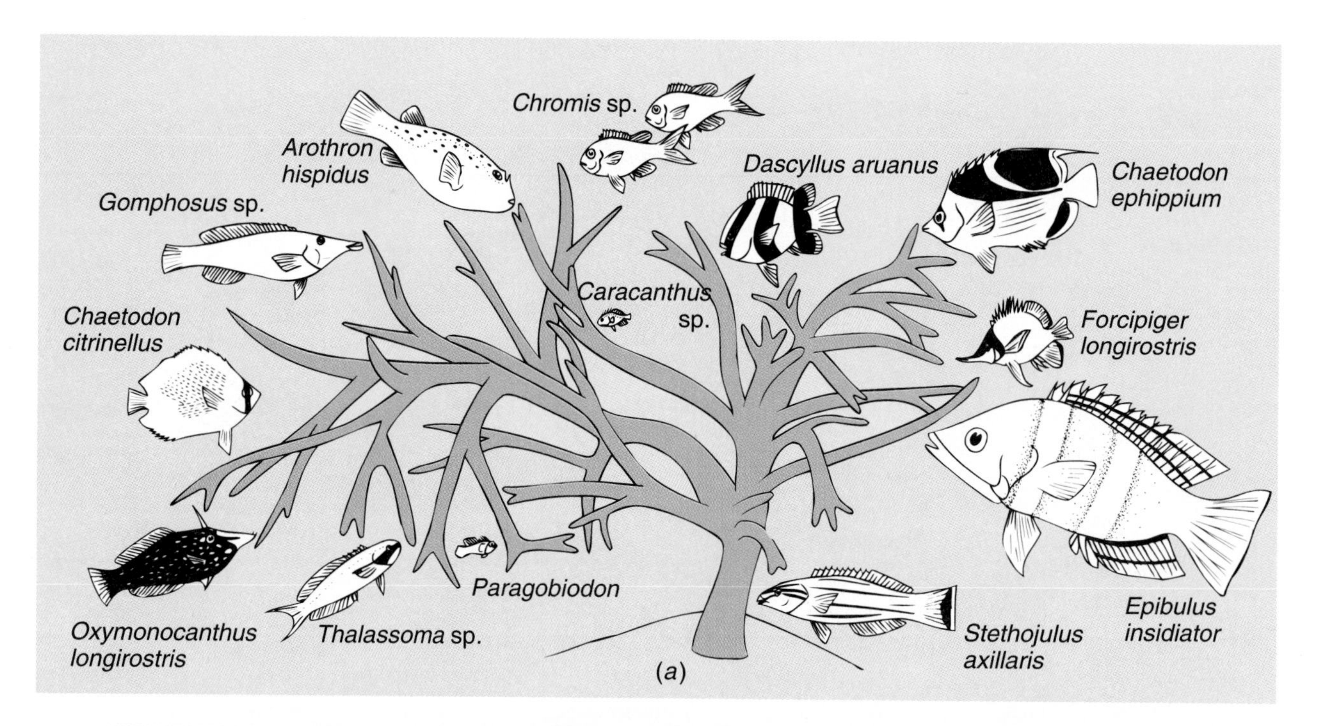

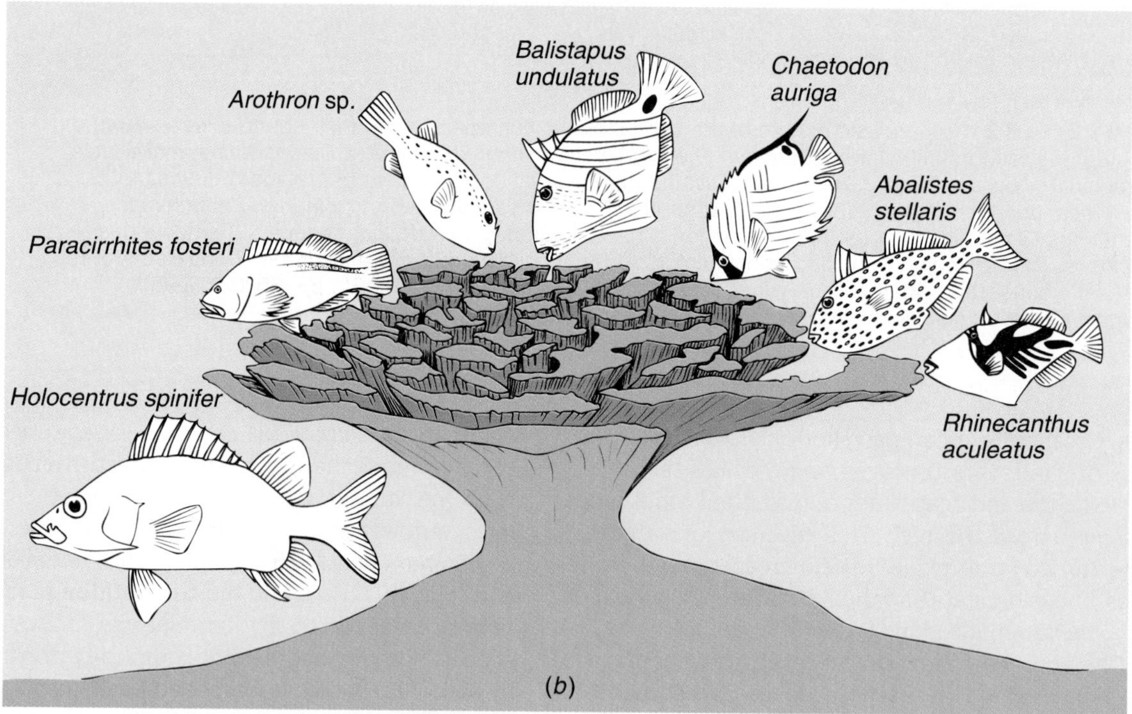

FIGURE 9.33 Fishes associated with the individual coral colonies of the (a) branching and (b) plate type. Butterfly fishes (Chaetodontidae), *Chaetodon*, *Forcipiger*; damselfishes (Pomacentridae), *Chromis*, *Dascyllus*; wrasses (Labridae), *Epibulus*, *Thalassoma*, *Stethojulus*, *Gomphusus*; velvet fishes (Caracanthidae), *Caracanthus*; gobies (Gobiidae), *Paragobiodon*; puffers (Tetraodontidae), *Arothron*; filefishes (Monocanthidae), *Oxymonocanthus*; hawk fishes (Cirrhitidae), *Paracirrhites*; triggerfishes (Balistidae), *Balistapus*, *Rhinecanthus*, *Abalistes*; squirrel fishes (Holocentridae), *Holocentrus*. (Not to scale.) (Modified from "Ecological Relationships of the Fish Fauna on Coral Reefs of the Marshall Islands' R. W. Hiatt & D. W. Strasburg, *Ecological Monographs*, Vol. 30, pp. 65-127, 1960. Copyright © 1960 Ecological Society of America. Reprinted by permission.)

the same requirements), and that there is active competition among the species. Local success and persistence result from chance as to which species of the planktonic larval pool occupies the vacant space. Here, competition is unimportant and recruitment dominates. The third view, by Talbot, Russell, and Anderson (1978), is that the fish populations do not reach equilibrium. Predation, catastrophe, and unpredictable recruitment ensure that populations never become large enough to undergo competitive exclusion, because they are kept below the numbers at which resources or food become limiting. This is the **predation-disturbance model**. The most recent explanation is the **recruitment limitation model**. This model, championed by Victor (1983) and Doherty (1982), argues that the larval supply is never sufficient for the adult population size to reach carrying capacity. The adult population reflects variation in the larval recruitment and not postrecruitment events. Doherty and Fowler (1994) have provided data from an empirical test of this hypothesis in which 90% of the variation in abundance of the damselfish *Pomacentrus moluccensis* was explained by recruitment. However, further data are needed for other fishes. Evidence is currently inconclusive as to which of these views, if any, is correct. We can say, however, that one consistent fact that has arisen in the 25 years of controversy is that fish recruitment is variable.

Perhaps as a result of the large numbers of species and the partitioning of the habitat, we find that most reef fishes, despite their obvious mobility, are restricted to certain areas of the reef and are very localized. Most do not migrate, and many of the smaller species, such as gobies, blennies, and damselfishes, defend territories. However, certain of the larger fish do make diel migrations to feeding areas. The distances covered may range from only a few meters to several kilometers and are often predictable (Figure 9.35).

	Postrecruitment (PR) competition: Intense	Postrecruitment (PR) competition: Weak
Recruitment modified by PR processes	1. Competition model (Smith and Tyler)	3. Predation disturbance models (e.g., Talbot et al.)
Recruitment not modified by PR processes	2. Lottery model (Sale)	4. Recruitment limitation model (e.g., Victor)

FIGURE 9.34 Classification of the four models offered to explain the diversity of species among coral reef fishes based on the importance of competition and recruitment processes. (From *The Ecology of Fishes on Coral Reefs*, edited by Peter F. Sale. Copyright © 1991 by Academic Press, reproduced by permission of the publisher.)

One of the most interesting discoveries concerning the fishes of coral reefs is the differences in fishes between day and night. Most people see reefs during the day when the majority of the fish species are visible. At night, these diurnal fishes seek shelter in the reef and are replaced by a smaller number of nocturnal species not seen during the day (Figures 9.36 and 9.37). Since some of the nocturnal species are ecologically similar to certain diurnal species (Apogonidae, replace

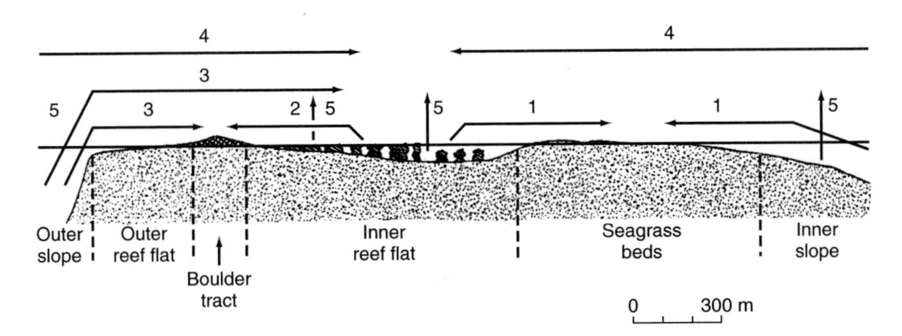

FIGURE 9.35 Movement patterns of different groups of fishes within a fringing reef during an incoming tide. 1 and 2: from the inner reef flat to seagrass beds and rubble bank, respectively, to feed; 3: from the outer slopes to feed in shallower waters; 4: movement of pelagic fishes from deeper waters to feed on inner reef flat; 5: general movement up into the water column. (From *The Ecology of Fishes on Coral Reefs*, edited by Peter F. Sale. Copyright © 1991 by Academic Press, reproduced by permission of the publisher. All rights of reproduction in any form reserved.)

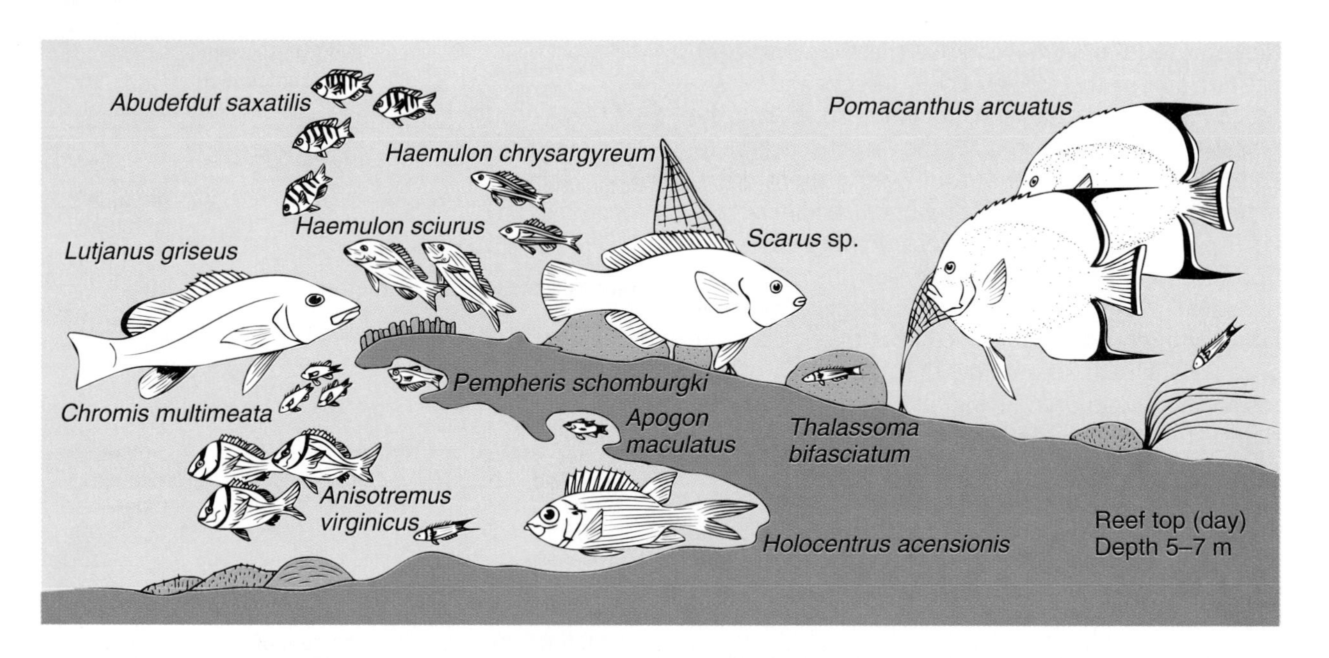

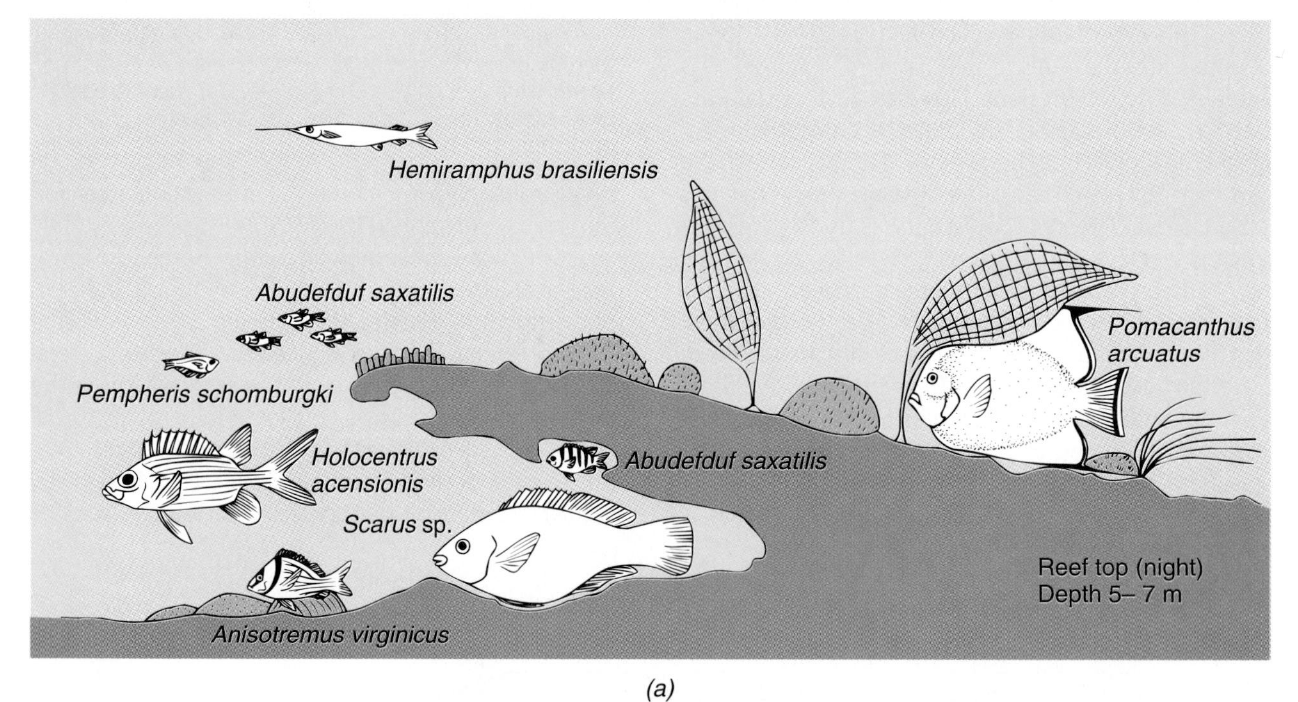

(a)

FIGURE 9.36 Day and night fish distribution on a Florida coral reef at two different depths. (a) Reef top at 5–7 m. (b) Deep reef at 25–30 m. Damselfishes (Pomacentridae), *Abudefduf*, *Chromis*; wrasses (Labridae), *Clepticus*, *Thalassoma*, *Lachnolaimus*; parrotfishes (Scaridae), *Scarus*; angelfishes (Chaetodontidae), *Pomacanthus*, *Holacanthus*; grunts (Pomadasyidae), *Haemulon*, *Anisotremus*; cardinal fishes (Apogonidae), *Apogon*; squirrel fishes (Holocentridae), *Holocentrus*; sweepers (Pempheridae), *Pempheris*; snappers (Lutjanidae), *Lutjanus*; halfbeaks (Hemiramphidae), *Hemiramphus*. (Data for the drawing based on W. A. Starck and W. P. Davis, 1966, Night habits of fishes of Alligator Reef, Florida. *Ichthyolog. Aquar. J.* (38) 41:313–356.)

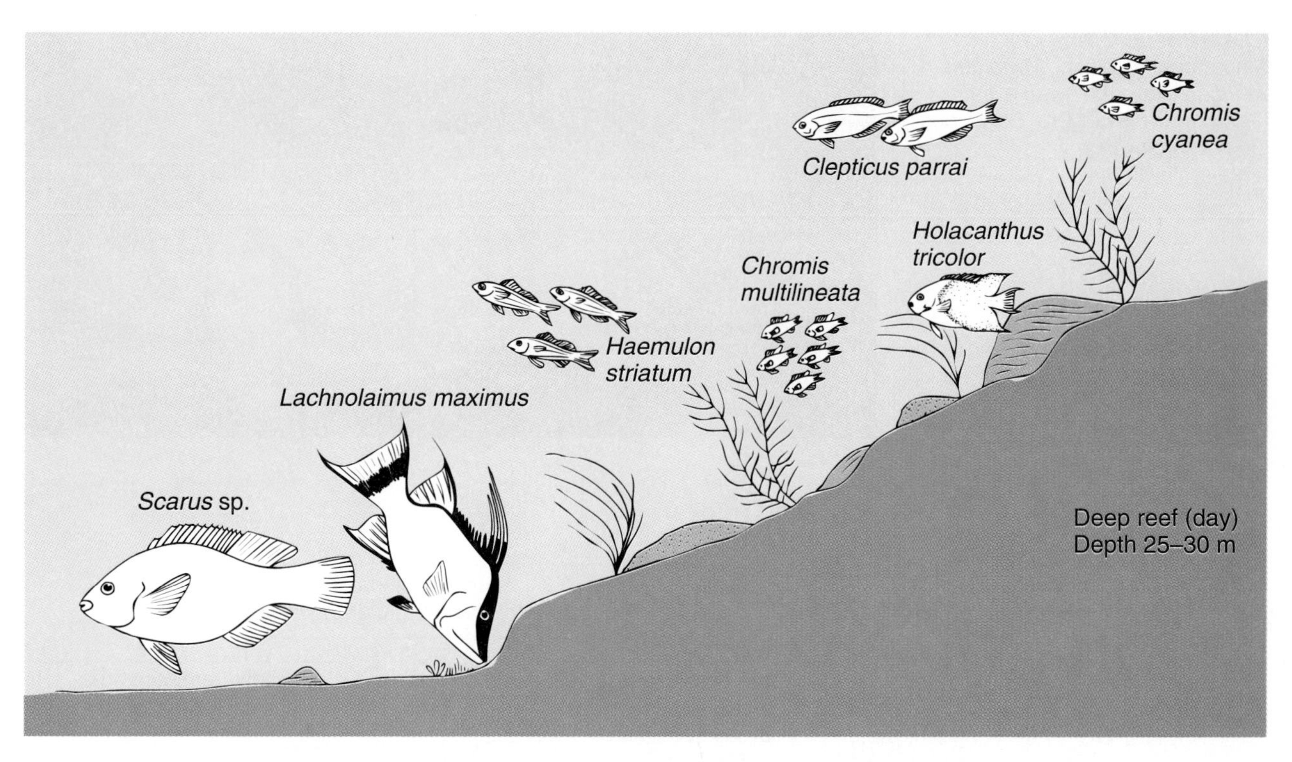

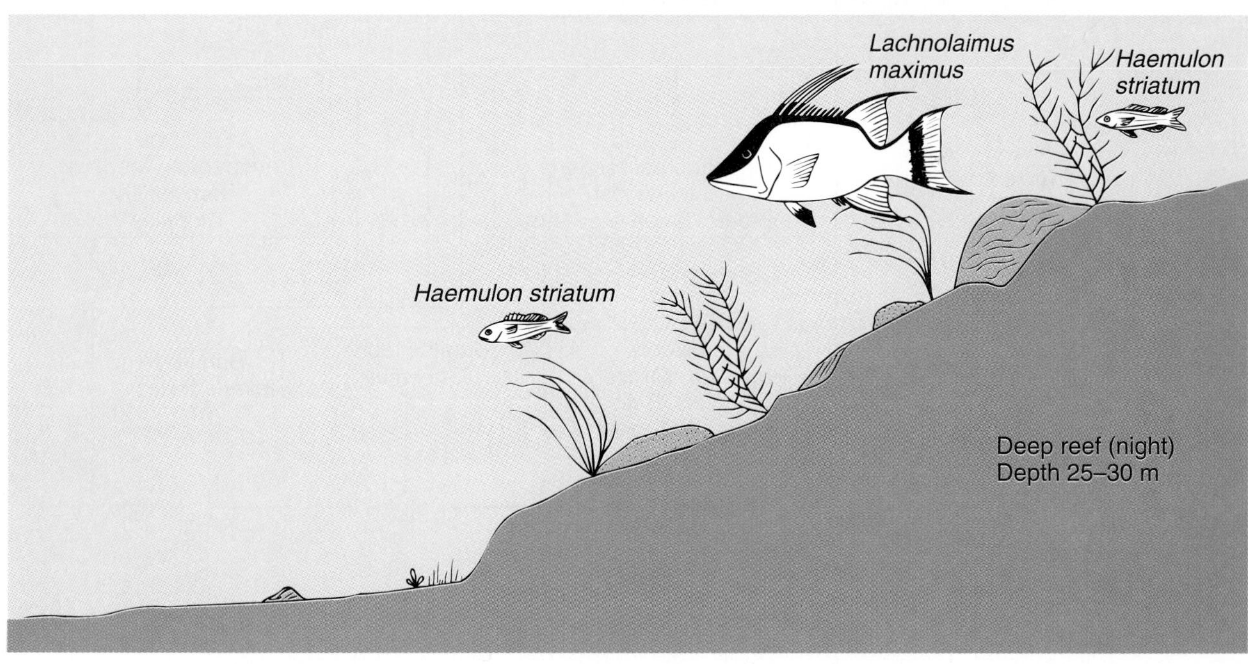

(*b*)

FIGURE 9.36 (CONTINUED)

FIGURE 9.37 A nocturnal reef fish, *Myripristis murdjan*. Note the large eyes. (Photo courtesy of Drs. Lovell and Libby Langstroth.)

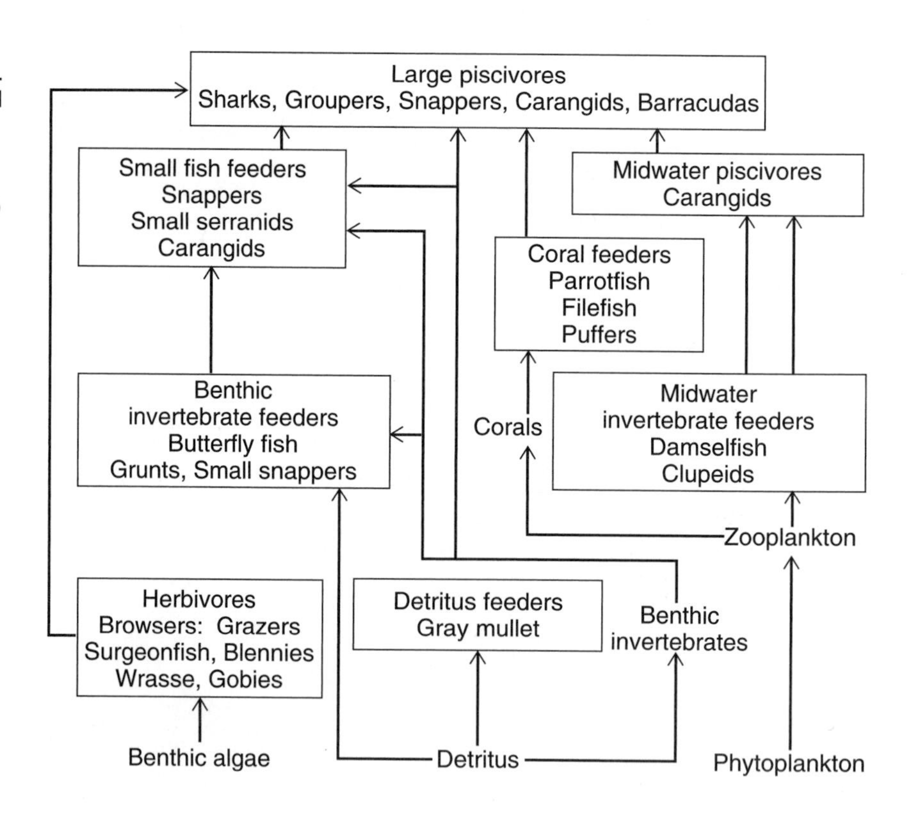

FIGURE 9.38 The trophic relationships of coral reef fisheries. (After "Ecology of Fishes in Tropical Waters" R. H. Lowe-McConnell, 1977, *Studies in Biology*, No. 76, 1977. Reprinted with the permission of Cambridge University Press.)

Pomacentridae), this is another way of permitting a greater number of species to exist on the reef without competing directly. According to Hobson (1968), all nocturnal species of reef fishes are predaceous, but the diurnal fishes span nearly all trophic categories (carnivores, planktivores, omnivores, etc.).

The feeding relationships of reef fishes are of some interest (Figure 9.38 and Table 9.3). The most abundant feeding type on reefs are the carnivores, which may constitute 50–70% of the fish species. Goldman and Talbot (1976) say that many of these carnivores are not specialized to a given food source but are opportunistic, taking whatever is available to them. They also feed on different prey at different stages in their life cycles. The fact that the majority of the fishes on reefs are generalized carnivores argues against the competition model explanation for the species richness of reefs. Indeed, this lack of specialization makes it difficult to explain satisfactorily the species richness. This is not to say that no

TABLE 9.3

Proportion of Fish Species in Different Trophic Categories from Seven Tropical Studies[a]

Trophic Category	H & S[b]	Rand[c]	G & T[d]	*Hobson*	W & H	*Sano*	T & C
Herbivores	26	13	22	7	15	18	20
Planktivores	4	12	15	18	20	15	38
Benthic invertebrate feeders	49	44	27	56	53	41	33
Coral feeders	6	1		9	5	9	
Sessile animal feeders	8	6		13	3		
Mobile invertebrate feeders	35	37		34	45		
Omnivores	13	7		10	4	19	
Piscivores	10	25	38	7	8	4	8
Others (e.g., cleaners)				2		2	1

[a] Figures given are percentages of the number of species surveyed unless otherwise noted. The category of benthic animal feeders is the sum of the three subcategories below and indented from it. The seven studies were: H & S, Hiatt and Strasburg (1960); Rand, Randall (1967); G & T, Boldman and Talbot (1976); Hobson, Hobson (1975); W & H, Williams and Hatcher (1983); Sano, Sano et al. (1946b); and T & C, Thresher and Colin (1986).

[b] Hiatt and Strasburg (1960) classified some species in more than one category.

[c]Randall (1967) indicated that the survey was biased toward important sport and commercial species.

[d]Goldman and Talbot (1976) used only the four categories shown. Herbivores were included by them into a category of grazers, which included coral feeders. Figures given are percent weight of all groups, averaged over three habitat types: fore reef, lagoon, and back reef.

Source: *The Ecology of Fishes on Coral Reefs*, edited by Peter F. Sale. Copyright © 1991 by Academic Press, reproduced by permission of the publisher.

specialized carnivores exist; there are some. Correlated with the generalized feeding habits of most carnivores, the numbers of true scavenger fishes are few because the carnivores also act as scavengers and pick up any recently dead organisms.

Herbivores and coral grazers make up the next largest group of fishes (about 15% of the species), and the most important of these are the families Scaridae and Acanthuridae (Figure 9.39). The remainder of the fishes are generally classed as omnivores or multivores and include representatives from virtually all families of fishes on the reef (Pomacentridae, Chaetodontidae, Pomacanthidae, Monocanthidae, Ostraciontidae, Tetraodontidae). Only a few fishes are zooplankton feeders, and these are mainly small, schooling fishes of the families Pomacentridae, Clupeidae, and Atherinidae.

A side effect of the feeding of some fishes, such as grunts (haemulids), is to enhance the nutrition and growth of corals. Meyer *et al.* (1983) have demonstrated that schools of certain fishes, which rest by day in coral heads, feed at night on seagrass beds and then return to defecate material rich in nitrogen and phosphorus into the coral heads. Whether this fecal material may result in faster coral growth has not as yet been demonstrated.

When one considers that most of the fishes on the reef are unspecialized carnivores, it is easy to understand why most invertebrate organisms are hidden from view, since to expose a soft body on a reef

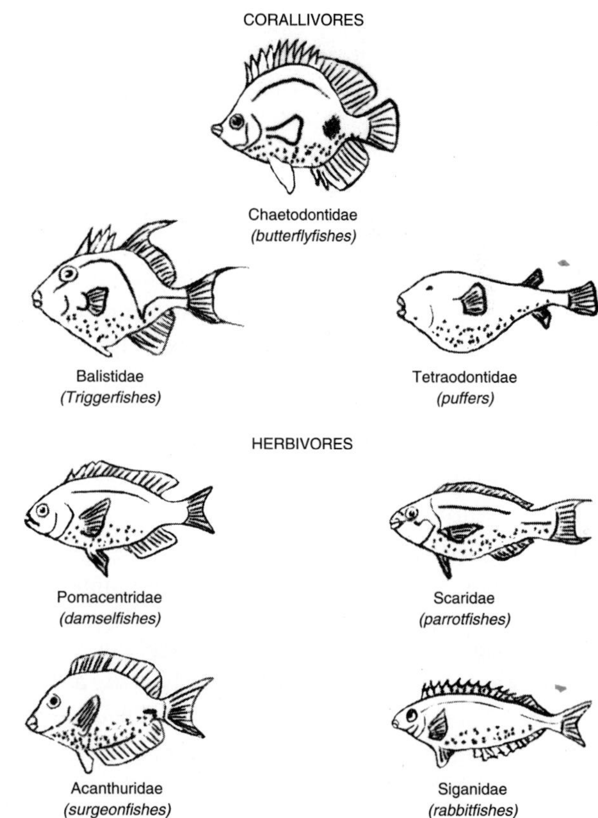

FIGURE 9.39 Families of reef fishes that include the most corallivorous and herbivorous species. (From *The Life and Death of Coral Reefs*, C. Birkeland, editor, © 1997 Chapman & Hall, p. 355, F 15.1, with kind permission from Kluwer Academic Publishers.)

would certainly invite instant attack by some carnivore. How, then, do a few invertebrates, notably the large sea cucumbers, remain abundant lying unprotected on reefs? The answer apparently is that those few that are visible have evolved defense mechanisms to deter potential predators. The most common is a poison. According to Bakus (1973), tropical sea cucumbers produce toxic substances that can kill fishes and have viscous, sticky strands (cuvierian organs), which can be extruded to literally tie up any potential predator.

The development of distasteful or toxic substances is not limited to invertebrates. Many of the fishes also produce toxic substances. These may take the form of venom associated with various spines or poisonous material extruded onto the body surface (**crinotoxin**), or the flesh and internal organs may be toxic. Truly venomous fishes are rare on reefs and are confined mainly to the stonefishes (Synanceiidae) and scorpion fishes (Scorpaenidae), but a large number of fishes have toxic secretions on their outer surfaces, including the abundant parrotfish (Scaridae), wrasses (Labridae), and surgeonfish (Acanthuridae). One explanation for the prevalence of these substances is that they deter predation by the abundant carnivores.

The final category, namely, toxic flesh or internal organs, has caused the most interest among humans, because eating tropical fishes with toxins in the flesh or organs produces a serious disease called **ciguatera**. Ciguatera has been one of the most mysterious and puzzling diseases found in the tropics. As noted by Banner (1976), the symptoms in humans are primarily neurological, including exhaustion, visual disturbance, inversion of senses (hot feels cold, and vice versa), paralysis, loss of reflexes, and finally, death by respiratory failure. The mystery came originally from several sources:

1. The symptoms are not consistent.
2. The fishes that have the toxin vary in toxicity and can include virtually any or all of the large reef fishes, ranging from carnivores to herbivores. A given species may be toxic on one reef but not on another, or the same species may be toxic at one time of the year on a given reef and not at another.
3. There seems to be no repeatable pattern to the occurrence of the toxin.

We now know somewhat more about the disease. Perhaps most significant was the discovery in 1977 of the causative organism, the dinoflagellate *Gambierdiscus toxicus*. This organism is scarce on reefs with large amounts of living coral, but it proliferates after disturbances have created dead coral surfaces covered by filamentous or calcareous algae. The toxin, which is a polyether, has also been isolated and partially characterized. An efficient and sensitive bioassay employing mosquitoes has also been developed. We are, however, still ignorant of most of the ecological aspects of G. *toxicus* and the factors that trigger a bloom, as well as of the mechanisms by which the toxin comes to reside in the flesh of certain carnivorous fish. Ciguatera, therefore, still remains of considerable importance to the people of the tropics, since the fisheries of many areas cannot be developed until this toxicity problem is solved.

Apparently, coral reef fishes are also subject to occasional epidemics that cause mass mortality. A Caribbean-wide mass die off of fishes was recorded by Williams and Williams (1987–88) in August and September of 1980 following the passage of a hurricane. However, this mass mortality event was not studied, so that the nature of the cause(s) is unknown.

Another of the outstanding characteristics of reef fishes is their color. Why, especially given the great predation pressure, should they be so brightly colored? Biologists cannot agree as yet, and it is likely that color and pattern serve several different functions. One explanation is that the bright colors advertise that the species is toxic or otherwise distasteful so predators will ignore it (warning coloration). They may also camouflage the species, either by breaking up the shape of the fish or else making it appear as something else. The latter explanation especially has been used to explain the dramatic banding patterns seen in some reef fishes (Figure 9.40; see also Figures 9.15 and 9.18). In addition to their role in various predator-prey interactions, color and pattern may be important in visual communication, since these fishes are highly dependent on vision. Color and pattern may serve in species and gender recognition and may be used in courtship and mating behaviors.

Despite the ubiquitous presence of fishes on coral reefs in seemingly large numbers, little is known about the population dynamics of reef fishes. We do know that the populations of individual species may fluctuate over time and in different areas of the reef, but we are ignorant of the exact causes. Much of our lack of knowledge comes from the presently almost impossible task of following the dynamics of the larval populations in the plankton and their settlement patterns on the reef.

A final aspect of reef fish ecology concerns the phenomenon of **cleaning behavior**. Cleaning behavior is a specialized form of predation in which certain small fishes (*Labroides* spp.) or shrimps remove various ectoparasites from other, usually larger, fish species. Cleaning behavior is widespread and apparently occurs on all reefs. In this process, the cleaner fishes (or shrimps) often set up "cleaning stations" where they appear to advertise their presence through bright, contrasting colors. The fish to be cleaned comes to the cleaning station area (often a prominent coral head or boulder) and remains motionless as the cleaner moves over its body removing the parasites (Figures 9.41 and 9.42). The cleaners may even enter the mouth and gill chambers of the fish. A sort of truce among predators prevails at the cleaning stations, as the fishes being cleaned are vulnerable to predation. Fishes even "line up" at these stations, awaiting their turn for cleaning.

The importance of this cleaning behavior to the fish population and economy of the reef has not been established. Some biologists believe that if cleaners are removed from a reef, the fish fauna decreases, moving away or otherwise showing signs of distress. Other biologists have removed cleaners with no noticeable effect on the reef or reef fishes.

■ We can summarize this section by stressing the importance of fishes to the reef ecosystem. Fishes are important in determining the zonation of various parts of the reef through their grazing activities, which keep the algae reduced to a short mat and prevent them from outcompeting the corals. At the same time, their grazing on corals is responsible for the exclusion of certain corals from areas of the reef, which also contributes to the zonation of reefs. Consumption of coral by fishes also leads to the breakdown of coral and contributes to the sediment budget of a reef. The tremendous number of carnivores among the fishes undoubtedly exerts a great predation pressure on all reef organisms and is the reason for the cryptic nature of most soft-bodied invertebrates and the development of various toxic passive defense mechanisms among fishes and invertebrates. ■

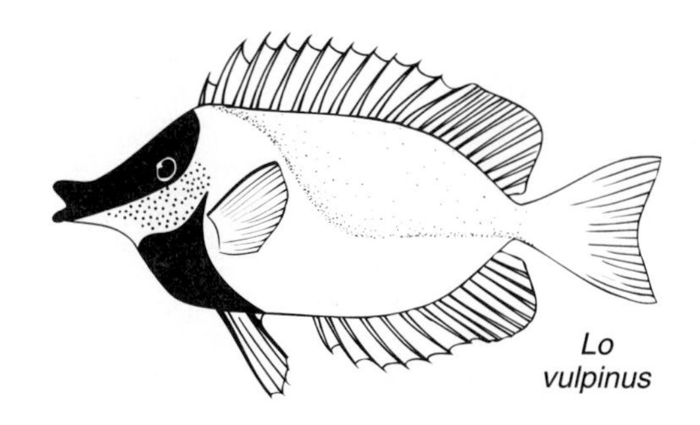

FIGURE 9.40 Reef fishes with dramatic banding patterns.

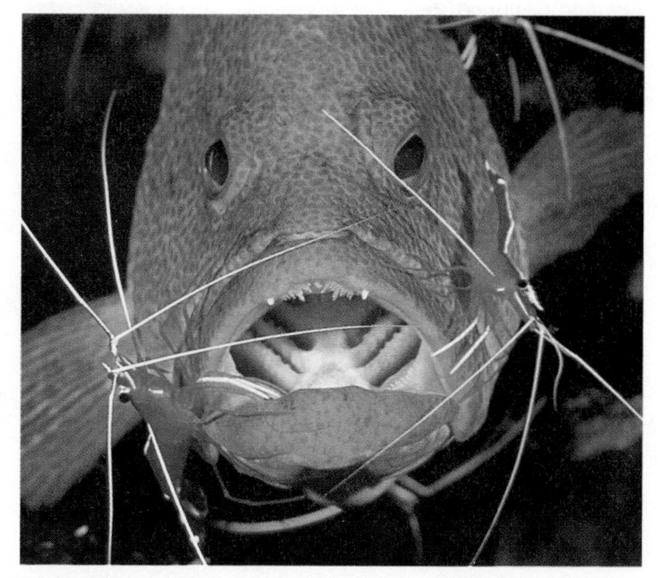

FIGURE 9.41 Humpback cleaner shrimp on a tomato cod (*Lysmata ambionesis*). (Photo courtesy of Steve Wulper/ DRK Photo).

FIGURE 9.42 A cleaning shrimp, *Periclimenes yucatanicus*, on the tentacles of the anemone *Condylactis gigantea*. (Photo courtesy of Dr. Diane Nelson.)

Coral Reef Cryptofauna and Bioerosion

The term **cryptofauna**, applied to coral reefs, refers to all invertebrates that inhabit the coral substrates, boring into the calcium carbonate, living in crevices and cracks, and occupying space under coral heads and boulders. According to Hutchings (1983), two groups exist: those that actually bore into the calcium carbonate—the true borers—and those that are unable to bore the coral rock but live in preexisting holes, cracks, and crevices—the "opportunists" (Figure 9.43). True borers of coral rock include sponges, bivalve mollusks, sipunculans, and some polychaete worms. The nestlers or opportunists represent most marine invertebrate phyla, but crustaceans and polychaete worms seem to be particularly common.

Colonial cryptofaunal organisms, including sponges, bryozoans, and ascidians, tend to dominate the surface fauna of the undersides of corals, caves, and overhangs. Their ability to reproduce asexually and to overgrow other potential space occupiers gives them a competitive edge in these areas (see discussion on p. 393).

FIGURE 9.43 Diagrammatic representation of a variety of external and internal bioeroders of coral skeletons. (From *The Life and Death of Coral Reefs*, C. Birkeland, editor, © 1997 Chapman and Hall, p. 69, F 4.1, with kind permission from Kluwer Academic Publishers.)

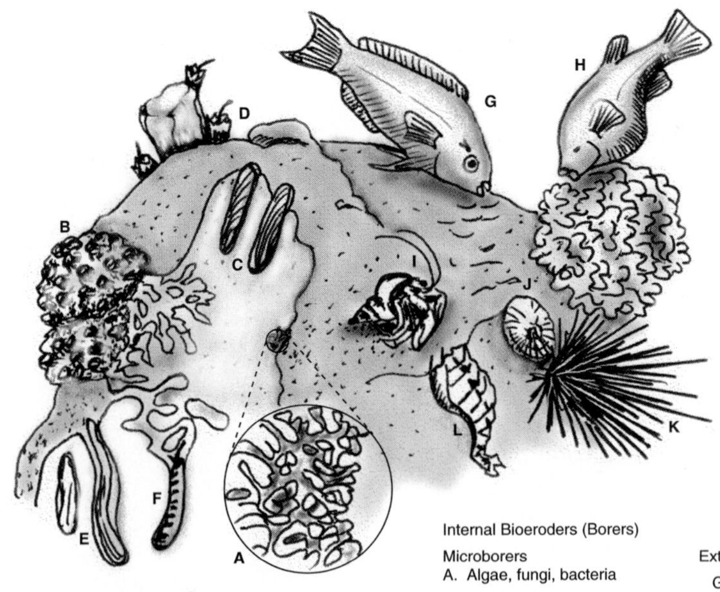

Very little is known about the structure, recruitment, or succession of these cryptofaunal communities. According to Hutchings (1983), the community is dominated by deposit feeders. These animals feed mainly on organic detritus that has been collected in mucus secreted by the coral and moved off the coral colony. They may play an important part in the coral reef system by using the large amounts of coral-produced mucus and in recycling the organic material and nutrients trapped therein. Cryptofaunal organisms are also an important food source for certain reef carnivores, such as holocentrid fishes and certain gastropod mollusks, such as *Conus* (Kohn and Nybakken, 1975).

Most certainly the cryptofauna is significant in bioerosion of the reef. The rates of bioerosion due to activities of sponges, echinoids, and grazing mollusks have been assessed and when added to that of various grazing fishes suggest that bioerosion may be the most important of the processes that break down coral reefs. For instance, Stearn and Scoffin (1977) reported that on Barbados corals and algae laid down about 160 metric tons of calcium carbonate annually over an area of about 1 hectare. In that same area, boring sponges, barnacles, and bivalves annually removed 1.5 metric tons from live corals and 23.5 metric tons from dead coral. Parrotfish removed an additional 1 metric ton, and the urchin *Diadema antillarum* removed 163 metric tons. This suggests that the reef is eroding faster than it can be built up. However, a second study by Scoffin *et al.* (1980) on the same reef suggested a higher calcification rate of 206 tons per year and a total bioerosion rate of 163 tons per year. This points up the variability in the data and suggests that few conclusions can yet be drawn about calcification and bioerosion rates without more study.

A review of bioerosion by Hutchings (1986) has further substantiated the earlier work on the significance of biological agents in reef destruction. Hutchings also noted that, in addition to the major groups of sponges, mollusks, echinoderms, and fishes, such erosive agents as fungi, endolithic algae, polychaetes, and sipunculan worms are present on reefs, but that little or nothing is known about their relative importance. More important, the rates at which any and all bioerosive organisms remove calcium carbonate from reefs are variable in time and space, compounding the problem of determining the role of these organisms in the overall ecology of coral reefs.

Succession and Stability in Coral Reefs

Ecological succession refers to the orderly change in community structure and composition over time (See pp. 21-22). Whereas early evidence suggested that succession did not occur in coral reefs, current studies, primarily from two long-term studies in the Pacific, have indicated predictable species replacements do occur in some areas. In Guam, the coral reefs were devastated by *Acanthaster* in 1967–69, losing up to 90% of the coral cover. Colgan (1982) followed the recovery of these reefs through the next 11 years and identified a definite successional sequence. At Heron Island on the Great Barrier Reef, Connell et al. (1997) have been studying the dynamics of the coral community continuously since 1962, giving us the longest continuous time record for any coral reef. Connell et al. (1997), too, report a predictive succession of species through time. However, similar long-term studies of reefs in Hawaii by Grigg (1983) and Dollar and Tribble (1993) did not reveal any predictable species replacements. These studies have suggested that local disturbance events can strongly influence successional events and that biological mechanisms such as competition, life history attributes, inhibition or facilitation of colonization, and predation on eggs and larvae all influence the direction and rate of successional changes such that the entire process is highly variable geographically. It seems obvious that much more work is needed before we can make any definitive statements about succession in coral reefs.

Related to the question of succession is that of stability of coral reefs. On short time scales reefs appear to be quite stable, but they are not necessarily so over the longer term. The structure and even the existence of reefs can be radically changed by natural catastrophic events such as hurricanes, El Niño induced weather and oceanographic changes, *Acanthaster* outbreaks, and diseases affecting corals and other reef organisms (see following section). As with succession, the stability of reefs appears to be quite variable in time and space.

Catastrophic Mortality and Recovery of Reefs

Although reefs appear to be large and very stable systems, existing, in the case of atolls, for millions of years, they do suffer large-scale destruction from

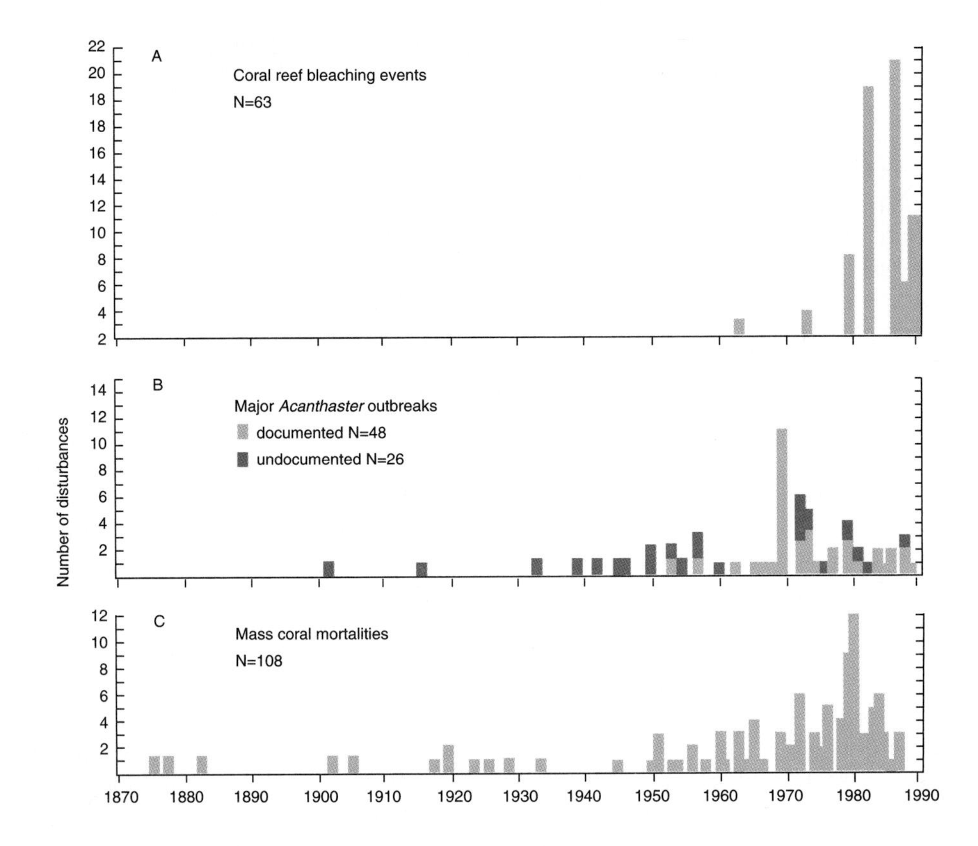

FIGURE 9.44 Reported coral bleaching events, *Acanthaster* outbreaks, and mass coral mortalities from 1870 to 1990. (From *The Life and Death of Coral Reefs*, C. Birkeland, editor, © 1997 Chapman and Hall, p. 355, F 15.1, with kind permission from Kluwer Academic Publishers.)

various forces. Perhaps the major physical source of massive reef mortality is mechanical destruction by severe tropical storms. Hurricanes or typhoons are intense enough to destroy large areas of reef as they pass over them. Since most coral reefs lie within the zone most frequently traversed by typhoons or hurricanes, the chance that a reef or a section of a reef will be destroyed or severely damaged is great. Damage in such storms usually is due to coral colonies being toppled or uprooted and carried off the reef, thus freeing space for new occupants.

The second major source of catastrophic mortality of reefs, at least in recent times, is the population explosion of the sea star *Acanthaster planci*. Since 1957, when the population explosions were first observed, *A. planci* has caused catastrophic mortality of reefs in a number of places in the western Pacific (Figure 9.44). This starfish's ability to destroy large areas of reef is tremendous. In Guam, Chesher (1969) estimated that 90% of the coral reef along 38 km of shoreline was destroyed in 2½ years. On the Great Barrier Reef, Endean (1973) indicated that the bulk of the corals on an 8 km^2 reef was destroyed in 12 months. More recent surveys (1985–86) reported in Walbran et al. (1989) indicated that 27% of the Great Barrier Reef has been affected by *Acanthaster*, while Endean and Cameron (1990) report that about one-third of the reefs of the Great Barrier Reef have been destroyed. When these rates are multiplied by the number of reefs in the Pacific that were infested in the 1960s (including 60% of the reefs of the Great Barrier Reef between 16 and 19°S), one can understand the concern for the survival of reefs.

What causes these population explosions, and are they natural phenomena? These are the two major questions that biologists have asked regarding the "*Acanthaster* problem," and their answers show how little we really know about how coral reefs function. Although there is no consensus concerning the answer to either question, there is evidence that the explosions are natural phenomena. That similar outbreaks have occurred in the past is suggested by the findings of aggregations of *Acanthaster* skeletal remains in sediment cores by Frankel (1977) and more recently by Walbran et al. (1989), by the mention in folklore and memories of certain island peoples according to Birkeland (1982), by their occurrences in areas remote from human interference and activities, and by their recurrence in several areas of the Pacific. Whether such explosions are natural to other reef systems is not known. Even if these earlier outbreaks were natural, the recent ones could still have been caused by human factors. Whether natural or human-induced, what is the trigger? Examination of the infested reefs has produced no consistent pattern of human-induced or natural activity that indicates a single cause. Many reasons have been proffered, such as various dredging activities opened new space for the A. *planci* juveniles; some chemical (e.g., pesticides) somehow freed A. *planci* from its normal population controls; humans removed an important predator, allowing increased survival of juveniles; or severe tropical storms opened space for juvenile settlement or caused adults to aggregate.

Although no one knows for certain what the cause is, of the above suggestions the two that have received most attention and coverage in the press are the two involving predators and the effects of storms. The predator-removal theory necessitates the presence of predators on adult *Acanthaster* that normally keep the population in check. Are there such predators? Yes, there are at least three predators in the Indo-Pacific that feed on *Acanthaster*: a fish, the hump-headed wrasse, *Cheilinus undulatus*; the puffer fish, *Arothron hispidus*; and the large gastropod, *Charonia tritonis*. In addition in the eastern Pacific, the polychaete worm *Pherecardia striata* and the shrimp *Hymenocera picta* appear to be important predators. Of course, there are undoubtedly predators on the planktonic larvae and newly settled juveniles of *Acanthaster* as well, but they are little known or studied. Of these predators on the adults, attention has been mainly focused on *Charonia* because it is a known specialist predator on starfish. It is also in demand by shell collectors. Endean and Cameron (1990) argue that the removal of this relatively uncommon gastropod from the reefs by collectors could have allowed the population explosion to begin, but currently there is little evidence to support this hypothesis. If this explanation is correct, C. *tritonis* is a keystone species in the same sense that *Pisaster ochraceus* was in the rocky intertidal. However, we know so little about the interactions on reefs that this certainly could not have been predicted ahead of time.

Another idea is the "runoff" hypothesis by Birkeland (1982) which suggests that juvenile recruitment of A. *planci* is enhanced by a combination of unusually low salinities, high nutrients, and high temperatures. In this scenario, the occurrence of a year or years of abnormally heavy rainfall in an area, especially if coupled with human destruction of the native vegetation on adjacent land areas, causes increased runoff. This runoff provides enough nutrients to stimulate phytoplankton blooms and provide food for large numbers of *Acanthaster* larvae. The increased survival of the larvae and their subsequent settlement on the reefs leads to outbreaks three years later. This model uses a natural explanation that fits the pattern of recent outbreaks, but currently there is little evidence to support it. Correlated with this theory is the adult-aggregation theory of Dana et al. (1972), supported by Potts (1981), which says that the destruction of reef areas by typhoons causes the starfish to condense into aggregations, which attack the remaining living coral en masse.

A third cause of catastrophic mortality is the phenomenon known as El Niño. El Niño is the name given to a naturally recurring periodic phenomenon manifested as a massive influx of nutrient-poor warm water into the usually cold, nutrient-rich surface waters off the coasts of Ecuador and Peru. In fact, El Niño is but part of an interacting set of atmospheric and oceanographic conditions called the El Niño-Southern Oscillation (ENSO) that periodically alters the atmospheric pressure, winds, rainfall patterns, ocean currents, and sea level over large areas of the tropical and subtropical Pacific Ocean. These natural, unpredictable ENSO events vary greatly in intensity, areal extent, and ecological effects. The greater the intensity, the greater the change in oceanographic and meteorological conditions, the larger the area affected, and the greater the effect on living organisms. The 1982–83 El Niño, according to Glynn (1988), was perhaps the strongest event of the twentieth century up to that date and caused unprecedented disturbances in marine populations and communities. It may have had a periodicity of 1 in 100 to 1 in 250 years, and resulted in a rise of ocean surface temperatures to 30–32°C in the eastern Pacific. This was 2–4°C above normal and remained for several months. This prolonged elevation of temperature, coupled with changes in sea level up to 44 cm below normal in the eastern Pacific and unusual hurricane activity and hurricane paths, killed large tracts of reef corals in the Gulf of Panama, Galápagos Islands, Tahiti, and the Society Islands, as well as in other places throughout much of the tropical Pacific (Lessios et al., 1983). Mortality rates for coral ranged from 50 to 98% overall according to Glynn (1988). Particularly devastated were the reefs in the Galápagos, where Glynn (1985) reported 95% of corals died, and in the Gulf of Panama, where 70–80% of the corals were destroyed. (The 1997–98 El Niño event had an intensity equal to the 1982–83 event, but studies of its effects are not yet available.)

FIGURE 9.45 Coral bleaching. (Photo courtesy of Dr. Gary Williams.)

The loss of the corals and coral reef framework in areas, such as the Society Islands and the Galápagos Islands, had profound secondary effects. On the one hand, it eliminated habitats for numerous reef-associated species, according to Glynn (1988). Whole fish communities disappeared from areas of dead coral. On the other hand, in the eastern Pacific the recruitment of sea urchins was enhanced and densities of *Diadema* in Panama and *Eucidaris* in the Galápagos rose. Their activities, in turn, caused increased bioerosion, which destroyed the dead reef framework. Even corals that survived the initial mortality by the change in physical factors were subjected to increased predation by surviving corallivorous sea stars and sea urchins when the protective shrimp and crab symbionts died or migrated. Hence, the eastern Pacific reefs have not shown much recovery in the more than 15 years since the 1982–83 El Niño event.

A puzzling event, which can cause catastrophic coral mortality, is the phenomenon of **coral bleaching** (Figure 9.45). Corals bleach when they expel the zooxanthellae that normally inhabit their tissues. Following the expulsion, corals become white, hence the name of the phenomenon. If the zooxanthellae are absent for a significant time, the coral will die. Coral bleaching events have been reported for many reefs in both the Caribbean and the Indo-Pacific in the 1980s and 1990s (Figure 9.44). Usually the bleaching affects most, but often not all, coral species and is most severe in shallow-water areas. Widespread and severe coral bleaching occurred in the Caribbean in 1987–88 and in the eastern Pacific following the previously described El Niño in 1982–83. More recent bleaching events have been reported by Goreau (1990) for Jamaica and by Brown (1997) for Thailand, French Polynesia, and the Andaman Sea.

The cause of bleaching is unclear, but most scientists attribute it to stress, particularly stress induced by water temperatures of 30°C or above. Bleaching events are growing more common, and there is concern that the pattern of bleaching events may be an early warning of the onset of a global warming (see Chapter 11). If bleachings continue to increase, there is concern for the integrity of many reefs because either the length of the events or the closeness of events will be such that corals will not have time to recover, and large reef areas may die.

The activities of humans can directly cause catastrophic mortality on reefs through dredging, pollution, and overfishing. The reefs in Kaneohe Bay, Hawaii, for example, were destroyed through sewage pollution. During World War II, the military closed off the lagoon at Palmyra Atoll, which resulted in the death of all lagoon reefs. In Jamaica, the combination of the spectacular die-off of the sea urchin *Diadema antillarum* coupled with the chronic overfishing of most fishes has caused the reefs there to change from coral dominated to algal dominated. In certain areas of India, coral is actively mined for use in building, and this is also causing catastrophic mortality. In the Philippines, dynamiting for fishes is a major cause of destruction. Oil pollution has also been shown to adversely affect corals, with the branching species, such as *Acropora*, being more heavily affected than the massive species.

Two diseases that affect corals may cause local die-offs but have not as yet caused catastrophic mortality. They are the **black-band disease** and **white-band disease**. Black-band disease has also been found in milleporinids and gorgonians. The cyanobacterium *Phormidium corallyticum* has been suggested as the cause of the disease, but it also provides a microhabitat for other bacteria that may be important. The cause of the coral tissue lysis as the band moves has yet to be determined. Not all corals are equally susceptible to black-band disease. Large brain corals are most susceptible, whereas the branching corals are resistant. In white-band disease, the cause is unknown. White-band disease seems more prevalent in branching corals.

When catastrophic mortality occurs, how long does it take a reef to recover? Again, we are ignorant, but the little information we have on recolonization of reefs destroyed by hurricanes suggests that it may take as long as 25–30 years before the reef has completely recovered. Estimates of recovery time from *Acanthaster* devastation range from 7 to 40 years. It may take even more time depending on whether the devastated reefs are reinfested with *Acanthaster* before they can recover. In many parts of the Great Barrier Reef, reefs attacked and devastated in the 1960s by *Acanthaster* were reinfested in the 1980s, suggesting to Endean and Cameron (1990) that complete recovery may take considerably more than 40 years. In the case of the 1982–83 El Niño, Glynn (1985) feels that some areas may take 100 or more years to recover. The reasons for the widely varying estimates of recovery time are because recovery depends on several factors, including the extent of initial destruction, the nearness of a source of larvae for recolonization, favorable water currents and conditions for establishment of the larvae, time between major disturbances, and biological interactions. Recovery of some reefs may be inhibited by persistent grazing by large numbers of sea urchins or by the mats of algae and soft corals, which quickly take over the dead coral skeletons; thus, the skeletons do not offer suitable substrates for the recolonizing reef corals.

Long-Term Dynamics

Coral reefs throughout the world are suffering from an onslaught of both natural and human disturbances that in concert over the last few decades have degraded or destroyed many reefs. Extensive assessment by 80 countries and the Global Coral Reef Monitoring Network has raised the percentage of reefs destroyed throughout the world by late 2000 to 27%. Reefs in the Caribbean seem to have suffered disproportionately and are also the best documented. It seems appropriate to end this section on coral reefs with a brief discussion of the current concern for the long-term survival of reefs.

The island of Jamaica lies in the center of the coral reef diversity in the Atlantic, and more than 60 species of reef-building corals occur along its reef-lined coast. The Jamaican reefs have also been the subject of intensive study going back to the 1950s, making them among the best understood reef systems in the world. With this background, Hughes (1994) has documented what 40 years of human and natural disturbances have done to these reefs. It can serve here as an example of the kinds of changes that we may see in other parts of the coral reef areas of the world if we are not careful.

As Hughes points out, the amount of damage to the Jamaican coral reefs is enormous. Censuses taken from 1977 to 1993 along 300 kilometers of coastline reveal a decline in coral cover from an average of 52% to 3% and a concomitant increase in fleshy algae from a mean of 4% to 92%! In other words, there has been a phase shift from coral domination to algae domination in shallow waters. The classic zonation pattern of coral reefs first described by Goreau in the 1950s no longer exists.

The beginning of this phase change probably began in the 1960s when the exponentially rising Jamaican human population greatly increased fishing pressure on the reefs. By the end of the decade, fishing had reduced the fish biomass by as much as 80% along the north coast. By 1973, the fishing pressure was perhaps two to three times the sustainable yield, and large predatory fish species were rare. Species composition changed, indicating severe overfishing. This paucity of fish, particularly herbivorous fish that compete with sea urchins, such as scarids and acanthurids, and predatory fish that eat sea urchins, such as balistids, permitted the increase in the sea urchin *Diadema antillarum* such that populations on overfished reefs were much higher than on pristine reefs. During this time, the reefs appeared normal and no large algal blooms occurred as a result of the demise of the herbivorous fish because the abundant *Diadema* replaced the fish.

In 1980, Hurricane Allen, a class 5 storm, hit Jamaica after almost 40 years in which the island had been spared a major storm. The storm smashed the shallow-water reef corals, especially the dominant branching *Acropora cervicornis* and *A. palmata*. Immediately after the storm, there was a short-lived algal bloom, but it was not sustained, probably due to urchin grazing, and corals began to recolonize.

The recolonization was cut short by another natural disaster. Starting in 1982, a water-borne pathogen began to attack the *Diadema* populations, eventually causing mass mortality of the urchins throughout the entire Caribbean. *Diadema* populations were reduced by 99%. In the decade since the die-off, there has been no significant recovery.

Without either the herbivorous fishes or the *Diadema* to keep the algae in check, all reefs have undergone an explosive algal bloom that has preempted virtually all the space on the former reefs. The algae have overgrown and killed the coral colonies that survived the hurricane. The takeover by the algae was aided by a series of bleaching events in 1987, 1989, and 1990 that further reduced the remaining corals.

It now appears that this new pattern will persist even in the face of future hurricanes, because while such future storms will destroy the algae, they will also destroy the remaining corals, and the algae regenerate

TABLE 9.4

The Genera of Mangroves

Avicennia	*Aegiceras*
Suaeda	*Aegialitis*
Laguncularia	*Rhizophora*
Lumnitzera	*Bruguiera*
Conocarpus	*Ceriops*
Xylocarpus	*Sonneratia*

Source: After Lugo and Snedaker, 1974.

or recruit much more rapidly than the corals. The ultimate recovery of the corals depends on the return of significant numbers of herbivores to the reefs, probably both the fishes and the urchins. As Hughes concludes, to save the reefs for the future, it will be necessary to control overfishing and build the populations of the herbivores. If this is not done, the prime coral reef area of the Atlantic faces a gloomy future.

That this scenario for Jamaica is not unique is supported by work by Porter and Meier (1992), who monitored six coral reefs between Miami and Key West from 1984 to 1992. All six areas lost coral species during this time, and coral cover was reduced in five areas. Net losses were from 7.3 to 43.9% of live coral cover, and they did not record any recruitment by frame-building coral species. Losses of this magnitude are usually only associated with hurricane damage. In this case, there were no such storms, and losses were due to bleaching and black-band disease. Porter and Meier concluded that loss rates of the magnitude measured could not be sustained if the coral reefs were to persist in their current configuration.

MANGROVE FORESTS

Mangrove forest or **mangal** is a general term used to describe a variety of tropical inshore communities dominated by several species of trees or shrubs that grow in salt water. These "mangroves" are flowering terrestrial plants that have reinvaded the fringes of the sea. The term **mangrove** refers to the individual plants, whereas mangrove forest, mangrove swamp, tidal forest, and mangal refer to the whole community or association dominated by these plants. Walsh (1974) reports that from 60 to 75% of the coastline of the earth's tropical regions is lined with mangroves, and so their ecological importance is clear. Recognizable descriptions of mangal associations were made by the Greeks as early as 325 B.C. In this section, we shall cover the basic ecological factors governing the existence of these communities.

Structure and Adaptations

The mangrove forest comprises trees and shrubs belonging to some 12 genera of flowering plants in eight different families (Table 9.4). The most important or dominant genera appear to be *Rhizophora*, *Avicennia*, *Bruguiera*, and *Sonneratia*. Mangroves share a number of characteristic features that allow them to live in shallow marine waters. They are, first of all, shallow rooted, with roots that spread widely or with peculiar prop roots that stem from the trunk or branches (Figure 9.46). The shallow roots often send up extensions, called **pneumatophores**, to the surface of the substrate that allow the roots to receive oxygen in the otherwise anoxic mud in which these trees grow (Figure 9.47). The leaves are tough and succulent and have internal water storage tissue. Some mangroves have elevated salt concentrations and salt glands, which help them maintain the osmotic balance by secreting salt (see also salt glands in seabirds in Chapter 3, pp. 121-122). Others exclude salt and separate fresh water from salt water at the roots by a reverse osmosis process. Nearly all mangroves share two reproductive strategies, according to Rabinowitz (1978). These are vivipary, or uninterrupted development of the embryo following fertilization such that there is no "resting" or true "seed" stage; and dispersal by water. The latter is particularly well developed in the genus *Rhizophora* and may serve as a model. The seed, while still on the parent plant, germinates and begins to grow into a seedling without a resting stage. During this time, the seedling elongates and the weight distribution changes, so it becomes heavier at the outer, free end (Figure 9.48). Eventually, this seedling is dropped from the parent plant and, because of the weight distribution, floats upright in the water. It then is carried by the water currents until it enters water shallow enough for its root end to strike the bottom. When this happens, it puts out roots to anchor itself and continues growth into a tree. The advantages of this system of reproduction are obvious for a plant living on the fringe of the sea. Having seeds able to float means it can be dispersed by water currents, while the fact that it floats upright with much of the seed below water means that, when water shallow enough for the mangrove to grow in is reached, the seed will ground itself (Figure 9.49).

Distribution

Mangrove forests are distributed throughout the tropical and subtropical oceans of the world. They can grow only on shores sheltered from wave action; otherwise, the seeds would never be able to ground

FIGURE 9.46 Mangrove forest, showing typical root system and roots descending from the branches. (Photo by Dr. Roger Seapy.)

FIGURE 9.47 Pneumatophores on the roots of a mangrove tree. (Photo courtesy of Wards Natural Science Establishment.)

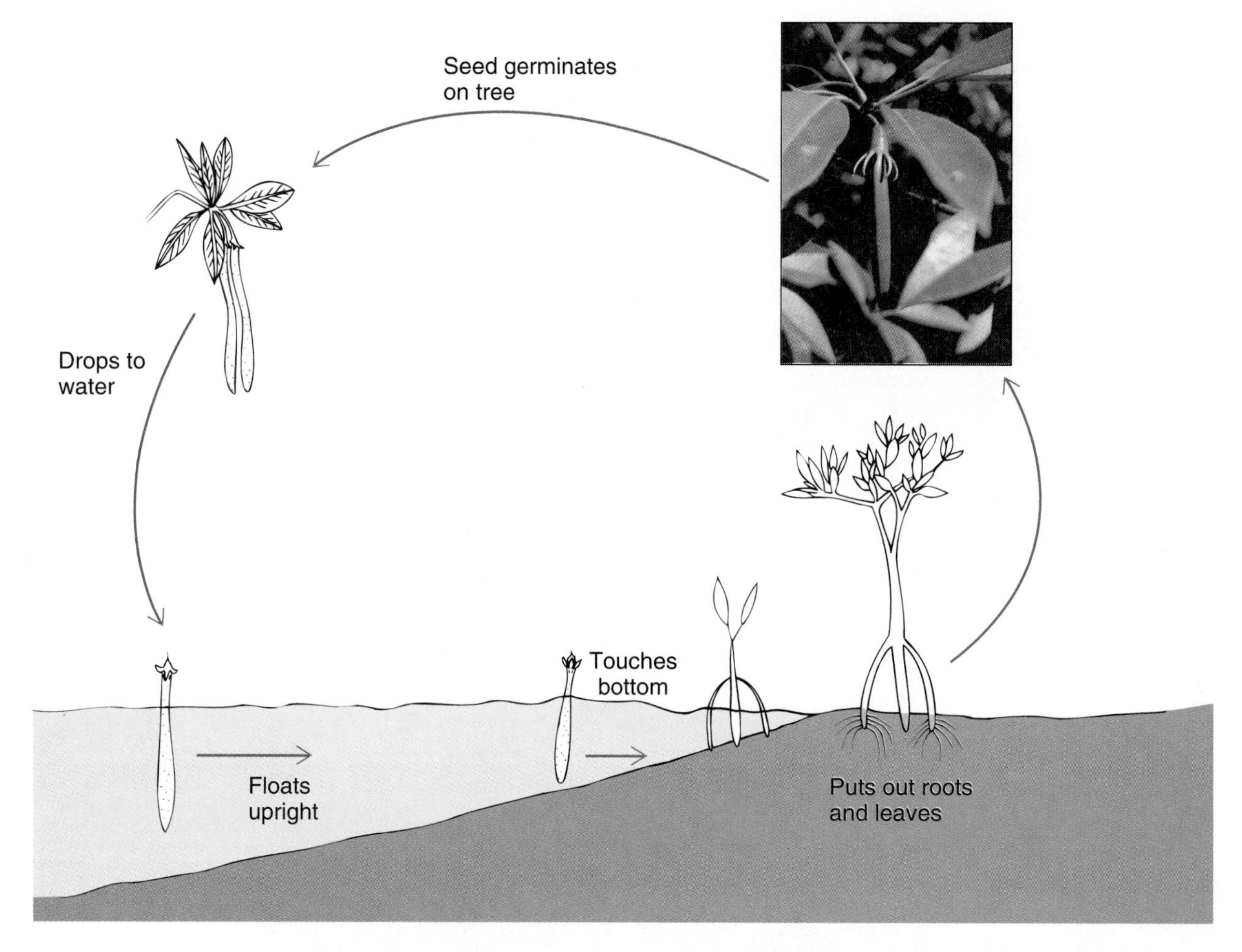

FIGURE 9.48 The life cycle of a typical mangrove tree.

FIGURE 9.49 Mangrove seedlings coming up among the stilt roots of the adult trees. Lizard Island, Australia. (Photo courtesy of Dr. Roger Seapy.)

properly and put down roots. These shores may be directly along the lee sides of islands or island chains or on islands or landmasses behind protecting offshore coral reefs. They are particularly well developed in estuarine portions of the tropics, where they reach their greatest areal extent.

Mangroves occur over a larger geographical area than coral reefs and may be found in areas well outside the tropics, such as along the west coast of central and northern South America and Africa, where reefs are rare or absent, and as far south as the north island of New Zealand and nearly to the mouth of the Rio de la Plata in Argentina (Figure 9.50). These mangrove forests may also penetrate some distance upstream along the banks of rivers (as far as 300 km along the Fly River in New Guinea; see McNae, 1968). Mangroves are usually absent from atolls and isolated high islands like Hawaii. (Mangroves were introduced to Hawaii in 1902.)

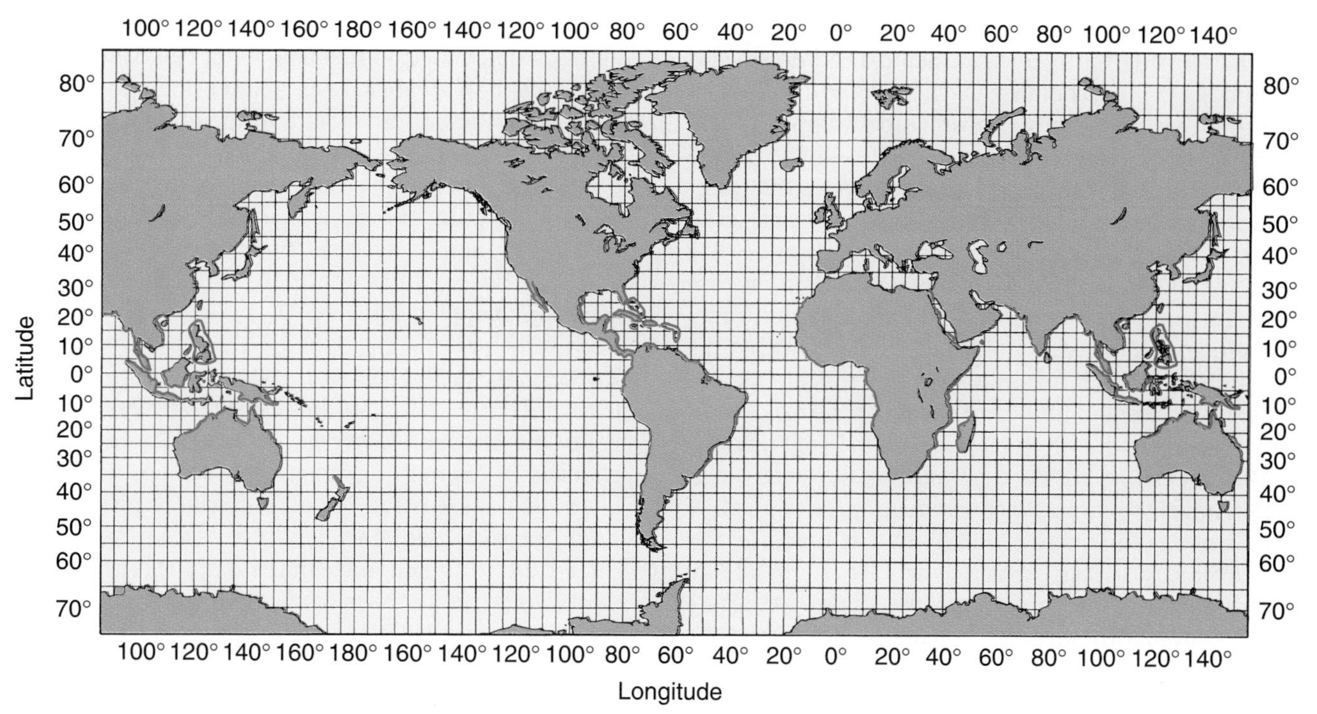

FIGURE 9.50 Distribution of mangrove forests in the world.

TABLE 9.5

Species Diversity of Mangroves in Different Geographical Regions

Region	*Number of Genera*	*Number of Species*
Australia/New Guinea	16	35
Asia/Indonesia	17	39
Esat Africa/Madagascar	8	9
West Africa	3	5
West Atlantic/Caribbean	3	6
Eastern Pacific	4	7

Source: Species Diversity, Vol. 1: *Ecological Communities* by Riclefs and Schluter, p. 218. Copyright © 1993 University of Chicago Press. Reprinted by permission.

As with corals, the diversity of mangroves varies geographically. The largest number of species is in the Indonesia-Australia-New Guinea area; the eastern Pacific and the tropical Atlantic are represented by only a few species (Table 9.5).

Physical Conditions of Mangrove Forests

Since mangroves can establish themselves only where there is no significant wave action, the first physical condition noted in mangrove areas is that the water motion is minimal. This lack of vigorous water motion has significant effects. Slow water movement means that fine sediment particles settle out and accumulate on the bottom. The result is an accumulation of mud; hence, the substrate in mangrove swamps is usually mud. In this sense, then, these areas are much like the previously discussed muddy shores, where poor interstitial circulation and high bacterial numbers lead to anoxic conditions. Perhaps this is also why mangroves have such shallow roots and pneumatophores. We should note that mangroves can grow on many types of substrates, so that it is not unusual to see small islands of mangroves even on coral rock or sand. The extensive mangrove forests, however, flourish only on muds or fine-grained sediments (Odum et al., 1982).

The initial slow water movement in mangrove forests is further enhanced by the mangroves themselves. As we noted above, many mangroves have peculiar prop (stilt) roots, which extend downward from the trunk and branches. These roots are often so numerous that they form an impenetrable tangle between the surface of the mud and the surface of the water. These dense root systems decrease the water movement even more, and the finest particles settle out around the roots of the mangroves, creating ever accumulating layers of sediment. Once dropped, the sediment is usually not picked up again and transported out. Sediment rapidly accumulates within the mangrove association. Mangrove forests are significant producers of new coastal land at the expense of the sea. The rate at which this occurs is remarkable. McNae (1968) notes, for example, that the great Venetian traveler Marco Polo visited a port city called Palembang in Sumatra in 1292. At that time, the city was directly on the sea. At present it lies 50 km inland! This translates into an accumulation of land at the rate of 73 m/year over the last 675 years. In other areas, the rate may be even faster (200 m/year in the Bodri Delta of east Java). This creation of land means that, as one progresses through a mangrove association from seaward side to landward edge, one will see a progressive change from marine to terrestrial conditions. This is probably also the main reason for the zonation observed in mangrove forests. The soils of mangrove forests are low in oxygen and high in salt content. They are fine grained, with a high organic content.

Although mangroves seem to be facultative halophytes and are capable of growing in freshwater areas, mangrove forests do not develop in strictly freshwater environments. This is probably because, as Kuenzler (1974) notes, they are not good competitors and the salinity is important in reducing competition from freshwater and terrestrial vascular plants.

The final physical factor to be considered is the tide. The tidal range and type vary across the geographical range of the mangroves. Mangrove forests develop only in shallow water and intertidal areas and are thus strongly influenced by the tides. Whereas tidal action is not a direct requirement for the mangrove according to Odum et al. (1982), it is indirectly important in a number of ways. First, it helps exclude other competitively superior vascular plants, permitting mangroves to exist. Second, tides bring salt water up the estuaries, permitting mangroves to penetrate inland. Tides also transport nutrients into and export material out of mangrove forests. Finally, the action of the tides prevents soil salinities from reaching lethal levels in areas of high evaporation, and they aid in dispersing propagules. Perhaps these are the reasons mangrove communities reach their greatest development where there are large tidal fluctuations. It is the tide and its vertical range that determine the periodicity of the inundation of the forests. This periodicity of inundation is of considerable importance in determining what kinds of mangrove associations develop in an area and may be responsible for the different types of zonation observed.

Zonation

Until recently, most studies on mangrove forests have been descriptive works with an emphasis on the changes in vegetation across the associations, from seaward edge to true terrestrial communities. As might be expected, each of these studies has produced differing schemes that reflect the differing underlying conditions, such as tidal factors. No one has suggested a truly universal scheme, such as that proposed for rocky shores (see Chapter 6), but a general scheme of wide applicability for mangrove forests of the Indo-Pacific region may be suggested.

The seaward area of most Pacific mangrove forests is dominated by one or more species of *Avicennia*. This *Avicennia* fringe is usually narrow because *Avicennia* seedlings do not grow well under conditions of shade or heavy siltation, which prevail in the interior of the forest. Associates in this zone and growing seaward of it are trees of the genus *Sonneratia*, which grow where they experience daily wetting (Figure 9.51).

Behind the *Avicennia* fringe lies the *Rhizophora* zone, dominated by one or more species of *Rhizophora*. These are the trees that most characterize mangrove communities, because they have the arching stilt roots that make these areas such an impenetrable maze for humans. Species of *Rhizophora* are often quite tall and develop across a broad area of the intertidal, from levels flooded at all high tides to areas flooded only at highest spring tides.

As one progresses landward, the next zone is the *Bruguiera* zone. Trees of the genus *Bruguiera* develop on more heavy sediment (clays) deposited at the high water level. The final zone of the mangrove forest, which is sometimes present, is the *Ceriops* zone, an association of small shrubs. This is a variable zone and may be merged with the trees of the *Bruguiera* zone.

In the Americas, mangrove zonation is less pronounced, in part due to the lesser number of mangrove species. For example, many Indo-Pacific regions have 30–39 mangrove species present, but Florida has just

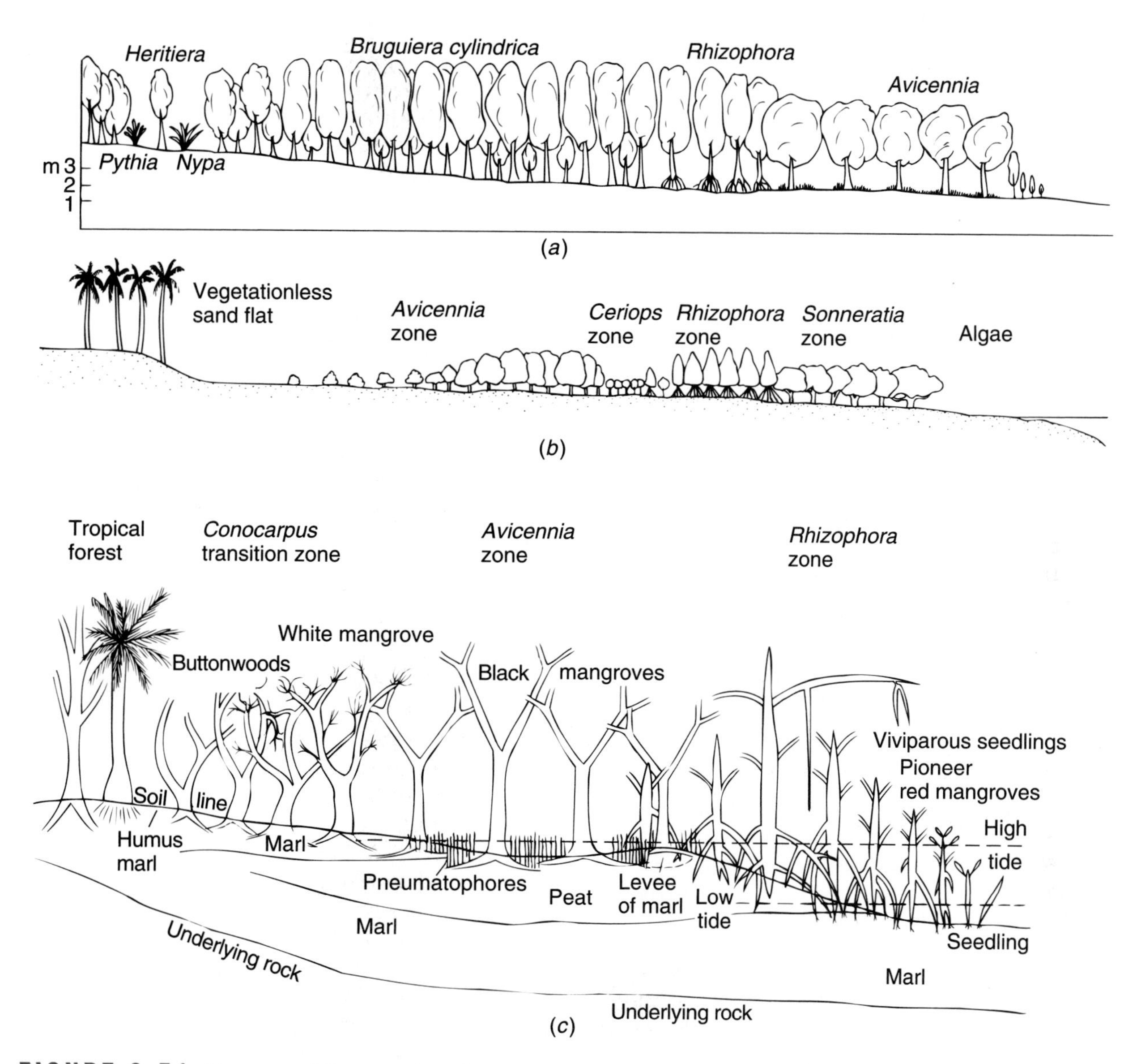

FIGURE 9.51 Diagrams of the zonation of mangroves in several areas. (a) Malaya. (b) East Africa. (c) Florida. (a, b, from *Ecosystems of The World*, Vol. 1, ed., V. J. Chapman, 1977 c, from "The Ecology and Geologic Role of Mangroves in Florida," J. H. Davis, Jr., 1940, Carnegie Institute of Washington, Publication 517, Tortugas Lab, Paper 32. Reprinted by permission.)

three: red mangrove (*Rhizophora mangle*), black mangrove (*Avicennia germinans*), and white mangrove (*Laguncularia racemosa*). Another species, buttonwood (*Conocarpus erectus*), though not a mangrove, often occurs between mangroves and terrestrial vegetation. Zonation in the Americas, according to West (1977), places the *Rhizophora* mangal zone at the seaward edge. This zone is followed inland by a broad band of *Avicennia* and then a belt of *Laguncularia* or *Conocarpus* (Figure 9.51c).

It should also be remembered that the above are general schemes and not all mangrove forests will correspond to them. In fact, in some areas, the association may be considerably abbreviated or represented only by a few individuals. This is particularly true at the limits of the distribution of the mangroves. The zonation may also be interrupted where local conditions are such that evaporation of water from the soils makes them hypersaline. Hypersalinity kills mangroves, creating bare areas. Full development of mangrove forests is found in areas of high rainfall or in areas where rivers furnish enough fresh water to preclude the development of hypersaline conditions.

Zonation may also be limited by tidal action. Wherever the tidal range is small, the intertidal zone is also restricted, as are the mangrove forests. The most extensive forests are developed on shores that have a substantial vertical tidal range.

Associated Organisms

Mangrove communities are unique. Due to the height of the trees, true terrestrial organisms can occupy the portions of the tree that are never in water, while true marine animals occupy the bases that are in the water. Mangrove forests, then, form a strange mixture of marine and terrestrial organisms and have been suggested as a pathway from land to sea, and vice versa.

Terrestrial organisms show no special adaptations for life in the mangrove forest, since they spend their lives out of reach of the marine waters in the upper reaches of the trees, though they may forage on the marine animals at low tide. There is a large variety of terrestrial arthropods in the mangrove forest. It includes herbivorous and carnivorous insects, but their significance to the mangrove trees is little known.

The marine organisms are of two types: those that inhabit the hard substrate of the numerous stilt roots of the mangroves and those that occupy the mud. Mangrove associations are different from those of muddy shores primarily because of the vast, hard surface area of roots available to organisms, which is absent from typical mud shores (see pp. 292–293).

The dominant groups of marine animals in the mangrove forests are mollusks, certain crustaceans, and some peculiar fishes. Mollusks are represented by a number of snails, one group of which generally lives on the roots and trunks of the mangrove trees (Littorinidae) and the other on the mud at the base of the roots, comprising mainly detritus feeders (Ellobiidae and Potamididae). Little is known of the contribution of these snails to the mangrove forest. A second group of mollusks includes the bivalves, especially oysters, which attach to the mangrove roots, forming a significant biomass (Figure 9.52).

Mangrove forests are also inhabited by numerous large-sized crabs and shrimps. These animals excavate burrows in the soft substrate and include such common genera as *Uca*, the fiddler crabs; *Cardisoma*, the tropical land crabs; and various ghost crabs (*Dotilia*, *Cleistostoma*). The mangrove crab *Aratus pisonii* is extremely abundant in Caribbean mangroves, where it lives on the roots just above the waterline. Also abundant in Caribbean mangroves is *Goniopsis cruentata*, which lives among the roots. It does not construct burrows but may take refuge in the burrows dug by other crabs. These crabs usually are specialized to feed on the detrital particles found in the mud. Generally, they separate the organic detrital particles from the nonorganic matter by filtering the substrate through a set of fine hairs around the mouth. These crabs also show varying degrees of adaptation toward a more terrestrial mode of life. This usually expresses itself in a vascularization of the walls of the gill chambers so that they become more "lunglike." As we noted previously, mangrove associations are transition areas between land and sea, and the crabs living in mangroves certainly reflect this in their partial adaptation to air breathing.

The burrows of these crabs, as well as those of mud shrimps, such as *Upogebia* and *Thalassina*, serve several functions. They provide refuges from predation, a breeding place, and an aid to feeding. They also serve the mangrove community in that they allow oxygen to enter more deeply into the substrate; thus, they ameliorate anoxic conditions. The mangrove crab *Aratus pisonii* is not a particle feeder. It appears to be an omnivore with a diet that includes mangrove leaves. Thus, it is a potentially significant herbivore. According to Sousa and Mitchell (1999), some of the mangrove crabs are consumers of mangrove seedlings and may be significant in maintaining mangrove zonation by differential consumption of different species' seedlings. Crabs are, however, not the only organisms that feed on mangroves. Devlin (1999) has reported that the most significant consumer of mangrove propagules and seedlings in Florida is the beetle *Cocotrypes rhizophorae*, and there is even a report of mangrove foliage being consumed by green sea turtles!

The large numbers of marine invertebrates inhabiting the prop roots form an association that is poorly understood and little studied. The few studies done suggest that some intriguing interactions occur among the epifaunal organisms and that they have a corresponding significance to the mangroves. For example, Perry (1988), working on the red mangrove *Rhizophora mangle* in Costa Rica, experimentally demonstrated some interesting ecological interactions among the mangrove, the epifauna on the mangrove roots, and various benthic and water column

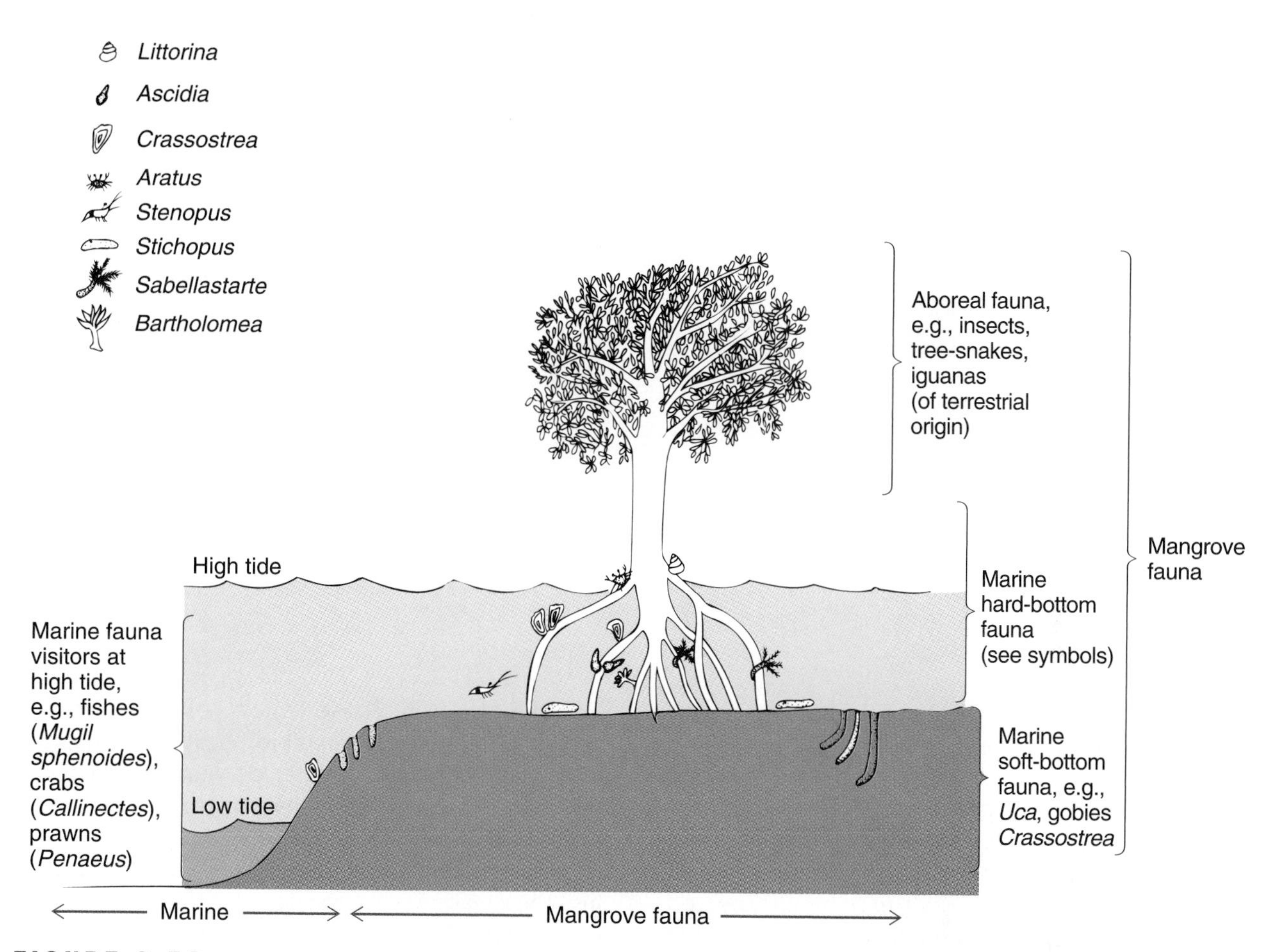

FIGURE 9.52 Representation of the macrofauna of a mangrove forest showing the vertical distribution and ecological relationships. (Modified from *Marine Biology*, H. Friedrich, Sedwick & Jackson.)

predators. In this system, the isopod *Sphaeroma peruvianum* bores into the roots of the mangrove, while barnacles (*Balanus* spp.) settle on the surface of the roots. Both cause a decrease in root growth rate, of 50% and 30%, respectively, as well as a decrease in root production of over 50% in both cases. However, not all roots are equally affected. Those roots not in contact with the substratum are most affected, whereas those in contact suffer considerably less. The reason for this discrepancy is that those roots in contact with the substrate are accessible to a benthic predator, the hermit crab *Clibarnarius panamaensis*, which climbs the roots at high tide to feed on barnacles. According to Perry, the absence of the isopod from certain roots can also be attributed to predation, as demonstrated by studies of predator exclusion. The exact predator, however, remains unknown. A similar situation may occur with respect to algae. Taylor et al. (1986) have found that in mangroves in Belize, fleshy algae grow on roots that do not intersect the bottom but not on those that do. Their experiment demonstrated that the hanging roots offered protection from benthic herbivores, such as urchins.

According to Sutherland (1980), who studied the same red mangrove species in Venezuela, these epifaunal organisms that occur on mangrove roots seem to be long-lived and resistant to displacement by other organisms. They also show little or no evidence of succession, a given root remaining with the same species array for many years. In contrast to Perry, Sutherland found predation had little effect on community development. These conflicting studies suggest that our understanding of how species interact in mangroves is still very poor.

FIGURE 9.53
A fish of the genus *Periophthalmus* in the mud of a mangrove swamp. (Photo courtesy of Dr. Roger seapy.)

Mangrove roots are not the only parts of the plant subject to consumption. Wada and Wower (1989) demonstrated that the ocypodid crab *Macrophthalmus quadratus* grazes on the pneumatophores of the mangrove *Sonneratia alba* and removes epiphytic algae and bark. Onuf et al. (1977) found that mangroves in Florida responded to nutrient enrichment derived from the guano of a bird-breeding colony by higher growth rates compared with nonenriched areas, but that the nutrient-enriched mangroves also had increased herbivory on the leaves by terrestrial herbivores. Underwood and Barrett (1990) have experimentally demonstrated a relationship between the root-dwelling oyster *Crassostrea commercialis*, which resides on the roots of the mangrove *Avicennia marina*, and the presence of the littorinid snail *Bembicium auratum*; the oysters increase the distribution, abundance, and size of the *Bembicium*. Underwood and Barrett ascribe this positive relationship to the ability of the oysters to provide a refuge from predation for the littorinid, thus protecting it from crustaceans and fish.

Mangrove areas also are nursery grounds for penaeid shrimps and such fishes as mullet, which may spend the early part of their life cycles in these areas before moving offshore. The prop root area of mangroves is known to be a nursery for juvenile spiny lobsters (*Panulirus argus*) in the Caribbean. They may spend up to two years in this area.

Conspicuous by their size and abundance in the water and on the mud in Indo-Pacific mangrove forests are the small, big-eyed fishes of the genus *Periophthalmus* and relatives. These fishes, collectively called mud-skippers, are remarkable because they spend most of their time out of water, crawling around on the mud flats or even climbing the mangrove roots (Figure 9.53)! In fact, they act much like frogs or toads. These fishes create burrows in the mud for refuge and breeding. They are quite fast, moving over the exposed mud flats by "walking" on their strong pectoral fins or by using a series of "skipping" motions or "bounds," in which the tail and caudal fins provide the thrust.

Aside from their peculiar unfishlike habit of crawling around on the mud flats, the other characteristic feature of these fishes is the modification of the eyes. The eyes are set high on the head and are arranged so that they focus best in air, not in the water. When mudskippers are swimming, the eyes protrude above the surface; hence, these fishes look for all the world like frogs. A final noteworthy adaptation is in the respiratory system. The gills are often reduced, and aerial respiration is accomplished by vascularized sacs in the mouth cavity and gill chambers.

Succession and Mortality

Mangroves are subject to mortality from a number of different natural and human-induced causes. They exist under delicately balanced conditions involving a somewhat predictable steady sedimentation rate, minimal water movement, a certain tidal regime, and water and soil of a certain salinity. Any change that upsets this balance produces corresponding changes in the mangrove community. If

FIGURE 9.54
Photo of dead mangroves destroyed in Everglades National Park by the hurricane of 1960. (Photo by the author.)

the changes are slow enough, there may be a gradual change or succession.

Davis (1940), working in Florida, was the first to suggest a connection between zonation, succession, and changes in physical factors. In this now classical view, Davis argued that the mangrove zonal patterns were equivalent to seral stages in succession. In his scheme, the seaward zone of red mangroves is the "pioneer" or initial seral state. Farther landward, the black and white mangrove zones were regarded as progressively later stages of the successional sequence, leading eventually to terrestrial communities. This view, however, is not universally accepted. Other workers, such as Thom (1975), have demonstrated that mangrove zonation is mostly a response to external physical factors rather than a successional sequence induced by the plants themselves. This view is supported by Rabinowitz (1975), who demonstrated in transplant experiments that various species of mangroves could grow well within any zone and that physical factors alone were not responsible for their absence. It now appears that zonation of mangroves is not controlled directly by physical factors but by an interplay of these factors with interspecific competition and propagule dispersion and sorting. Unless there is heavy sedimentation, once a mangrove area reaches an equilibrium state, change occurs only if there is some sort of disturbance. That is, there is no further "succession" induced simply by the plants. Disturbances may be small (lightning strikes, fires, or droughts) or they may be large (sea level changes or hurricanes). Such perturbations set zonation back to an earlier state. Evidence from Florida, according to Odum et al. (1982), suggests that hurricane perturbations are frequent and regular enough to keep mangrove systems in a cyclical pattern of succession rather than in the classic facilitation sequence that would continually produce additional terrestrial forests as the climax stage. In this newer view, the role of mangroves in building land at the expense of the sea is seen as more passive than active, and various physical processes are the dominant forces determining the rate of growth or recession of mangrove shorelines. Mangroves in this scenario have the role of stabilizing sediments, the origin and major deposition of which are due to physical processes.

Perhaps the greatest cause of large-scale mortality is typhoons and hurricanes. These violent storms destroy large areas of mangrove forests by uprooting trees, by massive sedimentation, or by altered salinity of water and soil. Hurricane Donna, which struck Everglades National Park in 1960, caused a mortality ranging from 25 to 75% of 100,000 acres of mangroves in the park. Recovery from such massive mortality has a similar time span for mangroves as for coral reefs, and estimates of recovery time are about 20–25 years (Figure 9.54).

Another suggested source of mortality is the small isopod *Sphaeroma terebrans*. This isopod burrows into the prop roots of red mangroves, and the boring, plus secondary decomposition by bacteria and fungi, usually severs the prop root at the level of mean high tide. Rehm and Humm (1973) were the first to describe this phenomenon and suggested that this prop root destruction left the underlying sediment free to erode, allowing the shallow-rooted trees to fall into the water. In this way, the whole forest would slowly be cut down and destroyed. Since that time other workers have taken issue with this "eco-catastrophe" view. For example, Estevez and Simon (1975) have shown that the infestations of the isopod are patchy and limited in extent. They also demonstrated that burrowing was related to salinity changes. Simberloff et al. (1978) suggested that since the boring stimulated the mangrove to produce more root growth and branching, which is beneficial to the tree, the isopod has a positive effect. Thus, we find ourselves in the peculiar position of having dramatically opposed views concerning this unresolved problem.

Most other massive mortality is due to human activity. Perhaps the greatest destruction was as a result of herbicide spraying on the mangrove forests of Vietnam during the war. According to Tschirley (1969), perhaps 100,000 hectares of these forests were defoliated and destroyed by herbicides during the war. It is not known how long recovery of these forests will take, or whether they will regenerate at all in view of the extreme sensitivity of mangroves to residual herbicides.

Mangrove forests also have been destroyed by filling, dredging, and channelizing waters. Historically, mangrove forests probably occurred along the Persian Gulf, from which they are now absent, due to cutting by early humans for firewood and boat building. In the tropics today, the mangroves are exploited by humans mainly for firewood, but they may be partially destroyed in order to build ponds for fish and shrimp culture or for salt production; mangroves have also been felled for construction.

Mangrove forests have demonstrably important value to humans. They are the nurseries for penaeid shrimp, spiny lobsters, blue crabs, and various fishes, such as the mullet. The whole Tortugas pink shrimp industry of Florida, an 11-million-pound-per-year harvest, is dependent on the mangrove system. Mangroves also shelter a number of endangered species, such as the American crocodile, the brown pelican, and the Atlantic Ridley sea turtle. Finally, their ability to stabilize land in the face of tropical storms may be important in protecting human habitations further inland.

SUMMARY OF KEY CONCEPTS

- Coral reefs are estimated to occupy 0.17% of the area of the planet.
- Coral reefs have the greatest higher taxon diversity of any marine ecosystem, comprising perhaps 4–5% of all species.
- Coral reefs are among the most productive systems in the marine environment and have existed for hundreds of millions of years, yet the living coral is sensitive to a number of natural and human-made disturbances.
- Coral reefs are unique among marine communities in that they are built up entirely by biological activity.
- Hermatypic corals have symbiotic zooxanthellae algae in their tissues and are the corals that produce reefs; ahermatypic corals lack zooxanthellae and do not produce reefs.
- Six major physical factors limit coral reef development: temperature, depth, light, salinity, sedimentation, and emergence into air.
- Coral reefs are limited to waters of full salinity, that are not deeper than 50–70 m, do not fall below 18°C, and do not have excessive sediment.
- Corals are members of the phylum Cnidaria and most hermatypic corals are colonial, the various individual polyps occupying cups or corallites in the massive skeletal mass.
- Coral reefs may be grouped into three categories: fringing, barrier, and atoll.
- The subsidence or compensation theory of Darwin explains the existence of atolls and links fringing, barrier, and atoll reefs together in an evolutionary sequence, but it is not an explanation for all fringing and barrier reefs.
- Although the corals are the major organisms that form the basic reef structure, members of practically all phyla and classes may be found on coral reefs.
- Other organisms contributing strongly to the calcium carbonate budget of reefs include the coralline algae, the green algal genus *Halimeda*, the hydrocoral *Millepora*, and various mollusks including the giant clams.
- Other prominent groups on coral reefs include the soft corals, gorgonians, echinoderms, sponges, crustaceans, polychaete worms, and fishes.
- The largest number of reef coral species and genera occurs in the Indo-Pacific in an area that includes the Philippines, Indonesia, New Guinea, and northern Australia; the number of species declines in all directions away from that area.
- The tropical Atlantic reefs have many fewer coral genera and species than the Indo-Pacific.

- Coral reefs are complex associations of organisms that have a high diversity of habitat types, which is one reason for the high diversity of species.
- The zonation pattern across an atoll beginning at the windward side and progressing across the atoll to the leeward side includes an outer seaward slope with a windward reef margin, spur and groove or buttress zone, algal ridge, reef flat zone, seaward beach of the atoll island, island, lagoon with lagoon reefs and seagrass beds, leeward reef flat, and leeward reef margin.
- There are many differences between Atlantic and Pacific coral reefs, including structure and zonation features as well as presence, absence, and abundances of various faunal and floral components.
- Coral reefs maintain a high primary productivity in nutrient-poor tropical ocean waters, but there is little agreement among scientists as to the underlying sources of the nutrients that sustain the high productivity.
- Other factors that help sustain high productivity include the large amount of plant tissue present on reefs, close coupling of the photosynthetic organisms with the animals or matrix of the reef, and rapid cycling of nutrients by the large bacteria populations.
- Although reef corals are known to be carnivores similar to other cnidarians, it now appears that these animals are phototropic, depending upon the symbiotic zooxanthellae for much of their nutrition.
- Coral growth and calcification depend upon sufficient light for the zooxanthellae. The zooxanthellae increase the rate of calcification and the growth of the coral colonies.
- The rate at which coral colonies grow differs with different species, with age of the colonies, and in different areas of the reef.
- We are ill informed about the ages of corals, and what little information we have suggests that most coral colonies are ten years or younger, but that some colonies can reach ages of 100 years or more.
- Corals reproduce both sexually and asexually. Asexual reproduction is usually by budding off new polyps from the parent and increases the size of the colony, but fragmentation of the parent colony can also lead to new colonies. Sexual reproduction results in a planula larvae, which exits the parent to initiate a new colony.
- Most corals are hermaphroditic and may either brood the fertilized eggs to the planula stage or release the gametes to be fertilized externally in seawater.
- The rate of recruitment of new coral colonies to the reef system is variable in time and in space.
- Corals compete with each other for light mainly by overtopping and shading other colonies, a process that is a type of exploitative competition. Still other species maintain space on the reef by killing adjacent competing species with their nematocyst-studded mesenterial filaments, a process called interference competition.
- Corals capable of using their mesenterial filaments in interference competition can be arranged in a specific order so that when different species are paired, the winner could be predicted. This aggressive pecking order was originally suggested as a means to maintain coral diversity, but the competitive interaction is now known to be more complicated and there are also other mechanisms employed by corals and other coral reef invertebrates in competitive interactions.
- Another group of organisms in which space competition seems common is the cryptofauna.
- The role of predation in determining the structure and composition of coral reefs is not well studied or understood. Taken together, most predators of corals do not appear to have a significant effect on coral colonies nor do they seem to be able to affect the community structure. Most of these predators are small relative to the coral colony and do not remove enough coral tissue to kill the colony.
- Two taxa of predators are capable of destroying whole coral colonies and modifying the structure of reefs. The first is the starfish *Acanthaster planci*, which, when present in large numbers, can consume whole reefs. The second is the various corallivorous fishes.
- Algae are the major space competitors with corals on coral reefs and can outcompete them, but they are kept from doing so by the intense grazing pressure of herbivorous fishes and urchins.
- Grazing can also produce shifts in the abundance of different groups of algae. Fleshy filamentous algae grow faster and may eliminate the slow-growing coralline algae, but the fleshy algae are preferred by the grazers and thus under heavy grazing pressure the coralline algae survive.
- Algae deter herbivores by being less palatable through deposition of calcium carbonate, by producing noxious chemicals, and by settling into habitats that provide spatial or temporal refuges from grazers.
- Moderate grazing prevents monopolization of the substrate by algae and increases diversity on reefs, whereas heavy grazing destroys not only the algae but corals as well and decreases diversity.
- Damselfish defend territories against other grazing and corallivorous fishes and in so doing permit the algae to overgrow, shade, and kill young slow-growing massive coral colonies while enhancing the growth of fast-growing branching corals, so that the latter dominate such territories.
- Herbivory and predation are more intense on Pacific reefs than on Atlantic reefs.
- Large numbers of bacteria exist on reefs, representing a huge energy source and suggesting the presence of a microbial loop system similar to that found in the plankton, but the existence of such a loop is still being debated.
- Algae are important to the reef in cementing calcium carbonate pieces together, in the production of sand, in erosion of reefs, as primary producers, and in furnishing habitat.
- Coral reefs harbor more species and diverse fish communities than any environment on earth, but only a fraction

of this diversity can be explained by the large variety of habitats on the reef.

- Four competing hypotheses exist to explain the high local fish diversity of reefs. These are the competition model, the lottery hypothesis, the predator disturbance model, and the recruitment limitation model.
- There are marked differences in the fish species active on the reef between day and night.
- The largest trophic group of fishes on reefs are predators, constituting 50–70% of the species; most are unspecialized opportunists.
- The second largest trophic group of fishes are the herbivores, while the remainder of the fishes are generally classed as omnivores with a few planktivores.
- Venomous fishes are rare on reefs, but many fishes have toxic secretions on their body surfaces.
- Ciguatera is a disease of humans caused by eating certain fishes whose flesh contains toxins produced by the dinoflagellate *Gambierdiscus toxicus*.
- Mass mortalities of reef fishes are uncommon.
- Little is known of the population dynamics of reef fishes.
- Cleaning behavior is a specialized form of predation in which small fishes or shrimps remove ectoparasites from other, usually larger, fishes.
- The cryptofauna refers to those invertebrates that inhabit the coral substrates living in cracks, crevices, and caves and under and in coral heads and rock. They are important bioerosive forces on reefs.
- Present data give conflicting evidence as to whether or not ecological succession occurs on coral reefs.
- On short time scales reefs appear to be stable entities, but not on longer time scales.
- There are three known natural causes of mass mortality on reefs. One is hurricanes or typhoons, a second is population outbreaks of the corallivorous sea star *Acanthaster planci*, and the last is the meteorological and oceanographic phenomenon known as El Niño. Another puzzling event called coral bleaching may also cause significant coral mortality.
- Human activity may also cause mass mortality of reefs by dredging, mining, overfishing, and pollution.
- White-band and black-band disease may cause local die-offs, but as yet they have not caused mass mortality.
- Recovery times for reefs that have undergone mass mortality range from 7 to 100 years.
- Coral reefs throughout the world are suffering from an onslaught of both natural and human disturbances that have degraded or destroyed many reefs. Many have not recovered in the decade or more since devastation, leading to considerable concern among humans about their survival.
- Mangrove forest or mangal is the term used to describe a variety of tropical inshore communities dominated by species of trees or shrubs that grow in salt water. They are found on 60–75% of the tropical coastlines of the planet.
- Mangroves are shallow-rooted plants with widely spread roots and often with peculiar prop roots that stem from the trunk or branches, commonly with pneumatophores, and with tough leaves and internal water storage tissue.
- Nearly all mangroves have vivipary and dispersal by water.
- Mangroves occur over a larger geographical area than coral reefs but are restricted to shores protected from wave action.
- Mangrove forests are areas of slow water movement, further slowed by the presence of the numerous roots of the mangroves themselves, thus accumulating sediment.
- The soils in mangrove forests are fine grained, low in oxygen, and high in organic content and salt content.
- Mangrove forests are strongly influenced by the tides, which exclude other potential plant competitors, bring salt water further up estuaries permitting mangroves to penetrate further inland, prevent soil salinities from reaching lethal levels, and disperse the mangrove propagules.
- The general zonation pattern for Pacific mangrove forests is to have a seaward area dominated by *Avicennia*, behind which is a zone of *Rhizophora*, then a *Bruguiera* zone, and most landward a *Ceriops* zone.
- Zonation is less pronounced in the Americas, where the *Rhizophora* zone is the most seaward, followed by *Avicennia* and then *Laguncularia* or *Conocarpus*.
- Zonation patterns may be interrupted by local conditions, mainly hypersalinity, which kills mangroves.
- Mangrove forests consist of organisms derived both from the marine environment and the terrestrial environment. Marine organisms include those that attach to the hard substrate of the roots and those that occupy the mud.
- Invertebrates inhabiting the mud often dig burrows that allow oxygen to enter more deeply into the substrate, ameliorating the anoxic conditions.
- Invertebrates inhabiting the prop roots have some intriguing interactions, but in general the prop root association is poorly understood and studied.
- Mangroves are also nursery grounds for shrimp, lobsters, and fishes that live as adults offshore.
- Mangroves are subject to massive natural mortality, primarily by hurricanes and typhoons, and recovery times are in the same time range as for reef recovery.
- Another suggested agent of large-scale mortality by some scientists is the isopod *Sphaeroma terebrans*, which bores into the prop roots. However, other scientists think the isopod has a positive effect.
- Most other massive mortality is due to human activity. Historically, the mangroves have been destroyed by filling, dredging, cutting for firewood, and building ponds for aquaculture.
- The greatest destruction of mangroves in recent times was the result of herbicide spraying on the mangrove forests of Vietnam during the war.

REVIEW QUESTIONS

ESSAY: Develop complete answers to these questions.

1. Other than in the composition of coral species, how do coral reefs of the Indo-Pacific and Atlantic oceans differ? Discuss.
2. Examine the impact that multivorous coral predators have on coral reefs in terms of both destruction and stabilization.
3. Compare and contrast exploitative and interference competition, providing appropriate examples of each to support your essay.
4. Explore the arguments that attempt to explain the high diversity of coral reef fish. Present evidence that supports each idea.
5. Explore how mangals have both horizontal and vertical zonation patterns. How does the tide affect both?
6. What role does the crown-of-thorns starfish (*Acanthaster*) play in coral reef stability (structural and biological)?
7. Compare the primary productivity values of atoll reefs with those of open tropical oceans of the same latitude. Why do the open tropical oceans have such a low productivity? In contrast, why are the coral reefs so productive, especially in light of the fact that the waters of the coral reef are nutrient poor?
8. What environmental factors cause destruction of mangals? What other forces are at work that destroy mangals?

BIBLIOGRAPHY

Adey, W. H., I. G. MacIntyre, and R. Stuckenrath. 1977. Relict barrier system off St. Croix: Its implications with respect to late Cenozoic coral reef development in the Western Atlantic. *Proc. 3rd Int. Coral Reef Symp.* 2:15–22.

Anderson, I. 1990. Darwin may founder on the Great Barrier Reef. *New Scien.* (October):15.

Atkinson, M. J. 1992. Productivity of Enewetak Atoll reef flats predicted from mass transfer relationships. *Cont. Shelf Res.* 12(7/8):799–807.

Babcock, R. C. 1985. Growth and mortality in juvenile corals (*Goniastrea, Platygyra* and *Acropora*): The first year. *Proc. 5th Int. Coral Reef Symp.* 4:355–360.

Bakus, G. 1964. The effects of fish-grazing on invertebrate evolution in shallow tropical waters. Occ. paper no. 27, Allan Hancock Foundation.

Bakus, G. 1973. The biology and ecology of tropical holothurians. In *Biology and geology of coral reefs*, edited by O. A. Jones and R. Endean. Vol. II, *Biology* 1. New York: Academic Press, 325–367.

Banner, A. H. 1976. Ciguatera: A disease from coral reef fish. In *Biology and geology of coral reefs*, edited by O. A. Jones and R. Endean. Vol. 3, *Biology* 2. New York: Academic Press, 177–213.

Birkeland, C. 1977. The importance of rate of biomass accumulation in early successional stages of benthic communities to the survival of coral recruits. Vol. 1, *Biology. Proc. 3rd Int. Coral Reef Symp.* 15–21.

Birkeland, C. 1982. Terrestrial runoff as a cause of outbreaks of *Acanthaster planci* (Echinodermata: Asteroidea). *Mar. Biol.* 69:175–185.

Birkeland, C. 1989. The influence of echinoderms on coral reef communities. In *Echinoderm studies*, Vol. 3, edited by M. Jangoux and J. M. Lawrence. Rotterdam: A. A. Balkema, 1–80.

Birkeland, C., ed. 1997. *Life and death of coral reefs*. New York: Chapman and Hall.

Borowitzka, M. A. and A. W. D. Larkum. 1986 Reef algae. *Oceanus* 29(2)j 52.

Brock, R. E. 1979. An experimental study on the effects of grazing by parrot fishes and the role of refuges in benthic community structure. *Mar. Biol.* 51:381–388.

Brown, B. E. 1997. Disturbances to reefs in recent times. In *Life and death of coral reefs*, edited by C. Birkeland. New York: Chapman and Hall.

Brown, B. E., and J. C. Ogden. 1993. Coral bleaching. *Sci. Amer.* 268(1):64–70.

Buddemeir, R. W., and R. A. Kinzie III. 1976. Coral growth. *Oceanog. Mar. Biol. Ann. Rev.* 14:183–225.

Chapman, V. J., ed. 1977. *Ecostystems of the world*. Vol. 1, *Wet coastal ecosystems*. New York: Elsevier.

Chesher, R. H. 1969. Destruction of Pacific corals by the sea star *Acanthaster planci*. *Science* 165:280–283.

Colgan, M. W. 1982. Succession and recovery of a coral reef after predation by *Acanthaster planci*. *Proc. 4th Int. Coral Reef Symp.* 2:333–338.

Connell, J. 1973. Population ecology of reef building coral. In *Biology and geology of coral reefs*, edited by O. A. Jones and R. Endean. Vol. II. *Biology* 1. New York: Academic Press, 205–245.

Connell, J. H., T. P. Hughes, and C. C. Wallace. 1997. A 30-year study of coral abundance, recruitment and disturbance in space and time. *Ecol. Monogr.* 67:461–488.

Crossland, C. 1952. Madreporaria, Hydrocorallinae, *Heliopora*, and *Tubipora*. *Sci. Repts. Great Barrier Reef Exped. 1928–29* 6:85–257.

Dana, T. F., W. A. Newman, and E. W. Fager. 1972. *Acanthaster* aggregations: Interpreted as primarily responses to natural phenomena. *Pac. Sci.* 26:355–372.

Davies, P. J. 1983. Reef growth. In *Perspectives on coral reefs*, edited by D. J. Barnes. AIMS *Contribution*, no. 200:69–106.

Davis, J. H., Jr. 1940. The ecology and geologic role of mangroves in Florida. Carnegie Inst. Wash. Publ. 517, Tortugas Lab. Pap. 32:303–412.

Devlin, D. J. 1999. A field experiment of predator strategies and mangrove resistance: The relationship between the red mangrove (*Rhizophora mangle*) and the scolytid beetle *Cocotrypes rhizophorae*. *Gulf Res. Rep.* 10:73.

Doherty, P. J. 1982. Some effects of density on the juveniles of two species of tropical, territorial damselfishes. *J. Exp. Mar. Biol. Ecol.* 65:249–261.

Doherty, P., and T. Fowler. 1994. An empirical test of recruitment limitation in a coral reef fish. *Science* 263:935–939.

Dollar, S. J., and G. W. Tribble. 1993. Recurrent storm disturbance and recovery: A long-term study of coral communities in Hawaii. *Coral Reefs* 12:223–233.

Ducklow, H. W. 1990. The biomass, production and fate of bacteria in coral reefs. In *Ecosystems of the world* 25, *Coral reefs*, edited by Z. Dubinsky. New York: Elsevier, 265–289.

Endean, R. 1973. Population explosions of *Acanthaster planci* and associated destruction of hermatypic corals in the Indo-West Pacific region. In *Biology and geology of coral reefs*, edited by O. A. Jones and R. Endean. Vol. II, *Biology* 1. New York: Academic Press, 390–438.

Endean, R., and A. M. Cameron. 1990. *Acanthaster planci* population outbreaks. In *Ecosystems of the world* 25, *Coral reefs*, edited by Z. Dubinsky. New York: Elsevier, 419–437.

Estevez, E. D., and J. L. Simon. 1975. Systematics and ecology of *Sphaeroma* (Crustacea, Isopoda) in the mangrove habitat of Florida. In *Proceedings of the international symposium on the biology and management of mangroves*, edited by G. Walsh, S. Snedaker, and H. Teas. Gainesville: University of Florida, 286–304.

Fadlallah, Y. H. 1983. Sexual reproduction, development and larval biology in scleractinian corals. A review. *Coral Reefs* 2:129–150.

Frankel, E. 1977. Evidence from the Great Barrier Reef of ancient *Acanthaster* aggregations. *Atoll Res. Bull.* 220:75–93.

Franzisket, L, 1969. Riffcorallen konnen autotroph leben. *Naturwissenschaften* 56:144.

Ghiold, J. 1990. White death—the fate of a deserted coral. *New Scien.* 126 (1719):46.

Glynn, P. W. 1972. Observations on the ecology of the Caribbean and Pacific coasts of Panama. In *The Panamic biota: Some observations prior to a sea level canal*, edited by M. L. Jones. *Bull. Biol. Soc. Wash.* 2:13–30.

Glynn, P. W. 1973. Aspects of the ecology of coral reefs in the western Atlantic region. In *Biology and geology of coral reefs*, edited by O. A. Jones, and R. Endean. Vol. II, *Biology* 1. New York: Academic Press, 271–324.

Glynn, P. W. 1974. Rolling stones among the Scleractinia: Mobile coraliths in the Gulf of Panama. *Proc. 2nd Int. Coral Reef Symp.* 2:183–198.

Glynn, P. W. 1976. Some physical and biological determinants of coral community structure in the eastern Pacific. *Ecol. Monogr.* 46:431–456.

Glynn, P. W. 1984. Widespread coral mortality and the 1982–83 El Niño warming event. *Envir. Conserv.* 11:133–146.

Glynn, P. W. 1985. Corallivore population sizes and feeding effects following El Niño (1982–1983) associated coral mortality in Panama. *Proc. 5th Int. Coral Reef Symp.* 2:149.

Glynn, P. W. 1988. El Niño-Southern Oscillation 1982–1983: Nearshore population, community and ecosystem responses. *Ann. Rev. Ecol. Syst.* 19:309–345.

Glynn, P. W. 1990. Feeding ecology of selected coral reef macroconsumers: Patterns and effects on coral reef community structure. In *Ecosystems of the world* 25, *Coral reefs*, edited by Z. Dubinsky. New York: Elsevier, 365–400.

Glynn, P. W. 1997. Bioerosion and coral reef growth: A dynamic balance. In *The life and death of coral reefs*, edited by C. Birkeland. New York: Chapman and Hall.

Glynn, P. W., and G. M. Wellington. 1983. Corals and coral reefs of the Galápagos Islands. Berkeley: University of California Press.

Glynn, P. W., G. M. Wellington, and C. Birkeland. 1979. Coral reef growth in the Galápagos: Limitations by sea urchins. *Science* 203:47–48.

Goldman, B., and F. Talbot. 1976. Aspects of the ecology of coral reef fishes. In *Biology and geology of coral reefs*, edited by O. A. Jones and R. Endean. Vol. III, *Biology* 2. New York: Academic Press, 125–154.

Goreau, T. F. 1961. On the relation of calcification to primary production in reef building organisms. In *The biology of Hydra, etc.*, edited by H. M. Lenhoff and W. F. Loomis. Coral Gables, FL: University of Miami Press, 269–285.

Goreau, T. F., and J. W. Wells. 1967. The shallow water Scleractinia of Jamaica: Revised list of species and their vertical distribution range. *Bull. Mar. Sci.* 17(2):442–453.

Goreau, T. F., N. I. Goreau, and T. J. Goreau. 1979. Corals and coral reefs. *Sci. Amer.* 241(2):124–136.

Goreau, T. J. 1990. Coral bleaching in Jamaica. *Nature* 343:417.

Grigg, R. W. 1983. Community structure, succession and development of coral reefs in Hawaii. *Mar. Ecol. Progr. Ser.* 11:1–14.

Hatcher, B. G. 1983. Grazing in coral reef ecosystems. In *Perspectives on coral reefs*, edited by D. J. Barnes. Townsville, Australia: AIMS, 169–179.

Hay, M. E. 1984. Patterns of fish and urchin grazing on Caribbean coral reefs: Are previous results typical? *Ecology* 65:446–454.

Hay, M. E. 1991. Fish-seaweed interactions on coral reefs: Effects of herbivorous fishes and adaptations of their prey. In *The ecology of fishes on coral reefs*, edited by P. Sale. New York: Academic Press, 96–119.

Hiatt, R. W., and D. W. Strasburg. 1960. Ecological relationship of the fish fauna on coral reefs of the Marshall Islands. *Ecol. Monogr.* 30:65–127.

Highsmith, R. C. 1982. Reproduction by fragmentation in corals. *Mar. Ecol. Progr. Ser.* 7:207–226.

Hobson, E. S. 1968. *Predatory behavior of some shore fishes in the Gulf of California*. Bur. Sports Fish. and Wildlife (U.S.), Res. Rept. 73.

Hogarth, P. J. 1999. *The biology of mangroves*. Oxford: Oxford University Press.

Hughes, T. P. 1994. Catastrophic phase shifts and large scale degradation of a Caribbean coral reef. *Science* 265:1547–1551.

Huston, M. A. 1985. Patterns of species diversity on coral reefs. *Ann. Rev. Ecol. Syst.* 16:149–177.

Hutchings, P. A. 1983. Cryptofaunal communities of coral reefs. In *Perspectives on coral reefs*, edited by D. J. Barnes. Townsville, Australia: AIMS, 200–208.

Hutchings, P. A. 1986. Biological destruction of coral reefs: A review. *Coral Reefs* 4:239–252.

Jackson, J. B. C. 1977. Competition on marine hard substrata: The adaptive significance of solitary and colonial strategies. *Amer. Nat.* 111:743–767.

Jackson, J. B. C. 1979. Overgrowth competition between encrusting cheilostome ectoprocts in a Jamaican cryptic reef environment. *J. Anim. Ecol.* 48:805–823.

Jackson, J. B. C., and T. P. Hughes. 1985. Adaptive strategies of coral-reef invertebrates. *Amer. Sci.* 73:265–274.

Johannes, R. E. 1974. Sources of nutritional energy for reef corals. *Proc. 2nd Int. Coral Reef Symp.* 1:133–137.

Johannes, R. E., N. T. Kuenzel, and S. L. Coles. 1970. The role of zooplankton in the nutrition of some scleractian corals. *Limn. Oceanog.* 15:579–586.

Johannes, R. E., et al. 1972. The metabolism of some coral reef communities: A team study of nutrient and energy flux at Eniwetok Atoll. *Bioscience* 22:541–543.

Jones, O. A., and R. Endean, eds. 1973. *Biology and geology of coral reefs*. Vol. II, *Biology* 1. New York: Academic Press.

Jones, O. A., and R. Endean, eds. 1976. *Biology and geology of coral reefs*. Vol. III, *Biology* 2. New York: Academic Press.

Karlson, R. H. 1980. Alternative competitive strategies in a periodically disturbed habitat. *Bull. Mar. Sci.* 30:118–131.

Karlson, R. H. 1999. *Dynamics of coral communities*. Boston: Kluwer Academic Publishers.

Kaufman, L. 1977. The three spot damselfish: Effects on benthic biota of Caribbean coral reefs. *Proc. 3rd Int. Coral Reef Symp.* 1:559–564.

Knowlton, N., J. C. Lang, M. C. Rooney, and P. Clifford. 1981. Evidence for delayed mortality in hurricane damaged Jamaican staghorn corals. *Nature* 294:251–252.

Kohn, A., and P. Helfrich. 1957. Primary productivity of a Hawaiian coral reef. *Limn. Oceanog.* 2(3):241–251.

Kohn, A., and J. Nybakken. 1975. Ecology of *Conus* on eastern Indian Ocean fringing reefs: Diversity of species and resource utilization. *Mar. Biol.* 29:211–234.

Kuenzler, E. J. 1974. Mangrove swamp systems. In *Coastal ecological systems*, edited by H. T. Odum, B. J. Copeland, and E. A. McMahon. Vol. 1. Washington, DC: Conservation Foundation, 346–371.

Ladd, H. S., E. Ingerson, R. C. Townshend, M. Russell, and H. K. Stephenson. 1953. Drilling on Eniwetok Atoll, Marshall Islands. *Bull. Amer. Assoc. Petrol. Geol.* 37:2257–2280.

Lang, J. C. 1973. Interspecific aggression by scleractinian corals, 2. Why the race is not only to the swift. *Bull. Mar. Sci.* 23(2):260–279.

Lang, J. C., and E. A. Chornesky. 1990. Competition between scleractinian reef corals—a review of mechanisms and effects. In *Ecosystems of the world* 25, *Coral reefs*, edited by Z. Dubinsky. New York: Elsevier, 209–252.

Larkham, A. W. D. 1983. The primary productivity of plant communities on coral reefs. In *Perspectives on coral reefs*, edited by D. J. Batches. Townsville, Australia: AIMS, 221–230.

Lessios, H. A. 1988. Mass mortality of *Diadema antillarum* in the Caribbean: What have we learned? *Ann. Rev. Ecol. Syst.* 19:371–394.

Lessios, H. A., P. W. Glynn, and D. R. Robertson. 1983. Mass mortalities of coral reef organisms. *Science* 222:715.

Lobel, P. S. 1980. Herbivory by damselfishes and their role in coral reef community ecology. *Bull. Mar. Sci.* 30:273–289.

Lowe-McConnell, R. H. 1977. *Ecology of fishes in tropical waters*. Studies in biology no. 76. London: Edward Arnold.

Loya, Y. 1976. Recolonization of Red Sea corals affected by natural catastrophes and man-made perturbations. *Ecology* 57(2):278–289.

Luckhurst, B. E., and K. Luckhurst. 1978. Analysis of the influence of substrate variables on coral reef fish communities. *Mar. Biol.* 49:317–323.

Lugo, A. E., and S.C. Snedaker. 1974. The ecology of mangroves. *Ann. Rev. Ecol. Syst.* 5:39–64.

McMichael, D. F. 1974. Growth rate, population size and mantle coloration in the small giant clam *Tridacna maxima* (Roding) at One Tree Island, Capricorn Group, Queensland. *Proc. 2nd Int. Coral Reef Symp.* 1:241–254.

McNae, W. 1968. A general account of the fauna and flora of mangrove swamps and forests in the IndoWest Pacific region. *Adv. Mar. Biol.* 6:73–270.

Meyer, J. L., E. T. Schultz, and G. S. Helfman. 1983. Schools: An asset to corals. *Science* 220:1047–1048.

Motoda, S. 1940. The environment and life of massive reef coral, *Goniastrea aspera* Verrill, inhabiting the reef flat in Palao. *Palao Tropical Biol. Sta. Studies* 2:61–104.

Muscatine, L., and E. Cernichiari. 1969. Assimilation of photosynthetic products of zooxanthellae by a reef coral. *Biol. Bull.* 137:506–523.

Neudecker, S. 1977. Transplant experiments to test the effect of fish grazing on coral distribution. *Proc. 3rd. Int. Coral Reef Symp.* 1:317–323.

Newel, N. D. 1972. The evolution of reefs. *Sci. Amer.* 226(6):54–65.

Odum, H. T., and E. P. Odum. 1955. Trophic structure and productivity of a windward coral reef community on Eniwetok Atoll. *Ecol. Monogr.* 25:291–320.

Odum, H. T., B. J. Copeland, and E. A. McMahan, eds. 1974. *Coastal ecological systems of the United States*. Vol. I: B, *Natural tropical ecosystems of high diversity*. Washington, DC: Conservation Foundation, 346–514.

Odum, W. E., C. C. McIvor, and T. J. Smith III. 1982. *The ecology of the mangroves of South Florida: A community profile*. U.S. Fish and Wildlife Service, Office of Biological Services FWS/OBS-81/24.

Ogden, J. C., and P. S. Lobel. 1978. The role of herbivorous fishes and urchins in coral reef communities. *Envir. Biol. Fishes* 3:49–63.

Onuf, C. P., J. M. Teal, and I. Valiela. 1977. Interactions of nutrients, plant growth and herbivory in a mangrove ecosystem. *Ecology* 58:514–526.

Pearson, R. G. 1975. Coral reefs, unpredictable climatic factors and *Acanthaster*. Crown-of-Thorns starfish seminar proceedings. *Aust. Govt. Pub. Ser. Canberra*, 131–134.

Perry, D. M. 1984. Direct and indirect regulatory effects by root fauna on the growth rate of prop roots in the red mangrove. *Amer. Zool.* 24(3):126.

Perry, D. M. 1988. Effects of associated fauna on growth and productivity in the red mangrove. *Ecology* 69(4):1064–1075.

Porter, J. 1972. Predation by *Acanthaster* and its effect on coral species diversity. *Amer. Nat.* 106(950):487–492.

Porter, J. 1974. Zooplankton feeding by the Caribbean reef-building coral *Montastrea cavernosa*. *Proc. 2nd Int. Coral Reef Symp.* 1:111–125.

Porter, J. 1976. Autotrophy, heterotrophy and resource partitioning in Caribbean reef building corals. *Amer. Nat.* 110:731–742.

Porter, J. W., and O. W. Meier. 1992. Quantification of loss and change in Floridian reef coral populations. *Amer. Zoo.* 32:625–640.

Potts, D. C. 1977. Suppression of coral populations by filamentous algae within damselfish territories. *J. Exp. Mar. Biol. Ecol.* 28:207–216.

Potts, D. C. 1981. Crown-of-Thorns starfish—man induced pest or natural phenomenon? In *The ecology of pests*, edited by R. L. Kitching and R. E. Jones. Melbourne: CSIRO Publications Service, 55–86.

Rabinowitz, D. 1975. Planting experiments in mangrove swamps of Panama. In *Proceedings of the international symposium on the biology and management of mangroves*, edited by G. Walsh, S. Snedaker, and H. Teas. Gainesville: University of Florida, 385–393.

Rabinowitz, D. 1978. Dispersal properties of mangrove propagules. *Biotropica* 10:47–57.

Rehm, A. E., and H. J. Humm. 1973. *Sphaeroma terebrans*: A threat to the mangroves of southeastern Florida. *Science* 182:173–174.

Richardson, C. A., P. Dustan, and J. C. Lang. 1979. Maintenance of living space by sweeper tentacles of *Montastrea cavernosa*, a Caribbean reef coral. *Mar. Biol.* 55:181–186.

Risk, M. J. 1972. Fish diversity on a coral reef in the Virgin Islands. *Atoll Res. Bull.* 153:1–6.

Rougerie, F. J., A. Fagerstrom, and C. Andrie. 1992. Geothermal endo-upwelling: A solution to the reef nutrient paradox? *Cont. Shelf Res.* 12(7/8):785–798.

Sale, P. F. 1977. Maintenance of high diversity in coral reef fish communities. *Amer. Nat.* 111:337–359.

Sale, P. F. 1980. The ecology of fishes on coral reefs. *Oceanog. Mar. Biol. Ann. Rev.* 18:367–421.

Sale, P. F., ed. 1991. *The ecology of fishes on coral reefs*. New York: Academic Press.

Salvat, B., and G. Richard. 1985. Takapoto atoll, Tuamotu archipelago. In *Fifth international coral reef congress, Tahiti 27 May–1 June, 1985*. Vol 1: *French Polynesian coral reef*, edited by B. Delesalle, R. Galzin, and B. Salvat. Antenne Museum Ephe: Moorea, French Polynesia, 323–378.

Sammarco, P. W. 1980. *Diadema* and its relationship to coral spat mortality: Grazing, competition and biological disturbance. *J. Exp. Mar. Biol. Ecol.* 45:245–272.

Sammarco, P. W. 1985. The Great Barrier Reef versus the Caribbean: Comparisons of grazers, coral recruitment patterns, and reef recovery. *Proc. 5th Int. Coral Reef Symp.* 4:391–397.

Sammarco, P. W. 1987. A comparison of some ecological processes on coral reefs of the Caribbean and the Great Barrier Reef. *Unesco Rept. Mar. Sci.*, no. 46: 127–166.

Sammarco, P. W., and J. C. Coll. 1992. Chemical adaptations in the Octocorallia: Evolutionary considerations. *Mar. Ecol. Prog. Ser.* 88:93–104.

Sammarco, P. W., J. S. Levinton, and J. C. Ogden. 1974. Grazing and control of coral reef community structure by *Diadema antillarum* Philippi (Echinodermata: Echinoidea): A preliminary study. *J. Mar. Res.* 32(1):47–53.

Scoffin, T. P., C. W. Stearn, D. Boucher, P. Frydl, C. M. Hawkins, I. G. Hunter, and J. K. MacGeachy. 1980. Calcium carbonate budget of a fringing reef on the west coast of Barbados. II, Erosion sediments and internal structure. *Bull. Mar. Sci.* 30:475–528.

Simberloff, D., B. J. Brown, and S. Lowrie. 1978. Isopod and insect root borers may benefit Florida mangroves. *Science* 201:630–632.

Smith, C. L., and J. C. Tyler. 1972. Space resource sharing in a coral reef fish community. *Bull. Nat. Hist. Mus. Los Angeles County* 14:125–170.

Smith, S. V., and J. A. Marsh. 1973. Organic carbon production and consumption on the windward reef flat of Eniwetok Atoll. *Limn. Oceanog.* 18:953–961.

Sorokin, Yu. I. 1990. Aspects of trophic relations, productivity and energy balance in coral reef ecosystems. In *Ecosystems of the world* 25, *Coral reefs*, edited by Z. Dubinsky. New York: Elsevier, 401–410.

Sorokin, Yu. I. 1993. *Coral reef ecology*. Ecological studies, vol. 102. New York: Springer.

Sousa, W. P., and B. J. Mitchell. 1999. The effect of seed predators on plant distributions: Is there a general pattern in mangroves? *Oikos* 86:55–66.

Stearn, C. W., and T. P. Scoffin. 1977. Carbonate budget of a fringing reef, Barbados. *Proc. 3rd Int. Coral Reef Symp.* 2:471–476.

Stehli, F. G., and J. W. Wells. 1971. Diversity and age patterns in hermatypic corals. *Sys. Zool.* 20:115–126.

Steneck, R. S., and W. H. Adey. 1976. The role of environment in control of morphology in *Lithophyllum congestum*, a Caribbean algal ridge builder. *Botan. Mar.* 19: 197–215.

Stoddard, D. R. 1969. Ecology and morphology of recent coral reefs. *Biol. Rev.* 44(4):433–498.

Sutherland, J. P. 1980. Dynamics of the epibenthic community of roots of the mangle, *Rhizophora mangle*, at Bahia de Buche, *Venezuela. Mar. Biol.* 58:75–84.

Szmant-Froelich, A. 1985. Reproduction and recruitment of corals: Conclusion. *Proc. 5th Int. Coral Reef Symp.* 4:399–400.

Taylor, P., M. Litter, and D. Litter. 1986. Escapes from herbivory in relation to the structure of mangrove island macroalgael communities. *Oceologia* 69:481–490.

Talbot, F. H., B. C. Russell, and G. R. V. Anderson. 1978. Coral fish communities: Unstable, high diversity systems? *Ecol. Monogr.* 48:425–440.

Thom, B. G. 1975. Mangrove ecology from a geomorphic viewpoint. In *Proceedings of the international symposium on the biology and management of mangroves*, edited by G. Walsh, S. Snedaker, and H. Teas. Gainesville: University of Florida, 469–481.

Tschirley, F. H. 1969. Defoliation in Vietnam. *Science* 163:779–786.

Underwood, A. J., and G. Barrett. 1990. Experiments on the influence of oysters on the distribution, abundance and sizes of the gastropod *Bembicum auraturn* in a mangrove swamp in New South Wales, Australia. *J. Exp. Mar. Biol. Ecol.* 137(1):25–45.

Vaughn, T. W. 1915. The geologic significance of the growth rate of the Floridian and Bahamian shoalwater corals. *J. Wash. Acad. Sci.* 5:591–600.

Vermeij, G. J. 1987. Interocean differences in architecture and ecology: The effects of history and productivity. *Unesco Rept. Mar. Sci.*, no. 46:105–126.

Veron, J. E. N. 1986. Distribution of reef-building corals. *Oceanus* 29:27–31.

Victor, B. C. 1983. Recruitment and population dynamics of coral reef fish. *Science* 219:419–420.

Wada, K., and D. Wower. 1989. Foraging on mangrove pneumatophores by ocypodid crabs. *J. Exp. Mar. Biol. Ecol.* 134(2):89–100.

Wahle, C. M. 1980. Detection, pursuit and overgrowth of tropical gorgonians by milleporid hydrocorals: Perseus and Medusa revisited. *Science* 209:689–691.

Walbran, P. D., R. A. Henderson, A. J. Timothy Jull, and M. J. Head. 1989. Evidence from sediments of long-term *Acanthaster planci* predation on corals of the Great Barrier Reef. *Science* 245:847–850.

Walsh, G. E. 1974. Mangroves, a review. In *Ecology of halophytes*, edited by R. J. Reimold and W. J. Queen. New York: Academic Press, 51–174.

Wellington, G. W. 1980. Reversal of digestive interactions between Pacific reef corals: Mediation by sweeper tentacles. *Oecologia* 47:340–343.

Wellington, G. W. 1982a. Depth zonation of corals in the Gulf of Panama: Control and facilitation by resident reef fishes. *Ecol. Monogr.* 52:223–241.

Wellington, G. W. 1982b. An experimental analysis of the effects of light and zooplankton on coral zonation. *Oecologia* 52:311–320.

Wells, J. W. 1954. *Recent corals of the Marshall Islands, Bikini and nearby atolls.* U.S. Geol. Survey Paper 260-I:385–486.

Wells, J. W. 1957. Coral reefs. In *The treatise on marine ecology and paleoecology*. Vol. I, *Ecology*. Geol. Soc. of Amer. Memoir 67, 609–631.

West, R. C. 1977. Tidal salt marsh and mangal formations of middle and south America. In *Wet coastal ecosystems*, edited by V. J. Chapman. New York: Elsevier, 193–213.

Wilkinson, C. 1983. Net primary productivity in coral reef sponges. *Science* 219:410–411.

Wilkinson, C. 1987a. Interocean differences in size and nutrition of coral reef sponges. *Science* 236:1654–1657.

Wilkinson, C. 1987b. Sponge biomass as an indication of reef productivity in two oceans. *Unesco Rept. Mar. Sci.*, no. 46:99–104.

Williams, E. H., and L. B. Williams. 1987–88. Caribbean marine mass mortality: A problem with a solution. *Oceanus* 30(4):69–75.

Yentsch, C. S. 1966. Primary production. In *The encyclopedia of oceanography*, edited by R. W. Fairbridge. New York: Van Nostrand, 722–725.

Yonge, C. M. 1963. The biology of coral reefs. *Adv. Mar. Biol.* 1:209–260.

SYMBIOTIC RELATIONSHIPS

One of the most striking features of marine organisms is the large number of close associations among many different species. These are a series of associations that are not predator-prey or herbivore-plant relationships in which one member consumes the other. These close relationships between unlike species generally are either harmless to either member or, more likely, beneficial to one or both. ***Symbiosis****, the name given to such associations, means an interrelationship between two different species. Symbiotic relationships are found in the terrestrial environment as well as in the aquatic, but they appear disproportionately common, extensive, and well developed in the marine environment, such that they merit separate coverage in a book such as this.*

Coverage and Definitions

Symbiotic relationships cover a broad spectrum of associations from random, casual, or facultative associations through more and more obligatory groupings that benefit one or both members, finally to those that are parasitic. Such differences in the degree of association have led to the subdivision of symbiosis into more narrowly defined groupings. The term **commensalism** is often used to refer to an association that is clearly to the advantage of one member while not harming the other member. **Inquilinism** is a special subdivision of commensalism in which an animal lives in the home of another, or in its digestive tract, without being parasitic. **Mutualism** is that form of symbiosis in which two species associate for their mutual benefit. In a mutualistic relationship, the partners are often called **symbionts.** In a commensal or inquiline relationship, however, the partner gaining advantage is called the **commensal** and the other the **host**.

Parasitism generally refers to an association in which one species lives in or on another and draws nourishment from that species at its expense—that is, to its detriment. In other words, it is an association in which the advantage is solely to one member at the expense of the other. Parasitism is an extremely common association in both marine and terrestrial environments. It will not be discussed in this chapter, since it contains few features unique to the marine environment.

There are two broad groupings of nonparasitic symbiotic associations in the sea: those between algal cells and various invertebrate animals and those between various animals, both vertebrate and invertebrate. We shall consider these groupings in that order. A final category is that between various bacteria and animals, which has become particularly

TABLE 10.1

Summary of the Types of Associations Between Algae and Marine Invertebrates

Algal Symbiont Group	*Algal Taxon*	*Invertebrate Animal Host Taxa*
Zooxanthellae	Dinophyceae	Prostista, Porifera, Cnidaria, Platyhelminthes, Mollusca
	Haptophyceae	Protista
	Chrysophyceae	Protista
	Bacillariophyceae	Platyhelminthes (*Convoluta convoluta*)
	Cryptophyceae	Protozoa
Zoochlorellae	Prasinophyceae	Platyhelminthes (*Convoluta roscoffensis*), Protista
	(?) Chlorophyceae	Cnidaria, Protista
Cyanellae	Cyanophyceae	Porfera, Protista, Echiura
Prochlorophyta	*Prochloron*	Urochordata
Chloroplasts	Chlorophyceae	Mollusca (Sacoglossa)

Source: Modified from "Symbiosis of Algae with Invertebrates" by D.C. Smith in *Oxford Biology Reader*, No. 43, p. 16..

well publicized through the discoveries of organisms living in deep-sea thermal-vent systems. These symbioses were covered in Chapter 4.

SYMBIOSES OF ALGAE AND ANIMALS

All known symbiotic relationships in the sea between plants and animals are between unicellular algae or their chloroplasts and a wide variety of marine invertebrate animals. These symbiotic relationships are most common in tropical waters, but they also are prevalent in temperate oceans. They are, however, virtually absent from polar waters. Furthermore, the associations are, for obvious reasons, restricted to the shallow subtidal or intertidal areas or to the uppermost layers of the pelagic realm where sufficient light for photosynthesis is present.

Types and Composition of the Associations

There are basically two types of symbiotic association between algae and invertebrates. The more common is to have the entire functioning algal cell associated with the invertebrate animal. The second is to have only the functioning chloroplasts from the algal cells incorporated into the tissues of the invertebrate body.

The algal cell symbionts have been typically classified into groups on the basis of their color. **Zooxanthellae** is the name given to brown, golden, or brownish-yellow cells, and **zoochlorellae** to those that are green. A third, smaller group is blue or bluish-green and has been called **cyanellae** (Table 10.1). These color groups, however, do not distinguish among the actual algal species involved. Zooxanthellae, the most common in the seas of the world, are primarily species of dinoflagellates (Figure 10.1), but include a few diatoms and cryptophyceans, haptophyceans, and chrysophyceans. Schoenberg and Trench (1976) reported the most common zooxanthellae species to be the dinoflagellate (Dinophyceae) *Symbiodinium* (*Gymnodinium*) *microadriaticum*. The difficulty in identifying these symbiotic algae with traditional methods has led researchers in the past to embrace concepts as disparate as that of a single zooxanthellae species in all hosts to one that says there is a different species in each host. By the use of modern molecular techniques, however, primarily RNA sequencing, Rowan and Powers (1991) have been able to recognize 10 algal taxa among 22 hosts sampled. The zoochlorellae are much less common in the sea but dominate symbiotic associations in fresh water. Among the marine associations, zoochlorellae include the algae *Platymonas* (*Tetraselmis*) *convolutae*, *Pedinomonas* of the Prasinophyceae, and *Chlorella* and *Chlamydomonas* of the Chlorophyceae. Cyanellae are all cyanobacteria (formerly called blue-green algae) and include the genera *Aphanocapsa* and *Phormidium*, and probably others as yet unidentified

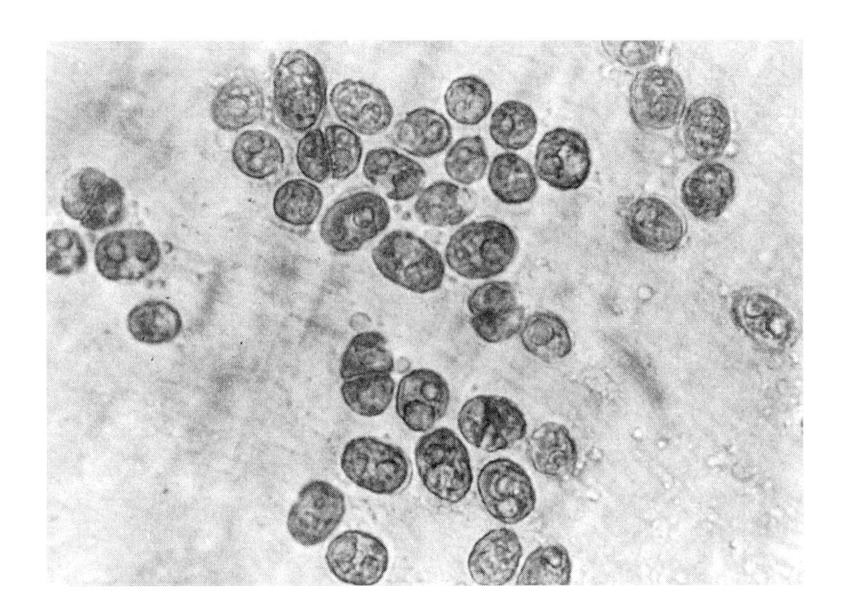

FIGURE 10.1 Symbiotic zooxanthellae (dinoflagellates) from the anemone *Anthopleura xanthogrammica*. (Photomicrograph by the author.)

(Table 10.1). They are most common as symbionts in sponges and planktonic diatoms.

The algae all occur inside the bodies of the animals, and most are found either within vacuoles inside individual tissue cells or in various body spaces between or within tissue layers. Generally, the algal cells are restricted to certain tissues or areas of the host, and they grow and reproduce within the invertebrate without being digested.

Among those associations that involve only the chloroplasts and the invertebrate body, the chloroplasts are usually derived from the cells of larger green algae (*Codium*, *Caulerpa*, *Cladophora*, *Bryopsis*) first ingested by the animal in feeding and subsequently transferred from the digestive tract of the animal to other tissues.

Algal symbiosis, either cellular or chloroplast, is widespread among various invertebrate groups. Symbiotic associations occur in Protista, Porifera, Cnidaria, Platyhelminthes, Mollusca, and Echiura. Within these phyla, McLaughlin and Zahl (1966) report about 130 genera having algal symbionts. It is most common among Protista and Cnidaria.

Although most algal symbionts can be readily identified in their hosts, it has been recently discovered that a number of protistans previously considered photosynthetic, such as the ciliate *Mesodinium rubrum* and the dinoflagellates *Peridinium balticum* and *Kryptoperidinium foliaceum* are, in fact, symbiotic associations in which the algal cells have lost their cellular integrity and do not appear as separate from the host body. These associations are examples of **cryptic symbiosis**.

Origin of the Association

Although we will probably never know the exact origin of algal-invertebrate symbiotic relationships, it is possible to outline some suggestions. In all the phyla mentioned that contain symbiotic algae or chloroplasts, the final phase of digestion is intracellular. Individual cells take up particles from the stomach. It is thus possible to conceive of the origin of this association through the ingestion by cells of the digestive tract of either intact algal cells or chloroplasts, taken in by the animal during feeding. If, then, the algal cells were resistant to digestive action or the animal lacked enzymes to digest plant cellulose, the basis for a new association would be laid. This association could then evolve into a more obligatory relationship, provided that the new association conveyed an enhanced survival value to both symbionts. Since these associations are common, they must have some selective value (see pp. 441–442).

Distribution of Algae-Invertebrate Associations

Although there is neither space nor time to list and discuss all the species in which algal symbionts occur, it is of use to discuss briefly the various groups of organisms in which this association is prevalent.

Beginning with protozoans, symbiotic relationships are found in all the epipelagic planktonic

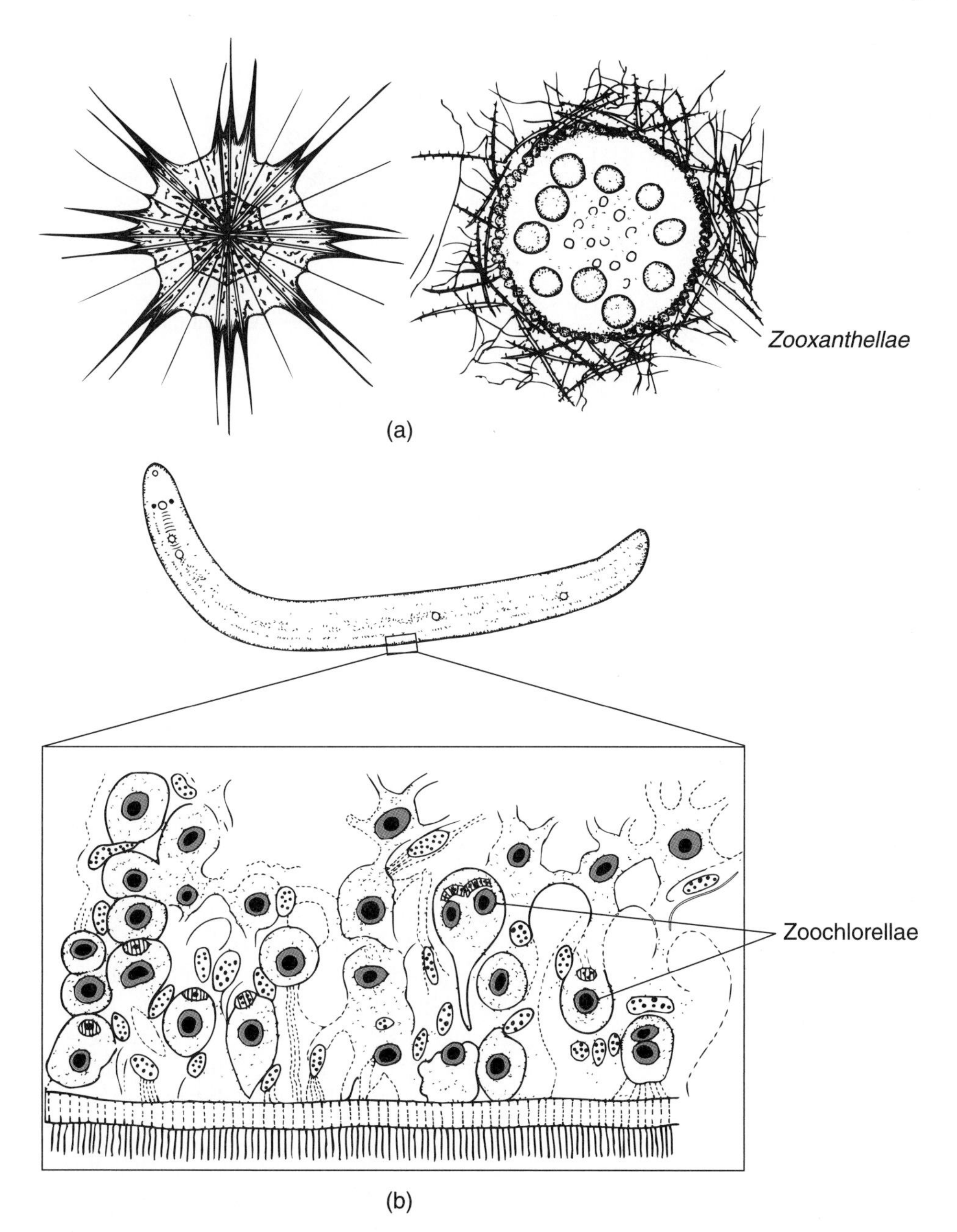

FIGURE 10.2 Zooxanthellae and zoochlorellae in various invertebrate taxa. (a) Zooxanthellae in the radiolarian *Acanthometra pellucida* (left) and *Sphaerozoum acuferum* (right). (b) Zoochlorellae in the acoel flatworm *Convoluta roscoffensis*. (After *The Biology of Marine Animals*, J. A. Colin Nicol, 1960, Pittman Books.)

Radiolaria, where the zooxanthellae occur in the outer frothy layer (Figure 10.2). They also are found in a number of planktonic Foraminifera (*Globigerinoides*), in all the larger benthic species (*Elphidium*, *Nonion*, *Heterostregina*), and even in marine ciliates (*Paraeuplotes*, *Trichodina*).

Symbiosis in sponges is particularly important in species that inhabit coral reefs. The association is usually with cyanobacteria. According to Wilkinson (1983), sponges with such symbionts constitute 80% of the sponge species and biomass on certain reefs of the Great Barrier Reef. Zooxanthellae occur in *Cliona* and cyanellae in Demospongia.

Among the Cnidaria, symbiosis becomes extremely common, especially among tropical species. Virtually all tropical, shallow-water anemones, soft corals, sea fans, whips, and stony corals have symbiotic zooxanthellae in their tissues. Even certain tropical

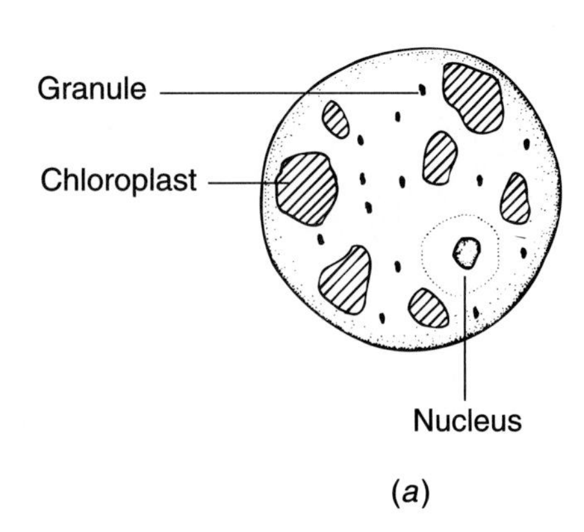

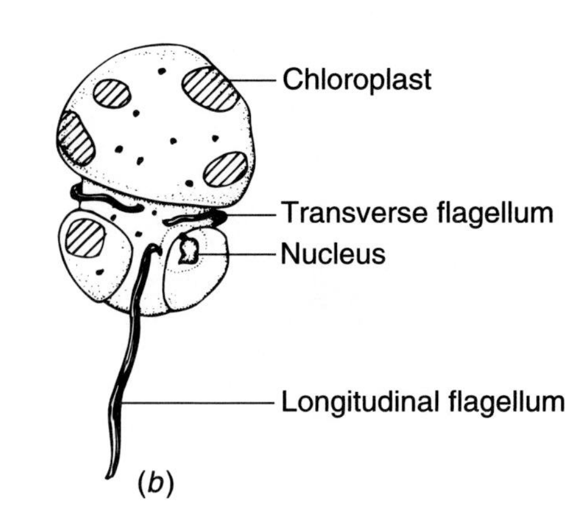

FIGURE 10.3 Symbiotic (a) and free-living (b) zooxanthellae cells.

jellyfish (*Cassiopeia*, *Mastigius*) contain zooxanthellae. In temperate seas, the incidence is not as great, though many anemones, such as *Anthopleura*, retain the symbionts.

A peculiar symbiosis occurs in some compound ascidians (*Didemnum*, *Tridedmnum*) where an unusual prokaryotic unicellular alga called *Prochloron*, which combines features of both cyanobacteria and green algae, occurs embedded in the test or in the cloacal system.

Whereas a few symbiotic relationships occur in Ctenophora (*Beröe*), Annelida (*Eunice*), and Echinodermata (*Ophioglypha*), the remaining major reservoirs of symbiotic relationships are found in the Platyhelminthes and Mollusca. Among marine flatworms, the classic case is that of the genus *Convoluta*, which has been extensively investigated. Symbiotic relationships in the Mollusca are generally restricted to certain marine gastropods of the order Sacoglossa, where the animals retain only the chloroplasts, and to the bivalve family Tridacnidae and the genera *Corculum* and *Fragum*.

Modifications Resulting from the Association

The symbiotic association between the algae and the invertebrates is generally a very close one (mutualistic), which has resulted in significant anatomical and physiological changes in both the algal cells and the various invertebrate hosts.

Perhaps the most profound changes that occur as a result of the association are found in the algal cells. Most marine symbiotic algae are dinoflagellates. The symbionts, however, have lost their locomotory flagellae and the characteristic grooves around their body. Furthermore, their cell walls are much reduced in thickness. The zoochlorellae in the flatworm *Convoluta* have lost even more, in that the cell wall disappears, as do the light-sensitive stigmata. These cells become little more than bags containing chloroplasts.

If the algal cells are removed from the host and grown outside the animal in culture, they will develop the characteristic flagellae, cell walls, and other organs of a typical free-living form. It is apparent that the changes observed are a direct result of the symbiotic association (Figure 10.3).

In the case of the animal hosts, the changes vary with the type of organism and with the degree of interdependence established between the symbionts. The single universal modification is that all of these invertebrates live in shallow water, where they can obtain adequate light, so the algae can carry on photosynthesis. Perhaps the least amount of modification occurs in the lower invertebrates, such as protozoans and Porifera, where the algae occur as simple inclusions in the cytoplasm of the animal. Among sponges, those with symbiotic cyanobacteria enhance their light-gathering ability according to Wilkinson (1983) by morphological flattening, thereby increasing the area available to intercept light.

Among the Cnidaria, definite modifications begin to become apparent. The algae occur in marine Cnidaria in the innermost of the cell layers, the gastrodermis (endoderm). The numbers of zooxanthellae in different corals vary, and those that have the most seem to have reduced tentacle size, indicating they are less able to capture zooplankton food. Among certain soft corals of the family Xeniidae, the digestive regions of the animal are reduced and the animals are not responsive to animal food. In the jellyfish *Cassiopeia*, a striking modification is behavioral. These animals, rather than swimming in open waters as do most jellyfishes, lie upside down on the bottom in shallow tropical waters, exposing their oral arms to the light to illuminate their algae (Figure

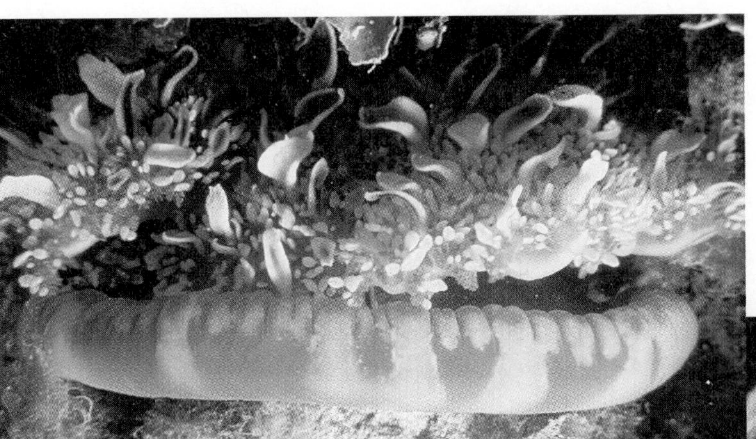

FIGURE 10.4
Cassiopeia jellyfish in normal resting position upside down on the bottom. The brownish color is due to the symbiotic zooxanthellae. (Photo courtesy of Diane Nelson.)

10.4). The oral arms are also much enlarged and expanded to provide more area for habitation by the algal cells. Truly a strange way of life for a jellyfish!

Among flatworms, the most studied case of symbiosis is that of *Convoluta roscoffensis*, a small animal inhabiting sand beaches on the Brittany coast of France. The alga inhabiting *Convoluta* is *Platymonas* (*Tetraselmis*) *convolutae*, which also occurs free-living in the same area. *Convoluta* exhibit perhaps the most profound behavioral and life history modifications in response to the alga, but little in the way of anatomical changes (Figure 10.5).

Convoluta roscoffensis live burrowed in the sand of beaches near the upper reaches of the tidal zone. Whenever the tide is in, the animals are found buried. When the tide recedes, the animals move up onto the surface of the sand, where they spread out to expose their symbionts to the light. When the tide begins to return, the vibrations trigger a burrowing response, so the animals are safely under the sand again before the rising tidal waters can sweep them out to sea. Even more significantly, as Smith (1973) notes, these animals apparently do not feed as adults. Furthermore, if the young worms do not ingest the algal symbionts upon hatching, they will not complete development and will die even if they feed.

It is among the mollusks, however, that we find the most dramatic changes of all resulting from the association. Although there are thousands of species of mollusks in the seas of the world (the second largest phylum after arthropods), symbiotic associations with algal cells have developed in only nine species. All are bivalve mollusks, seven in a single family, the Tridacnidae, which includes the giant clams. Six of the species are in the genus *Tridacna* itself, and the remaining one is in the genus *Hippopus*. The remaining two species, *Corculum cardissa* and *Fragum fragum*, are in the family Cardiidae.

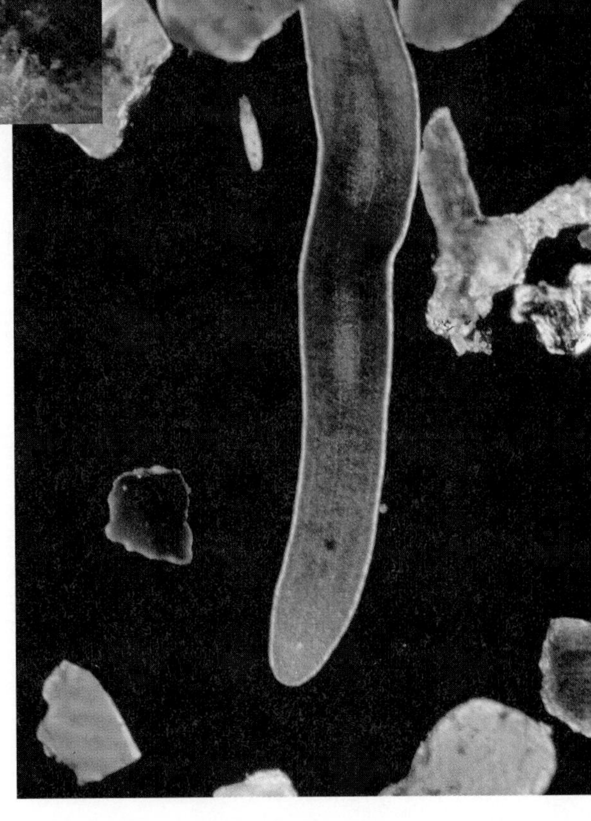

FIGURE 10.5 *Convoluta roscoffensis*, showing the green color of the symbiotic zoochlorellae. (Photo courtesy of D. P. Wilson/Science Source/Photo Researchers Inc.)

All tridacnid clams are distributed only in the Old World tropics of the Indo-Pacific. They also include

FIGURE 10.6
A giant clam, *Tridacna gigas*, with siphonal tissue expanded. (Photo courtesy of R. Shimek.)

the largest bivalve mollusk in existence: *Tridacna gigas*, which has been recorded by Rosewater (1965) to reach 1.2 m in length and 263 kg in weight. The other species are more modest in size, but all are still large in comparison with most other bivalve species.

These tridacnid clams are inhabitants of coral reef areas, where they are found abundantly in the shallow, sunlit water (Figure 10.6). They usually have an uncharacteristic position for a bivalve mollusk, in that they either lie on the surface of the bottom or bore into coral or coral rock with the opening between the valves facing up toward the surface of the water. The valves usually gape widely, and within this opening is an extensive, brightly colored tissue layer. It is in this tissue layer that the symbiotic zooxanthellae are found. The brightly colored tissue exposed within the gape is siphonal tissue; the color results from the interaction of various pigments deposited there. The bright colors protect the tissues of the clam from the damaging effects of sunlight, while they pass enough light to allow the zooxanthellae to photosynthesize.

What is truly remarkable about tridacnids, however, is the tremendous change their bodies have undergone from that of a typical clam to accommodate this symbiotic association. Bivalves generally rest with their foot either embedded in the substrate or held against it. In this position, a normal clam has its hinge uppermost. Obviously, this would not do for a tridacnid, since such a position would not permit the tissue and zooxanthellae to be illuminated. The result, as Yonge (1975) has described, is that the entire tridacnid has undergone a tremendous rotation with respect to the foot, so the hinge comes to lie on the underside next to the foot and the opening of the shell faces upward (Figure 10.7). At the same time, the siphons and siphonal tissue underwent an expansion. The siphons grew and extended themselves, covering the length of the upward-facing opening and providing the expanded area for occupation by the zooxanthellae. As a result of this rotation and expansion, one of the tridacnid's shell-closing muscles was lost. Because of these profound anatomical changes, the tridacnids are markedly different in body orientation from any other bivalves. These changes can be attributed only to the association with the symbiotic algae.

Symbiotic relationships with chloroplasts have been reported by Greene (1970) in the marine gastropods of the order Sacoglossa (*Elysia*, *Placida*, *Placobranchus*; Figure 10.8). Animals of this order regularly consume the contents of algal cells, usually of the order Caulerpales. Algae of this order are particularly suitable for these suctorian molluscan feeders, because they have partially syncytial (no cell walls) construction. Some species of sacoglossans have evolved the ability to retain the chloroplasts in a functioning position on the dorsal surface. The only modifications so far observed are that they do not feed very often, and the tropical forms provide a screen of light-absorbing material above the chloroplasts to cut down the light intensity.

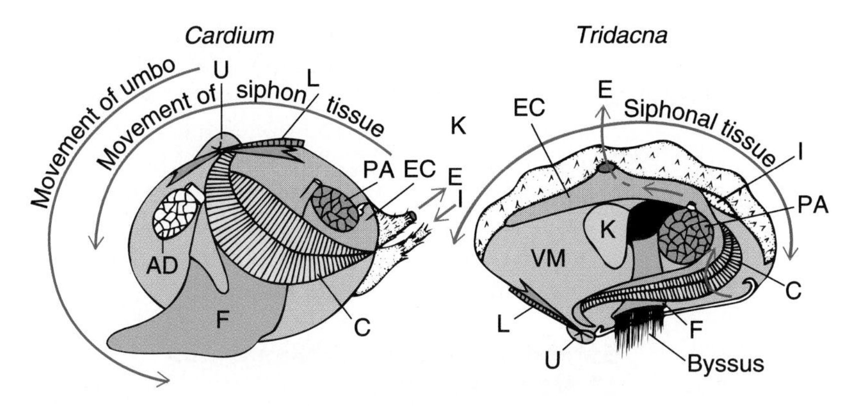

FIGURE 10.7 Comparison of the body orientation and internal anatomy of a typical clam, *Cardium*, with that of *Tridacna*. (AD) Anterior adductor; (C) ctenidium; (E) exhalent current; (EC) exhalent chamber; (F) foot; (I) inhalent current entering inhalent aperture; (K) kidney; (L) ligament; (PA) posterior adductor; (U) umbo; (VM) visceral mass. (Modified from "Symbiosis" *The Treatise on Marine Ecology and Paleontology*, Vol. 1, G. Thorson, 1957, eds. J. Hedgpeth, © 1957 Geological Society of America.)

FIGURE 10.8 *Elysia hedgpethi*, a sacoglossan mollusk that contains chloroplasts obtained from the green algae *Bryopsis corticulans* or *Codium fragile*. (Photo by author.)

Value of the Association

Since symbiotic associations with algae are so common among invertebrates, and since they often result in profound anatomical or behavioral changes in the partners, it only seems natural to assume they must have some positive value to each of the partners. What might be the value of such associations?

We have already partially answered this question with respect to corals in Chapter 9, where we saw that the corals obtain food materials (see p. 387) from the zooxanthellae and also that the zooxanthellae enhance the ability of corals to lay down calcium carbonate (see p. 388). In turn, the zooxanthellae in corals receive nutrients in the form of nitrates and phosphates produced in the metabolic processes of the coral, nutrients that are rare in external waters.

Among most invertebrates that have a symbiotic association with algal cells, the cells retain their integrity and are not digested by the animal to obtain nutrients. The nutrients that pass are in the form of chemical compounds. Energy-containing molecules, such as glycerol, produced by the zooxanthellae in photosynthesis pass to the animal. Nitrates and phosphates, needed nutrients for the algae, pass from the invertebrate to the algal cell.

A similar situation also occurs between the sacoglossan mollusks and the symbiotic chloroplasts. Greene (1970) has shown translocation of organic material from chloroplast to animal. In the sacoglossans so far investigated, however, the duration of life of the chloroplasts is much shorter, and the animals must periodically replenish their supply by ingesting the contents of algal cells.

With respect to the giant clams, however, Fankboner (1971) has suggested that, in addition to reciprocal transference of nutrients, the animals may also digest the zooxanthellae cells. This would seem counterproductive, especially considering the great anatomical changes these animals have undergone to provide properly for their algal guests. Apparently, the clams can discern between healthy zooxanthellae cells and those that are degenerating or senile. As a result, the clam transports only the senile or degenerating cells from the blood spaces of the outer mantle, where they photosynthesize, into the deeper tissues, where they are consumed by phagocytic blood cells. In this manner, the clams retain the symbiotic relationship. At the same time, they cull the unfit algal cells to obtain additional nutrients otherwise lost when such cells die.

Other values may result from the association. As we have noted, most of these associations are tropical, and these waters are typically low in plant nutrients. It may be that the algae benefit from the association through access to a larger and more reliable source of nutrients in the metabolic products of the animal (NO_2^{2-}, PO_4^{3-}, CO_2) than they would obtain from the open water. Among the benthic Foraminifera, the partnership probably accounts for the 20- to 100-fold increase in the calcification rate observed between those with symbionts and those without them.

Tropical waters also are lower in oxygen than temperate waters, because water at higher temperatures holds less oxygen. Since the photosynthetic process produces oxygen, it may be that the animals, especially in the crowded conditions of the shallow waters of the tropics, gain through additional amounts of oxygen produced by the symbiotic algae. Zooxanthellae also are protected from the actinic damage of the intense solar radiation of the tropics by living in the tissues of various marine animals.

In the symbiotic relationship between cyanobacteria and sponges, Wilkinson (1983) has demonstrated that the cyanobacteria are photosynthetically active and has estimated that 5–12% of the photosynthetically produced carbon is translocated to the host sponge. That this additional energy source is significant has been demonstrated by Wilkinson and Vacelet (1979), who have shown faster growth rates of sponges in light than in the dark. It has also been suggested that the cyanobacteria may protect host-sponge cells from high light intensity.

Establishment and Transmission of Algal Symbionts

Whereas each generation of sacoglossans must feed on algae to obtain its supply of symbiotic chloroplasts, perpetuation of the symbiotic association with zooxanthellae from generation to generation of hosts is accomplished in one of two ways. Either the zooxanthellae cells must be passed on directly from the parent to the eggs or larvae, or each new generation must reinfect itself with algal cells found in the surrounding environment.

Passage of algal symbionts directly to the next generation via the egg or larvae is the method used by most Cnidaria. In brooding corals, for example, the zooxanthellae enter the eggs or larvae at some time prior to their release from the parent. Among broadcast spawners, however, the zooxanthellae do not make their appearance, according to Babcock (1985), until 5–10 days after the larvae have settled. It is also assumed, though without much evidence, that the giant clams transmit the zooxanthellae through the eggs. It has been suggested that the clams receive them from the surrounding corals as well.

For other hosts, the perpetuation of a symbiotic relationship depends on the reinfection of each generation with algal symbionts obtained from the environment. Among protistans and Porifera, this is the common means of transmission. Each new generation obtains the algal cells through ingestion of the free-living form of the symbiont. Surprisingly, this is also the case in *Convoluta roscoffensis*, where the association is a highly dependent one. In this case, the algae-free larvae ingest the free-living *Platymonas* (*Tetraselmis*) *convolutae*. To ensure the larvae will be infected, however, a chemical has evolved that attracts the alga to the egg cases of *Convoluta*, so when the young hatch and begin to feed, the algae are present. Under such conditions, the new generation is virtually assured of reinfection.

A final possibility for transmission is the ingestion by the potential host of food organisms themselves containing symbiotic algae. Although this has not yet been proved important, it remains an area requiring study.

Little is known about the transmission of cyanobacteria to succeeding generations of sponges, but cyanobacteria have been found in the eggs of some sponges, suggesting that these algae may be transmitted directly.

SYMBIOSES AMONG ANIMALS

We are concerned here with those special interactions in which members of different species are regularly associated with each other in nonparasitic relationships. Such symbiotic associations are widespread in the sea, primarily in the crowded reaches of the epipelagic, intertidal, and shallow subtidal zones. As with algal symbiosis, such relationships are somewhat more common or spectacular in the tropics, but they are also common in temperate seas.

Types of Associations

Symbiotic relationships among marine animals cover a broader spectrum than the strictly mutualistic associations we have seen among algae or chloroplasts and marine invertebrates. The simplest type of associations are commensal, wherein the "guest" lives on another "host" organism or in or on some construction of that organism, such as a tube or burrow. Such associations are similar to epiphytic relationships among terrestrial plants, wherein the epiphyte uses the other plant as a substrate without actually taking sustenance from it. In these cases, the commensal usually gains in some measurable way from the association, and the host is not seriously inconvenienced.

Marine commensals that live on other invertebrates are called **epizoites**. Those that live inside other animals but are not parasites are called **endozoites**. Epizoites are abundant in marine waters, and many are probably not true commensals; the relationship is the result of organisms that normally settle on the substrate but instead settle at random on the outside of a slow-moving or sessile invertebrate. These will not be considered further here. Other epizoites are highly specific, and a symbiotic relationship is certainly the case, with the "guest" somehow seeking out the correct "host." These symbiotic epizoites are the most abundant group of commensals and are spread among the protozoans and the phyla Cnidaria, Entoprocta, Annelida, Arthropoda, and Mollusca.

Many invertebrates harbor specialized ciliate protistans, either on their external surfaces, internally in the digestive tract, or among the gills. For example, the vorticellid ciliate *Ellobiophyra donacis* is restricted to the gills of the clam *Donax vittatus* (Figure 10.9), and the collared ciliate *Lobochona prorates* is found on the telson of the isopod crustacean *Limnoria tripunctata*. Many other examples are also known. Endozoic commensal ciliates are common as well, especially in the digestive tracts of larger animals. For example, Beers (1961) found that the large green sea urchin of the cold temperate waters of North America, *Strongylocentrotus droebachiensis*, has as many as seven species of ciliate protozoans in its gut, which are found nowhere else and die if removed.

Among the Cnidaria, the class Hydrozoa furnishes most of the cases of epizoites. The best studied is the case of two species of *Proboscidactyla*, two-tentacled hydroids, which always occur on the rim of the tubes of polychaete worms of the genera *Potamilla* and *Schizobranchia* in the North Pacific Ocean. Similarly, along the Pacific, Atlantic, and Gulf coasts, the hydroid genus *Clytia* occurs on the clams of the genus *Donax*. Hydroids of the genera *Hydractinia* and *Podocoryne* are well-known ectocommensals on shells occupied by hermit crabs and on the snail *Cantharus*, respectively. Other hydroids are epizoic on gorgonians, pennatulids, and ascidians, but the closeness of the association is not always known.

Turbellarian flatworms are often endozoic in the digestive tracts of larger marine invertebrates and in the mantle cavities of various mollusks. The polyclad *Notoplana ovalis* occurs in the mantle cavity of the limpet *Patella oculis*, while the triclad *Nexilis epichitonius* occurs in the mantle cavity of the common Pacific coast chiton *Mopalia hindsii*. The rhabdocoels *Syndesmis dendrastorum* and *S. franciscanus* are endozoites in the guts of the sand dollar, *Dendraster excentricus*, and the sea urchins *Strongylocentrotus franciscanus*, *S. purpuratus*, and *S. droebachiensis*, respectively. Bryozoans offer two remarkable examples of symbiosis. Larvae of *Hipporida* seek out a tiny shell occupied by a young hermit crab. The bryozoan colony grows in such a way that it forms a chamber that enlarges continuously with the aperture of the snail shell and keeps pace with the size increase of the hermit crab, so the crab never needs to change shells. The growth forms of the bryozoan take on peculiar spindle shapes that are

FIGURE 10.9
Some epizoites. (a) *Ellobiophyra donacis* on the gills of the bivalve mollusk *Donax.* (b) *Proboscidactyla stellata* on the tube of a polychaete worm. (a from Chatton, E., and Lwoff, A. (1929). Contribution à l'étude l'adaptation *Ellobiophrya donacis* Ch. Et Lw., péritriche vivant sur les de l'aephale *Donax viinius* da Costa, *Bull. Biol.* 63, pp 321-349. b after *The Biology of Marine Animals*, J. A. Colin Nicol, 1960 Pittman Books.)

(*a*)

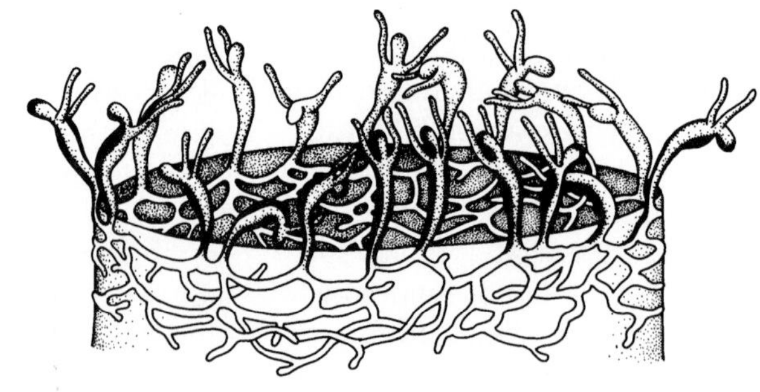

(*b*)

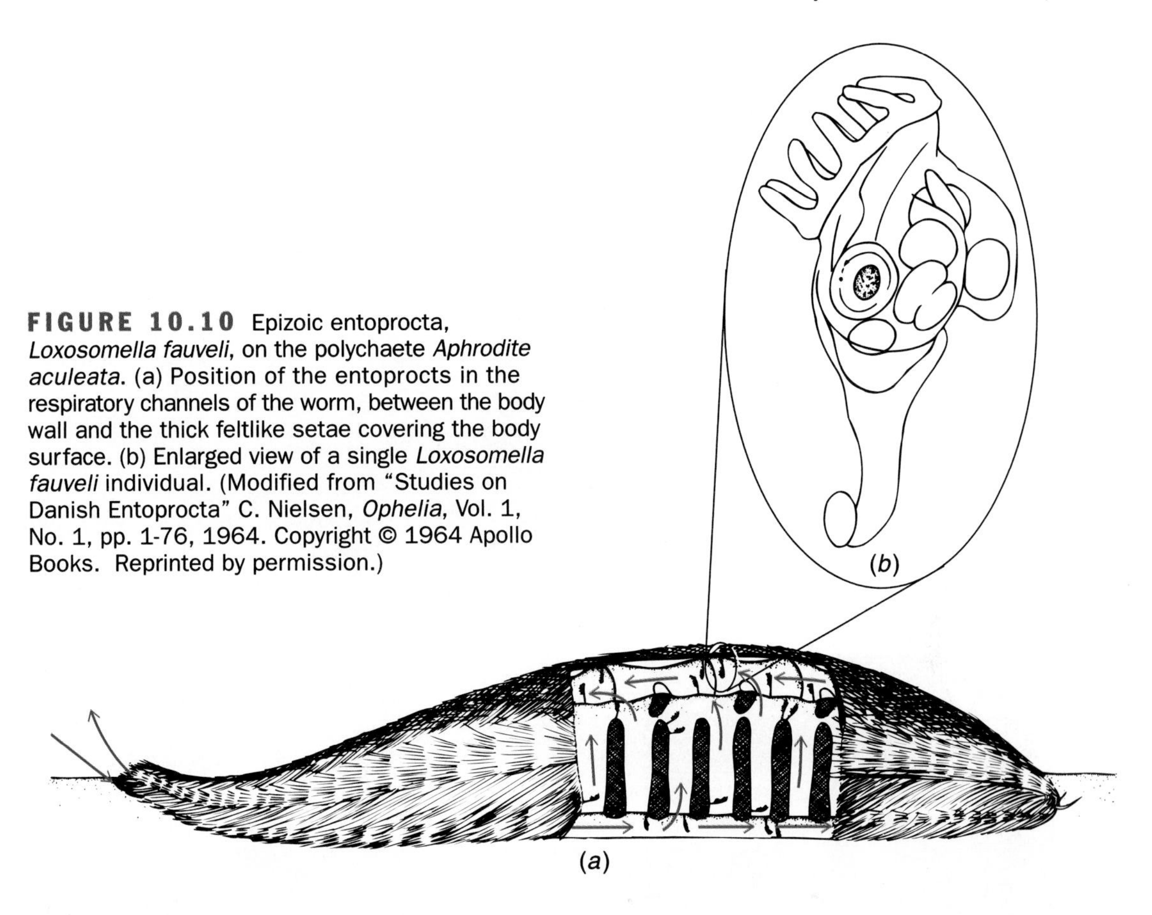

FIGURE 10.10 Epizoic entoprocta, *Loxosomella fauveli*, on the polychaete *Aphrodite aculeata*. (a) Position of the entoprocts in the respiratory channels of the worm, between the body wall and the thick feltlike setae covering the body surface. (b) Enlarged view of a single *Loxosomella fauveli* individual. (Modified from "Studies on Danish Entoprocta" C. Nielsen, *Ophelia*, Vol. 1, No. 1, pp. 1-76, 1964. Copyright © 1964 Apollo Books. Reprinted by permission.)

known to Gulf coast beachcombers as "Texas longhorns"). A second bryozoan, *Triticella elongata*, is commonly found on the crab *Pinnixa*, which is commensal in the tubes of the polychaete *Chaetopterus*. The phylum Entoprocta is a small, little-known group, most of which, as Nielsen (1964) has documented, are epizoic commensals on other marine invertebrates. For example, they inhabit respiratory current areas of various polychaete worms (Figure 10.10).

Among the Annelida, a number of polychaete worms are epizoites with definite host specificity. The best studied is the hesionid *Ophiodromus pugettensis*, which occurs on the ambulacral grooves of the bat star, *Asterina* (*Patiria*) *miniata*, on the Pacific coast (Figure 10.11). However, the whole group of so-called "scale worms" (Polynoidae) are all very common epizoites on various other echinoderms and mollusks.

Among the class Crustacea, the epizoic forms occur mainly within three groups: copepods, amphipods, and decapod crabs. Copepods associated with other marine animals are usually highly evolved and modified parasites, but there are a few that are simply commensal, scampering over the outer surfaces of various invertebrates. Such examples include various species of the genus *Hemicyclops*, which move on the surface of other, larger crustaceans. Others, such as *Paranthessius*, are found in the mantle cavity of various bivalves (*Tresus*, *Protothaca*, *Saxidomus*). One of the most specialized epizoic associations is that among several species of barnacles and various large whales. Two families of barnacles are found only on whales, and several are species specific. *Cryptolepas rachianectes* occurs only on the California gray whale, *Eschrichtius robustus*, and *Conchoderma auritum* attaches to *Coronula diadema*, which, in turn, is attached to the skin of whales (Figure 10.12). A curious form of epizoic relationship occurs between the amphipods of the suborder Hyperiidea and the large jellyfish. Hyperiids are often epizoic or endozoic on jellyfish. Also epizoic on

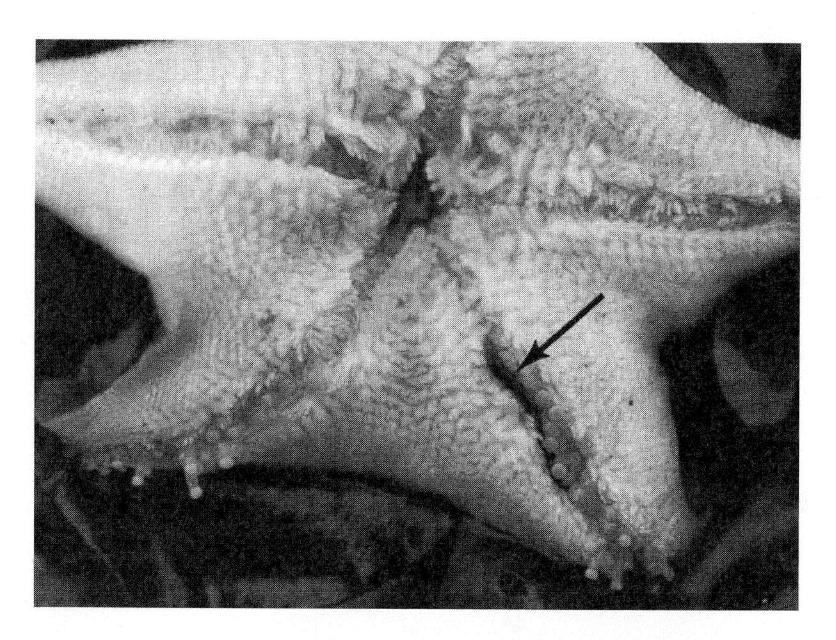

FIGURE 10.11 The scale worm *Ophiodromus pugettensis* on the bat star *Asterina miniata*. Arrow points to worm. (Photo by the author.)

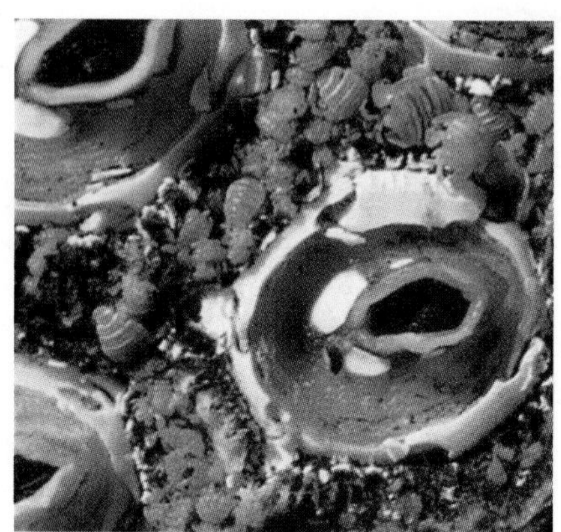

FIGURE 10.12 The whale barnacle *Cryptolepas rachianectes* and the cyamid amphipod *Cyamus scammoni* on the skin of the California gray whale, *Eschrichtius robustus*. (Photo by the author.)

FIGURE 10.13 Commensal slipper shells, *Crepidula adunca*, adhering to the black shell of the predatory gastropod *Mitra idae*. (Photo courtesy of Drs. Lovell and Libby Langstroth.)

jellyfish are certain juveniles of various benthic crab species (e.g., *Cancer jordani*).

Most mollusk epizoites are found among certain groups of clams. Generally, these are small clams that attach via a foot or byssus to the outside of other invertebrates. Thus, *Montacuta ferruginosa* is found on the heart urchin, *Echinocardium cordatum*; *Orobitella rugifera*, attached via a byssus, is found on the underside of the burrowing shrimp *Upogebia stellata*; and *Mysella pedroana* is seen on the legs of the sand crab *Blepharipoda occidentalis*. Among the gastropods found on other invertebrates, the genus *Crepidula* is one of the most common (Figure 10.13).

The above is certainly not an exhaustive list of the various epizoic and endozoic symbiotic associations, but it gives an idea of the extent and diversity of such associations. Many, many more could be listed but would not serve the purpose of this text.

A second type of symbiotic association is that of organisms associated with animals forming a

FIGURE 10.14 The echiuran worm *Urechis caupo*, to the left, and some of the symbionts that inhabit its burrow. Shown are the pea crab *Scleroplax granulata*, the scale worm *Hesperonoe adventor*, and the long-fingered shrimp *Betaeus longidactylus*. (Photo by Dr. Richard Mariscal.)

tube or burrow. In this case, the commensal or guest occupies the tube or burrow constructed by the host but is not necessarily in close association with the body of the host. These commensals use the tube or burrow as a refuge from predation. They may or may not also tap the food source of the host. This type of symbiotic relationship ranges from one simply fortuitous, as in the case of an animal diving into any convenient burrow to avoid a predator, to examples of obligatory associations, where the commensals are not found other than in such tubes or burrows and are dependent on them for existence. In the former category are certain small gobiid fishes, such as *Clevelandia ios*, which take refuge in any available tube or burrow but emerge to feed. Obligate tube dwellers include many small crabs of the family Pinnotheridae, known as "pea" crabs. Many species of this family are known throughout the world, where they inhabit the tubes and burrows of polychaete annelids and other invertebrates, sometimes in species-specific associations.

One of the best known examples of symbiosis with tube dwellers is that of the echiurid worm *Urechis caupo* on the Pacific coast of North America. This animal creates a permanent U-shaped burrow in which can be found four or more commensals (Figure 10.14). At the upper end is the fish *Clevelandia ios*, which merely uses the tube as a refuge. Further in the tube and closely associated with the worm are the scale worm *Hesperonoe adventor* and the pinnotherid crab *Scleroplax granulata*. *Hesperonoe adventor* seems restricted to *Urechis* burrows and lives against the worm, snatching food from it. Although the crabs are restricted to tubes, they may be found in the burrows of other animals as well. Another commensal is the small clam *Cryptomya californica*, which has very short siphons and inserts them into the burrow of *Urechis*, enabling it to live lower in the substrate than the length of its siphons would normally permit. Occasionally, the shrimp *Betaeus longidactylus* may inhabit the burrow.

A similar situation prevails with the tube-building polychaete *Chaetopterus*. This genus is virtually cosmopolitan, building tough parchment tubes in the muddy intertidal. These tubes are also inhabited by various species of pinnotherid crabs and scale worms, the species differing depending on the geographical location.

Another type of association concerns those invertebrates that live in the mantle cavities of various mollusks. This is a very common site for commensals to occupy; thus, the nemertine *Malacobdella grossa* lives in the mantle cavity of various clams on the Pacific coast. Pinnotherid crabs occupy the man-

FIGURE 10.15 Symbiotic relationships among crabs and anemones. a) The anemone *Dardanus gemmatus* on a hermit crab. b) The crab *Lybia tessellata* with anemones of the genus *Triactis* in its chelae. (Photos courtesy of a, Mike Severns/Tom Stuck & Associates. b, Andrew J. Martinez/Photo Researchers Inc.)

FIGURE 10.16 An anemone fish of the genus *Amphiprion* nestled in the tentacles of its anemone. (Photo courtesy of Drs. Lovell and Libby Langstroth.)

tle cavity of numerous bivalves, including oysters, where they are considered a pest by oyster fishers. Polychaete worms inhabit mantle cavities of chitons and limpets. In the tropics, shrimps of the family Palaemonidae are often found in mantle cavities of large bivalves such as *Pinna*, *Atrina*, and *Tridacna*.

The remaining types of associations are mainly mutualistic rather than strictly commensal, as aforementioned. The organisms involved display varying levels of behavioral and physiological modifications for the association.

One of the most obvious, but not well understood, associations occurs between various crabs and sea anemones. Many hermit crabs all over the world have anemones attached to their shells, and in some genera, such as *Dardanus*, the presence of anemones seems to be universal. Still other crabs, such as *Hepatus*, *Munidopagurus macrocheles*, and *Stenocionops furcata*, are found with anemones attached directly to their backs. The ultimate symbiotic relationship appears to be that evolved among a few crabs (*Lybia*, *Polydactylus*), which carry anemones on their chelae and use them for defense or food capture (Figure 10.15). Interestingly, most of the anemones involved in these associations are of three genera, *Calliactis*, *Paracalliactis*, and *Adamsia*,

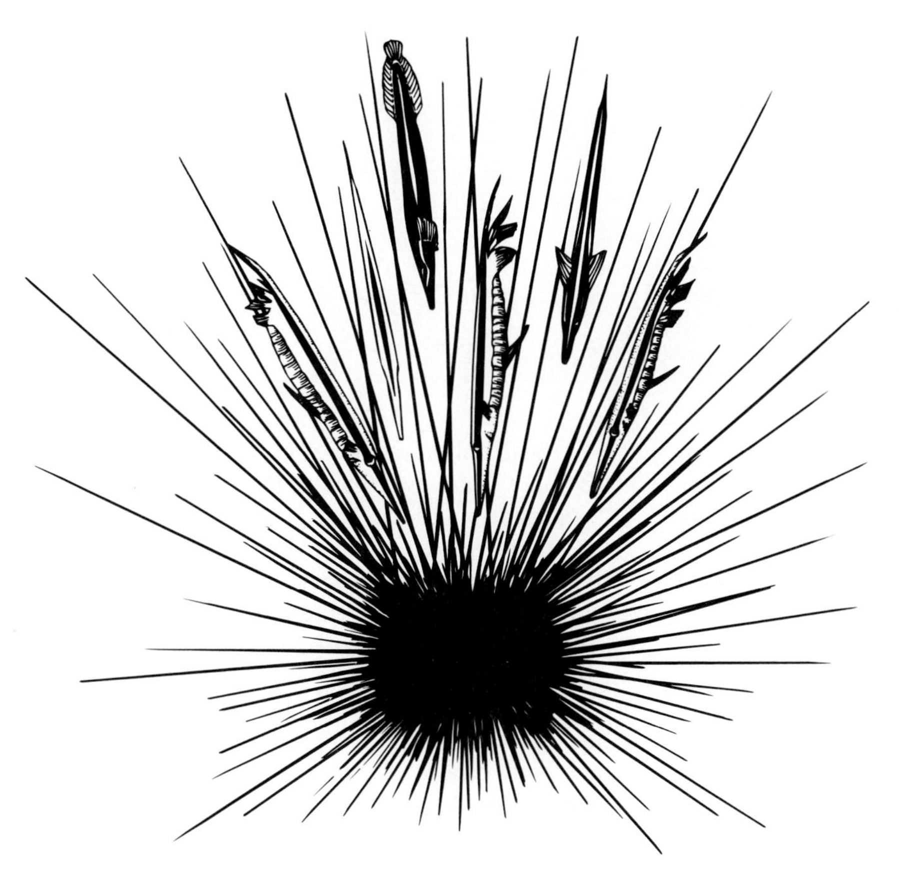

FIGURE 10.17
Diadema, the long-spined sea urchin, with three shrimp fish, *Aeoliscus strigatus*, and two cling fish, *Diademichthys deversor*, sheltering among the spines. (After *Marine Animals, Partnerships & Other Associations*, R. V. Gotto, 1969.)

species of which are rarely found elsewhere than with crabs. These associations must be mutualistic, because whenever the crab changes shells or molts its carapace, the anemone is usually also transferred. This is effected by the manipulatory movements of the crab but requires a degree of cooperation on the part of the anemone in loosening its grip on the shell or carapace and allowing itself to be physically moved.

Another type of association is that between other Cnidaria and fishes. Throughout the Indo-West Pacific in the coral reef areas, there is a striking association of small fishes of the genera *Amphiprion*, *Premnas*, and *Dascyllus* with various large anemones. These anemone fishes are able to live nestled among the tentacles of the anemones by preventing discharge of the formidable nematocysts of these tentacles (Figure 10.16). Other fishes cannot prevent this discharge, and the anemones can kill and eat fishes of a similar size. A similar association seems to exist between the dangerous siphonophore, the Portuguese man-of-war (*Physalia*), and the little fish *Nomeus gronovii*, which swims among the tentacles bearing the powerful nematocysts (see Figure 10.20). Still other juvenile fishes often congregate under the bells of the large scyphozoan jellyfishes (*Cyanea*, *Chrysaora*), where they presumably obtain protection from predation.

A similar situation prevails between fishes and the long-spined tropical sea urchin, *Diadema*. As we noted in Chapter 9, these urchins are circumtropical in shallow water and have enormously long, thin spines, that easily penetrate flesh. At least two tropical fishes, *Aeoliscus strigatus* and *Diademichthys deversor*, have adapted themselves to live among these spines, presumably for protection (Figure 10.17). The fish also look like sea urchin spines and so are camouflaged.

FIGURE 10.18 A remora, *Echeneis* attached to a manta ray. (Photo courtesy of Jeff Rothmann Photography.)

There is also the association of fishes with other fishes, usually large, predaceous fishes. Pilot fishes (*Naucrates ductor*) and remoras (*Echeneis remora*) are always found with larger fishes; marine vertebrates, such as turtles; or even moving, inanimate objects (Figure 10.18).

The final type of association is that known as cleaning behavior. As noted in Chapter 9, this is an association in which various species of fishes and shrimps actively attract large fishes to clean them of various ectoparasites (see p. 409).

Origin of the Association

With such a diverse array of relationships, it seems unlikely that there is any single explanation for the origin of all the different types of associations. However, one factor common to all associations is that they have arisen in areas of the ocean that are the most crowded with life. It would, therefore, appear reasonable that epizoites in particular could have become established due to the great competition for space among the settling invertebrate larvae. As a result, many settled on other invertebrates, and this later evolved, in some cases, into more obligatory relationships in which commensal individuals gained some advantage over the free-living forms. This is borne out by noting that many epizoites or burrow-inhabiting species are not strongly attracted to any given host species, while others are highly specific. Similarly, for many of the tube and burrow inhabitants, and those found in the mantle cavities of various mollusks, we can probably trace the origin of the relationship to the search for space. After all, tubes, burrows, and mantle cavities really represent a certain amount of potentially occupiable, semienclosed space.

The initial association of crabs and anemones may also have begun when anemones began to reside on shells already occupied by hermit crabs. Alternatively, the association could have arisen out of the habit of various crabs that pick up materials from the bottom and apply them to their carapaces as camouflage.

Many of the commensals that are epizoites or burrow or tube dwellers, especially the polychaete worms, tend to be from groups that show strong **thigmotaxis**; that is, they respond positively to stimulation by contact with other, solid bodies. Such a behavioral response would naturally preadapt them to move into tubes or burrows or into various cavities in the bodies of larger animals. Once established, selection would favor the development of chemosensory clues enabling these animals to find such burrows, tubes, or cavities, and more specific associations would follow.

Among the fishes and Cnidarians, the various types of symbiotic relationships seem to have their origin in the protection from larger predators that such associations confer. It does not appear that this relationship stems from crowding, especially between epipelagic fishes and Cnidarians. Provided that the symbionts can avoid having the nematocysts discharged, the tentacles of anemones and jellyfishes represent a secure, formidable barrier to a large predator. However, in the case of large sharks, pilot fishes, and remoras, the association probably originated in the search for food in the epipelagic. Pilot fishes and remoras probably were originally attracted by the amounts of food dropped in feeding by the larger fishes, and they could have established themselves either by avoiding falling prey to the sharks, or by being small enough not to be recognized as food by the larger fishes.

Cleaning behavior is a complex relationship involving many fish species, and its origin and ecological functions have been confounded by alternative hypotheses and factual inaccuracies, as Gorlick et al. (1978) have discussed. More recently, Grutter (1999) has demonstrated that for at least one cleaner, *Labroides dimidiatus*, the association is a mutualistic one.

Modifications Due to the Association

As with the algal-animal symbiosis, the animal-animal relationships can result in anatomical, physiological, and behavioral modifications to one or both of the partners.

The fewest modifications occur among the epizoites, especially those that are facultative epizoites and also are found in nonsymbiotic situations. Often, these animals have no anatomical modifications attributable to the symbiotic association, but those that are found with specific hosts usually have developed the means for recognition of the host from a distance. Such ability to recognize hosts is even more highly developed among obligate symbionts of various types. What is the basis for such recognition? Investigators such as Davenport (1950) have shown that it is most likely a chemical, unique to the host, that is released into the water. The commensal has, in turn, developed chemosensory receptors that detect the chemical and enable it to home in on its host. If such symbionts are tested in the laboratory in an apparatus designed to give them a "choice," they show a remarkable ability to choose the host on which they are found from among other closely related forms. Such chemical attraction has been demonstrated for several scale worms (Polynoidae) commensal with various starfish species; for the peculiar fishes, *Carapus*, which live in sea cucumbers; and for the polynoid *Hesperonoe adventor* and its host, *Urechis caupo*. More recently, VandenSpiegel *et al.* (1998) have suggested that for the shrimp *Synalpheus stimpsoni*, which lives on tropical commatulid crinoids, vision is the first sense that allows the shrimp to find a commatulid and then olfaction acts, permitting selection of the right commatulid species.

Another modification of commensals, and one that also aids in host recognition, is the ability of the larvae or young to select the appropriate substrate. In many epizoite species that settle onto their hosts from the plankton, metamorphosis and subsequent development do not occur unless the larva is in contact with the appropriate host. The substrate discrimination abilities of some of these symbionts are truly remarkable. Some can recognize substrate surface texture at the molecular level! It is not surprising, then, that whale barnacles can settle not only on whale skin, but only on certain species of whales. In this case, however, it is likely that a chemical unique to the host is also present. Similarly, the clam *Modiolaria* is able to recognize the structure of tunicin, the major chemical component of the external layer of the sea squirts with which it lives.

A basic modification of tube-dwelling symbionts is that they are often smaller or thinner and flatter than their free-living relatives. Thus, the pinnotherids, or "pea" crabs, are among the smallest of the marine crabs. The shrimps that inhabit the mantle cavities of bivalve mollusks are also among the smallest of the shrimps. Similarly, the scale worms are often very flattened in comparison with other polychaetes. The fishes, such as *Clevelandia ios*, commensal in tubes, and *Apogonichthys* and *Carapus*, in the internal spaces of clams and sea cucumbers, respectively, are very thin so they may fit into these restricted areas. The pearl fishes, *Carapus*, show the greatest modifications in that they have lost their scales and pelvic fins and have shifted the anal opening far forward under the head. This latter change apparently is to ensure that defecation will occur outside the body of the sea cucumber. Furthermore, as Arnold (1957) has shown, these fishes show behavioral modification in that they enter the cucumber tail first (Figure 10.19)! Such extensive modifications suggest a long history of adaptation.

Among anemones associated with crabs, a number of behavioral modifications occur. First, the anemones must respond positively to manipulations by the crab when it rushes to transfer the anemone to another shell or from its shed carapace to a new one. It does this by releasing its grip on the old shell, so the crab can remove it with its claws. This requires a higher degree of nervous integration than we have believed could be accomplished by the extremely primitive, brainless nervous system of the anemone. We do not understand this process among crabs that

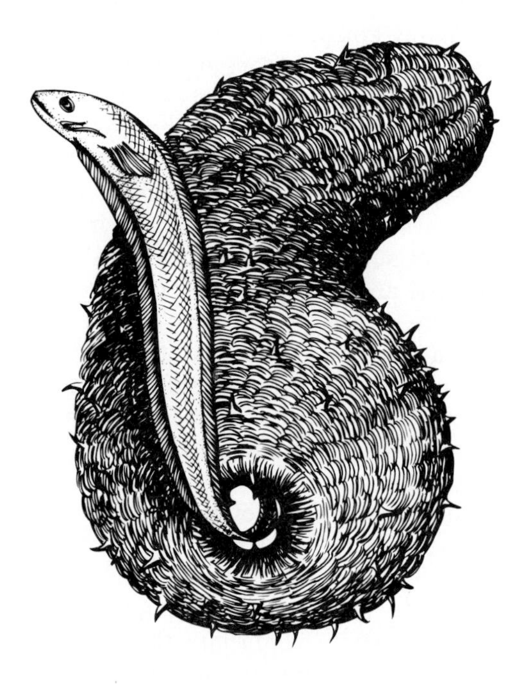

FIGURE 10.19 A pearl fish, *Carapus*, entering tail first into the cloacal opening of its holothurian host. (After *Marine Animals, Partnerships and Other Associations*, R. V. Gotto, 1969.)

carry anemones in their chelae. The anemones again show a behavioral adaptation in that they allow themselves to be carried in the chelae and do not close up. In these cases, however, the crab also shows changes in that the chelae are modified to carry the anemones rather than to gather food. The crab extends the anemones when threatened or when gathering food. It uses its second pair of legs to transport food to the mouth, rather than the chelae, since the latter carry the anemones. Perhaps the most drastic changes undergone by the anemone member of these crab-anemone systems is that exhibited by *Adamsia palliata*, which lives with the hermit crab, *Eupagurus prideauxi*. Here, the anemone steadily expands its basal disk to completely encircle the vulnerable abdomen of the hermit crab; thus, it eliminates the need for the crab to change its shell as it grows. The anemone then expands to allow for the growth of the crab and orients itself so the oral disk and mouth are ventral to the crab; thus, it can obtain food from the crab.

The greatest morphological changes among fishes associated with other fishes occur in the remoras. Remoras have their dorsal fin modified to form a large sucker, enabling them to remain attached to their hosts (see Figure 10.19). Pilot fishes (*Naucrates*) and the anemone fishes (*Amphiprion*) have no special morphological modifications. Among anemone fishes, however, there are definite behavioral and biochemical modifications. Studies have shown that anemone fishes of the genus *Amphiprion* must undergo a period of acclimatization to an anemone before they can dive into its tentacles with impunity. During this period, the fish is not capable of preventing nematocyst discharge by the anemone. During the acclimatization, Schlichter (1976) demonstrated that the fish goes through a series of first brief and then longer encounters with the anemone. During this time, the fish coats itself with anemone mucus; thus, it tricks the anemone into discerning the fish as itself. As a result, there is no nematocyst discharge from the anemone. When this coating is complete, the acclimatization period is over and the fish can dive in among the tentacles without causing nematocyst discharge.

In addition to this fascinating coating process, anemone fishes have additional behavioral modifications. In contrast to most fishes, which swim away when approached by humans or other large potential predators, anemone fishes swim out of their anemones toward the potential predator! This attracts the predator to them, at which time, they do an abrupt about-face and dive into the anemone. Presumably, the behavior would make an unwary potential predator chase them and be caught and consumed by the anemone. To further attract attention to themselves, anemone fish also have vivid color patterns, contrasting markedly with their background (see Figure 10.17).

In contrast to the anemone fishes, which develop immunity to the nematocysts of their host, the fishes (*Nomeus gronovii*) associated with the Portuguese man-of-war do not acclimate, nor do they ever obtain complete immunity from the nematocysts (Figure 10.20). They instead appear to spend their lives playing a perpetual game of Russian roulette with the tentacles of the siphonophore, constantly moving to avoid contact with the lethal nematocyst batteries! The same is true for the juveniles of various fishes that shelter below the umbrellas of the large jellyfish.

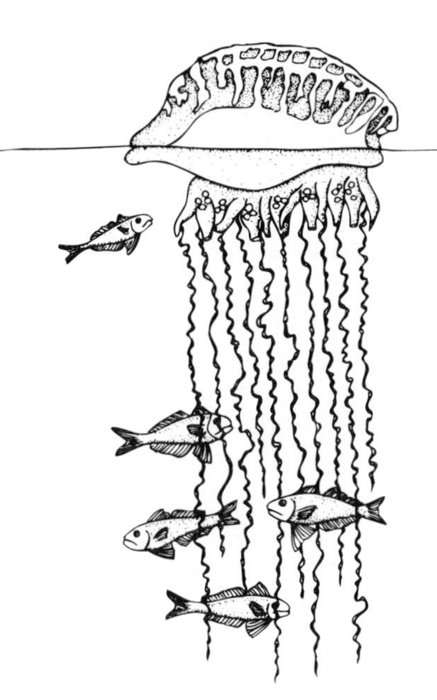

FIGURE 10.20
Portuguese man-of-war, *Physalia*, with the fish *Nomeus gronovii* sheltering among the tentacles. (After *The Biology of Marine Animals*, J. A. Colin Nicol, 1960 Pittman Books.)

Among cleaning fishes, certain morphological modifications can be noted, such as pointed, narrow snouts and forcepslike teeth, adaptations for picking up the small ectoparasites. Both cleaning shrimps and fishes also tend to have very bright and contrasting colors, which make them stand out against the background and advertise their presence. Certain cleaning shrimps, notably the Pederson shrimp, *Periclimenes pedersoni*, have extraordinarily long antennae, which they wave about to further advertise their presence to the fishes.

Cleaning behavior relationships also impose behavioral modifications on the fishes that are being cleaned. These animals must seek out and present themselves at the "cleaning station" and then remain motionless while the cleaner moves over their bodies. They must also open their mouths and spread their gills and opercula to allow the cleaner to enter the mouth and gill area. Of course, at the same time, the larger fish being cleaned must refrain from sudden inhalations, which might accidentally ingest the small cleaner!

Value of the Associations

These symbiotic associations have obvious value to one or both of the members, and the relative value of the relationship to each varies depending on the association.

The major value of the symbiotic relationship to the epizoites would seem to be with the epizoite rather than the host organism. For some epizoites, the value of the association lies in the fact that the host represents the only available suitable substrate in an otherwise unsuitable area. For example, the hydroid *Clytia bakeri*, which is epizoic on the clam *Donax gouldi* in sand areas in southern California, could not live in the area unless it had the clam to settle on, because it cannot exist on sand. There is, however, no obvious value to the clam. Other epizoites fix themselves on the host to take advantage of feeding or respiratory currents produced by the host. This seems particularly common. The two-tentacled hydroids of the genus *Proboscidactyla* are fixed on the rims of the tubes of their annelid hosts to take advantage of the feeding tentacles and to remove food particles from the host. Similarly, the various epizoic entoprocts are usually found where there are respiratory currents. In this latter case, they do not steal food directly from the host, but filter the food from the moving current. Perhaps most epizoites benefit from the protection afforded them by the larger host; thus, the various flatworms occurring in the guts and mantle cavities of their larger hosts are protected from predation and, in the case of intertidal forms, from possible desiccation. Surely, the hyperiid amphipods found on the large scyphozoan jellyfish would be much more subject to predation if free-living in the plankton. Finally, the epizoite may also gain the advantage of movement. This advantage may be particularly significant if the epizoite is sessile and the host freely moving.

Commensals that inhabit tubes primarily gain protection from predation. This, for example, is the main benefit to the goby *Clevelandia ios* that uses the tubes of *Urechis caupo*. Other commensals, however, such as the scale worms (Polynoidae), gain not only protection but food as they snatch morsels from their tube-dwelling hosts. Commensals in the mantle cavities of bivalve mollusks benefit from the protection of the bivalve shells, but they can also use the feeding and respiratory currents of the clam for their own needs. In a few cases, these symbionts may tend toward parasitism in that they may consume pieces of the gills or mantle of their bivalve host.

Among the fishes associated with large jellyfishes, the Portuguese man-of-war, and the large sharks, the whole value of the association is to the fishes in the form of protection from predation in an area (the epipelagic zone) notorious for great predation pressure. Pilot fishes and remoras are also protected from predation by their association with the large predaceous sharks, but they may also obtain food when the shark is feeding. The fishes associated with sea urchins and living in the respiratory trees of sea cucumbers are protected by their association, but their presence is no advantage to their host.

Among the mutualistic associations, the values of the association are more pronounced and, as the term implies, extend to the host as well. In the various crab and anemone associations, the crabs, as hosts, gain the advantage of the nematocyst batteries of the anemones, either as protection against predators or else as an offensive weapon to capture food. The anemones, while still sessile, gain the advantage of movement; some, such as those on hermit crabs, may also be able to obtain food fragments lost during the feeding process of the crab. A similar situation prevails between anemone fishes and their anemones. The fishes gain protection, but they benefit their partners by attracting prey within reach of the tentacles and also by picking up pieces of food outside the reach of the anemone and depositing them in the anemone's mouth. Fautin (1991) has demonstrated that the fish protect the anemone from predatory fishes. The fish may also benefit the anemone by producing nutrients in the form of wastes that the anemone takes up.

In the case of cleaning behavior, the cleaner organisms' gain is in the form of food, whereas the large fishes cleaned presumably enjoy an increased measure of health due to the removal of the various ectoparasites.

Luminescent Bacteria

A final category of symbiosis is the curious relationship that has developed between various marine animals and luminescent bacteria. In these relationships, the bacteria are usually confined to a cavity in the body of the larger animal near the outer surface. This cavity is usually connected to the outside through an opening of some sort.

Such symbiotic relationships are confined to three groups of marine animals: fishes, squids, and pelagic tunicates. The relationship is most common among fishes and squids inhabiting the mesopelagic zone (see pp. 152–154), but it is still less common among these groups than is bioluminescence produced intrinsically without bacteria. The symbiotic bacteria are species of the genera *Vibrio* or *Photobacterium*. The fish or squid of each generation apparently acquire their luminescent bacteria from the environment, where the bacteria are widespread. The various suggested values for the production of light by mesopelagic animals have been discussed in Chapter 4.

The relationship of bacteria and the fishes or squids is a mutualistic one in which the bacteria obtain food from the larger animal. In turn, the fishes or squid use the light produced by the bacteria for various defensive or offensive purposes, as discussed in Chapter 4.

The light produced by bacteria is usually continuous. As a result, the fishes and squids often develop elaborate modifications to control the light. These anatomical developments can include the production of a reflecting layer behind the cavity to reflect the light outward, lenses to focus and concentrate the light, and most often, a screen or shade on the outside that can be raised or lowered to either "turn on" or "turn off" the light. The bacteria show no corresponding changes in behavior or morphology as a result of the association.

The origin of these elaborate associations is not well understood, and the bacteria apparently are not passed from one generation of fishes or squid through the eggs. It would appear that each generation somehow must reinfect its light organs from the external environment.

SUMMARY OF KEY CONCEPTS

- Symbiosis is the name given to close relationships between two different species.
- Symbiotic relationships cover a broad spectrum of associations, including commensalism, inquilinism, and mutualism.
- Basically there are two groupings of nonparasitic symbiotic associations in the oceans: between algae and invertebrates, and between various animals.
- All symbiotic relationships between plants and animals are between unicellular algae or their chloroplasts and a wide variety of invertebrates.
- Symbiotic relationships are most common in the tropics, less common in temperate waters, and absent from polar waters.
- There are three groups of algal cell symbionts: brown zooxanthellae, green zoochlorellae, and blue-green cyanellae.
- Algal symbionts are generally restricted to certain tissues or areas of the host.
- In cryptic symbiosis, the algal symbionts have lost their cellular integrity and do not appear separate from the host body.
- Origin of the symbiotic association is likely through ingestion of the algal cells or chloroplasts by cells of the digestive tract of the host.
- Algal-invertebrate associations are found in protozoans, sponges, cnidarians, ascidians, flatworms, and mollusks most commonly and rarely in a few other higher taxa.
- Modifications in the algal symbionts include loss of locomotory organs and reduced cell walls or loss of cell walls.
- Among the animal hosts a wide variety of anatomical and behavioral modifications are found, of which the most dramatic occur in the jellyfish *Cassiopeia*, the flatworm *Convoluta*, and the bivalves of the genus *Tridacna*.
- The value of the symbiotic association to the algae seems to be in obtaining nutrients from the host, while the host benefits by obtaining food molecules produced by photosynthesis in the algal cell and perhaps additional oxygen.
- Each new generation of host either receives the algal symbiont directly from the parent by passage via egg or larvae or must be reinfected from the environment.
- Symbiotic relationships among animals cover a broader spectrum than the strictly mutualistic associations seen among plants and animals.
- The simplest type of animal symbiosis is the commensal wherein the "guest" lives on the host organism or in some construction of the host, usually a tube or burrow, gains in some way, and does not seriously inconvenience the host.
- Commensals that live on other animals are called epizoites; those that live inside other animals but are not parasitic are called endozoites.
- Epizoites are the most abundant group of commensals and are found among protozoans, Cnidaria, Entoprocta, Annelida, Arthropoda, and Mollusca.
- Associations between hosts and commensals in which the latter occupy tubes or burrows constructed by the host range from simply fortuitous to obligatory.
- Most other types of associations among animals are mutualistic and include associations between anemones and crabs and between cnidarians and fishes.
- Cleaning behavior is an association in which various species of shrimp or small fishes clean larger fishes of ectoparasites.
- Although there is not a single explanation for the origin of all the different animal-animal associations, the one factor common to all the associations is that they have arisen in areas of the ocean most crowded with life; hence, epizoites might well have become established due to competition for space.
- Associations between crabs and anemones may have begun when anemones began to reside on shells already occupied by hermit crabs.
- Many tube- or burrow-inhabiting commensals show strong thigmotaxis, which would naturally preadapt them move into tubes or burrows.
- Symbiotic relationships among fishes and cnidarians seem to have their origin in the protection from larger predators that such relationships confer.
- Cleaning behavior is a complex relationship and its origin is still debatable.
- Anatomical and behavioral modifications due to the association of animals with other animals vary dramatically with the fewest modifications, oftentimes none, occurring among the epizoites.
- A common modification of obligate commensals is their ability to recognize the host from a distance, usually through chemosensory means.
- Some commensals have great powers of substrate recognition.
- A basic modification of tube- or burrow-dwelling symbionts is that they are often smaller or thinner and flatter than their free-living relatives.
- Anemones associated with crabs show a number of primarily behavioral modifications that allow the crab host to manipulate and move the anemone.
- The greatest morphological changes among fishes associated with other fishes occurs in the remoras, where the dorsal fin is modified to a sucker.
- Anemone fishes lack morphological modifications but exhibit a definite set of behavioral and biochemical modifications.
- Cleaning fishes have long, pointed snouts and forceps-like teeth, and both cleaning fishes and shrimps have

contrasting colors that make them stand out against the background.
- Fishes being cleaned must remain motionless during cleaning, with mouths and opercula open to permit the cleaner to enter the mouth and gill area.
- The major value of the symbiotic relationship between epizoites and host would seem to be with the epizoites.
- Commensals inhabiting tubes or burrows primarily gain protection from predation, as do fishes associated with jellyfish, whereas pilot fish and remoras also are protected from predators but may also obtain food when the host is feeding.
- In mutualistic associations such as between crabs and anemones and between anemone fish and anemones, both partners benefit.
- In cleaning behavior, the cleaners gain food while the cleaned fishes enjoy better health due to removal of the parasites.
- Symbiotic relationships between luminescent bacteria and fishes, squids, and tunicates are mutualistic, with the bacteria obtaining food from the host and the host using the light for various defensive and offensive purposes.

REVIEW QUESTIONS

ESSAY: Develop complete answers to these questions.

1. Compare and contrast chloroplast symbiosis with the endosymbiotic theory of the evolution of eukaryotic organelles.
2. Develop a table that presents the benefits accrued by host and guest in algal-invertebrate symbiosis. Evaluate the evidence that supports one of these benefits.
3. Speculate on why symbiosis is such a recurring and dominant theme in marine systems.
4. There appear to be many more symbiotic associations (e.g., corals and their algae, and the *Tridacna* and its algae) in marine systems than in freshwater systems. Speculate on why this might be so.
5. Discuss more than one evolutionary scenario for the development of the cleaning symbiosis exhibited by cleaner shrimp and fish.

BIBLIOGRAPHY

Arnold, D. C. 1957. Further studies on the behavior of the fish *Carapus acus*. *Staz. Zool., Naples* 23:91.

Babcock, R. C. 1985. Growth and mortality in juvenile corals (*Goniastrea*, *Platygyra*, and *Acropora*): The first year. *Proc. 5th Int. Coral Reef Symp.* 4:355–360.

Beers, C. D. 1961. The obligate commensal ciliates of *Strongylocentrotus droebachiensis*. Occurrence and division in urchins of diverse ages; survival in sea water in relation to infectivity. *Biol. Bull.* 121:69–81.

Cheng, T., ed. 1971. *Aspects of the biology of symbiosis*. Baltimore: University Park Press.

Cote, I. M. 2000. Evolution and ecology of cleaning symbiosis in the sea. *Oceanography and Marine Biology, An Annual Review* 38: 311–355.

Dales, R. P. 1957. Commensalism. In *The treatise on marine ecology and paleoecology*, edited by J. E. Hedgpeth. Vol. I, *Ecology*. Geol. Soc. of Amer. Memoir 67, 391–412.

Davenport, D. 1950. Studies in the physiology of commensalism. I, The polynoid genus *Arctonoe*. *Biol. Bull.* 98(2):81–93.

Fankboner, P. V. 1971. Intracellular digestion of symbiotic zooxanthellae by host amoebocytes in giant clams (Bivalvia: Tridacnidae) with a note on the nutritional role of the hypertrophied siphonal epidermis. *Biol. Bull.* 141:222–234.

Fautin, D. G. 1991. The anemone fish symbiosis: What is known and what is not. *Symbiosis* 10:23–46.

Gorlick, D. L., P. D. Atkins, and G. S. Losey. 1978. Cleaning stations as water holes, garbage dumps and sites for evolution of reciprocal altruism. *Amer. Nat.* 112 (984):341–353.

Gotto, R. V. 1969. *Marine animals, partnerships and other associations*. New York: Elsevier.

Greene, R. W. 1970. Symbiosis in sacoglossan opisthobranchs: Symbiosis with algal chloroplasts; translocation of photosynthetic products from chloroplast to host tissue. *Malacologia* 10(2):357–380.

Grutter, A. S. 1999. Cleaner fish really do clean. *Nature* 398:672–673.

Henry, S. M. 1966. *Symbiosis*. Vol. I, *Associations of microorganisms, plants and marine organisms*. New York: Academic Press.

McLaughlin, J. J. A., and P. Zahl. 1966. Endozoic algae. In *Symbiosis*, edited by S. M. Henry. Vol. I, *Associations of microorganisms, plants and marine organisms*. New York: Academic Press, 257–297.

Nicol, J. A. C. 1960. *The biology of marine animals*. New York: Wiley.

Nielsen, C. 1964. Studies on Danish Entoprocta. *Ophelia* 1(1):1–76.

Rosewater, J. 1965. The family Tridacnidae in the Indo-Pacific. *Indo-Pacific Mollusca* I 6:347–396.

Rowan, R., and D. Powers. 1991. A molecular genetic classification of zooxanthellae and the evolution of animal-algal symbiosis. *Science* 251:1348–1351.

Schlichter, D. 1976. Macromolecular mimicry: Substances released by sea anemones and their role in the protection of anemone fish. In *Coelenterate ecology and behavior*, edited by G. O. Mackie. New York: Plenum Press, 438–441.

Schoenberg, D. A., and R. K. Trench. 1976. Specificity of symbioses between marine cnidarians and zooxanthellae. In *Coelenterate ecology and behavior*, edited by G. O. Mackie. New York: Plenum Press, 423–432.

Smith, D. C. 1973. *Symbiosis of algae with invertebrates*. Oxford Biology Reader, no. 43. London: Oxford University Press.

Smith, D. C., and A. E. Douglas. 1987. *The biology of symbiosis*. Baltimore: Edward Arnold.

VandenSpiegel, D., I. Eekhaut and M. Jangoux. 1998. Host selection by *Synalpheus stimpsoni*, an ectosymbiotic shrimp of comatulid crinoids, inferred by a field survey and laboratory experiments. *J. Exp. Mar. Bio. Ecol.* 225:185–196.

Vernberg, W. B. 1974. *Symbiosis in the sea*. The Belle W. Baruch Library in Marine Science, no. 2. Columbia: University of South Carolina Press.

Wilkinson, C. R. 1983. Phylogeny of bacterial and cyanobacterial symbionts. In *Endocytobiology, endosymbiosis and cell biology*, edited by W. Schwemmler and H. E. A. Schenk. Vol. II. Berlin: Walter de Gruyter, 993–1002.

Wilkinson, C. R., and J. Vacelet. 1979. Transplantation of marine sponges to different conditions of light and current. *J. Exp. Mar. Biol. Ecol.* 37:91–104.

Yonge, C. M. 1957. Symbiosis. In *The treatise on marine ecology and paleoecology*, edited by J. E. Hedgpeth. Vol. I, *Ecology*. Geol. Soc. of Amer. Memoir 67, 429–442.

Yonge, C. M. 1975. Giant clams. *Sci. Amer.* 232:96–105.

HUMAN IMPACT ON THE SEA

The oceans are truly the last frontier for human exploration and exploitation on this planet. For years, the vastness and volume of the oceans, coupled with the fact that humans are terrestrial organisms naturally ill equipped to enter into water more than superficially, kept the oceans and their communities safe from significant human interference or effects. As the human population of the world increased through the centuries, large land areas were significantly altered, but the oceans remained relatively untouched. The major usage of the sea during these centuries was as a source of food, but the primitive nature of gear and the limited range of vessels combined to make the effects of fishing insignificant.

As we enter the twenty-first century, all this has changed. The explosion of technology has enabled human beings to penetrate all parts of the oceans. The rapidly increasing human population, coupled with sophisticated technology, has had significant effects on the ecology of the world's oceans in a few short decades. The purpose of this chapter is to discuss and evaluate some of these effects in light of our knowledge of the functioning of the oceans' ecosystems.

FISHERIES

The earliest use of the oceans by humans was probably for food. Early human populations living along the oceans captured various shore fishes, marine mammals, and shellfishes for consumption. This is recorded in the various large "shell mounds" excavated by archaeologists in different parts of the world. With the advent of vessels to venture onto the sea and the development of more refined nets, fishing became an important source of food to people living near the oceans. However, the gear and the vessels were so inefficient through much of human history that, even as late as the 1880s, such scientists as T. H. Huxley believed that the major fisheries were inexhaustible. In the last decades of the twentieth century, however, the old ships and gear were replaced by much larger and more powerful vessels, more effective nets and traps, and electronic devices for detecting fish schools. The result has been a significant reduction in many fish populations and the disappearance or overexploitation of others at a time when increasing human populations are demanding more food. Since many believe the oceans are the major food reservoir for upcoming generations of humans, the impact of fishing and its potential to sustain future generations bear consideration here.

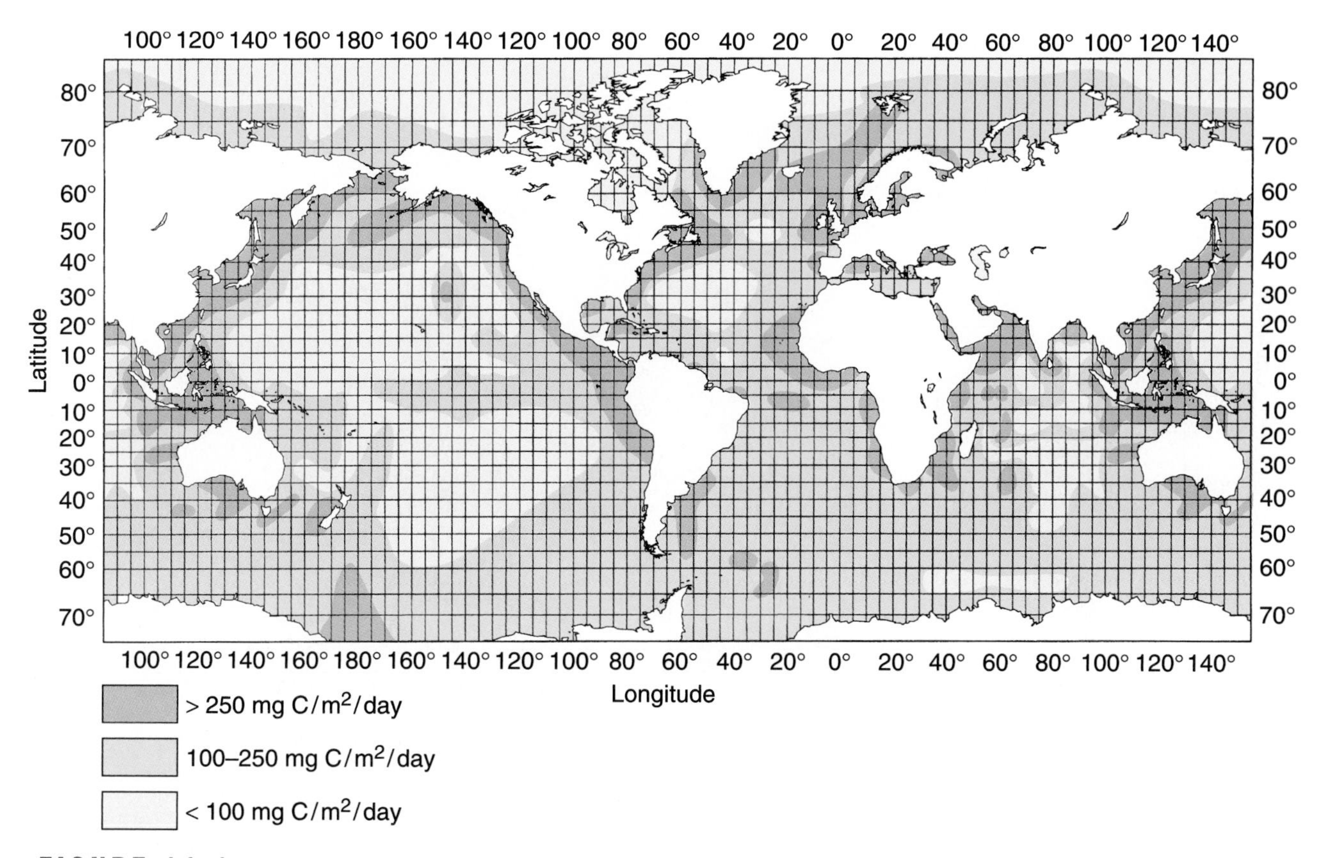

FIGURE 11.1 Geographical variations in the primary productivity of the world's oceans. White areas are areas of lowest productivity (less than 100 mg C/m^2/day); light blue indicates medium productivity (100–250 mg C/m^2/day); dark blue indicates areas of highest productivity (more than 250 mg C/m^2/day). (After *FAO atlas of the living resources of the sea*, 1972, Food and Agriculture Organization of the United Nations.)

Major Fishing Areas

The major fisheries are concentrated in the waters overlying the continental shelves around the world. The only major fisheries that operate in the open oceanic regions of the world are those for tuna and whales. There are several reasons for the concentration of fisheries in neritic waters. First, the inshore waters have a much higher primary productivity than most open ocean waters (see pp. 68–69) and, therefore, support larger populations of fishes at all trophic levels (Figure 11.1). Second, the bottom is fairly shallow on the shelf and is accessible to the various nets and traps used by humans to capture fishes. The deep ocean floor, on the other hand, is so far from the surface that even with our current sophisticated capture mechanisms, we cannot commercially fish there. Finally, the lack of food in abyssal depths precludes any large fish population from existing there. Any fishes there could not sustain a fishery for any length of time.

Although the fisheries are concentrated in the shelf regions, there is a very unequal distribution of the tonnage caught in various areas. The shelf areas of northwest Europe, along the upwelling coast of western South America, and off Japan produce the largest catches of fishes (Figure 11.2). The southern oceans and the tropics, with the exception of the west coast of South America, contribute much less to world fisheries.

Major Commercial Species

The commercially important marine animals are drawn from four groups: bony and cartilaginous fishes, marine mammals, mollusks, and crustaceans.

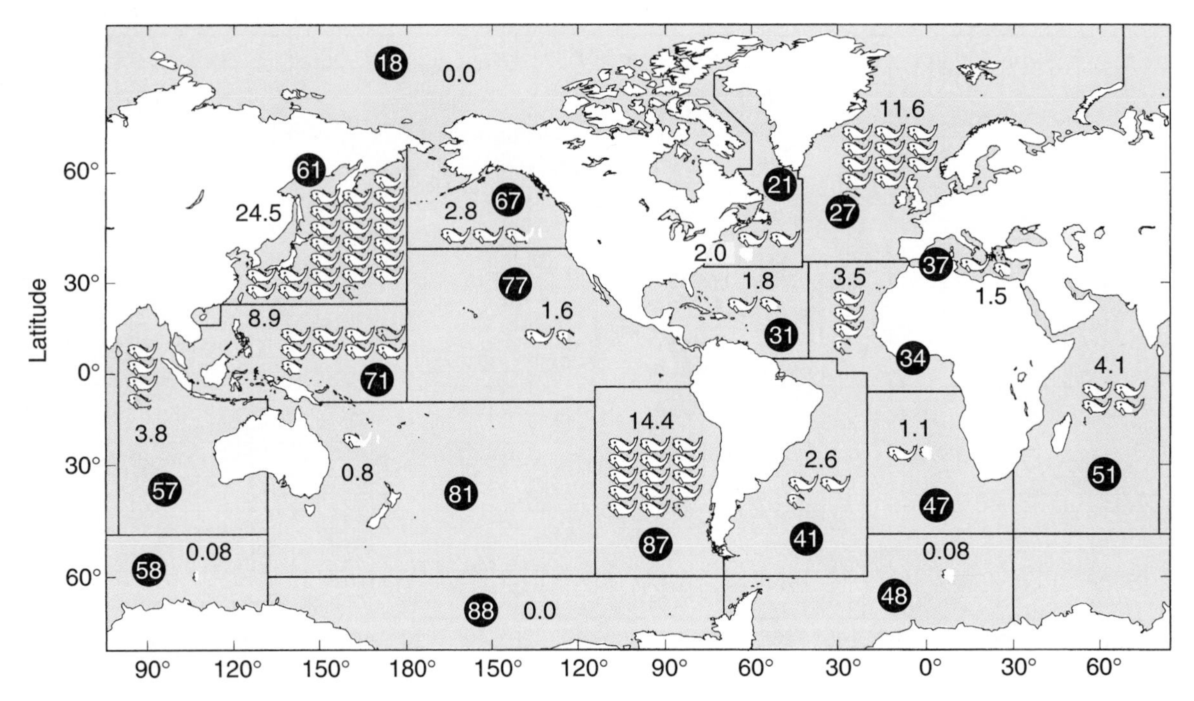

FIGURE 11.2 The major marine fishing regions of the world and their respective 1997 landings. Each symbol represents 1 million metric tons. The circled numbers refer to the region in the FAO division of the oceans. (Modified from the *FAO 1997 yearbook of fishery statistics*, Food and Agriculture Organization of the United Nations.)

Of these groups, the various fishes constitute by far the greatest tonnage. Among the thousands of species of marine fishes, only a very few make up the majority of catches in fisheries throughout the world. These can be assembled into a few major groups (Table 11.1). The herrings, sardines, and anchovies account for the largest tonnage of fishes. Of these, a single species, the Peruvian **anchoveta** (*Engraulis ringens*), in the past has accounted for almost half the catch and was the basis of the largest single fishery in the world, off Peru (Figure 11.3). Significant other fisheries in this group are for menhaden off the Atlantic and Gulf coasts of the United States and the herring fishery of northwestern Europe. Before its collapse in the 1940s and 1950s, the California sardine fishery was an important contributor to tonnage in this group. Herrings, anchovies, and anchovetas, commonly called clupeoid fishes, are small, pelagic, and feed on a low trophic level, either directly on phytoplankton (*Engraulis*) or on herbivores, such as copepods. Most are not used as food for humans but are reduced to protein meal for animal food.

The second largest group of fishes landed are the gadoids, comprising various cod, haddock, pollock, and hake. These are bottom-dwelling (demersal) fishes, and the major fisheries are on the shallow banks of the North Atlantic and North Pacific oceans (Figure 11.3).

The third largest group are the scombroid fishes, generally known by the common name mackerel. Fishes of this group are common in both temperate and tropical waters, where they are fast-moving pelagic carnivores in shallow water.

Closely related to mackerels are the various tunas. Tunas are among the largest fishes in commercial fisheries. They are also the basis of the only major open ocean fishery. They are widespread, fast-swimming carnivores of tropical and warm temperate seas (see Chapter 3 for a discussion of their biology). Although the tonnage is not as great as for other fishes, they are significant because they command a high price both as a canned fish product and as a fresh fish.

Redfishes, rockfishes, and sea basses are demersal, cold-water fishes, primarily used as human food.

The flatfishes, such as halibut, sole, plaice, and flounder, are all well-known fishes of considerable economic importance because they are premium

TABLE 11.1

Worldwide Catches of Different Groups of Fishes and Shellfishes (in Millions of Metric Tons)

	1989	1990	1991	1992	1993	1994	1995	1996	1997
Herrings, sardines, anchovies	24.57	22.28	21.71	20.39	22.01	25.91	22.02	22.31	21.58
Cod, hake, haddock	12.83	11.79	10.39	10.54	9.93	9.73	10.72	10.74	10.23
Jacks, mullets, sauries	9.24	9.72	10.31	10.49	9.96	9.88	10.81	11.36	10.73
Miscellaneous marine fishes	10.12	9.91	9.76	10.33	10.30	9.69	9.79	10.25	10.06
Redfish, basses, congers	5.90	5.74	5.94	5.85	5.69	6.29	6.89	6.74	7.24
Mackerels, snoeks, cutlass fishes	3.82	3.51	3.44	3.30	4.02	4.53	4.72	5.15	5.26
Tunas, bonitos, bill fishes	4.00	4.37	4.45	4.39	4.57	4.65	4.73	4.63	4.85
Shrimps, prawns	2.44	2.60	2.80	2.91	2.08	2.25	2.30	2.45	2.54
Squid, cuttlefishes, octopus	2.53	2.34	2.61	2.75	2.71	2.77	2.88	3.06	3.32
Clams, cockles, ark shells	1.40	1.46	1.52	1.69	1.05	0.93	0.95	0.90	0.83
Flounder, halibut, sole	1.19	1.22	1.10	1.17	1.10	0.98	0.92	0.93	0.99
Oysters	1.01	1.00	0.99	1.06	0.16	0.17	0.19	0.19	0.19
Crabs	1.14	1.27	1.51	1.58	0.99	1.17	1.15	1.21	1.18
Mussels	1.24	1.32	1.30	1.33	0.26	0.28	0.24	0.20	0.22
Sharks, rays, chimeras	0.68	0.68	0.69	0.69	0.74	0.76	0.75	0.80	0.79
Scallops, pectens, etc.	0.84	0.87	0.84	1.04	0.49	0.61	0.50	0.48	0.48
Lobsters, squat lobsters	0.29	0.21	0.21	0.20	0.21	0.22	0.23	0.22	0.25
Krill, planktonic crustacea	0.39	0.37	0.23	0.29	0.09	0.08	0.12	0.10	0.08
Salmon, trout, smelt	1.43	1.45	1.64	1.44	0.98	0.99	1.15	1.03	0.92

Source: From *United Nations Food and Agriculture Organization Yearbook of Fishery Statistics*, 1997.

food fishes. All are caught in fairly shallow waters, and many have been heavily exploited for years.

The sharks and other cartilaginous fishes have recently been more extensively exploited for food, and some have been demonstrated to have excellent flesh.

The salmons, trouts, and smelts produce a respectable tonnage, but since the trouts are freshwater fishes, it is not possible to determine what fraction of the landings are various salmons and which are trouts. Also, salmon are currently being raised commercially in mariculture.

The major crustacean group in world fisheries is shrimp, various species being caught in both warm and cool waters in all oceans. The other significant crustaceans are various species of crabs and lobsters. All are economically valuable and command high prices for direct human consumption.

The major mollusk group in terms of tonnage is the squid. Significant numbers of squids are caught in Japan, Europe, and California. The other major mollusk group commercially harvested includes the bivalves, such as oysters, mussels, and clams.

The major fishery for marine mammals has been that for various species of whales. Persistent overexploitation of the stocks in the twentieth century has, however, virtually doomed this fishery. Other harvestable mammals are mainly seals and sea lions.

Other minor fisheries exist for various species of algae. Some are taken in Japan for human consumption, but most are collected to extract various products for use in industry (algin derivatives).

Sustainable Yield and the Future

In 1997, the last year for which published FAO data are available, the total world fishery catch was 93.3 million metric tons. Of the 93.3 million metric tons, 85.6 million metric tons were from marine fisheries. Weber (1994) shows that 1989 was the peak year for marine fisheries landings. Since that time, the landings have actually dropped or at best remained the same (Figure 11.4). The total fishes caught for human consumption represents perhaps 1% of all human food, but a significant 10% of the protein intake; hence, it is important.

Throughout the 1970s and 1980s the total marine landings increased steadily, finally peaking in 1989 (Figure 11.4). This was due primarily to the heavy fishing for certain unexploited fishes, such as pollock and hake, that are used mainly for fish meal, and the expansion of fisheries in the Southern Hemisphere oceans. Since the human population is still increasing and is expected to reach nearly 7 billion early in the twenty-first century and since many are anticipating that much of the food needed to

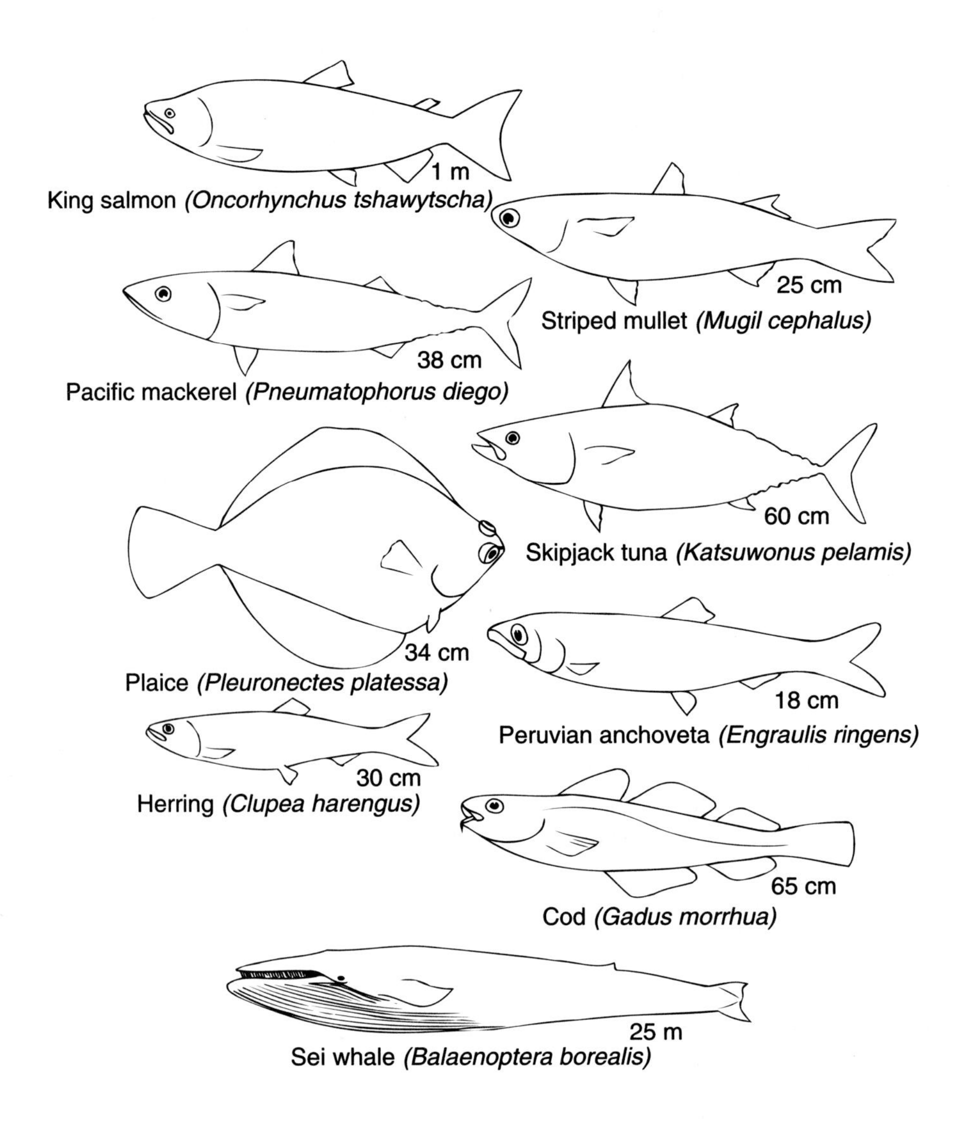

FIGURE 11.3
Some of the major commercial fishes of the world. (Not to scale.)

feed these increased numbers will come from the sea, does this current decline or stagnation mean we have reached the limit of the amount of fish protein we can harvest from the seas to meet the increased needs of a burgeoning human population? What are the facts with respect to current and potential amounts of food from the sea?

To answer these questions, it is first necessary to look briefly at the situation of current fisheries. Of the present fishing effort, 90% is concentrated on the 7% of the world ocean area represented by the continental shelves. These shallow areas are the most productive ocean areas.

The great growth of the world fisheries in the 1950s and 1960s was due to two factors. First was the discovery and exploitation of the anchoveta fishery of Peru, which sent Peru from having less than 1% of the world fisheries landings in 1950 to 17% in 1967 and to a position as the top fishing nation in the world. The second factor was the introduction of modern fish-capturing and -handling gear, including factory ships, larger nets, and more accurate methods of locating fishes. Most of this effort was expended on the continental shelf regions around the world with the result that, at the present time, this limited area has been fully exploited for its existing fish stocks and many, if not most, have been overfished (see pp. 465–468). It would appear, therefore, that little or no increase can be expected from these areas.

What about the remaining 93% of the world's oceans? Currently, the major oceanic fisheries of the world are for various tunas, swordfish, bill fish, and so forth. In 1997, the landings of this group were 4.85 million metric tons, or only 5.6% of the total marine fish-

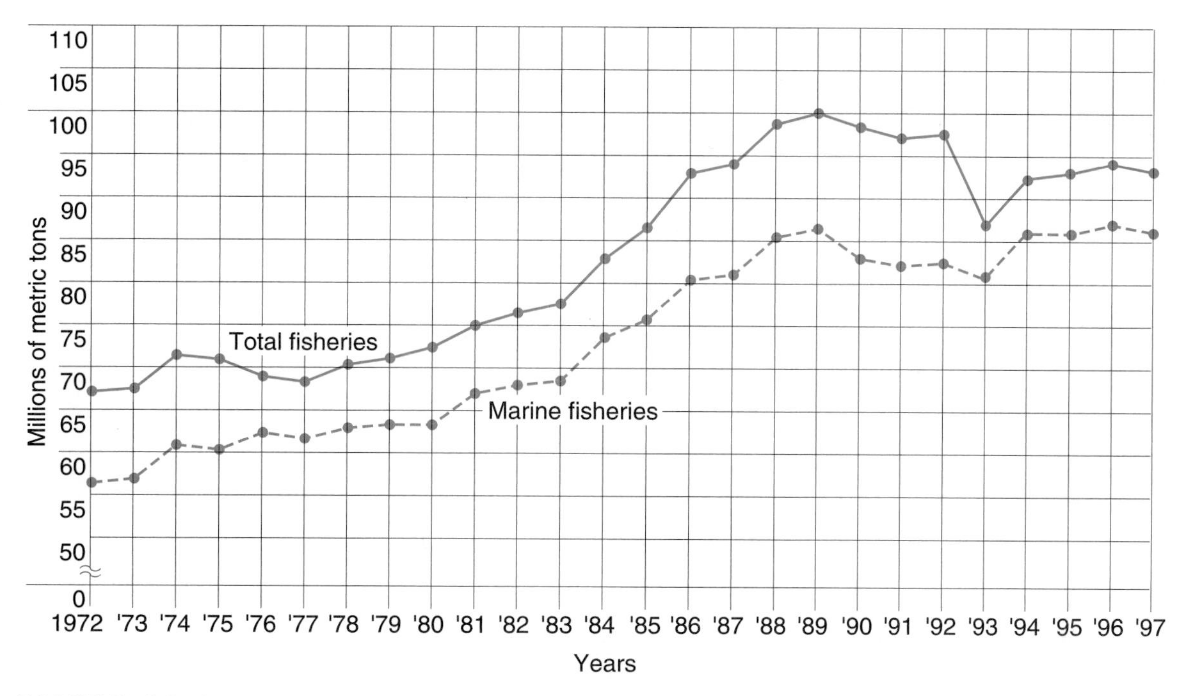

FIGURE 11.4 Changes in the annual world fisheries catch in recent years.

ery landings for 1997. However, this 4.85 million tons is not a significant increase in the landings over previous years and suggests that we are approaching the limit in tunas that may be caught without depleting the stocks. Whereas 25 years ago the tuna stocks were considered nearly impossible to overfish, the advent of purse seines dramatically changed that fact in a few years. Today, there is considerable concern about the continued ability of these fish stocks to sustain the current fishing pressure. Indeed, the bluefin tuna of the Atlantic, perhaps the world's most valuable fish bringing $260 per kilogram in Japan, has had a population decline of 90% in the western Atlantic and has been suggested for listing as an endangered species.

Bottom fishing on the abyssal plains under the oceanic area is not feasible, since the diminished amount of energy reaching the abyssal floor does not allow the existence of substantial fish stocks (see p. 142).

Furthermore, as we noted in Chapter 2, the surface waters of these open ocean areas are impoverished with respect to nutrients and phytoplankton. They cannot sustain the populations of fishes, even in the surface waters, that the continental shelf waters are able to. They are "watery deserts." Even though this is a vast area, it cannot match the production of the shallow coastal waters. In fact, we are approaching the limit of what we can obtain from the sea.

What, then, is a reasonable estimate for the maximum sustainable yield of food that we can expect from the seas? By **maximum sustainable yield** we mean the largest number of fishes that can be harvested year after year without diminishing the stocks. Perhaps the best estimate is based on total primary productivity of the oceans. As we have already discussed, most food chains and trophic levels depend on photosynthesis. The energy fixed in photosynthesis is finite in the world's oceans. By knowing the trophic level of the various fishes taken as food and the efficiency of energy transfer from level to level, it is possible to obtain a figure for maximum production. Such was the approach taken by Ryther (1969) to estimate potential yield.

In this estimate, he first divided the ocean into three areas, each representing waters with significantly different levels of primary productivity. The largest area (90%) was that of the open oceans of the world, but it had the lowest primary productivity (Table 11.2). The second largest area in size (9.9%) was the neritic zone, and the final—highest in productivity but smallest in size (0.1%)—was the restricted area where upwelling occurs (Table 11.2). Next, he estimated the efficiency of energy transfer among trophic levels in each division and, finally, the trophic level of the major fishes captured in each (Table 11.3). Combining the data, he arrived at a total world fish production of 24×10^7 tons (= 240 million tons).

TABLE 11.2

Division of the Oceans into Provinces Based on Level of Primary Productivity

Province	*Percentage of the Ocean*	*Area (km²)*	*Mean Productivity (grams of carbon per m² per year)*	*Total Productivity (10^9 tons of carbon per year)*
Open ocean	90.0	326.0×10^6	50	16.3
Coastal zone	9.9	36.0×10^6	100	3.6
Upwelling areas	0.1	3.6×10^5	300	0.1
Total				20.0

Source: After Ryther, 1969.

TABLE 11.3

Estimated Total Fish Production of the World's Oceans Based on the Three Provinces of Table 11.2

Province	*Primary Production (tons of organic carbon)*	*Trophic Levels*	*Efficiency (percent)*	*Fish Production (tons, fresh weight)*
Oceanic	16.3×10^9	5	10	16×10^5
Coastal	3.6×10^9	3	15	12×10^7
Upwelling areas	0.1×10^9	1.5	20	12×10^7
Total				24×10^7

Source: After Ryther, 1969.

TABLE 11.4

Change in Catch for Major Marine Fishing Regions, Peak Year to 1997

Region	*Peak Year*	*Peak Catch (million tons)*	*1997 Catch (million tons)*	*Change (percent)*
Atlantic Ocean				
Northwest	1973	4.4	2.05	–53
Northeast[a]	1976	13.2	11.66	–12
West Central	1984	2.6	1.8	–31
East Central	1990	4.1	3.6	–12
Southwest	1997	2.6	2.6	0
Southeast[a]	1973	3.1	1.1	–64
Mediterranean and Black Seas[a]	1988	2.1	1.5	–28
Pacific Ocean				
Northwest	1988	26.4	24.6	–7
Northeast[a]	1987	3.4	2.8	–17
West Central	1997	8.9	8.9	0
East Central	1981	1.9	1.6	–16
Southwest	1991	0.9	0.8	–12
Southeast	1994	20.2	14.4	–29
Indian Ocean				
Western	Still rising		4.1	+6[b]
Eastern	Still rising		3.9	+5[b]

[a]Rebounding from a larger decline. [b]Average annual growth since 1988.

Note: Percentages calculated before rounding off catch figures. The catch in the Antarctic is at 356,000 tons, down from a peak of 653,000 tons in 1982, primarily because of reduced interest in krill. The catch in the Arctic is zero.

Source: From Weber, 1994 and FAO *Yearbook*, 1997.

This figure, however, does not represent the total amount that may be taken. A significant fraction of this 240 million tons must be left as breeding stock to sustain future yields. Another large fraction is consumed by other carnivores in the sea. Ryther estimated these latter fractions to account for somewhat more than half of the total production, leaving about 100 million tons as the sustained yield for the world's oceans.

Other scientists, such as Cushing (1969) and Schaefer (1965), have also estimated the total harvestable amounts of fishes in the sea. These estimates have been made on the basis of primary productivity, extrapolating from current landings, and trophic level and ecological efficiencies. More recently, Pauly and Christensen (1995) estimated that in the most intensively fished continental shelf regions more than 20% of the primary productivity is needed to sustain the fisheries. Because scientists disagree on the proper figures to use for primary productivity, ecological efficiency, and average trophic level, these estimates have varied from a low of 60 million tons to a high of 290 million tons.

Given the current situation of world marine fisheries in which we are seeing continued declines or stagnation in landings, and few, if any, unexploited fisheries, it appears at this time (1999) that we are either at or above the point of sustained yield for the world's fisheries. That we may well have overshot the limit on the number of fish we can take is manifested in the decline of various fisheries.

Overexploitation

In recent years, there have been abundant examples of the decline of fish stocks of all types in all areas of the world. For example, Weber (1994) notes that with the exception of two areas, productivity has declined in all of the 15 major marine fishing areas of the planet. In four of the most significant areas—the northwest, west central, and southeast Atlantic and the northeast Pacific—the catch has declined by 12% to 64% (Table 11.4). If one looks at individual species of fishes, one finds that 18 have dropped by more than 100,000 tons since their peak landings (Figure 11.5). In the New England region alone, the total harvest of groundfish (cod, haddock, flounder, etc.) has declined from near 800,000 tons to about 100,000 tons. Many of these declines are clearly due to overexploitation, but others have more complex origins. All indicate that we are exceeding the maximum amount of fishes we can take from the oceans and have often mismanaged fisheries, so we have not achieved a maximum sustainable yield. The signs of overfishing are normally a decline in average size of fishes and an increase in the effort needed to land the same amount of fishes. Age structure also changes. In some cases, overexploitation has been compounded by certain natural changes in the environment that also reduced the stocks. In most situations, however, overexploitation is the direct result of human activity. Three examples are discussed here, one with environmental consequences affecting the fishery, one in which causes are uncertain, and one related directly to humans.

The small clupeid fish known as the Peruvian anchoveta (*Engraulis ringens*) is a small filter feeder that occurs in large schools in the upwelling water mass of the Peru-Chile Current off the west coast of South America, where it feeds on the abundant plankton. This small fish occurred in tremendous numbers in these cold waters before the advent of human fishing. The major predators were various species of marine birds, collectively called guano birds, which lived by the millions along these shores and produced at their nesting areas huge quantities of droppings or guano, which were harvested for fertilizer.

The anchoveta population depends on abundant plankton, which, in turn, depends on the nutrient-rich upwelling water. In certain years, the upwelling is blocked in this area by an influx of warm surface water known as the El Niño Current (see Chapter 9, pp. 411–414). When this occurs, plankton disappear and the population of anchoveta drops, as do the bird numbers; hence, there is a natural abundance cycle.

Direct human exploitation of the anchoveta began in the 1950s, and the fishery increased at a phenomenal rate until the late 1960s. The catch peaked in 1970 at 13 million metric tons, more than one-sixth of the entire world's catch!

This huge catch was accompanied by a decline in bird numbers, since they had fewer fishes to eat. The fishery experienced one drop in numbers in 1965 as a result of the occurrence of the El Niño, but this temporary setback was made up the following year.

Fishery experts had estimated a maximum sustainable anchoveta yield of 9.5 million metric tons. This was exceeded in 1967 when over 10 million tons were landed. By 1970, more than 13.1 million metric tons were landed, but the size of the fishes was decreasing, and it was apparent to experts that the fishery was in trouble and that too few fishes were escaping to provide the breeding stock. As could

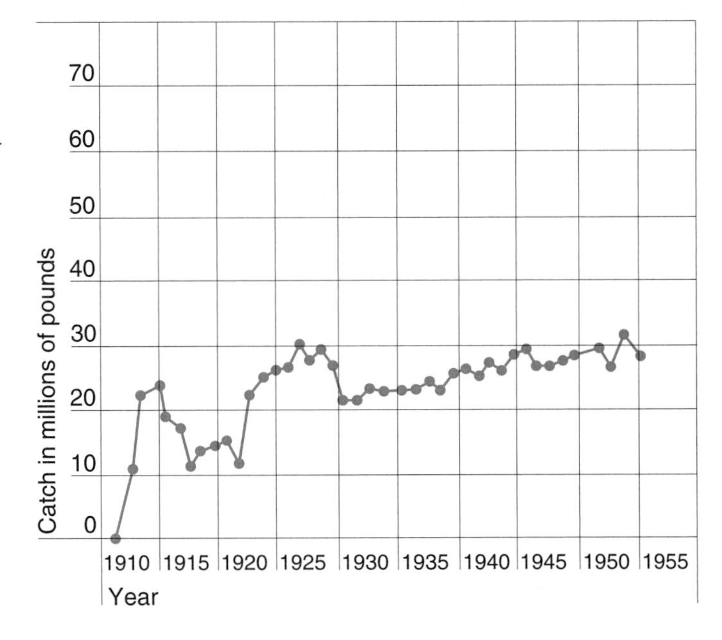

FIGURE 11.5
Pacific halibut fishery exploitation and recovery. (Modified from *Fisheries Resources of The Sea And Their Management*, D. H. Cushing, © Oxford University Press, 1975. Reprinted by permission.)

have been forecast, the crash came in 1972, aided by another El Niño, and landings dropped to 4.7 million tons. This was followed by an even worse year in 1973, when only 1.7 million tons were landed. Subsequently, this fishery has not fully recovered. Even in 1997, only 7.7 million tons were landed (up from 5.5 million tons in 1982) (Table 11.5). A once huge resource has, due to overfishing, been reduced far below its sustainable yield, and the human population of the world has been denied a significant protein supply.

Much more mysterious is the demise of the California sardine fishery. The California sardine, *Sardinops sagax*, is also a small, plankton-feeding, clupeid fish with a range from Mexico to southern Alaska. In the 1930s, this species was the basis of the greatest fishery in North America. During the 1936–37 season, fishermen landed 650,000 tons in California alone, the major fishing region. The decline began in the 1940s, when the age structure of captured sardines began to change to older age groups, reflecting the decimation of the younger fishes and their failure to recruit to the fishery. Despite the establishment of the California Cooperative Oceanic Fisheries Research Program in 1949, the fishery was commercially dead by the early 1950s. In Monterey, California, a center of the industry, the famed Cannery Row of Steinbeck closed up. The fishery has not recovered in the nearly 40 years since then, though since 1981 there has been an encouraging increase in the numbers of fish. The population now approaches commercial levels and some limited fishing has begun.

Although most people attributed the demise of the fishery over the 10-year period from 1940 to 1950 to overfishing, it seems apparent now that other factors were also at work. The decline of the fishery began with the onset of a 15-year period of unusually cold water in the sardine habitat. It is stated by Marr (1960) that this cold water affected the sardine population adversely, either directly through a reduction in breeding or through an effect on the food supply. The result was a catastrophic decline in numbers. If overfishing were the only reason for the demise, one would have expected that, in the more than 35 years during which there has been no fishing, the stocks would have recovered or at least returned to their known breeding grounds. This has occurred very slowly, hence the strong suspicion that other factors are involved.

Finally, Soutar and Isaacs (1969) have presented evidence, from the examination of the scales of sar-

TABLE 11.5

Fishery Declines of More than 100,000 Tons, Peak Year to 1992

Species	*Peak Year*	*Peak Catch (million tons)*	*1992 Catch (million tons)*	*Decline*	*Change (percent)*
Pacific herring	1964	0.7	0.2	0.5	−71
Atlantic herring	1966	4.1	1.4	2.6	−63
Atlantic cod	1968	3.9	1.2	2.7	−69
Southern African pilchard	1968	1.7	0.1	1.9	−94
Haddock	1969	1.0	0.2	0.8	−80
Peruvian anchovy[a]	1970	13.1	5.5	7.6	−58
Polar cod	1971	0.35	0.02	0.33	−94
Cape hake	1972	1.1	0.2	0.9	−82
Silver hake	1973	0.43	0.05	0.38	−88
Greater yellow croaker	1974	0.20	0.04	0.16	−80
Atlantic redfish	1976	0.7	0.3	0.4	−57
Cape horse mackerel	1977	0.7	0.4	0.3	−43
Chub mackerel	1978	3.4	0.9	2.5	−74
Blue whiting	1980	1.1	0.5	0.6	−55
South American pilchard	1985	6.5	3.1	3.4	−52
Asian pollock	1986	6.8	5.0	1.8	−26
North Pacific hake	1987	0.30	0.06	0.24	−80
Japanese pilchard	1988	5.4	2.5	2.9	−54
TOTALS		51.48	21.67	30.01	−58

[a]The catch of the Perusian anchovy hit a low of 94,000 tons in 1984, less than one percent of the 1970 level, before climbing up to the 1992 level.
Source: FAO. From Weber, 1994.

dines preserved in layers of sediments in certain anoxic basins, that the California sardine naturally undergoes huge cyclic changes in population numbers, and the cycles range from 500 to 1,700 years. If this is correct, the California sardine would be present in numbers for a certain period of years, then vanish for a longer period before reappearing. During its absence, the niche is filled by another clupeid, the anchovy. Currently, the increase in anchovy numbers would seem to substantiate this theory. If true, Cannery Row might again bustle with the activities of fish rendering in only 500 years or so!

The final example of the demise of a fishery is much simpler. It is the direct result of human overfishing. Here, we refer to the whaling industry. The ignominious end of the world's whale fishery is now being observed.

In contrast to most animals taken in various fisheries, whales are warm-blooded mammals that have a long life span and a very low reproductive rate. This means that they are very slow to build up their numbers once reduced and are readily susceptible to extinction.

Large-scale whaling began in the North Atlantic in the sixteenth century. At first, methods were primitive and refuges from whalers existed for most species. However, the advent of long-range vessels and, especially, the harpoon gun with explosive charge spelled the end of whale stocks in the North Atlantic and North Pacific by 1900.

Attention then turned to the large stocks of baleen whales in the Antarctic region, where most whaling of the twentieth century has been concentrated. Coincident with the exploitation of these southern stocks was the advent of huge factory ships, more speedy catcher boats, and airplanes and helicopters to find the whales. All these technological advancements led to a greater slaughter of whales in the first 60 years of the twentieth century than had been achieved by all whaling in all preceding centuries. In approximately 60 years of Antarctic whaling, Rounsefell (1975) notes that perhaps 1.3 million whales were taken!

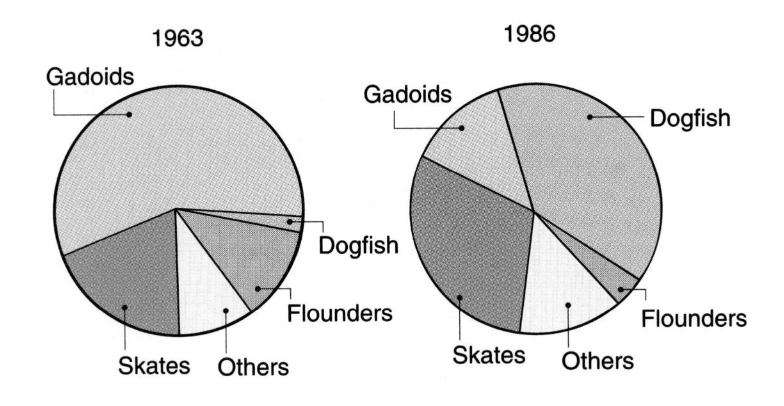

FIGURE 11.6
Percent composition of the demersal fish of Georges Bank in 1963 and 1986. (From *Advancements in Marine Biology*, Vol. 34, p. 281, Fig. 26, © Academic Press, London.)

In an attempt to regulate this fishery, the International Whaling Commission (IWC) was set up in 1948. It was to obtain data on whale stocks and to set catch quotas for the various species to conserve the stocks. Unfortunately, because the Commission had no enforcement powers, no independent means of obtaining data on the status of the various whale stocks outside of what the whalers reported, and was subject to intense political pressure by the whaling nations that comprised it, its attempts to conserve stocks has been a mockery. It has consistently set its quotas too high and refused to heed the advice of its own and outside experts. As a result, the stocks of whales have continued to decline. Today, many species are on the verge of extinction, and others are so reduced in numbers that they are commercially extinct. Along with the demise of the whales has been the demise of the whaling industry. When the IWC was set up in 1948, its 20 member nations were whalers. By 1979, only two remained as international whalers, Russia and Japan.

The dramatic decline in this fishery can be understood from a few examples. In the 1930–31 season in the Antarctic, 25,000 blue whales (*Balaenoptera musculus*) were taken, but by the 1963–64 season, only 112 were taken. In 1968, the IWC reluctantly gave complete protection to the species. Declining blue whale stocks forced whalers to change to fin whales (*Balaenoptera physalus*) after World War II, and from 1955–61, more than 25,000 were taken each season. By 1965, this number had dropped to less than 2,500, and by the 1974–75 season, it had dropped to less than 1,000. Sei whales (*Balaenoptera borealis*) went from 25,453 in 1964–65 to 3,859 in 1974–75.

Clearly, the unrealistically high quotas set by the IWC and the greed of the whaling nations have reduced the whale stocks to the point where it is, or soon will be, uneconomical to continue whaling. Even if all whaling is stopped now, as many countries have urged, it will take many species years, perhaps centuries, to recover because of their low fecundity and the long time necessary for them to reach reproductive size. That there is some hope for recovery of at least some of these leviathans lies in the recovery of the California gray whale (*Eschrichtius robustus*). This species was reduced to only a few individuals in the early decades of the twentieth century, but since complete protection in 1938, they have now increased in numbers to more than 8,000.

Collapse or decline of fisheries may be caused indirectly by other activities of humans. The collapse of the salmon fishery in Washington, Oregon, and British Columbia is one example. In this case, logging and dam building destroyed or silted up the spawning streams on which the stocks of salmon depended. The stocks may also have been overfished.

Ecosystem Changes

Overfishing may also result in changes in the larger ecosystems or communities in which the target fish reside. For example, the demise of the North Atlantic cod and haddock fishery resulted in the increase of dogfish sharks and skates that now fill that niche; this increase may contribute to additional declines in cod and haddock as these sharks feed on the young of these fishes (Figure 11.6). In the northeast Pacific, the heavy fishing of the Alaska pollock resulted in the decrease of the

Steller's sea lion (*Eumetopias jubatus*) population from 300,000 individuals in 1960 to 66,000 in 1990, most likely because the Alaska pollock was the chief food of the sea lions.

Fishing has certain other effects on marine communities, which may be changed directly or indirectly through removal of certain species or alteration of the habitat by the fishing gear. Fishing affects the community directly by removal of species. One of the effects of this is the reduction in species diversity, particularly on a local level. Species diversity decreases may also come about indirectly through alteration in predator-prey relationships and through removal of individuals of fished species. This loss of species richness has been best documented for coral reef fisheries. Although fishing reduces species diversity, it does not appear to cause species extinctions. Fishing may also alter the size structure, reproduction, and life history characteristics of the fished populations. Since most fishing selectively removes the larger specimens of a target species, this may affect the reproduction and subsequent recruitment if the major contributors to the reproductive pool are the larger individuals or if reproductive age is not reached until large size is attained. Such size selection may also alter the sex ratios in the target species and curtail reproductive life span, particularly for hermaphroditic species and gonochoristic species in which one sex is larger than the other. For example, Jennings and Kaiser (1998) note that about 50% of tropical reef fish are hermaphroditic, and these usually are sequentially hermaphroditic, meaning that they change from one sex to another as they grow. This means that if the larger individuals are either male or female and are systematically removed by fishing, the reproductive potential of the population is reduced leaving a preponderance of just one sex in the fished population. Similarly, Smith and Jamieson (1991) report that in a trap fishery for Dungeness crab in British Columbia the size of females is smaller than males and they are effectively barred from the fishery due to size limits on the traps. Thus males are rare in the population, hence reducing mating and thus egg production.

Fishing can also alter the structure of the fished communities themselves. Perhaps the best and most recent example is the change in the composition of the fish communities of the Georges Bank fishing grounds in the North Atlantic. As Sherman (1991) reported, during the period from 1963 to 1986 the total biomass of the demersal fish declined precipitously and the composition of the community changed as well, as the desirable gadoids and flounders were fished out and replaced by spiny dogfish and skates (Figure 11.6).

When overfishing is coupled with natural changes in the environment, the effects may extend well beyond the immediate area. A good example of this is the relationship between changes in oceanographic conditions in the Barents and Norwegian seas and the fisheries for cod, herring, and capelin as reported by Blindheim and Skjoldal (1993). In this system, intensive fishing of herring coupled with lower sea temperatures in the 1960s reduced the herring predation on juvenile capelin and led to heightened capelin recruitment. However, in the 1980s the sea temperatures rose again, this time favoring cod recruitment but not capelin. Capelin recruitment failed in 1984 and 1985, and this failure coupled with intensive fishing led to the collapse of the capelin stocks. Herring biomass was still low, and hence those other species which fed on herring and capelin had few alternative prey. These included cod, sea birds, and seals. As a result, cod, sea birds, and seals began to starve and mortality rates increased (Figure 11.7).

Another source of destruction in fisheries is the bycatch. **Bycatch** refers to nontarget organisms and undersized target organisms that are captured by the fishing gear. Most of this bycatch arrives on the fishing vessels dead or dying and is discarded back overboard, either because it is illegal to land it or because there is no market for it. Alverson et al. (1994) have estimated this bycatch at 27 million tons annually or about 31% of the total world marine fisheries catch at that time. This bycatch thus represents a significant destructive impact on marine communities. This bycatch is particularly devastating if it involves mammals, sea birds, turtles, and sharks, which have long life spans, experience long times before reaching reproductive age, and produce few young. The best known example of mammal bycatch is the tuna purse seine fishery in the eastern tropical Pacific. The tuna schools are often associated with porpoises, and when the nets are set porpoises are captured and drowned in the net. Dayton et al. (1995) note that up to 1987 perhaps 6 million porpoise were taken, substantially reducing the populations. Since that time various changes have been introduced to the fishery that have markedly reduced the porpoise deaths. Not all fisheries are equal in the amount of

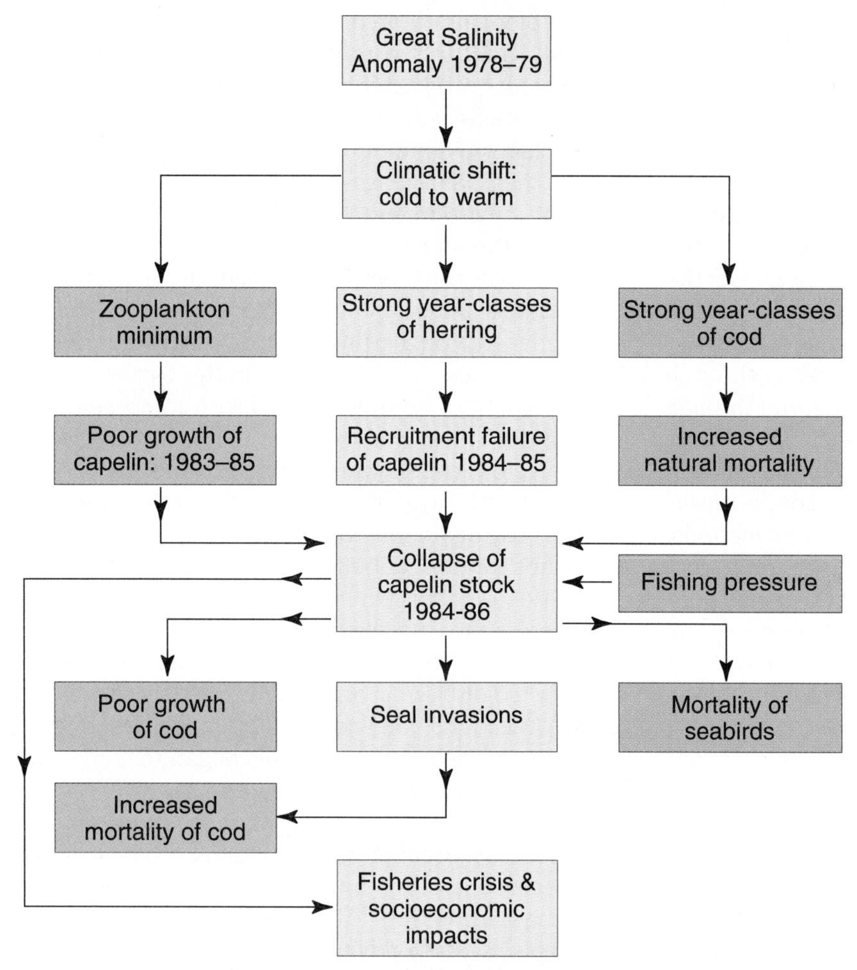

FIGURE 11.7 Interactions among the organisms and the environment in the Barents Sea and Norwegian Sea ecosystems from the late 1970s to the mortality of birds and seals in the 1980s. (Modified from *Advancements in Marine Biology*, Vol. 34, p. 267, Fig. 19, © Academic Press, London.)

bycatch; the ratio of discards to landings is the highest for the shrimp and prawn trawl fisheries (Figure 11.8). The extensive bycatch of marine birds and mammals by the high seas drift net fisheries, particularly those of the North Pacific, finally caused such a stir that the nets were banned.

In connection with the above, we must remember that the fishing gear itself has a deleterious effect on the habitat. Large trawling nets with heavy steel doors and bottom chains that dig into the bottom destroy large numbers of invertebrates that constitute the infaunal community structure. As a result the whole benthic community may change (Figure 11.9). The size of the area that can be so affected by the gear is enormous in some heavily fished areas. In the North Sea, for example, researchers have demonstrated that every square meter of the seabed is affected by fishing gear at least once a year.

Another type of human activity that is removing organisms from marine habitats and certainly contributing to community change is collecting of live marine organisms for the aquarium trade. This has the same effect as fishing, since it removes organ-

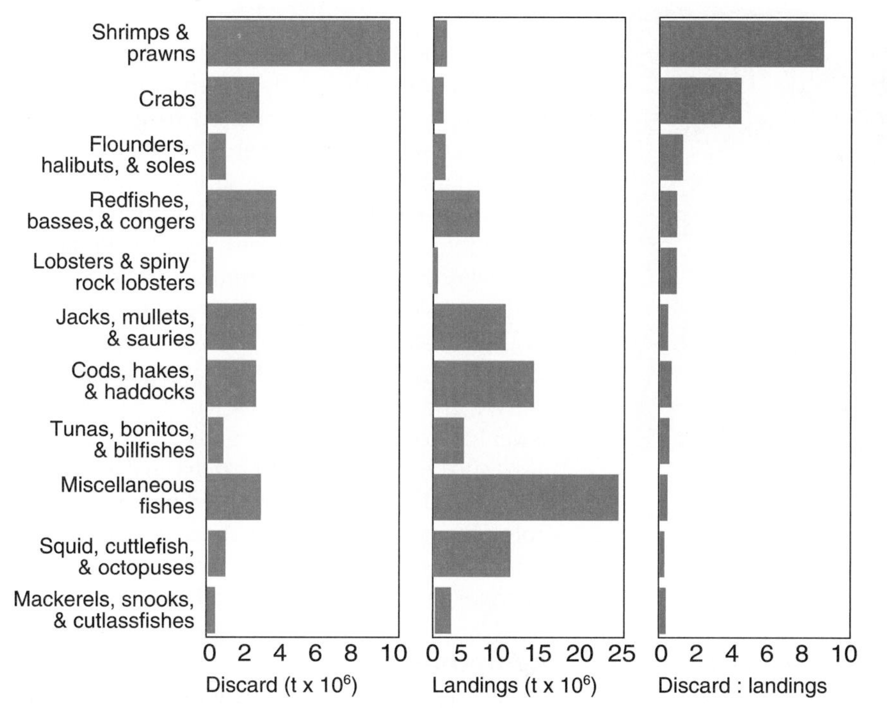

FIGURE 11.8
Ratio between bycatch and landed weight in 11 groups of fisheries. (Modified from *Advancements in Marine Biology*, Vol. 34, p. 268, Fig. 2m © Academic Press, London.)

FIGURE 11.9
Photos showing the difference in habitat before (a) and after (b) trawling gear has passed over an area. (From S. Simpson, 1998, Fishing trawlers scrape rock bottom. *Science News* 154:388. Photos courtesy of *Science News*.)

isms permanently. This fishery is concentrated primarily on coral reefs because of their colorful inhabitants. Wheeler (1996) has indicated that the numbers of corals, fishes, and invertebrates collected for the aquarium trade is massive and the United States is by far the largest consumer. According to Wheeler (1996) the worldwide trade in marine fishes alone is 18–60 million fish. While some of these fish are aquarium bred, most, probably between 90–99%, are wild collected. Another unfortunate aspect of this trade is that as many as 90% of these animals die in transit or within a year or two in captivity. While there are several methods employed to capture live fishes, one of the most common is to spray cyanide on the reef to stun the fishes. Unfortunately, cyanide also kills corals at concentrations hundreds of times lower than that used to collect fish, thus leading to high coral mortality. Cyanide is poisonous to fishes, often causing delayed mortality only after the live fish have been shipped. If this is taken into account, it has been estimated that only one fish out of ten collected survives. The effect of this collection on marine communities, especially coral reefs, is little known due to poor or nonexistent records, lack of enforcement of regulations (if any), and remoteness of collecting areas. The good news is that the aquarium trade is striving to improve survivability and working toward captive breeding and improved husbandry that will eventually reduce the numbers harvested from the wild.

FIGURE 11.10 Grey reef shark (*Carcharhinus amblyrhnchus*) dead after being entangled in an abandoned fishing net. (Photo courtesy of Norbent Wu/Peter Arnold Inc.)

Finally, there is the problem of lost gear. Every year fishermen lose gear. This, however, does not mean that the gear stop fishing. Lost nets or traps may continue to capture marine organisms for varying periods of time (Figure 11.10). This is known as **ghost fishing.** The extent of this problem can be appreciated by a few examples. Dayton *et al.* (1995) report that on Georges Bank 236 grapnel tows retrieved 341 actively fishing ghost nets while an ROV survey off California found 1% of the bottom was covered with lost gear, much of it still fishing. In Norway, nets that had been lost in 1983 were found to be still fishing in 1990. In the aforementioned high seas drift net fishery of the North Pacific, some 30,000–40,000 kilometers of nets were set per day and the daily lost rate was estimated to be 20%.

Tragedy of the Commons

In 1968, Garrett Hardin published an article in *Science* magazine entitled "The Tragedy of the Commons," in which he pointed out that freedom of access to a common resource ultimately brings ruin to the resource. The core of this argument is that whenever there is a resource held in common by a large number of individuals or political entities, such as countries, each will try to maximize its own gain from that resource. Not to do so will allow another to do so, and the negative aspects of acting selfishly do not devolve on the one acting alone, but on all who share the resource. Unfortunately, such an analysis is made by every entity sharing the resource. As a result, the resource is destroyed. Therein is the tragedy.

The plight of the whale fishery can be likened to such a tragedy. What happened to it may also be applicable to other open ocean fisheries, such as tuna, or potential fisheries, such as krill (see pp. 473–474).

Open ocean fisheries lie in international waters beyond the control of any one country. Presumably, these fisheries are held in common for the good of the world's people. Since they are owned by no one, they and their resources are accessible by all. Countries with the means to exploit these areas do so because it serves their own interests. They do not have to answer to another nation or regulatory agency, and the tragedy is that, if they do not catch the fishes, another nation will. This short-term greed has destroyed the whale fishery and may next do so for tuna. Recent interest of some countries in mining of deep ocean manganese nodules may also fall into this category.

Regulation

The decline of many world fisheries due to common access and the pressures of an increased demand for food by an ever increasing human population have led to friction among fishing nations and various attempts to regulate fishery resources. As long as the fishery exists on the continental shelf where individual nations have jurisdiction by international convention, the fishery can be controlled by a single nation. In the past, the area of an individual nation's jurisdiction was 12 miles (19.3 km). Some, however, have unilaterally acted to extend these limits out to 50 or 200 miles or beyond. This has caused several confrontations, such as the "cod war" between Iceland and Britain, the seizure and fining of many United States tuna boats by Ecuador and Peru, and more recently, the confrontation between Canada and Spain over fishing the depleted stocks on the Grand Banks. Following the third United Nations Conference on the Law of the Sea, a standard 200-mile-wide fishing and economic zone (Exclusive Economic Zone) was established for the waters of each coastal nation. This means that individual nations now have full control of all fishery activity within 200 miles of their shores.

International agreements among nations for the control of fisheries have proved more difficult to bring about and effectively implement, as we witnessed in the IWC. However, there are some outstanding examples that have worked. One of the oldest is the International Fur Seal Treaty, established in 1911 among the United States, Russia, Japan, and Canada, to regulate exploitation of the northern fur seal (*Callorhinus ursinus*). It has been an outstanding success in regulating fur seal numbers to achieve a sustained yield. (The more recent declines in the fur seal populations stem from large losses suffered by the adults in encountering the huge, miles-long gill nets set in the open Pacific for salmon and tuna.) Another success has been the International Halibut Commission, established in 1924 between Canada and the United States to regulate the then declining Pacific halibut fishery. This was the first fish species to be conserved internationally. After establishment of the Commission, the stocks of halibut recovered (see Figure 11.5). Since these early attempts, other international regulatory agencies have been established, including the Atlantic and Baltic commissions, the International Tuna Commission, and the International North Pacific Fisheries Commission (INPFC). Many of these more recent regulatory agencies have had a beneficial effect on certain fisheries, but problems still remain. The major one is that the various commissions have limited jurisdiction geographically or are limited to only certain species. For example, the waters around Japan are not in the jurisdiction of the INPFC, and the Halibut Commission regulates only that species.

It is well to remember that not all commissions and regulatory agencies are successful, witness the IWC. Perhaps the most dramatic recent failure of a regulatory agency is the International Commission for the Conservation of Atlantic Tuna (ICCAT). Despite the existence of this regulatory body, the Atlantic bluefin tuna populations have continued to drop, 90% since 1975 for those that breed in the western Atlantic and 50% for those that breed in the Mediterranean.

To effect a truly worldwide regulation of fisheries, the current commissions or others will have to expand their jurisdiction to include all species in their areas. Those areas, mainly in the Southern Hemisphere, where no regulatory agencies exist, will have to be brought under control. Also, regulatory agencies increasingly will have to take a whole ecosystem approach.

New Fisheries

Given the fact that the current world fisheries catch is either declining or plateauing in many long-established fisheries, are there any prospects for finding substantial new fisheries? The answer may be yes, but whether these new fisheries will be able to contribute significantly to the human food supply is much less certain.

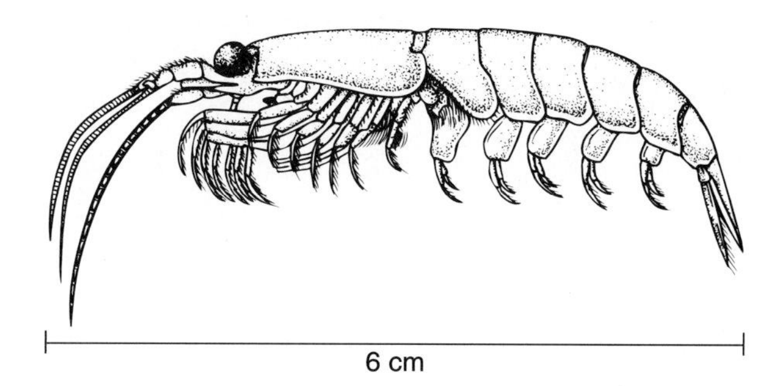

FIGURE 11.11 Krill, or *Euphausia superba*, of the southern ocean.

Now that the stocks of baleen whales have been reduced to the point of extinction in Antarctic waters, a new fishery has developed for the **krill** (euphausiids) that formerly constituted the food of the whales. Krill exist in large numbers in the seas around Antarctica, and they are large enough (6 cm) to be captured without resorting to very small mesh plankton nets (Figure 11.11). The most conservative estimates of stocks of krill are 50 million tons, but other estimates ranged up to 100 million tons, according to Idyll (1970). In 1977, the harvest of a krill fishery was estimated at as much as 150 million tons. The disparity in these early estimates is because we really do not have reliable estimates of krill population sizes, where they live, or their local distribution. We also lack information on the age structure of the populations and where they spend the winters. Because of this lack of basic biological knowledge, the most recent estimates of the annual production of krill reported by Nicol and de la Mare (1993) range from 75 million tons to 1.35 billion tons depending on what parameter is used to make the extrapolation (Figure 11.12). There is little doubt, however, that the potential for this fishery is enormous, certainly the largest of any unexploited fishery in the world today. Much remains to be learned about this fishery and what effect the taking of huge numbers of krill may do to the remainder of the fragile Antarctic ecosystem.

Another area of potential exploitation, until recently relatively untapped, is the area of the continental slopes at depths of 300–1,000 m. Recently, substantial stocks of blue whiting have been discovered in this zone off the British Isles. This area is now being exploited off the Pacific coast of North America for grenadiers (Pacific roughy) and "rock cod"

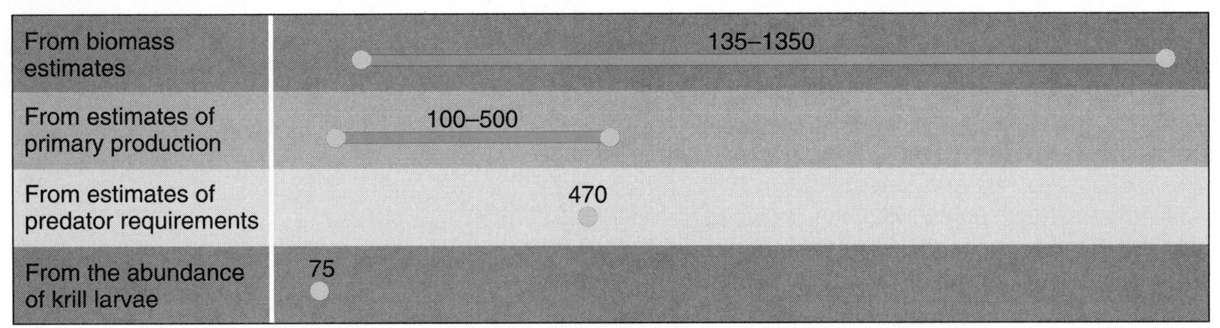

FIGURE 11.12 Various estimates of the annual production of krill, showing the widely varying numbers depending on the source. (After "Ecosystem Management and Antarctic Arill" S. Nicol and W. de la Mare, *American Scientist,* Vol. 81, No. 5, pp. 36–47. Copyright © 1993 American Scientist/Elyse Carter. Reprinted by permission.)

(*Sebastolobus*). The zone also contains substantial numbers of squids.

Rodhouse (1995) has suggested that there are significant underexploited squid resources in the southern oceans surrounding Antarctica. He estimated that the various higher vertebrate predators in these oceans take about 34 million tons of squid, and some of these squid would be suitable for human food if they are palatable and have behavior patterns that would allow them to be harvested commercially.

Finally, the entire Indian Ocean has a low level of fisheries development by current standards. It produces only about one-fifth the catch per unit area of the Atlantic and Pacific. It would appear that increased efforts in these waters would increase the catches without overexploiting the stocks. On the basis of primary production alone, Idyll (1970) estimated that this ocean can produce 10 million metric tons of fishes as opposed to 7.96 million tons harvested in 1997.

■ We can summarize this section by saying the landings of marine fisheries have peaked and are now stagnating or declining, and many world fisheries are showing signs of overexploitation due to unregulated access and overfishing. Furthermore, only limited amounts of increases can be expected from fishing stocks and underexploited areas, with the possible exception of the Antarctic krill. The estimate of a total sustainable yield of 100 million metric tons or less for the world's oceans appears reasonable. With catches now at about 87 million tons, it is unlikely that a significant increase in food can come from harvesting the wild populations in the sea. Therefore, the seas cannot and will not be the source of substantial amounts of food for an increasing human population. ■

MARICULTURE

As the limit is in sight for the harvesting of wild populations of marine animals and plants in fisheries, humans have begun to culture various marine and estuarine species in situations that are more analogous to our terrestrial-based agriculture, a field that is termed **aquaculture**. In aquaculture, humans rear selected aquatic plants and animals under controlled conditions to increase the amount of food available to humans compared with that obtained through traditional fishing. We are concerned here only with marine aquaculture, or **mariculture**.

History and Extent

Despite evidence that culturing of marine organisms existed at least as far back as ancient Rome, aquaculture through the centuries has never become a significant contributor to human food. Food from the seas remained the province of the hunters and gatherers, the fishermen. Only in the twentieth century has aquaculture received wide recognition and undergone significant development. Partly, this development is in response to the pressure for more human food, especially protein, and partly from the decline of wild stocks. Mariculture encompasses marine organisms raised not only for human food, but also for other products, such as pearls and terrestrial animal food.

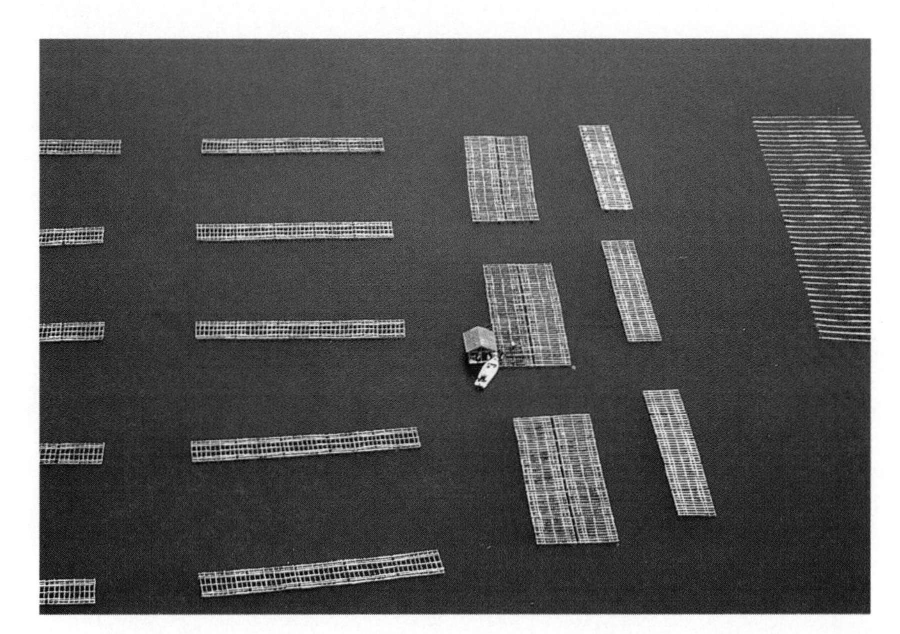

FIGURE 11.13
Pearl oyster culture racks in Japan. (Photo courtesy of George Gerster/ Photo Researchers Inc.)

Mariculture is practiced throughout the world in the tropical and temperate zones but is more technologically advanced in the industrialized nations of the Northern Hemisphere. At present, it is confined to shallow coastal embayments or artificial ponds.

Species Cultivated

Few marine organisms are currently cultivated, and most are either the type that fetch a premium price on the market, which economically justifies the high cost of their culture, or else the type that produce large amounts of biomass under intensive culture. In the former category, we could include certain species of shrimps and abalone cultured in Japan. In the latter category are oysters and mussels. As might be anticipated, estuarine and bottom-dwelling species are more commonly cultivated than pelagic, open ocean species.

Fishes that are successfully cultured commercially include several species of Pacific salmon (*Oncorhynchus*), Atlantic salmon (*Salmo*), mullets (*Mugil, Chanos*), yellowtail (*Seriola*), and flatfishes (*Pleuronectes, Solea*). Crustaceans cultured for market include various species of shrimps and prawns (*Penaeus, Metapenaeus, Leander*). Among the mollusks, oysters (*Crassostrea, Ostrea*) and mussels (*Mytilus*) are the subject of intensive commercial culturing (Figure 11.13). Other mollusks raised include several clams (*Anadara, Tapes, Mactra, Mercenaria, Meretrix*) and the abalone (*Haliotis*). Compared with the total list of fishes, crustaceans, and mollusks taken for food in various fisheries, this is a very short list. Other species have been cultured experimentally, including lobsters, scallops, and crabs, but at present, no commercial industry exists. This suggests there is considerable room for future expansion.

Yields

Even though aquaculture has lagged far behind agriculture in development, its potential for production of human food even in its current low state is considerable. For example, Bardach (1968) has noted that the unmanaged common oyster grounds along the East Coast of the United States yield only about 10 kg of oysters per hectare, while the intensive culture of oysters in hanging racks in Japan yields 58,000 kg/hectare! Even larger yields are possible with mussels. In Spain, an intensive culture on hanging racks produces 300,000 kg/hectare.

A similar situation occurs with fishes. In Indonesia, the milk fish (*Chanos chanos*) is cultured in brackish ponds, but is not subjected to such intensive care as in Taiwan. Yield in Indonesia is 400 kg/hectare, but 2,000 kg/hectare in Taiwan.

Presently, the world yield from aquacultural practices is probably over 4 million tons, but most is from fresh water. Indeed, in 1992, two aquacultured freshwater fish, silver carp and grass carp, were included among the top ten fish, by weight, landed in the world. This was the first time that any cultured fish had made that listing. While freshwater aquaculture is becoming significant, mariculture currently contributes little human food yield. However, the science of mariculture is as yet at a low level of development. If the previous examples are any indication of what can be expected, the potential future yield could be substantial; however, it is not great for reasons discussed in the next section.

Problems with and Restrictions to Mariculture Development

Mariculture is not a panacea for the problem of feeding the human race, nor is it likely to contribute significantly to food supplies in the immediate future. The reasons for this are numerous; space permits mentioning only the more significant.

First, marine animals require that the water in which they live has the proper physical and chemical characteristics. This is difficult to maintain in ponds or tanks and requires complex filtering and water treatment to remove wastes and other toxic materials. Attempts to raise marine organisms on a large scale mean an often economically crippling investment in equipment to maintain the water quality.

Another drawback is that many marine species go through a series of larval stages, each requiring different conditions and food, before attaining adult size. To rear these forms successfully is often too costly or simply impossible in captivity. Still other species are pelagic in the open ocean and will not survive when confined in smaller spaces. Since we cannot control animals in the pelagic areas, we cannot culture there.

There is also the problem of various diseases and parasites, which proliferate under captive or crowded conditions. Last, but certainly not least, there is in the United States the problem of government-imposed regulations and permits that are often extremely costly both in time and money. For example, in both Maine and California the permitting process may take up to six or seven years to complete and cost over $100,000 on average.

Present and future mariculture appears most practical for species with fewer or no larval stages, are nonpelagic, and live and reproduce well in crowded conditions. Furthermore, our inability to control conditions or organisms in the open ocean means future mariculture and increased yields must come from sheltered inshore bays, lagoons, and estuaries. Since the areas suitable for such culturing are a very minor part of the total world oceans, even at maximum culturing intensity, we cannot expect substantial quantities of food from mariculture in the near future.

POLLUTION

The seas have been considered the ultimate dumping grounds for the wastes of human societies. In the same way fishery stocks were once considered inexhaustible, so we have felt until recently that the immense volume of the world's oceans had an infinite capacity for absorbing all of our waste. Dilution, in other words, was the solution to pollution. In recent years, we have realized that however large, the oceans are not infinite in their capacity to absorb wastes. We have also learned that some of our wastes in very small amounts have significant effects on communities and species. It is the purpose of this section to discuss briefly a few of the more significant sources of marine pollution and their effects or potential effects on marine ecosystems. For more extensive information on pollution, see the bibliography at the end of this chapter.

Oil

Extensive media coverage over the last 25–30 years has made the oil pollution problem in the seas widely known to the public. Oil pollution in the seas results primarily from the spillage of crude oil from offshore drilling platforms or from accidents involving tankers. What is less well known is that there are a few marine environments subject to natural oil contamination (Coal Oil Point in Santa Barbara, California, for example). One of the largest natural seeps was that reported by Harvey et al. (1979). They discovered a seep 1.5 km wide, extending for hundreds of kilometers off the island of Trinidad in the Caribbean Sea. This seep was found at a depth of 100 m, had a thickness of 100 m, and contained in excess of 1 megaton of oil.

Crude oil released into the sea usually floats, although some components may sink, and after a time, evaporation of certain fractions may cause even more to sink. Away from land, the floating oil probably has little effect on the majority of planktonic and nektonic organisms. The exceptions are the marine birds and certain mammals, such as sea otters. When birds contact the oil, they become coated, and their feathers lose their insulating qualities. As a result, most will die of exposure in the water or may be unable to feed, with the same result. Oil is, therefore, devastating to marine birds, causing significant losses.

Inshore, the oil may coat the shallow subtidal and intertidal and smother the communities. The devastation is nearly complete initially, but recovery usually occurs with time. More serious than the oil itself have been the various chemicals, such as detergents, used to break up or disperse the oil in the water. As

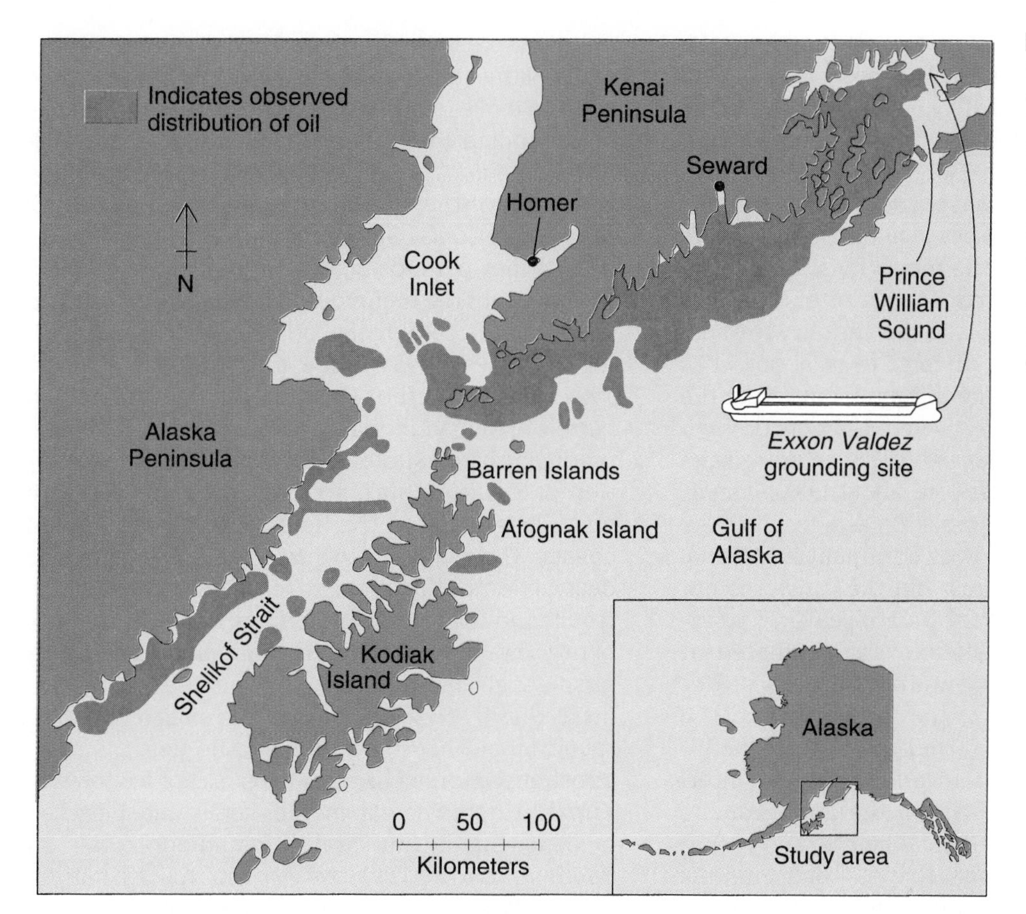

FIGURE 11.14 Observed distribution of oil from the *Exxon Valdez* oil spill between March 24 and May 20, 1989. (From "Immediate Impact of the Exxon Valdez Oil Spill on Marine Birds" J. F. Platt et al., *Auk*, Vol. 107, pp. 387-397, 1990 American Ornithologists' Union.)

Nelson-Smith (1973) notes, these were shown in the Torrey Canyon disaster in 1967 to have caused more mortality of marine organisms than the oil itself.

Oil disasters have become more numerous with the increased oil demand of the industrialized world, the necessity of oil transport from distant sources, and increased numbers of offshore drilling rigs.

The immensity of oil disasters involving the giant supertankers is revealed by the loss of the *Amoco Cadiz* off the Brittany coast of France in March 1978. In this case, 220,000 tons of oil spilled and one-third reached shore to contaminate 320 km (192 miles) of shoreline. The spill contaminated or destroyed major commercial marine products, including oyster beds, mussel beds, and lobster holding pens. As of 1985, many of the areas affected by the oil spill appeared to the casual observer to be completely recovered. However, investigations such as that done by a joint United States-France commission (NOAA-CNEXO) have shown that some differences still exist between communities in the spill area and those present in the same area prior to the spill. This is particularly true for the bays and estuaries where studies, such as those by Riaux-Gobin (1985) and Boucher (1985), suggest that the hydrocarbons have affected the long-term trends of community structure, particularly with respect to microalgae and nematodes.

The immensity of the area that can be affected by a single oil spill from a modern, large oil tanker was brought home to the world through the extensive media coverage of the disaster of the *Exxon Valdez*. Just past midnight on March 24, 1989, the tanker *Exxon Valdez* went aground in Prince William Sound, Alaska, spilling 260,000 barrels of Alaskan crude oil. This oil eventually was distributed by winds and currents over 30,000 km^2 of the coastal waters of Alaska (Figure 11.14). According to Platt et al. (1990), at the time of the spill, this area was occupied by about 1 million seabirds of which about 400,000 were killed. Of these, 70% were murres—certainly a significant loss. There was, however, no massive loss of fish.

Undoubtedly the greatest losses from the spill were among the birds and marine mammals. It has been hard, however, to pinpoint exactly how large the losses were because of the lack of recent census data on these animals immediately before the spill. In the case of killer whales (*Orcinus orca*), before the spill there were an estimated 297 individuals in Prince William Sound in 11 resident pods. Following the spill some pods showed decreases in numbers, but we are unsure of the source of the mortality since no carcasses were ever recovered. For sea otters (*Enhydra lutris*) our data are somewhat better. Prior to the spill the population size was estimated to be about 5,000 individuals in Prince William Sound. Following the spill, 878 carcasses were recovered; 123 were brought in alive but died in treatment, and 37 were treated and farmed out to aquaria. Therefore, about 1,000 animals were removed from the population as a result of the spill. What we do not know is how many additional animals may have been lost but the carcasses not recovered. In the case of the harbor seals (*Phoca vitulina*), the prespill population was estimated at 2,500–3,000 individuals and was declining about 10% annually. Following the spill the decline went to 35% in oiled sites, and about 200 individual were killed.

Among the algae and invertebrates, the intertidal forms suffered most from the spill, but the results varied among sites, tidal height, and organisms. High intertidal algae, limpets, and mussels were most affected. The losses from the smothering effects of the oil were compounded by the use of hot water and steam to clean the rocky intertidal. This use of hot water made the intertidal cosmetically more appealing, but the destruction it brought actually delayed the recolonization of these shores as compared with those that were left to cleaning by natural sources. Interestingly enough, there is little evidence that the subtidal areas and communities suffered much from the spill.

Until relatively recently, we had experienced no extensive spills from offshore wells, though they are potentially more devastating, since the amount of oil gushing out is not limited by the hold of a ship. Such a disaster happened in the Gulf of Mexico where, in 1979, the Ixtoc I well blew and spewed out from 450,000 to 1.4 million tons of oil in 295 days before being capped. The discrepancy in the amount spilled is due to the uncertainties of estimating the flow from the well. It was the world's greatest oil spill. It is also one of the least studied, so data on its effects are not available.

As if accidental oil spills were not enough of a problem, we were witness in 1991 to a massive deliberate oil spill in the northern Persian Gulf. It was the result of the Iraqi army releasing stored Kuwaiti crude oil from tanker loading piers. The extent and damage of this spill remains questionable because of few follow-up studies, as well as a lack of knowledge of the condition of the communities in the Persian Gulf prior to the spills. During the war, early reports suggested that perhaps 30,000 birds had been killed, about 20% of the mangroves had been oiled, and up to 50% of the coral reefs had been affected. According to Gerges (1993), follow-up studies by NOAA in 1992, a year after the war, indicated that 6–8 million barrels of oil had probably been spilled into the sea, producing an oil slick that contaminated a large portion of Saudi Arabia's northern coastline but with much less effect on the Iraqi, Kuwaiti, and Iranian coasts. The oil pollution was exacerbated by the destruction of the sewage treatment plants, which resulted in the release of perhaps 50,000 m^3 per day of raw sewage into the marine waters of Kuwait. Surprisingly, given this volume of pollutants, the follow-up studies by NOAA at 150 bottom sites a year later found oil at only three sites and no demonstrable direct effects on most coral reefs except for three in Kuwaiti waters where numerous corals were bleached or dead. Significant mortalities among corals off Saudi Arabia in 1992 were attributed to natural causes. What effect these pollutants may have on commercial fisheries or other communities in the Persian Gulf remains to be assessed.

Only a few of these larger disasters per year could add up to significant destruction, particularly of shallow-water communities. We can expect more such disasters in future years with increased oil shipping and increased offshore drilling. A more optimistic view comes from a recent study. According to a comprehensive review by the National Academy of Sciences (1985) of the fate of oil in the sea, biological impacts vary considerably. Low-energy environments, such as estuaries, salt marshes, and intertidal and subtidal sedimentary areas, appear most vulnerable and also show the slowest rates of recovery. There appears to be little detectable long-term effect on pelagic communities, save for birds, and in all cases examined, there appear to be mechanisms for oil degradation that should allow eventual recovery.

Sewage and Garbage

The discharge of human sewage and garbage into coastal waters is practiced throughout the world. The

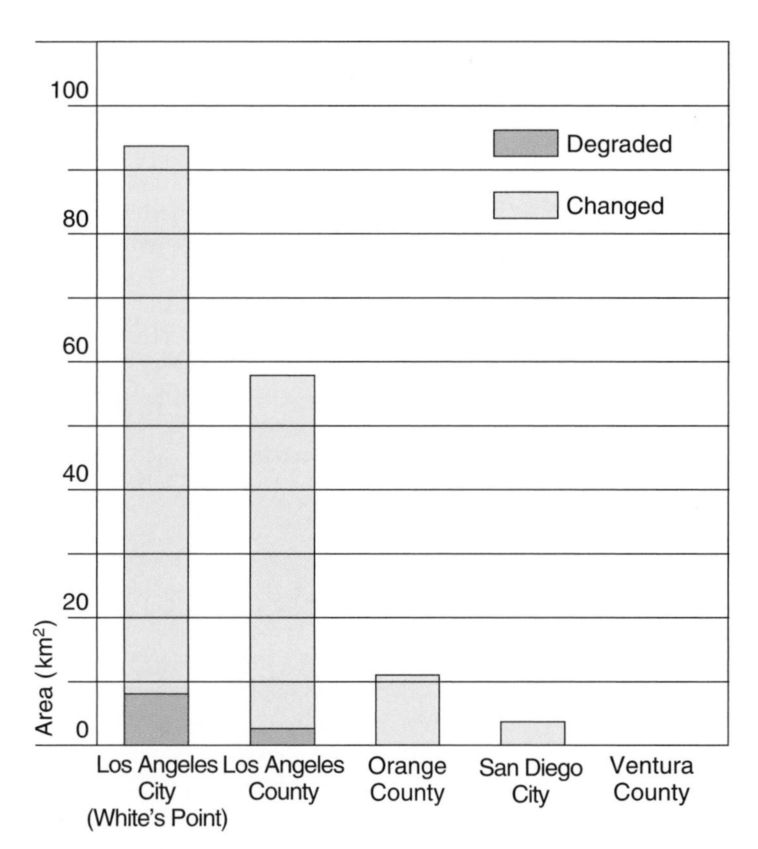

FIGURE 11.15
Changed or degraded areas near southern California's major sewer outfalls. (After W. Bascom ed., 1978, Coastal water research project annual report, Los Angeles.)

sewage may or may not have had some treatment before discharge. Sewage adds a large volume of small particles to the water and also large amounts of nutrients. In small volumes and with adequate diffusing pipes, it is difficult to detect any long-term effect on the communities of the open coast. In large volumes and in semienclosed embayments, the effect can be devastating. Two examples should suffice.

In southern California, the Los Angeles area discharges in excess of 330 million gallons of sewage per day at the White's Point outfall off the Palos Verdes Peninsula. Extensive studies in and around this outfall and others in the area over a period of years by personnel of the Southern California Coastal Water Research Project (Bascom, 1978) have revealed that sewage has caused significant degradation in benthic invertebrate communities in areas near the outfall as compared with similar areas some distance away. In addition, the kelp beds in the vicinity of the outfall have disappeared; sea urchins have markedly increased; and diseased fishes are more prevalent. About 4.6% or 168 km^2 of the 3,640-km^2 southern California mainland shelf has been changed or degraded as a result of sewage discharge from four major outfalls (Figure 11.15).

Kaneohe Bay on the Island of Oahu is a large, shallow embayment with restricted water circulation that was once known for its flourishing coral reefs. Since World War II, Banner (1974) reports that the area around the bay was subjected to a tenfold increase in population. As a direct result of this urbanization, the bay was subjected to massive domestic sewage discharges along with significant siltation from runoff during storms. The result was the total destruction of the coral reef communities in two-thirds of the bay and their replacement by a massive, smothering growth of the green alga *Dictyosphaeria carvernosa* (Figure 11.16). By 1978, the sewage discharges were eliminated from the bay, and in a resurvey of the area in 1983, Maragos et al. (1985) have reported remarkable recovery of corals and water clarity, suggesting an early recovery of the reefs in this bay.

In addition to sewage, large amounts of garbage are dumped into the sea each year from shore and from ships. What effect this material may have on the marine communities is unknown. Evidence of the

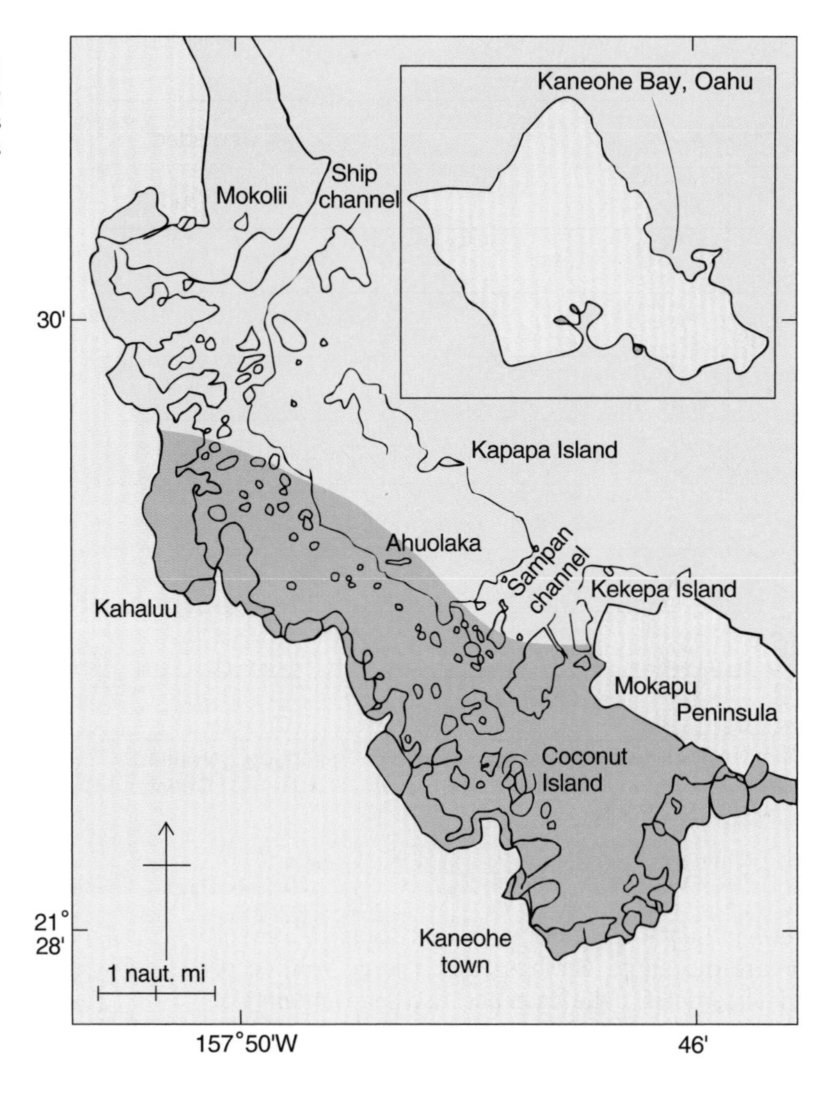

FIGURE 11.16 Kaneohe Bay, Hawaii, showing the extent of destruction of coral and overgrowth by the alga *Dictyosphaeria* at its maximum extent before cleanup was initiated. The darker blue portion indicates the area of destruction. (Modified from *Proceedings on the Second International Coral Reef Symposium*, Vol.2 , pp. 685-702)

extent of the garbage pollution is found in the occurrence of small bits of plastic in plankton tows taken in many parts of the North Atlantic and Pacific oceans and in the digestive tracts of fishes, diving seabirds, and marine mammals. As yet, there is little scientific evidence that these plastic bits cause mortality, though it has been suggested that they cause intestinal blockage. More ominous is that they do not readily decompose, and increasing use of plastics by modern human societies probably means increasing amounts in the oceans.

The whole spectrum of the offal of civilization characterizes the material dumped into the waters off New York City. It includes dredge spoils, sewage, chemicals, garbage, and various construction materials. The amounts of these materials poured into these open coastal waters is of such volume it is visible even from satellites, according to Gunnerson (1981). For sewage alone, 127 municipal discharges contribute 2.6×10^9 gallons per day, or more than 100 times that of the White's Point outfall in the Los Angeles area. The results of 35 years of dumping have been significant. Oxygen levels near zero have occurred over extensive bottom areas off New Jersey and have led, on numerous occasions, to mass fish and shellfish mortalities, as well as to chemical and bacteriological contamination of other fish and shellfish. Some fisheries have even had to be abandoned, according to Gunnerson (1981). Furthermore, even though most of the waste is dumped many miles offshore, certain materials, such as plastic, tar, grease balls, and even such medical wastes as hypodermic needles, have come ashore to contaminate bathing beaches throughout the area. Clearly, human activities have had and will continue to have a significant effect on the marine ecosystem.

Chemicals

More insidious than oil or sewage, which are at least visible, are the various invisible toxic chemicals produced by the industrialized nations that find their way into the oceans' ecosystems. These chemicals are often transferred through the food chains in the sea and exert their effects in animals and places removed in time and space from the source. Because of this, it is difficult to pin down the effects of a given chemical, especially if the effects turn up years later. Certain marine organisms also enhance the toxic effects of many chemicals because they accumulate the substances in their bodies far above levels found in the surrounding water. Another factor that increases the effects of chemicals on living systems is **biological magnification**, in which the chemical increases in concentration in the bodies of organisms with each succeeding trophic level. These chemicals are not metabolized in an organism; therefore, the amount accumulated in the tissues remains there. When several such individuals are consumed by a carnivore of the next trophic level, the carnivore gains the chemicals from all the individuals, which increases the concentration in its body. Continuation of this process can lead to significant levels in the top carnivore, if the food chains are long. As we saw in Chapter 1, marine food chains are long, and so top carnivores in marine systems often have heavy loads of these chemicals. Humans also consume marine organisms, mainly from the higher trophic levels. This has already had an effect on certain of our fisheries, such as swordfish and tuna, in which the levels of mercury have often been higher than safe levels established by the FDA for humans, causing fishes to be rejected for consumption. More tragic than that is the case of Minimata disease, documented by Goldberg (1974).

In the late 1930s, the Chisso Corporation of Japan established a factory on the shores of Minimata Bay to produce vinyl chloride and formaldehyde. By-products of this plant contained mercury and were discharged into the bay. Through biomagnification, the marine fishes and shellfish accumulated high concentrations of the toxic compound methylmercury chloride. The fishes and shellfish were, in turn, consumed by the human inhabitants of the area. About 15 years after the dumping of mercury into the bay began, a strange, permanently disabling neurological disorder began to appear among the inhabitants, especially among the children. This disorder was called **Minimata disease**. The cause was diagnosed as mercury poisoning in 1959, but it was not until the early 1960s that the active mercury compound was identified and the link to the factory discharge established. It took even longer, until the 1970s, for Japan to halt the discharge of mercury into the ocean and for other nations to blacklist mercury dumping and establish standards for acceptable levels in food.

A second, well-documented example is the case of the pesticide DDT. Dichlorodiphenyltrichloroethane (DDT) was the first of the synthetic pesticides of a class known as chlorinated hydrocarbons. It was put into worldwide use in 1945 and was hailed as a boon to humankind for its effectiveness in destroying a wide range of insect pests while remaining relatively nontoxic to humans. It has, however, a few attributes that were to bring about serious environmental consequences. First, it is a remarkably stable compound in natural systems. It, or its first breakdown product, DDE, persists for years; just how long is unknown. It is relatively insoluble in water, but soluble in fats or lipids, and it adheres strongly to particles.

Although DDT is not used in the marine environment, it has entered the marine food webs through runoff from land, precipitation, and dumping. As a result, by the 1960s, DDT had been found in marine organisms as remote as Antarctica. The insidious effects of DDT in marine food webs became apparent in the 1960s, when certain top carnivores were found to have high levels in their tissues. DDT enters the marine food chains at the level of plankton, where the chemical adsorbs on the surfaces of the plankton organisms or is dissolved in the lipid. Since it is not metabolized, biomagnification concentrates it through succeeding trophic levels to the top carnivore. Concentrations of DDT in certain marine fish, such as mackerel, exceeded the permissible 5 ppm established as safe for human consumption by the FDA, and many catches had to be destroyed.

The most serious effects, however, occurred among marine birds. Once DDT reached certain high levels in birds, it interfered with the calcium deposition in the eggshell. As a result, thin eggs were laid, which broke open when the birds incubated them. The greatest amount of attention was focused on the brown pelican (*Pelicanus occidentalis*), in which successful breeding virtually disappeared in the main rookeries on the islands off southern California in the late 1960s and early 1970s. As Goldberg (1976) notes, this was traced directly to the high levels of DDT in their tissues. What was curious here was that most of southern California is urbanized and had not used much DDT, certainly not as much as that used in agricultural areas. The waters around Los Angeles, however, had concentrations of DDT averaging 370 ppm as opposed to 1 ppm off Baja California. Where did it come from? The source was discovered to be the White's Point sewer outfall. It turned out that a single chemical plant that produced most of the world's DDT was located in Los Angeles and was dumping its wastes into this outfall at the rate of 100 tons a year, a figure ten times as high as the total amount of DDT carried per year by the Mississippi River into the Gulf of Mexico. The decline of the brown pelican was linked through the marine food web to a toxic chemical discharged from a single source. Subsequent cessation of DDT dumping in Los Angeles has been followed by the slow recovery of the pelican.

These are only two examples of the problems of toxic chemicals. They emphasize the insidious nature of the often much-delayed effects and the difficulty of tracing the source of a problem through complex food webs, about which we know relatively little. Since modern industrial states now produce thousands of actually or potentially toxic chemicals each year and many will enter the oceans, how many more of these disasters must we look forward to in the future?

Radioactive Wastes

Since 1944, the world's oceans have been receiving an input of radioactive wastes generated from production of nuclear weapons and electric power. Major sources of input have been fallout from nuclear weapons tests, releases or dumping of wastes from nuclear fuel cycle systems, mainly power or reprocessing plants, and certain accidents. In the latter category, one of the most significant contributors was the 1964 incident involving an aerospace nuclear generator that reentered the atmosphere when the launch malfunctioned and deposited into the oceans a quantity of plutonium 238 equal to half the total oceanic deposits of plutonium 238.

There are two main categories of radioactive wastes. **Low-level wastes** contain less than 1×10^{-8} curies of radioactivity per gram. **High-level wastes** contain more than 1×10^{-8} curies per gram.

Presently, the dumping of high-level radioactive wastes in the oceans is prohibited by international agreement, but the dumping of low-level wastes is permitted. Low-level ocean dumpsites have been designated and used at one site in the northeast Atlantic, off Europe at three sites, off the United States East Coast at one site, and at one off the Pacific coast. From 1946 to 1970, the United States dumped approximately 107,000 containers of low-level radioactive wastes at the three sites off our coasts, according to Park et al. (1983). Currently, the only active site is off the East Coast. Whereas low-level radioactive waste seems manageable, there is considerable concern about high-level wastes, such as those generated from discarded fuel assemblies from nuclear generators or that arise from various reprocessing operations. These wastes usually have long half-lives and powerful ionizing radiation. Such wastes must be safely isolated from the biosphere for periods of time that are geological in scale.

If a radioactive waste can be contained for ten times the half-life of the radionuclide, the radioactivity will be reduced by 1,024 times. Plutonium 239, a common radionuclide of spent nuclear fuel, has a half-life of 24,100 years. This means that it must be isolated from the biosphere for 241,000 years.

Currently, there is much debate over what to do about disposal of these high-level nuclear wastes. This debate and its resolution have been heightened by the proliferation of nuclear power plants and weapons, generating more wastes, and by the fact that the United States also needs to decommission and safely dispose of three or four nuclear submarines per year for the next 25 or more years. This problem has become more acute with the demise of the Cold War and the decommissioning of more weapon systems.

Since high-level radioactive wastes remain toxic to humans for many thousands of years, it is necessary to find a disposal site that will be stable for an equivalent time period and that also will be beyond the reach of human tampering for the same length of time. Given the ingenuity of humans and our present technological prowess, there are few places on the planet that will meet these requirements. One such place is the deep ocean, well away from the edges of the tectonic plates. The abyssal plains of the deep ocean lie under 4–5 km of water. This is beyond any regular current human activity in the ocean. Furthermore, the bottom is covered with a thick layer of sediment that has accumulated over a long time and has remained stable. If the radioactive waste could be placed in this sediment, there is a good chance that it would remain safely away from the realm of human activities for the required time. Even if wastes should leak from the containers after some years, the nuclides would

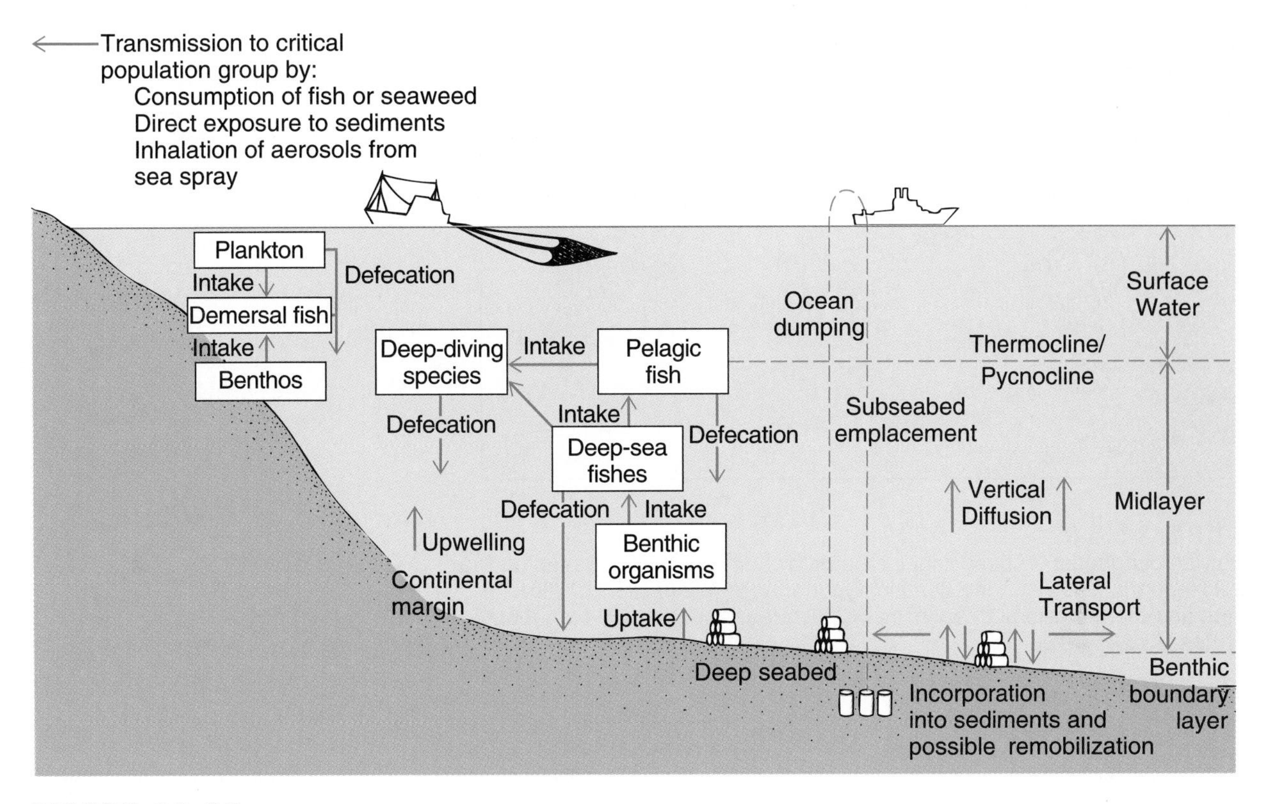

FIGURE 11.17
Physicochemical and biological transmission of waste radionuclides from deep-sea disposal sites. (Modified from *Wastes in the Ocean*, Vol. 3, P. K. Park et al. Copyright © 1983 John Wiley & Sons, Inc. Reprinted by permission.)

adhere to the sediment and probably remain in place (Figure 11.17).

Because of the foregoing advantages, the deep-sea bed has been suggested as the one area of the planet where we could safely isolate these toxic wastes for thousands of years. However, in order to evaluate the deep sea for feasibility of high-level waste disposal, we need to consider the consequences of leakage or failure of the containment vessels. It is, therefore, necessary to understand the fate of escaped radionuclides once they are out of the containment vessel. This requires a considerable knowledge of both the physical and biological mechanisms that will influence radionuclide movement. Since, as we saw in Chapter 4, the deep sea is the least understood of all the ocean areas of the planet, there is obviously a great deal of information that we need. As we grapple with the problem of safe disposal of radioactive wastes, we should encourage the research necessary to answer the important questions lest we again fail to make an intelligent choice and resort to expediency.

Miscellaneous Pollution Problems

To provide docking facilities for commercial and pleasure vessels, humans have engaged in a great deal of dredging of estuaries and bays. Such activities destroy large areas of productive habitat for marine organisms and may have effects beyond the immediate area if nursery grounds for commercial species caught offshore are destroyed. Not only are communities destroyed by physically being removed, but the increased load of silt suspended in the water reduces light, and hence photosynthesis, and clogs the feeding and respiratory surfaces of many invertebrates, leading to their deaths. Such activity has destroyed many productive areas worldwide, and the expanding human population suggests that we can look forward to even more destruction in the future.

The siting of electrical generating plants along the seacoasts to take advantage of the marine waters for cooling purposes has led to the perturbation of marine communities by heated water discharge. This **thermal pollution** has been little investigated for the marine environment; thus far, the effects appear restricted to

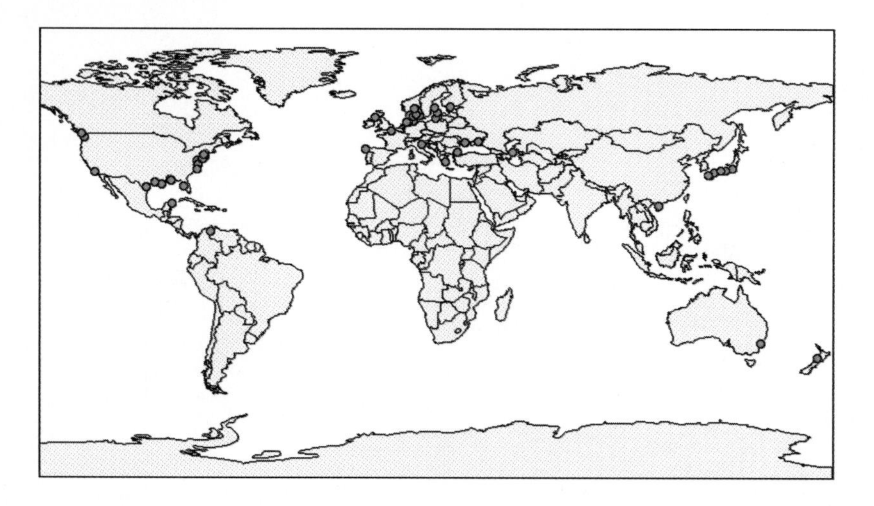

FIGURE 11.18
Worldwide distribution of dead zones created by human activity. (Based on Diaz and Rosenberg, 1995, Marine Benthic Hyporia: A Review of its Ecological Effects and the Behavioral Responses of Macrofauna. *Oceanogr. Mar. Biol. Ann. Rev.* 33:245–303.)

the communities immediately adjacent to the discharge. The damaging effects of thermal discharges are most pronounced in the tropics, where the organisms are living naturally at water temperatures close to their thermal maximum. It takes little increase in temperature to stress such communities. An example of the effects of a thermal discharge on tropical communities is described by Banus and Kolehmainen (1976) in Guayanilla on the south coast of Puerto Rico, where a power-generating station discharges heated water into a lagoon. The discharged water is 8–10°C higher than the ambient seawater and produces water temperatures in the lagoon of 39°C in the summer. This is high enough to affect reproduction of the dominant mangrove trees in the bay, and no young rooted and growing mangrove seedlings have been found in the lagoon.

Eutrophication refers to the release of excess nutrients into coastal waters. These nutrients come from the fertilizers spread on agricultural land that make their way to the sea via rivers and streams. The results of such increases are the red tides, various yellow and green foams, slimes, and slicks—all of which indicate excess phytoplankton growth. Although on a small scale certain of these blooms are natural, they have become more common and extensive in recent years. For example, in 1988 a yellow-brown slick off the Scandinavian countries threatened the entire Norwegian salmon mariculture industry and led to the most massive peacetime maritime rescue operation in the history of Norway, according to Cherfas (1990). In another incident in 1978, the German North Sea coast beaches were inundated by a meter-deep, foul-smelling foam created by surf beating on a massive outbreak of a gelatinous encapsulated alga of the genus *Phyaeocystis*. Aside from aesthetic considerations, many of these massive outbreaks are also increasingly toxic. This further contributes to the disaster.

The significance of eutrophication in the oceans is most dramatically demonstrated by the presence of some 50 oxygen-starved coastal dead zones in various parts of the oceans (Figure 11.18). **Dead zones** result when oxygen in ocean waters is reduced below the level necessary to sustain marine life. The largest such zone in the Western Hemisphere is located every summer off the Louisiana and Texas coasts and spreads with each advancing year. In 1989, according to Malakoff (1998), it covered 9,000 km^2, but by 1997 it had expanded to 16,000 km^2 (Figure 11.19). The Louisiana dead zone is the result of the massive amounts of nutrients carried off the agricultural lands of the Mississippi River drainage. During the spring and summer the nutrient-laden warm fresh water of the Mississippi

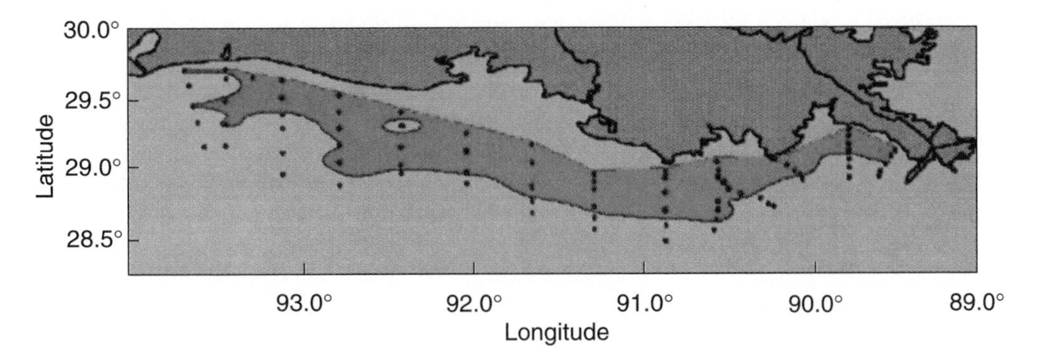

FIGURE 11.19
Extent of the dead zone off Louisiana and Texas. (From Rabalais, Turner, and Wiseman/Gulf Hypoxia Studies.)

flows out over the more dense Gulf of Mexico seawater, forming a lidlike layer. Fueled by the nutrients, the phytoplankton bloom and attract zooplankton grazers. Large amounts of dead phytoplankton, zooplankton, and fecal pellets then sink to the bottom, where proliferating oxygen-consuming bacteria consume them and reduce the oxygen below the level necessary to sustain multicellular organisms. The dead zone is generally confined to shallow water up to 60 meters deep, but it is not confined to just the water immediately adjacent to the bottom and may affect as much as 80% of the water column according to Malakoff (1998). The mobile organisms in the dead zone are driven away, but the sedentary and sessile species such as clams, starfishes, and worms are killed. One of the largest concerns is that as this dead zone spreads and/or becomes more persistent is that it will destroy one of the richest fishing grounds in the Gulf of Mexico. To solve this problem it will be necessary to somehow reduce the amount of fertilizers that leach from agricultural lands throughout some of the richest farmlands of the United States, a task which is not only daunting, but certainly cannot be achieved quickly.

How do we know that these blooms are not natural events and their apparent increases are not just the result of better observation and quick communication common to modern times? Studies in Japan and Europe, areas with good historical records, show that such phenomena have increased markedly in the past 20 years. The good side to this is that the increased numbers of algal cells may absorb more carbon dioxide and retard the greenhouse effect (see subsequent section). However, at present, no one knows whether this is true or not. Another unanswered problem concerns dimethyl sulfide (DMS), a natural compound of phytoplankton. When the phytoplankton die, DMS is released into the atmosphere, where it may be changed to sulfuric acid and contribute to acid rain.

It may seem strange to suggest that dams built on freshwater rivers and streams can affect the oceans, but by preventing the flow of fresh water into the oceans, these structures have had effects in certain areas. Probably the best example is the Aswan Dam on the Nile River in Egypt. Prior to its construction in 1965, the Nile carried into the Mediterranean Sea 43 billion cubic meters of water. It now carries only 3 billion. The result of this dam has been the erosion of the Nile Delta and the loss of the fisheries of the eastern Mediterranean. For example, the sardine fishery in the eastern Mediterranean fell from an average of 18,000 tons per year in the 1960s before the dam was built to 600 tons in 1969 and has remained low in subsequent years according to Weber (1994).

A final consideration concerns pollution by the introduction of **alien animals**. The devastating effects on native fauna and flora of alien animals and plants are well known and documented for the terrestrial environment. We know, for example, of the loss of the American chestnut through a disease imported from Europe and, more recently, the demise of our native elm trees through Dutch elm disease. Equally dramatic have been the devastations of European rabbits on the flora and fauna of Australia. What about introductions in marine waters?

Our knowledge of the effects of introducing an alien marine species is limited. The major intentional introductions of marine species have been those of commercial or sports fishing value. Thus, the eastern (*Crassostrea virginica*) and Japanese (*Crassostrea gigas*) oysters were introduced into the bays and estuaries of the Pacific coast of North America to provide the basis of an oyster industry. Similarly, the striped bass (*Roccus saxatilis*) was introduced to the Pacific coast from the Atlantic and has become the basis of a valuable sports fishery.

Along with these purposeful introductions, however, have come other, unintentional species. Other inadvertent introductions have occurred from the fouling organisms on ships and in their ballast water and through commercial bait and seafood importations. The number of these species and their cumulative effects on indigenous communities are little known but potentially great. That they may have profound effects is witnessed by studies of introduced species on the central California coast. Carlton (1975, 1989) has shown that 150–200 marine invertebrates from various parts of the world have become established in the bays and estuaries of the Pacific coast. In the case of San Francisco Bay, the majority of the present invertebrate fauna has been introduced, and it is now nearly impossible to know what the indigenous communities of this great bay were like.

How serious a threat these introductions are to marine communities around the world remains to be seen, but if the results we have observed in terrestrial communities are any guide, the threat could be considerable. As an example of the potential seriousness of such aliens we can mention the case of the ctenophore *Mnemiopsis leidyi*. In 1982, according to Travis (1993), this voracious planktonic carnivore was introduced into the Black Sea, probably from the ballast water of a ship that had taken on that water on the East Coast of North America where *Mnemiopsis* is native. *Mnemiopsis* underwent an almost unbelievable population explosion in the Black Sea. At times, it now constitutes as much as 95% of the entire biomass! This gelatinous carnivore feeds on zooplankton, including the eggs and larvae of fishes, and is the prime suspect in the precipitous decline of the Black Sea fisheries. The question of how to combat this alien is now under debate, but none of the options is without risk. Perhaps the best option is to introduce a predator or parasite of this animal. The problem is that we know so little about the species or its predators and parasites that we cannot be certain that they, too, might not become a problem if introduced. Time may be of the essence here, since *Mnemiopsis* has recently emigrated into the Mediterranean Sea and may threaten its fisheries next. The potential for such introductions in the future is great. If, for example, a sea level canal is put through in Central America, allowing free interchange of the tropical Atlantic and Pacific faunas, massive changes may occur. The giant offshore oil-drilling rigs are another source. They accumulate a whole marine fauna in one area during drilling. When they are moved, often to an entirely different area, the whole fauna is brought along, transferred intact.

MARINE DISEASES

Within the last few decades, there has been an increase in mortalities of many taxa of marine organisms apparently due to diseases. In addition, there seems to have been an increase not only in the frequency of epidemics but in the occurrence of new diseases as well. Many of these outbreaks have occurred in shallow waters heavily affected by humans, suggesting human activity as a factor. Unfortunately, we know so little about diseases in the marine environment that we are, in many cases, unable to assess whether these epidemics are natural or whether they have resulted from some human intervention or activity in the marine environment. We have also made little progress in identifying the causative factor for many of the diseases. For example, the massive die-off of the dominant Caribbean sea urchin, *Diadema antillarum*, in the 1980s (see pp. 415–416) was one of the best-studied epidemics and yet we have not identified the pathogen. Similarly, the near elimination of the two dominant species of the coral *Acropora* at many locations in the Caribbean at the same time was by an agent as yet unidentified. Currently, of the dozen or more coral diseases in the Caribbean, Harvell *et al.* (1999) indicate that the causative agents are known for only three. These mass mortalities are not limited to the Western tropical Atlantic; pathogen-caused epidemics have been reported in various parts of the temperate North Atlantic and the tropical Pacific and have affected a wide range of both animal and plant taxa. Concomitant with the apparent increase in diseases has been the increase in toxic algal blooms that have affected both marine mammal and fish populations. For example, the toxic dinoflagellate *Pfisteria piscicida* was determined as the cause of massive fish kills on the Atlantic coast of North America according to Burkholder et al. (1992).

The question to be asked is whether this apparent increase in disease is real or whether it is natural and results from our increased scrutiny of the oceans. Several lines of evidence, primarily from coral reefs, suggest that the increase is real. First, our monitoring of coral reefs in Florida indicates that we are seeing diseases that we have not seen before. Second, coral reefs are long-lived structures and if these diseases had been in existence before, causing changes in community structure, we would see them in the fossil record. Finally, because of the massive mortality, if they had occurred in previous decades, we certainly would have detected them.

If these diseases are increasing and some are new, where did they come from? From current evidence, it appears that most have come about not by new pathogenic organisms but by existing organisms shifting hosts. Evidence for this **host shift** comes from several sources. In the Antarctic the appearance of canine distemper virus in crabeater seals is thought be have come from contacts with sled dogs used in several Antarctic expeditions. In the Caribbean a fungus pathogen that has attacked various species of sea fans has been identified as *Aspergillus sydownii*, which has somehow spilled over from its usual terrestrial reservoir. Still other diseases appear to be new. For example a virus, phocine distemper virus, previously unrecognized, was determined to have caused mass mortality among harbor seals (*Phoca vitulina*) and gray seals (*Halichorus gryphus*) in Northwestern Europe in the 1980s, while two other new viruses, dolphin morbillivirus and porpoise morbillivirus, were found to be the cause of worldwide mortalities of dolphins, porpoises, and other cetaceans.

Why have we had these increasing outbreaks? Although we are uncertain as yet, correlative data suggest two conditions favoring outbreaks. The first is the climate-induced changes that have affected marine ecosystems. Although there is little or no direct proof, periodic climatic changes such as produced by El Niño-Southern Oscillation events cause extreme stress on many marine communities and some of the resultants, such as coral bleaching, may be accelerated by pathogens that are themselves enhanced by such events. Correlative proof of this has been found in the variation in Dermo, a disease of the eastern oyster *Crassostrea virginica* caused by the protozoan *Perkinsus marinus*. Dermo outbreaks occur when La Niña conditions prevail and drop during El Niño events. The other condition favoring disease outbreaks is direct and indirect human activity. Humans have transported many marine species out of their native regions and introduced them into other areas. This usually means transporting their pathogens as well. The best examples of transference of pathogens has been in the mariculture industry, where human-facilitated epidemics have been documented in the shrimp and bivalve mollusk culturing industries and have sometimes invaded the wild populations. Another example of disease outbreaks brought about by human activities is that of human-produced pollutants that enter the oceans. These pollutants, from pesticides to sewage, may compromise the immune systems of marine animals or facilitate the movement of terrestrial pathogens to marine organisms.

In closing this section, we should note that humans themselves are not immune to pathogens in seawater since many terrestrial pathogens can persist for varying amounts of time in marine environments and can infect humans exposed to seawater. Perhaps the best example of this is the human disease cholera. The organism responsible for cholera is *Vibrio cholerae*, which is associated with marine plankton. The incidence and severity of human epidemics of cholera have been strongly correlated with seawater temperature, turbidity, and plankton blooms.

DRUGS FROM THE SEA

Many marine organisms produce chemicals that are not found, or are rare, in terrestrial organisms. These chemicals are used mainly by the marine organisms to defend against predators. Certain of these compounds have biomedical properties that may be useful in combating human disease. Organisms especially productive of such biomedical compounds include sponges, cnidarians, tunicates, and opisthobranch mollusks. Among the useful chemicals are a shark-repelling saponin; ichthyotoxin, from the fish *Pardachirus marmoratus*; antitumor chemicals from the sponges *Haliclona* and *Halichondria*; and a potent non-protein toxin, palytoxin, from the soft coral *Palythoa*. Perhaps the most interesting, however, is a steroid called squalamine found in the dogfish shark (*Squalus acanthias*). Stone (1993) reports it as effective in killing fungi that cause dangerous infections in AIDS and cancer patients. More than 1,700 compounds with some biomedical property have been reported in the last ten years, and more are certain to come with additional testing of organisms. The main stumbling block thus far to commercial pharmaceutical development has been the inability to raise most of the marine organisms in large numbers in captivity.

GLOBAL WARMING AND SEA LEVEL CHANGE

Certainly, one of the most talked about subjects in science today, not only among scientists but among the public, politicians, and governments, is that of **global warming** or the **greenhouse effect** and how it will affect the climate of the planet. Global warming refers to the increase in the temperature of the planet primarily as a result of the increase in the atmosphere of certain gases, such as carbon dioxide, methane, nitrous oxide, and chlorofluorocarbons. These gases have the ability to trap the radiant energy from the sun, leading to increases in atmospheric temperatures. Increased temperatures, in turn, may cause changes in the climate, alter rainfall and storm patterns, change ocean current patterns, and increase the rate of sea level rise.

There seems little doubt that the "greenhouse" gases have increased over the last 100 years, and much of this increase is due to the activities of humans in burning fossil fuels and wood. Certainly, the increase in chlorofluorocarbons is directly attributable to humans, since this gas is a human invention that is used in refrigeration systems. The other gases are derived from natural sources. Exacerbating the problem has been the wholesale cutting of the world's forests, which would otherwise serve as a major sink to absorb excess carbon dioxide. The other major sink for carbon dioxide is the world's oceans (see the carbon cycle in Chapter 1). It is both the increase in sea level rise and the potential of the ocean to absorb greenhouse gases that directly concern us here. As global temperatures rise, the oceans of the earth will expand because of the melting of glaciers and ice caps, the most significant of which are those of Greenland and Antarctica, and because of the thermal expansion of the ocean water. Depending on the amount of warming, sea levels could rise as much as 5–7 ft over the next century. This is opposed to the rise of 0.5 ft in the last 100 years. If the "worst case" rise were to occur, there would be massive inundations of low-lying coastal areas, such as Louisiana, Cape Cod, Florida, Egypt, and Bangladesh, and many low-lying island archipelagos, such as the Maldives and Marshall islands, might be completely eliminated.

If these unprecedented disasters could occur, why is there such controversy concerning global warming? The answer stems from our own lack of knowledge. Whereas we can chart with relative accuracy the trends in atmospheric gases, sea level rise (or fall), and the advance or retreat of glaciers over the last 100–200 years, this time frame is geologically insignificant but significantly variable such that it cannot serve as a basis to predict long-term changes. Besides, we know that the planet has undergone various sea level and climatic changes in the past, but these changes have usually proceeded very slowly, over thousands or millions of years, not within a hundred or so years, as now seems to be the case. The real question, and the source of the controversy, is whether or not the changes we are now observing are simply natural variations in an otherwise stable climate since the end of the last ice age, or whether they are the forewarnings of a long-term change. In spite of our current lack of the necessary long-term data, there now seems to be a growing consensus among scientists, according to Kerr (1990), that global warming is a reality.

There are a few startling data sets from the marine environment that suggest long-term changes are happening and may be related to global warming. They also give an inkling as to what we might expect if trends continue. The first concerns the effects of the depletion of the ozone layer over Antarctica precipitated by the chlorofluorocarbons. Smith et al. (1992) have reported that the thinning of the ozone layer over the Antarctic led to an increase in the ultraviolet light impinging on the Antarctic Ocean. This resulted in photoinhibition of the phytoplankton and a decrease in primary productivity of from 6 to 12%, which, in turn, decreased the uptake of carbon dioxide. The results of this study suggest that should the ozone layer be depleted over much larger areas of the planet, even less carbon dioxide would be absorbed, thus intensifying global warming.

In the second example, Roemmich and McGowan (1995) report on a 40-year study of the zooplankton in the waters off southern California. During this time, the surface water temperature warmed by about 1.5°C. This resulted in increased stability for the thermocline such that wind-driven upwelling was displaced upward into shallower water, and there were less nutrients (NO_3, PO_4) for the phytoplankton and lower productivity. In turn, this led to decreases in the zooplankton biomass by 70% between 1951 and 1993. The future consequences of such an observed trend depend on whether or not it is part of a natural cycle. If natural, the trend may reverse itself. However, if it is a manifestation of the longer human-induced global warming and there is a further 1–2°C temperature rise in ocean waters around the world in the next 40–50 years causing stratification increases, the results for the marine ecosystems could be devastating.

Given the fact that greenhouse gases are continuing to increase and the known fact that if this trend continues there will be considerable sea level rise, is there anything that humans can do to forestall this? The answer is yes, but it would require massive action on the part of all nations on the planet. Forests must be prevented from being cut (indeed, more need to be planted to absorb the carbon dioxide), chlorofluorocarbons must be taken out of use, and fossil fuel usage must be reduced. There is, finally, the intriguing suggestion by Martin et al. (1990) that much of the excess carbon dioxide could be removed from the atmosphere to the ocean sink if a few hundred thousand tons of iron could be made available to the phytoplankton of the equatorial Pacific Ocean, southern Ocean, and subarctic Pacific Ocean. All of these areas have been puzzling to oceanographers because they have high nutrient content but low chlorophyll. This enigma has led scientists to ask why there is such low biomass in areas with abundant nutrients and light. If Martin is correct, these phytoplankton are limited in growth and photosynthesis by the lack of trace amounts of iron in the water. If iron were added, the increase in photosynthetic activity would be great enough to draw down the carbon dioxide from the atmosphere.

Initially, the hypothesis was experimentally supported by laboratory measurements with seawater samples that were uncontaminated with metals. Additional experiments were performed using deckboard bottle enrichments of surface waters from the subarctic Pacific, southern Ocean, and equatorial Pacific by Coale (1991) and Johnson et al. (1994), which also confirmed the hypothesis. However, these experiments were strongly criticized because of the effects of containment and the failure to take into account the complexities of the natural environment and the other members of the pelagic community.

Finally, in a unique series of experiments known as the IronEx experiments, scientists from the Moss Landing Marine Laboratories led two international expeditions to the equatorial Pacific to test the hypothesis in a series of open ocean enrichment experiments. In these experiments when approximately half a ton of iron was added to 100 km^2 of the nitrate-rich waters, phytoplankton biomass increased by a factor of 30, depleting carbon dioxide to 60% of its initial value, and the ocean turned green for miles (Coale et al., 1998). Whereas the confirmation of the iron hypothesis has suggested to some industrialists and politicians that we should use this to control global carbon dioxide, the scientific community is urging caution and that much more needs to be understood about the oceans before large-scale fertilization is undertaken.

CONCLUDING REMARKS

For too long, the seas of the world have been considered an inexhaustible source of food, as having an infinite capacity to absorb and purify our wastes, and as a source of all the raw materials needed to maintain an industrial society. It is now apparent that none of these assumptions is true, and the human population at the current level of technological development has the ability to inflict massive destruction on the seas, just as we have done on land. At present, the seas remain in good condition relative to the land, and we cannot afford to permit them to be degraded in the same way if we wish to continue our tenure as a species on this planet. We must employ ecological principles, such as those outlined in this text, to ensure that the potential of the oceans is realized without degradation or reliving yet another tragedy of the commons.

SUMMARY OF KEY CONCEPTS

- The explosion of technology and the rapidly increasing human population have had significant effects on the ecology of the world's oceans in recent years.
- Technological advances in fishing gear and ships have resulted in the significant reduction in and overexploitation of many fish stocks.
- The major fisheries are concentrated in the waters overlying the continental shelves, but there is a very unequal tonnage caught in different geographical areas.
- The commercially important marine animals are drawn from four groups: bony and cartilaginous fishes, marine mammals, mollusks, and crustaceans; the greatest tonnage comes from fishes.
- In 1997 the total marine fisheries landings were 85.6 million metric tons, but this figure is lower than the peak catch, which was reached in 1989.
- Currently most of the fish stocks on the continental shelf areas of the world are fully exploited or overexploited.

- Open ocean waters are impoverished with respect to production and cannot sustain the population of fishes that the continental shelves do.
- The maximum sustainable yield is the largest number of fishes that can be harvested year after year without diminishing stocks.
- The estimated sustainable yield from the world's oceans varies between 60 million tons and 290 million tons because of disagreements among scientists on the proper figures to use for primary productivity, ecological efficiency, and average trophic level.
- With few exceptions, productivity has declined in all of the major fishing areas of the planet due to overexploitation often compounded by certain natural changes in the environment.
- The decline of the Peruvian anchoveta fishery was due to overfishing combined with environmental changes known as El Niño, the demise of the whale stocks was due entirely to overfishing, and the failure of the California sardine fishery remains a mystery.
- Overfishing may also result in changes in the larger ecosystems or communities in which the target species reside.
- Fishing affects the community directly by removal of species, thus reducing the species diversity, and may also alter the size structure, reproduction, and life history characteristics of the fished populations.
- Fishing may also alter the species structure of the fished communities.
- Overfishing coupled with environmental changes may have effects extending well beyond the immediate area.
- Bycatch is also a source of destruction to marine communities.
- Fishing gear itself has a deleterious effect on the habitat.
- Capturing fishes for the aquarium trade has resulted in significant losses, particularly to coral reefs.
- Ghost fishing by lost gear can cause continued depletion of organisms.
- Open ocean fisheries resources, presumably held in common for the good of the world's people, are at risk of succumbing to the tragedy of the commons.
- International agreements among nations for the control of fisheries have proved difficult to bring about and effectively implement, but they will be necessary to effect a truly worldwide regulation of fisheries.
- Potential new fisheries include southern ocean krill and squid, the area of the continental slope at 300–1,000 meters, and many areas in the Indian Ocean.
- Mariculture is the rearing of marine plants and animals under controlled conditions to provide increased amounts of food for humans.
- Few marine organisms are cultivated, and most of these are either the type that fetch a premium price on the market or produce large amounts of biomass under culture.
- Although yields from mariculture can be very high, the science is currently at a low level of development, and thus mariculture presently contributes relatively little to human food.
- Mariculture development is restricted by several factors, including the difficulty of maintaining the proper physical and chemical characteristics of the water, the presence of many larval forms, diseases and parasites, and the cost in money and time to obtain permits.
- In recent years we have come to realize that the oceans are not infinite in their capacity to absorb human wastes and wastes generated by human activity.
- Oil pollution usually results from large tanker accidents, leading to an oil spill, or from blowouts occurring in offshore drilling rigs.
- Oil is devastating to marine birds and mammals and can suffocate intertidal and shallow subtidal communities.
- The area affected by a spill from a large tanker can be extremely large, as witnessed by the *Exxon Valdez* spill in Alaska.
- Oil seems to have little effect on pelagic communities, save for birds.
- Discharge of human sewage to the marine environment in small amounts and with adequate diffusing seems to cause no effects on marine communities on the open coast, but in large volumes and in semienclosed embayments such discharge may be devastating.
- Large amounts of human garbage are dumped into the ocean each year, but there is little scientific data regarding their effects on marine communities.
- Various chemicals produced by the industrialized nations find their way into marine ecosystems; these are often transferred through marine food chains and exert effects in animals and places removed in time and space from the source.
- Biological magnification is the process whereby the concentration of chemicals in the bodies of organisms increases with each succeeding trophic level.
- Major sources of radioactive materials in the oceans have been from fallout from nuclear testing, releases or dumping of wastes from nuclear fuel cycle systems, and certain accidents.
- Presently, the dumping of high-level radioactive wastes into the oceans is prohibited but dumping of low-level wastes is still permitted.
- High-level radioactive wastes must be isolated from the biosphere for geological periods of time; the deep ocean floor has been suggested as the safe place to put such wastes.
- Dredging of estuaries and bays for boating facilities destroys large areas of productive habitat for marine organisms.
- Thermal pollution by heated water discharges is most pronounced in the tropics and is usually restricted to the communities immediately adjacent to the discharge.
- Eutrophication, the release of excess nutrients into coastal waters, leads to excess phytoplankton growth

resulting in red tides, various foams and slicks, and most devastating, the depletion of oxygen, creating dead zones.

- Dams built on freshwater rivers affect the oceans by reducing the sediment load, leading to erosion of deltas and reductions in fisheries.
- Alien marine animals and plants have been introduced in various marine habitats around the world both intentionally and unintentionally. The number of the unintentional introductions and their effect on indigenous communities is little known but potentially great.
- In recent years there has been an increase in mortalities of many marine taxa from diseases. There has not only been an increase in the frequency of epidemics but also in the occurrence of new diseases.
- Some of the new diseases are a result of host shifts by the pathogens.
- Some of the recent epidemics have come about as a result of climate-induced changes in marine ecosystems, whereas others have been due to human activity.
- The incidence and severity of the human disease cholera have been strongly correlated with seawater temperature, turbidity, and plankton blooms.
- Certain marine organisms have been found to contain chemicals that have proved useful in combating human diseases.
- Global warming or the greenhouse effect has the potential to create significant sea level rise and also to change marine ecosystems.

REVIEW QUESTIONS

ESSAY: Develop complete answers to these questions.

1. What is meant by the term *maximum sustainable yield*? How does this relate to the biological potential of a fish? Elaborate.
2. Resolve: The solution to pollution is dilution. Present evidence to support your argument.
3. Describe the concept of the tragedy of the commons in terms of two different marine issues, two different terrestrial issues, two different atmospheric issues, and two different freshwater issues.
4. Why did the total catch in marine fisheries decline in the 1990s even though new fisheries resources, such as krill, are being exploited?
5. Discuss ways human efforts to increase food production on land have reduced the potential yields of food from the sea.

BIBLIOGRAPHY

Alverson, D. L., M. H. Freeberg, J. G. Pope, and S. A. Murawski. 1994. A global assessment of fisheries bycatch and discards. FAO Fisheries Technical Paper 339.

Banner, A. H. 1974. Kaneohe Bay, Hawaii: Urban pollution and a coral reef ecosystem. In *Proceedings of the 2nd Int. Coral Reef Symp*. 2:685–702.

Banus, M. D., and S. E. Kolehmainen. 1976. Rooting and growth of red mangrove seedlings from thermally stressed trees. In *Thermal ecology*, Vol. 2, edited by G. W. Esch and R. W. McFarlen. Washington Technical Information Center of the Energy Research and Development Administration, 46–53.

Bardach, J. E. 1968. Aquaculture. *Science* 161:1098–1106.

Bardach, J. E., J. H. Ryther, and W. O. McLarney. 1972. *Aquaculture*. New York: Wiley.

Bascom, W., ed. 1978. Coastal water research project annual report. Los Angeles.

Blindheim, J., and H. R. Skjoldal. 1993. Effects of climatic changes on the biomass yield of the Barent's Sea, Norwegian Sea and West Greenland large marine ecosystem. In *Large marine ecosytems: stress, mitigation and sustainability*, edited by L. M. Alexander and B. D. Gold. AAAS, 185–189.

Boucher, G. 1985. Long-term monitoring of meiofauna densities after the *Amoco Cadiz* oil spill. *Mar. Pollution Bull.* 16:328–333.

Burkholder, J. M., E. J. Noga, C. H. Hobbs, H. B. Glasgow, and S. A. Smith. 1992. New "phantom" dinoflagellate is the causative agent of major estuarine fish kills. *Nature* 358:407–410.

Busch, L. 1991. Science under wraps in Prince William Sound. *Science* 252:772–773.

Carlton, J. 1975. Introduced intertidal invertebrates. In *Light's manual: Intertidal invertebrates of the central California coast*, edited by R. I. Smith and J. Carlton. 3d ed. Berkeley: University of California Press, 17–25.

Carlton, J. 1989. Man's role in changing the face of the ocean: Biological invasions and implications for conservation of near-shore environments. *Cons. Biol.* 3(3):265–273.

Cherfas, J. 1990. The fringe of the ocean—under siege from the land. *Science* 248:163–165.

Coale, K. H. 1991. The effects of iron, manganese, copper and zinc on primary production and biomass in plankton of the subarctic Pacific. *Limn. Oceanog.* 36:1851–1864.

Coale, K. H., K. S. Johnson, S. E. Fitzwater, S. P. G. Blain, T. P. Stanton, and T. L. Coley. 1998. IronEx I, an in situ

iron-enrichment experiment: Experimental design, implementation and results. *Deep-Sea Res.* II 45:919–945.
Culley, M. 1971. *The pilchard, biology and exploitation*. III, *The California sardine*. New York: Pergamon, 143–176.
Cushing, D. H. 1969. Upwelling and fish production. FAO Fisheries Technical Paper 84, 1–40.
Cushing, D. H. 1975. *Fisheries resources of the sea and their management*. Oxford, UK: Oxford University Press.
Cushing, D. H. 1977. *Science and the fisheries*. Studies in Biology, no. 85. Baltimore: Edward Arnold.
Dayton, P. K., S. F. Thrush, M. T. Agardy, and R. J. Hoffman. 1995. Environmental effects of marine fishing. *Aquat. Cons. Mar. Freshwat. Ecosyst.* 5:202–232.
FAO. 1998. 1997 *Yearbook of fisheries statistics*, Vol. 85. Rome: Food and Agriculture Organization of the United Nations.
Fautin, D., ed. 1987. *Biomedical importance of marine organisms*. Memoirs of the California Academy of Sciences, no. 13.
Gerges, M. 1993. Rehabilitating ROPME's ecosystems. *Siren* 48:1–4.
Goldberg, E. D. 1974. Marine pollution: Action and reaction times. *Oceanus* 18(1):6–18.
Goldberg, E. D. 1976. *The health of the oceans*. Paris: UNESCO Press.
Gunnerson, C. 1981. The New York Bight ecosystem. In *Marine environmental pollution*, edited by R. A. Geyer. Vol. 2, *Dumping and mining*. New York: Elsevier, 313–378.
Hardin, G. 1968. The tragedy of the commons. *Science* 162(3859):1243–1248.
Harvell, C. D., K. Kim, J. M. Burkholder, R. R. Colwell, P. R. Epstein, D. J. Grimes, E. E. Hofman, E. K. Lipp, A. D. M. E. Osterhaus, R. M. Overstreet, J. W. Porter, G. W. Smith, and G. R. Vasta. 1999. Emerging marine diseases—climate links and anthropogenic factors. *Science* 285:1505–1510.
Harvey, G. R., A. G. Requejo, P. A. McGillivary, and J. M. Tokar. 1979. Observations of a subsurface oil-rich layer in the open ocean. *Science* 205:999–1001.
Hess, W. 1978. The *Amoco Cadiz* oil spill. A preliminary scientific report. NOAA/EPA.
Holmes, R. 1994. Biologists sort the lessons of fisheries collapse. *Science* 264:1252–1253.
Idyll, C. P. 1970. *The sea against hunger*. New York: Crowell.
Jennings, S., and M. J. Kaiser. 1998. The effects of fishing on marine ecosystems. In *Advances in marine biology*, edited by J. H. S. Blaxter, A. J. Southward, and P. A. Tyler. 201–352.
Johnson, K. S., K. H. Coale, V. A. Elrod, and N. Tindale. 1994. Iron photochemistry and bioavailability in the equatorial Pacific. *Mar. Chem.* 46:319–334.
Kerr, R. A. 1990. New greenhouse report puts down dissenters. *Science* 249:481–482.
Loughlin, T. R., ed. 1994. Marine mammals and the *Exxon Valdez*. San Diego: Academic Press.
Malakov, D. 1998. Death by suffocation in the Gulf of Mexico. *Science* 281:190–191.
Maragos, J., C. Evans, and P. Holthus. 1985. Reef corals in Kaneohe Bay six years before and six years after termination of sewage discharges. *Proc. 5th Int. Coral Reef Symp.* 2:236.
Marine pollution. 1974. A series of articles in *Oceanus* 18(1).
Marr, J. C. 1960. The causes of major variations in the catch of the Pacific sardine, *Sardinops caerulea*. Spec. Sci. Rept., U.S. Fish and Wildlife Ser. 208:108–125.
Martin, J. H., S. E. Fitzwater, and R. M. Gordon. 1990. Iron deficiency limits phytoplankton growth in Antarctic waters. *Global Biogeochem. Cycl.* 4:5–12.
McVay, S. 1966. The last of the great whales. *Sci. Amer.* 215:13–21.
Nelson-Smith, A. 1973. *Oil pollution and marine ecology*. New York: Plenum Press.
Nicol, S., and W. de la Mare. 1993. Ecosystem management and the Antarctic krill. *Amer. Sci.* 81:36–47.
National Academy of Sciences. 1985. *Oil in the sea, inputs, fates and effects*. Washington, DC: National Academy Press.
Park, P. K., D. R. Kester, I. W. Duedall, and B. Ketchum. 1983. *Wastes in the ocean*. Vol. 3, *Radioactive wastes and the ocean*. New York: Wiley.
Pauly, D., and V. Christensen. 1995. Primary production required to sustain global fisheries. *Nature* 374:255–257.
Platt, J. F., C. J. Lensink, W. Butler, M. Kendziorek, and D. R. Nysewander. 1990. Immediate impact of the *Exxon Valdez* oil spill on marine birds. *Auk* 107:387–397.
Riaux-Gobin, C. 1985. Long-term change in microphytobenthos in a Brittany estuary after the *Amoco Cadiz* oil spill. *Mar. Ecol. Progr. Ser.* 24:51–56.
Rodhouse, P. G. 1995. Southern ocean cephalopod resources. *Bull. Malacological Soc. London* 25:1, 3.
Roemmich, D., and J. McGowan. 1995. Climatic warming and the decline of zooplankton in the California Current. *Science* 267:1324–1326.
Rounsefell, B. A. 1975. *Ecology, utilization and management of marine fisheries*. St. Louis: Mosby.
Russell-Hunter, A. D. 1970. *Aquatic productivity*. New York: Macmillan.
Ryther, J. H. 1969. Photosynthesis and fish production in the sea. *Science* 166:72–76.
Schaefer, M. B. 1965. The potential harvest of the sea. *Trans. Amer. Fish. Soc.* 94:123–128.
Scheuer, P. J. 1990. Some marine ecological phenomena: Chemical basis and biochemical potential. *Science* 248:173–177.
Sherman, K. 1991. The large marine ecosystem concept: Research and management strategy for living marine resources. *Ecol. App.* 1:349–360.
Smith, B. D., and G. S. Jamieson. 1991. Possible consequences of intensive fishing formales on the mating

opportunities of Dungeness crabs. *Trans. Amer. Fish. Soc.* 120:650–653.

Smith, R. C., B. B. Prezelin, K. S. Baker, R. R. Bidigare, N. P. Boucher, T. Coley, D. Karentz, S. MacIntyre, H. A. Matlick, D. Menzies, M. Ondrusek, Z. Wan, and K. J. Waters. 1992. Ozone depletion: Ultraviolet radiation and phytoplankton biology in Antarctic waters. *Science* 255:952–959.

Soutar, A., and J. D. Isaacs. 1969. History of fish populations inferred from fish scales in anaerobic sediments off California. *Calif. Mar. Res. Comm.*, CalCOFI *Rept.* 13:63–70.

Stone, R. 1993. *Deja vu* guides the way to new antimicrobial steroid. *Science* 259:1125.

Travis, J. 1993. Invader threatens Black, Azov Seas. *Science* 262:1366–1367.

Thrush, S. F., J. E. Hewitt, V. J. Cummings, P. K. Dayton, M. Cryer, S. J. Turner, G. A. Funnel, R. G. Budd, C. J. Milburn, and M. R. Wilkinson. 1998. Disturbance of the marine benthic habitat by commercial fishing: Impacts at the scale of the fishery. *Ecol. App.* 8:866–879.

Weber, P. 1994. Net loss: Fish, jobs and the marine environment. Worldwatch paper 120.

Wheeler, J. A. 1996. The marine aquarium trade: A tool for coral reef conservation. A report for the Sustainable Development and Conservation Program. University of Maryland, College Park. 47 pp.

GLOSSARY

abyssal The bottom zone of the oceans at depths between 4000 and 6000 m.

abyssal gigantism Phenomenon observed among several crustacean groups in which general size increases with increasing depth.

abyssal plain That area of the deep ocean floor lying between 4000 and 6000 m.

abyssalpelagic Pelagic aphotic zone lying between the bathypelagic and hadalpelagic zones.

acidity A measure of the concentration of hydrogen ions in a solution.

adenosine triphosphate The energy storage molecule of most living systems.

advection Movement of seawater horizontally or vertically, as in a current.

aerenchyma Tissue of thin-walled cells and spaces found in stems and roots of certain marsh plants serving to transfer oxygen.

aerobic Condition in which oxygen is present.

ahermatypic coral A nonreef-producing coral without the symbiotic zooxanthellae in the tissues.

algal ridge Coral-free ridge of encrusting coralline algae lying immediately behind the buttress zone.

alien animal Animal that is not indigenous to the area.

alkalinity A measure of the concentration of hydroxyl ions in a solution.

allochthonous Of foreign origin; transported into the area from outside.

alternate stable states The occurrence in a single habitat of several different persistent communities in which the structure of each is the result of historical events.

altricial Refers to young birds that are hatched in an immature state requiring extended parental care.

American estuary An estuary dominated by extensive stands of emergent vegetation surrounding less extensive mud flats.

anaerobic Condition where oxygen is absent.

anchor ice Ice that forms around any convenient nucleus in the area below the permanent pack ice in Antarctica; it tends to carry organisms out of the area to be incorporated in the sea ice above.

anchoveta Common name for *Engraulis ringens*, the Peruvian anchovy.

anoxic Without oxygen.

anticyclonic Moving in a clockwise direction.

aphotic Without light; that area of the oceans without light.

apogee That point during the orbit of the moon around the earth when the moon is farthest from the earth.

aquaculture The culture of aquatic organisms.

aquatic Living or existing within or on water.

asexual In reproduction, without involving sex.

atoll A modified ringshaped coral reef arising from deep water far from continental landmasses and enclosing a shallow lagoon.

ATP *See* adenosine triphosphate.

autochthonous Formed or occurring in the place where it is found.

autotrophic Living organisms capable of producing their own energy resources.

auxospore In diatoms, a reproductive cell that reestablishes the initial size of the species.

bacterioplankton Bacteria that live in the plankton, usually of a size range of 0.2 to 2.0 μm (picoplankton).

backwash Swash water that flows back down the beach.

baleen Horny material growing in comblike, fringed units from the upper jaws of whales of the order Mysticeti.

barrier reef Coral reef adjacent to landmasses and separated from them by a lagoon or channel of variable extent.

bathyal Bottom zone encompassing the continental shelf down to about 4000 meters.

bathypelagic Pelagic aphotic zone lying between the mesopelagic and abyssalpelagic zones.

bathyscape A free-moving, deep submersible designed to carry human observers into the deep sea.

bay A partially enclosed inlet of the ocean.

bends Name of disease occurring in humans who breathe compressed air in deep water. It results from a rapid decrease in pressure, which causes nitrogen bubbles to form in the internal tissues and blood vessels.

benthic The area of the sea bottom; organisms that occur on the sea bottom.

benthos Those organisms living on or in the sea bottom.

bet hedging A theory of life-history evolution in which the life-history traits displayed by the organisms result from the fluctuations in juvenile survivorship probabilities relative to those of the adult; opposite of "r-K" selection theory.

binocular vision Type of vision providing depth-of-field focus due to overlap of the field of vision of two closely set eyes.

biochemistry A scientific discipline concerned with the study of the chemistry of living organisms and their products.

bioerosion The dissolution or breakdown of substrates, usually calcium carbonate, by a variety of living organisms.

biogeochemical cycle Cyclical movement of an element or compound through living organisms and the nonliving environment.

biological conditioning Excretion by organisms of certain organic compounds into the water column that make the water either more or less suitable for themselves while enhancing or inhibiting other organisms.

biological interaction In ecology, a general term in which organisms have a mutual or reciprocal action or influence, including predation, competition, and grazing.

biological magnification Increase in concentration of chemicals in the bodies of animals with increasing trophic level.

bioluminescence The production of light by living organisms.

bioturbator Organism that is capable of modifying the local environment to enhance or exclude other organisms.

blade In a kelp plant, the flattened part of the plant that terminates the stipe.

blubber The layer of lipid that serves as an insulating layer under the skin of whales and other marine mammals.

blue-green algae Group of prokaryotic photosynthetic organisms of the kingdom Monera.

boiling point Temperature at which a liquid changes to a gas.

bottom-up process Control of community structure by environmental factors or the prey.

bradycardia Slowing of the heart rate.

buffer A chemical solution that resists or dampens changes in pH with the addition of acids or bases.

bycatch Other non-target organisms that are capture by fishing gear.

^{14}C method Method of measuring primary productivity using the radioactive isotope ^{14}C.

carbohydrate Group of biochemical compounds composed of carbon, hydrogen, and oxygen, often in the ratio of 1 carbon to 2 hydrogen and 1 oxygen.

Carnivora An order of mammals adapted to feed on other animals.

carnivore An animal that consumes other animals as food.

caudal The tail or posterior end of an organism.

caudal peduncle In fishes, that part of the body immediately in front of the tail.

cellulose A long-chain carbohydrate composed of repeated units of the sugar glucose.

centrifugal force Force that pulls objects out from the center of rotation.

Cetacea The order of mammals that contains the whales and porpoises.

chemolithoautotrophic bacteria Those bacteria able to obtain energy and therefore synthesize organic material through oxidation of reduced sulfur compounds.

chitin Biochemically, a polymer of the carbohydrate glucosamine that forms the hard outer integument of crustaceans and other marine invertebrates.

chlorophyll maximum Depth in the ocean where the concentration of chlorophyll reaches its highest level on a per-volume basis.

ciguatera Disease of humans caused by consumption of tropical fishes with toxins in the flesh and organs.

cleaning behavior Special category of symbiosis in which large animals, usually fishes, permit themselves to be cleaned of various parasites by smaller fishes or invertebrates.

climax In ecology, the final stage of a successional sequence that is able to persist in the absence of environmental change.

clone A group of genetically identical individuals of a plant or animal species produced by asexual reproduction from a single sexually produced individual.

coastal front As water flows over the shallow continental shelf, friction with the bottom causes vertical turbulence, which transfers nutrients and changes water chemistry, leading to changes in the plankton.

coastal plain estuary Estuary formed by flooding of a low-lying coastal valley due to rising sea level.

coccolithophore Small, unicellular, flagellated algae usually with an external covering of small pieces of calcium carbonate.

cohesion Mutual attraction of similar molecules, which resist external forces that would break them.

cold seeps Areas in the deep sea where reduced carbon or sulfur compounds exit from the rocks at ambient temperatures.

commensal In a symbiotic relationship the name given to the partner that gains the advantage.

commensalism A symbiotic relationship between two different species in which one benefits and the other is neither benefited nor harmed.

compensation depth Depth at which the processes of photosynthesis and respiration are equal.

compensation intensity In the water column, that point at which light intensity is equal to 1% of surface intensity.

compensation theory *See* subsidence theory.

competition The interaction among organisms for a necessary resource that is in short supply.

competition model Hypothesis to explain the high diversity of coral reef fishes which derives the high diversity from intense competition leading to a high degree of niche specialization.

competitive exclusion The ecological principal that states that complete competitors cannot coexist.

competitive interference Exclusion of one species by another species through interruption of its normal activities.

continental shelf The shallow underwater extension of a continent; usually limited in depth to 200 m.

continental slope The steeply descending bottom between the edge of the continental shelf and the abyssal plain; the ocean bottom between the depths of 200 and 4000 m.

copepod Small crustacean of the order Copepoda; the dominant macroscopic planktonic herbivore.

copepodid The larval stages of a copepod that follow the nauplius and metanauplius stages and precede the adult.

coral bleaching Expelling of zooxanthellae from the tissues of reef-building corals.

coral reef Massive limestone structure built up through the constructional cementing and depositional activities of anthozoans of the order Scleractinia and certain other invertebrate and algal species.

coralline algae A group of red algae common on coral reefs and that lay down calcium carbonate in their tissues.

corallite The cup into which a coral polyp fits.

corallivore An animal that consumes corals.

Coriolis effect The deflection imparted to moving water masses due to spinning of the earth on its axis; the deflection is to the right in the Northern Hemisphere and to the left in the Southern Hemisphere.

cosmopolitan In biogeography, an organism that is distributed throughout the world in suitable habitats.

crinotoxin A poisonous material secreted onto the surface of an organism.

critical depth The depth at which photosynthesis for the water column is equal to respiration for the water column; the depth to which phytoplankton cells may be mixed and still spend sufficient time above the compensation depth so that respiration and photosynthesis are equal.

critical tide level Points on the shore where a small change in vertical distance results in a disproportionate change in exposure time to air.

cropper In the disturbance theory of Dayton and Hessler, the term applies to deep-sea animals that consume living and dead animals smaller than themselves.

cryptic coloration Color, hue, or pattern that mimics the background.

cryptic symbiosis Symbiotic association between an algal cell and a nonphotosynthetic organism in which the algal species' cellular integrity is so reduced that it appears as an integral part of the host.

cryptofauna General term referring to the fauna living on or in coral substrates.

current Water movements that result in the horizontal transport of water masses.

cyanellae Symbiotic blue-green algae.

cyanobacteria *See* blue-green algae.

Cyanophyceae A class of blue-green algae.

cyclonic Moving in a counterclockwise direction.

decomposer An organism that breaks down dead protoplasm, freeing simple chemical substances for use by other organisms.

deep scattering layer (DSL) Concentration of midwater organisms that reflect sound waves produced by sonar devices.

deep sea A general term referring to that area of the ocean beyond the continental shelf and below the lighted zone.

deep water mass One of several water masses below the thermocline.

demersal Living close to the bottom of the sea.

density The mass per unit volume of a substance (physics); the number of individuals per unit area (biology).

deposit feeder An animal that feeds by consuming particles on or in the substrate.

desiccation The process of losing water.

diatom Microscopic autotrophic organism of the algae class Bacillariophyceae characterized by being enclosed in a two-part siliceous capsule.

diel Daily or once every 24 hours.

diffuse predation Condition where the total predation is strong and capable of controlling the abundance of a competitively dominant species but in which the predation is spread over more than one predator.

dinoflagellate Microscopic organism of the class Dinophyceae possessing two locomotory flagellae.

disphotic The area of the ocean between the euphotic and aphotic zones; it has insufficient light for photosynthesis but sufficient light for detection and response by animals.

dissipative beach A flat beach, maximally eroded, with fine sediments and with wave energy consumed in a surf zone away from the beach proper.

disturbance theory A hypothesis for explaining the high diversity in the deep sea that suggests that extreme predation pressure prevents any species from becoming numerous enough to eliminate others by competition, therefore perpetuating a large number of species and high diversity.

diurnal Occurring daily or relating to daytime.

diurnal tide Tide with a single high and low each day.

divergence Surface horizontal flow of water away from a central area, permitting water to upwell.

diversity *See* species diversity.

DOM Abbreviation for dissolved organic matter.

dominant The numerically abundant species in a community.

drag Resistance to movement through a medium.

dysphotic *See* disphotic.

echolocation *See* sonar.

ecological succession An orderly process of community change controlled through modification of the physical environment.

ecology Discipline in biology that treats the spectrum of relationships existing between organisms and their environment and among organisms.

ecotone An area of intergradation between two biological communities or associations.

ectothermic Cold-blooded; unable to regulate body temperature.

eddy A circular movement of water.

Ekman spiral The change of direction and velocity of a current as one progresses downward in the water column.

electrical conductivity Relative ability of a material to allow the passage of electricity.

electron A negatively charged elementary particle in all matter.

element A substance composed of atoms having the same atomic number.

elutriation Means of separating interstitial fauna from the sand through constant stirring with water and subsequent filtering on a fine mesh.

endobenthic Meiofaunal-sized organisms that move within the sediments, displacing particles.

endothermic Warm-blooded; able to regulate internal body temperature.

endozoite An animal living symbiotically inside another animal.

energy The capacity to do work.

epibenthic Organisms living at the sediment-water interface.

epicercal tail A caudal fin of fishes with the dorsal lobe larger than the ventral lobe.

epifauna Benthic organisms that live on or move over the substrate surface.

epipelagic *See* photic.

epiphytic Plants that are not parasitic, but that are attached to other plants.

epizoite An animal that is not parasitic, but that lives attached to another animal.

equilibrium species A species that has a life history characterized by long life, long development time to reach maturity, low death rates, and few reproductive periods per year.

estuary A partially enclosed coastal embayment where fresh water and seawater meet and mix.

euphotic *See* photic.

European estuary An estuary dominated by large mudflats and relatively barren of large plants but has large populations of surface benthic diatoms.

euryhaline Able to tolerate wide fluctuations in salinity.

eutrophic Containing abundant nutrient material.

eutrophication The process of excessively increasing the nutrient levels in the oceans through natural or artificial means.

evaporite estuary *See* negative estuary.

exploitative competition Competition among different species or members of the same species for a necessary resource that is in short supply.

extinction coefficient Ratio between the intensity of light at a given depth and the intensity at the surface.

femtoplankton A size class of plankton ranging from 0.02 to 0.2 μm.

fetch The distance over which the wind can blow.

filter feeder Animals that obtain their food by filtering particles out of the water column.

fin A flattened appendage of an aquatic animal used in locomotion or maneuvering in the water.

fishery The place or act of taking fishes or other sea organisms for human use.

fjord Estuaries occurring in deep, drowned valleys, originally cut by glacial action; characterized by the presence of a shallow sill at the mouth, restricting water interchange between surface water and deeper fjord waters.

flipper The pectoral or pelvic appendage of a marine mammal or reptile.

flushing time Time required for a given mass of freshwater to be discharged from the estuary.

food chain The pathway that transfers energy from a given source autotrophic organism through a series of consumers.

food web The combinations of all food chains in a given community or ecosystem.

form resistance Condition in which drag is proportional to the cross-sectional area of the object in contact with the water.

frictional resistance Condition in which drag is proportional to the amount of surface area in contact with the water.

fringing reef A coral reef that forms immediately adjacent to a landmass.

fugitive species *See* opportunistic species.

fundamental niche The total of all the ranges of all the biological and physical factors within which a species can exist; a multidimensional or *n*-dimensional hyperspace.

gametophyte In algae, the haploid organism that produces gametes.

gas bladder Structure on dorsal side of body cavity of bony fishes that contains gas and is used by the fishes to regulate buoyancy.

geomorphology Discipline dealing with the form and configuration of the surface of the earth.

ghost fishing Continued capture of marine organisms by fishing gear which has been lost in the ocean.

global warming Increase in average global temperature due to build-up of gases such as CO_2, NO_2, and chlorofluorocarbons in the atmosphere.

gorgonians Anthozoans of the subclass Octocorallia that have spiculate internal skeletons; commonly called sea fans or sea whips.

grab A collecting device that picks up a volume of the bottom; usually deployed on a cable from a vessel.

gravity The attraction of terrestrial bodies toward the center of the earth (the earth's gravitational attraction).

grazer An animal that feeds on autotrophic organisms or other sessile animals.

grazing Feeding on vegetation or sessile colonial animals by either consuming the whole food organism or by cropping all or part of the surface growth.

greenhouse effect *See* global warming.

gross photosynthesis The total amount of photosynthesis before subtracting losses due to respiration.

gross primary production The total amount of organic material fixed in the primary production process.

guild Species that make their living in the same way.

guyot A submerged, isolated, flat-topped mountain.

gyre Circular motion of water in the major ocean basins.

habitat The place where an organism can be found.

hadal Parts of the ocean bottom below 6000 m.

hadalpelagic Deepest pelagic aphotic zone lying below the abyssalpelagic.

halophyte A plant adapted to grow in soils with high salt concentrations.

haptophytes Chlorophyll-bearing organisms of the class Haptophyceae.

harem In mammals, a group of females controlled by a single male during the breeding season.

heat Energy that moves from a higher to a lower temperature system.

heat capacity Amount of energy required to raise the temperature of 1 g of material 1°C.

heat of vaporization Amount of energy required to evaporate a unit mass of a substance at constant temperature and pressure.

herbivore An animal that consumes autotrophic organisms.

hermatypic coral A reef-building coral with symbiotic zooxanthellae in the tissues.

heteropods Pelagic molluscs of the prosobranch order Heteropoda.

heterotrophic Organisms that require an external source of food.

high-level radioactive wastes Those wastes generating greater than 1×10^{-8} curies of ionizing radiation.

holdfast In kelp, that part of the plant that fixes it to the substrate.

holoepipelagic Nektonic animals that spend their entire lives in the open ocean.

holoplankton Plankton organisms that spend their entire lives in the plankton.

homeothermic *See* endothermic.

homeoviscous adaptation An adaptation to high pressure by incorporation of more fluid lipids into biological membranes.

homogeneous estuary An estuary with complete mixing such that the salinity is similar from surface to bottom.

host The major provisioning partner of a symbiotic or commensal relationship.

host shift Movement of a pathogen from a natural host to another that was not previously a host.

hydrodynamic The physical features of water motion.

hydrogen bond Weak bond formed between a hydrogen atom in one molecule and an electronegative atom of the same or another molecule.

hydrographic The physical and chemical features of the oceans.

induced drag Disruption of laminar flow patterns around objects by throwing the water into vortices or eddies.

infauna Benthic organisms that live in or burrow through the bottom sediment.

infaunal predator A predator that lives within a sedimentary environment.

infralittoral fringe In the Stephenson Universal Zonation scheme, the lowest zone on the shore, bounded above by the upper limit of laminarians and below by the lowest tide level.

infralittoral zone *See* infralittoral fringe.

inquilinism Subcategory of commensalism in which the animal lives in the home or digestive tract of another without being parasitic.

insolation Being exposed directly to the sun's rays.

interference competition Condition where one species competes with another by directly interacting with the species in some way.

intermittent estuary *See* seasonal estuary.

interstitial The space between adjacent particles in a sedimentary bottom.

intertidal Benthic area lying between the extremes of high and low tides.

interzonal fauna Mesopelagic animals that migrate vertically into other zones.

ion A positively or negatively charged atom or molecule produced through loss or gain of one or more electrons.

isohaline A line joining points of equal salinity concentration.

isotherm A line joining points of equal temperature.

iteroparous Referring to organisms that reproduce more than once during their life spans.

keel A median longitudinal ventral ridge in certain nektonic fishes.

kelp Collective name for various large brown algae.

kelp bed A community dominated by large brown algae in which the plants do not form a surface canopy.

kelp forest A community dominated by large brown algae in which the plants form a surface canopy.

key industry species *See* keystone species.

keystone species A species that is disproportionately important in the maintenance of community integrity and without which drastic alterations of the community would occur.

krill Colloquial name for crustaceans of the order Euphausiacea.

lagoon (a) Type of estuary formed through the cutting off of inshore waters by the buildup of sandbars parallel to the shore; (b) a shallow stretch of water separated from the open ocean by a coral reef or island.

lagoon reef *See* patch reef.

Langmuir convection cell A localized area of water a few meters wide and hundreds of meters long in which there is a local circulation pattern created by the wind causing water to upwell and diverge from the center and converge and sink at the boundaries.

larva An independent morphologically different stage of an animal that develops from a fertilized egg and must undergo a greater or lesser amount of change before assuming adult features.

latent heat of fusion Increase in heat content that accompanies converting a unit mass of substance from a solid to a liquid state.

law of the minimum Law stating that of those essential substances that are necessary to the survival of an organism, the one that is present in the smallest quantities or is reduced to a minimum first will limit the growth and survival of the organisms, even if all others are plentiful.

lecithotrophic larvae Planktonic larvae that do not feed in the plankton.

leeward The side of an island of reef protected from winds and waves.

Leibig's law of the minimum *See* law of the minimum.

light Electromagnetic radiation with wavelengths between 400 and 700 μm.

light-dark bottle method Method of measuring primary productivity employing two identical bottles, one transparent and the other opaque. Photosynthetic organisms in the transparent bottle undergo both respiration and photosynthesis while those in the dark bottle undergo only respiration Subtracting oxygen contents of the two bottles give a measure of photosynthesis.

lithothamnion ridge Synonym for algal ridge.

littoral *See* intertidal.

local distribution *See* patchiness.

lottery hypothesis Hypothesis to explain the high divesity of coral reef fishes which derives the high diversity due to chance as to which species from the larval pool occupies the vacant space.

low-level radioactive wastes Those wastes that generate less than 1×10^{-8} curies of ionizing radiation.

luciferase The enzyme that catalyzes the production of bioluminescence in the presence of luciferin and ATP.

luciferin The protein that produces bioluminescence in the presence of ATP and luciferase.

mackerel Common name for several species of economically important fishes of the family Scombridae.

macrofauna A general term referring to benthic organisms more than 1 mm in size.

macroplankton Plankton with a size range from 2 to 20 cm.

Madreporaria The order of the class Anthozoa that contains the corals.

mangal *See* mangrove forest.

mangrove Common name for any of several species of inshore tropical trees or shrubs that dominate the mangal associations.

mangrove forest A variety of tropical inshore communities dominated by several species of shrubs or trees that grow in salt water.

mariculture The culture of marine organisms.

marine-dominated estuary *See* homogeneous estuary.

marine snow Name given to the various types of amorphous particulate matter found in the water column and derived from living organisms.

mass strandings When large numbers of whales beach themselves.

maximum sustainable yield The largest number of fishes that can be harvested year after year without diminishing the stock.

megaplankton Plankton with a size range above 20 cm.

meiobenthos A synonym for meiofauna.

meiofauna Benthic organisms between 0.5 mm and 62 μm in size, often used synonymously with interstitial fauna.

melon The large lipid-filled body in the head of whales that serves as an acoustical lens.

meroepipelagic Open-ocean nekton that spend part of their lives in the open ocean and part in other areas.

meroplankton Planktonic organisms that spend only part of their life cycles in the plankton.

mesobenthic Meiofaunal-sized organisms living and moving within the interstitial space.

mesopelagic Uppermost pelagic aphotic zone.

mesoplankton Plankton with a size range between 0.2 and 20 mm.

mesopsammon Organisms living between sand grains in fresh water.

metabolism A general term referring to all physical and chemical processes that occur in a living organism.

metabolite A biochemical produced by an organism and secreted into the surrounding medium.

metamorphosis The process of structural transformation in an animal, which changes the organism from larva to adult.

microbial loop Name given to that part of the planktonic food web made up of bacteria and various nanoplankton organisms that recovers and cycles dissolved and particulate carbon back into the larger planktonic food web.

microfauna Benthic organisms below the size of 0.1 mm.

micrograzer hypothesis Hypothesis explaining the persistently higher nutrient levels and lack of a spring bloom of phytoplankton in the North Pacific as opposed to the North Atlantic.

microplankton Plankton with a size range from 20 to 200 μm.

midlittoral zone In the Stephenson Scheme of Zonation, that area between the upper limits of laminarians and the upper limits of barnacles.

migration Periodic movement of animals from one place to another.

mimic An organism that assumes the shape, pattern, or behavior of a different species, usually for protection.

Minimata disease Crippling neurological disorder caused by mercury poisoning.

mixed tide Tides consisting of a mixture of diurnal and semidiurnal tides.

mixing and micrograzer hypothesis Explanation of the lack of a spring phytoplankton bloom in the subarctic Pacific Ocean due to the continued presence of protistan micrograzers.

multivore *See* omnivore.

mutualism That form of symbiosis in which two species associate to their mutual benefit.

myoglobin An iron-containing protein compound in muscle that acts as an oxygen carrier and reserve oxygen supply.

nanoplankton Plankton with a size range from 2 to 20 μm.

nauplius First larval form of most crustaceans; characterized by possessing three pairs of appendages.

negative estuary Estuary in which evaporation is high and freshwater input low, so that salt water enters at the surface, evaporates, becomes hypersaline, and sinks, forming an outflowing bottom current.

nekton Pelagic animals that are powerful enough swimmers to move at will in the water column.

neritic Refers to the water mass overlying the continental shelves.

net community photosynthesis The amount of photosynthesis in excess of respiration by both autotrophic and heterotrophic organisms.

net plankton Organisms caught in a standard zooplankton net (above 0.2 mm).

net production Amount of total production left after losses from respiration (or that amount left to support other trophic levels).

neutral In chemistry, having equal numbers of OH^- and H^+ ions; a pH of 7.

neutral estuary *See* homogeneous estuary.

new production *See* net community photosynthesis.

niche The role of an organism in a community.

Niño, El An episodic increased warming of the equatorial Pacific Ocean due to atmospheric changes; it can cause large-scale changes in oceanographic and atmospheric conditions that may lead to catastrophic mortality of many marine organisms.

nonpelagic development Development without a planktonic period.

nucleus The central, dense, positively charged part of an atom.

oceanic Refers to the open ocean water masses away from the continental shelves.

oligotrophic Containing little nutrient material.

omnivore An organism that consumes both plant and animal material (synonym = multivore).

ooze A deep-sea sediment composed, at least partially, of the skeletal remains of certain pelagic organisms.

opportunistic species A species that has a life history characterized by short life span, short development time to reach maturity, high death rate, and many reproductive periods per year.

osmoconformer An organism unable to regulate its internal fluid and salt balance, and therefore, one with a varying internal salt concentration.

osmoregulation Physiological activity within an organism that maintains the internal salt and fluid balance within a narrow, acceptable range.

osmoregulator An organism that has physiological mechanisms to control the salt content of its internal fluids.

osmosis The movement of water across a semipermeable membrane separating two solutions of differing solute concentrations, movement being from the more dilute solution to the more concentrated.

oval Special gas absorptive organ in physoclist fishes.

oxygen minimum zone An area in the ocean, usually between 500 and 1000 m, where oxygen values approach zero.

parallel bottom community Concept that similar sediment types at similar depths around the world contain similar communities of organisms in which the dominant animals are closely comparable taxonomically and ecologically.

parasitism Association in which one species lives on, or in, another, drawing nourishment from that species at the expense of, or to the detriment of, the other.

patch reef A small isolated reef structure existing in the lagoon behind the main coral reef.

patchiness An ecological condition in which organisms occur in isolated groups within a larger contiguous suitable habitat.

peat Organic matter in the soil of salt marshes accumulated from the root masses of marsh plants.

pectoral Pertaining to the chest region of the body.

pelagic The area of the open water of the world's oceans; organisms that occur in the water column.

pelvic Pertaining to the hip region of the body.

perigee That point during the orbit of the moon around the earth when the moon is closest to the earth.

period Time required for two wave crests to pass a fixed point.

permanent meiofauna Organisms that spend their entire lives in the interstices of sediment.

Petersen grab A quantitative grab designed by C. G. Joh. Petersen to take a fixed area of soft bottom, usually a tenth or one-half of a square meter.

pheromone A chemical secreted by a species that influences the behavior of others of the same species.

photic The lighted area of the ocean; the epipelagic zone.

photophore A complex organ of certain deep-sea animals that produces light and regulates its emission and use.

photosynthesis Process occurring in autotrophic organisms, whereby they use the energy of sunlight to synthesize energy-rich organic compounds from carbon dioxide and water.

phototaxis Movement of an organism in response to light.

phototrophic Name applied to organisms that use light as a source of energy or an animal dependent upon photosynthetic zooxanthellae symbionts for energy.

physoclist In bony fishes, those fishes with no open duct between the gas bladder and the digestive tract.

physostome Gas bladder type in fishes where there is an open duct between the gas bladder and the esophagus.

phytoplankton Planktonic autotrophic organisms.

picoplankton Plankton in the size range of 0.2 to 2.0 μm.

Pinnipedia A suborder of mammals containing the seals, sea lions, and walruses.

plankton Those organisms free-floating or drifting in the open water of the oceans having their lateral and vertical movements determined by water motion.

plankton net Fine-mesh conical nets dragged in the water to collect plankton.

planktotrophic larvae Larvae that feed in the plankton.

planula The sexually produced larvae of many cnidarians, including corals.

pneumatocyst The gas-filled floats on certain kelps.

pneumatophore Extensions above the substrate of the roots of mangroves to permit aeration.

POC Abbreviation for particulate organic carbon.

poikilothermic *See* ectothermic.

polar The Arctic or Antarctic zones.

pollution Degradation of the natural environment.

polyp A sessile or sedentary cnidarian individual with a cylindrical body and, usually, tentacles surrounding a mouth at the free end.

positive estuary Estuary with substantial freshwater in-put and reduced evaporation to the extent that fresh-water flows out at the surface and salt water flows in at the bottom; also known as a salt wedge estuary.

practical salinity unit (PSU) The total amount of dissolved material in seawater expressed as parts per thousand.

preadaptation Possession by a species or group of species of features that favor easy occupation of a new habitat.

predation Consumption of an individual of one species by an individual of another species.

predator An animal that feeds on another animal.

predator-disturbance model Hypothesis to explain the high diversity of coral reef fishes which derives the high diversity because the fish populations never reach equilibrium and therefore do not undergo competitive exclusion because of predation, catastrophe, and unpredictable recruitment.

primary productivity The rate of formation of energy-rich organic compounds from inorganic materials.

prochlorophytes Tiny 0.6-0.8 μm, chlorophyll-bearing organisms that resemble cyanobacteria but possess a different combination of pigments.

protein Complex biochemical compounds of large molecular weight composed of joined amino acids.

psammon An obsolete term referring to organisms living interstitially.

pteropods Pelagic molluscs of the opisthobranch orders Thecosomata and Gymnosomata.

pycnocline That area in the water column where the highest rate of change in density occurs for a given change in depth.

realized niche That portion of the fundamental niche that the species actually occupies or uses within its habitat.

recruitment limitation model Hypothesis to explain the high diversity of coral reef fishes which derives the high diversity from the variation in larval recruitment.

redox potential discontinuity (RDP) layer Zone in muddy shores characterized by a rapid change from positive to negative redox potential; the boundary layer between aerobic and anaerobic sediments.

red tide The name given to massive blooms of dinoflagellates in which the concentrations of the plants are such that the water becomes a red-brown in color. Toxins are often secreted in sufficient quantities to affect other marine organisms.

reflective beach Beach with no surf zone, a steep slope, and waves that break directly on the beach and that are then reflected back, producing large swashes on the beach.

refractive index For a medium, the ratio of the speed of light in a vacuum to that in the medium.

rete mirabile A large network of anastomosing small blood vessels serving a number of functions in marine vertebrates.

reverse migration Opposite of vertical migration; animals migrate down at night and toward the surface during the day.

rhodopsin A light-sensitive pigment found in the rods of the eye.

rings *See* eddy.

river-dominated estuary *See* salt wedge estuary.

rod The sensory body of the retina that is most responsive to low light levels.

salinity The total amount of dissolved material (salts) in seawater.

salt glands Special glands in plants and animals for eliminating excess salt.

salt marsh Communities of emergent vegetation rooted in soils alternately inundated and drained by tidal action.

salt pans Vegetation-free areas in salt marshes caused by high soil salinity.

salt wedge estuary Estuary with a pronounced difference in salinity between surface and deeper water, the deeper waters being more saline (*see* also positive estuary).

Scleractinia Synonym for Madreporaria.

sclerosepta The calcium carbonate septa in the skeletons of corals.

seagrass Collective name for marine flowering plants of the families Potamogetonaceae and Hydrocharitaceae.

seamount *See* guyot.

seasonal estuary Estuary having freshwater inflow and mixing with ocean water only for part of a year; the rest of the year it is a stagnant lagoon.

seasonal succession The change of dominant species in the phytoplankton with succeeding seasons of the year.

sedentary In reference to organisms, those that remain for long periods in one place or have limited movement.

sediment destabilizer Organisms whose activities cause the substrate to move, become resuspended, erode, or otherwise change.

sediment stabilizer Organisms whose activities bind the substrate particles or otherwise enhance substrate stability.

semelparous Breeding or reproducing only once during a lifetime.

semidiurnal tide A tide with two highs and two lows per lunar day.

semienclosed bay *See* lagoon (a).

sere One of the stages in a successional sequence leading to a climax community.

sessile An organism that is fixed in place.

shelf *See* sublittoral.

silicoflagellate Flagellated photosynthetic organism of the class Chrysophyta with an internal skeleton of silicon dioxide.

Sirenia The order of the class Mammalia that contains the sea cows (dugongs and manatees).

slough *See* lagoon.

sonar A method of using sound waves to locate objects in the water column.

species diversity In ecology, a numerical measure combining the number of species in an area with their relative abundance.

species richness The number of species in a given area.

spermaceti organ The melon of the sperm whale.

spermatophore A special packet that contains sperm.

sporophyte In algae, the diploid plant that produces spores.

spring bloom Dramatic increase in the population of photosynthetic organisms in the surface waters of the North Atlantic in the spring as a result of the abundant nutrients and lack of grazers.

stability-time hypothesis Theory *seeking* to explain the occurrence of high diversity among communities of the marine environment; the hypothesis states that high diversity occurs because highly stable environmental conditions have persisted over long periods of time, allowing numerous, highly specialized species to evolve.

standing crop The biomass of an organism or group of organisms present per unit volume or per unit area at a given point in time.

statocyst An organ that detects gravity.

stenohaline Able to tolerate only a narrow range of salinity changes.

stipe That part of a kelp plant between blade and holdfast.

stratified estuary *See* salt wedge estuary.

sublittoral Bottom zone of the continental shelf.

submarine ridge Submerged ranges of mountains usually marking the boundries of crustal plates.

subsidence theory Darwin's theory as to the origin of coral atolls whereby atolls begin as fringing reefs around volcanic islands; subsequent subsidence of the island, coupled with upward reef growth, then produces first a barrier reef and finally, with the disappearance of the volcano, an atoll.

succession The process of community change through time. The facilitation model of succession is the classical definition of succession, whereby there is an orderly sequence of seral stages, the organisms of each modifying the environment making it less suitable for them and more suitable for the next. The end point or climax perpetuates itself. The inhibition model of succession has no organism as competitively superior and those that first occupy the area remain until a disturbance removes them. A third model, the tolerance model, is intermediate between the other two.

succulence strategy Adaptation of vascular plants living in highly saline soils to maintain high tissue water concentrations to buffer against water loss by osmosis.

sulfur bacteria *See* chemolithoautotrophic bacteria.

supralittoral fringe In the Stephenson Universal Zonation scheme, the highest zone on the shore, bounded below by the upper limit of barnacles and above by the upper limit of *Littorina*.

surface tension With respect to water, the attraction of the molecules of water at the surface (air-water interface) so that the surface resists penetration by small objects.

surge channel Deep channels in the windward face of a coral reef through which water moves in and out of the reef.

suspension feeding Feeding by filtering particles out of the water column.

swash Water that runs up the face of a beach after a wave breaks.

swim bladder *See* gas bladder.

symbiont Name given to the partners in a mutualistic relationship.

symbiosis The interrelationship between two different species.

tectonic estuary Estuary formed when the sea reinvades the land due to subsidence of the land as a result of earthquake activity.

temporary meiofauna Meiofaunal-sized organisms that are the juveniles of macrofaunal organisms.

terrigenous Ocean sediments derived from the land.

territoriality A pattern of behavior of an individual or group that actively keeps other members of the same or different species out of an area.

territory An area occupied and protected by a species against others of its own species, usually for breeding or feeding purposes.

thalassopsammon The interstitial organisms in marine waters.

thermal pollution Degradation of the environment through increases in temperature.

thermal stratification Situation prevailing at all times in the tropics and during the summer in the temperate zones where the uppermost water mass decreases in density due to increased temperature such that it is isolated from the more dense, colder water below.

thermocline That portion of the water column where temperature changes most rapidly with each unit change in depth.

thigmotaxis Behavioral or movement response triggered by contact with a solid body.

thiobios Meiofauna that live in anoxic sediments.

tide The periodic rise and fall in the surface water of the oceans due to gravitational attraction of the sun and moon and the rotation of the earth.

tidepool Area in the rocky intertidal that retains some volume of water at low tide.

top down processes Control of community structure by predators.

toxin Any of a variety of chemical substances produced by organisms that are, in turn, poisonous to other organisms.

trenches The narrow, steep-sided depressions in the ocean floor usually lying between 6000 and 10,000 m in depth.

trophic group amensalism Exclusion of one group of organisms by modification of the environment by another.

trophic level An ecological system of classification of organisms according to their means of obtaining nutrition; the basic level is that of the autotrophs, the second is that of the herbivores, and the succeeding levels are carnivores.

trophic structure The total number of feeding levels through which energy passes in an ecosystem.

trophogenic zone The area in the ocean above the compensation point.

tropholytic zone The area in the ocean below the compensation point.

tsunami A very long period wave generated by an underwater earthquake, landslide, or volcanic eruption; also known as a tidal wave.

tubular eyes Eye in the form of a short cylinder and containing two retinas found in some deep sea fishes.

tuna The common name for any of several large pelagic fishes of the family Thunnidae.

turbidity Condition of reduced visibility in water due to the presence of suspended particles.

twilight zone *See* mesopelagic.

ultraabyssal *See* hadal.

ultraplankton Plankton less than 2 μm in size.

upper water mass The well-mixed water layer above the thermocline.

vernal The spring of the year.

vertical migration Diurnal vertical movement of pelagic organisms in the water column toward the surface at night and down to depth during the day.

viroplankton Planktonic organisms in the size range of 0.02-0.2 micrometers.

virus Infectious agent composed of a protein sheath around a nucleic acid core characterized by a lack of an independent metabolism making them totally dependent upon living organisms for reproduction.

viscosity That property of a liquid to resist movement through it.

vitamin An organic compound essential to certain metabolic processes in organisms.

wave With respect to the oceans, a disturbance that moves through the water but does not cause particles of the water to advance with it.

wavelength The horizontal distance between the tops of successive wave crests.

weak predation Condition where the total effect of predation on a competitive dominant prey is such that predation alone does not control the abundance of the competitive dominant.

whalebone *See* baleen.

windward The side of an island or reef that faces the prevailing wind.

zonation Prominent horizontal bands of organisms that succeed each other vertically.

zoochlorellae Symbiotic green algae.

zooplankton Planktonic animals.

zooxanthellae Dinoflagellates symbiotic in the tissues of various marine invertebrates; symbionts that are brownish in color.

INDEX

Note: Page references followed by *f* and *t* indicate figures and tables, respectively.